This online teaching and learning environment integrates the entire digital textbook with the most effective instructor and student resources to fit every learning style.

With **WileyPLUS:**

- Students achieve concept mastery in a rich, structured environment that's available 24/7.
- Instructors personalize and manage their course more effectively with assessment, assignments, grade tracking, and more.

From multiple study paths, to self-assessment, to a wealth of interactive visual and audio resources, *WileyPLUS* gives you everything you need to personalize the teaching and learning experience.

»Find out how to MAKE IT YOURS»

www.wiley**plus**.com

ALL THE HELP, RESOURCES, AND PERSONAL SUPPORT YOU AND YOUR STUDENTS NEED!

2-Minute Tutorials and all of the resources you & your students need to get started
www.wileyplus.com/firstday

Student support from an experienced student user. Ask your local representative for details!

Collaborate with your colleagues, find a mentor, attend virtual and live events, and view resources
www.WhereFacultyConnect.com

Pre-loaded, ready-to-use assignments and presentations
www.wiley.com/college/quickstart

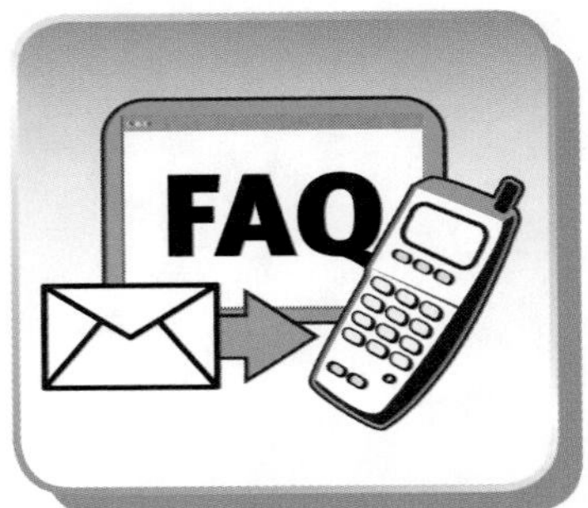

Technical Support 24/7
FAQs, online chat, and phone support
www.wileyplus.com/support

Your *WileyPLUS* Account Manager.
Training and implementation support
www.wileyplus.com/accountmanager

MAKE IT YOURS!

PART 3

9TH EDITION

FUNDAMENTALS OF PHYSICS

PART 3

9TH EDITION

HALLIDAY & RESNICK

FUNDAMENTALS OF PHYSICS

Jearl Walker

Cleveland State University

John Wiley & Sons, Inc.

SPONSORING EDITOR Geraldine Osnato
EXECUTIVE EDITOR Stuart Johnson
ASSISTANT EDITOR Aly Rentrop
ASSOCIATE MARKETING DIRECTOR Christine Kushner
SENIOR PRODUCTION EDITOR Elizabeth Swain
TEXT DESIGNER Madelyn Lesure
COVER DESIGNER M77 Design
DUMMY DESIGNER Lee Goldstein
PHOTO EDITOR Hilary Newman
EXECUTIVE MEDIA EDITOR Thomas Kulesa
COVER IMAGE ©Eric Heller/Photo Researchers, Inc.

This book was set in 10/12 Times Ten by Prepare and was printed and bound by R.R.Donnelley/Jefferson City. The cover was printed by R.R.Donnelley/Jefferson City.

This book is printed on acid free paper.

Copyright © 2011, 2008, 2005, 2003 John Wiley & Sons, Inc. All rights reserved.

No part of this publication may be reproduced, stored in a retrieval system or transmitted in any form or by any means, electronic, mechanical, photocopying, recording, scanning or otherwise, except as permitted under Sections 107 or 108 of the 1976 United States Copyright Act, without either the prior written permission of the Publisher, or authorization through payment of the appropriate per-copy fee to the Copyright Clearance Center, Inc. 222 Rosewood Drive, Danvers, MA 01923, website www.copyright.com. Requests to the Publisher for permission should be addressed to the Permissions Department, John Wiley & Sons, Inc., 111 River Street, Hoboken, NJ 07030-5774, (201)748-6011, fax (201)748-6008, or online at http://www.wiley.com/go/permissions.

Evaluation copies are provided to qualified academics and professionals for review purposes only, for use in their courses during the next academic year. These copies are licensed and may not be sold or transferred to a third party. Upon completion of the review period, please return the evaluation copy to Wiley. Return instructions and a free of charge return shipping label are available at www.wiley.com/go/returnlabel. Outside of the United States, please contact your local representative.

Library of Congress Cataloging-in-Publication Data

Halliday, David
Fundamentals of physics / David Halliday, Robert Resnick, Jearl Walker.—9th ed.
p. cm.
Includes index.
Part 3: ISBN 978-0-470-54793-9 (pbk.)
Also catalogued as
Extended version: ISBN 978-0-470-46908-8
1. Physics—Textbooks. I. Resnick, Robert II. Walker, Jearl III. Title.
QC21.3.H35 2011
530—dc22
2009033774

Printed in the United States of America

10 9 8 7 6 5 4 3 2 1

Volume 1

PART 1

PART 2

Volume 2

PART 3

PART 4

PART 5

CONTENTS

28 MAGNETIC FIELDS 735

29 MAGNETIC FIELDS DUE TO CURRENTS 764

30 INDUCTION AND INDUCTANCE 791

31 ELECTROMAGNETIC OSCILLATIONS AND ALTERNATING CURRENT 826

32 MAXWELL'S EQUATIONS; MAGNETISM OF MATTER 861

APPENDICES

ANSWERS

INDEX I-1

WHY I WROTE THIS BOOK

Fun with a big challenge. That is how I have regarded physics since the day when Sharon, one of the students in a class I taught as a graduate student, suddenly demanded of me, "What has any of this got to do with my life?" Of course I immediately responded, "Sharon, this has everything to do with your life—this is physics."

She asked me for an example. I thought and thought but could not come up with a single one. That night I began writing the book *The Flying Circus of Physics* (John Wiley & Sons Inc., 1975) for Sharon but also for me because I realized her complaint was mine. I had spent six years slugging my way through many dozens of physics textbooks that were carefully written with the best of pedagogical plans, but there was something missing. Physics is the most interesting subject in the world because it is about how the world works, and yet the textbooks had been thoroughly wrung of any connection with the real world. The fun was missing.

I have packed a lot of real-world physics into this HRW book, connecting it with the new edition of *The Flying Circus of Physics*. Much of the material comes from the HRW classes I teach, where I can judge from the faces and blunt comments what material and presentations work and what do not. The notes I make on my successes and failures there help form the basis of this book. My message here is the same as I had with every student I've met since Sharon so long ago: "Yes, you *can* reason from basic physics concepts all the way to valid conclusions about the real world, and that understanding of the real world is where the fun is."

I have many goals in writing this book but the overriding one is to provide instructors with tools by which they can teach students how to effectively read scientific material, identify fundamental concepts, reason through scientific questions, and solve quantitative problems. This process is not easy for either students or instructors. Indeed, the course associated with this book may be one of the most challenging of all the courses taken by a student. However, it can also be one of the most rewarding because it reveals the world's fundamental clockwork from which all scientific and engineering applications spring.

Many users of the eighth edition (both instructors and students) sent in comments and suggestions to improve the book.These improvements are now incorporated into the narrative and problems throughout the book. The publisher John Wiley & Sons and I regard the book as an ongoing project and encourage more input from users. You can send suggestions, corrections, and positive or negative comments to John Wiley & Sons or Jearl Walker (mail address: Physics Department, Cleveland State University, Cleveland, OH 44115 USA; or email address: physics@wiley.com; or the blog site at www.flyingcircusofphysics. com). We may not be able to respond to all suggestions, but we keep and study each of them.

LEARNINGS TOOLS

Because today's students have a wide range of learning styles, I have produced a wide range of learning tools, both in this new edition and online in *WileyPLUS*:

ANIMATIONS of one of the key figures in each chapter. Here in the book, those figures are flagged with the swirling icon. In the online chapter in *WileyPLUS*, a mouse click begins the animation. I have chosen the figures that are rich in information so that a student can see the physics in action and played out over a minute or two instead of just being flat on a printed page. Not only does this give life to the physics, but the animation can be repeated as many times as a student wants.

Animation

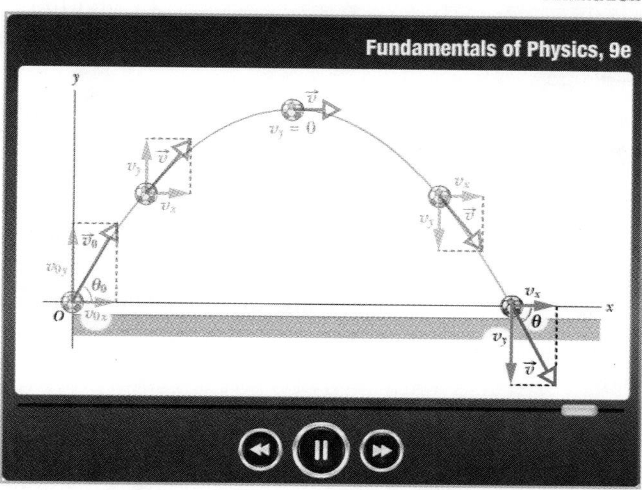

WILEY PLUS **VIDEOS** I have made well over 1000 instructional videos, with more coming each semester. Students can watch me draw or type on the screen as they hear me talk about a solution, tutorial, sample problem, or review, very much as they would experience were they sitting next to me in my office while I worked out something on a notepad. An instructor's lectures and tutoring will always be the most valuable learning tools, but my videos are available 24 hours a day, 7 days a week, and can be repeated indefinitely.

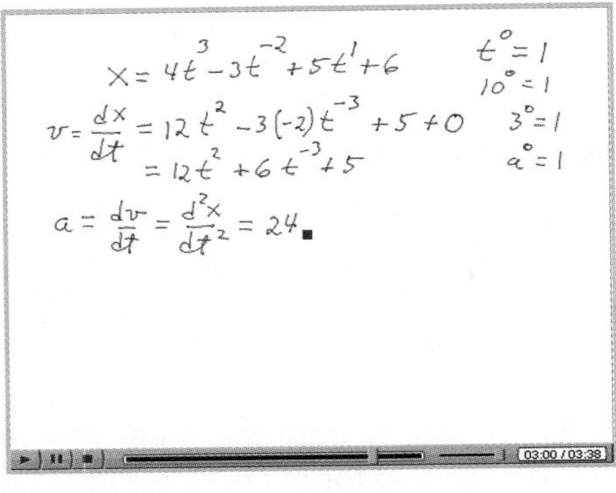

Video Review

- **Video tutorials on subjects in the chapters**. I chose the subjects that challenge the students the most, the ones that my students scratch their heads about.
- **Video reviews of high school math,** such as basic algebraic manipulations, trig functions, and simultaneous equations.
- **Video introductions to math,** such as vector multiplication, that will be new to the students.
- **Video presentations of every Sample Problem** in the textbook chapters (both 8e and 9e). My intent is to work out the physics, starting with the Key Ideas instead of just grabbing a formula. However, I also want to demonstrate how to read a sample problem, that is, how to read technical material to learn problem-solving procedures that can be transferred to other types of problems.
- **Video solutions to 20% of the end-of chapter problems.** The availability and timing of these solutions are controlled by the instructor. For example, they might be available after a homework deadline or a quiz. Each solution is not simply a plug-and-chug recipe. Rather I build a solution from the Key Ideas to the first step of reasoning and to a final solution. The student learns not just how to solve a particular problem but how to tackle any problem, even those that require *physics courage*.
- **Video examples of how to read data from graphs** (more than simply reading off a number with no comprehension of the physics).

GO Tutorial Close

This GO Tutorial will provide you with a step-by-step guide on how to approach this problem. When you are finished, go back and try the problem again on your own. To view the original question while you work, you can just drag this screen to the side.
(This GO Tutorial consists of 6 steps).

Step 1.1 : *Concept – Evaluate quantities*

KEY IDEAS:
(1) We want the net force on particle 1 to be a certain value (zero). Thus, we choose particle 1 as our system.
(2) Because particles 1 and 4 have the same sign of charge, they repel each other.
(3) In order for the net force on particle 1 to be zero, the forces on it due to particles 2 and 3 must be attractive, to counter the repulsive force.
(4) Because the net force on particle 1 is zero, the x component and the y component of that net force must each be zero.

GETTING STARTED: We need to consider each of the three forces acting on particle 1. We start with the force due to particle 4. What is the distance between particles 1 and 4?

Number ______ Units ______

the tolerance is +/-2%

Check Your Input

Step 1.2 : *Concept – Evaluate quantities*

What is the magnitude of the force on particle 1 due to particle 4?

Number ______ Units ______

the tolerance is +/-2%

Check Your Input

Step 1.3 : *Concept – Evaluate quantities*

Which of the vectors in Figure (1 or 2) best shows the force on particle 1 due to particle 4?

○ 2
○ 1

Check Your Input

Step 1.4 : *Concept – Evaluate quantities*

What is the magnitude of the *x* component that force?

Number ______ Units ______

the tolerance is +/-2%

Check Your Input

Step 1.5 : *Concept – Evaluate quantities*

To counter that *x* component, should the force on particle 1 due to particle 2 be leftward or rightward?

○ rightward
○ leftward

Check Your Input

Step 1.6 : *Concept – Evaluate quantities*

What is the magnitude of the force on particle 1 due to particle 2?

Number ______ Units ______

the tolerance is +/-2%

Check Your Input

Now that you know how to solve the problem, go back and try again on your own. Close

GO Tutorial

WILEY PLUS **READING MATERIAL** I have written a large number of reading resources for *WileyPLUS*.

- **Every sample problem in the textbook** (both 8e and 9e) is available online in both reading and video formats.
- **Hundreds of additional sample problems.** These are available as stand-alone resources but (at the discretion of the instructor) they are also linked out of the homework problems. So, if a homework problem deals with, say, forces on a block on a ramp, a link to a related sample problem is provided. However, the sample problem is not just a replica of the homework problem and thus does not provide a solution that can be merely duplicated without comprehension.
- GO **GO Tutorials** for 10% of the end-of-chapter homework problems. In multiple steps, I lead a student through a homework problem, starting with the Key Ideas and giving hints when wrong answers are submitted. However, I purposely leave the last step (for the final answer) to the student so that they are responsible at the end. Some online tutorial systems trap a student when wrong answers are given, which can generate a lot of frustration. My GO Tutorials are not traps, because at any step along the way, a student can return to the main problem.
- **Hints on every end-of-chapter homework problem** are available online (at the discretion of the instructor). I wrote these as true hints about the main ideas and the general procedure for a solution, not as recipes that provide an answer without any comprehension.

WILEY PLUS **EVALUATION MATERIALS** Both self-evaluations and instructor evaluations are available.

- **Reading questions are available within each online section.** I wrote these so that they do not require analysis or any deep understanding; rather they simply test whether a student has read the section. When a student opens up a section, a randomly chosen reading question (from a bank of questions) appears at the end. The instructor can decide whether the question is part of the grading for that section or whether it is just for the benefit of the student.

- **Checkpoints are available within most sections.** I wrote these so that they require analysis and decisions about the physics in the section. *Answers to all checkpoints are in the back of the book.*

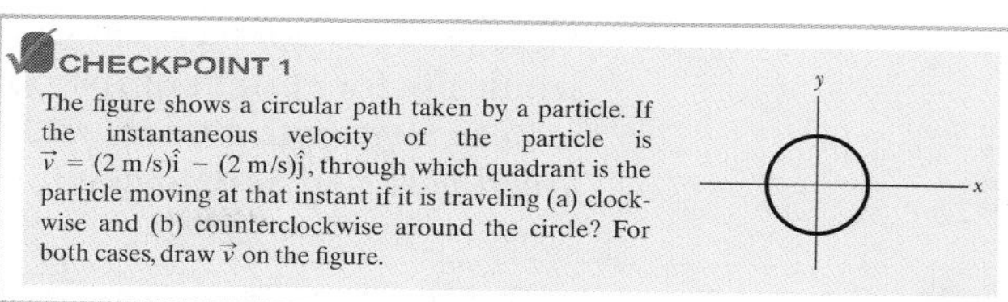

CHECKPOINT 1

The figure shows a circular path taken by a particle. If the instantaneous velocity of the particle is $\vec{v} = (2\text{ m/s})\hat{i} - (2\text{ m/s})\hat{j}$, through which quadrant is the particle moving at that instant if it is traveling (a) clockwise and (b) counterclockwise around the circle? For both cases, draw $\vec{v}$ on the figure.

Checkpoint

- **All end-of-chapter homework questions and problems** in the book (and many more problems) are available in *WileyPLUS*. The instructor can construct a homework assignment and control how it is graded when the answers are submitted online. For example, the instructor controls the deadline for submission and how many attempts a student is allowed on an answer. The instructor also controls which, if any, learning aids are available with each homework problem. Such links can include hints, sample problems, in-chapter reading materials, video tutorials, video math reviews, and even video solutions (which can be made available to the students after, say, a homework deadline).

- **Symbolic notation problems** are available in every chapter and require algebraic answers.

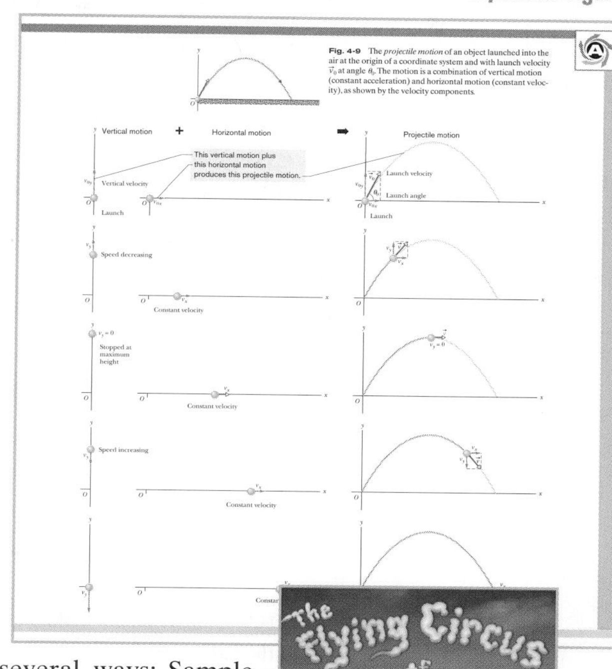

Expanded Figure

WileyPLUS DEMONSTRATIONS AND INTERACTIVE SIMULATIONS

These have been produced by a number of instructors, to provide the experience of a computerized lab and lecture-room demonstrations.

ART PROGRAM

- Many of the figures in the book have been modified to make the physics ideas more pronounced.
- At least one key figure per chapter has been greatly expanded so that its message is conveyed in steps.

FLYING CIRCUS OF PHYSICS

- Flying Circus material has been incorporated into the text in several ways: Sample Problems, text examples, and end-of-chapter Problems. The purpose of this is two-fold: (1) make the subject more interesting and engaging, (2) show the student that the world around them can be examined and understood using the fundamental principles of physics.
- Links to *The Flying Circus of Physics* are shown throughout the text material and end-of-chapter problems with a biplane icon. In the electronic version of this book, clicking on the icon takes you to the corresponding item in *Flying Circus*. The bibliography of *Flying Circus* (over 11 000 references to scientific and engineering journals) is located at www.flyingcircusofphysics.com.

SAMPLE PROBLEMS are chosen to demonstrate how problems can be solved with reasoned solutions rather than quick and simplistic plugging of numbers into an equation with no regard for what the equation means.

KEY IDEAS in the sample problems focus a student on the basic concepts at the root of the solution to a problem. In effect, these key ideas say, "We start our solution by using this basic concept, a procedure that prepares us for solving many other problems. We don't start by grabbing an equation for a quick plug-and-chug, a procedure that prepares us for nothing."

WHAT IS PHYSICS? The narrative of every chapter begins with this question, and with an answer that pertains to the subject of the chapter. (A plumber once asked me, "What do you do for a living?" I replied, "I teach physics." He thought for several minutes and then asked, "What is physics?" The plumber's career was entirely based on physics, yet he did not even know what physics is. Many students in introductory physics do not know what physics is but assume that it is irrelevant to their chosen career.)

ICONS FOR ADDITIONAL HELP. When worked-out solutions are provided either in print or electronically for certain of the odd-numbered problems, the statements for those problems include an icon to alert both student and instructor as to where the solutions are located. An icon guide is provided here and at the beginning of each set of problems

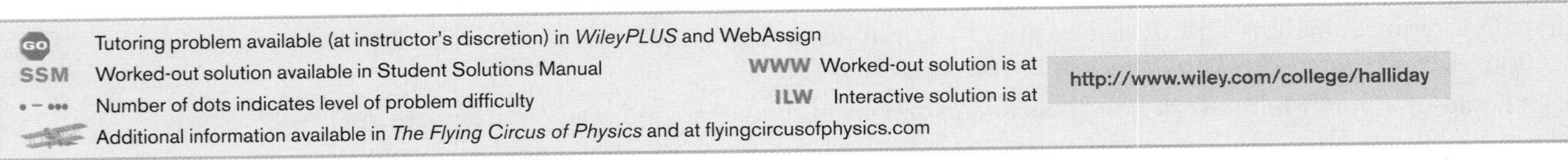

Icon Guide

VERSIONS OF THE TEXT

To accommodate the individual needs of instructors and students, the ninth edition of *Fundamentals of Physics* is available in a number of different versions.

The **Regular Edition** consists of Chapters 1 through 37 (ISBN 978-0-470-04472-8).

The **Extended Edition** contains seven additional chapters on quantum physics and cosmology, Chapters 1–44 (ISBN 978-0-471-75801-3).

Both editions are available as single, hardcover books, or in the following alternative versions:

Volume 1 - Chapters 1–20 (Mechanics and Thermodynamics), hardcover, ISBN 978-0-47004473-5

Volume 2 - Chapters 21–44 (E&M, Optics, and Quantum Physics), hardcover, ISBN 978-0-470-04474-2

INSTRUCTOR SUPPLEMENTS

INSTRUCTOR'S SOLUTIONS MANUAL by Sen-Ben Liao, Lawrence Livermore National Laboratory. This manual provides worked-out solutions for all problems found at the end of each chapter.

INSTRUCTOR COMPANION SITE http://www.wiley.com/college/halliday

- **Instructor's Manual** This resource contains lecture notes outlining the most important topics of each chapter; demonstration experiments; laboratory and computer projects; film and video sources; answers to all Questions, Exercises, Problems, and Checkpoints; and a correlation guide to the Questions, Exercises, and Problems in the previous edition. It also contains a complete list of all problems for which solutions are available to students (SSM,WWW, and ILW).
- **Lecture PowerPoint Slides** by Sudipa Kirtley of The Rose Hulman Institute. These PowerPoint slides serve as a helpful starter pack for instructors, outlining key concepts and incorporating figures and equations from the text.
- **Classroom Response Systems ("Clicker") Questions** by David Marx, Illinois State University. There are two sets of questions available: Reading Quiz questions and Interactive Lecture questions. The Reading Quiz questions are intended to be relatively straightforward for any student who reads the assigned material. The Interactive Lecture questions are intended for use in an interactive lecture setting.
- **Wiley Physics Simulations** by Andrew Duffy, Boston University. This is a collection of 50 interactive simulations (Java applets) that can be used for classroom demonstrations.
- **Wiley Physics Demonstrations** by David Maiullo, Rutgers University. This is a collection of digital videos of 80 standard physics demonstrations. They can be shown in class or accessed from the Student Companion site. There is an accompanying Instructor's Guide that includes "clicker" questions.
- **Test Bank** The Test Bank includes more than 2200 multiple-choice questions. These items are also available in the Computerized Test Bank which provides full editing features to help you customize tests (available in both IBM and Macintosh versions). The Computerized Test Bank is offered in both Diploma and Respondus.
- *Instructor's Solutions Manual,* in both MSWord and PDF files.
- All text illustrations, suitable for both classroom projection and printing.

ONLINE HOMEWORK AND QUIZZING. In addition to *WileyPLUS*, *Fundamentals of Physics,* ninth edition, also supports WebAssignPLUS and LON-CAPA, which are other programs that give instructors the ability to deliver and grade homework and quizzes online. WebAssign PLUS also offers students an online version of the text.

STUDENT SUPPLEMENTS

STUDENT COMPANION SITE. The web site http://www.wiley.com/college/halliday was developed specifically for *Fundamentals of Physics*, ninth edition, and is designed to further assist students in the study of physics. It includes solutions to selected end-of-chapter problems (which are identified with a www icon in the text); self-quizzes; simulation exercises; tips on how to make best use of a programmable calculator; and the Interactive LearningWare tutorials that are described below.

STUDENT STUDY GUIDE by Thomas Barrett of Ohio State University. The Student Study Guide consists of an overview of the chapter's important concepts, problem solving techniques and detailed examples.

STUDENT SOLUTIONS MANUAL by Sen-Ben Liao, Lawrence Livermore National Laboratory. This manual provides students with complete worked-out solutions to 15 percent of the problems found at the end of each chapter within the text. The Student Solutions Manual for the ninth edition is written using an innovative approach called TEAL which stands for Think, Express, Analyze, and Learn. This learning strategy was originally developed at the Massachusetts Institute of Technology and has proven to be an effective learning tool for students. These problems with TEAL solutions are indicated with an SSM icon in the text.

INTERACTIVE LEARNINGWARE. This software guides students through solutions to 200 of the end-of-chapter problems. These problems are indicated with an ILW icon in the text. The solutions process is developed interactively, with appropriate feedback and access to error-specific help for the most common mistakes.

INTRODUCTORY PHYSICS WITH CALCULUS AS A SECOND LANGUAGE: *Mastering Problem Solving* by Thomas Barrett of Ohio State University. This brief paperback teaches the student how to approach problems more efficiently and effectively. The student will learn how to recognize common patterns in physics problems, break problems down into manageable steps, and apply appropriate techniques. The book takes the student step by step through the solutions to numerous examples.

ACKNOWLEDGMENTS

A great many people have contributed to this book. J. Richard Christman, of the U.S. Coast Guard Academy, has once again created many fine supplements; his recommendations to this book have been invaluable. Sen-Ben Liao of Lawrence Livermore National Laboratory, James Whitenton of Southern Polytechnic State University, and Jerry Shi, of Pasadena City College, performed the Herculean task of working out solutions for every one of the homework problems in the book. At John Wiley publishers, the book received support from Stuart Johnson and Geraldine Osnato, the editors who oversaw the entire project from start to finish, and Tom Kulesa, who coordinated the state-of-the-art media package. We thank Elizabeth Swain, the production editor, for pulling all the pieces together during the complex production process. We also thank Maddy Lesure for her design of the text and art direction of the cover; Lee Goldstein for her page make-up; and Lilian Brady for her proofreading. Hilary Newman was inspired in the search for unusual and interesting photographs. Both the publisher John Wiley & Sons, Inc. and Jearl Walker would like to thank the following for comments and ideas about the 8th edition: Jonathan Abramson, Portland State University; Omar Adawi, Parkland College; Edward Adelson, The Ohio State University; Steven R. Baker, Naval Postgraduate School; George Caplan, Wellesley College; Richard Kass, The Ohio State University; M. R. Khoshbin-e-Khoshnazar, Research Institution for Curriculum Development & Educational Innovations (Tehran); Stuart Loucks, American River College; Laurence Lurio, Northern Illinois University; Ponn Maheswaranathan, Winthrop University; Joe McCullough, Cabrillo College; Don N. Page, University of Alberta; Elie Riachi, Fort Scott Community College; Andrew G. Rinzler, University of Florida; Dubravka Rupnik, Louisiana State University; Robert Schabinger, Rutgers University; Ruth

Schwartz, Milwaukee School of Engineering; Nora Thornber, Raritan Valley Community College; Frank Wang, LaGuardia Community College; Graham W. Wilson, University of Kansas; Roland Winkler, Northern Illinois University; Ulrich Zurcher, Cleveland State University. Finally, our external reviewers have been outstanding and we acknowledge here our debt to each member of that team.

Maris A. Abolins, *Michigan State University*
Edward Adelson, *Ohio State University*
Nural Akchurin, *Texas Tech*
Yildirim Aktas, *University of North Carolina- Charlotte*
Barbara Andereck, *Ohio Wesleyan University*
Tetyana Antimirova, *Ryerson University*
Mark Arnett, *Kirkwood Community College*
Arun Bansil, *Northeastern University*
Richard Barber, *Santa Clara University*
Neil Basecu, *Westchester Community College*
Anand Batra, *Howard University*
Richard Bone, *Florida International University*
Michael E. Browne, *University of Idaho*
Timothy J. Burns, *Leeward Community College*
Joseph Buschi, *Manhattan College*
Philip A. Casabella, *Rensselaer Polytechnic Institute*
Randall Caton, *Christopher Newport College*
Roger Clapp, *University of South Florida*
W. R. Conkie, *Queen's University*
Renate Crawford, *University of Massachusetts-Dartmouth*
Mike Crivello, *San Diego State University*
Robert N. Davie, Jr., *St. Petersburg Junior College*
Cheryl K. Dellai, *Glendale Community College*
Eric R. Dietz, *California State University at Chico*
N. John DiNardo, *Drexel University*
Eugene Dunnam, *University of Florida*
Robert Endorf, *University of Cincinnati*
F. Paul Esposito, *University of Cincinnati*
Jerry Finkelstein, *San Jose State University*
Robert H. Good, *California State University-Hayward*
Michael Gorman, *University of Houston*
Benjamin Grinstein, *University of California, San Diego*
John B. Gruber, *San Jose State University*
Ann Hanks, *American River College*
Randy Harris, *University of California-Davis*
Samuel Harris, *Purdue University*
Harold B. Hart, *Western Illinois University*
Rebecca Hartzler, *Seattle Central Community College*
John Hubisz, *North Carolina State University*
Joey Huston, *Michigan State University*
David Ingram, *Ohio University*
Shawn Jackson, *University of Tulsa*
Hector Jimenez, *University of Puerto Rico*
Sudhakar B. Joshi, *York University*
Leonard M. Kahn, *University of Rhode Island*
Sudipa Kirtley, *Rose-Hulman Institute*
Leonard Kleinman, *University of Texas at Austin*
Craig Kletzing, *University of Iowa*
Peter F. Koehler, *University of Pittsburgh*
Arthur Z. Kovacs, *Rochester Institute of Technology*
Kenneth Krane, *Oregon State University*
Priscilla Laws, *Dickinson College*
Edbertho Leal, *Polytechnic University of Puerto Rico*
Vern Lindberg, *Rochester Institute of Technology*
Peter Loly, *University of Manitoba*
James MacLaren, *Tulane University*
Andreas Mandelis, *University of Toronto*
Robert R. Marchini, *Memphis State University*
Andrea Markelz, *University at Buffalo, SUNY*
Paul Marquard, *Caspar College*
David Marx, *Illinois State University*
Dan Mazilu, *Washington and Lee University*
James H. McGuire, *Tulane University*
David M. McKinstry, *Eastern Washington University*
Jordon Morelli, *Queen's University*
Eugene Mosca, *United States Naval Academy*
Eric R. Murray, *Georgia Institute of Technology, School of Physics*
James Napolitano, *Rensselaer Polytechnic Institute*
Blaine Norum, *University of Virginia*
Michael O'Shea, *Kansas State University*
Patrick Papin, *San Diego State University*
Kiumars Parvin, *San Jose State University*
Robert Pelcovits, *Brown University*
Oren P. Quist, *South Dakota State University*
Joe Redish, *University of Maryland*
Timothy M. Ritter, *University of North Carolina at Pembroke*
Dan Styer, *Oberlin College*
Frank Wang, *LaGuardia Community College*

CHAPTER

ELECTRIC CHARGE

21

21-1 WHAT IS PHYSICS?

You are surrounded by devices that depend on the physics of electromagnetism, which is the combination of electric and magnetic phenomena. This physics is at the root of computers, television, radio, telecommunications, household lighting, and even the ability of food wrap to cling to a container. This physics is also the basis of the natural world. Not only does it hold together all the atoms and molecules in the world, it also produces lightning, auroras, and rainbows.

The physics of electromagnetism was first studied by the early Greek philosophers, who discovered that if a piece of amber is rubbed and then brought near bits of straw, the straw will jump to the amber. We now know that the attraction between amber and straw is due to an electric force. The Greek philosophers also discovered that if a certain type of stone (a naturally occurring magnet) is brought near bits of iron, the iron will jump to the stone. We now know that the attraction between magnet and iron is due to a magnetic force.

From these modest origins with the Greek philosophers, the sciences of electricity and magnetism developed separately for centuries—until 1820, in fact, when Hans Christian Oersted found a connection between them: an electric current in a wire can deflect a magnetic compass needle. Interestingly enough, Oersted made this discovery, a big surprise, while preparing a lecture demonstration for his physics students.

The new science of electromagnetism was developed further by workers in many countries. One of the best was Michael Faraday, a truly gifted experimenter with a talent for physical intuition and visualization. That talent is attested to by the fact that his collected laboratory notebooks do not contain a single equation. In the mid-nineteenth century, James Clerk Maxwell put Faraday's ideas into mathematical form, introduced many new ideas of his own, and put electromagnetism on a sound theoretical basis.

Our discussion of electromagnetism is spread through the next 16 chapters. We begin with electrical phenomena, and our first step is to discuss the nature of electric charge and electric force.

21-2 Electric Charge

In dry weather, you can produce a spark by walking across certain types of carpet and then bringing one of your fingers near a metal doorknob, metal faucet, or even a friend. You can also produce multiple sparks when you pull, say, a sweater from your body or clothes from a dryer. Sparks and the "static cling" of clothing (similar to what is seen in Fig. 21-1) are usually just annoying. However, if you happen to pull off a sweater and then spark to a computer, the results are more than just annoying.

Fig. 21-1 Static cling, an electrical phenomenon that accompanies dry weather, causes these pieces of paper to stick to one another and to the plastic comb, and your clothing to stick to your body. *(Fundamental Photographs)*

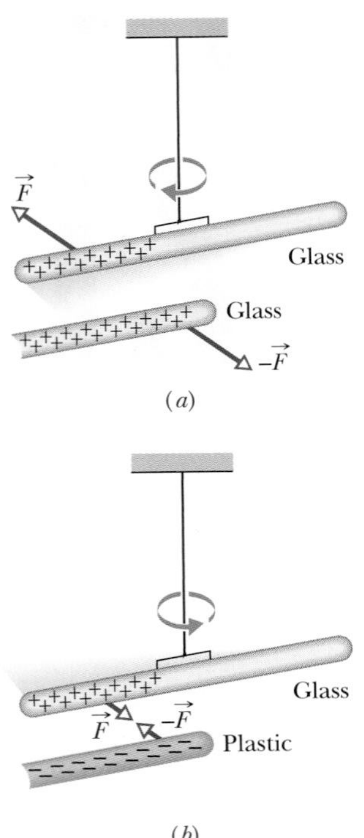

Fig. 21-2 (*a*) Two charged rods of the same sign repel each other. (*b*) Two charged rods of opposite signs attract each other. Plus signs indicate a positive net charge, and minus signs indicate a negative net charge.

These examples reveal that we have electric charge in our bodies, sweaters, carpets, doorknobs, faucets, and computers. In fact, every object contains a vast amount of electric charge. **Electric charge** is an intrinsic characteristic of the fundamental particles making up those objects; that is, it is a property that comes automatically with those particles wherever they exist.

The vast amount of charge in an everyday object is usually hidden because the object contains equal amounts of the two kinds of charge: *positive charge* and *negative charge.* With such an equality—or *balance*—of charge, the object is said to be *electrically neutral;* that is, it contains no *net* charge. If the two types of charge are not in balance, then there *is* a net charge. We say that an object is *charged* to indicate that it has a charge imbalance, or net charge. The imbalance is always much smaller than the total amounts of positive charge and negative charge contained in the object.

Charged objects interact by exerting forces on one another. To show this, we first charge a glass rod by rubbing one end with silk. At points of contact between the rod and the silk, tiny amounts of charge are transferred from one to the other, slightly upsetting the electrical neutrality of each. (We *rub* the silk over the rod to increase the number of contact points and thus the amount, still tiny, of transferred charge.)

Suppose we now suspend the charged rod from a thread to *electrically isolate* it from its surroundings so that its charge cannot change. If we bring a second, similarly charged, glass rod nearby (Fig. 21-2*a*), the two rods *repel* each other; that is, each rod experiences a force directed away from the other rod. However, if we rub a *plastic* rod with fur and then bring the rod near the suspended glass rod (Fig. 21-2*b*), the two rods *attract* each other; that is, each rod experiences a force directed toward the other rod.

We can understand these two demonstrations in terms of positive and negative charges. When a glass rod is rubbed with silk, the glass loses some of its negative charge and then has a small unbalanced positive charge (represented by the plus signs in Fig. 21-2*a*). When the plastic rod is rubbed with fur, the plastic gains a small unbalanced negative charge (represented by the minus signs in Fig. 21-2*b*). Our two demonstrations reveal the following:

> Charges with the same electrical sign repel each other, and charges with opposite electrical signs attract each other.

In Section 21-4, we shall put this rule into quantitative form as Coulomb's law of *electrostatic force* (or *electric force*) between charges. The term *electrostatic* is used to emphasize that, relative to each other, the charges are either stationary or moving only very slowly.

The "positive" and "negative" labels and signs for electric charge were chosen arbitrarily by Benjamin Franklin. He could easily have interchanged the labels or used some other pair of opposites to distinguish the two kinds of charge. (Franklin was a scientist of international reputation. It has even been said that Franklin's triumphs in diplomacy in France during the American War of Independence were facilitated, and perhaps even made possible, because he was so highly regarded as a scientist.)

The attraction and repulsion between charged bodies have many industrial applications, including electrostatic paint spraying and powder coating, fly-ash collection in chimneys, nonimpact ink-jet printing, and photocopying. Figure 21-3 shows a tiny carrier bead in a photocopying machine, covered with particles of black powder called *toner*, which stick to it by means of electrostatic forces. The negatively charged toner particles are eventually attracted from the carrier bead to a rotating drum, where a positively charged image of the document being copied has formed. A charged sheet of paper then attracts the toner particles from the drum to itself, after which they are heat-fused permanently in place to produce the copy.

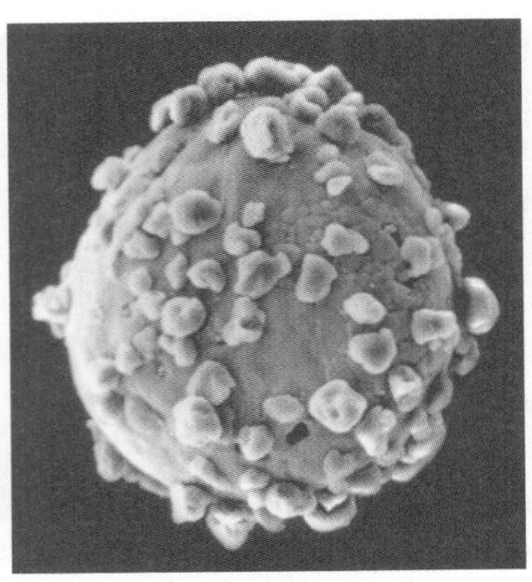

Fig. 21-3 A carrier bead from a photocopying machine; the bead is covered with toner particles that cling to it by electrostatic attraction. The diameter of the bead is about 0.3 mm. *(Courtesy Xerox)*

21-3 Conductors and Insulators

We can classify materials generally according to the ability of charge to move through them. **Conductors** are materials through which charge can move rather freely; examples include metals (such as copper in common lamp wire), the human body, and tap water. **Nonconductors**—also called **insulators**—are materials through which charge cannot move freely; examples include rubber (such as the insulation on common lamp wire), plastic, glass, and chemically pure water. **Semiconductors** are materials that are intermediate between conductors and insulators; examples include silicon and germanium in computer chips. **Superconductors** are materials that are *perfect* conductors, allowing charge to move without *any* hindrance. In these chapters we discuss only conductors and insulators.

Here is an example of how conduction can eliminate excess charge on an object. If you rub a copper rod with wool, charge is transferred from the wool to the rod. However, if you are holding the rod while also touching a faucet, you cannot charge the rod in spite of the transfer. The reason is that you, the rod, and the faucet are all conductors connected, via the plumbing, to Earth's surface, which is a huge conductor. Because the excess charges put on the rod by the wool repel one another, they move away from one another by moving first through the rod, then through you, and then through the faucet and plumbing to reach Earth's surface, where they can spread out. The process leaves the rod electrically neutral.

In thus setting up a pathway of conductors between an object and Earth's surface, we are said to *ground* the object, and in neutralizing the object (by eliminating an unbalanced positive or negative charge), we are said to *discharge* the object. If instead of holding the copper rod in your hand, you hold it by an insulating handle, you eliminate the conducting path to Earth, and the rod can then be charged by rubbing (the charge remains on the rod), as long as you do not touch it directly with your hand.

The properties of conductors and insulators are due to the structure and electrical nature of atoms. Atoms consist of positively charged *protons*, negatively charged *electrons*, and electrically neutral *neutrons*. The protons and neutrons are packed tightly together in a central *nucleus*.

The charge of a single electron and that of a single proton have the same magnitude but are opposite in sign. Hence, an electrically neutral atom contains equal numbers of electrons and protons. Electrons are held near the nucleus because they have the electrical sign opposite that of the protons in the nucleus and thus are attracted to the nucleus.

When atoms of a conductor like copper come together to form the solid, some of their outermost (and so most loosely held) electrons become free to wander about within the solid, leaving behind positively charged atoms (*positive ions*). We call the mobile electrons *conduction electrons*. There are few (if any) free electrons in a nonconductor.

The experiment of Fig. 21-4 demonstrates the mobility of charge in a conductor. A negatively charged plastic rod will attract either end of an isolated neutral

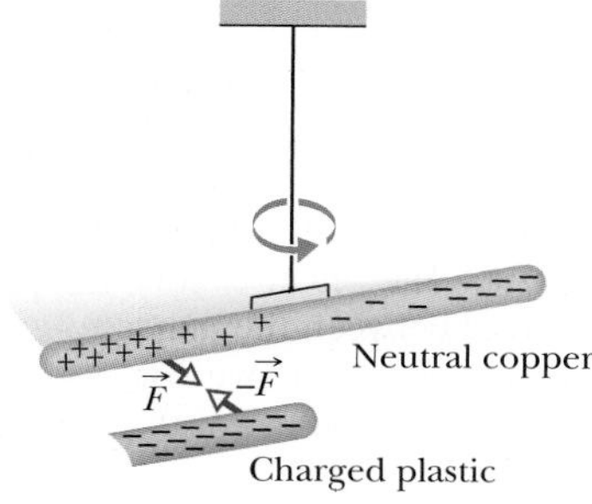

Fig. 21-4 A neutral copper rod is electrically isolated from its surroundings by being suspended on a nonconducting thread. Either end of the copper rod will be attracted by a charged rod. Here, conduction electrons in the copper rod are repelled to the far end of that rod by the negative charge on the plastic rod. Then that negative charge attracts the remaining positive charge on the near end of the copper rod, rotating the copper rod to bring that near end closer to the plastic rod.

Fig. 21-5 Two pieces of a wintergreen LifeSaver candy as they fall away from each other. Electrons jumping from the negative surface of piece A to the positive surface of piece B collide with nitrogen (N_2) molecules in the air.

copper rod. What happens is that many of the conduction electrons in the closer end of the copper rod are repelled by the negative charge on the plastic rod. Some of the conduction electrons move to the far end of the copper rod, leaving the near end depleted in electrons and thus with an unbalanced positive charge. This positive charge is attracted to the negative charge in the plastic rod. Although the copper rod is still neutral, it is said to have an *induced charge*, which means that some of its positive and negative charges have been separated due to the presence of a nearby charge.

Similarly, if a positively charged glass rod is brought near one end of a neutral copper rod, conduction electrons in the copper rod are attracted to that end. That end becomes negatively charged and the other end positively charged, so again an induced charge is set up in the copper rod. Although the copper rod is still neutral, it and the glass rod attract each other.

Note that only conduction electrons, with their negative charges, can move; positive ions are fixed in place. Thus, an object becomes positively charged only through the *removal of negative charges.*

Blue Flashes from a Wintergreen LifeSaver

Indirect evidence for the attraction of charges with opposite signs can be seen with a wintergreen LifeSaver (the candy shaped in the form of a marine lifesaver). If you adapt your eyes to darkness for about 15 minutes and then have a friend chomp on a piece of the candy in the darkness, you will see a faint blue flash from your friend's mouth with each chomp. Whenever a chomp breaks a sugar crystal into pieces, each piece will probably end up with a different number of electrons. Suppose a crystal breaks into pieces A and B, with A ending up with more electrons on its surface than B (Fig. 21-5). This means that B has positive ions (atoms that lost electrons to A) on its surface. Because the electrons on A are strongly attracted to the positive ions on B, some of those electrons jump across the gap between the pieces.

As A and B fall away from each other, air (primarily nitrogen, N_2) flows into the gap, and many of the jumping electrons collide with nitrogen molecules in the air, causing the molecules to emit ultraviolet light. You cannot see this type of light. However, the wintergreen molecules on the surfaces of the candy pieces absorb the ultraviolet light and then emit blue light, which you *can* see—it is the blue light coming from your friend's mouth.

CHECKPOINT 1

The figure shows five pairs of plates: A, B, and D are charged plastic plates and C is an electrically neutral copper plate. The electrostatic forces between the pairs of plates are shown for three of the pairs. For the remaining two pairs, do the plates repel or attract each other?

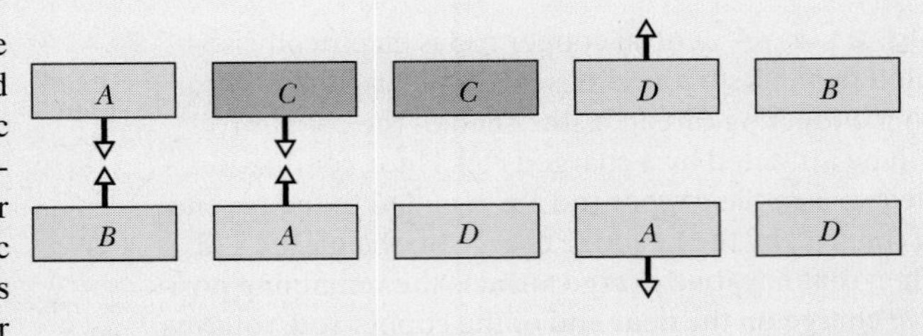

21-4 Coulomb's Law

If two charged particles are brought near each other, they each exert a force on the other. If the particles have the same sign of charge, they repel each other (Figs. 21-6*a* and *b*). That is, the force on each particle is directed away from the other particle, and if the particles can move, they move away from each other. If, instead, the particles have opposite signs of charge, they attract each other (Fig. 21-6*c*) and, if free to move, they move closer to each other.

This force of repulsion or attraction due to the charge properties of objects is called an **electrostatic force.** The equation giving the force for charged *particles* is called **Coulomb's law** after Charles-Augustin de Coulomb, whose experiments in 1785 led him to it. In terms of the particles in Fig. 21-7, where particle 1 has charge q_1 and particle 2 has charge q_2, the force on particle 1 is

$$\vec{F} = k\frac{q_1 q_2}{r^2}\,\hat{r} \qquad \text{(Coulomb's law)}, \qquad (21\text{-}1)$$

in which $\hat{r}$ is a unit vector along an axis extending through the two particles, r is the distance between them, and k is a constant. (As with other unit vectors, $\hat{r}$ has a magnitude of exactly 1 and no dimension or unit; its purpose is to point.) If the particles have the same signs of charge, the force on particle 1 is in the direction of $\hat{r}$; if they have opposite signs, the force is opposite $\hat{r}$.

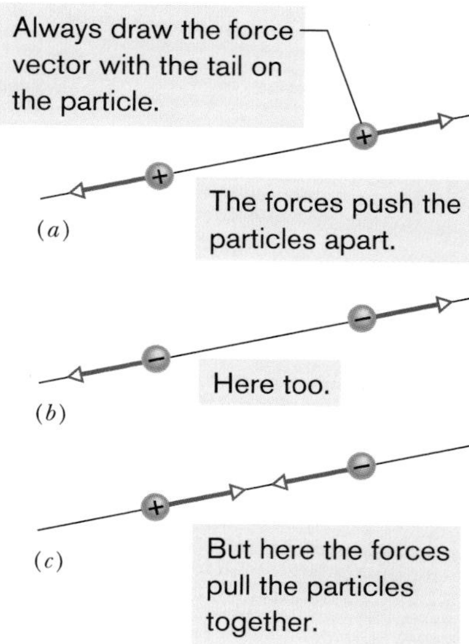

Fig. 21-6 Two charged particles repel each other if they have the same sign of charge, either (*a*) both positive or (*b*) both negative. (*c*) They attract each other if they have opposite signs of charge.

Curiously, the form of Eq. 21-1 is the same as that of Newton's equation (Eq. 13-3) for the gravitational force between two particles with masses m_1 and m_2 that are separated by a distance r:

$$\vec{F} = G\frac{m_1 m_2}{r^2}\,\hat{r} \qquad \text{(Newton's law)}, \qquad (21\text{-}2)$$

in which G is the gravitational constant.

The constant k in Eq. 21-1, by analogy with the gravitational constant G in Eq. 21-2, may be called the *electrostatic constant.* Both equations describe inverse square laws that involve a property of the interacting particles—the mass in one case and the charge in the other. The laws differ in that gravitational forces are always attractive but electrostatic forces may be either attractive or repulsive, depending on the signs of the two charges. This difference arises from the fact that, although there is only one kind of mass, there are two kinds of charge.

Coulomb's law has survived every experimental test; no exceptions to it have ever been found. It holds even within the atom, correctly describing the force between the positively charged nucleus and each of the negatively charged electrons, even though classical Newtonian mechanics fails in that realm and is replaced there by quantum physics. This simple law also correctly accounts for the forces that bind atoms together to form molecules, and for the forces that bind atoms and molecules together to form solids and liquids.

The SI unit of charge is the **coulomb.** For practical reasons having to do with the accuracy of measurements, the coulomb unit is derived from the SI unit *ampere* for *electric current i.* Current is the rate dq/dt at which charge moves past a point or through a region. In Chapter 26 we shall discuss current in detail. Until then we shall use the relation

$$i = \frac{dq}{dt} \qquad \text{(electric current)}, \qquad (21\text{-}3)$$

in which i is the current (in amperes) and dq (in coulombs) is the amount of charge moving past a point or through a region in time dt (in seconds). Rearranging Eq. 21-3 tells us that

$$1\ \text{C} = (1\ \text{A})(1\ \text{s}).$$

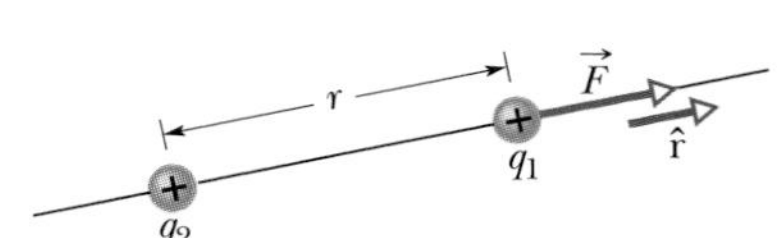

Fig. 21-7 The electrostatic force on particle 1 can be described in terms of a unit vector $\hat{r}$ along an axis through the two particles.

For historical reasons (and because doing so simplifies many other formulas), the electrostatic constant k of Eq. 21-1 is usually written $1/4\pi\varepsilon_0$. Then the magnitude of the force in Coulomb's law becomes

$$F = \frac{1}{4\pi\varepsilon_0}\frac{|q_1||q_2|}{r^2} \quad \text{(Coulomb's law)}. \tag{21-4}$$

The constants in Eqs. 21-1 and 21-4 have the value

$$k = \frac{1}{4\pi\varepsilon_0} = 8.99 \times 10^9\,\text{N}\cdot\text{m}^2/\text{C}^2. \tag{21-5}$$

The quantity ε_0, called the **permittivity constant,** sometimes appears separately in equations and is

$$\varepsilon_0 = 8.85 \times 10^{-12}\,\text{C}^2/\text{N}\cdot\text{m}^2. \tag{21-6}$$

Still another parallel between the gravitational force and the electrostatic force is that both obey the principle of superposition. If we have n charged particles, they interact independently in pairs, and the force on any one of them, let us say particle 1, is given by the vector sum

$$\vec{F}_{1,\text{net}} = \vec{F}_{12} + \vec{F}_{13} + \vec{F}_{14} + \vec{F}_{15} + \cdots + \vec{F}_{1n}, \tag{21-7}$$

in which, for example, $\vec{F}_{14}$ is the force acting on particle 1 due to the presence of particle 4. An identical formula holds for the gravitational force.

Finally, the shell theorem that we found so useful in our study of gravitation has analogs in electrostatics:

A shell of uniform charge attracts or repels a charged particle that is outside the shell as if all the shell's charge were concentrated at its center.

If a charged particle is located inside a shell of uniform charge, there is no net electrostatic force on the particle from the shell.

(In the first theorem, we assume that the charge on the shell is much greater than that of the particle. Then any redistribution of the charge on the shell due to the presence of the particle's charge can be neglected.)

Spherical Conductors

If excess charge is placed on a spherical shell that is made of conducting material, the excess charge spreads uniformly over the (external) surface. For example, if we place excess electrons on a spherical metal shell, those electrons repel one another and tend to move apart, spreading over the available surface until they are uniformly distributed. That arrangement maximizes the distances between all pairs of the excess electrons. According to the first shell theorem, the shell then will attract or repel an external charge as if all the excess charge on the shell were concentrated at its center.

If we remove negative charge from a spherical metal shell, the resulting positive charge of the shell is also spread uniformly over the surface of the shell. For example, if we remove n electrons, there are then n sites of positive charge (sites missing an electron) that are spread uniformly over the shell. According to the first shell theorem, the shell will again attract or repel an external charge as if all the shell's excess charge were concentrated at its center.

CHECKPOINT 2

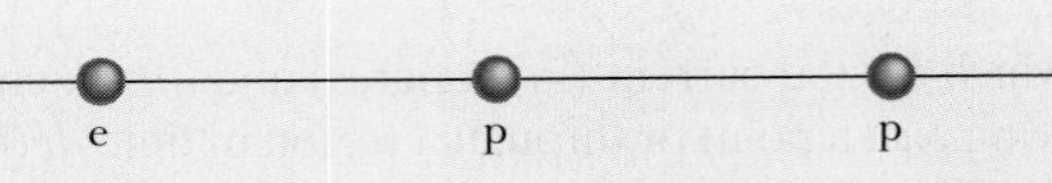

The figure shows two protons (symbol p) and one electron (symbol e) on an axis. What is the direction of (a) the electrostatic force on the central proton due to the electron, (b) the electrostatic force on the central proton due to the other proton, and (c) the net electrostatic force on the central proton?

Sample Problem

Finding the net force due to two other particles

(a) Figure 21-8*a* shows two positively charged particles fixed in place on an *x* axis. The charges are $q_1 = 1.60 \times 10^{-19}$ C and $q_2 = 3.20 \times 10^{-19}$ C, and the particle separation is $R = 0.0200$ m. What are the magnitude and direction of the electrostatic force $\vec{F}_{12}$ on particle 1 from particle 2?

KEY IDEAS

Because both particles are positively charged, particle 1 is repelled by particle 2, with a force magnitude given by Eq. 21-4. Thus, the direction of force $\vec{F}_{12}$ on particle 1 is *away from* particle 2, in the negative direction of the *x* axis, as indicated in the free-body diagram of Fig. 21-8*b*.

Two particles: Using Eq. 21-4 with separation *R* substituted for *r*, we can write the magnitude F_{12} of this force as

$$F_{12} = \frac{1}{4\pi\varepsilon_0}\frac{|q_1||q_2|}{R^2}$$

$$= (8.99 \times 10^9\ \text{N}\cdot\text{m}^2/\text{C}^2)$$

$$\times \frac{(1.60 \times 10^{-19}\ \text{C})(3.20 \times 10^{-19}\ \text{C})}{(0.0200\ \text{m})^2}$$

$$= 1.15 \times 10^{-24}\ \text{N}.$$

Thus, force $\vec{F}_{12}$ has the following magnitude and direction (relative to the positive direction of the *x* axis):

$$1.15 \times 10^{-24}\ \text{N} \quad \text{and} \quad 180°. \qquad \text{(Answer)}$$

We can also write $\vec{F}_{12}$ in unit-vector notation as

$$\vec{F}_{12} = -(1.15 \times 10^{-24}\ \text{N})\hat{\text{i}}. \qquad \text{(Answer)}$$

(b) Figure 21-8*c* is identical to Fig. 21-8*a* except that particle 3 now lies on the *x* axis between particles 1 and 2. Particle 3 has charge $q_3 = -3.20 \times 10^{-19}$ C and is at a distance $\frac{3}{4}R$ from particle 1. What is the net electrostatic force $\vec{F}_{1,\text{net}}$ on particle 1 due to particles 2 and 3?

KEY IDEA

The presence of particle 3 does not alter the electrostatic force on particle 1 from particle 2. Thus, force $\vec{F}_{12}$ still acts on particle 1. Similarly, the force $\vec{F}_{13}$ that acts on particle 1 due to particle 3 is not affected by the presence of particle 2. Because particles 1 and 3 have charge of opposite signs, particle 1 is attracted to particle 3. Thus, force $\vec{F}_{13}$ is directed *toward* particle 3, as indicated in the free-body diagram of Fig. 21-8*d*.

Three particles: To find the magnitude of $\vec{F}_{13}$, we can rewrite Eq. 21-4 as

$$F_{13} = \frac{1}{4\pi\varepsilon_0}\frac{|q_1||q_3|}{(\frac{3}{4}R)^2}$$

$$= (8.99 \times 10^9\ \text{N}\cdot\text{m}^2/\text{C}^2)$$

$$\times \frac{(1.60 \times 10^{-19}\ \text{C})(3.20 \times 10^{-19}\ \text{C})}{(\frac{3}{4})^2(0.0200\ \text{m})^2}$$

$$= 2.05 \times 10^{-24}\ \text{N}.$$

We can also write $\vec{F}_{13}$ in unit-vector notation:

$$\vec{F}_{13} = (2.05 \times 10^{-24}\ \text{N})\hat{\text{i}}.$$

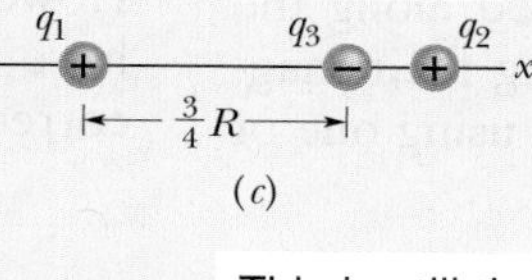

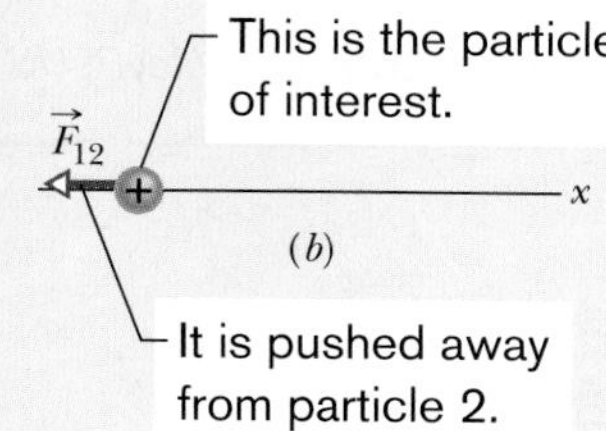

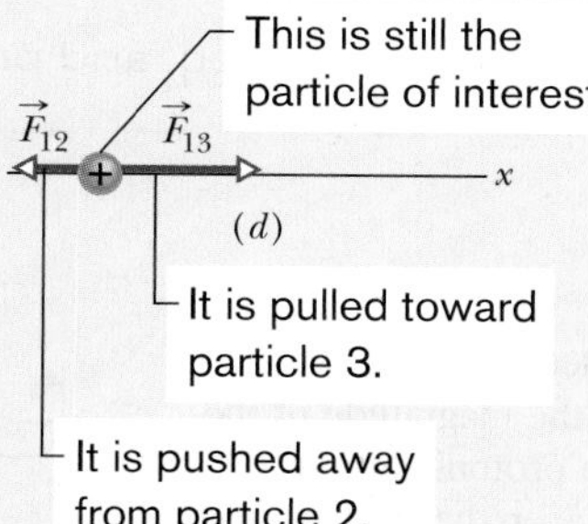

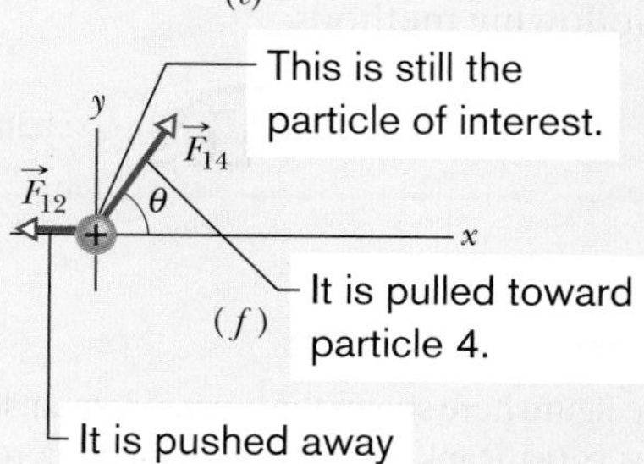

Fig. 21-8 (*a*) Two charged particles of charges q_1 and q_2 are fixed in place on an *x* axis. (*b*) The free-body diagram for particle 1, showing the electrostatic force on it from particle 2. (*c*) Particle 3 included. (*d*) Free-body diagram for particle 1. (*e*) Particle 4 included. (*f*) Free-body diagram for particle 1.

The net force $\vec{F}_{1,\text{net}}$ on particle 1 is the vector sum of $\vec{F}_{12}$ and $\vec{F}_{13}$; that is, from Eq. 21-7, we can write the net force $\vec{F}_{1,\text{net}}$ on particle 1 in unit-vector notation as

$$\begin{aligned}\vec{F}_{1,\text{net}} &= \vec{F}_{12} + \vec{F}_{13}\\ &= -(1.15 \times 10^{-24}\,\text{N})\hat{\text{i}} + (2.05 \times 10^{-24}\,\text{N})\hat{\text{i}}\\ &= (9.00 \times 10^{-25}\,\text{N})\hat{\text{i}}. \qquad \text{(Answer)}\end{aligned}$$

Thus, $\vec{F}_{1,\text{net}}$ has the following magnitude and direction (relative to the positive direction of the x axis):

$$9.00 \times 10^{-25}\,\text{N} \quad \text{and} \quad 0°. \qquad \text{(Answer)}$$

(c) Figure 21-8*e* is identical to Fig. 21-8*a* except that particle 4 is now included. It has charge $q_4 = -3.20 \times 10^{-19}$ C, is at a distance $\frac{3}{4}R$ from particle 1, and lies on a line that makes an angle $\theta = 60°$ with the x axis. What is the net electrostatic force $\vec{F}_{1,\text{net}}$ on particle 1 due to particles 2 and 4?

KEY IDEA

The net force $\vec{F}_{1,\text{net}}$ is the vector sum of $\vec{F}_{12}$ and a new force $\vec{F}_{14}$ acting on particle 1 due to particle 4. Because particles 1 and 4 have charge of opposite signs, particle 1 is attracted to particle 4. Thus, force $\vec{F}_{14}$ on particle 1 is directed *toward* particle 4, at angle $\theta = 60°$, as indicated in the free-body diagram of Fig. 21-8*f*.

Four particles: We can rewrite Eq. 21-4 as

$$\begin{aligned}F_{14} &= \frac{1}{4\pi\varepsilon_0}\frac{|q_1||q_4|}{(\frac{3}{4}R)^2}\\ &= (8.99 \times 10^9\,\text{N}\cdot\text{m}^2/\text{C}^2)\\ &\quad\times \frac{(1.60 \times 10^{-19}\,\text{C})(3.20 \times 10^{-19}\,\text{C})}{(\frac{3}{4})^2(0.0200\,\text{m})^2}\\ &= 2.05 \times 10^{-24}\,\text{N}.\end{aligned}$$

Then from Eq. 21-7, we can write the net force $\vec{F}_{1,\text{net}}$ on particle 1 as

$$\vec{F}_{1,\text{net}} = \vec{F}_{12} + \vec{F}_{14}.$$

Because the forces $\vec{F}_{12}$ and $\vec{F}_{14}$ are not directed along the same axis, we *cannot* sum simply by combining their magnitudes. Instead, we must add them as vectors, using one of the following methods.

Method 1. *Summing directly on a vector-capable calculator.* For $\vec{F}_{12}$, we enter the magnitude 1.15×10^{-24} and the angle 180°. For $\vec{F}_{14}$, we enter the magnitude 2.05×10^{-24} and the angle 60°. Then we add the vectors.

Method 2. *Summing in unit-vector notation.* First we rewrite $\vec{F}_{14}$ as

$$\vec{F}_{14} = (F_{14}\cos\theta)\hat{\text{i}} + (F_{14}\sin\theta)\hat{\text{j}}.$$

Substituting 2.05×10^{-24} N for F_{14} and 60° for θ, this becomes

$$\vec{F}_{14} = (1.025 \times 10^{-24}\,\text{N})\hat{\text{i}} + (1.775 \times 10^{-24}\,\text{N})\hat{\text{j}}.$$

Then we sum:

$$\begin{aligned}\vec{F}_{1,\text{net}} &= \vec{F}_{12} + \vec{F}_{14}\\ &= -(1.15 \times 10^{-24}\,\text{N})\hat{\text{i}}\\ &\quad + (1.025 \times 10^{-24}\,\text{N})\hat{\text{i}} + (1.775 \times 10^{-24}\,\text{N})\hat{\text{j}}\\ &\approx (-1.25 \times 10^{-25}\,\text{N})\hat{\text{i}} + (1.78 \times 10^{-24}\,\text{N})\hat{\text{j}}.\end{aligned}$$

(Answer)

Method 3. *Summing components axis by axis.* The sum of the x components gives us

$$\begin{aligned}F_{1,\text{net},x} &= F_{12,x} + F_{14,x} = F_{12} + F_{14}\cos 60°\\ &= -1.15 \times 10^{-24}\,\text{N} + (2.05 \times 10^{-24}\,\text{N})(\cos 60°)\\ &= -1.25 \times 10^{-25}\,\text{N}.\end{aligned}$$

The sum of the y components gives us

$$\begin{aligned}F_{1,\text{net},y} &= F_{12,y} + F_{14,y} = 0 + F_{14}\sin 60°\\ &= (2.05 \times 10^{-24}\,\text{N})(\sin 60°)\\ &= 1.78 \times 10^{-24}\,\text{N}.\end{aligned}$$

The net force $\vec{F}_{1,\text{net}}$ has the magnitude

$$F_{1,\text{net}} = \sqrt{F^2_{1,\text{net},x} + F^2_{1,\text{net},y}} = 1.78 \times 10^{-24}\,\text{N}. \qquad \text{(Answer)}$$

To find the direction of $\vec{F}_{1,\text{net}}$, we take

$$\theta = \tan^{-1}\frac{F_{1,\text{net},y}}{F_{1,\text{net},x}} = -86.0°.$$

However, this is an unreasonable result because $\vec{F}_{1,\text{net}}$ must have a direction between the directions of $\vec{F}_{12}$ and $\vec{F}_{14}$. To correct θ, we add 180°, obtaining

$$-86.0° + 180° = 94.0°. \qquad \text{(Answer)}$$

Additional examples, video, and practice available at *WileyPLUS*

CHECKPOINT 3

The figure here shows three arrangements of an electron e and two protons p. (a) Rank the arrangements according to the magnitude of the net electrostatic force on the electron due to the protons, largest first. (b) In situation *c*, is the angle between the net force on the electron and the line labeled *d* less than or more than 45°?

(*a*) (*b*) (*c*)

Sample Problem

Equilibrium of two forces on a particle

Figure 21-9*a* shows two particles fixed in place: a particle of charge $q_1 = +8q$ at the origin and a particle of charge $q_2 = -2q$ at $x = L$. At what point (other than infinitely far away) can a proton be placed so that it is in *equilibrium* (the net force on it is zero)? Is that equilibrium *stable* or *unstable*? (That is, if the proton is displaced, do the forces drive it back to the point of equilibrium or drive it farther away?)

KEY IDEA

If $\vec{F}_1$ is the force on the proton due to charge q_1 and $\vec{F}_2$ is the force on the proton due to charge q_2, then the point we seek is where $\vec{F}_1 + \vec{F}_2 = 0$. Thus,

$$\vec{F}_1 = -\vec{F}_2. \qquad (21\text{-}8)$$

This tells us that at the point we seek, the forces acting on the proton due to the other two particles must be of equal magnitudes,

$$F_1 = F_2, \qquad (21\text{-}9)$$

and that the forces must have opposite directions.

Reasoning: Because a proton has a positive charge, the proton and the particle of charge q_1 are of the same sign, and force $\vec{F}_1$ on the proton must point away from q_1. Also, the proton and the particle of charge q_2 are of opposite signs, so force $\vec{F}_2$ on the proton must point toward q_2. "Away from q_1" and "toward q_2" can be in opposite directions only if the proton is located on the x axis.

If the proton is on the x axis at any point between q_1 and q_2, such as point P in Fig. 21-9*b*, then $\vec{F}_1$ and $\vec{F}_2$ are in the same direction and not in opposite directions as required. If the proton is at any point on the x axis to the left of q_1, such as point S in Fig. 21-9*c*, then $\vec{F}_1$ and $\vec{F}_2$ are in opposite directions. However, Eq. 21-4 tells us that $\vec{F}_1$ and $\vec{F}_2$ cannot have equal magnitudes there: F_1 must be greater than F_2, because F_1 is produced by a closer charge (with lesser r) of greater magnitude ($8q$ versus $2q$).

Finally, if the proton is at any point on the x axis to the right of q_2, such as point R in Fig. 21-9*d*, then $\vec{F}_1$ and $\vec{F}_2$ are again in opposite directions. However, because now the charge of greater magnitude (q_1) is *farther* away from the proton than the charge of lesser magnitude, there is a point at which F_1 is equal to F_2. Let x be the coordinate of this point, and let q_p be the charge of the proton.

Calculations: With the aid of Eq. 21-4, we can now rewrite Eq. 21-9 (which says that the forces have equal magnitudes):

$$\frac{1}{4\pi\varepsilon_0}\frac{8qq_p}{x^2} = \frac{1}{4\pi\varepsilon_0}\frac{2qq_p}{(x-L)^2}. \qquad (21\text{-}10)$$

(Note that only the charge magnitudes appear in Eq. 21-10. We already decided about the directions of the forces in drawing Fig. 21-9*d* and do not want to include any positive or negative signs here.) Rearranging Eq. 21-10 gives us

$$\left(\frac{x-L}{x}\right)^2 = \frac{1}{4}.$$

After taking the square roots of both sides, we have

$$\frac{x-L}{x} = \frac{1}{2},$$

which gives us

$$x = 2L. \qquad \text{(Answer)}$$

The equilibrium at $x = 2L$ is unstable; that is, if the proton is displaced leftward from point R, then F_1 and F_2 both increase but F_2 increases more (because q_2 is closer than q_1), and a net force will drive the proton farther leftward. If the proton is displaced rightward, both F_1 and F_2 decrease but F_2 decreases more, and a net force will then drive the proton farther rightward. In a stable equilibrium, if the proton is displaced slightly, it returns to the equilibrium position.

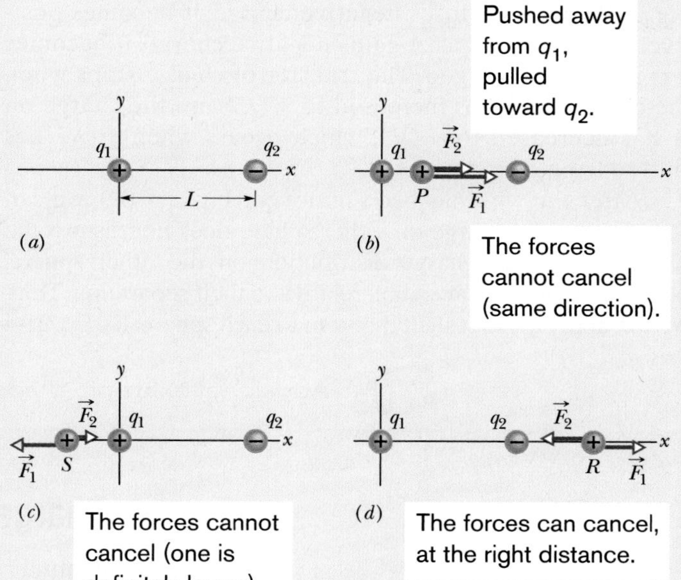

Fig. 21-9 (*a*) Two particles of charges q_1 and q_2 are fixed in place on an x axis, with separation L. (*b*)–(*d*) Three possible locations P, S, and R for a proton. At each location, $\vec{F}_1$ is the force on the proton from particle 1 and $\vec{F}_2$ is the force on the proton from particle 2.

Additional examples, video, and practice available at *WileyPLUS*

Sample Problem

Charge sharing by two identical conducting spheres

In Fig. 21-10*a*, two identical, electrically isolated conducting spheres *A* and *B* are separated by a (center-to-center) distance *a* that is large compared to the spheres. Sphere *A* has a positive charge of $+Q$, and sphere *B* is electrically neutral. Initially, there is no electrostatic force between the spheres. (Assume that there is no induced charge on the spheres because of their large separation.)

(a) Suppose the spheres are connected for a moment by a conducting wire. The wire is thin enough so that any net charge on it is negligible. What is the electrostatic force between the spheres after the wire is removed?

KEY IDEAS

(1) Because the spheres are identical, connecting them means that they end up with identical charges (same sign and same amount). (2) The initial sum of the charges (including the signs of the charges) must equal the final sum of the charges.

Reasoning: When the spheres are wired together, the (negative) conduction electrons on *B*, which repel one another, have a way to move away from one another (along the wire to positively charged *A*, which attracts them—Fig. 21-10*b*.) As *B* loses negative charge, it becomes positively charged, and as *A* gains negative charge, it becomes *less* positively charged. The transfer of charge stops when the charge on *B* has increased to $+Q/2$ and the charge on *A* has decreased to $+Q/2$, which occurs when $-Q/2$ has shifted from *B* to *A*.

After the wire has been removed (Fig. 21-10*c*), we can assume that the charge on either sphere does not disturb the uniformity of the charge distribution on the other sphere, because the spheres are small relative to their separation. Thus, we can apply the first shell theorem to each sphere. By Eq. 21-4

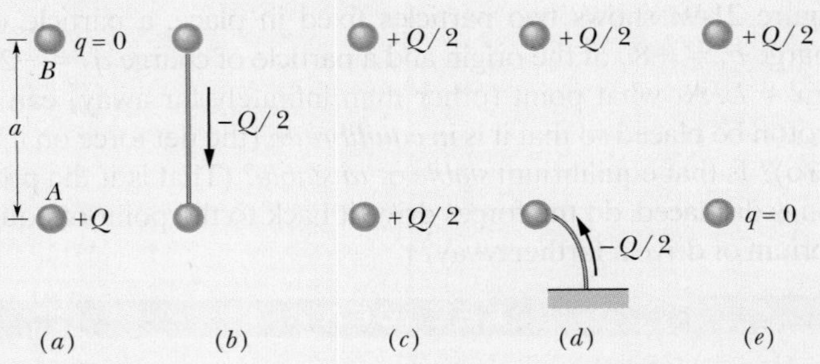

Fig. 21-10 Two small conducting spheres *A* and *B*. (*a*) To start, sphere *A* is charged positively. (*b*) Negative charge is transferred from *B* to *A* through a connecting wire. (*c*) Both spheres are then charged positively. (*d*) Negative charge is transferred through a grounding wire to sphere *A*. (*e*) Sphere *A* is then neutral.

with $q_1 = q_2 = Q/2$ and $r = a$,

$$F = \frac{1}{4\pi\varepsilon_0}\frac{(Q/2)(Q/2)}{a^2} = \frac{1}{16\pi\varepsilon_0}\left(\frac{Q}{a}\right)^2. \quad \text{(Answer)}$$

The spheres, now positively charged, repel each other.

(b) Next, suppose sphere *A* is grounded momentarily, and then the ground connection is removed. What now is the electrostatic force between the spheres?

Reasoning: When we provide a conducting path between a charged object and the ground (which is a huge conductor), we neutralize the object. Were sphere *A* negatively charged, the mutual repulsion between the excess electrons would cause them to move from the sphere to the ground. However, because sphere *A* is positively charged, electrons with a total charge of $-Q/2$ move *from* the ground up onto the sphere (Fig. 21-10*d*), leaving the sphere with a charge of 0 (Fig. 21-10*e*). Thus, there is (again) no electrostatic force between the two spheres.

Additional examples, video, and practice available at *WileyPLUS*

21-5 Charge Is Quantized

In Benjamin Franklin's day, electric charge was thought to be a continuous fluid—an idea that was useful for many purposes. However, we now know that fluids themselves, such as air and water, are not continuous but are made up of atoms and molecules; matter is discrete. Experiment shows that "electrical fluid" is also not continuous but is made up of multiples of a certain elementary charge. Any positive or negative charge *q* that can be detected can be written as

$$q = ne, \qquad n = \pm 1, \pm 2, \pm 3, \ldots, \tag{21-11}$$

in which *e*, the **elementary charge,** has the approximate value

$$e = 1.602 \times 10^{-19}\ \text{C}. \tag{21-12}$$

The elementary charge e is one of the important constants of nature. The electron and proton both have a charge of magnitude e (Table 21-1). (Quarks, the constituent particles of protons and neutrons, have charges of $\pm e/3$ or $\pm 2e/3$, but they apparently cannot be detected individually. For this and for historical reasons, we do not take their charges to be the elementary charge.)

You often see phrases—such as "the charge on a sphere," "the amount of charge transferred," and "the charge carried by the electron"—that suggest that charge is a substance. (Indeed, such statements have already appeared in this chapter.) You should, however, keep in mind what is intended: *Particles* are the substance and charge happens to be one of their properties, just as mass is.

When a physical quantity such as charge can have only discrete values rather than any value, we say that the quantity is **quantized.** It is possible, for example, to find a particle that has no charge at all or a charge of $+10e$ or $-6e$, but not a particle with a charge of, say, $3.57e$.

The quantum of charge is small. In an ordinary 100 W lightbulb, for example, about 10^{19} elementary charges enter the bulb every second and just as many leave. However, the graininess of electricity does not show up in such large-scale phenomena (the bulb does not flicker with each electron), just as you cannot feel the individual molecules of water with your hand.

Table 21-1

The Charges of Three Particles

Particle	Symbol	Charge
Electron	e or e^-	$-e$
Proton	p	$+e$
Neutron	n	0

CHECKPOINT 4

Initially, sphere A has a charge of $-50e$ and sphere B has a charge of $+20e$. The spheres are made of conducting material and are identical in size. If the spheres then touch, what is the resulting charge on sphere A?

Sample Problem

Mutual electric repulsion in a nucleus

The nucleus in an iron atom has a radius of about 4.0×10^{-15} m and contains 26 protons.

(a) What is the magnitude of the repulsive electrostatic force between two of the protons that are separated by 4.0×10^{-15} m?

KEY IDEA

The protons can be treated as charged particles, so the magnitude of the electrostatic force on one from the other is given by Coulomb's law.

Calculation: Table 21-1 tells us that the charge of a proton is $+e$. Thus, Eq. 21-4 gives us

$$F = \frac{1}{4\pi\varepsilon_0}\frac{e^2}{r^2}$$

$$= \frac{(8.99 \times 10^9\ \mathrm{N\cdot m^2/C^2})(1.602 \times 10^{-19}\ \mathrm{C})^2}{(4.0 \times 10^{-15}\ \mathrm{m})^2}$$

$$= 14\ \mathrm{N}. \qquad \text{(Answer)}$$

No explosion: This is a small force to be acting on a macroscopic object like a cantaloupe, but an enormous force to be acting on a proton. Such forces should explode the nucleus of any element but hydrogen (which has only one proton in its nucleus). However, they don't, not even in nuclei with a great many protons. Therefore, there must be some enormous attractive force to counter this enormous repulsive electrostatic force.

(b) What is the magnitude of the gravitational force between those same two protons?

KEY IDEA

Because the protons are particles, the magnitude of the gravitational force on one from the other is given by Newton's equation for the gravitational force (Eq. 21-2).

Calculation: With m_p ($= 1.67 \times 10^{-27}$ kg) representing the mass of a proton, Eq. 21-2 gives us

$$F = G\frac{m_p^2}{r^2}$$

$$= \frac{(6.67 \times 10^{-11}\ \mathrm{N\cdot m^2/kg^2})(1.67 \times 10^{-27}\ \mathrm{kg})^2}{(4.0 \times 10^{-15}\ \mathrm{m})^2}$$

$$= 1.2 \times 10^{-35}\ \mathrm{N}. \qquad \text{(Answer)}$$

Weak versus strong: This result tells us that the (attractive) gravitational force is far too weak to counter the repulsive electrostatic forces between protons in a nucleus. Instead, the protons are bound together by an enormous force called (aptly) the *strong nuclear force*—a force that acts between protons (and neutrons) when they are close together, as in a nucleus.

Although the gravitational force is many times weaker than the electrostatic force, it is more important in large-scale situations because it is always attractive. This means that it can collect many small bodies into huge bodies with huge masses, such as planets and stars, that then exert large gravitational forces. The electrostatic force, on the other hand, is repulsive for charges of the same sign, so it is unable to collect either positive charge or negative charge into large concentrations that would then exert large electrostatic forces.

Additional examples, video, and practice available at *WileyPLUS*

21-6 Charge Is Conserved

If you rub a glass rod with silk, a positive charge appears on the rod. Measurement shows that a negative charge of equal magnitude appears on the silk. This suggests that rubbing does not create charge but only transfers it from one body to another, upsetting the electrical neutrality of each body during the process. This hypothesis of **conservation of charge,** first put forward by Benjamin Franklin, has stood up under close examination, both for large-scale charged bodies and for atoms, nuclei, and elementary particles. No exceptions have ever been found. Thus, we add electric charge to our list of quantities—including energy and both linear and angular momentum—that obey a conservation law.

Important examples of the conservation of charge occur in the *radioactive decay* of nuclei, in which a nucleus transforms into (becomes) a different type of nucleus. For example, a uranium-238 nucleus (^{238}U) transforms into a thorium-234 nucleus (^{234}Th) by emitting an *alpha particle.* Because that particle has the same makeup as a helium-4 nucleus, it has the symbol ^{4}He. The number used in the name of a nucleus and as a superscript in the symbol for the nucleus is called the *mass number* and is the total number of the protons and neutrons in the nucleus. For example, the total number in ^{238}U is 238. The number of protons in a nucleus is the *atomic number Z*, which is listed for all the elements in Appendix F. From that list we find that in the decay

$$^{238}\text{U} \rightarrow {}^{234}\text{Th} + {}^{4}\text{He}, \tag{21-13}$$

the *parent* nucleus ^{238}U contains 92 protons (a charge of $+92e$), the *daughter* nucleus ^{234}Th contains 90 protons (a charge of $+90e$), and the emitted alpha particle ^{4}He contains 2 protons (a charge of $+2e$). We see that the total charge is $+92e$ before and after the decay; thus, charge is conserved. (The total number of protons and neutrons is also conserved: 238 before the decay and $234 + 4 = 238$ after the decay.)

Another example of charge conservation occurs when an electron e^- (charge $-e$) and its antiparticle, the *positron* e^+ (charge $+e$), undergo an *annihilation process,* transforming into two *gamma rays* (high-energy light):

$$e^- + e^+ \rightarrow \gamma + \gamma \qquad \text{(annihilation)}. \tag{21-14}$$

In applying the conservation-of-charge principle, we must add the charges algebraically, with due regard for their signs. In the annihilation process of Eq. 21-14 then, the net charge of the system is zero both before and after the event. Charge is conserved.

In *pair production*, the converse of annihilation, charge is also conserved. In this process a gamma ray transforms into an electron and a positron:

$$\gamma \rightarrow e^- + e^+ \qquad \text{(pair production)}. \tag{21-15}$$

Figure 21-11 shows such a pair-production event that occurred in a bubble cham-

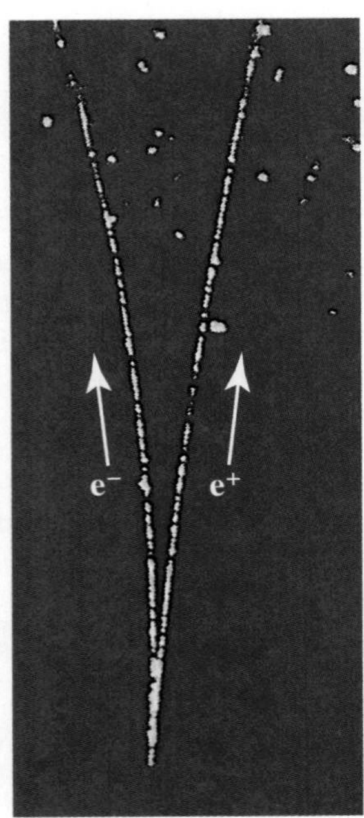

Fig. 21-11 A photograph of trails of bubbles left in a bubble chamber by an electron and a positron. The pair of particles was produced by a gamma ray that entered the chamber directly from the bottom. Being electrically neutral, the gamma ray did not generate a telltale trail of bubbles along its path, as the electron and positron did. *(Courtesy Lawrence Berkeley Laboratory)*

ber. A gamma ray entered the chamber from the bottom and at one point transformed into an electron and a positron. Because those new particles were charged and moving, each left a trail of tiny bubbles. (The trails were curved because a magnetic field had been set up in the chamber.) The gamma ray, being electrically neutral, left no trail. Still, you can tell exactly where it underwent pair production—at the tip of the curved V, which is where the trails of the electron and positron begin.

REVIEW & SUMMARY

Electric Charge The strength of a particle's electrical interaction with objects around it depends on its **electric charge,** which can be either positive or negative. Charges with the same sign repel each other, and charges with opposite signs attract each other. An object with equal amounts of the two kinds of charge is electrically neutral, whereas one with an imbalance is electrically charged.

Conductors are materials in which a significant number of charged particles (electrons in metals) are free to move. The charged particles in **nonconductors,** or **insulators,** are not free to move.

The Coulomb and Ampere The SI unit of charge is the **coulomb** (C). It is defined in terms of the unit of current, the ampere (A), as the charge passing a particular point in 1 second when there is a current of 1 ampere at that point:

$$1\text{ C} = (1\text{ A})(1\text{ s}).$$

This is based on the relation between current i and the rate dq/dt at which charge passes a point:

$$i = \frac{dq}{dt} \quad \text{(electric current).} \qquad (21\text{-}3)$$

Coulomb's Law *Coulomb's law* describes the **electrostatic force** between small (point) electric charges q_1 and q_2 at rest (or nearly at rest) and separated by a distance r:

$$F = \frac{1}{4\pi\varepsilon_0}\frac{|q_1||q_2|}{r^2} \quad \text{(Coulomb's law).} \qquad (21\text{-}4)$$

Here $\varepsilon_0 = 8.85 \times 10^{-12}\ \text{C}^2/\text{N}\cdot\text{m}^2$ is the **permittivity constant,** and $1/4\pi\varepsilon_0 = k = 8.99 \times 10^9\ \text{N}\cdot\text{m}^2/\text{C}^2$.

The force of attraction or repulsion between point charges at rest acts along the line joining the two charges. If more than two charges are present, Eq. 21-4 holds for each pair of charges. The net force on each charge is then found, using the superposition principle, as the vector sum of the forces exerted on the charge by all the others.

The two shell theorems for electrostatics are

A shell of uniform charge attracts or repels a charged particle that is outside the shell as if all the shell's charge were concentrated at its center.

If a charged particle is located inside a shell of uniform charge, there is no net electrostatic force on the particle from the shell.

The Elementary Charge Electric charge is **quantized:** any charge can be written as ne, where n is a positive or negative integer and e is a constant of nature called the **elementary charge** ($\approx 1.602 \times 10^{-19}$ C). Electric charge is **conserved:** the net charge of any isolated system cannot change.

QUESTIONS

1 Figure 21-12 shows four situations in which five charged particles are evenly spaced along an axis. The charge values are indicated except for the central particle, which has the same charge in all four situations. Rank the situations according to the magnitude of the net electrostatic force on the central particle, greatest first.

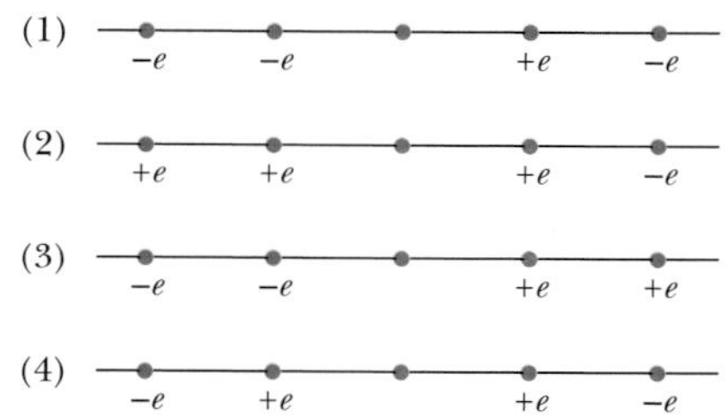

Fig. 21-12 Question 1.

2 Figure 21-13 shows three pairs of identical spheres that are to be touched together and then separated. The initial charges on them are indicated. Rank the pairs according to (a) the magnitude of the charge transferred during touching and (b) the charge left on the positively charged sphere, greatest first.

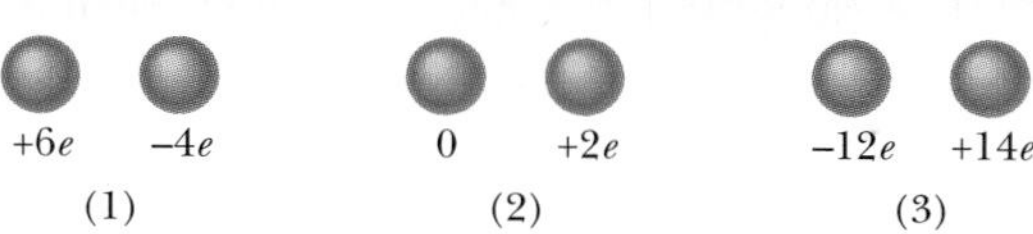

Fig. 21-13 Question 2.

3 Figure 21-14 shows four situations in which charged particles are fixed in place on an axis. In which situations is there a point to the left of the particles where an electron will be in equilibrium?

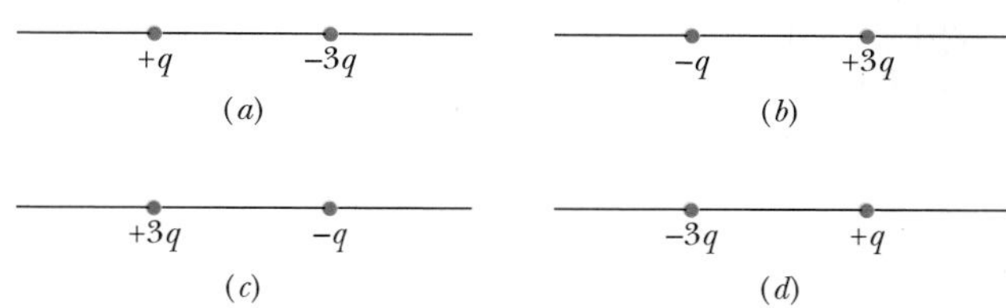

Fig. 21-14 Question 3.

4 Figure 21-15 shows two charged particles on an axis. The charges are free to move. However, a third charged particle can be placed at a certain point such that all three particles are then in equilibrium. (a) Is that point to the left of the first two particles, to their right, or between them? (b) Should the third particle be positively or negatively charged? (c) Is the equilibrium stable or unstable?

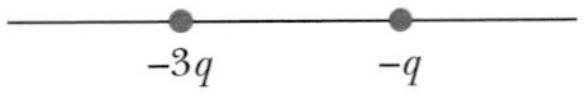

Fig. 21-15 Question 4.

5 In Fig. 21-16, a central particle of charge $-q$ is surrounded by two circular rings of charged particles. What are the magnitude and direction of the net electrostatic force on the central particle due to the other particles? (*Hint:* Consider symmetry.)

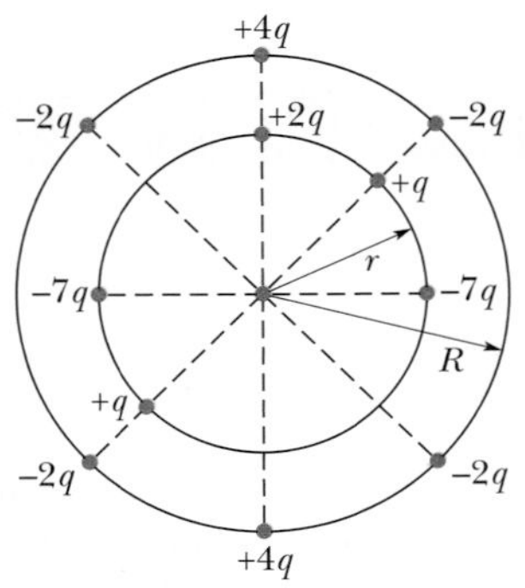

Fig. 21-16 Question 5.

6 A positively charged ball is brought close to an electrically neutral isolated conductor. The conductor is then grounded while the ball is kept close. Is the conductor charged positively, charged negatively, or neutral if (a) the ball is first taken away and then the ground connection is removed and (b) the ground connection is first removed and then the ball is taken away?

7 Figure 21-17 shows three situations involving a charged particle and a uniformly charged spherical shell. The charges are given, and the radii of the shells are indicated. Rank the situations according to the magnitude of the force on the particle due to the presence of the shell, greatest first.

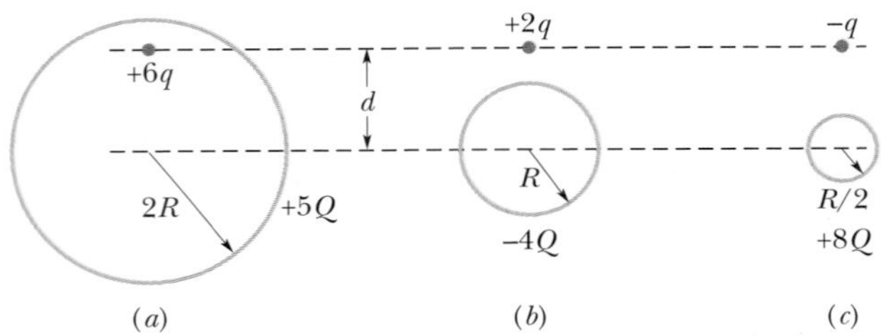

Fig. 21-17 Question 7.

8 Figure 21-18 shows four arrangements of charged particles. Rank the arrangements according to the magnitude of the net electrostatic force on the particle with charge $+Q$, greatest first.

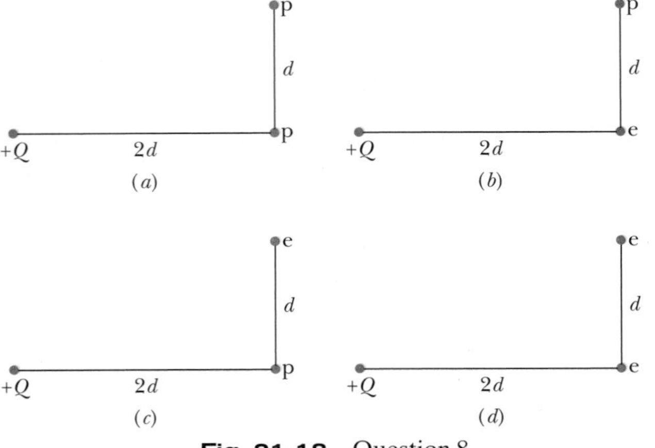

Fig. 21-18 Question 8.

9 Figure 21-19 shows four situations in which particles of charge $+q$ or $-q$ are fixed in place. In each situation, the particles on the x axis are equidistant from the y axis. First, consider the middle particle in situation 1; the middle particle experiences an electrostatic force from each of the other two particles. (a) Are the magnitudes F of those forces the same or different? (b) Is the magnitude of the net force on the middle particle equal to, greater than, or less than $2F$? (c) Do the x components of the two forces add or cancel? (d) Do their y components add or cancel? (e) Is the direction of the net force on the middle particle that of the canceling components or the adding components? (f) What is the direction of that net force? Now consider the remaining situations: What is the direction of the net force on the middle particle in (g) situation 2, (h) situation 3, and (i) situation 4? (In each situation, consider the symmetry of the charge distribution and determine the canceling components and the adding components.)

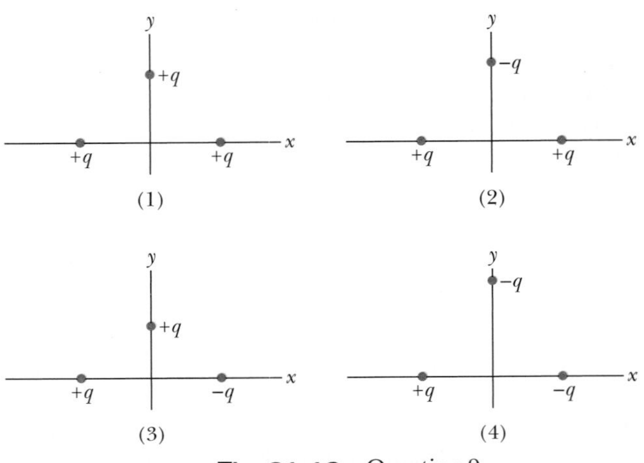

Fig. 21-19 Question 9.

10 In Fig. 21-20, a central particle of charge $-2q$ is surrounded by a square array of charged particles, separated by either distance d or $d/2$ along the perimeter of the square. What are the magnitude and direction of the net electrostatic force on the central particle due to the other particles? (*Hint:* Consideration of symmetry can greatly reduce the amount of work required here.)

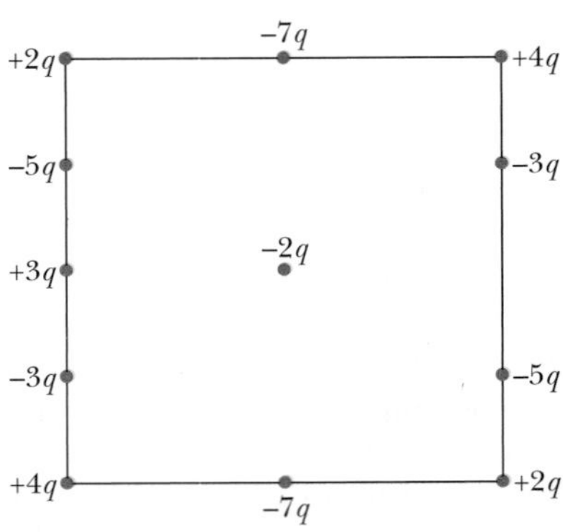

Fig. 21-20 Question 10.

PROBLEMS

Tutoring problem available (at instructor's discretion) in *WileyPLUS* and WebAssign

SSM Worked-out solution available in Student Solutions Manual

WWW Worked-out solution is at

ILW Interactive solution is at

http://www.wiley.com/college/halliday

• – ••• Number of dots indicates level of problem difficulty

Additional information available in *The Flying Circus of Physics* and at flyingcircusofphysics.com

sec. 21-4 Coulomb's Law

•1 SSM ILW Of the charge Q initially on a tiny sphere, a portion q is to be transferred to a second, nearby sphere. Both spheres can be treated as particles. For what value of q/Q will the electrostatic force between the two spheres be maximized?

•2 Identical isolated conducting spheres 1 and 2 have equal charges and are separated by a distance that is large compared with their diameters (Fig. 21-21*a*). The electrostatic force acting on sphere 2 due to sphere 1 is $\vec{F}$. Suppose now that a third identical sphere 3, having an insulating handle and initially neutral, is touched first to sphere 1 (Fig. 21-21*b*), then to sphere 2 (Fig. 21-21*c*), and finally removed (Fig. 21-21*d*). The electrostatic force that now acts on sphere 2 has magnitude F'. What is the ratio F'/F?

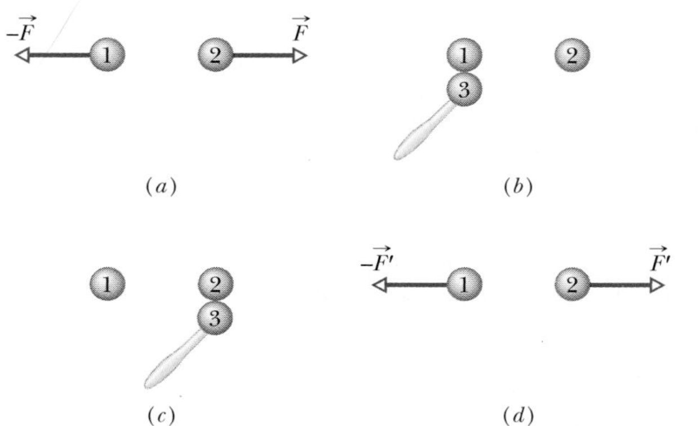

Fig. 21-21 Problem 2.

•3 SSM What must be the distance between point charge $q_1 = 26.0\ \mu C$ and point charge $q_2 = -47.0\ \mu C$ for the electrostatic force between them to have a magnitude of 5.70 N?

•4 In the return stroke of a typical lightning bolt, a current of 2.5×10^4 A exists for 20 μs. How much charge is transferred in this event?

•5 A particle of charge $+3.00 \times 10^{-6}$ C is 12.0 cm distant from a second particle of charge -1.50×10^{-6} C. Calculate the magnitude of the electrostatic force between the particles.

•6 ILW Two equally charged particles are held 3.2×10^{-3} m apart and then released from rest. The initial acceleration of the first particle is observed to be 7.0 m/s² and that of the second to be 9.0 m/s². If the mass of the first particle is 6.3×10^{-7} kg, what are (a) the mass of the second particle and (b) the magnitude of the charge of each particle?

••7 In Fig. 21-22, three charged particles lie on an x axis. Particles 1 and 2 are fixed in place. Particle 3 is free to move, but the net electrostatic force on it from particles 1 and 2 happens to be zero. If $L_{23} = L_{12}$, what is the ratio q_1/q_2?

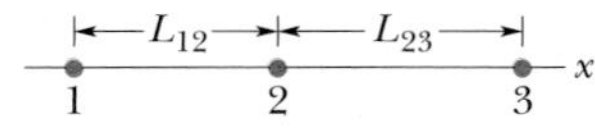

Fig. 21-22 Problems 7 and 40.

••8 In Fig. 21-23, three identical conducting spheres initially have the following charges: sphere A, $4Q$; sphere B, $-6Q$; and sphere C, 0. Spheres A and B are fixed in place, with a center-to-center separation that is much larger than the spheres. Two experiments are conducted. In experiment 1, sphere C is touched to sphere A and then (separately) to sphere B, and then it is removed. In experiment 2, starting with the same initial states, the procedure is reversed: Sphere C is touched to sphere B and then (separately) to sphere A, and then it is removed. What is the ratio of the electrostatic force between A and B at the end of experiment 2 to that at the end of experiment 1?

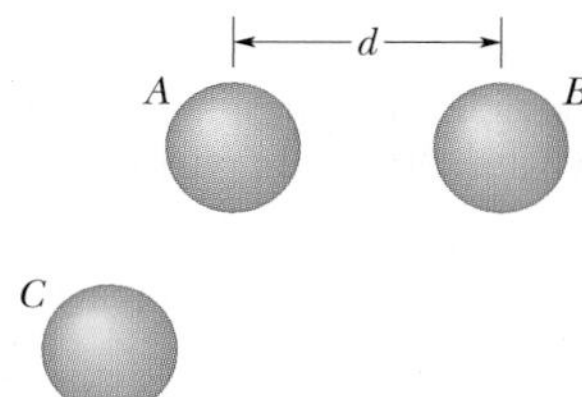

Fig. 21-23 Problems 8 and 65.

••9 SSM WWW Two identical conducting spheres, fixed in place, attract each other with an electrostatic force of 0.108 N when their center-to-center separation is 50.0 cm. The spheres are then connected by a thin conducting wire. When the wire is removed, the spheres repel each other with an electrostatic force of 0.0360 N. Of the initial charges on the spheres, with a positive net charge, what was (a) the negative charge on one of them and (b) the positive charge on the other?

••10 GO In Fig. 21-24, four particles form a square. The charges are $q_1 = q_4 = Q$ and $q_2 = q_3 = q$. (a) What is Q/q if the net electrostatic force on particles 1 and 4 is zero? (b) Is there any value of q that makes the net electrostatic force on each of the four particles zero? Explain.

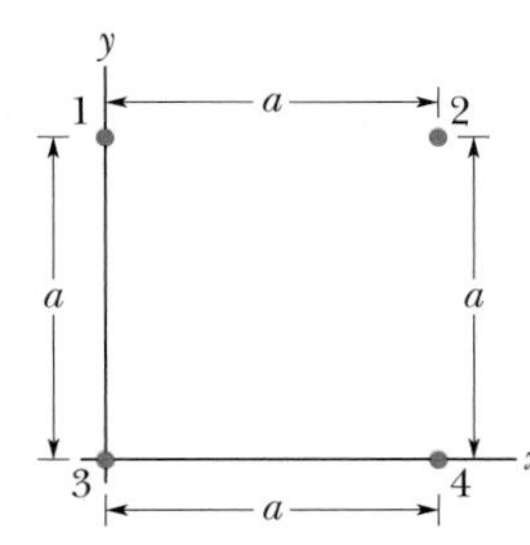

Fig. 21-24 Problems 10, 11, and 70.

••11 ILW In Fig. 21-24, the particles have charges $q_1 = -q_2 = 100$ nC and $q_3 = -q_4 = 200$ nC, and distance $a = 5.0$ cm. What are the (a) x and (b) y components of the net electrostatic force on particle 3?

••12 Two particles are fixed on an x axis. Particle 1 of charge 40 μC is located at $x = -2.0$ cm; particle 2 of charge Q is located at $x = 3.0$ cm. Particle 3 of charge magnitude 20 μC is released from rest on the y axis at $y = 2.0$ cm. What is the value of Q if the initial acceleration of particle 3 is in the positive direction of (a) the x axis and (b) the y axis?

••13 GO In Fig. 21-25, particle 1 of charge $+1.0\ \mu C$ and particle 2 of charge $-3.0\ \mu C$ are held at separation $L = 10.0$ cm on an x axis. If particle 3 of unknown charge q_3 is to be located such that the net electrostatic force on it from particles 1 and 2 is zero, what must be the (a) x and (b) y coordinates of particle 3?

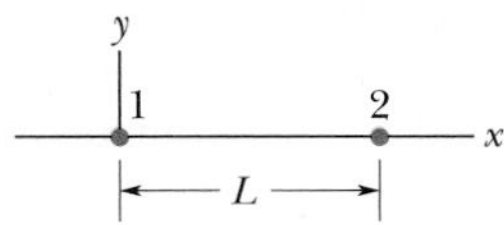

Fig. 21-25 Problems 13, 19, 30, 58, and 67.

••14 Three particles are fixed on an x axis. Particle 1 of charge q_1 is at $x = -a$, and particle 2 of charge q_2 is at $x = +a$. If their net electrostatic force on particle 3 of charge $+Q$ is to be zero, what must be the ratio q_1/q_2 when particle 3 is at (a) $x = +0.500a$ and (b) $x = +1.50a$?

••15 GO The charges and coordinates of two charged particles held fixed in an xy plane are $q_1 = +3.0\ \mu C$, $x_1 = 3.5$ cm, $y_1 = 0.50$ cm, and $q_2 = -4.0\ \mu C$, $x_2 = -2.0$ cm, $y_2 = 1.5$ cm. Find the (a) magnitude and (b) direction of the electrostatic force on particle 2 due to particle 1. At what (c) x and (d) y coordinates should a third particle of charge $q_3 = +4.0\ \mu C$ be placed such that the net electrostatic force on particle 2 due to particles 1 and 3 is zero?

••16 GO In Fig. 21-26a, particle 1 (of charge q_1) and particle 2 (of charge q_2) are fixed in place on an x axis, 8.00 cm apart. Particle 3 (of charge $q_3 = +8.00 \times 10^{-19}$ C) is to be placed on the line between particles 1 and 2 so that they produce a net electrostatic force $\vec{F}_{3,net}$ on it. Figure 21-26b gives the x component of that force versus the coordinate x at which particle 3 is placed. The scale of the x axis is set by $x_s = 8.0$ cm. What are (a) the sign of charge q_1 and (b) the ratio q_2/q_1?

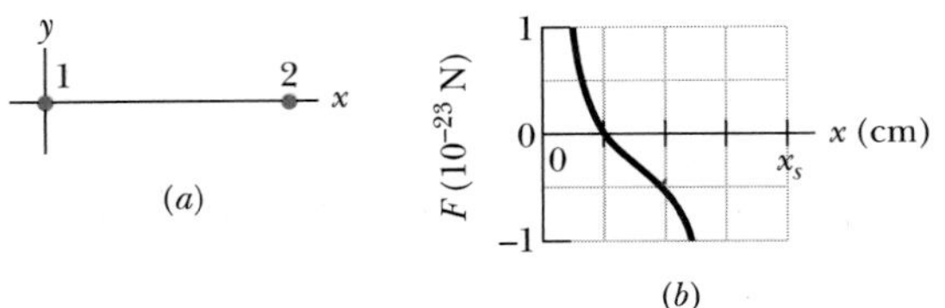

Fig. 21-26 Problem 16.

••17 In Fig. 21-27a, particles 1 and 2 have charge 20.0 μC each and are held at separation distance $d = 1.50$ m. (a) What is the magnitude of the electrostatic force on particle 1 due to particle 2? In Fig. 21-27b, particle 3 of charge 20.0 μC is positioned so as to complete an equilateral triangle. (b) What is the magnitude of the net electrostatic force on particle 1 due to particles 2 and 3?

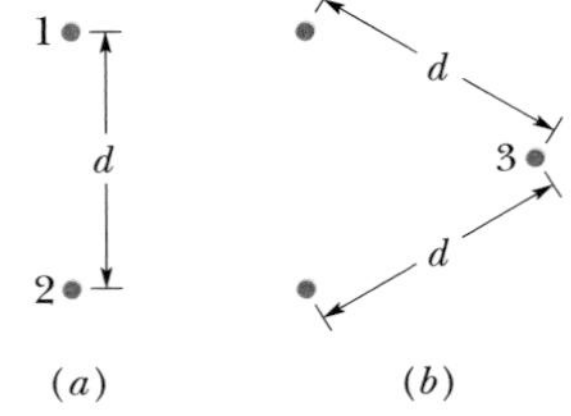

Fig. 21-27 Problem 17.

••18 In Fig. 21-28a, three positively charged particles are fixed on an x axis. Particles B and C are so close to each other that they can be considered to be at the same distance from particle A. The net force on particle A due to particles B and C is 2.014×10^{-23} N in the negative direction of the x axis. In Fig. 21-28b, particle B has been moved to the opposite side of A but is still at the same distance from it. The net force on A is now 2.877×10^{-24} N in the negative direction of the x axis. What is the ratio q_C/q_B?

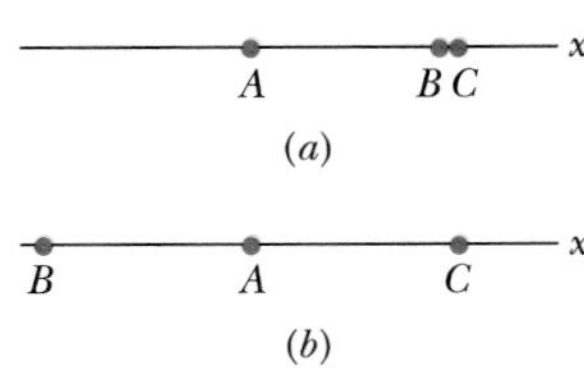

Fig. 21-28 Problem 18.

••19 SSM WWW In Fig. 21-25, particle 1 of charge $+q$ and particle 2 of charge $+4.00q$ are held at separation $L = 9.00$ cm on an x axis. If particle 3 of charge q_3 is to be located such that the three particles remain in place when released, what must be the (a) x and (b) y coordinates of particle 3, and (c) the ratio q_3/q?

•••20 Figure 21-29a shows an arrangement of three charged particles separated by distance d. Particles A and C are fixed on the x axis, but particle B can be moved along a circle centered on particle A. During the movement, a radial line between A and B makes an angle θ relative to the positive direction of the x axis (Fig. 21-29b). The curves in Fig. 21-29c give, for two situations, the magnitude F_{net} of the net electrostatic force on particle A due to the other particles. That net force is given as a function of angle θ and as a multiple of a basic amount F_0. For example on curve 1, at $\theta = 180°$, we see that $F_{net} = 2F_0$. (a) For the situation corresponding to curve 1, what is the ratio of the charge of particle C to that of particle B (including sign)? (b) For the situation corresponding to curve 2, what is that ratio?

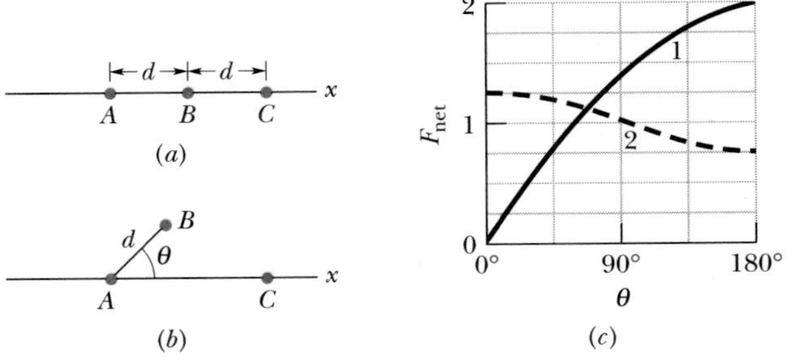

Fig. 21-29 Problem 20.

•••21 A nonconducting spherical shell, with an inner radius of 4.0 cm and an outer radius of 6.0 cm, has charge spread nonuniformly through its volume between its inner and outer surfaces. The *volume charge density* ρ is the charge per unit volume, with the unit coulomb per cubic meter. For this shell $\rho = b/r$, where r is the distance in meters from the center of the shell and $b = 3.0\ \mu C/m^2$. What is the net charge in the shell?

•••22 GO Figure 21-30 shows an arrangement of four charged particles, with angle $\theta = 30.0°$ and distance $d = 2.00$ cm. Particle 2 has charge $q_2 = +8.00 \times 10^{-19}$ C; particles 3 and 4 have charges $q_3 = q_4 = -1.60 \times 10^{-19}$ C. (a) What is distance D between the origin and particle 2 if the net electrostatic force on particle 1 due to the other particles is zero? (b) If particles 3 and 4 were moved closer to the x axis but maintained their symmetry about that axis, would the required value of D be greater than, less than, or the same as in part (a)?

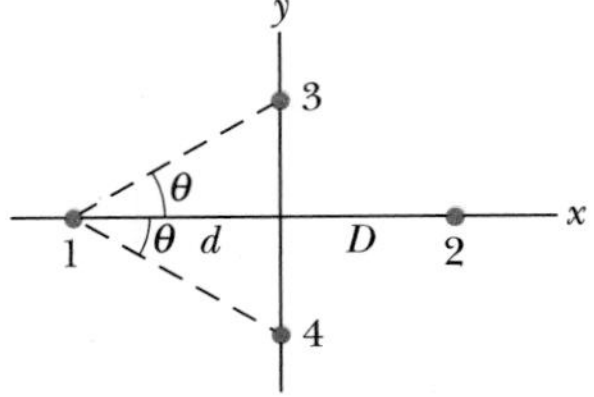

Fig. 21-30 Problem 22.

•••23 In Fig. 21-31, particles 1 and 2 of charge $q_1 = q_2 = +3.20 \times 10^{-19}$ C are on a y axis at distance $d = 17.0$ cm from the origin. Particle 3 of charge $q_3 = +6.40 \times 10^{-19}$ C is moved gradually along the x axis from $x = 0$ to $x = +5.0$ m. At what values of x will the magnitude of the electrostatic force on the third particle from the other two particles be (a) minimum and (b) maximum? What are the (c) minimum and (d) maximum magnitudes?

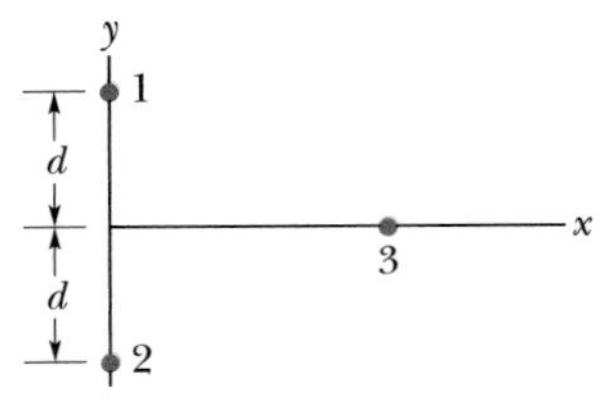

Fig. 21-31 Problem 23.

sec. 21-5 Charge Is Quantized

•24 Two tiny, spherical water drops, with identical charges of -1.00×10^{-16} C, have a center-to-center separation of 1.00 cm. (a) What is the magnitude of the electrostatic force acting between them? (b) How many excess electrons are on each drop, giving it its charge imbalance?

•25 ILW How many electrons would have to be removed from a coin to leave it with a charge of $+1.0 \times 10^{-7}$ C?

•26 What is the magnitude of the electrostatic force between a singly charged sodium ion (Na^+, of charge $+e$) and an adjacent singly charged chlorine ion (Cl^-, of charge $-e$) in a salt crystal if their separation is 2.82×10^{-10} m?

•27 SSM The magnitude of the electrostatic force between two identical ions that are separated by a distance of 5.0×10^{-10} m is 3.7×10^{-9} N. (a) What is the charge of each ion? (b) How many electrons are "missing" from each ion (thus giving the ion its charge imbalance)?

•28 A current of 0.300 A through your chest can send your heart into fibrillation, ruining the normal rhythm of heartbeat and disrupting the flow of blood (and thus oxygen) to your brain. If that current persists for 2.00 min, how many conduction electrons pass through your chest?

••29 GO In Fig. 21-32, particles 2 and 4, of charge $-e$, are fixed in place on a y axis, at $y_2 = -10.0$ cm and $y_4 = 5.00$ cm. Particles 1 and 3, of charge $-e$, can be moved along the x axis. Particle 5, of charge $+e$, is fixed at the origin. Initially particle 1 is at $x_1 = -10.0$ cm and particle 3 is at $x_3 = 10.0$ cm. (a) To what x value must particle 1 be moved to rotate the direction of the net electric force $\vec{F}_{net}$ on particle 5 by 30° counterclockwise? (b) With particle 1 fixed at its new position, to what x value must you move particle 3 to rotate $\vec{F}_{net}$ back to its original direction?

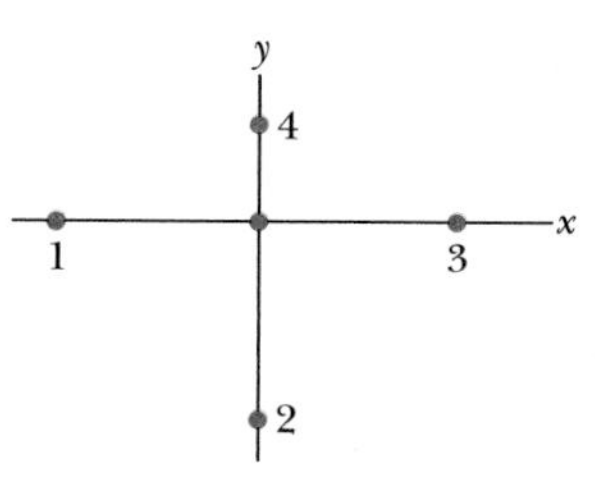

Fig. 21-32 Problem 29.

••30 In Fig. 21-25, particles 1 and 2 are fixed in place on an x axis, at a separation of $L = 8.00$ cm. Their charges are $q_1 = +e$ and $q_2 = -27e$. Particle 3 with charge $q_3 = +4e$ is to be placed on the line between particles 1 and 2, so that they produce a net electrostatic force $\vec{F}_{3,net}$ on it. (a) At what coordinate should particle 3 be placed to minimize the magnitude of that force? (b) What is that minimum magnitude?

••31 ILW Earth's atmosphere is constantly bombarded by *cosmic ray protons* that originate somewhere in space. If the protons all passed through the atmosphere, each square meter of Earth's surface would intercept protons at the average rate of 1500 protons per second. What would be the electric current intercepted by the total surface area of the planet?

••32 Figure 21-33a shows charged particles 1 and 2 that are fixed in place on an x axis. Particle 1 has a charge with a magnitude of $|q_1| = 8.00e$. Particle 3 of charge $q_3 = +8.00e$ is initially on the x axis near particle 2. Then particle 3 is gradually moved in the positive direction of the x axis. As a result, the magnitude of the net electrostatic force $\vec{F}_{2,net}$ on particle 2 due to particles 1 and 3 changes. Figure 21-33b gives the x component of that net force as a function of the position x of particle 3. The scale of the x axis is set by $x_s = 0.80$ m. The plot has an asymptote of $F_{2,net} = 1.5 \times 10^{-25}$ N as $x \to \infty$. As a multiple of e and including the sign, what is the charge q_2 of particle 2?

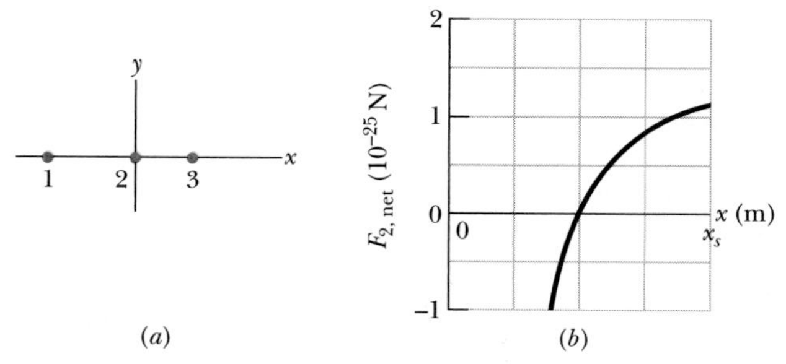

Fig. 21-33 Problem 32.

••33 Calculate the number of coulombs of positive charge in 250 cm^3 of (neutral) water. (*Hint:* A hydrogen atom contains one proton; an oxygen atom contains eight protons.)

•••34 Figure 21-34 shows electrons 1 and 2 on an x axis and charged ions 3 and 4 of identical charge $-q$ and at identical angles θ. Electron 2 is free to move; the other three particles are fixed in place at horizontal distances R from electron 2 and are intended to hold electron 2 in place. For physically possible values of $q \leq 5e$, what are the (a) smallest, (b) second smallest, and (c) third smallest values of θ for which electron 2 is held in place?

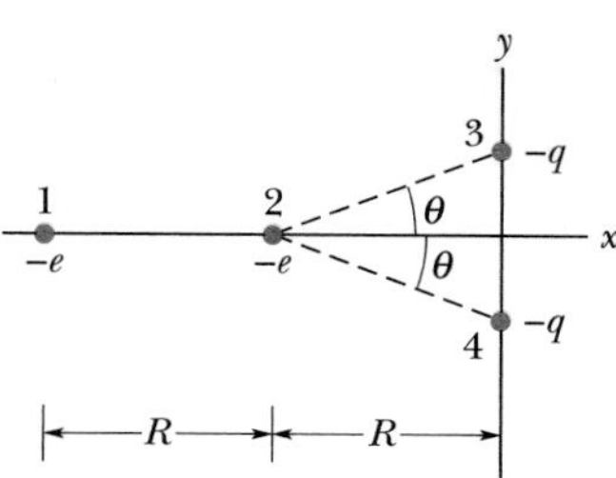

Fig. 21-34 Problem 34.

•••35 SSM In crystals of the salt cesium chloride, cesium ions Cs^+ form the eight corners of a cube and a chlorine ion Cl^- is at the cube's center (Fig. 21-35). The edge length of the cube is 0.40 nm. The Cs^+ ions are each deficient by one electron (and thus each has a charge of $+e$), and the Cl^- ion has one excess electron (and thus has a charge of $-e$). (a) What is the magnitude of the net electrostatic force exerted on the Cl^- ion by the eight Cs^+ ions at the corners of the cube? (b) If one of the Cs^+ ions is missing, the crystal is said to have a *defect*; what is the magnitude of the net electrostatic force exerted on the Cl^- ion by the seven remaining Cs^+ ions?

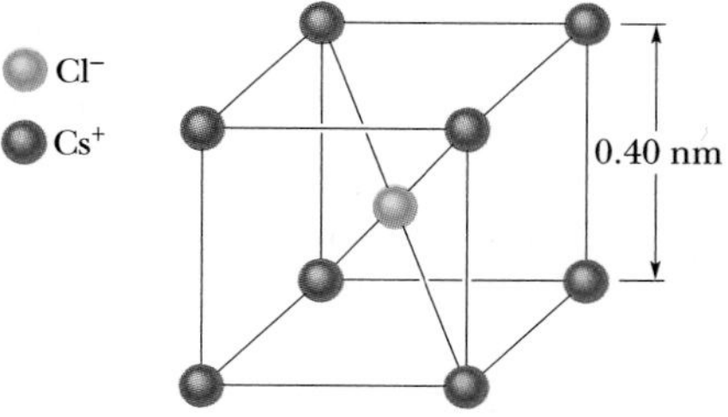

Fig. 21-35 Problem 35.

sec. 21-6 Charge Is Conserved

•36 Electrons and positrons are produced by the nuclear transformations of protons and neutrons known as *beta decay.* (a) If a proton transforms into a neutron, is an electron or a positron produced? (b) If a neutron transforms into a proton, is an electron or a positron produced?

•37 SSM Identify X in the following nuclear reactions: (a) $^1H + {}^9Be \to X + n$; (b) $^{12}C + {}^1H \to X$; (c) $^{15}N + {}^1H \to {}^4He + X$. Appendix F will help.

Additional Problems

38 GO Figure 21-36 shows four identical conducting spheres that are actually well separated from one another. Sphere W (with an initial charge of zero) is touched to sphere

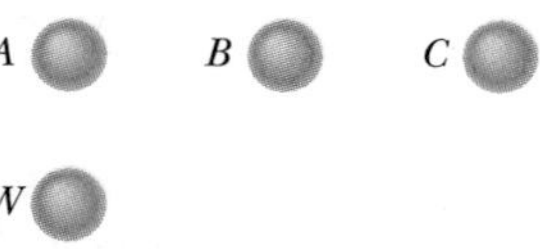

Fig. 21-36 Problem 38.

A and then they are separated. Next, sphere W is touched to sphere B (with an initial charge of $-32e$) and then they are separated. Finally, sphere W is touched to sphere C (with an initial charge of $+48e$), and then they are separated. The final charge on sphere W is $+18e$. What was the initial charge on sphere A?

39 SSM In Fig. 21-37, particle 1 of charge $+4e$ is above a floor by distance $d_1 = 2.00$ mm and particle 2 of charge $+6e$ is on the floor, at distance $d_2 = 6.00$ mm horizontally from particle 1. What is the x component of the electrostatic force on particle 2 due to particle 1?

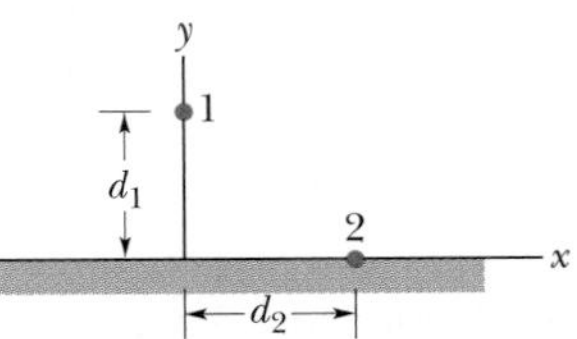

Fig. 21-37 Problem 39.

40 In Fig. 21-22, particles 1 and 2 are fixed in place, but particle 3 is free to move. If the net electrostatic force on particle 3 due to particles 1 and 2 is zero and $L_{23} = 2.00L_{12}$, what is the ratio q_1/q_2?

41 (a) What equal positive charges would have to be placed on Earth and on the Moon to neutralize their gravitational attraction? (b) Why don't you need to know the lunar distance to solve this problem? (c) How many kilograms of hydrogen ions (that is, protons) would be needed to provide the positive charge calculated in (a)?

42 In Fig. 21-38, two tiny conducting balls of identical mass m and identical charge q hang from nonconducting threads of length L. Assume that θ is so small that tan θ can be replaced by its approximate equal, sin θ. (a) Show that

$$x = \left(\frac{q^2L}{2\pi\varepsilon_0 mg}\right)^{1/3}$$

gives the equilibrium separation x of the balls. (b) If $L = 120$ cm, $m = 10$ g, and $x = 5.0$ cm, what is $|q|$?

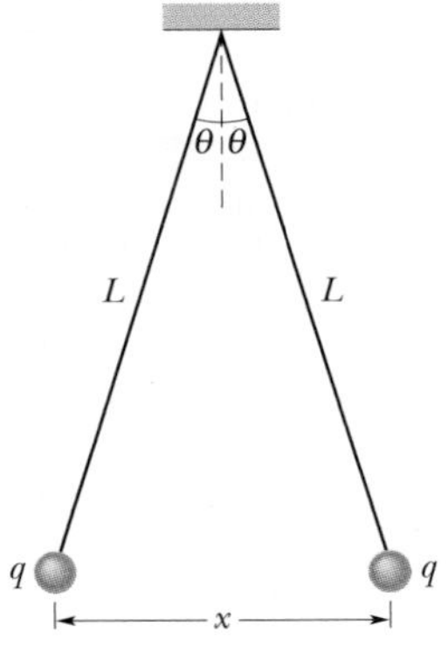

Fig. 21-38 Problems 42 and 43.

43 (a) Explain what happens to the balls of Problem 42 if one of them is discharged (loses its charge q to, say, the ground). (b) Find the new equilibrium separation x, using the given values of L and m and the computed value of $|q|$.

44 SSM How far apart must two protons be if the magnitude of the electrostatic force acting on either one due to the other is equal to the magnitude of the gravitational force on a proton at Earth's surface?

45 How many megacoulombs of positive charge are in 1.00 mol of neutral molecular-hydrogen gas (H_2)?

46 In Fig. 21-39, four particles are fixed along an x axis, separated by distances $d = 2.00$ cm. The charges are $q_1 = +2e$, $q_2 = -e$, $q_3 = +e$, and $q_4 = +4e$, with $e = 1.60 \times 10^{-19}$ C. In unit-vector notation, what is the net electrostatic force on (a) particle 1 and (b) particle 2 due to the other particles?

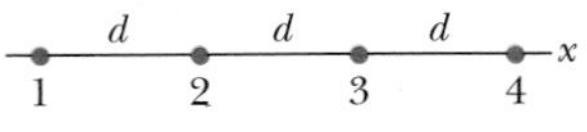

Fig. 21-39 Problem 46.

47 GO Point charges of $+6.0\ \mu$C and $-4.0\ \mu$C are placed on an x axis, at $x = 8.0$ m and $x = 16$ m, respectively. What charge must be placed at $x = 24$ m so that any charge placed at the origin would experience no electrostatic force?

48 In Fig. 21-40, three identical conducting spheres form an equilateral triangle of side length $d = 20.0$ cm. The sphere radii are much smaller than d, and the sphere charges are $q_A = -2.00$ nC, $q_B = -4.00$ nC, and $q_C = +8.00$ nC. (a) What is the magnitude of the electrostatic force between spheres A and C? The following steps are then taken: A and B are connected by a thin wire and then disconnected; B is grounded by the wire, and the wire is then removed; B and C are connected by the wire and then disconnected. What now are the magnitudes of the electrostatic force (b) between spheres A and C and (c) between spheres B and C?

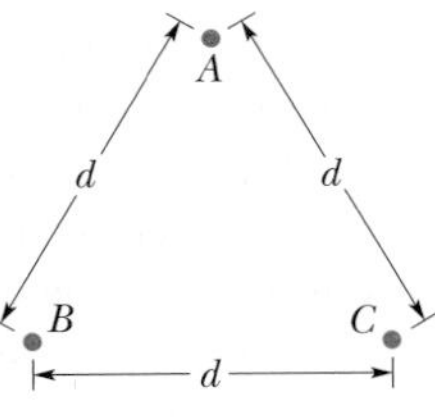

Fig. 21-40 Problem 48.

49 A neutron consists of one "up" quark of charge $+2e/3$ and two "down" quarks each having charge $-e/3$. If we assume that the down quarks are 2.6×10^{-15} m apart inside the neutron, what is the magnitude of the electrostatic force between them?

50 Figure 21-41 shows a long, nonconducting, massless rod of length L, pivoted at its center and balanced with a block of weight W at a distance x from the left end. At the left and right ends of the rod are attached small conducting spheres with positive charges q and $2q$, respectively. A distance h directly beneath each of these spheres is a fixed sphere with positive charge Q. (a) Find the distance x when the rod is horizontal and balanced. (b) What value should h have so that the rod exerts no vertical force on the bearing when the rod is horizontal and balanced?

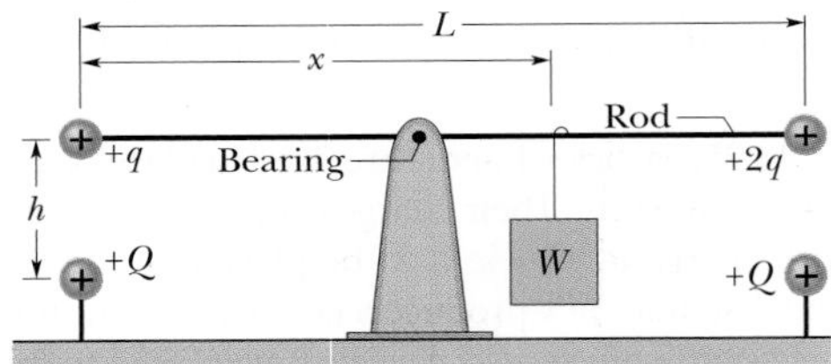

Fig. 21-41 Problem 50.

51 A charged nonconducting rod, with a length of 2.00 m and a cross-sectional area of 4.00 cm^2, lies along the positive side of an x axis with one end at the origin. The *volume charge density* ρ is charge per unit volume in coulombs per cubic meter. How many excess electrons are on the rod if ρ is (a) uniform, with a value of $-4.00\ \mu$C/m^3, and (b) nonuniform, with a value given by $\rho = bx^2$, where $b = -2.00\ \mu$C/m^5?

52 A particle of charge Q is fixed at the origin of an xy coordinate system. At $t = 0$ a particle ($m = 0.800$ g, $q = 4.00\ \mu$C) is located on the x axis at $x = 20.0$ cm, moving with a speed of 50.0 m/s in the positive y direction. For what value of Q will the moving particle execute circular motion? (Neglect the gravitational force on the particle.)

53 What would be the magnitude of the electrostatic force between two 1.00 C point charges separated by a distance of (a) 1.00 m and (b) 1.00 km if such point charges existed (they do not) and this configuration could be set up?

54 A charge of 6.0 μC is to be split into two parts that are then separated by 3.0 mm. What is the maximum possible magnitude of the electrostatic force between those two parts?

55 Of the charge Q on a tiny sphere, a fraction α is to be transferred to a second, nearby sphere. The spheres can be treated as particles. (a) What value of α maximizes the magnitude F of the electrostatic force between the two spheres? What are the (b) smaller and (c) larger values of α that put F at half the maximum magnitude?

56 If a cat repeatedly rubs against your cotton slacks on a dry day, the charge transfer between the cat hair and the cotton can leave you with an excess charge of $-2.00\ \mu C$. (a) How many electrons are transferred between you and the cat?

You will gradually discharge via the floor, but if instead of waiting, you immediately reach toward a faucet, a painful spark can suddenly appear as your fingers near the faucet. (b) In that spark, do electrons flow from you to the faucet or vice versa? (c) Just before the spark appears, do you induce positive or negative charge in the faucet? (d) If, instead, the cat reaches a paw toward the faucet, which way do electrons flow in the resulting spark? (e) If you stroke a cat with a bare hand on a dry day, you should take care not to bring your fingers near the cat's nose or you will hurt it with a spark. Considering that cat hair is an insulator, explain how the spark can appear.

57 We know that the negative charge on the electron and the positive charge on the proton are equal. Suppose, however, that these magnitudes differ from each other by 0.00010%. With what force would two copper coins, placed 1.0 m apart, repel each other? Assume that each coin contains 3×10^{22} copper atoms. (*Hint:* A neutral copper atom contains 29 protons and 29 electrons.) What do you conclude?

58 In Fig. 21-25, particle 1 of charge $-80.0\ \mu C$ and particle 2 of charge $+40.0\ \mu C$ are held at separation $L = 20.0$ cm on an x axis. In unit-vector notation, what is the net electrostatic force on particle 3, of charge $q_3 = 20.0\ \mu C$, if particle 3 is placed at (a) $x = 40.0$ cm and (b) $x = 80.0$ cm? What should be the (c) x and (d) y coordinates of particle 3 if the net electrostatic force on it due to particles 1 and 2 is zero?

59 What is the total charge in coulombs of 75.0 kg of electrons?

60 In Fig. 21-42, six charged particles surround particle 7 at radial distances of either $d = 1.0$ cm or $2d$, as drawn. The charges are $q_1 = +2e$, $q_2 = +4e$, $q_3 = +e$, $q_4 = +4e$, $q_5 = +2e$, $q_6 = +8e$, $q_7 = +6e$, with $e = 1.60 \times 10^{-19}$ C. What is the magnitude of the net electrostatic force on particle 7?

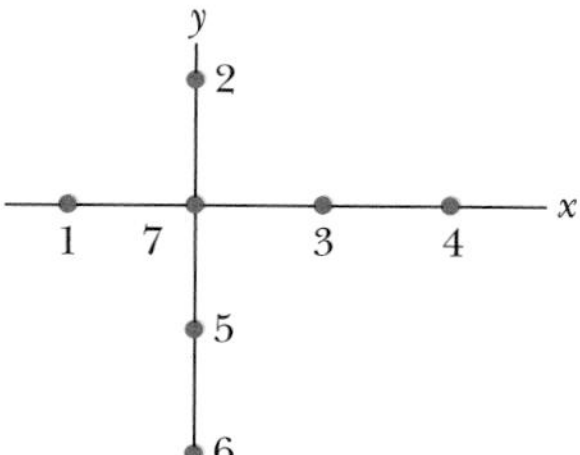

Fig. 21-42 Problem 60.

61 Three charged particles form a triangle: particle 1 with charge $Q_1 = 80.0$ nC is at xy coordinates (0, 3.00 mm), particle 2 with charge Q_2 is at (0, -3.00 mm), and particle 3 with charge $q = 18.0$ nC is at (4.00 mm, 0). In unit-vector notation, what is the electrostatic force on particle 3 due to the other two particles if Q_2 is equal to (a) 80.0 nC and (b) -80.0 nC?

62 SSM In Fig. 21-43, what are the (a) magnitude and (b) direction of the net electrostatic force on particle 4 due to the other three particles? All four particles are fixed in the xy plane, and $q_1 = -3.20 \times 10^{-19}$ C, $q_2 = +3.20 \times 10^{-19}$ C, $q_3 = +6.40 \times 10^{-19}$ C, $q_4 = +3.20 \times 10^{-19}$ C, $\theta_1 = 35.0°$, $d_1 = 3.00$ cm, and $d_2 = d_3 = 2.00$ cm.

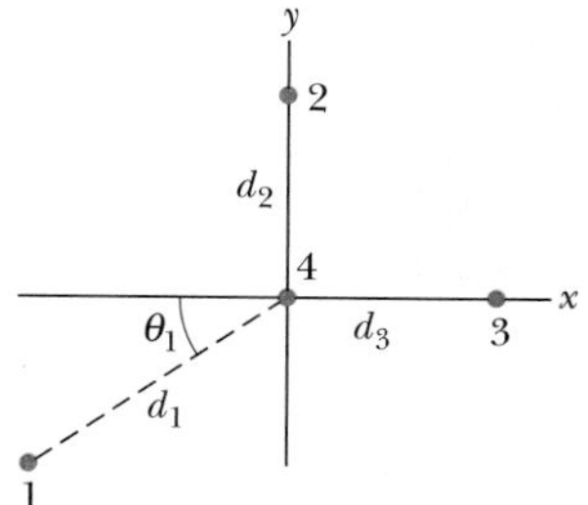

Fig. 21-43 Problem 62.

63 Two point charges of 30 nC and -40 nC are held fixed on an x axis, at the origin and at $x = 72$ cm, respectively. A particle with a charge of 42 μC is released from rest at $x = 28$ cm. If the initial acceleration of the particle has a magnitude of 100 km/s^2, what is the particle's mass?

64 Two small, positively charged spheres have a combined charge of 5.0×10^{-5} C. If each sphere is repelled from the other by an electrostatic force of 1.0 N when the spheres are 2.0 m apart, what is the charge on the sphere with the smaller charge?

65 The initial charges on the three identical metal spheres in Fig. 21-23 are the following: sphere A, Q; sphere B, $-Q/4$; and sphere C, $Q/2$, where $Q = 2.00 \times 10^{-14}$ C. Spheres A and B are fixed in place, with a center-to-center separation of $d = 1.20$ m, which is much larger than the spheres. Sphere C is touched first to sphere A and then to sphere B and is then removed. What then is the magnitude of the electrostatic force between spheres A and B?

66 An electron is in a vacuum near Earth's surface and located at $y = 0$ on a vertical y axis. At what value of y should a second electron be placed such that its electrostatic force on the first electron balances the gravitational force on the first electron?

67 SSM In Fig. 21-25, particle 1 of charge $-5.00q$ and particle 2 of charge $+2.00q$ are held at separation L on an x axis. If particle 3 of unknown charge q_3 is to be located such that the net electrostatic force on it from particles 1 and 2 is zero, what must be the (a) x and (b) y coordinates of particle 3?

68 Two engineering students, John with a mass of 90 kg and Mary with a mass of 45 kg, are 30 m apart. Suppose each has a 0.01% imbalance in the amount of positive and negative charge, one student being positive and the other negative. Find the order of magnitude of the electrostatic force of attraction between them by replacing each student with a sphere of water having the same mass as the student.

69 In the radioactive decay of Eq. 21-13, a ^{238}U nucleus transforms to ^{234}Th and an ejected ^{4}He. (These are nuclei, not atoms, and thus electrons are not involved.) When the separation between ^{234}Th and ^{4}He is 9.0×10^{-15} m, what are the magnitudes of (a) the electrostatic force between them and (b) the acceleration of the ^{4}He particle?

70 In Fig. 21-24, four particles form a square. The charges are $q_1 = +Q$, $q_2 = q_3 = q$, and $q_4 = -2.00Q$. What is q/Q if the net electrostatic force on particle 1 is zero?

CHAPTER

22 ELECTRIC FIELDS

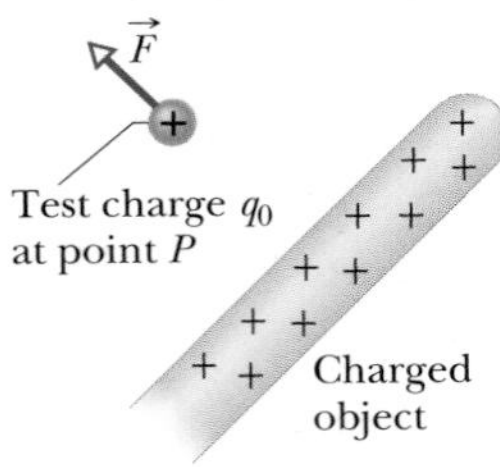

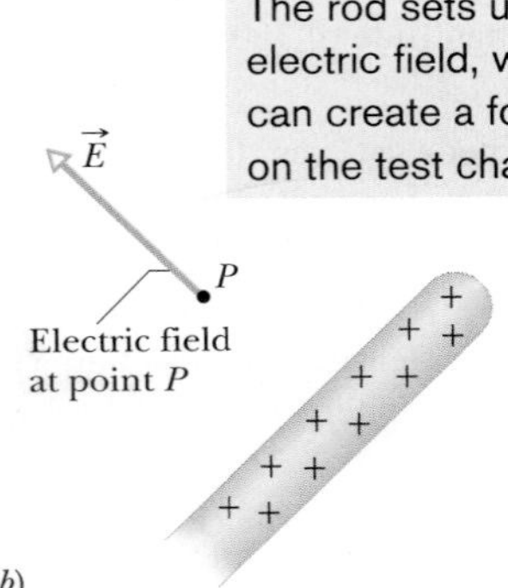

Fig. 22-1 (*a*) A positive test charge q_0 placed at point P near a charged object. An electrostatic force $\vec{F}$ acts on the test charge. (*b*) The electric field $\vec{E}$ at point P produced by the charged object.

Table 22-1

Some Electric Fields

Field Location or Situation	Value (N/C)
At the surface of a uranium nucleus	3×10^{21}
Within a hydrogen atom, at a radius of 5.29×10^{-11} m	5×10^{11}
Electric breakdown occurs in air	3×10^{6}
Near the charged drum of a photocopier	10^{5}
Near a charged comb	10^{3}
In the lower atmosphere	10^{2}
Inside the copper wire of household circuits	10^{-2}

22-1 WHAT IS PHYSICS?

The physics of the preceding chapter tells us how to find the electric force on a particle 1 of charge $+q_1$ when the particle is placed near a particle 2 of charge $+q_2$. A nagging question remains: How does particle 1 "know" of the presence of particle 2? That is, since the particles do not touch, how can particle 2 push on particle 1—how can there be such an *action at a distance*?

One purpose of physics is to record observations about our world, such as the magnitude and direction of the push on particle 1. Another purpose is to provide a deeper explanation of what is recorded. One purpose of this chapter is to provide such a deeper explanation to our nagging questions about electric force at a distance. We can answer those questions by saying that particle 2 sets up an **electric field** in the space surrounding itself. If we place particle 1 at any given point in that space, the particle "knows" of the presence of particle 2 because it is affected by the electric field that particle 2 has already set up at that point. Thus, particle 2 pushes on particle 1 not by touching it but by means of the electric field produced by particle 2.

Our goal in this chapter is to define electric field and discuss how to calculate it for various arrangements of charged particles.

22-2 The Electric Field

The temperature at every point in a room has a definite value. You can measure the temperature at any given point or combination of points by putting a thermometer there. We call the resulting distribution of temperatures a *temperature field.* In much the same way, you can imagine a *pressure field* in the atmosphere; it consists of the distribution of air pressure values, one for each point in the atmosphere. These two examples are of *scalar fields* because temperature and air pressure are scalar quantities.

The electric field is a *vector field;* it consists of a distribution of *vectors,* one for each point in the region around a charged object, such as a charged rod. In principle, we can define the electric field at some point near the charged object, such as point P in Fig. 22-1*a*, as follows: We first place a *positive* charge q_0, called a *test charge,* at the point. We then measure the electrostatic force $\vec{F}$ that acts on the test charge. Finally, we define the electric field $\vec{E}$ at point P due to the charged object as

$$\vec{E} = \frac{\vec{F}}{q_0} \quad \text{(electric field).} \tag{22-1}$$

Thus, the magnitude of the electric field $\vec{E}$ at point P is $E = F/q_0$, and the direction of $\vec{E}$ is that of the force $\vec{F}$ that acts on the *positive* test charge. As shown in Fig. 22-1*b*, we represent the electric field at P with a vector whose tail is at P. To define the electric field within some region, we must similarly define it at all points in the region.

The SI unit for the electric field is the newton per coulomb (N/C). Table 22-1 shows the electric fields that occur in a few physical situations.

Although we use a positive test charge to define the electric field of a charged object, that field exists independently of the test charge. The field at point P in Figure 22-1b existed both before and after the test charge of Fig. 22-1a was put there. (We assume that in our defining procedure, the presence of the test charge does not affect the charge distribution on the charged object, and thus does not alter the electric field we are defining.)

To examine the role of an electric field in the interaction between charged objects, we have two tasks: (1) calculating the electric field produced by a given distribution of charge and (2) calculating the force that a given field exerts on a charge placed in it. We perform the first task in Sections 22-4 through 22-7 for several charge distributions. We perform the second task in Sections 22-8 and 22-9 by considering a point charge and a pair of point charges in an electric field. First, however, we discuss a way to visualize electric fields.

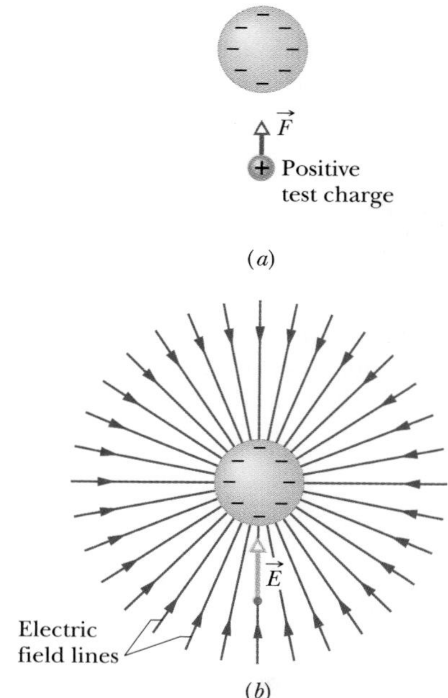

Fig. 22-2 (a) The electrostatic force $\vec{F}$ acting on a positive test charge near a sphere of uniform negative charge. (b) The electric field vector $\vec{E}$ at the location of the test charge, and the electric field lines in the space near the sphere. The field lines extend *toward* the negatively charged sphere. (They originate on distant positive charges.)

22-3 Electric Field Lines

Michael Faraday, who introduced the idea of electric fields in the 19th century, thought of the space around a charged body as filled with *lines of force*. Although we no longer attach much reality to these lines, now usually called **electric field lines,** they still provide a nice way to visualize patterns in electric fields.

The relation between the field lines and electric field vectors is this: (1) At any point, the direction of a straight field line or the direction of the tangent to a curved field line gives the direction of $\vec{E}$ at that point, and (2) the field lines are drawn so that the number of lines per unit area, measured in a plane that is perpendicular to the lines, is proportional to the *magnitude* of $\vec{E}$. Thus, E is large where field lines are close together and small where they are far apart.

Figure 22-2a shows a sphere of uniform negative charge. If we place a *positive* test charge anywhere near the sphere, an electrostatic force pointing *toward* the center of the sphere will act on the test charge as shown. In other words, the electric field vectors at all points near the sphere are directed radially toward the sphere. This pattern of vectors is neatly displayed by the field lines in Fig. 22-2b, which point in the same directions as the force and field vectors. Moreover, the spreading of the field lines with distance from the sphere tells us that the magnitude of the electric field decreases with distance from the sphere.

If the sphere of Fig. 22-2 were of uniform *positive* charge, the electric field vectors at all points near the sphere would be directed radially *away from* the sphere. Thus, the electric field lines would also extend radially away from the sphere. We then have the following rule:

> Electric field lines extend away from positive charge (where they originate) and toward negative charge (where they terminate).

Figure 22-3a shows part of an infinitely large, nonconducting *sheet* (or plane) with a uniform distribution of positive charge on one side. If we were to place a

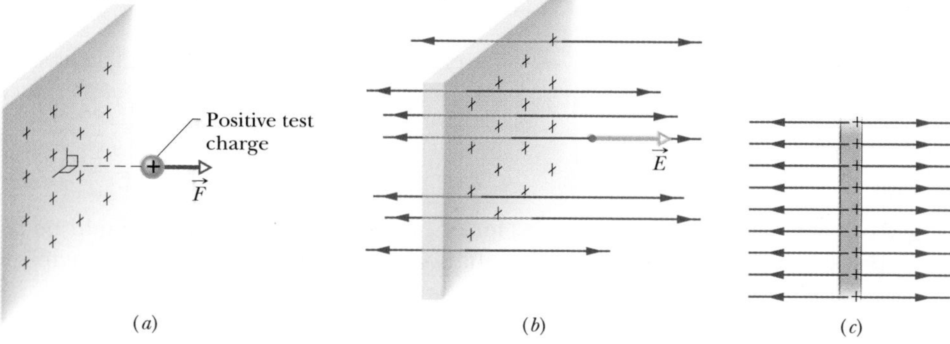

Fig. 22-3 (a) The electrostatic force $\vec{F}$ on a positive test charge near a very large, nonconducting sheet with uniformly distributed positive charge on one side. (b) The electric field vector $\vec{E}$ at the location of the test charge, and the electric field lines in the space near the sheet. The field lines extend *away from* the positively charged sheet. (c) Side view of (b).

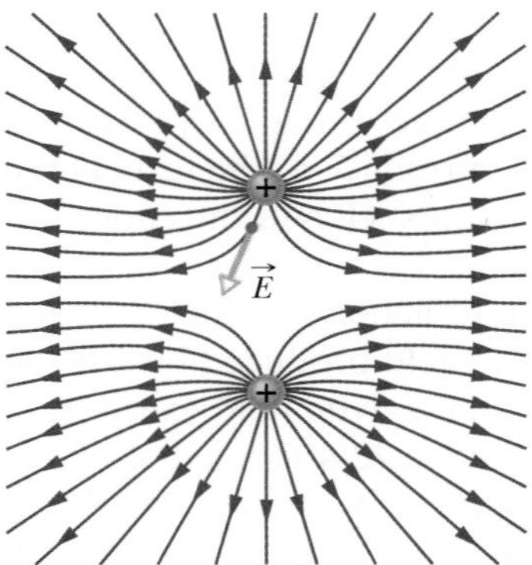

Fig. 22-4 Field lines for two equal positive point charges. The charges repel each other. (The lines terminate on distant negative charges.) To "see" the actual three-dimensional pattern of field lines, mentally rotate the pattern shown here about an axis passing through both charges in the plane of the page. The three-dimensional pattern and the electric field it represents are said to have *rotational symmetry* about that axis. The electric field vector at one point is shown; note that it is tangent to the field line through that point.

positive test charge at any point near the sheet of Fig. 22-3*a*, the net electrostatic force acting on the test charge would be perpendicular to the sheet, because forces acting in all other directions would cancel one another as a result of the symmetry. Moreover, the net force on the test charge would point away from the sheet as shown. Thus, the electric field vector at any point in the space on either side of the sheet is also perpendicular to the sheet and directed away from it (Figs. 22-3*b* and *c*). Because the charge is uniformly distributed along the sheet, all the field vectors have the same magnitude. Such an electric field, with the same magnitude and direction at every point, is a *uniform electric field.*

Of course, no real nonconducting sheet (such as a flat expanse of plastic) is infinitely large, but if we consider a region that is near the middle of a real sheet and not near its edges, the field lines through that region are arranged as in Figs. 22-3*b* and *c*.

Figure 22-4 shows the field lines for two equal positive charges. Figure 22-5 shows the pattern for two charges that are equal in magnitude but of opposite sign, a configuration that we call an **electric dipole.** Although we do not often use field lines quantitatively, they are very useful to visualize what is going on.

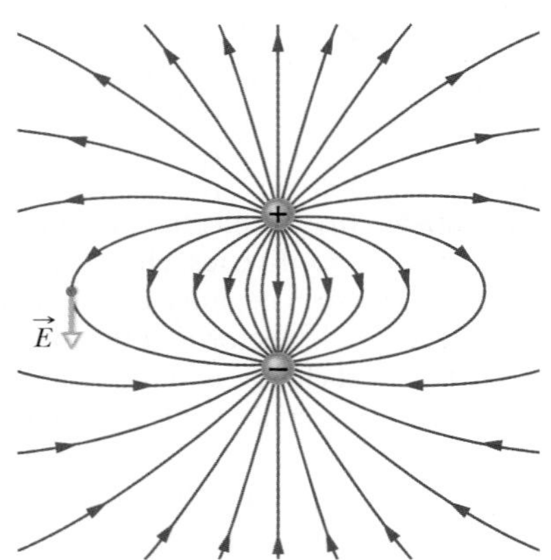

Fig. 22-5 Field lines for a positive point charge and a nearby negative point charge that are equal in magnitude. The charges attract each other. The pattern of field lines and the electric field it represents have rotational symmetry about an axis passing through both charges in the plane of the page. The electric field vector at one point is shown; the vector is tangent to the field line through the point.

22-4 The Electric Field Due to a Point Charge

To find the electric field due to a point charge q (or charged particle) at any point a distance r from the point charge, we put a positive test charge q_0 at that point. From Coulomb's law (Eq. 21-1), the electrostatic force acting on q_0 is

$$\vec{F} = \frac{1}{4\pi\varepsilon_0}\frac{qq_0}{r^2}\,\hat{\mathrm{r}}. \tag{22-2}$$

The direction of $\vec{F}$ is directly away from the point charge if q is positive, and directly toward the point charge if q is negative. The electric field vector is, from Eq. 22-1,

$$\vec{E} = \frac{\vec{F}}{q_0} = \frac{1}{4\pi\varepsilon_0}\frac{q}{r^2}\,\hat{\mathrm{r}} \quad \text{(point charge)}. \tag{22-3}$$

The direction of $\vec{E}$ is the same as that of the force on the positive test charge: directly away from the point charge if q is positive, and toward it if q is negative.

Because there is nothing special about the point we chose for q_0, Eq. 22-3 gives the field at every point around the point charge q. The field for a positive point charge is shown in Fig. 22-6 in vector form (not as field lines).

We can quickly find the net, or resultant, electric field due to more than one point charge. If we place a positive test charge q_0 near n point charges $q_1, q_2, \ldots, q_n$, then, from Eq. 21-7, the net force $\vec{F}_0$ from the n point charges acting on the test charge is

$$\vec{F}_0 = \vec{F}_{01} + \vec{F}_{02} + \cdots + \vec{F}_{0n}.$$

Therefore, from Eq. 22-1, the net electric field at the position of the test charge is

$$\begin{aligned}\vec{E} &= \frac{\vec{F}_0}{q_0} = \frac{\vec{F}_{01}}{q_0} + \frac{\vec{F}_{02}}{q_0} + \cdots + \frac{\vec{F}_{0n}}{q_0} \\ &= \vec{E}_1 + \vec{E}_2 + \cdots + \vec{E}_n.\end{aligned} \tag{22-4}$$

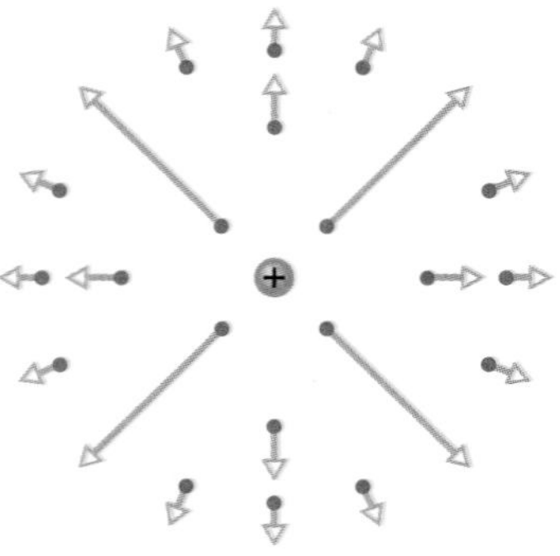

Fig. 22-6 The electric field vectors at various points around a positive point charge.

Here $\vec{E}_i$ is the electric field that would be set up by point charge i acting alone. Equation 22-4 shows us that the principle of superposition applies to electric fields as well as to electrostatic forces.

CHECKPOINT 1

The figure here shows a proton p and an electron e on an x axis. What is the direction of the electric field due to the electron at (a) point S and (b) point R? What is the direction of the net electric field at (c) point R and (d) point S?

Sample Problem

Net electric field due to three charged particles

Figure 22-7a shows three particles with charges $q_1 = +2Q$, $q_2 = -2Q$, and $q_3 = -4Q$, each a distance d from the origin. What net electric field $\vec{E}$ is produced at the origin?

KEY IDEA

Charges q_1, q_2, and q_3 produce electric field vectors $\vec{E}_1$, $\vec{E}_2$, and $\vec{E}_3$, respectively, at the origin, and the net electric field is the vector sum $\vec{E} = \vec{E}_1 + \vec{E}_2 + \vec{E}_3$. To find this sum, we first must find the magnitudes and orientations of the three field vectors.

Magnitudes and directions: To find the magnitude of $\vec{E}_1$, which is due to q_1, we use Eq. 22-3, substituting d for r and $2Q$ for q and obtaining

$$E_1 = \frac{1}{4\pi\varepsilon_0}\frac{2Q}{d^2}.$$

Similarly, we find the magnitudes of $\vec{E}_2$ and $\vec{E}_3$ to be

$$E_2 = \frac{1}{4\pi\varepsilon_0}\frac{2Q}{d^2} \quad \text{and} \quad E_3 = \frac{1}{4\pi\varepsilon_0}\frac{4Q}{d^2}.$$

We next must find the orientations of the three electric field vectors at the origin. Because q_1 is a positive charge, the field vector it produces points directly *away* from it, and because q_2 and q_3 are both negative, the field vectors they produce point directly *toward* each of them. Thus, the three electric fields produced at the origin by the three charged particles are oriented as in Fig. 22-7b. (*Caution:* Note that we have placed the tails of the vectors at the point where the fields are to be evaluated; doing so decreases the chance of error. Error becomes very probable if the tails of the field vectors are placed on the particles creating the fields.)

Adding the fields: We can now add the fields vectorially just as we added force vectors in Chapter 21. However, here we can use symmetry to simplify the procedure. From Fig. 22-7b, we see that electric fields $\vec{E}_1$ and $\vec{E}_2$ have the same direction. Hence, their vector sum has that direction and has the magnitude

$$E_1 + E_2 = \frac{1}{4\pi\varepsilon_0}\frac{2Q}{d^2} + \frac{1}{4\pi\varepsilon_0}\frac{2Q}{d^2} = \frac{1}{4\pi\varepsilon_0}\frac{4Q}{d^2},$$

which happens to equal the magnitude of field $\vec{E}_3$.

We must now combine two vectors, $\vec{E}_3$ and the vector sum $\vec{E}_1 + \vec{E}_2$, that have the same magnitude and that are oriented symmetrically about the x axis, as shown in Fig. 22-7c. From the symmetry of Fig. 22-7c, we realize that the equal y components of our two vectors cancel (one is upward and the other is downward) and the equal x components add (both are rightward). Thus, the net electric field $\vec{E}$ at the origin is in the positive direction of the x axis and has the magnitude

$$E = 2E_{3x} = 2E_3 \cos 30^\circ = (2)\frac{1}{4\pi\varepsilon_0}\frac{4Q}{d^2}(0.866) = \frac{6.93Q}{4\pi\varepsilon_0 d^2}. \quad \text{(Answer)}$$

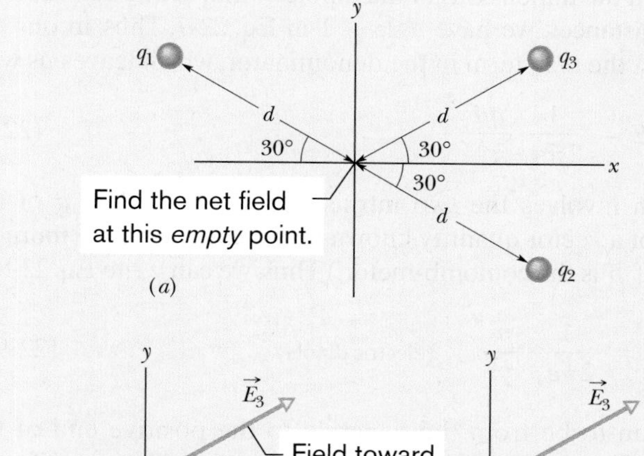

Fig. 22-7 (*a*) Three particles with charges q_1, q_2, and q_3 are at the same distance d from the origin. (*b*) The electric field vectors $\vec{E}_1$, $\vec{E}_2$, and $\vec{E}_3$, at the origin due to the three particles. (*c*) The electric field vector $\vec{E}_3$ and the vector sum $\vec{E}_1 + \vec{E}_2$ at the origin.

WILEY PLUS Additional examples, video, and practice available at *WileyPLUS*

22-5 The Electric Field Due to an Electric Dipole

Figure 22-8*a* shows two charged particles of magnitude q but of opposite sign, separated by a distance d. As was noted in connection with Fig. 22-5, we call this configuration an *electric dipole*. Let us find the electric field due to the dipole of Fig. 22-8*a* at a point P, a distance z from the midpoint of the dipole and on the axis through the particles, which is called the *dipole axis*.

From symmetry, the electric field $\vec{E}$ at point P—and also the fields $\vec{E}_{(+)}$ and $\vec{E}_{(-)}$ due to the separate charges that make up the dipole—must lie along the dipole axis, which we have taken to be a z axis. Applying the superposition principle for electric fields, we find that the magnitude E of the electric field at P is

$$\begin{aligned} E &= E_{(+)} - E_{(-)} \\ &= \frac{1}{4\pi\varepsilon_0}\frac{q}{r_{(+)}^2} - \frac{1}{4\pi\varepsilon_0}\frac{q}{r_{(-)}^2} \\ &= \frac{q}{4\pi\varepsilon_0(z - \frac{1}{2}d)^2} - \frac{q}{4\pi\varepsilon_0(z + \frac{1}{2}d)^2}. \end{aligned} \tag{22-5}$$

After a little algebra, we can rewrite this equation as

$$E = \frac{q}{4\pi\varepsilon_0 z^2}\left(\frac{1}{\left(1 - \frac{d}{2z}\right)^2} - \frac{1}{\left(1 + \frac{d}{2z}\right)^2}\right). \tag{22-6}$$

After forming a common denominator and multiplying its terms, we come to

$$E = \frac{q}{4\pi\varepsilon_0 z^2}\frac{2d/z}{\left(1 - \left(\frac{d}{2z}\right)^2\right)^2} = \frac{q}{2\pi\varepsilon_0 z^3}\frac{d}{\left(1 - \left(\frac{d}{2z}\right)^2\right)^2}. \tag{22-7}$$

We are usually interested in the electrical effect of a dipole only at distances that are large compared with the dimensions of the dipole—that is, at distances such that $z \gg d$. At such large distances, we have $d/2z \ll 1$ in Eq. 22-7. Thus, in our approximation, we can neglect the $d/2z$ term in the denominator, which leaves us with

$$E = \frac{1}{2\pi\varepsilon_0}\frac{qd}{z^3}. \tag{22-8}$$

The product qd, which involves the two intrinsic properties q and d of the dipole, is the magnitude p of a vector quantity known as the **electric dipole moment** $\vec{p}$ of the dipole. (The unit of $\vec{p}$ is the coulomb-meter.) Thus, we can write Eq. 22-8 as

$$E = \frac{1}{2\pi\varepsilon_0}\frac{p}{z^3} \quad \text{(electric dipole).} \tag{22-9}$$

The direction of $\vec{p}$ is taken to be from the negative to the positive end of the dipole, as indicated in Fig. 22-8*b*. We can use the direction of $\vec{p}$ to specify the orientation of a dipole.

Equation 22-9 shows that, if we measure the electric field of a dipole only at distant points, we can never find q and d separately; instead, we can find only their product. The field at distant points would be unchanged if, for example, q were doubled and d simultaneously halved. Although Eq. 22-9 holds only for distant points along the dipole axis, it turns out that E for a dipole varies as $1/r^3$ for *all* distant points, regardless of whether they lie on the dipole axis; here r is the distance between the point in question and the dipole center.

Inspection of Fig. 22-8 and of the field lines in Fig. 22-5 shows that the direction of $\vec{E}$ for distant points on the dipole axis is always the direction of the dipole

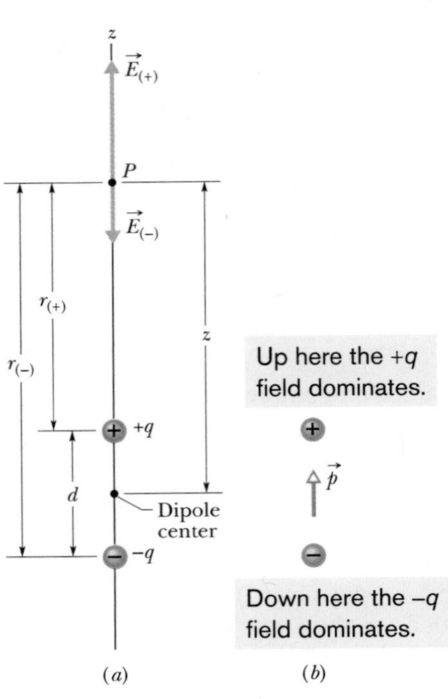

Fig. 22-8 (*a*) An electric dipole. The electric field vectors $\vec{E}_{(+)}$ and $\vec{E}_{(-)}$ at point P on the dipole axis result from the dipole's two charges. Point P is at distances $r_{(+)}$ and $r_{(-)}$ from the individual charges that make up the dipole. (*b*) The dipole moment $\vec{p}$ of the dipole points from the negative charge to the positive charge.

moment vector $\vec{p}$. This is true whether point P in Fig. 22-8*a* is on the upper or the lower part of the dipole axis.

Inspection of Eq. 22-9 shows that if you double the distance of a point from a dipole, the electric field at the point drops by a factor of 8. If you double the distance from a single point charge, however (see Eq. 22-3), the electric field drops only by a factor of 4. Thus the electric field of a dipole decreases more rapidly with distance than does the electric field of a single charge. The physical reason for this rapid decrease in electric field for a dipole is that from distant points a dipole looks like two equal but opposite charges that almost—but not quite—coincide. Thus, their electric fields at distant points almost—but not quite—cancel each other.

Sample Problem

Electric dipole and atmospheric sprites

Sprites (Fig. 22-9*a*) are huge flashes that occur far above a large thunderstorm. They were seen for decades by pilots flying at night, but they were so brief and dim that most pilots figured they were just illusions. Then in the 1990s sprites were captured on video. They are still not well understood but are believed to be produced when especially powerful lightning occurs between the ground and storm clouds, particularly when the lightning transfers a huge amount of negative charge $-q$ from the ground to the base of the clouds (Fig. 22-9*b*).

Just after such a transfer, the ground has a complicated distribution of positive charge. However, we can model the electric field due to the charges in the clouds and the ground by assuming a vertical electric dipole that has charge $-q$ at cloud height h and charge $+q$ at below-ground depth h (Fig. 22-9*c*). If $q = 200$ C and $h = 6.0$ km, what is the magnitude of the dipole's electric field at altitude $z_1 = 30$ km somewhat above the clouds and altitude $z_2 = 60$ km somewhat above the stratosphere?

KEY IDEA

We can approximate the magnitude E of an electric dipole's electric field on the dipole axis with Eq. 22-8.

Calculations: We write that equation as

$$E = \frac{1}{2\pi\varepsilon_0}\frac{q(2h)}{z^3},$$

where $2h$ is the separation between $-q$ and $+q$ in Fig. 22-9*c*. For the electric field at altitude $z_1 = 30$ km, we find

$$E = \frac{1}{2\pi\varepsilon_0}\frac{(200\text{ C})(2)(6.0\times10^3\text{ m})}{(30\times10^3\text{ m})^3}$$

$$= 1.6\times10^3\text{ N/C}. \qquad \text{(Answer)}$$

Similarly, for altitude $z_2 = 60$ km, we find

$$E = 2.0\times10^2\text{ N/C}. \qquad \text{(Answer)}$$

As we discuss in Section 22-8, when the magnitude of an electric field exceeds a certain critical value E_c, the field can pull electrons out of atoms (ionize the atoms), and then the freed electrons can run into other atoms, causing those atoms to emit light. The value of E_c depends on the density of the air in which the electric field exists. At altitude $z_2 = 60$ km the density of the air is so low that $E = 2.0\times10^2$ N/C exceeds E_c, and thus light is emitted by the atoms in the air. That light forms sprites. Lower down, just above the clouds at $z_1 = 30$ km, the density of the air is much higher, $E = 1.6\times10^3$ N/C does not exceed E_c, and no light is emitted. Hence, sprites occur only far above storm clouds.

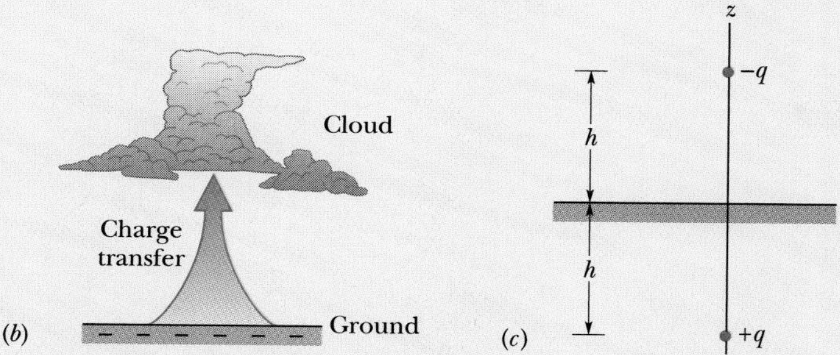

Fig. 22-9 (*a*) Photograph of a sprite. (*Courtesy NASA*) (*b*) Lightning in which a large amount of negative charge is transferred from ground to cloud base. (*c*) The cloud–ground system modeled as a vertical electric dipole.

Additional examples, video, and practice available at *WileyPLUS*

Table 22-2

Some Measures of Electric Charge

Name	Symbol	SI Unit
Charge	q	C
Linear charge density	λ	C/m
Surface charge density	σ	C/m^2
Volume charge density	ρ	C/m^3

22-6 The Electric Field Due to a Line of Charge

We now consider charge distributions that consist of a great many closely spaced point charges (perhaps billions) that are spread along a line, over a surface, or within a volume. Such distributions are said to be **continuous** rather than discrete. Since these distributions can include an enormous number of point charges, we find the electric fields that they produce by means of calculus rather than by considering the point charges one by one. In this section we discuss the electric field caused by a line of charge. We consider a charged surface in the next section. In the next chapter, we shall find the field inside a uniformly charged sphere.

When we deal with continuous charge distributions, it is most convenient to express the charge on an object as a *charge density* rather than as a total charge. For a line of charge, for example, we would report the *linear charge density* (or charge per unit length) λ, whose SI unit is the coulomb per meter. Table 22-2 shows the other charge densities we shall be using.

Figure 22-10 shows a thin ring of radius R with a uniform positive linear charge density λ around its circumference. We may imagine the ring to be made of plastic or some other insulator, so that the charges can be regarded as fixed in place. What is the electric field $\vec{E}$ at point P, a distance z from the plane of the ring along its central axis?

To answer, we cannot just apply Eq. 22-3, which gives the electric field set up by a point charge, because the ring is obviously not a point charge. However, we can mentally divide the ring into differential elements of charge that are so small that they are like point charges, and then we can apply Eq. 22-3 to each of them. Next, we can add the electric fields set up at P by all the differential elements. The vector sum of the fields gives us the field set up at P by the ring.

Let ds be the (arc) length of any differential element of the ring. Since λ is the charge per unit (arc) length, the element has a charge of magnitude

$$dq = \lambda\, ds. \tag{22-10}$$

This differential charge sets up a differential electric field $d\vec{E}$ at point P, which is a distance r from the element. Treating the element as a point charge and using Eq. 22-10, we can rewrite Eq. 22-3 to express the magnitude of $d\vec{E}$ as

$$dE = \frac{1}{4\pi\varepsilon_0}\frac{dq}{r^2} = \frac{1}{4\pi\varepsilon_0}\frac{\lambda\, ds}{r^2}. \tag{22-11}$$

From Fig. 22-10, we can rewrite Eq. 22-11 as

$$dE = \frac{1}{4\pi\varepsilon_0}\frac{\lambda\, ds}{(z^2 + R^2)}. \tag{22-12}$$

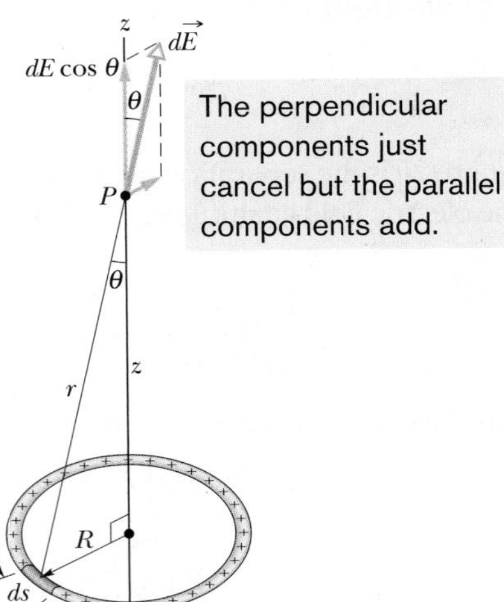

Fig. 22-10 A ring of uniform positive charge. A differential element of charge occupies a length ds (greatly exaggerated for clarity). This element sets up an electric field $d\vec{E}$ at point P. The component of $d\vec{E}$ along the central axis of the ring is $dE \cos\theta$.

Figure 22-10 shows that $d\vec{E}$ is at angle θ to the central axis (which we have taken to be a z axis) and has components perpendicular to and parallel to that axis.

Every charge element in the ring sets up a differential field $d\vec{E}$ at P, with magnitude given by Eq. 22-12. All the $d\vec{E}$ vectors have identical components parallel to the central axis, in both magnitude and direction. All these $d\vec{E}$ vectors have components perpendicular to the central axis as well; these perpendicular components are identical in magnitude but point in different directions. In fact, for any perpendicular component that points in a given direction, there is another one that points in the opposite direction. The sum of this pair of components, like the sum of all other pairs of oppositely directed components, is zero.

Thus, the perpendicular components cancel and we need not consider them further. This leaves the parallel components; they all have the same direction, so the net electric field at P is their sum.

The parallel component of $d\vec{E}$ shown in Fig. 22-10 has magnitude $dE \cos\theta$. The figure also shows us that

$$\cos\theta = \frac{z}{r} = \frac{z}{(z^2 + R^2)^{1/2}}. \tag{22-13}$$

Then multiplying Eq. 22-12 by Eq. 22-13 gives us, for the parallel component of $d\vec{E}$,

$$dE\cos\theta = \frac{z\lambda}{4\pi\varepsilon_0(z^2 + R^2)^{3/2}}\,ds. \tag{22-14}$$

To add the parallel components $dE \cos\theta$ produced by all the elements, we integrate Eq. 22-14 around the circumference of the ring, from $s = 0$ to $s = 2\pi R$. Since the only quantity in Eq. 22-14 that varies during the integration is s, the other quantities can be moved outside the integral sign. The integration then gives us

$$E = \int dE\cos\theta = \frac{z\lambda}{4\pi\varepsilon_0(z^2 + R^2)^{3/2}}\int_0^{2\pi R} ds$$

$$= \frac{z\lambda(2\pi R)}{4\pi\varepsilon_0(z^2 + R^2)^{3/2}}. \tag{22-15}$$

Since λ is the charge per length of the ring, the term $\lambda(2\pi R)$ in Eq. 22-15 is q, the total charge on the ring. We then can rewrite Eq. 22-15 as

$$E = \frac{qz}{4\pi\varepsilon_0(z^2 + R^2)^{3/2}} \quad \text{(charged ring)}. \tag{22-16}$$

If the charge on the ring is negative, instead of positive as we have assumed, the magnitude of the field at P is still given by Eq. 22-16. However, the electric field vector then points toward the ring instead of away from it.

Let us check Eq. 22-16 for a point on the central axis that is so far away that $z \gg R$. For such a point, the expression $z^2 + R^2$ in Eq. 22-16 can be approximated as z^2, and Eq. 22-16 becomes

$$E = \frac{1}{4\pi\varepsilon_0}\frac{q}{z^2} \quad \text{(charged ring at large distance)}. \tag{22-17}$$

This is a reasonable result because from a large distance, the ring "looks like" a point charge. If we replace z with r in Eq. 22-17, we indeed do have Eq. 22-3, the magnitude of the electric field due to a point charge.

Let us next check Eq. 22-16 for a point at the center of the ring—that is, for $z = 0$. At that point, Eq. 22-16 tells us that $E = 0$. This is a reasonable result because if we were to place a test charge at the center of the ring, there would be no net electrostatic force acting on it; the force due to any element of the ring would be canceled by the force due to the element on the opposite side of the ring. By Eq. 22-1, if the force at the center of the ring were zero, the electric field there would also have to be zero.

Sample Problem

Electric field of a charged circular rod

Figure 22-11*a* shows a plastic rod having a uniformly distributed charge $-Q$. The rod has been bent in a 120° circular arc of radius r. We place coordinate axes such that the axis of symmetry of the rod lies along the x axis and the origin is at the center of curvature P of the rod. In terms of Q and r, what is the electric field $\vec{E}$ due to the rod at point P?

KEY IDEA

Because the rod has a continuous charge distribution, we must find an expression for the electric fields due to differential elements of the rod and then sum those fields via calculus.

An element: Consider a differential element having arc length ds and located at an angle θ above the x axis (Figs. 22-11*b* and *c*). If we let λ represent the linear charge density of the rod, our element ds has a differential charge of magnitude

$$dq = \lambda\, ds. \tag{22-18}$$

The element's field: Our element produces a differential electric field $d\vec{E}$ at point P, which is a distance r from the element. Treating the element as a point charge, we can rewrite Eq. 22-3 to express the magnitude of $d\vec{E}$ as

$$dE = \frac{1}{4\pi\varepsilon_0}\frac{dq}{r^2} = \frac{1}{4\pi\varepsilon_0}\frac{\lambda\, ds}{r^2}. \tag{22-19}$$

The direction of $d\vec{E}$ is toward ds because charge dq is negative.

Symmetric partner: Our element has a symmetrically located (mirror image) element ds' in the bottom half of the rod. The electric field $d\vec{E}'$ set up at P by ds' also has the magnitude given by Eq. 22-19, but the field vector points toward ds' as shown in Fig. 22-11*d*. If we resolve the electric field vectors of ds and ds' into x and y components as shown in Figs. 22-11*e* and *f*, we see that their y components cancel (because they have equal magnitudes and are in opposite directions). We also see that their x components have equal magnitudes and are in the same direction.

Summing: Thus, to find the electric field set up by the rod, we need sum (via integration) only the x components of the differential electric fields set up by all the differential elements of the rod. From Fig. 22-11*f* and Eq. 22-19, we can write the component dE_x set up by ds as

$$dE_x = dE\cos\theta = \frac{1}{4\pi\varepsilon_0}\frac{\lambda}{r^2}\cos\theta\, ds. \tag{22-20}$$

Equation 22-20 has two variables, θ and s. Before we can integrate it, we must eliminate one variable. We do so by replacing ds, using the relation

$$ds = r\, d\theta,$$

in which $d\theta$ is the angle at P that includes arc length ds (Fig. 22-11*g*). With this replacement, we can integrate Eq. 22-20 over the angle made by the rod at P, from $\theta = -60°$ to $\theta = 60°$; that will give us the magnitude of the electric field at P due to the rod:

$$E = \int dE_x = \int_{-60°}^{60°} \frac{1}{4\pi\varepsilon_0}\frac{\lambda}{r^2}\cos\theta\, r\, d\theta$$

$$= \frac{\lambda}{4\pi\varepsilon_0 r}\int_{-60°}^{60°}\cos\theta\, d\theta = \frac{\lambda}{4\pi\varepsilon_0 r}\Big[\sin\theta\Big]_{-60°}^{60°}$$

$$= \frac{\lambda}{4\pi\varepsilon_0 r}[\sin 60° - \sin(-60°)]$$

$$= \frac{1.73\lambda}{4\pi\varepsilon_0 r}. \tag{22-21}$$

(If we had reversed the limits on the integration, we would have gotten the same result but with a minus sign. Since the integration gives only the magnitude of $\vec{E}$, we would then have discarded the minus sign.)

Charge density: To evaluate λ, we note that the rod subtends an angle of 120° and so is one-third of a full circle. Its arc length is then $2\pi r/3$, and its linear charge density must be

$$\lambda = \frac{\text{charge}}{\text{length}} = \frac{Q}{2\pi r/3} = \frac{0.477Q}{r}.$$

Substituting this into Eq. 22-21 and simplifying give us

$$E = \frac{(1.73)(0.477Q)}{4\pi\varepsilon_0 r^2}$$

$$= \frac{0.83Q}{4\pi\varepsilon_0 r^2}. \qquad \text{(Answer)}$$

The direction of $\vec{E}$ is toward the rod, along the axis of symmetry of the charge distribution. We can write $\vec{E}$ in unit-vector notation as

$$\vec{E} = \frac{0.83Q}{4\pi\varepsilon_0 r^2}\hat{\text{i}}.$$

Additional examples, video, and practice available at *WileyPLUS*

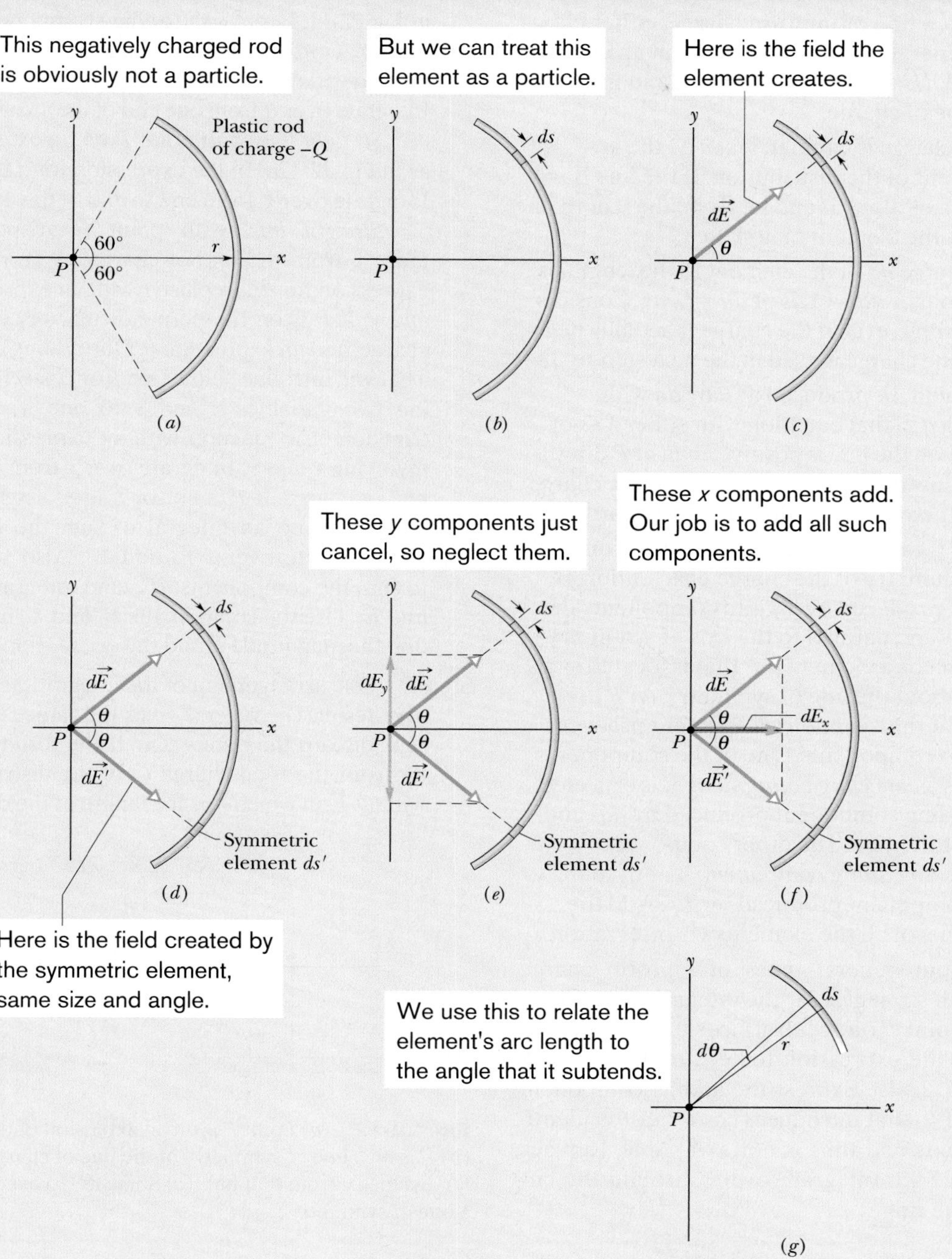

Fig. 22-11 (*a*) A plastic rod of charge $-Q$ is a circular section of radius r and central angle 120°; point P is the center of curvature of the rod. (*b*)–(*c*) A differential element in the top half of the rod, at an angle θ to the x axis and of arc length ds, sets up a differential electric field $d\vec{E}$ at P. (*d*) An element ds', symmetric to ds about the x axis, sets up a field $d\vec{E}'$ at P with the same magnitude. (*e*)–(*f*) The field components. (*g*) Arc length ds makes an angle $d\theta$ about point P.

Problem-Solving Tactics

A Field Guide for Lines of Charge

Here is a generic guide for finding the electric field $\vec{E}$ produced at a point P by a line of uniform charge, either circular or straight. The general strategy is to pick out an element dq of the charge, find $d\vec{E}$ due to that element, and integrate $d\vec{E}$ over the entire line of charge.

Step 1. If the line of charge is circular, let ds be the arc length of an element of the distribution. If the line is straight, run an x axis along it and let dx be the length of an element. Mark the element on a sketch.

Step 2. Relate the charge dq of the element to the length of the element with either $dq = \lambda\, ds$ or $dq = \lambda\, dx$. Consider dq and λ to be positive, even if the charge is actually negative. (The sign of the charge is used in the next step.)

Step 3. Express the field $d\vec{E}$ produced at P by dq with Eq. 22-3, replacing q in that equation with either $\lambda\, ds$ or $\lambda\, dx$. If the charge on the line is positive, then at P draw a vector $d\vec{E}$ that points directly away from dq. If the charge is negative, draw the vector pointing directly toward dq.

Step 4. Always look for any symmetry in the situation. If P is on an axis of symmetry of the charge distribution, resolve the field $d\vec{E}$ produced by dq into components that are perpendicular and parallel to the axis of symmetry. Then consider a second element dq' that is located symmetrically to dq about the line of symmetry. At P draw the vector $d\vec{E}'$ that this symmetrical element produces and resolve it into components. One of the components produced by dq is a *canceling component;* it is canceled by the corresponding component produced by dq' and needs no further attention. The other component produced by dq is an *adding component;* it adds to the corresponding component produced by dq'. Add the adding components of all the elements via integration.

Step 5. Here are four general types of uniform charge distributions, with strategies for the integral of step 4.

Ring, with point P on (central) axis of symmetry, as in Fig. 22-10. In the expression for dE, replace r^2 with $z^2 + R^2$, as in Eq. 22-12. Express the adding component of $d\vec{E}$ in terms of θ. That introduces $\cos\theta$, but θ is identical for all elements and thus is not a variable. Replace $\cos\theta$ as in Eq. 22-13. Integrate over s, around the circumference of the ring.

Circular arc, with point P at the center of curvature, as in Fig. 22-11. Express the adding component of $d\vec{E}$ in terms of θ. That introduces either $\sin\theta$ or $\cos\theta$. Reduce the resulting two variables s and θ to one, θ, by replacing ds with $r\, d\theta$. Integrate over θ from one end of the arc to the other end.

Straight line, with point P on an extension of the line, as in Fig. 22-12a. In the expression for dE, replace r with x. Integrate over x, from end to end of the line of charge.

Straight line, with point P at perpendicular distance y from the line of charge, as in Fig. 22-12b. In the expression for dE, replace r with an expression involving x and y. If P is on the perpendicular bisector of the line of charge, find an expression for the adding component of $d\vec{E}$. That will introduce either $\sin\theta$ or $\cos\theta$. Reduce the resulting two variables x and θ to one, x, by replacing the trigonometric function with an expression (its definition) involving x and y. Integrate over x from end to end of the line of charge. If P is not on a line of symmetry, as in Fig. 22-12c, set up an integral to sum the components dE_x, and integrate over x to find E_x. Also set up an integral to sum the components dE_y, and integrate over x again to find E_y. Use the components E_x and E_y in the usual way to find the magnitude E and the orientation of $\vec{E}$.

Step 6. One arrangement of the integration limits gives a positive result. The reverse gives the same result with a minus sign; discard the minus sign. If the result is to be stated in terms of the total charge Q of the distribution, replace λ with Q/L, in which L is the length of the distribution.

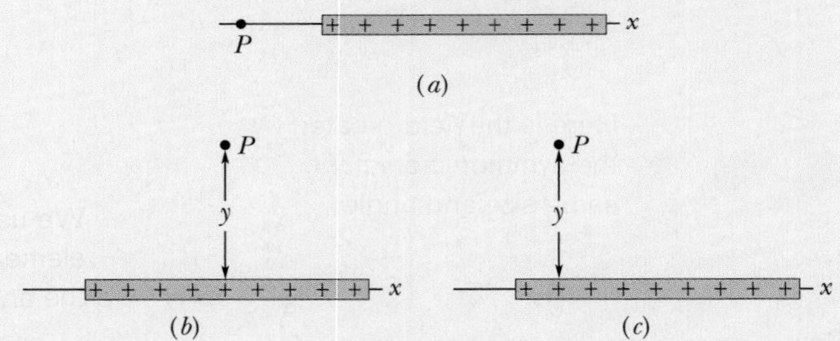

Fig. 22-12 (a) Point P is on an extension of the line of charge. (b) P is on a line of symmetry of the line of charge, at perpendicular distance y from that line. (c) Same as (b) except that P is not on a line of symmetry.

CHECKPOINT 2

The figure here shows three nonconducting rods, one circular and two straight. Each has a uniform charge of magnitude Q along its top half and another along its bottom half. For each rod, what is the direction of the net electric field at point P?

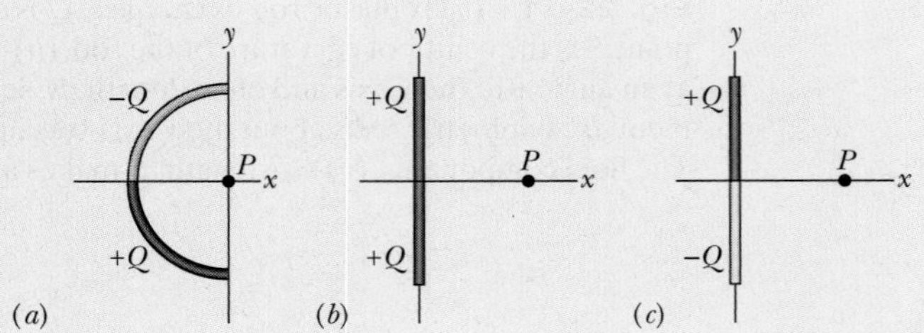

22-7 The Electric Field Due to a Charged Disk

Figure 22-13 shows a circular plastic disk of radius R that has a positive surface charge of uniform density σ on its upper surface (see Table 22-2). What is the electric field at point P, a distance z from the disk along its central axis?

Our plan is to divide the disk into concentric flat rings and then to calculate the electric field at point P by adding up (that is, by integrating) the contributions of all the rings. Figure 22-13 shows one such ring, with radius r and radial width dr. Since σ is the charge per unit area, the charge on the ring is

$$dq = \sigma\, dA = \sigma(2\pi r\, dr), \tag{22-22}$$

where dA is the differential area of the ring.

We have already solved the problem of the electric field due to a ring of charge. Substituting dq from Eq. 22-22 for q in Eq. 22-16, and replacing R in Eq. 22-16 with r, we obtain an expression for the electric field dE at P due to the arbitrarily chosen flat ring of charge shown in Fig. 22-13:

$$dE = \frac{z\sigma 2\pi r\, dr}{4\pi\varepsilon_0(z^2 + r^2)^{3/2}},$$

which we may write as

$$dE = \frac{\sigma z}{4\varepsilon_0}\frac{2r\, dr}{(z^2 + r^2)^{3/2}}. \tag{22-23}$$

We can now find E by integrating Eq. 22-23 over the surface of the disk—that is, by integrating with respect to the variable r from $r = 0$ to $r = R$. Note that z remains constant during this process. We get

$$E = \int dE = \frac{\sigma z}{4\varepsilon_0}\int_0^R (z^2 + r^2)^{-3/2}(2r)\, dr. \tag{22-24}$$

To solve this integral, we cast it in the form $\int X^m\, dX$ by setting $X = (z^2 + r^2)$, $m = -\frac{3}{2}$, and $dX = (2r)\, dr$. For the recast integral we have

$$\int X^m\, dX = \frac{X^{m+1}}{m+1},$$

and so Eq. 22-24 becomes

$$E = \frac{\sigma z}{4\varepsilon_0}\left[\frac{(z^2 + r^2)^{-1/2}}{-\frac{1}{2}}\right]_0^R. \tag{22-25}$$

Taking the limits in Eq. 22-25 and rearranging, we find

$$E = \frac{\sigma}{2\varepsilon_0}\left(1 - \frac{z}{\sqrt{z^2 + R^2}}\right) \quad \text{(charged disk)} \tag{22-26}$$

as the magnitude of the electric field produced by a flat, circular, charged disk at points on its central axis. (In carrying out the integration, we assumed that $z \geq 0$.)

If we let $R \to \infty$ while keeping z finite, the second term in the parentheses in Eq. 22-26 approaches zero, and this equation reduces to

$$E = \frac{\sigma}{2\varepsilon_0} \quad \text{(infinite sheet).} \tag{22-27}$$

This is the electric field produced by an infinite sheet of uniform charge located on one side of a nonconductor such as plastic. The electric field lines for such a situation are shown in Fig. 22-3.

We also get Eq. 22-27 if we let $z \to 0$ in Eq. 22-26 while keeping R finite. This shows that at points very close to the disk, the electric field set up by the disk is the same as if the disk were infinite in extent.

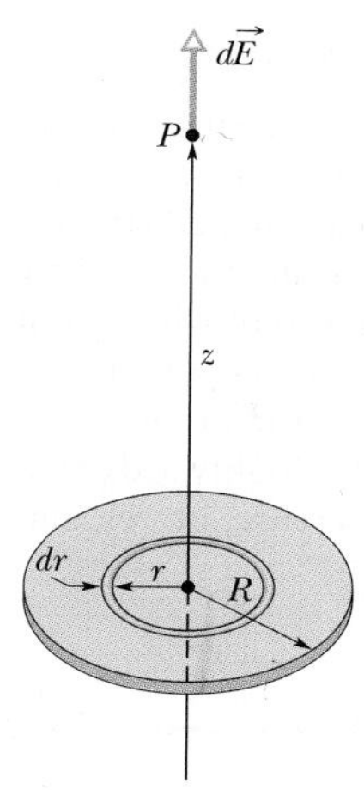

Fig. 22-13 A disk of radius R and uniform positive charge. The ring shown has radius r and radial width dr. It sets up a differential electric field $d\vec{E}$ at point P on its central axis.

CHECKPOINT 3

(a) In the figure, what is the direction of the electrostatic force on the electron due to the external electric field shown? (b) In which direction will the electron accelerate if it is moving parallel to the y axis before it encounters the external field? (c) If, instead, the electron is initially moving rightward, will its speed increase, decrease, or remain constant?

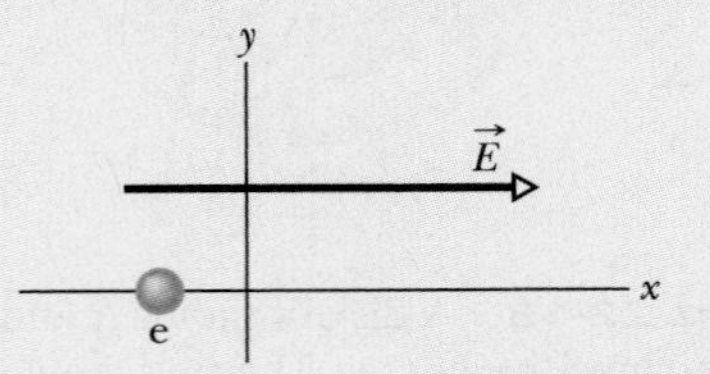

22-8 A Point Charge in an Electric Field

In the preceding four sections we worked at the first of our two tasks: given a charge distribution, to find the electric field it produces in the surrounding space. Here we begin the second task: to determine what happens to a charged particle when it is in an electric field set up by other stationary or slowly moving charges.

What happens is that an electrostatic force acts on the particle, as given by

$$\vec{F} = q\vec{E}, \tag{22-28}$$

in which q is the charge of the particle (including its sign) and $\vec{E}$ is the electric field that other charges have produced at the location of the particle. (The field is *not* the field set up by the particle itself; to distinguish the two fields, the field acting on the particle in Eq. 22-28 is often called the *external field.* A charged particle or object is not affected by its own electric field.) Equation 22-28 tells us

The electrostatic force $\vec{F}$ acting on a charged particle located in an external electric field $\vec{E}$ has the direction of $\vec{E}$ if the charge q of the particle is positive and has the opposite direction if q is negative.

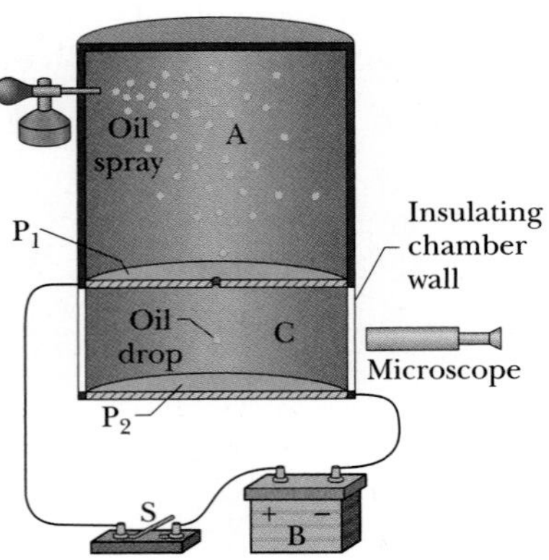

Fig. 22-14 The Millikan oil-drop apparatus for measuring the elementary charge e. When a charged oil drop drifted into chamber C through the hole in plate P_1, its motion could be controlled by closing and opening switch S and thereby setting up or eliminating an electric field in chamber C. The microscope was used to view the drop, to permit timing of its motion.

Measuring the Elementary Charge

Equation 22-28 played a role in the measurement of the elementary charge e by American physicist Robert A. Millikan in 1910–1913. Figure 22-14 is a representation of his apparatus. When tiny oil drops are sprayed into chamber A, some of them become charged, either positively or negatively, in the process. Consider a drop that drifts downward through the small hole in plate P_1 and into chamber C. Let us assume that this drop has a negative charge q.

If switch S in Fig. 22-14 is open as shown, battery B has no electrical effect on chamber C. If the switch is closed (the connection between chamber C and the positive terminal of the battery is then complete), the battery causes an excess positive charge on conducting plate P_1 and an excess negative charge on conducting plate P_2. The charged plates set up a downward-directed electric field $\vec{E}$ in chamber C. According to Eq. 22-28, this field exerts an electrostatic force on any charged drop that happens to be in the chamber and affects its motion. In particular, our negatively charged drop will tend to drift upward.

By timing the motion of oil drops with the switch opened and with it closed and thus determining the effect of the charge q, Millikan discovered that the values of q were always given by

$$q = ne, \qquad \text{for } n = 0, \pm 1, \pm 2, \pm 3, \ldots, \tag{22-29}$$

in which e turned out to be the fundamental constant we call the *elementary charge,* 1.60×10^{-19} C. Millikan's experiment is convincing proof that charge is quantized, and he earned the 1923 Nobel Prize in physics in part for this work. Modern measurements of the elementary charge rely on a variety of interlocking experiments, all more precise than the pioneering experiment of Millikan.

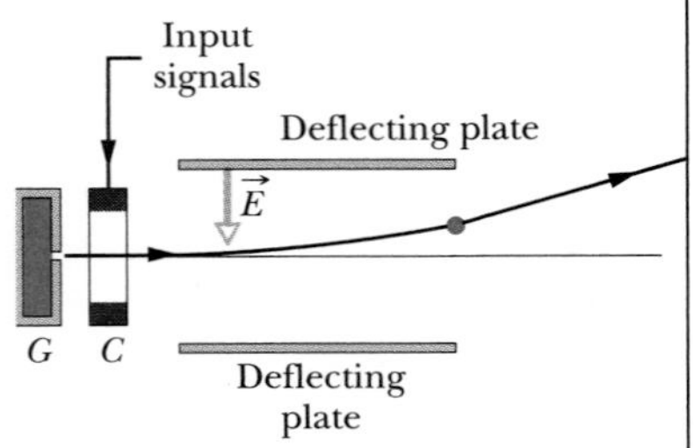

Fig. 22-15 Ink-jet printer. Drops shot from generator G receive a charge in charging unit C. An input signal from a computer controls the charge and thus the effect of field $\vec{E}$ on where the drop lands on the paper.

Ink-Jet Printing

The need for high-quality, high-speed printing has caused a search for an alternative to impact printing, such as occurs in a standard typewriter. Building up letters by squirting tiny drops of ink at the paper is one such alternative.

Figure 22-15 shows a negatively charged drop moving between two conducting deflecting plates, between which a uniform, downward-directed electric field $\vec{E}$ has been set up. The drop is deflected upward according to Eq. 22-28 and then strikes the paper at a position that is determined by the magnitudes of $\vec{E}$ and the charge q of the drop.

In practice, E is held constant and the position of the drop is determined by the charge q delivered to the drop in the charging unit, through which the drop must pass before entering the deflecting system. The charging unit, in turn, is activated by electronic signals that encode the material to be printed.

Electrical Breakdown and Sparking

If the magnitude of an electric field in air exceeds a certain critical value E_c, the air undergoes *electrical breakdown,* a process whereby the field removes electrons from the atoms in the air. The air then begins to conduct electric current because the freed electrons are propelled into motion by the field. As they move, they collide with any atoms in their path, causing those atoms to emit light. We can see the paths, commonly called sparks, taken by the freed electrons because of that emitted light. Figure 22-16 shows sparks above charged metal wires where the electric fields due to the wires cause electrical breakdown of the air.

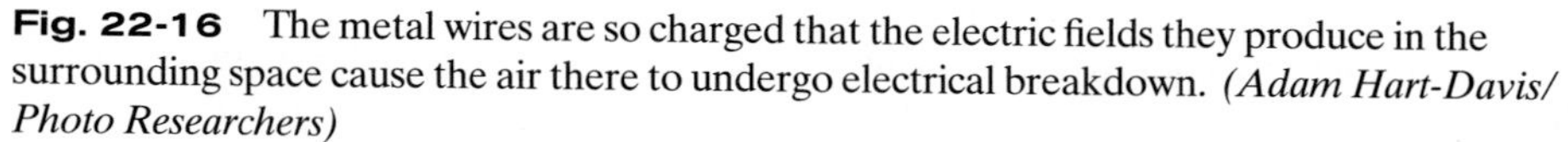

Fig. 22-16 The metal wires are so charged that the electric fields they produce in the surrounding space cause the air there to undergo electrical breakdown. *(Adam Hart-Davis/ Photo Researchers)*

Sample Problem

Motion of a charged particle in an electric field

Figure 22-17 shows the deflecting plates of an ink-jet printer, with superimposed coordinate axes. An ink drop with a mass m of 1.3×10^{-10} kg and a negative charge of magnitude $Q = 1.5 \times 10^{-13}$ C enters the region between the plates, initially moving along the x axis with speed $v_x = 18$ m/s. The length L of each plate is 1.6 cm. The plates are charged and thus produce an electric field at all points between them. Assume that field $\vec{E}$ is downward directed, is uniform, and has a magnitude of 1.4×10^6 N/C. What is the vertical deflection of the drop at the far edge of the plates? (The gravitational force on the drop is small relative to the electrostatic force acting on the drop and can be neglected.)

KEY IDEA

The drop is negatively charged and the electric field is directed *downward.* From Eq. 22-28, a constant electrostatic force of magnitude QE acts *upward* on the charged drop. Thus, as the drop travels parallel to the x axis at constant speed v_x, it accelerates upward with some constant acceleration a_y.

Calculations: Applying Newton's second law ($F = ma$) for components along the y axis, we find that

$$a_y = \frac{F}{m} = \frac{QE}{m}. \tag{22-30}$$

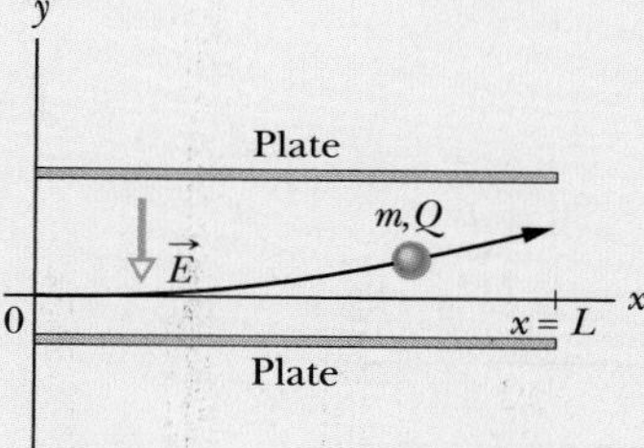

Fig. 22-17 An ink drop of mass m and charge magnitude Q is deflected in the electric field of an ink-jet printer.

Let t represent the time required for the drop to pass through the region between the plates. During t the vertical and horizontal displacements of the drop are

$$y = \tfrac{1}{2}a_y t^2 \quad \text{and} \quad L = v_x t, \tag{22-31}$$

respectively. Eliminating t between these two equations and substituting Eq. 22-30 for a_y, we find

$$y = \frac{QEL^2}{2mv_x^2}$$

$$= \frac{(1.5 \times 10^{-13}\ \text{C})(1.4 \times 10^6\ \text{N/C})(1.6 \times 10^{-2}\ \text{m})^2}{(2)(1.3 \times 10^{-10}\ \text{kg})(18\ \text{m/s})^2}$$

$$= 6.4 \times 10^{-4}\ \text{m}$$

$$= 0.64\ \text{mm}. \qquad \text{(Answer)}$$

Additional examples, video, and practice available at *WileyPLUS*

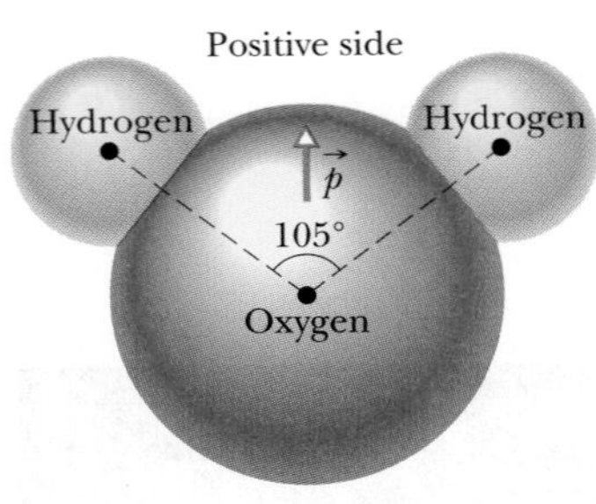

Fig. 22-18 A molecule of H_2O, showing the three nuclei (represented by dots) and the regions in which the electrons can be located. The electric dipole moment $\vec{p}$ points from the (negative) oxygen side to the (positive) hydrogen side of the molecule.

22-9 A Dipole in an Electric Field

We have defined the electric dipole moment $\vec{p}$ of an electric dipole to be a vector that points from the negative to the positive end of the dipole. As you will see, the behavior of a dipole in a uniform external electric field $\vec{E}$ can be described completely in terms of the two vectors $\vec{E}$ and $\vec{p}$, with no need of any details about the dipole's structure.

A molecule of water (H_2O) is an electric dipole; Fig. 22-18 shows why. There the black dots represent the oxygen nucleus (having eight protons) and the two hydrogen nuclei (having one proton each). The colored enclosed areas represent the regions in which electrons can be located around the nuclei.

In a water molecule, the two hydrogen atoms and the oxygen atom do not lie on a straight line but form an angle of about 105°, as shown in Fig. 22-18. As a result, the molecule has a definite "oxygen side" and "hydrogen side." Moreover, the 10 electrons of the molecule tend to remain closer to the oxygen nucleus than to the hydrogen nuclei. This makes the oxygen side of the molecule slightly more negative than the hydrogen side and creates an electric dipole moment $\vec{p}$ that points along the symmetry axis of the molecule as shown. If the water molecule is placed in an external electric field, it behaves as would be expected of the more abstract electric dipole of Fig. 22-8.

To examine this behavior, we now consider such an abstract dipole in a uniform external electric field $\vec{E}$, as shown in Fig. 22-19*a*. We assume that the dipole is a rigid structure that consists of two centers of opposite charge, each of magnitude q, separated by a distance d. The dipole moment $\vec{p}$ makes an angle θ with field $\vec{E}$.

Electrostatic forces act on the charged ends of the dipole. Because the electric field is uniform, those forces act in opposite directions (as shown in Fig. 22-19*a*) and with the same magnitude $F = qE$. Thus, *because the field is uniform,* the net force on the dipole from the field is zero and the center of mass of the dipole does not move. However, the forces on the charged ends do produce a net torque $\vec{\tau}$ on the dipole about its center of mass. The center of mass lies on the line connecting the charged ends, at some distance x from one end and thus a distance $d - x$ from the other end. From Eq. 10-39 ($\tau = rF \sin \phi$), we can write the magnitude of the net torque $\vec{\tau}$ as

$$\tau = Fx \sin \theta + F(d - x) \sin \theta = Fd \sin \theta. \tag{22-32}$$

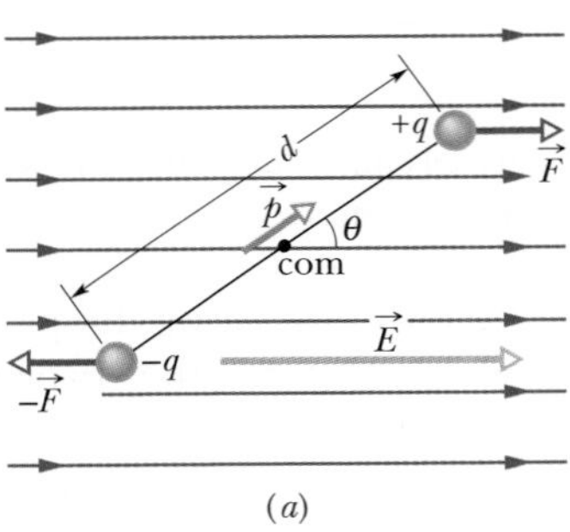

We can also write the magnitude of $\vec{\tau}$ in terms of the magnitudes of the electric field E and the dipole moment $p = qd$. To do so, we substitute qE for F and p/q for d in Eq. 22-32, finding that the magnitude of $\vec{\tau}$ is

$$\tau = pE \sin \theta. \tag{22-33}$$

We can generalize this equation to vector form as

$$\vec{\tau} = \vec{p} \times \vec{E} \quad \text{(torque on a dipole).} \tag{22-34}$$

The dipole is being torqued into alignment.

$\vec{p}$ θ $\vec{E}$ $\vec{\tau}$ ⊗

(b)

Fig. 22-19 (*a*) An electric dipole in a uniform external electric field $\vec{E}$. Two centers of equal but opposite charge are separated by distance d. The line between them represents their rigid connection. (*b*) Field $\vec{E}$ causes a torque $\vec{\tau}$ on the dipole. The direction of $\vec{\tau}$ is into the page, as represented by the symbol ⊗.

Vectors $\vec{p}$ and $\vec{E}$ are shown in Fig. 22-19*b*. The torque acting on a dipole tends to rotate $\vec{p}$ (hence the dipole) into the direction of field $\vec{E}$, thereby reducing θ. In Fig. 22-19, such rotation is clockwise. As we discussed in Chapter 10, we can represent a torque that gives rise to a clockwise rotation by including a minus sign with the magnitude of the torque. With that notation, the torque of Fig. 22-19 is

$$\tau = -pE \sin \theta. \tag{22-35}$$

Potential Energy of an Electric Dipole

Potential energy can be associated with the orientation of an electric dipole in an electric field. The dipole has its least potential energy when it is in its equilibrium orientation, which is when its moment $\vec{p}$ is lined up with the field $\vec{E}$ (then

$\vec{\tau} = \vec{p} \times \vec{E} = 0$). It has greater potential energy in all other orientations. Thus the dipole is like a pendulum, which has *its* least gravitational potential energy in *its* equilibrium orientation—at its lowest point. To rotate the dipole or the pendulum to any other orientation requires work by some external agent.

In any situation involving potential energy, we are free to define the zero-potential-energy configuration in a perfectly arbitrary way because only differences in potential energy have physical meaning. It turns out that the expression for the potential energy of an electric dipole in an external electric field is simplest if we choose the potential energy to be zero when the angle θ in Fig. 22-19 is 90°. We then can find the potential energy U of the dipole at any other value of θ with Eq. 8-1 ($\Delta U = -W$) by calculating the work W done by the field on the dipole when the dipole is rotated to that value of θ from 90°. With the aid of Eq. 10-53 ($W = \int \tau \, d\theta$) and Eq. 22-35, we find that the potential energy U at any angle θ is

$$U = -W = -\int_{90^\circ}^{\theta} \tau \, d\theta = \int_{90^\circ}^{\theta} pE \sin\theta \, d\theta. \tag{22-36}$$

Evaluating the integral leads to

$$U = -pE\cos\theta. \tag{22-37}$$

We can generalize this equation to vector form as

$$U = -\vec{p} \cdot \vec{E} \quad \text{(potential energy of a dipole).} \tag{22-38}$$

Equations 22-37 and 22-38 show us that the potential energy of the dipole is least ($U = -pE$) when $\theta = 0$ ($\vec{p}$ and $\vec{E}$ are in the same direction); the potential energy is greatest ($U = pE$) when $\theta = 180°$ ($\vec{p}$ and $\vec{E}$ are in opposite directions).

When a dipole rotates from an initial orientation θ_i to another orientation θ_f, the work W done on the dipole by the electric field is

$$W = -\Delta U = -(U_f - U_i), \tag{22-39}$$

where U_f and U_i are calculated with Eq. 22-38. If the change in orientation is caused by an applied torque (commonly said to be due to an external agent), then the work W_a done on the dipole by the applied torque is the negative of the work done on the dipole by the field; that is,

$$W_a = -W = (U_f - U_i). \tag{22-40}$$

CHECKPOINT 4

The figure shows four orientations of an electric dipole in an external electric field. Rank the orientations according to (a) the magnitude of the torque on the dipole and (b) the potential energy of the dipole, greatest first.

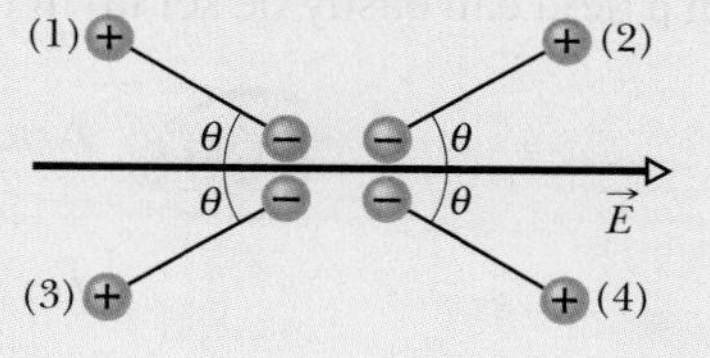

Microwave Cooking

Food can be warmed and cooked in a microwave oven if the food contains water because water molecules are electric dipoles. When you turn on the oven, the microwave source sets up a rapidly oscillating electric field $\vec{E}$ within the oven and thus also within the food. From Eq. 22-34, we see that any electric field $\vec{E}$ produces a torque on an electric dipole moment $\vec{p}$ to align $\vec{p}$ with $\vec{E}$. Because the oven's $\vec{E}$ oscillates, the water molecules continuously flip-flop in a frustrated attempt to align with $\vec{E}$.

Energy is transferred from the electric field to the thermal energy of the water (and thus of the food) where three water molecules happened to have bonded together to form a group. The flip-flop breaks some of the bonds. When the molecules reform the bonds, energy is transferred to the random motion of the group and then to the surrounding molecules. Soon, the thermal energy of the water is enough to cook the food. Sometimes the heating is surprising. If you heat a jelly donut, for example, the jelly (which holds a lot of water) heats far more than the donut material (which holds much less water). Although the exterior of the donut may not be hot, biting into the jelly can burn you. If water molecules were not electric dipoles, we would not have microwave ovens.

Sample Problem

Torque and energy of an electric dipole in an electric field

A neutral water molecule (H_2O) in its vapor state has an electric dipole moment of magnitude 6.2×10^{-30} C·m.

(a) How far apart are the molecule's centers of positive and negative charge?

KEY IDEA

A molecule's dipole moment depends on the magnitude q of the molecule's positive or negative charge and the charge separation d.

Calculations: There are 10 electrons and 10 protons in a neutral water molecule; so the magnitude of its dipole moment is

$$p = qd = (10e)(d),$$

in which d is the separation we are seeking and e is the elementary charge. Thus,

$$d = \frac{p}{10e} = \frac{6.2 \times 10^{-30}\ \text{C}\cdot\text{m}}{(10)(1.60 \times 10^{-19}\ \text{C})}$$
$$= 3.9 \times 10^{-12}\ \text{m} = 3.9\ \text{pm}. \qquad \text{(Answer)}$$

This distance is not only small, but it is also actually smaller than the radius of a hydrogen atom.

(b) If the molecule is placed in an electric field of 1.5×10^4 N/C, what maximum torque can the field exert on it? (Such a field can easily be set up in the laboratory.)

KEY IDEA

The torque on a dipole is maximum when the angle θ between $\vec{p}$ and $\vec{E}$ is 90°.

Calculation: Substituting $\theta = 90°$ in Eq. 22-33 yields

$$\tau = pE \sin\theta$$
$$= (6.2 \times 10^{-30}\ \text{C}\cdot\text{m})(1.5 \times 10^4\ \text{N/C})(\sin 90°)$$
$$= 9.3 \times 10^{-26}\ \text{N}\cdot\text{m}. \qquad \text{(Answer)}$$

(c) How much work must an *external agent* do to rotate this molecule by 180° in this field, starting from its fully aligned position, for which $\theta = 0$?

KEY IDEA

The work done by an external agent (by means of a torque applied to the molecule) is equal to the change in the molecule's potential energy due to the change in orientation.

Calculation: From Eq. 22-40, we find

$$W_a = U_{180°} - U_0$$
$$= (-pE\cos 180°) - (-pE\cos 0)$$
$$= 2pE = (2)(6.2 \times 10^{-30}\ \text{C}\cdot\text{m})(1.5 \times 10^4\ \text{N/C})$$
$$= 1.9 \times 10^{-25}\ \text{J}. \qquad \text{(Answer)}$$

Additional examples, video, and practice available at *WileyPLUS*

REVIEW & SUMMARY

Electric Field To explain the electrostatic force between two charges, we assume that each charge sets up an electric field in the space around it. The force acting on each charge is then due to the electric field set up at its location by the other charge.

Definition of Electric Field The *electric field* $\vec{E}$ at any point is defined in terms of the electrostatic force $\vec{F}$ that would be exerted on a positive test charge q_0 placed there:

$$\vec{E} = \frac{\vec{F}}{q_0}. \qquad (22\text{-}1)$$

Electric Field Lines *Electric field lines* provide a means for visualizing the direction and magnitude of electric fields. The electric field vector at any point is tangent to a field line through that point. The density of field lines in any region is proportional to the magnitude of the electric field in that region. Field lines originate on positive charges and terminate on negative charges.

Field Due to a Point Charge The magnitude of the electric field $\vec{E}$ set up by a point charge q at a distance r from the charge is

$$\vec{E} = \frac{1}{4\pi\varepsilon_0}\frac{q}{r^2}\hat{r} \qquad (22\text{-}3)$$

The direction of $\vec{E}$ is away from the point charge if the charge is positive and toward it if the charge is negative.

Field Due to an Electric Dipole An *electric dipole* consists of two particles with charges of equal magnitude q but opposite sign, separated by a small distance d. Their **electric dipole moment** $\vec{p}$ has magnitude qd and points from the negative charge to the positive charge. The magnitude of the electric field set up by the dipole at a distant point on the dipole axis (which runs through both charges) is

$$E = \frac{1}{2\pi\varepsilon_0}\frac{p}{z^3}, \qquad (22\text{-}9)$$

where z is the distance between the point and the center of the dipole.

Field Due to a Continuous Charge Distribution The electric field due to a *continuous charge distribution* is found by treating charge elements as point charges and then summing, via integration, the electric field vectors produced by all the charge elements to find the net vector.

Force on a Point Charge in an Electric Field When a point charge q is placed in an external electric field $\vec{E}$, the electrostatic force $\vec{F}$ that acts on the point charge is

$$\vec{F} = q\vec{E}. \tag{22-28}$$

Force $\vec{F}$ has the same direction as $\vec{E}$ if q is positive and the opposite direction if q is negative.

Dipole in an Electric Field When an electric dipole of dipole moment $\vec{p}$ is placed in an electric field $\vec{E}$, the field exerts a torque $\vec{\tau}$ on the dipole:

$$\vec{\tau} = \vec{p} \times \vec{E}. \tag{22-34}$$

The dipole has a potential energy U associated with its orientation in the field:

$$U = -\vec{p} \cdot \vec{E}. \tag{22-38}$$

This potential energy is defined to be zero when $\vec{p}$ is perpendicular to $\vec{E}$; it is least ($U = -pE$) when $\vec{p}$ is aligned with $\vec{E}$ and greatest ($U = pE$) when $\vec{p}$ is directed opposite $\vec{E}$.

QUESTIONS

1 Figure 22-20 shows three arrangements of electric field lines. In each arrangement, a proton is released from rest at point A and is then accelerated through point B by the electric field. Points A and B have equal separations in the three arrangements. Rank the arrangements according to the linear momentum of the proton at point B, greatest first.

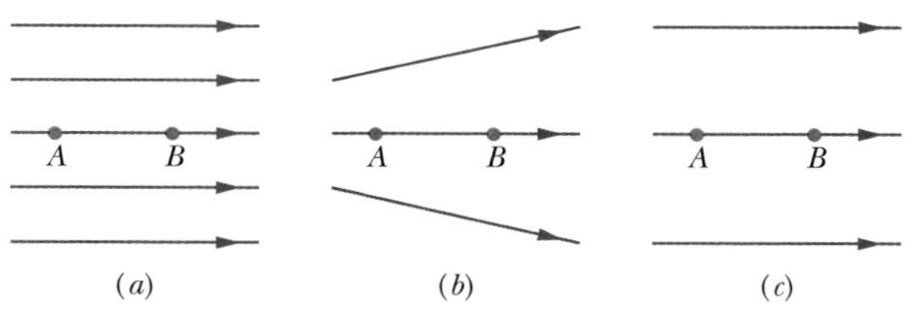

Fig. 22-20 Question 1.

2 Figure 22-21 shows two square arrays of charged particles. The squares, which are centered on point P, are misaligned. The particles are separated by either d or $d/2$ along the perimeters of the squares. What are the magnitude and direction of the net electric field at P?

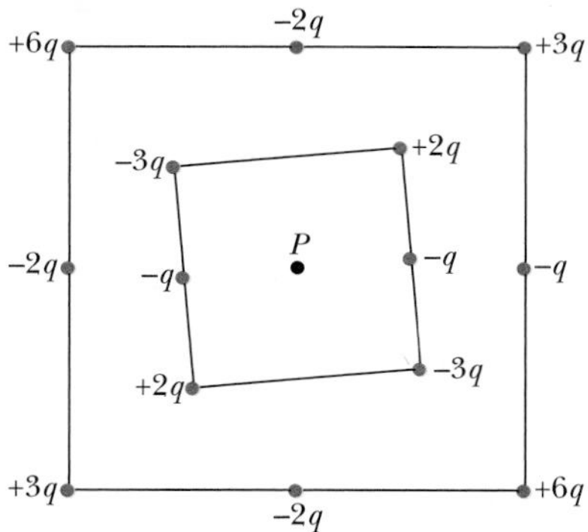

Fig. 22-21 Question 2.

3 In Fig. 22-22, two particles of charge $-q$ are arranged symmetrically about the y axis; each produces an electric field at point P on that axis. (a) Are the magnitudes of the fields at P equal? (b) Is each electric field directed toward or away from the charge producing it? (c) Is the magnitude of the net electric field at P equal to the sum of the magnitudes E of the two field vectors (is it equal to $2E$)? (d) Do the x components of those two field vectors add or cancel? (e) Do their y components add or cancel? (f) Is the direction of the net field at P that of the canceling components or the adding components? (g) What is the direction of the net field?

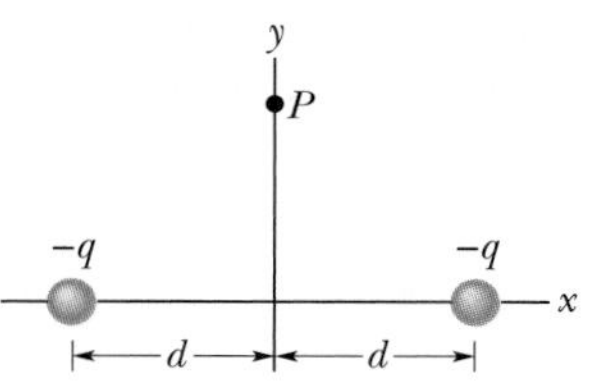

Fig. 22-22 Question 3.

4 Figure 22-23 shows four situations in which four charged particles are evenly spaced to the left and right of a central point. The charge values are indicated. Rank the situations according to the magnitude of the net electric field at the central point, greatest first.

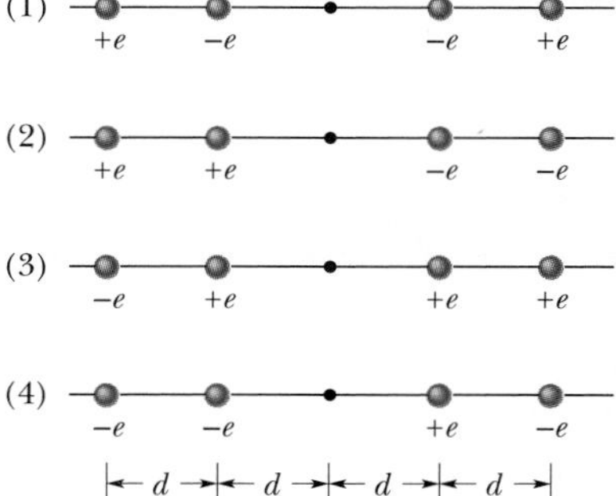

Fig. 22-23 Question 4.

5 Figure 22-24 shows two charged particles fixed in place on an axis. (a) Where on the axis (other than at an infinite distance) is there a point at which their net electric field is zero: between the charges, to their left, or to their right? (b) Is there a point of zero net electric field anywhere *off* the axis (other than at an infinite distance)?

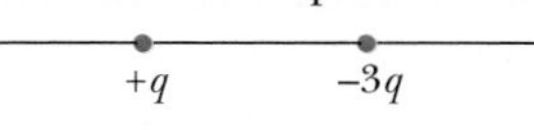

Fig. 22-24 Question 5.

6 In Fig. 22-25, two identical circular nonconducting rings are cen-

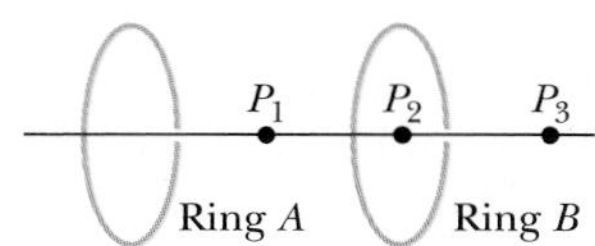

Fig. 22-25 Question 6.

tered on the same line. For three situations, the uniform charges on rings A and B are, respectively, (1) q_0 and q_0, (2) $-q_0$ and $-q_0$, and (3) $-q_0$ and q_0. Rank the situations according to the magnitude of the net electric field at (a) point P_1 midway between the rings, (b) point P_2 at the center of ring B, and (c) point P_3 to the right of ring B, greatest first.

7 The potential energies associated with four orientations of an electric dipole in an electric field are (1) $-5U_0$, (2) $-7U_0$, (3) $3U_0$, and (4) $5U_0$, where U_0 is positive. Rank the orientations according to (a) the angle between the electric dipole moment $\vec{p}$ and the electric field $\vec{E}$ and (b) the magnitude of the torque on the electric dipole, greatest first.

8 (a) In the Checkpoint of Section 22-9, if the dipole rotates from orientation 1 to orientation 2, is the work done on the dipole by the field positive, negative, or zero? (b) If, instead, the dipole rotates from orientation 1 to orientation 4, is the work done by the field more than, less than, or the same as in (a)?

9 Figure 22-26 shows two disks and a flat ring, each with the same uniform charge Q. Rank the objects according to the magnitude of the electric field they create at points P (which are at the same vertical heights), greatest first.

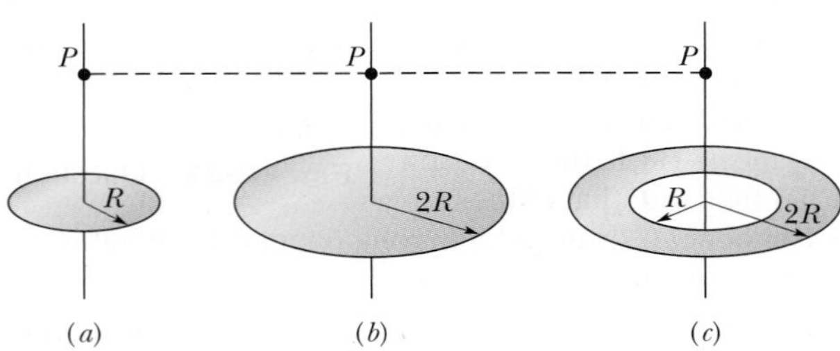

Fig. 22-26 Question 9.

10 In Fig. 22-27, an electron e travels through a small hole in plate A and then toward plate B. A uniform electric field in the region between the plates then slows the electron without deflecting it. (a) What is the direction of the field? (b) Four other particles similarly travel through small holes in either plate A or plate B and then into the region between the plates. Three have charges $+q_1$, $+q_2$, and $-q_3$. The fourth (labeled n) is a neutron, which is electrically neutral. Does the speed of each of those four other particles increase, decrease, or remain the same in the region between the plates?

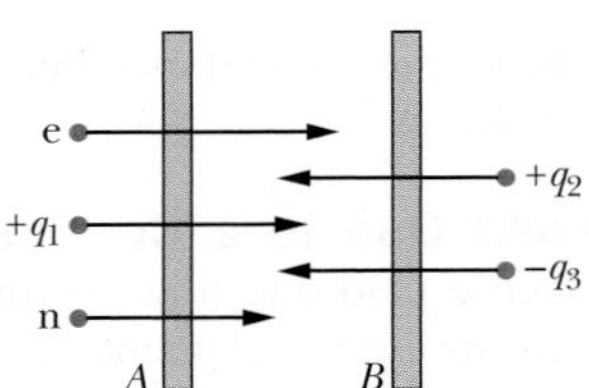

Fig. 22-27 Question 10.

11 In Fig. 22-28a, a circular plastic rod with uniform charge $+Q$ produces an electric field of magnitude E at the center of curvature (at the origin). In Figs. 22-28b, c, and d, more circular rods, each with identical uniform charges $+Q$, are added until the circle is complete. A fifth arrangement (which would be labeled e) is like that in d except the rod in the fourth quadrant has charge $-Q$. Rank the five arrangements according to the magnitude of the electric field at the center of curvature, greatest first.

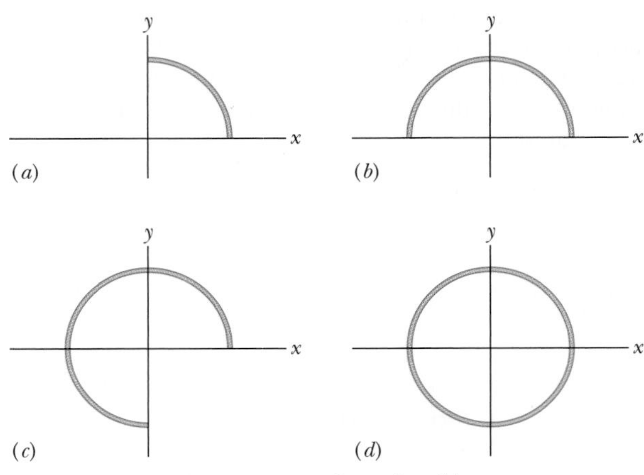

Fig. 22-28 Question 11.

PROBLEMS

GO Tutoring problem available (at instructor's discretion) in *WileyPLUS* and WebAssign
SSM Worked-out solution available in Student Solutions Manual
• – ••• Number of dots indicates level of problem difficulty
Additional information available in *The Flying Circus of Physics* and at flyingcircusofphysics.com
WWW Worked-out solution is at
ILW Interactive solution is at
http://www.wiley.com/college/halliday

sec. 22-3 Electric Field Lines

•1 Sketch qualitatively the electric field lines both between and outside two concentric conducting spherical shells when a uniform positive charge q_1 is on the inner shell and a uniform negative charge $-q_2$ is on the outer. Consider the cases $q_1 > q_2$, $q_1 = q_2$, and $q_1 < q_2$.

•2 In Fig. 22-29 the electric field lines on the left have twice the separation of those on the right. (a) If the magnitude of the field at A is 40 N/C, what is the magnitude of the force on a proton at A? (b) What is the magnitude of the field at B?

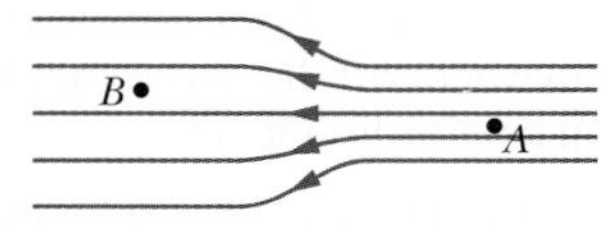

Fig. 22-29 Problem 2.

sec. 22-4 The Electric Field Due to a Point Charge

•3 SSM The nucleus of a plutonium-239 atom contains 94 protons. Assume that the nucleus is a sphere with radius 6.64 fm and with the charge of the protons uniformly spread through the sphere. At the nucleus surface, what are the (a) magnitude and (b) direction (radially inward or outward) of the electric field produced by the protons?

•4 Two particles are attached to an x axis: particle 1 of charge -2.00×10^{-7} C at $x = 6.00$ cm, particle 2 of charge $+2.00 \times 10^{-7}$ C at $x = 21.0$ cm. Midway between the particles, what is their net electric field in unit-vector notation?

•5 SSM What is the magnitude of a point charge whose electric field 50 cm away has the magnitude 2.0 N/C?

•6 What is the magnitude of a point charge that would create an electric field of 1.00 N/C at points 1.00 m away?

••7 SSM ILW WWW In Fig. 22-30, the four particles form a square of edge length $a = 5.00$ cm and have charges $q_1 = +10.0$ nC, $q_2 = -20.0$ nC, $q_3 = +20.0$ nC, and $q_4 = -10.0$ nC. In unit-vector notation, what net electric field do the particles produce at the square's center?

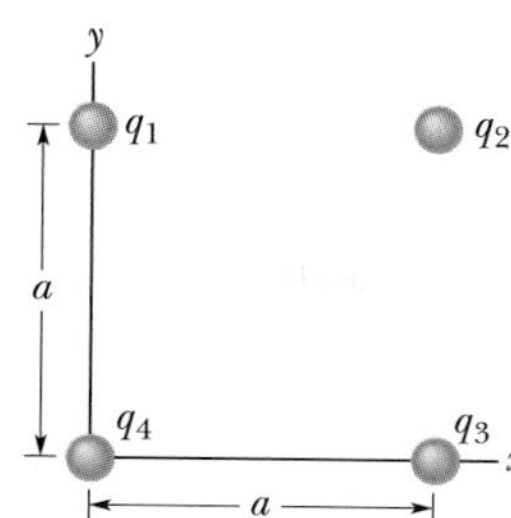

Fig. 22-30 Problem 7.

••8 GO In Fig. 22-31, the four particles are fixed in place and have charges $q_1 = q_2 = +5e$, $q_3 = +3e$, and $q_4 = -12e$. Distance $d = 5.0\ \mu$m. What is the magnitude of the net electric field at point P due to the particles?

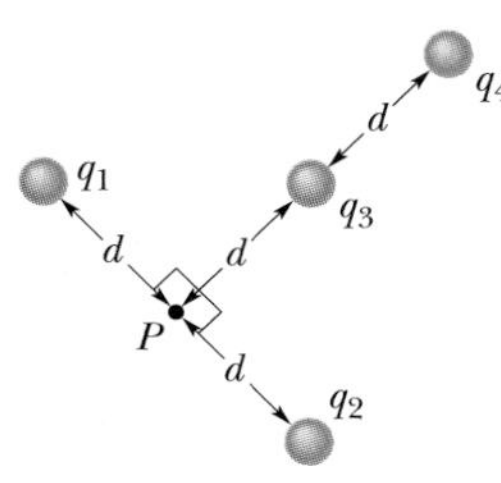

Fig. 22-31 Problem 8.

••9 GO Figure 22-32 shows two charged particles on an x axis: $-q = -3.20 \times 10^{-19}$ C at $x = -3.00$ m and $q = 3.20 \times 10^{-19}$ C at $x = +3.00$ m. What are the (a) magnitude and (b) direction (relative to the positive direction of the x axis) of the net electric field produced at point P at $y = 4.00$ m?

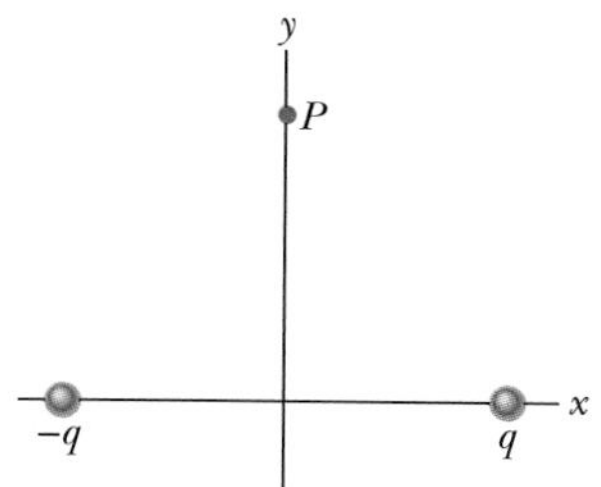

Fig. 22-32 Problem 9.

••10 GO Figure 22-33a shows two charged particles fixed in place on an x axis with separation L. The ratio q_1/q_2 of their charge magnitudes is 4.00. Figure 22-33b shows the x component $E_{\text{net},x}$ of their net electric field along the x axis just to the right of particle 2. The x axis scale is set by $x_s = 30.0$ cm. (a) At what value of $x > 0$ is $E_{\text{net},x}$ maximum? (b) If particle 2 has charge $-q_2 = -3e$, what is the value of that maximum?

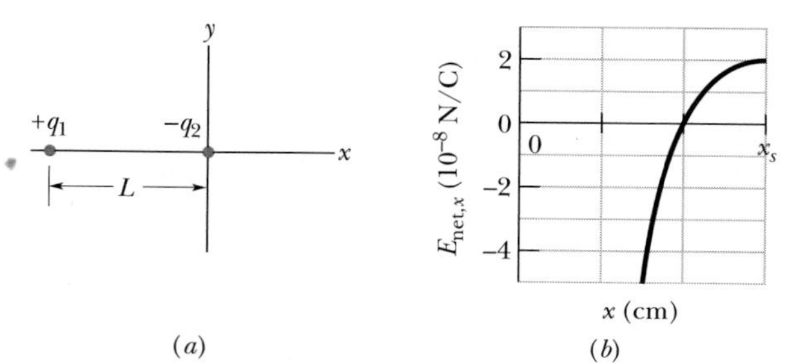

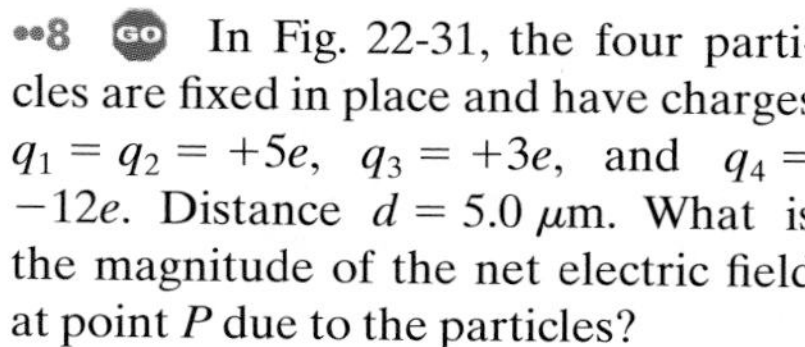

Fig. 22-33 Problem 10.

••11 SSM Two particles are fixed to an x axis: particle 1 of charge $q_1 = 2.1 \times 10^{-8}$ C at $x = 20$ cm and particle 2 of charge $q_2 = -4.00q_1$ at $x = 70$ cm. At what coordinate on the axis is the net electric field produced by the particles equal to zero?

••12 GO Figure 22-34 shows an uneven arrangement of electrons (e) and protons (p) on a circular arc of radius $r = 2.00$ cm, with angles $\theta_1 = 30.0°$, $\theta_2 = 50.0°$, $\theta_3 = 30.0°$, and $\theta_4 = 20.0°$. What are the (a) magnitude and (b) direction (relative to the positive direction of the x axis) of the net electric field produced at the center of the arc?

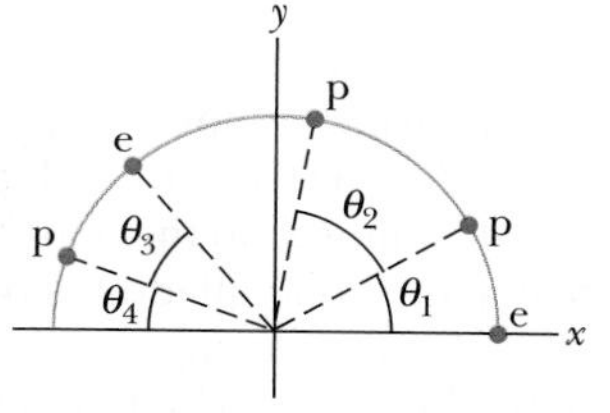

Fig. 22-34 Problem 12.

••13 Figure 22-35 shows a proton (p) on the central axis through a disk with a uniform charge density due to excess electrons. Three of those electrons are shown: electron e_c at the disk center and electrons e_s at opposite sides of the disk, at radius R from the center. The proton is initially at distance $z = R = 2.00$ cm from the disk. At that location, what are the magnitudes of (a) the electric field $\vec{E}_c$ due to electron e_c and (b) the *net* electric field $\vec{E}_{s,\text{net}}$ due to electrons e_s? The proton is then moved to $z = R/10.0$. What then are the magnitudes of (c) $\vec{E}_c$ and (d) $\vec{E}_{s,\text{net}}$ at the proton's location? (e) From (a) and (c) we see that as the proton gets nearer to the disk, the magnitude of $\vec{E}_c$ increases. Why does the magnitude of $\vec{E}_{s,\text{net}}$ decrease, as we see from (b) and (d)?

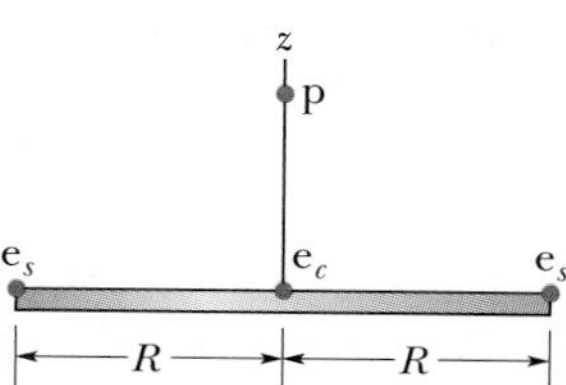

Fig. 22-35 Problem 13.

••14 In Fig. 22-36, particle 1 of charge $q_1 = -5.00q$ and particle 2 of charge $q_2 = +2.00q$ are fixed to an x axis. (a) As a multiple of distance L, at what coordinate on the axis is the net electric field of the particles zero? (b) Sketch the net electric field lines.

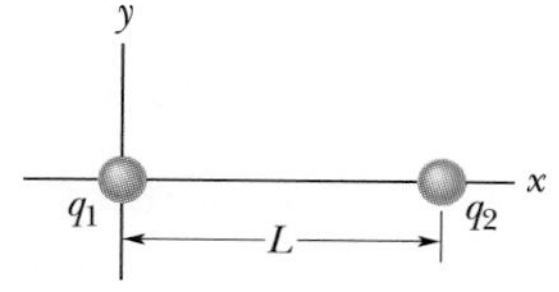

Fig. 22-36 Problem 14.

••15 In Fig. 22-37, the three particles are fixed in place and have charges $q_1 = q_2 = +e$ and $q_3 = +2e$. Distance $a = 6.00\ \mu$m. What are the (a) magnitude and (b) direction of the net electric field at point P due to the particles?

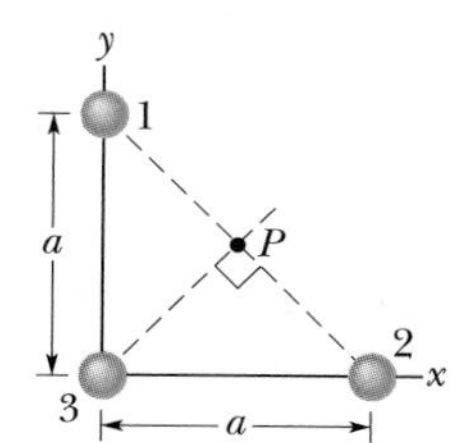

Fig. 22-37 Problem 15.

•••16 Figure 22-38 shows a plastic ring of radius $R = 50.0$ cm. Two small charged beads are on the ring: Bead 1 of charge $+2.00\ \mu$C is fixed in place at the left side; bead 2 of charge $+6.00\ \mu$C can be moved along the ring. The two beads produce a net electric field of magni-

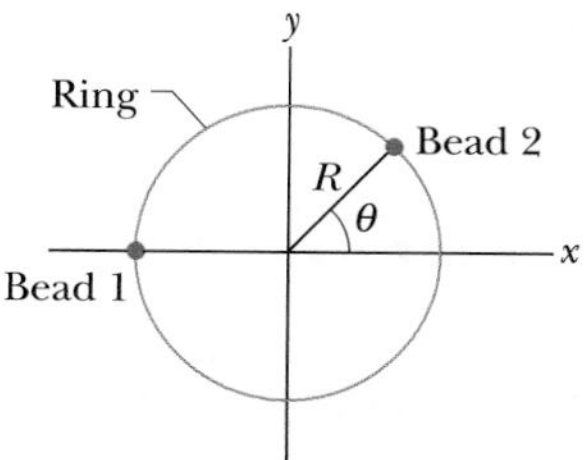

Fig. 22-38 Problem 16.

tude E at the center of the ring. At what (a) positive and (b) negative value of angle θ should bead 2 be positioned such that $E = 2.00 \times 10^5$ N/C?

•••17 Two charged beads are on the plastic ring in Fig. 22-39*a*. Bead 2, which is not shown, is fixed in place on the ring, which has radius $R = 60.0$ cm. Bead 1 is initially on the x axis at angle $\theta = 0°$. It is then moved to the opposite side, at angle $\theta = 180°$, through the first and second quadrants of the xy coordinate system. Figure 22-39*b* gives the x component of the net electric field produced at the origin by the two beads as a function of θ, and Fig. 22-39*c* gives the y component. The vertical axis scales are set by $E_{xs} = 5.0 \times 10^4$ N/C and $E_{ys} = -9.0 \times 10^4$ N/C. (a) At what angle θ is bead 2 located? What are the charges of (b) bead 1 and (c) bead 2?

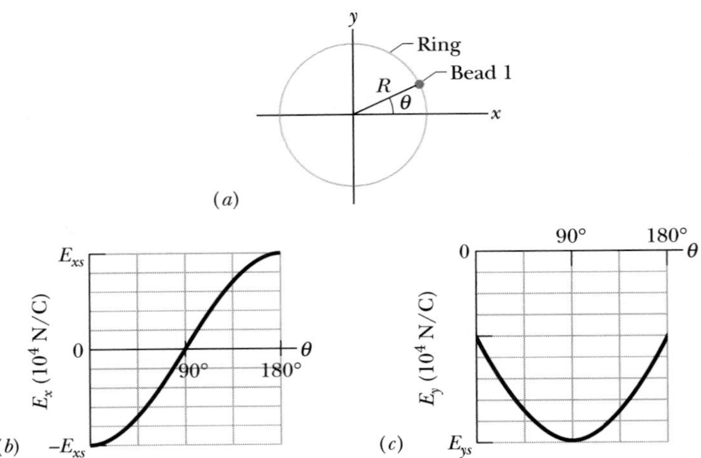

Fig. 22-39 Problem 17.

sec. 22-5 The Electric Field Due to an Electric Dipole

••18 The electric field of an electric dipole along the dipole axis is approximated by Eqs. 22-8 and 22-9. If a binomial expansion is made of Eq. 22-7, what is the next term in the expression for the dipole's electric field along the dipole axis? That is, what is E_{next} in the expression

$$E = \frac{1}{2\pi\varepsilon_0}\frac{qd}{z^3} + E_{next}?$$

••19 Figure 22-40 shows an electric dipole. What are the (a) magnitude and (b) direction (relative to the positive direction of the x axis) of the dipole's electric field at point P, located at distance $r \gg d$?

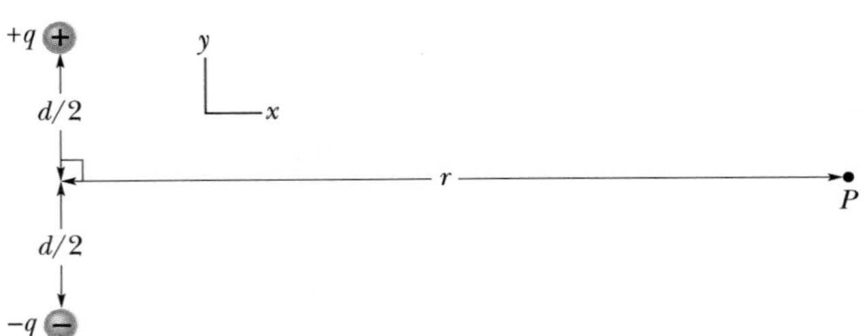

Fig. 22-40 Problem 19.

••20 Equations 22-8 and 22-9 are approximations of the magnitude of the electric field of an electric dipole, at points along the dipole axis. Consider a point P on that axis at distance $z = 5.00d$ from the dipole center (d is the separation distance between the particles of the dipole). Let E_{appr} be the magnitude of the field at point P as approximated by Eqs. 22-8 and 22-9. Let E_{act} be the actual magnitude. What is the ratio E_{appr}/E_{act}?

•••21 SSM *Electric quadrupole.* Figure 22-41 shows an electric quadrupole. It consists of two dipoles with dipole moments that are equal in magnitude but opposite in direction. Show that the value of E on the axis of the quadrupole for a point P a distance z from its center (assume $z \gg d$) is given by

$$E = \frac{3Q}{4\pi\varepsilon_0 z^4},$$

in which Q $(= 2qd^2)$ is known as the *quadrupole moment* of the charge distribution.

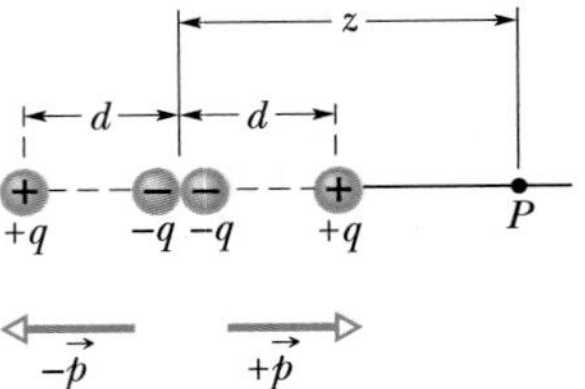

Fig. 22-41 Problem 21.

sec. 22-6 The Electric Field Due to a Line of Charge

•22 *Density, density, density.* (a) A charge $-300e$ is uniformly distributed along a circular arc of radius 4.00 cm, which subtends an angle of 40°. What is the linear charge density along the arc? (b) A charge $-300e$ is uniformly distributed over one face of a circular disk of radius 2.00 cm. What is the surface charge density over that face? (c) A charge $-300e$ is uniformly distributed over the surface of a sphere of radius 2.00 cm. What is the surface charge density over that surface? (d) A charge $-300e$ is uniformly spread through the volume of a sphere of radius 2.00 cm. What is the volume charge density in that sphere?

•23 Figure 22-42 shows two parallel nonconducting rings with their central axes along a common line. Ring 1 has uniform charge q_1 and radius R; ring 2 has uniform charge q_2 and the same radius R. The rings are separated by distance $d = 3.00R$. The net electric field at point P on the common line, at distance R from ring 1, is zero. What is the ratio q_1/q_2?

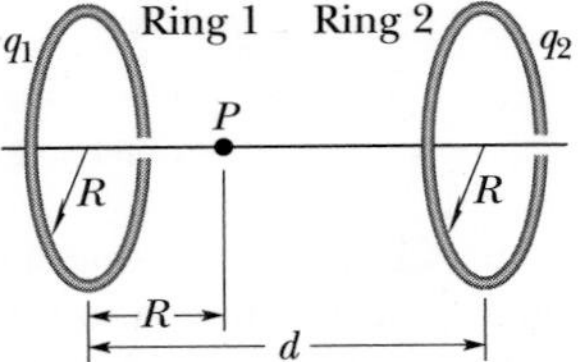

Fig. 22-42 Problem 23.

••24 A thin nonconducting rod with a uniform distribution of positive charge Q is bent into a circle of radius R (Fig. 22-43). The central perpendicular axis through the ring is a z axis, with the origin at the center of the ring. What is the magnitude of the electric field due to the rod at (a) $z = 0$ and (b) $z = \infty$? (c) In terms of R, at what positive value of z is that magnitude maximum? (d) If $R = 2.00$ cm and $Q = 4.00$ μC, what is the maximum magnitude?

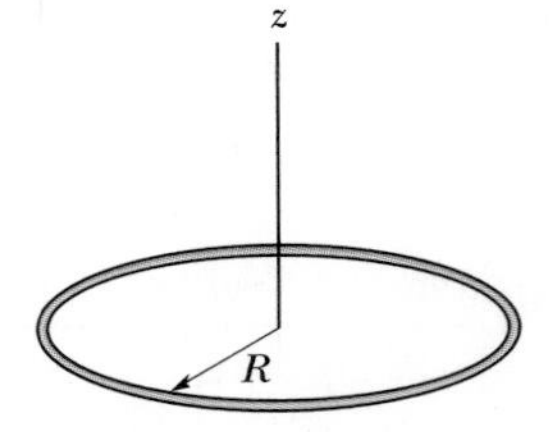

Fig. 22-43 Problem 24.

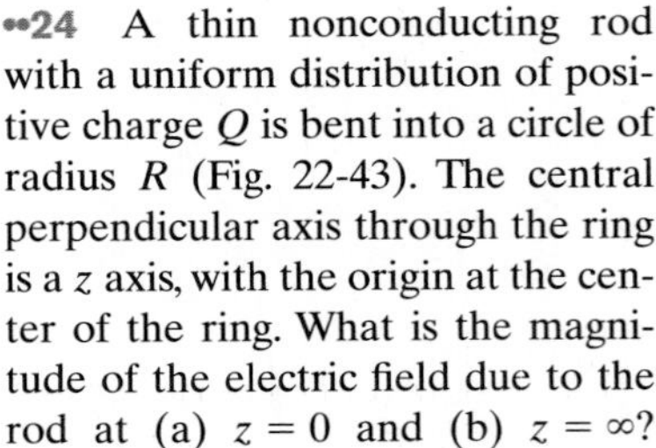

••25 Figure 22-44 shows three circular arcs centered on the origin of a coordinate system. On each arc, the uniformly distributed charge is given in terms of $Q = 2.00$ μC. The radii are given in terms of

$R = 10.0$ cm. What are the (a) magnitude and (b) direction (relative to the positive x direction) of the net electric field at the origin due to the arcs?

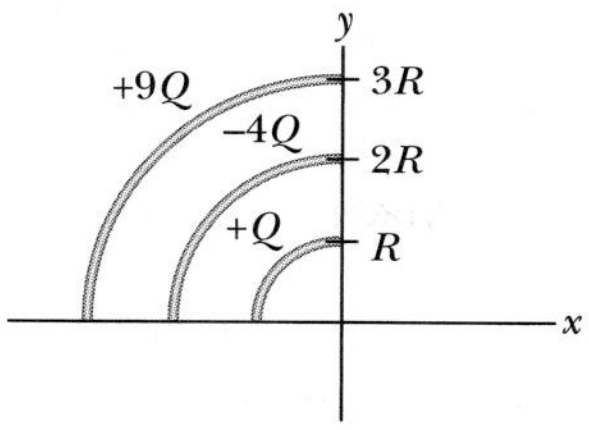

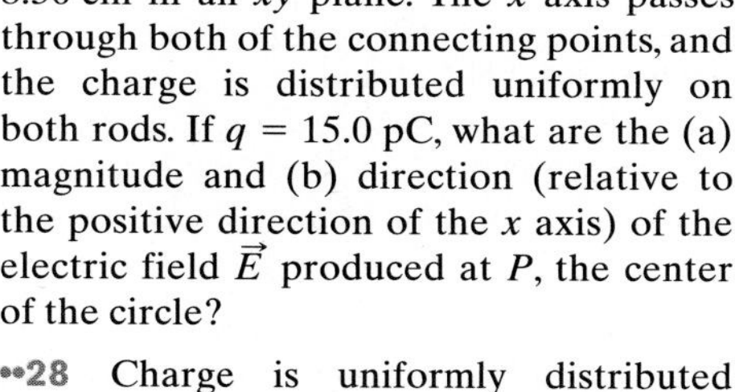

Fig. 22-44 Problem 25.

••26 ILW In Fig. 22-45, a thin glass rod forms a semicircle of radius $r =$ 5.00 cm. Charge is uniformly distributed along the rod, with $+q = 4.50$ pC in the upper half and $-q = -4.50$ pC in the lower half. What are the (a) magnitude and (b) direction (relative to the positive direction of the x axis) of the electric field $\vec{E}$ at P, the center of the semicircle?

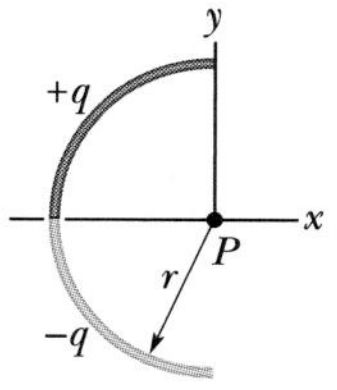

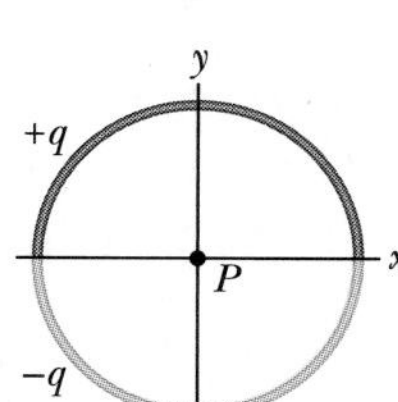

Fig. 22-45 Problem 26.

••27 In Fig. 22-46, two curved plastic rods, one of charge $+q$ and the other of charge $-q$, form a circle of radius $R =$ 8.50 cm in an xy plane. The x axis passes through both of the connecting points, and the charge is distributed uniformly on both rods. If $q = 15.0$ pC, what are the (a) magnitude and (b) direction (relative to the positive direction of the x axis) of the electric field $\vec{E}$ produced at P, the center of the circle?

Fig. 22-46 Problem 27.

••28 Charge is uniformly distributed around a ring of radius $R = 2.40$ cm, and the resulting electric field magnitude E is measured along the ring's central axis (perpendicular to the plane of the ring). At what distance from the ring's center is E maximum?

••29 Figure 22-47a shows a nonconducting rod with a uniformly distributed charge $+Q$. The rod forms a half-circle with radius R and produces an electric field of magnitude E_{arc} at its center of curvature P. If the arc is collapsed to a point at distance R from P (Fig. 22-47b), by what factor is the magnitude of the electric field at P multiplied?

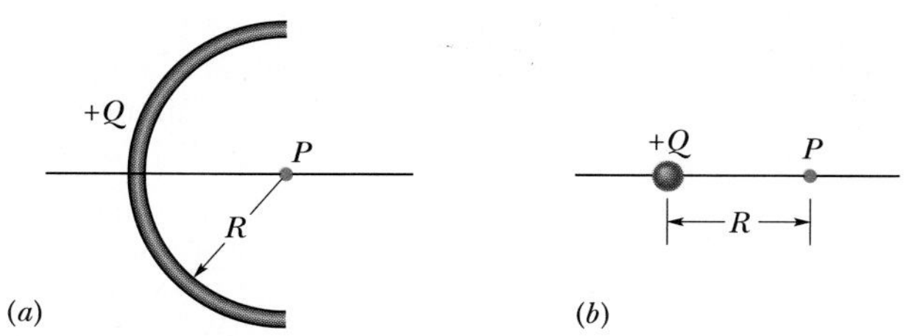

Fig. 22-47 Problem 29.

••30 Figure 22-48 shows two concentric rings, of radii R and $R' =$ 3.00R, that lie on the same plane. Point P lies on the central z axis, at distance $D = 2.00R$ from the center of the rings. The smaller ring has uniformly distributed charge $+Q$. In terms of Q, what is the uniformly distributed charge on the larger ring if the net electric field at P is zero?

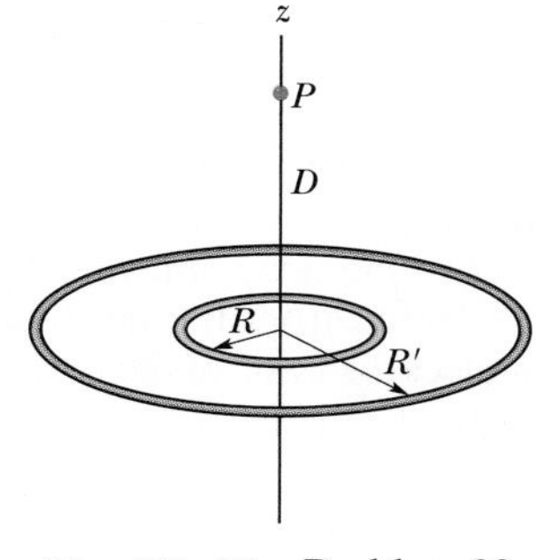

Fig. 22-48 Problem 30.

••31 SSM ILW WWW In Fig. 22-49, a nonconducting rod of length $L = 8.15$ cm has a charge $-q = -4.23$ fC uniformly distributed along its length. (a) What is the linear charge density of the rod? What are the (b) magnitude and (c) direction (relative to the positive direction of the x axis) of the electric field produced at point P, at distance $a = 12.0$ cm from the rod? What is the electric field magnitude produced at distance $a = 50$ m by (d) the rod and (e) a particle of charge $-q = -4.23$ fC that replaces the rod?

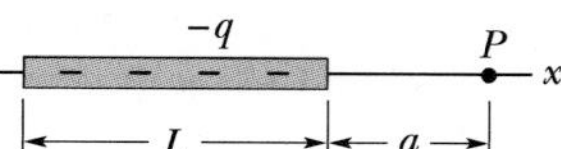

Fig. 22-49 Problem 31.

•••32 GO In Fig. 22-50, positive charge $q = 7.81$ pC is spread uniformly along a thin nonconducting rod of length $L = 14.5$ cm. What are the (a) magnitude and (b) direction (relative to the positive direction of the x axis) of the electric field produced at point P, at distance $R = 6.00$ cm from the rod along its perpendicular bisector?

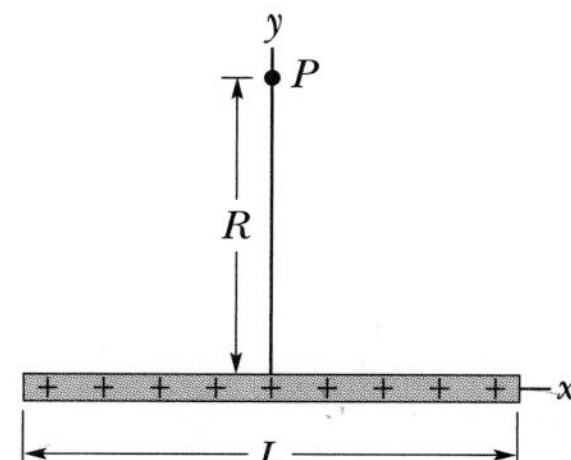

Fig. 22-50 Problem 32.

•••33 In Fig. 22-51, a "semi-infinite" nonconducting rod (that is, infinite in one direction only) has uniform linear charge density λ. Show that the electric field $\vec{E}_p$ at point P makes an angle of 45° with the rod and that this result is independent of the distance R. (*Hint:* Separately find the component of $\vec{E}_p$ parallel to the rod and the component perpendicular to the rod.)

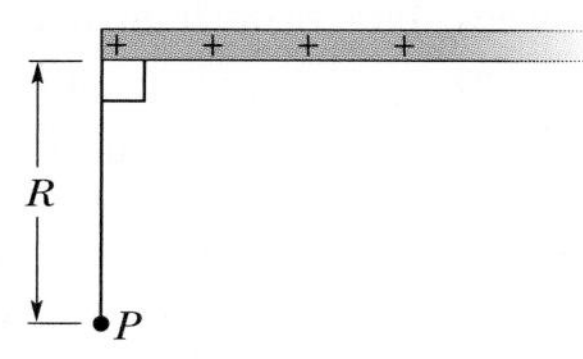

Fig. 22-51 Problem 33.

sec. 22-7 The Electric Field Due to a Charged Disk

•34 A disk of radius 2.5 cm has a surface charge density of 5.3 $\mu C/m^2$ on its upper face. What is the magnitude of the electric field produced by the disk at a point on its central axis at distance $z =$ 12 cm from the disk?

•35 SSM WWW At what distance along the central perpendicular axis of a uniformly charged plastic disk of radius 0.600 m is the magnitude of the electric field equal to one-half the magnitude of the field at the center of the surface of the disk?

••36 A circular plastic disk with radius $R = 2.00$ cm has a uniformly distributed charge $Q = +(2.00 \times 10^6)e$ on one face. A circular ring of width 30 μm is centered on that face, with the center of that width at radius $r = 0.50$ cm. In coulombs, what charge is contained within the width of the ring?

••37 Suppose you design an apparatus in which a uniformly charged disk of radius R is to produce an electric field. The field magnitude is most important along the central perpendicular axis of the disk, at a point P at distance 2.00R from the disk (Fig. 22-52a). Cost analysis suggests that you switch to a ring of the same outer radius R but with inner radius R/2.00 (Fig. 22-52b). Assume that the ring will have the same

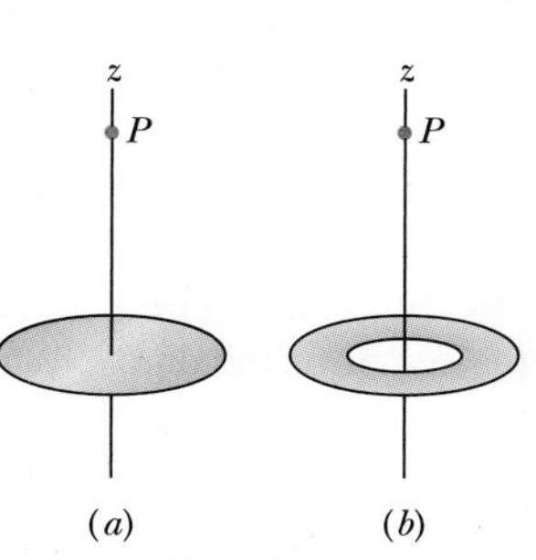

Fig. 22-52 Problem 37.

surface charge density as the original disk. If you switch to the ring, by what percentage will you decrease the electric field magnitude at P?

••38 Figure 22-53a shows a circular disk that is uniformly charged. The central z axis is perpendicular to the disk face, with the origin at the disk. Figure 22-53b gives the magnitude of the electric field along that axis in terms of the maximum magnitude E_m at the disk surface. The z axis scale is set by z_s = 8.0 cm. What is the radius of the disk?

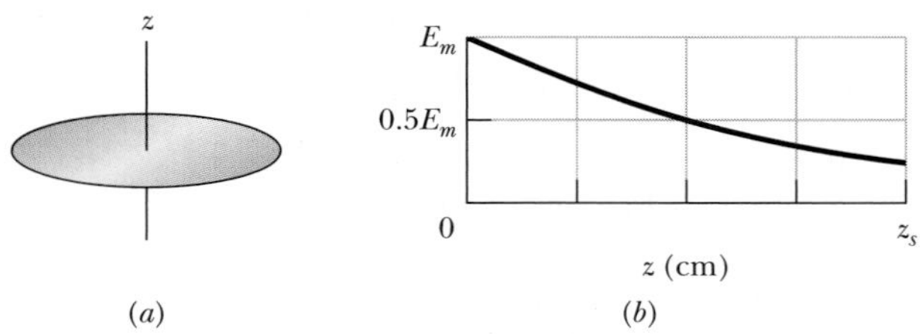

Fig. 22-53 Problem 38.

sec. 22-8 A Point Charge in an Electric Field

•39 In Millikan's experiment, an oil drop of radius 1.64 μm and density 0.851 g/cm^3 is suspended in chamber C (Fig. 22-14) when a downward electric field of 1.92×10^5 N/C is applied. Find the charge on the drop, in terms of e.

•40 GO An electron with a speed of 5.00×10^8 cm/s enters an electric field of magnitude 1.00×10^3 N/C, traveling along a field line in the direction that retards its motion. (a) How far will the electron travel in the field before stopping momentarily, and (b) how much time will have elapsed? (c) If the region containing the electric field is 8.00 mm long (too short for the electron to stop within it), what fraction of the electron's initial kinetic energy will be lost in that region?

•41 SSM A charged cloud system produces an electric field in the air near Earth's surface. A particle of charge -2.0×10^{-9} C is acted on by a downward electrostatic force of 3.0×10^{-6} N when placed in this field. (a) What is the magnitude of the electric field? What are the (b) magnitude and (c) direction of the electrostatic force $\vec{F}_{el}$ on the proton placed in this field? (d) What is the magnitude of the gravitational force $\vec{F}_g$ on the proton? (e) What is the ratio F_{el}/F_g in this case?

•42 Humid air breaks down (its molecules become ionized) in an electric field of 3.0×10^6 N/C. In that field, what is the magnitude of the electrostatic force on (a) an electron and (b) an ion with a single electron missing?

•43 SSM An electron is released from rest in a uniform electric field of magnitude 2.00×10^4 N/C. Calculate the acceleration of the electron. (Ignore gravitation.)

•44 An alpha particle (the nucleus of a helium atom) has a mass of 6.64×10^{-27} kg and a charge of $+2e$. What are the (a) magnitude and (b) direction of the electric field that will balance the gravitational force on the particle?

•45 ILW An electron on the axis of an electric dipole is 25 nm from the center of the dipole. What is the magnitude of the electrostatic force on the electron if the dipole moment is 3.6×10^{-29} C·m? Assume that 25 nm is much larger than the dipole charge separation.

•46 An electron is accelerated eastward at 1.80×10^9 m/s^2 by an electric field. Determine the field (a) magnitude and (b) direction.

•47 SSM Beams of high-speed protons can be produced in "guns" using electric fields to accelerate the protons. (a) What acceleration would a proton experience if the gun's electric field were 2.00×10^4 N/C? (b) What speed would the proton attain if the field accelerated the proton through a distance of 1.00 cm?

••48 In Fig. 22-54, an electron (e) is to be released from rest on the central axis of a uniformly charged disk of radius R. The surface charge density on the disk is $+4.00$ μC/m^2. What is the magnitude of the electron's initial acceleration if it is released at a distance (a) R, (b) $R/100$, and (c) $R/1000$ from the center of the disk? (d) Why does the acceleration magnitude increase only slightly as the release point is moved closer to the disk?

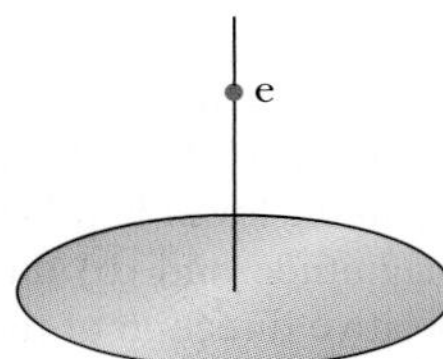

Fig. 22-54 Problem 48.

••49 A 10.0 g block with a charge of $+8.00 \times 10^{-5}$ C is placed in an electric field $\vec{E} = (3000\hat{i} - 600\hat{j})$ N/C. What are the (a) magnitude and (b) direction (relative to the positive direction of the x axis) of the electrostatic force on the block? If the block is released from rest at the origin at time $t = 0$, what are its (c) x and (d) y coordinates at $t = 3.00$ s?

••50 At some instant the velocity components of an electron moving between two charged parallel plates are $v_x = 1.5 \times 10^5$ m/s and $v_y = 3.0 \times 10^3$ m/s. Suppose the electric field between the plates is given by $\vec{E} = (120 \text{ N/C})\hat{j}$. In unit-vector notation, what are (a) the electron's acceleration in that field and (b) the electron's velocity when its x coordinate has changed by 2.0 cm?

••51 Assume that a honeybee is a sphere of diameter 1.000 cm with a charge of +45.0 pC uniformly spread over its surface. Assume also that a spherical pollen grain of diameter 40.0 μm is electrically held on the surface of the sphere because the bee's charge induces a charge of −1.00 pC on the near side of the sphere and a charge of +1.00 pC on the far side. (a) What is the magnitude of the net electrostatic force on the grain due to the bee? Next, assume that the bee brings the grain to a distance of 1.000 mm from the tip of a flower's stigma and that the tip is a particle of charge −45.0 pC. (b) What is the magnitude of the net electrostatic force on the grain due to the stigma? (c) Does the grain remain on the bee or does it move to the stigma?

••52 An electron enters a region of uniform electric field with an initial velocity of 40 km/s in the same direction as the electric field, which has magnitude E = 50 N/C. (a) What is the speed of the electron 1.5 ns after entering this region? (b) How far does the electron travel during the 1.5 ns interval?

••53 GO Two large parallel copper plates are 5.0 cm apart and have a uniform electric field between them as depicted in Fig. 22-55. An electron is released from the negative plate at the same time that a proton is released from the positive plate. Neglect the force of the particles on each other and find their distance from the positive plate when they pass each other. (Does it surprise you that you need not know the electric field to solve this problem?)

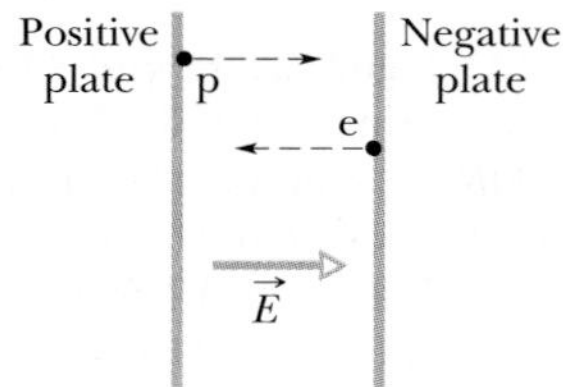

Fig. 22-55 Problem 53.

••54 GO In Fig. 22-56, an electron is shot at an initial speed of $v_0 = 2.00 \times 10^6$ m/s, at angle $\theta_0 = 40.0°$ from an x axis. It moves through a uniform electric field $\vec{E} = (5.00 \text{ N/C})\hat{j}$. A screen for detecting electrons is positioned parallel to the y axis, at distance $x = 3.00$ m. In unit-vector notation, what is the velocity of the electron when it hits the screen?

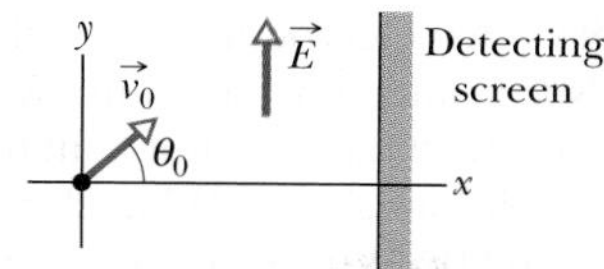

Fig. 22-56 Problem 54.

••55 ILW A uniform electric field exists in a region between two oppositely charged plates. An electron is released from rest at the surface of the negatively charged plate and strikes the surface of the opposite plate, 2.0 cm away, in a time 1.5×10^{-8} s. (a) What is the speed of the electron as it strikes the second plate? (b) What is the magnitude of the electric field $\vec{E}$?

sec. 22-9 A Dipole in an Electric Field

•56 An electric dipole consists of charges $+2e$ and $-2e$ separated by 0.78 nm. It is in an electric field of strength 3.4×10^6 N/C. Calculate the magnitude of the torque on the dipole when the dipole moment is (a) parallel to, (b) perpendicular to, and (c) antiparallel to the electric field.

•57 SSM An electric dipole consisting of charges of magnitude 1.50 nC separated by 6.20 μm is in an electric field of strength 1100 N/C. What are (a) the magnitude of the electric dipole moment and (b) the difference between the potential energies for dipole orientations parallel and antiparallel to $\vec{E}$?

••58 A certain electric dipole is placed in a uniform electric field $\vec{E}$ of magnitude 20 N/C. Figure 22-57 gives the potential energy U of the dipole versus the angle θ between $\vec{E}$ and the dipole moment $\vec{p}$. The vertical axis scale is set by $U_s = 100 \times 10^{-28}$ J. What is the magnitude of $\vec{p}$?

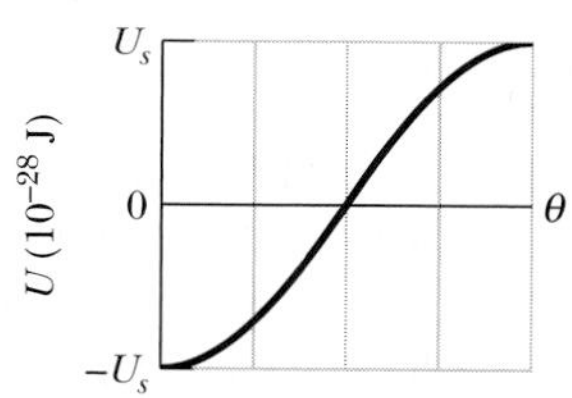

Fig. 22-57 Problem 58.

••59 How much work is required to turn an electric dipole 180° in a uniform electric field of magnitude $E = 46.0$ N/C if $p = 3.02 \times 10^{-25}$ C·m and the initial angle is 64°?

••60 A certain electric dipole is placed in a uniform electric field $\vec{E}$ of magnitude 40 N/C. Figure 22-58 gives the magnitude τ of the torque on the dipole versus the angle θ between field $\vec{E}$ and the dipole moment $\vec{p}$. The vertical axis scale is set by $\tau_s = 100 \times 10^{-28}$ N·m. What is the magnitude of $\vec{p}$?

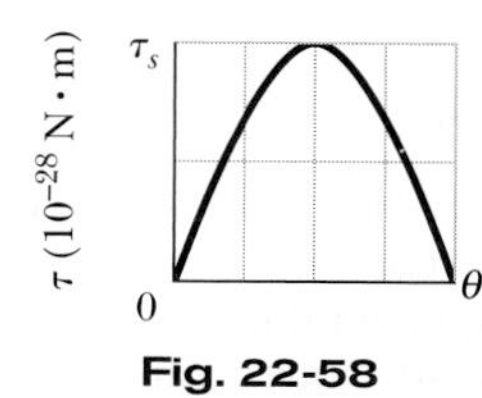

Fig. 22-58 Problem 60.

••61 Find an expression for the oscillation frequency of an electric dipole of dipole moment $\vec{p}$ and rotational inertia I for small amplitudes of oscillation about its equilibrium position in a uniform electric field of magnitude E.

Additional Problems

62 (a) What is the magnitude of an electron's acceleration in a uniform electric field of magnitude 1.40×10^6 N/C? (b) How long would the electron take, starting from rest, to attain one-tenth the speed of light? (c) How far would it travel in that time?

63 A spherical water drop 1.20 μm in diameter is suspended in calm air due to a downward-directed atmospheric electric field of magnitude $E = 462$ N/C. (a) What is the magnitude of the gravitational force on the drop? (b) How many excess electrons does it have?

64 Three particles, each with positive charge Q, form an equilateral triangle, with each side of length d. What is the magnitude of the electric field produced by the particles at the midpoint of any side?

65 In Fig. 22-59a, a particle of charge $+Q$ produces an electric field of magnitude E_{part} at point P, at distance R from the particle. In Fig. 22-59b, that same amount of charge is spread uniformly along a circular arc that has radius R and subtends an angle θ. The charge on the arc produces an electric field of magnitude E_{arc} at its center of curvature P. For what value of θ does $E_{arc} = 0.500E_{part}$? (*Hint:* You will probably resort to a graphical solution.)

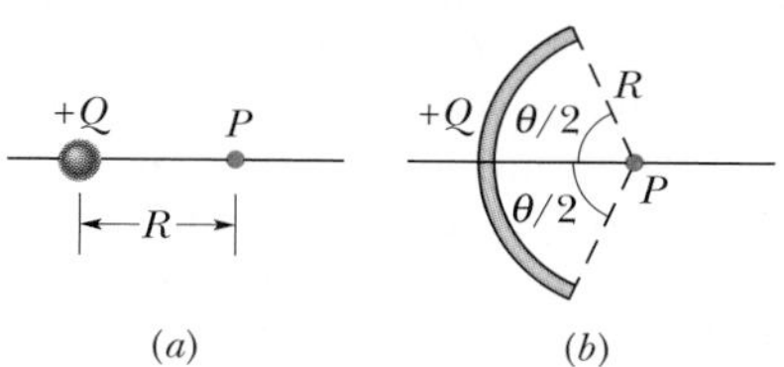

Fig. 22-59 Problem 65.

66 A proton and an electron form two corners of an equilateral triangle of side length 2.0×10^{-6} m. What is the magnitude of the net electric field these two particles produce at the third corner?

67 A charge (uniform linear density = 9.0 nC/m) lies on a string that is stretched along an x axis from $x = 0$ to $x = 3.0$ m. Determine the magnitude of the electric field at $x = 4.0$ m on the x axis.

68 In Fig. 22-60, eight particles form a square in which distance $d = 2.0$ cm. The charges are $q_1 = +3e$, $q_2 = +e$, $q_3 = -5e$, $q_4 = -2e$, $q_5 = +3e$, $q_6 = +e$, $q_7 = -5e$, and $q_8 = +e$. In unit-vector notation, what is the net electric field at the square's center?

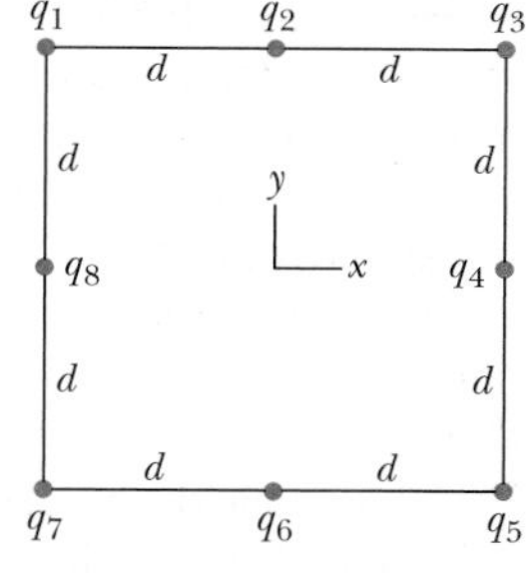

Fig. 22-60 Problem 68.

69 Two particles, each with a charge of magnitude 12 nC, are at two of the vertices of an equilateral triangle with edge length 2.0 m. What is the magnitude of the electric field at the third vertex if (a) both charges are positive and (b) one charge is positive and the other is negative?

70 In one of his experiments, Millikan observed that the following measured charges, among others, appeared at different times on a single drop:

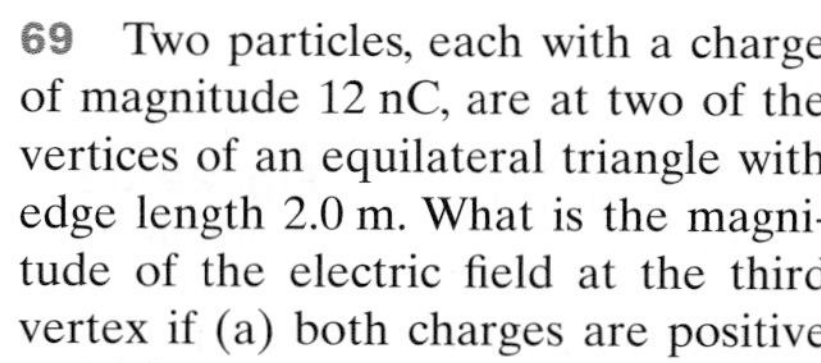

6.563×10^{-19} C	13.13×10^{-19} C	19.71×10^{-19} C
8.204×10^{-19} C	16.48×10^{-19} C	22.89×10^{-19} C
11.50×10^{-19} C	18.08×10^{-19} C	26.13×10^{-19} C

What value for the elementary charge e can be deduced from these data?

71 A charge of 20 nC is uniformly distributed along a straight rod of length 4.0 m that is bent into a circular arc with a radius of 2.0 m. What is the magnitude of the electric field at the center of curvature of the arc?

72 An electron is constrained to the central axis of the ring of charge of radius R in Fig. 22-10, with $z \ll R$. Show that the electrostatic force on the electron can cause it to oscillate through the ring center with an angular frequency

$$\omega = \sqrt{\frac{eq}{4\pi\varepsilon_0 mR^3}},$$

where q is the ring's charge and m is the electron's mass.

73 SSM The electric field in an xy plane produced by a positively charged particle is $7.2(4.0\hat{i} + 3.0\hat{j})$ N/C at the point (3.0, 3.0) cm and $100\hat{i}$ N/C at the point (2.0, 0) cm. What are the (a) x and (b) y coordinates of the particle? (c) What is the charge of the particle?

74 (a) What total (excess) charge q must the disk in Fig. 22-13 have for the electric field on the surface of the disk at its center to have magnitude 3.0×10^6 N/C, the E value at which air breaks down electrically, producing sparks? Take the disk radius as 2.5 cm, and use the listing for air in Table 22-1. (b) Suppose each surface atom has an effective cross-sectional area of 0.015 nm^2. How many atoms are needed to make up the disk surface? (c) The charge calculated in (a) results from some of the surface atoms having one excess electron. What fraction of these atoms must be so charged?

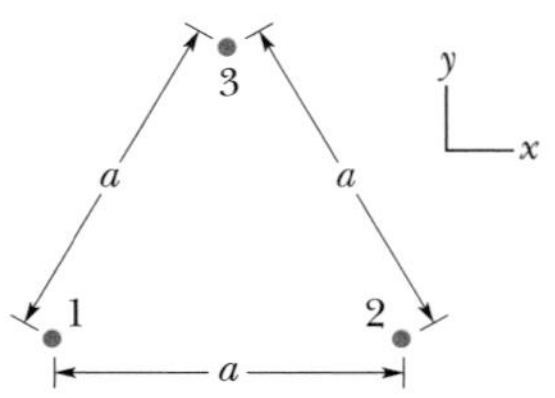

Fig. 22-61 Problems 75 and 86.

75 In Fig. 22-61, particle 1 (of charge $+1.00\ \mu$C), particle 2 (of charge $+1.00\ \mu$C), and particle 3 (of charge Q) form an equilateral triangle of edge length a. For what value of Q (both sign and magnitude) does the net electric field produced by the particles at the center of the triangle vanish?

76 In Fig. 22-62, an electric dipole swings from an initial orientation i ($\theta_i = 20.0°$) to a final orientation f ($\theta_f = 20.0°$) in a uniform external electric field $\vec{E}$. The electric dipole moment is 1.60×10^{-27} C·m; the field magnitude is 3.00×10^6 N/C. What is the change in the dipole's potential energy?

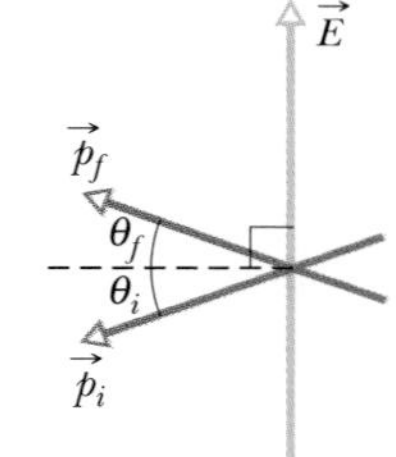

Fig. 22-62 Problem 76.

77 A particle of charge $-q_1$ is at the origin of an x axis. (a) At what location on the axis should a particle of charge $-4q_1$ be placed so that the net electric field is zero at $x = 2.0$ mm on the axis? (b) If, instead, a particle of charge $+4q_1$ is placed at that location, what is the direction (relative to the positive direction of the x axis) of the net electric field at $x = 2.0$ mm?

78 Two particles, each of positive charge q, are fixed in place on a y axis, one at $y = d$ and the other at $y = -d$. (a) Write an expression that gives the magnitude E of the net electric field at points on the x axis given by $x = \alpha d$. (b) Graph E versus α for the range $0 < \alpha < 4$. From the graph, determine the values of α that give (c) the maximum value of E and (d) half the maximum value of E.

79 A clock face has negative point charges $-q, -2q, -3q, \ldots, -12q$ fixed at the positions of the corresponding numerals. The clock hands do not perturb the net field due to the point charges. At what time does the hour hand point in the same direction as the electric field vector at the center of the dial? (*Hint:* Use symmetry.)

80 Calculate the electric dipole moment of an electron and a proton 4.30 nm apart.

81 An electric field $\vec{E}$ with an average magnitude of about 150 N/C points downward in the atmosphere near Earth's surface. We wish to "float" a sulfur sphere weighing 4.4 N in this field by charging the sphere. (a) What charge (both sign and magnitude) must be used? (b) Why is the experiment impractical?

82 A circular rod has a radius of curvature $R = 9.00$ cm and a uniformly distributed positive charge $Q = 6.25$ pC and subtends an angle $\theta = 2.40$ rad. What is the magnitude of the electric field that Q produces at the center of curvature?

83 SSM An electric dipole with dipole moment

$$\vec{p} = (3.00\hat{i} + 4.00\hat{j})(1.24 \times 10^{-30}\ \text{C}\cdot\text{m})$$

is in an electric field $\vec{E} = (4000\ \text{N/C})\hat{i}$. (a) What is the potential energy of the electric dipole? (b) What is the torque acting on it? (c) If an external agent turns the dipole until its electric dipole moment is

$$\vec{p} = (-4.00\hat{i} + 3.00\hat{j})(1.24 \times 10^{-30}\ \text{C}\cdot\text{m}),$$

how much work is done by the agent?

84 In Fig. 22-63, a uniform, upward electric field $\vec{E}$ of magnitude 2.00×10^3 N/C has been set up between two horizontal plates by charging the lower plate positively and the upper plate negatively. The plates have length $L = 10.0$ cm and separation $d = 2.00$ cm. An electron is then shot between the plates from the left edge of the lower plate. The initial velocity $\vec{v}_0$ of the electron makes an angle $\theta = 45.0°$ with the lower plate and has a magnitude of 6.00×10^6 m/s. (a) Will the electron strike one of the plates? (b) If so, which plate and how far horizontally from the left edge will the electron strike?

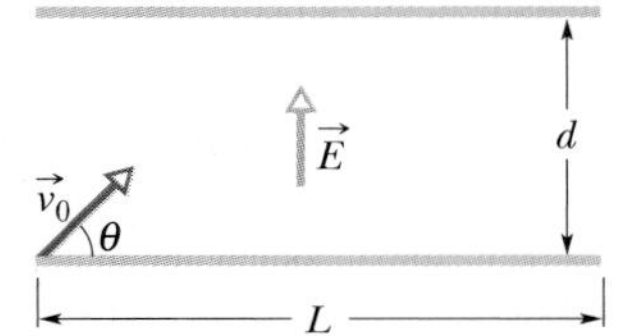

Fig. 22-63 Problem 84.

85 For the data of Problem 70, assume that the charge q on the drop is given by $q = ne$, where n is an integer and e is the elementary charge. (a) Find n for each given value of q. (b) Do a linear regression fit of the values of q versus the values of n and then use that fit to find e.

86 In Fig. 22-61, particle 1 (of charge +2.00 pC), particle 2 (of charge −2.00 pC), and particle 3 (of charge +5.00 pC) form an equilateral triangle of edge length $a = 9.50$ cm. (a) Relative to the positive direction of the x axis, determine the direction of the force $\vec{F}_3$ on particle 3 due to the other particles by sketching electric field lines of the other particles. (b) Calculate the magnitude of $\vec{F}_3$.

87 In Fig. 22-64, particle 1 of charge $q_1 = 1.00$ pC and particle 2 of charge $q_2 = -2.00$ pC are fixed at a distance $d = 5.00$ cm apart. In unit-vector notation, what is the net electric field at points (a) A, (b) B, and (c) C? (d) Sketch the electric field lines.

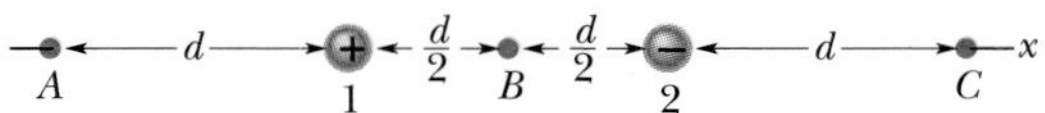

Fig. 22-64 Problem 87.

88 In Fig. 22-8, let both charges be positive. Assuming $z \gg d$, show that E at point P in that figure is then given by

$$E = \frac{1}{4\pi\varepsilon_0}\frac{2q}{z^2}.$$

CHAPTER 23

GAUSS' LAW

23-1 WHAT IS PHYSICS?

One of the primary goals of physics is to find simple ways of solving seemingly complex problems. One of the main tools of physics in attaining this goal is the use of symmetry. For example, in finding the electric field $\vec{E}$ of the charged ring of Fig. 22-10 and the charged rod of Fig. 22-11, we considered the fields $d\vec{E}$ $(= k\, dq/r^2)$ of charge elements in the ring and rod. Then we simplified the calculation of $\vec{E}$ by using symmetry to discard the perpendicular components of the $d\vec{E}$ vectors. That saved us some work.

For certain charge distributions involving symmetry, we can save far more work by using a law called Gauss' law, developed by German mathematician and physicist Carl Friedrich Gauss (1777–1855). Instead of considering the fields $d\vec{E}$ of charge elements in a given charge distribution, Gauss' law considers a hypothetical (imaginary) closed surface enclosing the charge distribution. This **Gaussian surface,** as it is called, can have any shape, but the shape that minimizes our calculations of the electric field is one that mimics the symmetry of the charge distribution. For example, if the charge is spread uniformly over a sphere, we enclose the sphere with a spherical Gaussian surface, such as the one in Fig. 23-1, and then, as we discuss in this chapter, find the electric field on the surface by using the fact that

> Gauss' law relates the electric fields at points on a (closed) Gaussian surface to the net charge enclosed by that surface.

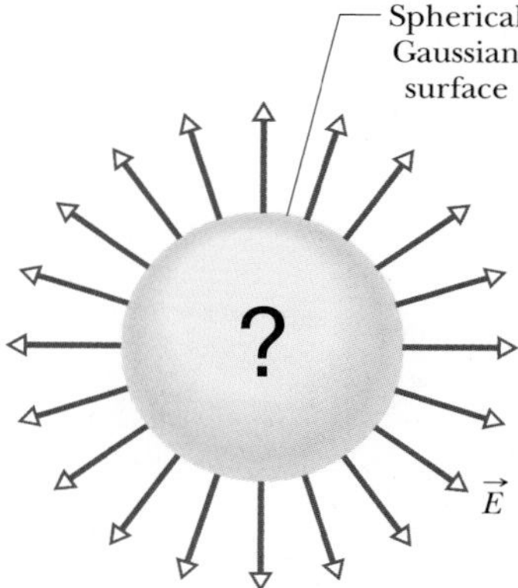

Fig. 23-1 A spherical Gaussian surface. If the electric field vectors are of uniform magnitude and point radially outward at all surface points, you can conclude that a net positive distribution of charge must lie within the surface and have spherical symmetry.

We can also use Gauss' law in reverse: If we know the electric field on a Gaussian surface, we can find the net charge enclosed by the surface. As a limited example, suppose that the electric field vectors in Fig. 23-1 all point radially outward from the center of the sphere and have equal magnitude. Gauss' law immediately tells us that the spherical surface must enclose a net positive charge that is either a particle or distributed spherically. However, to calculate how *much* charge is enclosed, we need a way of calculating how much electric field is intercepted by the Gaussian surface in Fig. 23-1. This measure of intercepted field is called *flux*, which we discuss next.

23-2 Flux

Suppose that, as in Fig. 23-2*a*, you aim a wide airstream of uniform velocity $\vec{v}$ at a small square loop of area A. Let Φ represent the *volume flow rate* (volume per unit time) at which air flows through the loop. This rate depends on the angle between $\vec{v}$ and the plane of the loop. If $\vec{v}$ is perpendicular to the plane, the rate Φ is equal to vA.

If $\vec{v}$ is parallel to the plane of the loop, no air moves through the loop, so Φ is zero. For an intermediate angle θ, the rate Φ depends on the component of $\vec{v}$ normal to the plane (Fig. 23-2*b*). Since that component is $v \cos \theta$, the rate of volume flow through the loop is

$$\Phi = (v \cos \theta)A. \tag{23-1}$$

This rate of flow through an area is an example of a **flux**—a *volume flux* in this situation.

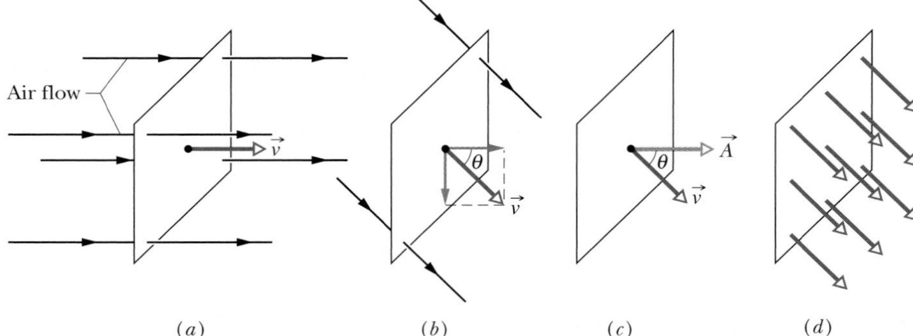

Fig. 23-2 (*a*) A uniform airstream of velocity $\vec{v}$ is perpendicular to the plane of a square loop of area A. (*b*) The component of $\vec{v}$ perpendicular to the plane of the loop is $v \cos \theta$, where θ is the angle between $\vec{v}$ and a normal to the plane. (*c*) The area vector $\vec{A}$ is perpendicular to the plane of the loop and makes an angle θ with $\vec{v}$. (*d*) The velocity field intercepted by the area of the loop.

Before we discuss a flux involved in electrostatics, we need to rewrite Eq. 23-1 in terms of vectors. To do this, we first define an *area vector* $\vec{A}$ as being a vector whose magnitude is equal to an area (here the area of the loop) and whose direction is normal to the plane of the area (Fig. 23-2*c*). We then rewrite Eq. 23-1 as the scalar (or dot) product of the velocity vector $\vec{v}$ of the airstream and the area vector $\vec{A}$ of the loop:

$$\Phi = vA \cos \theta = \vec{v} \cdot \vec{A}, \tag{23-2}$$

where θ is the angle between $\vec{v}$ and $\vec{A}$.

The word "flux" comes from the Latin word meaning "to flow." That meaning makes sense if we talk about the flow of air volume through the loop. However, Eq. 23-2 can be regarded in a more abstract way. To see this different way, note that we can assign a velocity vector to each point in the airstream passing through the loop (Fig. 23-2*d*). Because the composite of all those vectors is a *velocity field*, we can interpret Eq. 23-2 as giving the *flux of the velocity field through the loop.* With this interpretation, flux no longer means the actual flow of something through an area—rather it means the product of an area and the field across that area.

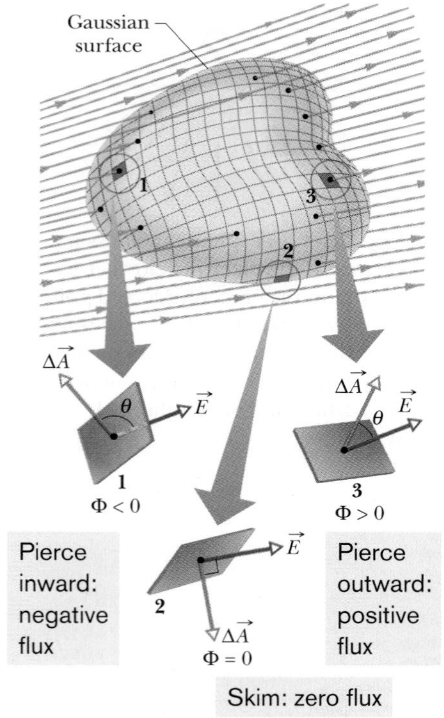

23-3 Flux of an Electric Field

To define the flux of an electric field, consider Fig. 23-3, which shows an arbitrary (asymmetric) Gaussian surface immersed in a nonuniform electric field. Let us divide the surface into small squares of area ΔA, each square being small enough to permit us to neglect any curvature and to consider the individual square to be flat. We represent each such element of area with an area vector $\Delta\vec{A}$, whose magnitude is the area ΔA. Each vector $\Delta\vec{A}$ is perpendicular to the Gaussian surface and directed away from the interior of the surface.

Because the squares have been taken to be arbitrarily small, the electric field $\vec{E}$ may be taken as constant over any given square. The vectors $\Delta\vec{A}$ and $\vec{E}$ for each square then make some angle θ with each other. Figure 23-3 shows an enlarged view of three squares on the Gaussian surface and the angle θ for each.

A provisional definition for the flux of the electric field for the Gaussian surface of Fig. 23-3 is

$$\Phi = \sum \vec{E} \cdot \Delta\vec{A}. \tag{23-3}$$

This equation instructs us to visit each square on the Gaussian surface, evaluate the scalar product $\vec{E} \cdot \Delta\vec{A}$ for the two vectors $\vec{E}$ and $\Delta\vec{A}$ we find there, and sum the results algebraically (that is, with signs included) for all the squares that make up the surface. The value of each scalar product (positive, negative, or zero) determines whether the flux through its square is positive, negative, or zero. Squares like square 1 in Fig. 23-3, in which $\vec{E}$ points inward, make a negative contribution to the sum of Eq. 23-3. Squares like 2, in which $\vec{E}$ lies in the surface, make zero contribution. Squares like 3, in which $\vec{E}$ points outward, make a positive contribution.

Fig. 23-3 A Gaussian surface of arbitrary shape immersed in an electric field. The surface is divided into small squares of area ΔA. The electric field vectors $\vec{E}$ and the area vectors $\Delta\vec{A}$ for three representative squares, marked 1, 2, and 3, are shown.

The exact definition of the flux of the electric field through a closed surface is found by allowing the area of the squares shown in Fig. 23-3 to become smaller and smaller, approaching a differential limit dA. The area vectors then approach a differential limit $d\vec{A}$. The sum of Eq. 23-3 then becomes an integral:

$$\Phi = \oint \vec{E} \cdot d\vec{A} \quad \text{(electric flux through a Gaussian surface).} \tag{23-4}$$

The loop on the integral sign indicates that the integration is to be taken over the entire (closed) surface. The flux of the electric field is a scalar, and its SI unit is the newton–square-meter per coulomb ($N \cdot m^2/C$).

We can interpret Eq. 23-4 in the following way: First recall that we can use the density of electric field lines passing through an area as a proportional measure of the magnitude of the electric field $\vec{E}$ there. Specifically, the magnitude E is proportional to the number of electric field lines per unit area. Thus, the scalar product $\vec{E} \cdot d\vec{A}$ in Eq. 23-4 is proportional to the number of electric field lines passing through area $d\vec{A}$. Then, because the integration in Eq. 23-4 is carried out over a Gaussian surface, which is closed, we see that

The electric flux Φ through a Gaussian surface is proportional to the net number of electric field lines passing through that surface.

CHECKPOINT 1

The figure here shows a Gaussian cube of face area A immersed in a uniform electric field $\vec{E}$ that has the positive direction of the z axis. In terms of E and A, what is the flux through (a) the front face (which is in the xy plane), (b) the rear face, (c) the top face, and (d) the whole cube?

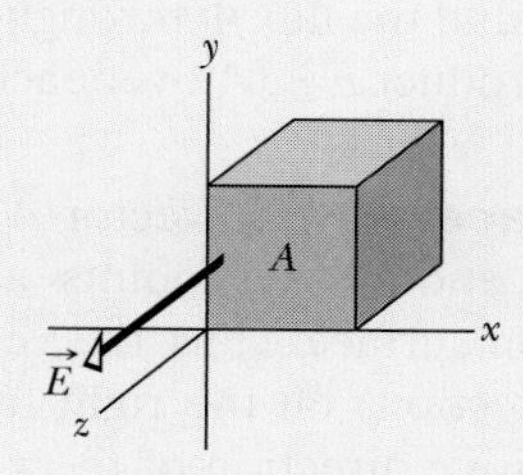

Sample Problem

Flux through a closed cylinder, uniform field

Figure 23-4 shows a Gaussian surface in the form of a cylinder of radius R immersed in a uniform electric field $\vec{E}$, with the cylinder axis parallel to the field. What is the flux Φ of the electric field through this closed surface?

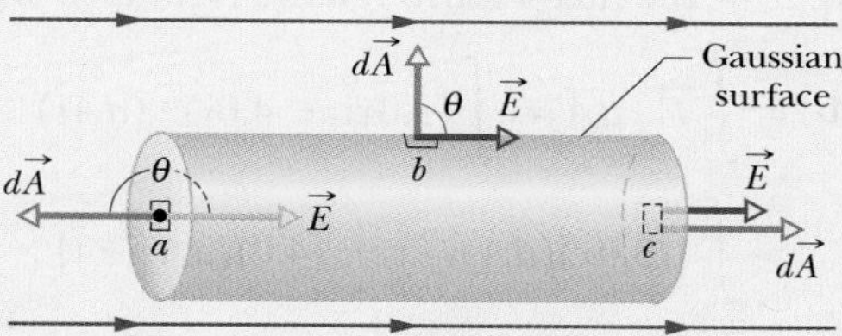

Fig. 23-4 A cylindrical Gaussian surface, closed by end caps, is immersed in a uniform electric field. The cylinder axis is parallel to the field direction.

KEY IDEA

We can find the flux Φ through the Gaussian surface by integrating the scalar product $\vec{E} \cdot d\vec{A}$ over that surface.

Calculations: We can do the integration by writing the flux as the sum of three terms: integrals over the left cylinder cap a, the cylindrical surface b, and the right cap c. Thus, from Eq. 23-4,

$$\Phi = \oint \vec{E} \cdot d\vec{A}$$

$$= \int_a \vec{E} \cdot d\vec{A} + \int_b \vec{E} \cdot d\vec{A} + \int_c \vec{E} \cdot d\vec{A}. \tag{23-5}$$

For all points on the left cap, the angle θ between $\vec{E}$ and $d\vec{A}$ is 180° and the magnitude E of the field is uniform. Thus,

$$\int_a \vec{E} \cdot d\vec{A} = \int E(\cos 180°)\, dA = -E \int dA = -EA,$$

where $\int dA$ gives the cap's area $A\ (= \pi R^2)$. Similarly, for the right cap, where $\theta = 0$ for all points,

$$\int_c \vec{E} \cdot d\vec{A} = \int E(\cos 0)\, dA = EA.$$

Finally, for the cylindrical surface, where the angle θ is 90° at all points,

$$\int_b \vec{E} \cdot d\vec{A} = \int E(\cos 90°)\, dA = 0.$$

Substituting these results into Eq. 23-5 leads us to

$$\Phi = -EA + 0 + EA = 0. \quad \text{(Answer)}$$

The net flux is zero because the field lines that represent the electric field all pass entirely through the Gaussian surface, from the left to the right.

Additional examples, video, and practice available at *WileyPLUS*

Sample Problem

Flux through a closed cube, nonuniform field

A *nonuniform* electric field given by $\vec{E} = 3.0x\hat{i} + 4.0\hat{j}$ pierces the Gaussian cube shown in Fig. 23-5*a*. (E is in newtons per coulomb and x is in meters.) What is the electric flux through the right face, the left face, and the top face? (We consider the other faces in another sample problem.)

KEY IDEA

We can find the flux Φ through the surface by integrating the scalar product $\vec{E} \cdot d\vec{A}$ over each face.

Right face: An area vector $\vec{A}$ is always perpendicular to its surface and always points away from the interior of a Gaussian surface. Thus, the vector $d\vec{A}$ for any area element (small section) on the right face of the cube must point in the positive direction of the x axis. An example of such an element is shown in Figs. 23-5*b* and *c*, but we would have an identical vector for any other choice of an area element on that face. The most convenient way to express the vector is in unit-vector notation,

$$d\vec{A} = dA\hat{i}.$$

From Eq. 23-4, the flux Φ_r through the right face is then

$$\begin{aligned}\Phi_r &= \int \vec{E} \cdot d\vec{A} = \int (3.0x\hat{i} + 4.0\hat{j}) \cdot (dA\hat{i}) \\ &= \int [(3.0x)(dA)\hat{i} \cdot \hat{i} + (4.0)(dA)\hat{j} \cdot \hat{i}] \\ &= \int (3.0x\, dA + 0) = 3.0 \int x\, dA.\end{aligned}$$

We are about to integrate over the right face, but we note that x has the same value everywhere on that face—namely, $x = 3.0$ m. This means we can substitute that constant value for x. This can be a confusing argument. Although x is certainly a variable as we move left to right across the figure, because the right face is perpendicular to the x axis, every point on the face has the same x coordinate. (The y and z coordinates do not matter in our integral.) Thus, we have

$$\Phi_r = 3.0 \int (3.0)\, dA = 9.0 \int dA.$$

The integral $\int dA$ merely gives us the area $A = 4.0\ \text{m}^2$ of the right face; so

$$\Phi_r = (9.0\ \text{N/C})(4.0\ \text{m}^2) = 36\ \text{N} \cdot \text{m}^2/\text{C}. \qquad \text{(Answer)}$$

Left face: The procedure for finding the flux through the left face is the same as that for the right face. However, two factors change. (1) The differential area vector $d\vec{A}$ points in the negative direction of the x axis, and thus $d\vec{A} = -dA\hat{i}$ (Fig. 23-5*d*). (2) The term x again appears in our integration, and it is again constant over the face being considered. However, on the left face, $x = 1.0$ m. With these two changes, we find that the flux Φ_l through the left face is

$$\Phi_l = -12\ \text{N} \cdot \text{m}^2/\text{C}. \qquad \text{(Answer)}$$

Top face: The differential area vector $d\vec{A}$ points in the positive direction of the y axis, and thus $d\vec{A} = dA\hat{j}$ (Fig. 23-5*e*). The flux Φ_t through the top face is then

$$\begin{aligned}\Phi_t &= \int (3.0x\hat{i} + 4.0\hat{j}) \cdot (dA\hat{j}) \\ &= \int [(3.0x)(dA)\hat{i} \cdot \hat{j} + (4.0)(dA)\hat{j} \cdot \hat{j}] \\ &= \int (0 + 4.0\, dA) = 4.0 \int dA \\ &= 16\ \text{N} \cdot \text{m}^2/\text{C}. \qquad \text{(Answer)}\end{aligned}$$

Additional examples, video, and practice available at *WileyPLUS*

23-4 Gauss' Law

Gauss' law relates the net flux Φ of an electric field through a closed surface (a Gaussian surface) to the *net* charge q_{enc} that is *enclosed* by that surface. It tells us that

$$\varepsilon_0 \Phi = q_{enc} \qquad \text{(Gauss' law).} \qquad (23\text{-}6)$$

By substituting Eq. 23-4, the definition of flux, we can also write Gauss' law as

$$\varepsilon_0 \oint \vec{E} \cdot d\vec{A} = q_{enc} \qquad \text{(Gauss' law).} \qquad (23\text{-}7)$$

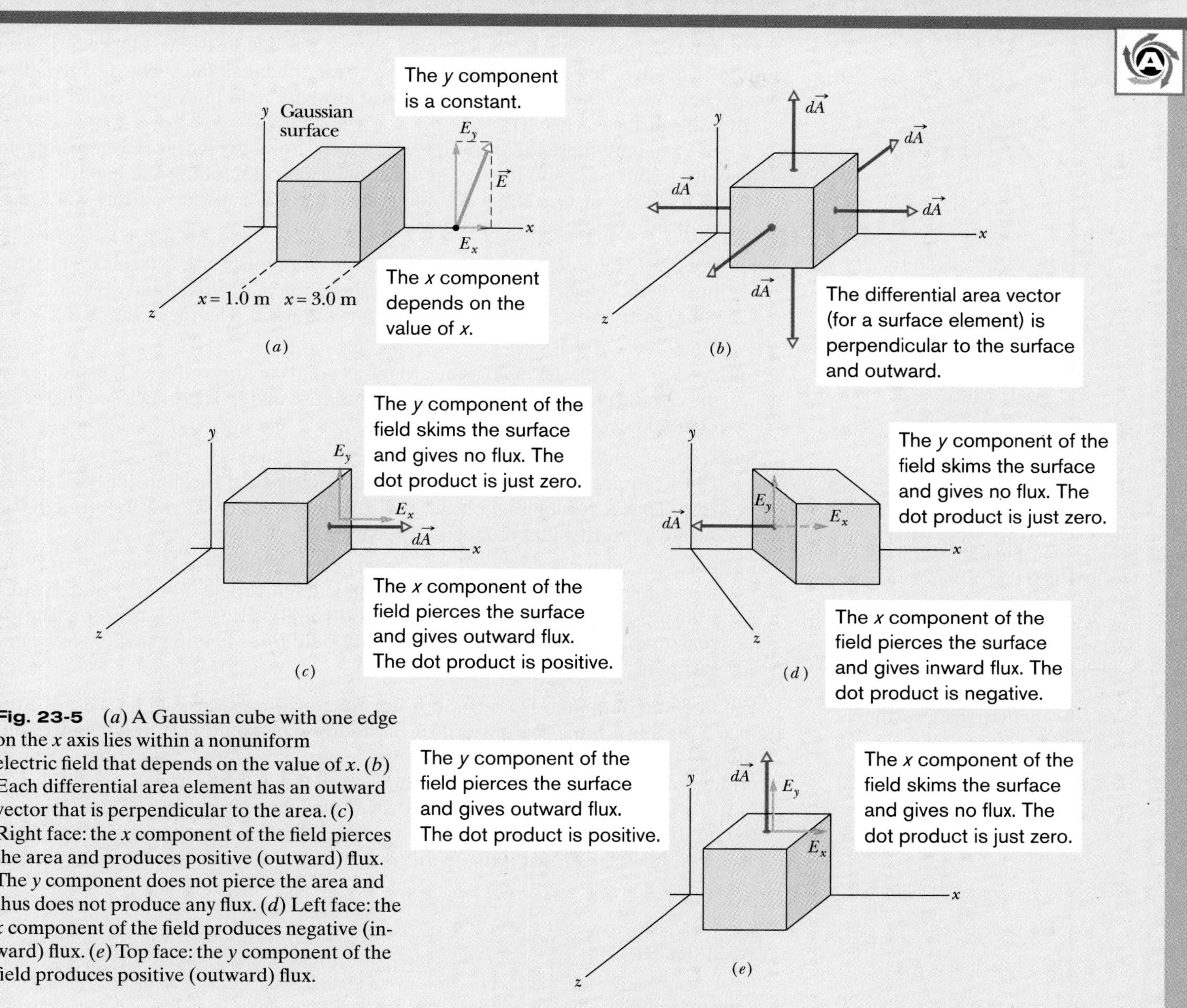

Fig. 23-5 (*a*) A Gaussian cube with one edge on the x axis lies within a nonuniform electric field that depends on the value of x. (*b*) Each differential area element has an outward vector that is perpendicular to the area. (*c*) Right face: the x component of the field pierces the area and produces positive (outward) flux. The y component does not pierce the area and thus does not produce any flux. (*d*) Left face: the x component of the field produces negative (inward) flux. (*e*) Top face: the y component of the field produces positive (outward) flux.

Equations 23-6 and 23-7 hold only when the net charge is located in a vacuum or (what is the same for most practical purposes) in air. In Chapter 25, we modify Gauss' law to include situations in which a material such as mica, oil, or glass is present.

In Eqs. 23-6 and 23-7, the net charge q_{enc} is the algebraic sum of all the *enclosed* positive and negative charges, and it can be positive, negative, or zero. We include the sign, rather than just use the magnitude of the enclosed charge, because the sign tells us something about the net flux through the Gaussian surface: If q_{enc} is positive, the net flux is *outward*; if q_{enc} is negative, the net flux is *inward*.

Charge outside the surface, no matter how large or how close it may be, is not included in the term q_{enc} in Gauss' law. The exact form and location of the charges inside the Gaussian surface are also of no concern; the only things that

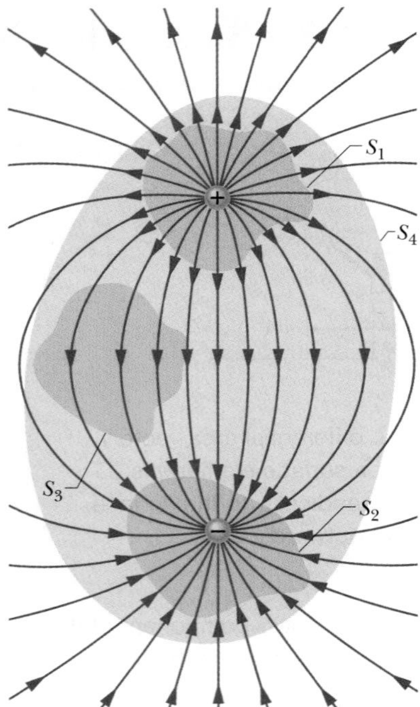

Fig. 23-6 Two point charges, equal in magnitude but opposite in sign, and the field lines that represent their net electric field. Four Gaussian surfaces are shown in cross section. Surface S_1 encloses the positive charge. Surface S_2 encloses the negative charge. Surface S_3 encloses no charge. Surface S_4 encloses both charges and thus no net charge.

matter on the right side of Eqs. 23-6 and 23-7 are the magnitude and sign of the net enclosed charge. The quantity $\vec{E}$ on the left side of Eq. 23-7, however, is the electric field resulting from *all* charges, both those inside and those outside the Gaussian surface. This statement may seem to be inconsistent, but keep this in mind: The electric field due to a charge outside the Gaussian surface contributes zero net flux *through* the surface, because as many field lines due to that charge enter the surface as leave it.

Let us apply these ideas to Fig. 23-6, which shows two point charges, equal in magnitude but opposite in sign, and the field lines describing the electric fields the charges set up in the surrounding space. Four Gaussian surfaces are also shown, in cross section. Let us consider each in turn.

Surface S_1. The electric field is outward for all points on this surface. Thus, the flux of the electric field through this surface is positive, and so is the net charge within the surface, as Gauss' law requires. (That is, in Eq. 23-6, if Φ is positive, q_{enc} must be also.)

Surface S_2. The electric field is inward for all points on this surface. Thus, the flux of the electric field through this surface is negative and so is the enclosed charge, as Gauss' law requires.

Surface S_3. This surface encloses no charge, and thus $q_{enc} = 0$. Gauss' law (Eq. 23-6) requires that the net flux of the electric field through this surface be zero. That is reasonable because all the field lines pass entirely through the surface, entering it at the top and leaving at the bottom.

Surface S_4. This surface encloses no *net* charge, because the enclosed positive and negative charges have equal magnitudes. Gauss' law requires that the net flux of the electric field through this surface be zero. That is reasonable because there are as many field lines leaving surface S_4 as entering it.

What would happen if we were to bring an enormous charge Q up close to surface S_4 in Fig. 23-6? The pattern of the field lines would certainly change, but the net flux for each of the four Gaussian surfaces would not change. We can understand this because the field lines associated with the added Q would pass entirely through each of the four Gaussian surfaces, making no contribution to the net flux through any of them. The value of Q would not enter Gauss' law in any way, because Q lies outside all four of the Gaussian surfaces that we are considering.

CHECKPOINT 2

The figure shows three situations in which a Gaussian cube sits in an electric field. The arrows and the values indicate the directions of the field lines and the magnitudes (in $N \cdot m^2/C$) of the flux through the six sides of each cube. (The lighter arrows are for the hidden faces.) In which situation does the cube enclose (a) a positive net charge, (b) a negative net charge, and (c) zero net charge?

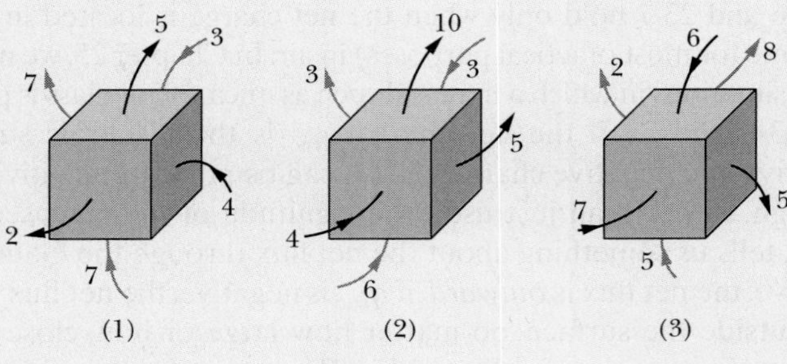

Sample Problem

Relating the net enclosed charge and the net flux

Figure 23-7 shows five charged lumps of plastic and an electrically neutral coin. The cross section of a Gaussian surface S is indicated. What is the net electric flux through the surface if $q_1 = q_4 = +3.1$ nC, $q_2 = q_5 = -5.9$ nC, and $q_3 = -3.1$ nC?

KEY IDEA

The *net* flux Φ through the surface depends on the *net* charge q_{enc} enclosed by surface S.

Calculation: The coin does not contribute to Φ because it is neutral and thus contains equal amounts of positive and negative charge. We could include those equal amounts, but they would simply sum to be zero when we calculate the *net* charge enclosed by the surface. So, let's not bother. Charges q_4 and q_5 do not contribute because they are outside surface S. They certainly send electric field lines through the surface, but as much enters as leaves and no net flux is contributed. Thus, q_{enc} is only the sum $q_1 + q_2 + q_3$ and Eq. 23-6 gives us

$$\Phi = \frac{q_{enc}}{\varepsilon_0} = \frac{q_1 + q_2 + q_3}{\varepsilon_0}$$

$$= \frac{+3.1 \times 10^{-9}\text{ C} - 5.9 \times 10^{-9}\text{ C} - 3.1 \times 10^{-9}\text{ C}}{8.85 \times 10^{-12}\text{ C}^2/\text{N}\cdot\text{m}^2}$$

$$= -670\text{ N}\cdot\text{m}^2/\text{C}. \qquad \text{(Answer)}$$

The minus sign shows that the net flux through the surface is inward and thus that the net charge within the surface is negative.

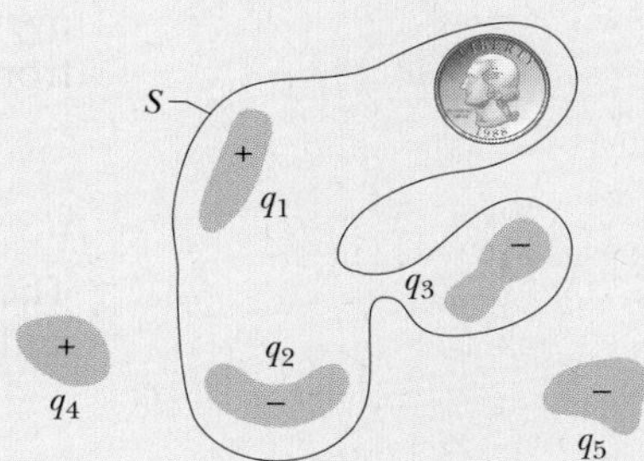

Fig. 23-7 Five plastic objects, each with an electric charge, and a coin, which has no net charge. A Gaussian surface, shown in cross section, encloses three of the plastic objects and the coin.

Sample Problem

Enclosed charge in a nonuniform field

What is the net charge enclosed by the Gaussian cube of Fig. 23-5, which lies in the electric field $\vec{E} = 3.0x\hat{i} + 4.0\hat{j}$? ($E$ is in newtons per coulomb and x is in meters.)

KEY IDEA

The net charge enclosed by a (real or mathematical) closed surface is related to the total electric flux through the surface by Gauss' law as given by Eq. 23-6 ($\varepsilon_0\Phi = q_{enc}$).

Flux: To use Eq. 23-6, we need to know the flux through all six faces of the cube. We already know the flux through the right face ($\Phi_r = 36$ N·m²/C), the left face ($\Phi_l = -12$ N·m²/C), and the top face ($\Phi_t = 16$ N·m²/C).

For the bottom face, our calculation is just like that for the top face *except* that the differential area vector $d\vec{A}$ is now directed downward along the y axis (recall, it must be *outward* from the Gaussian enclosure). Thus, we have $d\vec{A} = -dA\hat{j}$, and we find

$$\Phi_b = -16\text{ N}\cdot\text{m}^2/\text{C}.$$

For the front face we have $d\vec{A} = dA\hat{k}$, and for the back face, $d\vec{A} = -dA\hat{k}$. When we take the dot product of the given electric field $\vec{E} = 3.0x\hat{i} + 4.0\hat{j}$ with either of these expressions for $d\vec{A}$, we get 0 and thus there is no flux through those faces. We can now find the total flux through the six sides of the cube:

$$\Phi = (36 - 12 + 16 - 16 + 0 + 0)\text{ N}\cdot\text{m}^2/\text{C}$$
$$= 24\text{ N}\cdot\text{m}^2/\text{C}.$$

Enclosed charge: Next, we use Gauss' law to find the charge q_{enc} enclosed by the cube:

$$q_{enc} = \varepsilon_0\Phi = (8.85 \times 10^{-12}\text{ C}^2/\text{N}\cdot\text{m}^2)(24\text{ N}\cdot\text{m}^2/\text{C})$$
$$= 2.1 \times 10^{-10}\text{ C}. \qquad \text{(Answer)}$$

Thus, the cube encloses a *net* positive charge.

Additional examples, video, and practice available at *WileyPLUS*

23-5 Gauss' Law and Coulomb's Law

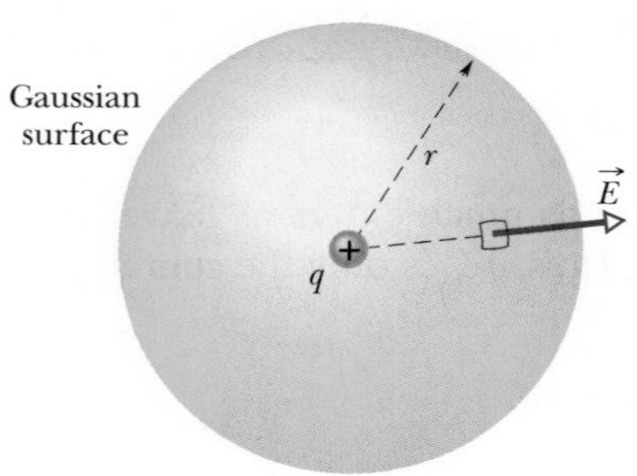

Fig. 23-8 A spherical Gaussian surface centered on a point charge q.

Because Gauss' law and Coulomb's law are different ways of describing the relation between electric charge and electric field in static situations, we should be able to derive each from the other. Here we derive Coulomb's law from Gauss' law and some symmetry considerations.

Figure 23-8 shows a positive point charge q, around which we have drawn a concentric spherical Gaussian surface of radius r. Let us divide this surface into differential areas dA. By definition, the area vector $d\vec{A}$ at any point is perpendicular to the surface and directed outward from the interior. From the symmetry of the situation, we know that at any point the electric field $\vec{E}$ is also perpendicular to the surface and directed outward from the interior. Thus, since the angle θ between $\vec{E}$ and $d\vec{A}$ is zero, we can rewrite Eq. 23-7 for Gauss' law as

$$\varepsilon_0 \oint \vec{E} \cdot d\vec{A} = \varepsilon_0 \oint E\, dA = q_{\text{enc}}. \tag{23-8}$$

Here $q_{\text{enc}} = q$. Although E varies radially with distance from q, it has the same value everywhere on the spherical surface. Since the integral in Eq. 23-8 is taken over that surface, E is a constant in the integration and can be brought out in front of the integral sign. That gives us

$$\varepsilon_0 E \oint dA = q. \tag{23-9}$$

The integral is now merely the sum of all the differential areas dA on the sphere and thus is just the surface area, $4\pi r^2$. Substituting this, we have

$$\varepsilon_0 E(4\pi r^2) = q$$

or

$$E = \frac{1}{4\pi\varepsilon_0} \frac{q}{r^2}. \tag{23-10}$$

This is exactly Eq. 22-3, which we found using Coulomb's law.

CHECKPOINT 3

There is a certain net flux Φ_i through a Gaussian sphere of radius r enclosing an isolated charged particle. Suppose the enclosing Gaussian surface is changed to (a) a larger Gaussian sphere, (b) a Gaussian cube with edge length equal to r, and (c) a Gaussian cube with edge length equal to $2r$. In each case, is the net flux through the new Gaussian surface greater than, less than, or equal to Φ_i?

23-6 A Charged Isolated Conductor

Gauss' law permits us to prove an important theorem about conductors:

> If an excess charge is placed on an isolated conductor, that amount of charge will move entirely to the surface of the conductor. None of the excess charge will be found within the body of the conductor.

This might seem reasonable, considering that charges with the same sign repel one another. You might imagine that, by moving to the surface, the added charges are getting as far away from one another as they can. We turn to Gauss' law for verification of this speculation.

Figure 23-9*a* shows, in cross section, an isolated lump of copper hanging from an insulating thread and having an excess charge q. We place a Gaussian surface just inside the actual surface of the conductor.

Copper surface
Gaussian surface
(*a*)

Gaussian surface
Copper surface
(*b*)

Fig. 23-9 (*a*) A lump of copper with a charge q hangs from an insulating thread. A Gaussian surface is placed within the metal, just inside the actual surface. (*b*) The lump of copper now has a cavity within it. A Gaussian surface lies within the metal, close to the cavity surface.

The electric field inside this conductor must be zero. If this were not so, the field would exert forces on the conduction (free) electrons, which are always present in a conductor, and thus current would always exist within a conductor. (That is, charge would flow from place to place within the conductor.) Of course, there is no such perpetual current in an isolated conductor, and so the internal electric field is zero.

(An internal electric field *does* appear as a conductor is being charged. However, the added charge quickly distributes itself in such a way that the net internal electric field—the vector sum of the electric fields due to all the charges, both inside and outside—is zero. The movement of charge then ceases, because the net force on each charge is zero; the charges are then in *electrostatic equilibrium.*)

If $\vec{E}$ is zero everywhere inside our copper conductor, it must be zero for all points on the Gaussian surface because that surface, though close to the surface of the conductor, is definitely inside the conductor. This means that the flux through the Gaussian surface must be zero. Gauss' law then tells us that the net charge inside the Gaussian surface must also be zero. Then because the excess charge is not inside the Gaussian surface, it must be outside that surface, which means it must lie on the actual surface of the conductor.

An Isolated Conductor with a Cavity

Figure 23-9*b* shows the same hanging conductor, but now with a cavity that is totally within the conductor. It is perhaps reasonable to suppose that when we scoop out the electrically neutral material to form the cavity, we do not change the distribution of charge or the pattern of the electric field that exists in Fig. 23-9*a*. Again, we must turn to Gauss' law for a quantitative proof.

We draw a Gaussian surface surrounding the cavity, close to its surface but inside the conducting body. Because $\vec{E} = 0$ inside the conductor, there can be no flux through this new Gaussian surface. Therefore, from Gauss' law, that surface can enclose no net charge. We conclude that there is no net charge on the cavity walls; all the excess charge remains on the outer surface of the conductor, as in Fig. 23-9*a*.

The Conductor Removed

Suppose that, by some magic, the excess charges could be "frozen" into position on the conductor's surface, perhaps by embedding them in a thin plastic coating, and suppose that then the conductor could be removed completely. This is equivalent to enlarging the cavity of Fig. 23-9*b* until it consumes the entire conductor, leaving only the charges. The electric field would not change at all; it would remain zero inside the thin shell of charge and would remain unchanged for all external points. This shows us that the electric field is set up by the charges and not by the conductor. The conductor simply provides an initial pathway for the charges to take up their positions.

The External Electric Field

You have seen that the excess charge on an isolated conductor moves entirely to the conductor's surface. However, unless the conductor is spherical, the charge does not distribute itself uniformly. Put another way, the surface charge density σ (charge per unit area) varies over the surface of any nonspherical conductor. Generally, this variation makes the determination of the electric field set up by the surface charges very difficult.

However, the electric field just outside the surface of a conductor is easy to determine using Gauss' law. To do this, we consider a section of the surface that is small enough to permit us to neglect any curvature and thus to take the section

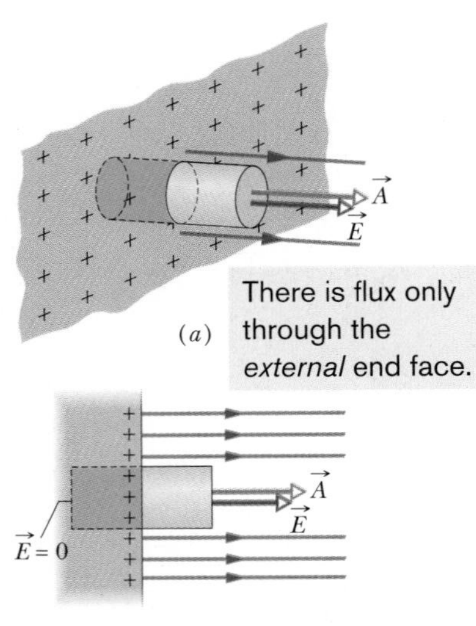

Fig. 23-10 (*a*) Perspective view and (*b*) side view of a tiny portion of a large, isolated conductor with excess positive charge on its surface. A (closed) cylindrical Gaussian surface, embedded perpendicularly in the conductor, encloses some of the charge. Electric field lines pierce the external end cap of the cylinder, but not the internal end cap. The external end cap has area A and area vector $\vec{A}$.

to be flat. We then imagine a tiny cylindrical Gaussian surface to be embedded in the section as in Fig. 23-10: One end cap is fully inside the conductor, the other is fully outside, and the cylinder is perpendicular to the conductor's surface.

The electric field $\vec{E}$ at and just outside the conductor's surface must also be perpendicular to that surface. If it were not, then it would have a component along the conductor's surface that would exert forces on the surface charges, causing them to move. However, such motion would violate our implicit assumption that we are dealing with electrostatic equilibrium. Therefore, $\vec{E}$ is perpendicular to the conductor's surface.

We now sum the flux through the Gaussian surface. There is no flux through the internal end cap, because the electric field within the conductor is zero. There is no flux through the curved surface of the cylinder, because internally (in the conductor) there is no electric field and externally the electric field is parallel to the curved portion of the Gaussian surface. The only flux through the Gaussian surface is that through the external end cap, where $\vec{E}$ is perpendicular to the plane of the cap. We assume that the cap area A is small enough that the field magnitude E is constant over the cap. Then the flux through the cap is EA, and that is the net flux Φ through the Gaussian surface.

The charge q_{enc} enclosed by the Gaussian surface lies on the conductor's surface in an area A. If σ is the charge per unit area, then q_{enc} is equal to σA. When we substitute σA for q_{enc} and EA for Φ, Gauss' law (Eq. 23-6) becomes

$$\varepsilon_0 EA = \sigma A,$$

from which we find

$$E = \frac{\sigma}{\varepsilon_0} \quad \text{(conducting surface).} \tag{23-11}$$

Thus, the magnitude of the electric field just outside a conductor is proportional to the surface charge density on the conductor. If the charge on the conductor is positive, the electric field is directed away from the conductor as in Fig. 23-10. It is directed toward the conductor if the charge is negative.

The field lines in Fig. 23-10 must terminate on negative charges somewhere in the environment. If we bring those charges near the conductor, the charge density at any given location on the conductor's surface changes, and so does the magnitude of the electric field. However, the relation between σ and E is still given by Eq. 23-11.

Sample Problem

Spherical metal shell, electric field and enclosed charge

Figure 23-11*a* shows a cross section of a spherical metal shell of inner radius R. A point charge of $-5.0\ \mu C$ is located at a distance $R/2$ from the center of the shell. If the shell is electrically neutral, what are the (induced) charges on its inner and outer surfaces? Are those charges uniformly distributed? What is the field pattern inside and outside the shell?

KEY IDEAS

Figure 23-11*b* shows a cross section of a spherical Gaussian surface within the metal, just outside the inner wall of the shell. The electric field must be zero inside the metal (and thus on the Gaussian surface inside the metal). This means that the electric flux through the Gaussian surface must also be zero. Gauss' law then tells us that the *net* charge enclosed by the Gaussian surface must be zero.

Reasoning: With a point charge of $-5.0\ \mu C$ within the shell, a charge of $+5.0\ \mu C$ must lie on the inner wall of the shell in order that the net enclosed charge be zero. If the point charge were centered, this positive charge would be uniformly distributed along the inner wall. However, since the point charge is off-center, the distribution of positive charge is skewed, as suggested by Fig. 23-11*b*, because the positive charge tends to collect on the section of the inner wall nearest the (negative) point charge.

Because the shell is electrically neutral, its inner wall can have a charge of $+5.0\ \mu C$ only if electrons, with a total

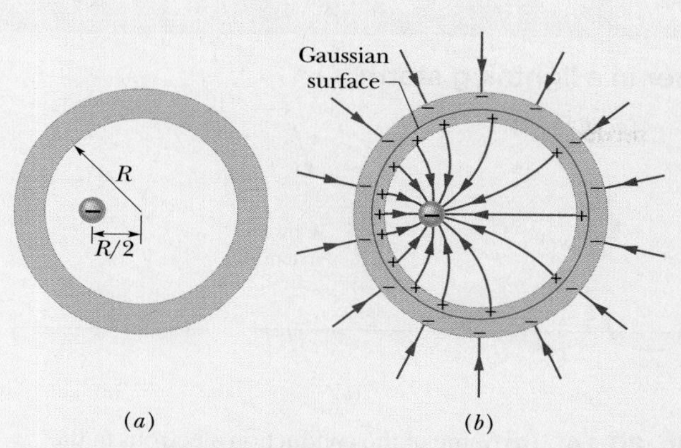

Fig. 23-11 (*a*) A negative point charge is located within a spherical metal shell that is electrically neutral. (*b*) As a result, positive charge is nonuniformly distributed on the inner wall of the shell, and an equal amount of negative charge is uniformly distributed on the outer wall.

charge of $-5.0\ \mu C$, leave the inner wall and move to the outer wall. There they spread out uniformly, as is also suggested by Fig. 23-11*b*. This distribution of negative charge is uniform because the shell is spherical and because the skewed distribution of positive charge on the inner wall cannot produce an electric field in the shell to affect the distribution of charge on the outer wall. Furthermore, these negative charges repel one another.

The field lines inside and outside the shell are shown approximately in Fig. 23-11*b*. All the field lines intersect the shell and the point charge perpendicularly. Inside the shell the pattern of field lines is skewed because of the skew of the positive charge distribution. Outside the shell the pattern is the same as if the point charge were centered and the shell were missing. In fact, this would be true no matter where inside the shell the point charge happened to be located.

WILEY PLUS Additional examples, video, and practice available at *WileyPLUS*

23-7 Applying Gauss' Law: Cylindrical Symmetry

Figure 23-12 shows a section of an infinitely long cylindrical plastic rod with a uniform positive linear charge density λ. Let us find an expression for the magnitude of the electric field $\vec{E}$ at a distance r from the axis of the rod.

Our Gaussian surface should match the symmetry of the problem, which is cylindrical. We choose a circular cylinder of radius r and length h, coaxial with the rod. Because the Gaussian surface must be closed, we include two end caps as part of the surface.

Imagine now that, while you are not watching, someone rotates the plastic rod about its longitudinal axis or turns it end for end. When you look again at the rod, you will not be able to detect any change. We conclude from this symmetry that the only uniquely specified direction in this problem is along a radial line. Thus, at every point on the cylindrical part of the Gaussian surface, $\vec{E}$ must have the same magnitude E and (for a positively charged rod) must be directed radially outward.

Since $2\pi r$ is the cylinder's circumference and h is its height, the area A of the cylindrical surface is $2\pi rh$. The flux of $\vec{E}$ through this cylindrical surface is then

$$\Phi = EA\cos\theta = E(2\pi rh)\cos 0 = E(2\pi rh).$$

There is no flux through the end caps because $\vec{E}$, being radially directed, is parallel to the end caps at every point.

The charge enclosed by the surface is λh, which means Gauss' law,

$$\varepsilon_0 \Phi = q_{enc},$$

reduces to

$$\varepsilon_0 E(2\pi rh) = \lambda h,$$

yielding

$$E = \frac{\lambda}{2\pi\varepsilon_0 r} \quad \text{(line of charge).} \tag{23-12}$$

This is the electric field due to an infinitely long, straight line of charge, at a point that is a radial distance r from the line. The direction of $\vec{E}$ is radially outward from the line of charge if the charge is positive, and radially inward if it is negative. Equation 23-12 also approximates the field of a *finite* line of charge at points that are not too near the ends (compared with the distance from the line).

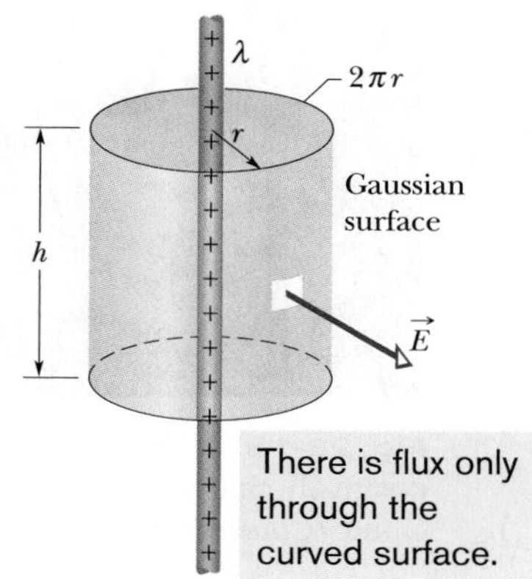

Fig. 23-12 A Gaussian surface in the form of a closed cylinder surrounds a section of a very long, uniformly charged, cylindrical plastic rod.

Sample Problem

Gauss' law and an upward streamer in a lightning storm

Upward streamer in a lightning storm. The woman in Fig. 23-13 was standing on a lookout platform in the Sequoia National Park when a large storm cloud moved overhead. Some of the conduction electrons in her body were driven into the ground by the cloud's negatively charged base (Fig. 23-14*a*), leaving her positively charged. You can tell she was highly charged because her hair strands repelled one another and extended away from her along the electric field lines produced by the charge on her.

Lightning did not strike the woman, but she was in extreme danger because that electric field was on the verge of causing electrical breakdown in the surrounding air. Such a breakdown would have occurred along a path extending away from her in what is called an *upward streamer.* An upward streamer is dangerous because the resulting ionization of molecules in the air suddenly frees a tremendous number of electrons from those molecules. Had the woman in Fig. 23-13 developed an upward streamer, the free electrons in the air would have moved to neutralize her (Fig. 23-14*b*), producing a large, perhaps fatal, charge flow through her body. That charge flow is dangerous because it could have interfered with or even stopped her breathing (which is obviously necessary for oxygen) and the steady beat of her heart (which is obviously necessary for the blood flow that carries the oxygen). The charge flow could also have caused burns.

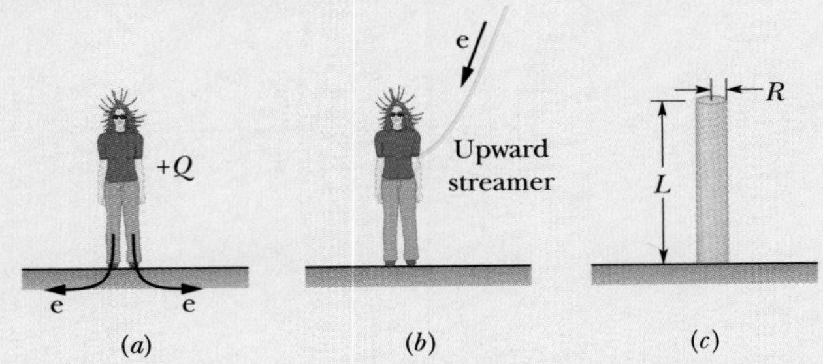

Fig. 23-14 (*a*) Some of the conduction electrons in the woman's body are driven into the ground, leaving her positively charged. (*b*) An upward streamer develops if the air undergoes electrical breakdown, which provides a path for electrons freed from molecules in the air to move to the woman. (*c*) A cylinder represents the woman.

Let's model her body as a narrow vertical cylinder of height $L = 1.8$ m and radius $R = 0.10$ m (Fig. 23-14*c*). Assume that charge Q was uniformly distributed along the cylinder and that electrical breakdown would have occurred if the electric field magnitude along her body had exceeded the critical value $E_c = 2.4$ MN/C. What value of Q would have put the air along her body on the verge of breakdown?

Fig. 23-13 This woman has become positively charged by an overhead storm cloud. *(Courtesy NOAA)*

KEY IDEA

Because $R \ll L$, we can approximate the charge distribution as a long line of charge. Further, because we assume that the charge is uniformly distributed along this line, we can approximate the magnitude of the electric field along the side of her body with Eq. 23-12 ($E = \lambda/2\pi\varepsilon_0 r$).

Calculations: Substituting the critical value E_c for E, the cylinder radius R for radial distance r, and the ratio Q/L for linear charge density λ, we have

$$E_c = \frac{Q/L}{2\pi\varepsilon_0 R},$$

or

$$Q = 2\pi\varepsilon_0 RLE_c.$$

Substituting given data then gives us

$$Q = (2\pi)(8.85 \times 10^{-12}\ \mathrm{C^2/N\cdot m^2})(0.10\ \mathrm{m}) \times (1.8\ \mathrm{m})(2.4 \times 10^6\ \mathrm{N/C}) = 2.402 \times 10^{-5}\ \mathrm{C} \approx 24\ \mu\mathrm{C}. \quad \text{(Answer)}$$

Additional examples, video, and practice available at *WileyPLUS*

23-8 Applying Gauss' Law: Planar Symmetry

Nonconducting Sheet

Figure 23-15 shows a portion of a thin, infinite, nonconducting sheet with a uniform (positive) surface charge density σ. A sheet of thin plastic wrap, uniformly charged on one side, can serve as a simple model. Let us find the electric field $\vec{E}$ a distance r in front of the sheet.

A useful Gaussian surface is a closed cylinder with end caps of area A, arranged to pierce the sheet perpendicularly as shown. From symmetry, $\vec{E}$ must be perpendicular to the sheet and hence to the end caps. Furthermore, since the charge is positive, $\vec{E}$ is directed *away* from the sheet, and thus the electric field lines pierce the two Gaussian end caps in an outward direction. Because the field lines do not pierce the curved surface, there is no flux through this portion of the Gaussian surface. Thus $\vec{E} \cdot d\vec{A}$ is simply $E\,dA$; then Gauss' law,

$$\varepsilon_0 \oint \vec{E} \cdot d\vec{A} = q_{\text{enc}},$$

becomes

$$\varepsilon_0(EA + EA) = \sigma A,$$

where σA is the charge enclosed by the Gaussian surface. This gives

$$E = \frac{\sigma}{2\varepsilon_0} \quad \text{(sheet of charge).} \tag{23-13}$$

Since we are considering an infinite sheet with uniform charge density, this result holds for any point at a finite distance from the sheet. Equation 23-13 agrees with Eq. 22-27, which we found by integration of electric field components.

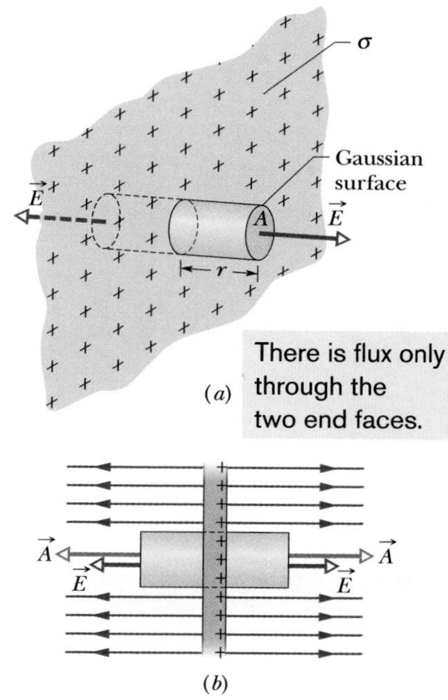

Fig. 23-15 (*a*) Perspective view and (*b*) side view of a portion of a very large, thin plastic sheet, uniformly charged on one side to surface charge density σ. A closed cylindrical Gaussian surface passes through the sheet and is perpendicular to it.

Two Conducting Plates

Figure 23-16*a* shows a cross section of a thin, infinite conducting plate with excess positive charge. From Section 23-6 we know that this excess charge lies on the surface of the plate. Since the plate is thin and very large, we can assume that essentially all the excess charge is on the two large faces of the plate.

If there is no external electric field to force the positive charge into some particular distribution, it will spread out on the two faces with a uniform surface charge density of magnitude σ_1. From Eq. 23-11 we know that just outside the plate this charge sets up an electric field of magnitude $E = \sigma_1/\varepsilon_0$. Because the excess charge is positive, the field is directed away from the plate.

Figure 23-16*b* shows an identical plate with excess negative charge having the same magnitude of surface charge density σ_1. The only difference is that now the electric field is directed toward the plate.

Suppose we arrange for the plates of Figs. 23-16*a* and *b* to be close to each other and parallel (Fig. 23-16*c*). Since the plates are conductors, when we bring them into this arrangement, the excess charge on one plate attracts the excess charge on the other plate, and all the excess charge moves onto the inner faces of the plates as in Fig. 23-16*c*. With twice as much charge now on each inner face, the new surface charge density (call it σ) on each inner face is twice σ_1. Thus, the electric field at any point between the plates has the magnitude

$$E = \frac{2\sigma_1}{\varepsilon_0} = \frac{\sigma}{\varepsilon_0}. \tag{23-14}$$

This field is directed away from the positively charged plate and toward the negatively charged plate. Since no excess charge is left on the outer faces, the electric field to the left and right of the plates is zero.

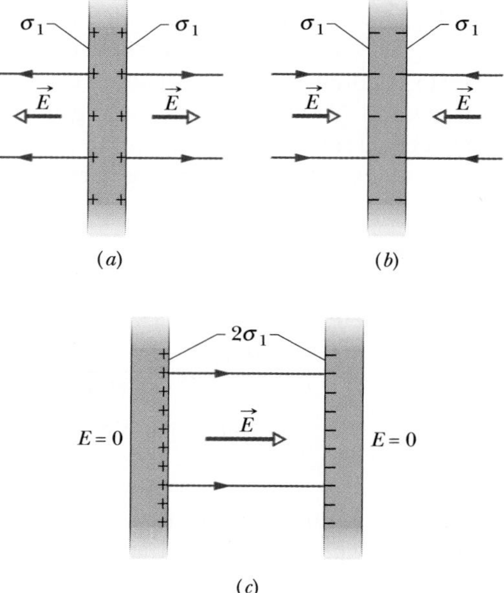

Fig. 23-16 (*a*) A thin, very large conducting plate with excess positive charge. (*b*) An identical plate with excess negative charge. (*c*) The two plates arranged so they are parallel and close.

Because the charges on the plates moved when we brought the plates close to each other, Fig. 23-16*c* is *not* the superposition of Figs. 23-16*a* and *b*; that is, the charge distribution of the two-plate system is not merely the sum of the charge distributions of the individual plates.

You may wonder why we discuss such seemingly unrealistic situations as the field set up by an infinite line of charge, an infinite sheet of charge, or a pair of infinite plates of charge. One reason is that analyzing such situations with Gauss' law is easy. More important is that analyses for "infinite" situations yield good approximations to many real-world problems. Thus, Eq. 23-13 holds well for a finite nonconducting sheet as long as we are dealing with points close to the sheet and not too near its edges. Equation 23-14 holds well for a pair of finite conducting plates as long as we consider points that are not too close to their edges.

The trouble with the edges of a sheet or a plate, and the reason we take care not to deal with them, is that near an edge we can no longer use planar symmetry to find expressions for the fields. In fact, the field lines there are curved (said to be an *edge effect* or *fringing*), and the fields can be very difficult to express algebraically.

Sample Problem

Electric field near two parallel charged metal plates

Figure 23-17*a* shows portions of two large, parallel, nonconducting sheets, each with a fixed uniform charge on one side. The magnitudes of the surface charge densities are $\sigma_{(+)} = 6.8\ \mu\text{C/m}^2$ for the positively charged sheet and $\sigma_{(-)} = 4.3\ \mu\text{C/m}^2$ for the negatively charged sheet.

Find the electric field $\vec{E}$ (a) to the left of the sheets, (b) between the sheets, and (c) to the right of the sheets.

KEY IDEA

With the charges fixed in place (they are on nonconductors), we can find the electric field of the sheets in Fig. 23-17*a* by (1) finding the field of each sheet as if that sheet were isolated and (2) algebraically adding the fields of the isolated sheets via the superposition principle. (We can add the fields algebraically because they are parallel to each other.)

Calculations: At any point, the electric field $\vec{E}_{(+)}$ due to the positive sheet is directed *away* from the sheet and, from Eq. 23-13, has the magnitude

$$E_{(+)} = \frac{\sigma_{(+)}}{2\varepsilon_0} = \frac{6.8\times10^{-6}\ \text{C/m}^2}{(2)(8.85\times10^{-12}\ \text{C}^2/\text{N}\cdot\text{m}^2)} = 3.84\times10^5\ \text{N/C}.$$

Similarly, at any point, the electric field $\vec{E}_{(-)}$ due to the negative sheet is directed *toward* that sheet and has the magnitude

$$E_{(-)} = \frac{\sigma_{(-)}}{2\varepsilon_0} = \frac{4.3\times10^{-6}\ \text{C/m}^2}{(2)(8.85\times10^{-12}\ \text{C}^2/\text{N}\cdot\text{m}^2)} = 2.43\times10^5\ \text{N/C}.$$

Figure 23-17*b* shows the fields set up by the sheets to the left of the sheets (L), between them (B), and to their right (R).

The resultant fields in these three regions follow from the superposition principle. To the left, the field magnitude is

$$\begin{aligned}E_L &= E_{(+)} - E_{(-)}\\ &= 3.84\times10^5\ \text{N/C} - 2.43\times10^5\ \text{N/C}\\ &= 1.4\times10^5\ \text{N/C}. \qquad \text{(Answer)}\end{aligned}$$

Because $E_{(+)}$ is larger than $E_{(-)}$, the net electric field $\vec{E}_L$ in this region is directed to the left, as Fig. 23-17*c* shows. To the right of the sheets, the electric field has the same magnitude but is directed to the right, as Fig. 23-17*c* shows.

Between the sheets, the two fields add and we have

$$\begin{aligned}E_B &= E_{(+)} + E_{(-)}\\ &= 3.84\times10^5\ \text{N/C} + 2.43\times10^5\ \text{N/C}\\ &= 6.3\times10^5\ \text{N/C}. \qquad \text{(Answer)}\end{aligned}$$

The electric field $\vec{E}_B$ is directed to the right.

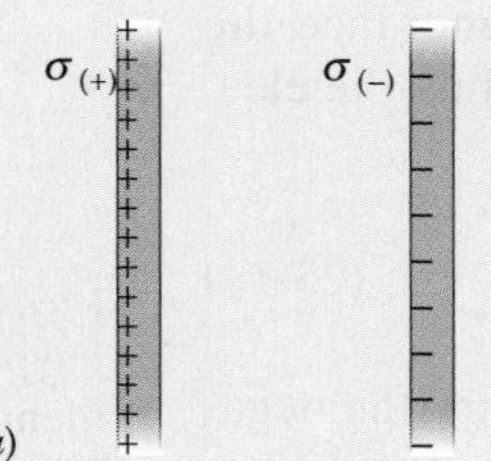

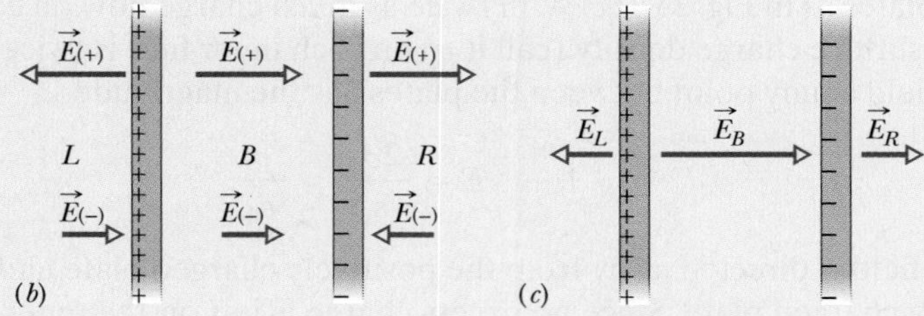

Fig. 23-17 (*a*) Two large, parallel sheets, uniformly charged on one side. (*b*) The individual electric fields resulting from the two charged sheets. (*c*) The net field due to both charged sheets, found by superposition.

Additional examples, video, and practice available at *WileyPLUS*

23-9 Applying Gauss' Law: Spherical Symmetry

Here we use Gauss' law to prove the two shell theorems presented without proof in Section 21-4:

> A shell of uniform charge attracts or repels a charged particle that is outside the shell as if all the shell's charge were concentrated at the center of the shell.

> If a charged particle is located inside a shell of uniform charge, there is no electrostatic force on the particle from the shell.

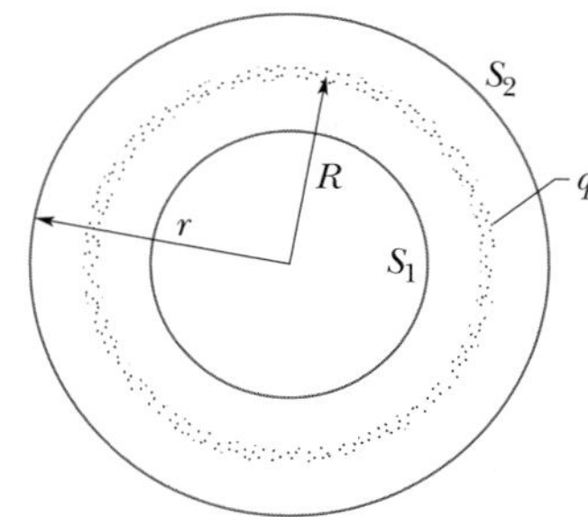

Fig. 23-18 A thin, uniformly charged, spherical shell with total charge q, in cross section. Two Gaussian surfaces S_1 and S_2 are also shown in cross section. Surface S_2 encloses the shell, and S_1 encloses only the empty interior of the shell.

Figure 23-18 shows a charged spherical shell of total charge q and radius R and two concentric spherical Gaussian surfaces, S_1 and S_2. If we followed the procedure of Section 23-5 as we applied Gauss' law to surface S_2, for which $r \geq R$, we would find that

$$E = \frac{1}{4\pi\varepsilon_0}\frac{q}{r^2} \quad \text{(spherical shell, field at } r \geq R). \tag{23-15}$$

This field is the same as one set up by a point charge q at the center of the shell of charge. Thus, the force produced by a shell of charge q on a charged particle placed outside the shell is the same as the force produced by a point charge q located at the center of the shell. This proves the first shell theorem.

Applying Gauss' law to surface S_1, for which $r < R$, leads directly to

$$E = 0 \quad \text{(spherical shell, field at } r < R), \tag{23-16}$$

because this Gaussian surface encloses no charge. Thus, if a charged particle were enclosed by the shell, the shell would exert no net electrostatic force on the particle. This proves the second shell theorem.

Any spherically symmetric charge distribution, such as that of Fig. 23-19, can be constructed with a nest of concentric spherical shells. For purposes of applying the two shell theorems, the volume charge density ρ should have a single value for each shell but need not be the same from shell to shell. Thus, for the charge distribution as a whole, ρ can vary, but only with r, the radial distance from the center. We can then examine the effect of the charge distribution "shell by shell."

In Fig. 23-19*a*, the entire charge lies within a Gaussian surface with $r > R$. The charge produces an electric field on the Gaussian surface as if the charge were a point charge located at the center, and Eq. 23-15 holds.

Figure 23-19*b* shows a Gaussian surface with $r < R$. To find the electric field at points on this Gaussian surface, we consider two sets of charged shells—one set inside the Gaussian surface and one set outside. Equation 23-16 says that the charge lying *outside* the Gaussian surface does not set up a net electric field on the Gaussian surface. Equation 23-15 says that the charge *enclosed* by the surface sets up an electric field as if that enclosed charge were concentrated at the center. Letting q' represent that enclosed charge, we can then rewrite Eq. 23-15 as

$$E = \frac{1}{4\pi\varepsilon_0}\frac{q'}{r^2} \quad \text{(spherical distribution, field at } r \leq R). \tag{23-17}$$

If the full charge q enclosed within radius R is uniform, then q' enclosed within radius r in Fig. 23-19*b* is proportional to q:

$$\frac{\begin{pmatrix}\text{charge enclosed by}\\ \text{sphere of radius } r\end{pmatrix}}{\begin{pmatrix}\text{volume enclosed by}\\ \text{sphere of radius } r\end{pmatrix}} = \frac{\text{full charge}}{\text{full volume}}$$

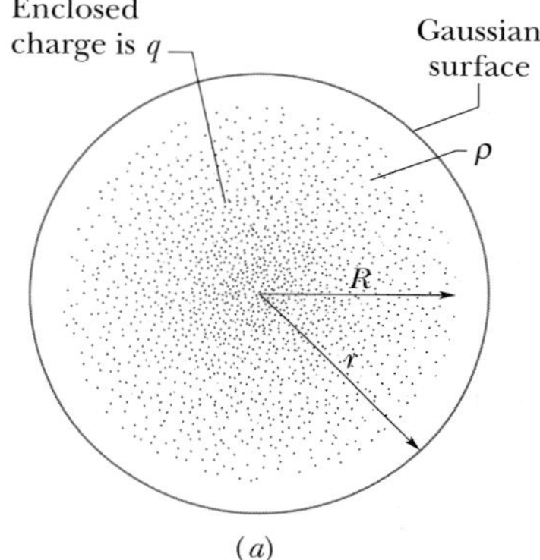

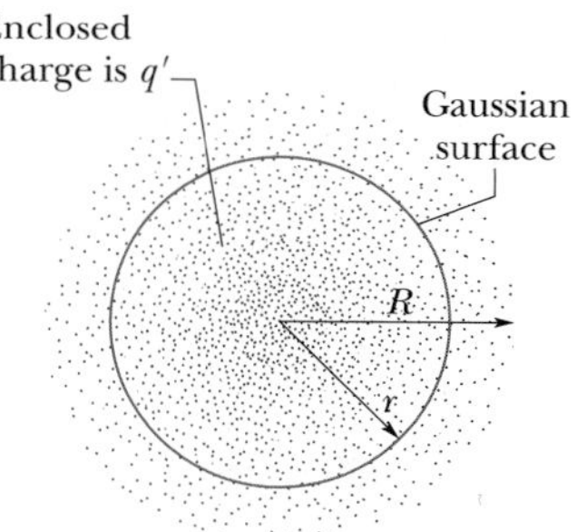

The flux through the surface depends on only the *enclosed* charge.

Fig. 23-19 The dots represent a spherically symmetric distribution of charge of radius R, whose volume charge density ρ is a function only of distance from the center. The charged object is not a conductor, and therefore the charge is assumed to be fixed in position. A concentric spherical Gaussian surface with $r > R$ is shown in (*a*). A similar Gaussian surface with $r < R$ is shown in (*b*).

or

$$\frac{q'}{\frac{4}{3}\pi r^3} = \frac{q}{\frac{4}{3}\pi R^3}. \tag{23-18}$$

This gives us

$$q' = q\frac{r^3}{R^3}. \tag{23-19}$$

Substituting this into Eq. 23-17 yields

$$E = \left(\frac{q}{4\pi\varepsilon_0 R^3}\right)r \quad \text{(uniform charge, field at } r \le R). \tag{23-20}$$

CHECKPOINT 4

The figure shows two large, parallel, nonconducting sheets with identical (positive) uniform surface charge densities, and a sphere with a uniform (positive) volume charge density. Rank the four numbered points according to the magnitude of the net electric field there, greatest first.

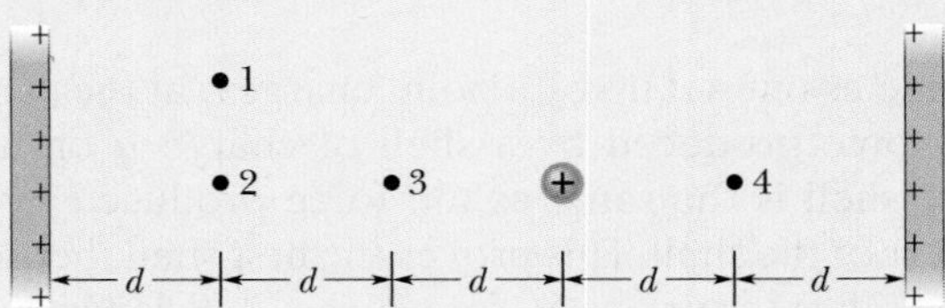

REVIEW & SUMMARY

Gauss' Law *Gauss' law* and Coulomb's law are different ways of describing the relation between charge and electric field in static situations. Gauss' law is

$$\varepsilon_0\Phi = q_{enc} \quad \text{(Gauss' law)}, \tag{23-6}$$

in which q_{enc} is the net charge inside an imaginary closed surface (a *Gaussian surface*) and Φ is the net *flux* of the electric field through the surface:

$$\Phi = \oint \vec{E}\cdot d\vec{A} \quad \text{(electric flux through a Gaussian surface)}. \tag{23-4}$$

Coulomb's law can be derived from Gauss' law.

Applications of Gauss' Law Using Gauss' law and, in some cases, symmetry arguments, we can derive several important results in electrostatic situations. Among these are:

1. An excess charge on an isolated *conductor* is located entirely on the outer surface of the conductor.
2. The external electric field near the *surface of a charged conductor* is perpendicular to the surface and has magnitude

$$E = \frac{\sigma}{\varepsilon_0} \quad \text{(conducting surface)}. \tag{23-11}$$

Within the conductor, $E = 0$.

3. The electric field at any point due to an infinite *line of charge* with uniform linear charge density λ is perpendicular to the line of charge and has magnitude

$$E = \frac{\lambda}{2\pi\varepsilon_0 r} \quad \text{(line of charge)}, \tag{23-12}$$

where r is the perpendicular distance from the line of charge to the point.

4. The electric field due to an *infinite nonconducting sheet* with uniform surface charge density σ is perpendicular to the plane of the sheet and has magnitude

$$E = \frac{\sigma}{2\varepsilon_0} \quad \text{(sheet of charge)}. \tag{23-13}$$

5. The electric field *outside a spherical shell of charge* with radius R and total charge q is directed radially and has magnitude

$$E = \frac{1}{4\pi\varepsilon_0}\frac{q}{r^2} \quad \text{(spherical shell, for } r \ge R). \tag{23-15}$$

Here r is the distance from the center of the shell to the point at which E is measured. (The charge behaves, for external points, as if it were all located at the center of the sphere.) The field *inside* a uniform spherical shell of charge is exactly zero:

$$E = 0 \quad \text{(spherical shell, for } r < R). \tag{23-16}$$

6. The electric field *inside a uniform sphere of charge* is directed radially and has magnitude

$$E = \left(\frac{q}{4\pi\varepsilon_0 R^3}\right)r. \tag{23-20}$$

QUESTIONS

1 A surface has the area vector $\vec{A} = (2\hat{i} + 3\hat{j})\ m^2$. What is the flux of a uniform electric field through the area if the field is (a) $\vec{E} = 4\hat{i}$ N/C and (b) $\vec{E} = 4\hat{k}$ N/C?

2 Figure 23-20 shows, in cross section, three solid cylinders, each of length L and uniform charge Q. Concentric with each cylinder is a cylindrical Gaussian surface, with all three surfaces having the same radius. Rank the Gaussian surfaces according to the electric field at any point on the surface, greatest first.

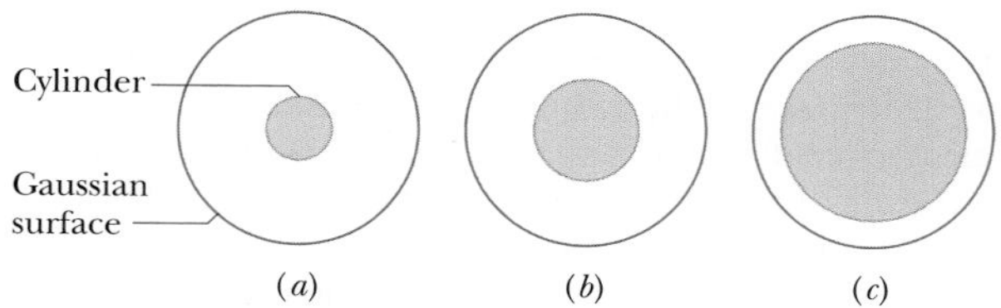

Fig. 23-20 Question 2.

3 Figure 23-21 shows, in cross section, a central metal ball, two spherical metal shells, and three spherical Gaussian surfaces of radii R, $2R$, and $3R$, all with the same center. The uniform charges on the three objects are: ball, Q; smaller shell, $3Q$; larger shell, $5Q$. Rank the Gaussian surfaces according to the magnitude of the electric field at any point on the surface, greatest first.

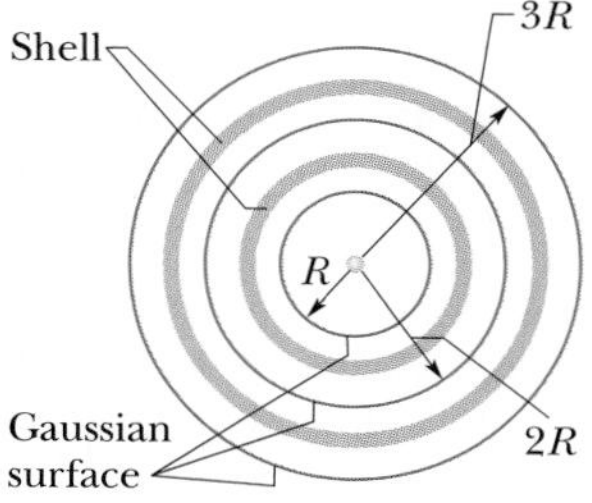

Fig. 23-21 Question 3.

4 Figure 23-22 shows, in cross section, two Gaussian spheres and two Gaussian cubes that are centered on a positively charged particle. (a) Rank the net flux through the four Gaussian surfaces, greatest first. (b) Rank the magnitudes of the electric fields on the surfaces, greatest first, and indicate whether the magnitudes are uniform or variable along each surface.

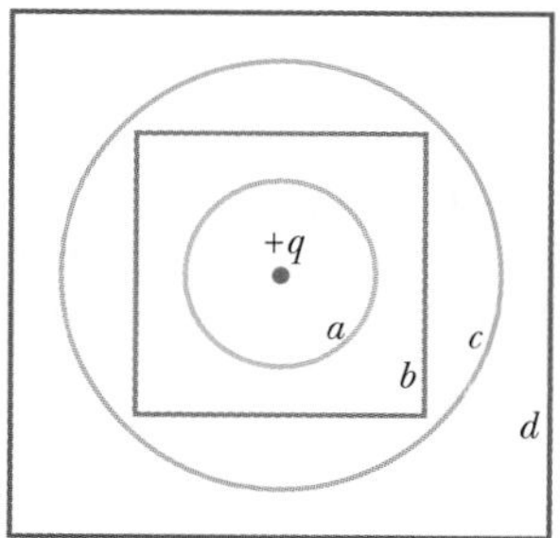

Fig. 23-22 Question 4.

5 In Fig. 23-23, an electron is released between two infinite nonconducting sheets that are horizontal and have uniform surface charge densities $\sigma_{(+)}$ and $\sigma_{(-)}$, as indicated. The electron is subjected to the following three situations involving surface charge densities and sheet separations. Rank the magnitudes of the electron's acceleration, greatest first.

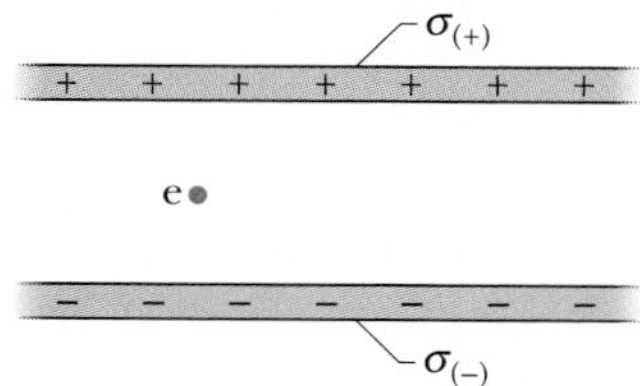

Fig. 23-23 Question 5.

Situation	$\sigma_{(+)}$	$\sigma_{(-)}$	Separation
1	$+4\sigma$	-4σ	d
2	$+7\sigma$	$-\sigma$	$4d$
3	$+3\sigma$	-5σ	$9d$

6 Three infinite nonconducting sheets, with uniform positive surface charge densities σ, 2σ, and 3σ, are arranged to be parallel like the two sheets in Fig. 23-17*a*. What is their order, from left to right, if the electric field $\vec{E}$ produced by the arrangement has magnitude $E = 0$ in one region and $E = 2\sigma/\varepsilon_0$ in another region?

7 Figure 23-24 shows four situations in which four very long rods extend into and out of the page (we see only their cross sections). The value below each cross section gives that particular rod's uniform charge density in microcoulombs per meter. The rods are separated by either d or $2d$ as drawn, and a central point is shown midway between the inner rods. Rank the situations according to the magnitude of the net electric field at that central point, greatest first.

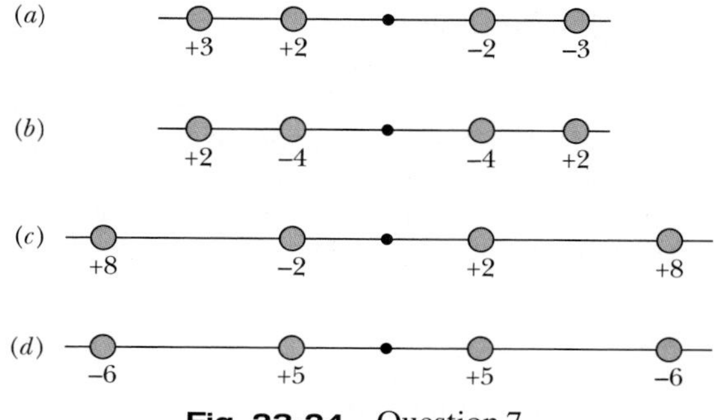

Fig. 23-24 Question 7.

8 Figure 23-25 shows four solid spheres, each with charge Q uniformly distributed through its volume. (a) Rank the spheres according to their volume charge density, greatest first. The figure also shows a point P for each sphere, all at the same distance from the center of the sphere. (b) Rank the spheres according to the magnitude of the electric field they produce at point P, greatest first.

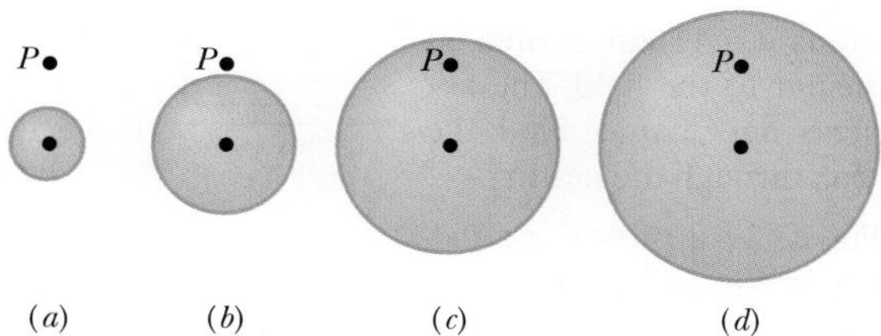

Fig. 23-25 Question 8.

9 A small charged ball lies within the hollow of a metallic spherical shell of radius R. For three situations, the net charges on the ball and shell, respectively, are (1) $+4q$, 0; (2) $-6q$, $+10q$; (3) $+16q$, $-12q$. Rank the situations according to the charge on (a) the inner surface of the shell and (b) the outer surface, most positive first.

10 Rank the situations of Question 9 according to the magnitude of the electric field (a) halfway through the shell and (b) at a point $2R$ from the center of the shell, greatest first.

PROBLEMS

GO Tutoring problem available (at instructor's discretion) in *WileyPLUS* and WebAssign
SSM Worked-out solution available in Student Solutions Manual
• – ••• Number of dots indicates level of problem difficulty
WWW Worked-out solution is at
ILW Interactive solution is at
http://www.wiley.com/college/halliday
Additional information available in *The Flying Circus of Physics* and at flyingcircusofphysics.com

sec. 23-3 Flux of an Electric Field

•1 SSM The square surface shown in Fig. 23-26 measures 3.2 mm on each side. It is immersed in a uniform electric field with magnitude $E = 1800$ N/C and with field lines at an angle of $\theta = 35°$ with a normal to the surface, as shown. Take that normal to be directed "outward," as though the surface were one face of a box. Calculate the electric flux through the surface.

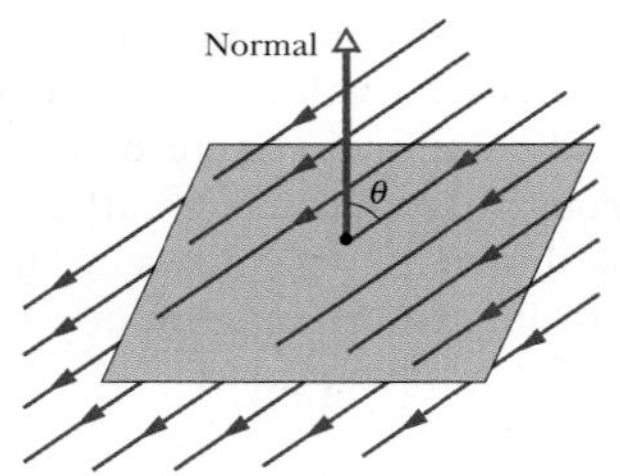

Fig. 23-26 Problem 1.

••2 An electric field given by $\vec{E} = 4.0\hat{i} - 3.0(y^2 + 2.0)\hat{j}$ pierces a Gaussian cube of edge length 2.0 m and positioned as shown in Fig. 23-5. (The magnitude E is in newtons per coulomb and the position x is in meters.) What is the electric flux through the (a) top face, (b) bottom face, (c) left face, and (d) back face? (e) What is the net electric flux through the cube?

••3 The cube in Fig. 23-27 has edge length 1.40 m and is oriented as shown in a region of uniform electric field. Find the electric flux through the right face if the electric field, in newtons per coulomb, is given by (a) $6.00\hat{i}$, (b) $-2.00\hat{j}$, and (c) $-3.00\hat{i} + 4.00\hat{k}$. (d) What is the total flux through the cube for each field?

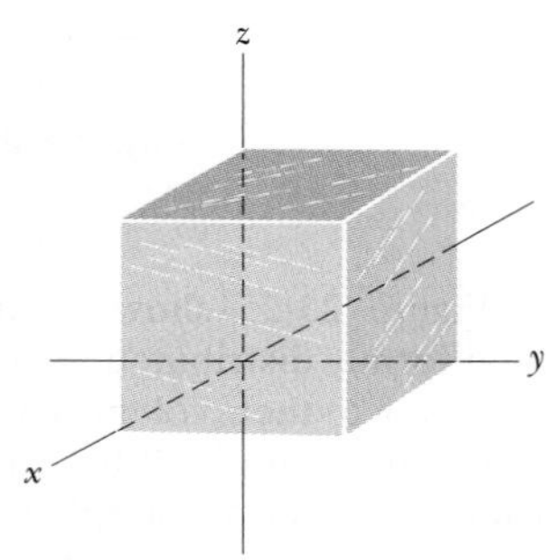

Fig. 23-27 Problems 3, 6, and 9.

sec. 23-4 Gauss' Law

•4 In Fig. 23-28, a butterfly net is in a uniform electric field of magnitude $E = 3.0$ mN/C. The rim, a circle of radius $a = 11$ cm, is aligned perpendicular to the field. The net contains no net charge. Find the electric flux through the netting.

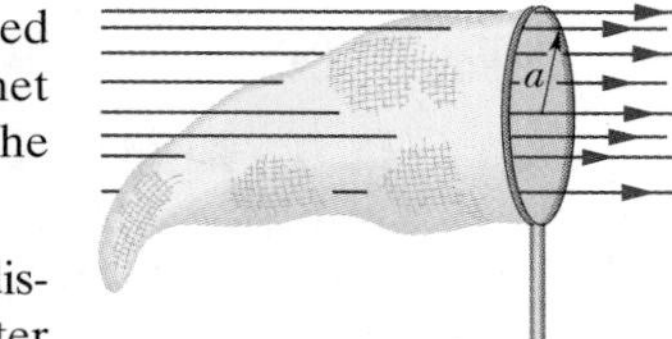

Fig. 23-28 Problem 4.

•5 In Fig. 23-29, a proton is a distance $d/2$ directly above the center of a square of side d. What is the magnitude of the electric flux through the square? (*Hint:* Think of the square as one face of a cube with edge d.)

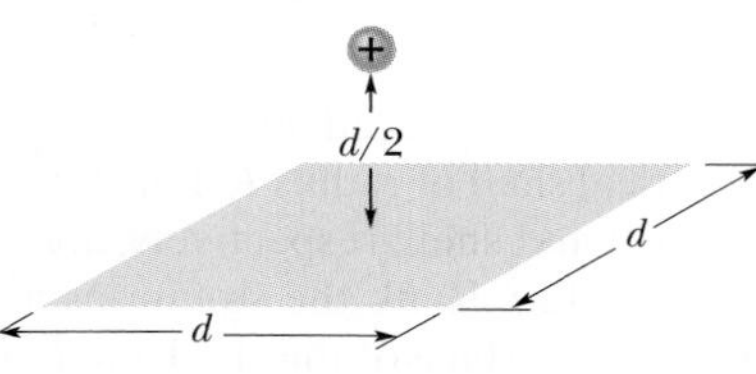

Fig. 23-29 Problem 5.

•6 At each point on the surface of the cube shown in Fig. 23-27, the electric field is parallel to the z axis. The length of each edge of the cube is 3.0 m. On the top face of the cube the field is $\vec{E} = -34\hat{k}$ N/C, and on the bottom face it is $\vec{E} = +20\hat{k}$ N/C. Determine the net charge contained within the cube.

•7 A point charge of 1.8 μC is at the center of a Gaussian cube 55 cm on edge. What is the net electric flux through the surface?

••8 When a shower is turned on in a closed bathroom, the splashing of the water on the bare tub can fill the room's air with negatively charged ions and produce an electric field in the air as great as 1000 N/C. Consider a bathroom with dimensions 2.5 m × 3.0 m × 2.0 m. Along the ceiling, floor, and four walls, approximate the electric field in the air as being directed perpendicular to the surface and as having a uniform magnitude of 600 N/C. Also, treat those surfaces as forming a closed Gaussian surface around the room's air. What are (a) the volume charge density ρ and (b) the number of excess elementary charges e per cubic meter in the room's air?

••9 ILW Fig. 23-27 shows a Gaussian surface in the shape of a cube with edge length 1.40 m. What are (a) the net flux Φ through the surface and (b) the net charge q_{enc} enclosed by the surface if $\vec{E} = (3.00y\hat{j})$ N/C, with y in meters? What are (c) Φ and (d) q_{enc} if $\vec{E} = [-4.00\hat{i} + (6.00 + 3.00y)\hat{j}]$ N/C?

••10 Figure 23-30 shows a closed Gaussian surface in the shape of a cube of edge length 2.00 m. It lies in a region where the nonuniform electric field is given by $\vec{E} = (3.00x + 4.00)\hat{i} + 6.00\hat{j} + 7.00\hat{k}$ N/C, with x in meters. What is the net charge contained by the cube?

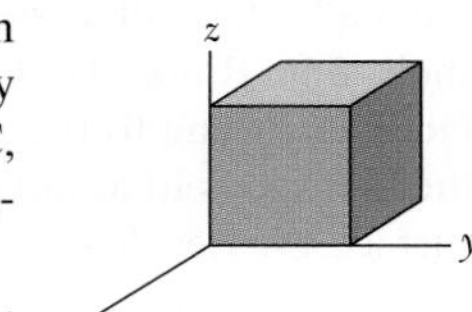

Fig. 23-30 Problem 10.

••11 GO Figure 23-31 shows a closed Gaussian surface in the shape of a cube of edge length 2.00 m, with one corner at $x_1 = 5.00$ m, $y_1 = 4.00$ m. The cube lies in a region where the electric field vector is given by $\vec{E} = -3.00\hat{i} - 4.00y^2\hat{j} + 3.00\hat{k}$ N/C, with y in meters. What is the net charge contained by the cube?

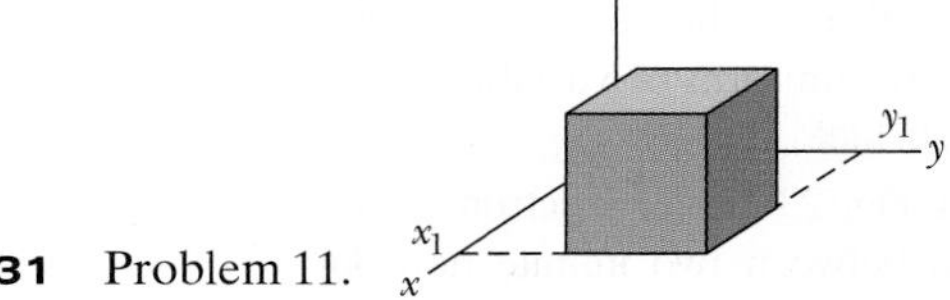

Fig. 23-31 Problem 11.

••12 Figure 23-32 shows two nonconducting spherical shells fixed in place. Shell 1 has uniform surface charge density +6.0 μC/m^2 on its outer surface and radius 3.0 cm; shell 2 has uniform surface charge density +4.0 μC/m^2 on its outer surface and radius 2.0 cm; the shell centers are separated by $L = 10$ cm. In unit-vector notation, what is the net electric field at $x = 2.0$ cm?

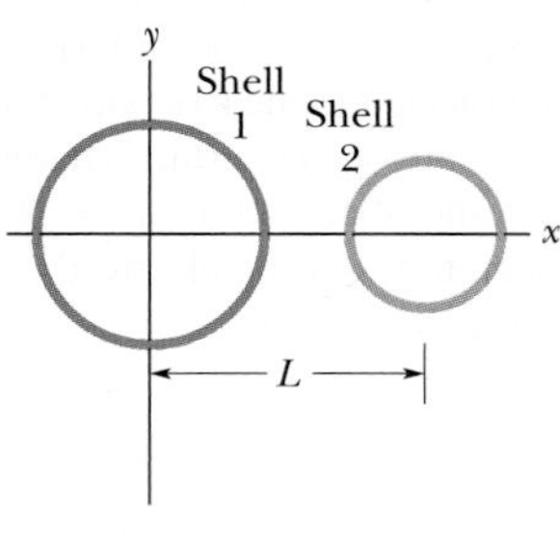

Fig. 23-32 Problem 12.

••13 SSM The electric field in a certain region of Earth's atmosphere is directed vertically down. At an altitude of 300 m the field has magnitude 60.0 N/C; at an altitude of 200 m, the magnitude is 100 N/C. Find the net amount of charge contained in a cube 100 m on edge, with horizontal faces at altitudes of 200 and 300 m.

••14 *Flux and nonconducting shells.* A charged particle is suspended at the center of two concentric spherical shells that are very thin and made of nonconducting material. Figure 23-33*a* shows a cross section. Figure 23-33*b* gives the net flux Φ through a Gaussian sphere centered on the particle, as a function of the radius r of the sphere. The scale of the vertical axis is set by $\Phi_s = 5.0 \times 10^5$ N·m²/C. (a) What is the charge of the central particle? What are the net charges of (b) shell A and (c) shell B?

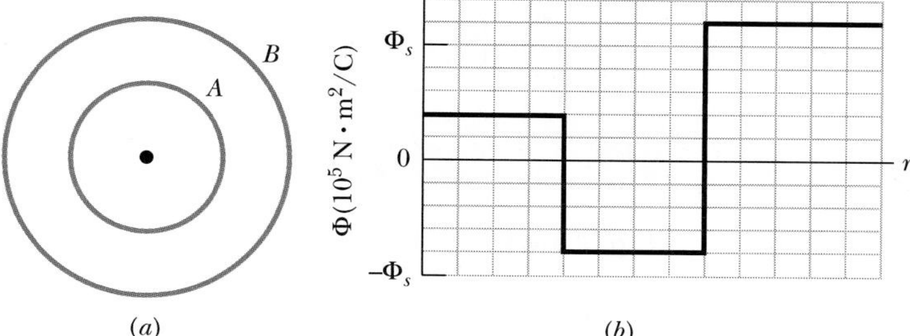

Fig. 23-33 Problem 14.

••15 A particle of charge $+q$ is placed at one corner of a Gaussian cube. What multiple of q/ε_0 gives the flux through (a) each cube face forming that corner and (b) each of the other cube faces?

•••16 GO The box-like Gaussian surface shown in Fig. 23-34 encloses a net charge of $+24.0\varepsilon_0$ C and lies in an electric field given by $\vec{E} = [(10.0 + 2.00x)\hat{i} - 3.00\hat{j} + bz\hat{k}]$ N/C, with x and z in meters and b a constant. The bottom face is in the xz plane; the top face is in the horizontal plane passing through $y_2 = 1.00$ m. For $x_1 = 1.00$ m, $x_2 = 4.00$ m, $z_1 = 1.00$ m, and $z_2 = 3.00$ m, what is b?

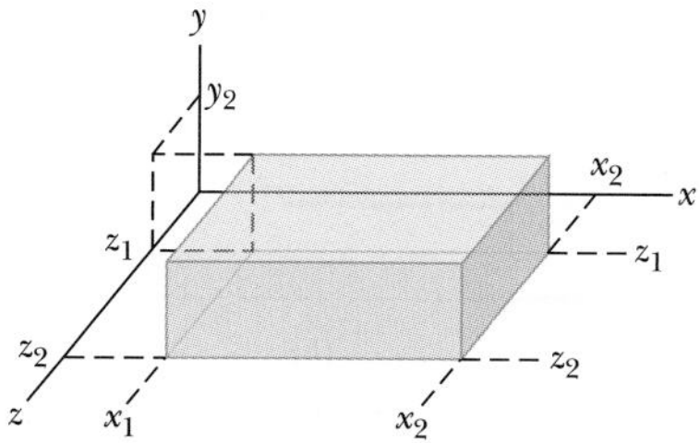

Fig. 23-34 Problem 16.

sec. 23-6 A Charged Isolated Conductor

•17 SSM A uniformly charged conducting sphere of 1.2 m diameter has a surface charge density of 8.1 μC/m². (a) Find the net charge on the sphere. (b) What is the total electric flux leaving the surface of the sphere?

•18 The electric field just above the surface of the charged conducting drum of a photocopying machine has a magnitude E of 2.3×10^5 N/C. What is the surface charge density on the drum?

•19 Space vehicles traveling through Earth's radiation belts can intercept a significant number of electrons. The resulting charge buildup can damage electronic components and disrupt operations. Suppose a spherical metal satellite 1.3 m in diameter accumulates 2.4 μC of charge in one orbital revolution. (a) Find the resulting surface charge density. (b) Calculate the magnitude of the electric field just outside the surface of the satellite, due to the surface charge.

•20 GO *Flux and conducting shells.* A charged particle is held at the center of two concentric conducting spherical shells. Figure 23-35*a* shows a cross section. Figure 23-35*b* gives the net flux Φ through a Gaussian sphere centered on the particle, as a function of the radius r of the sphere. The scale of the vertical axis is set by $\Phi_s = 5.0 \times 10^5$ N·m²/C. What are (a) the charge of the central particle and the net charges of (b) shell A and (c) shell B?

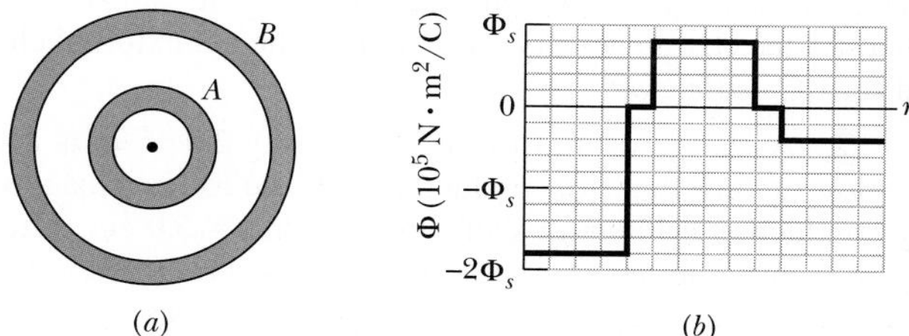

Fig. 23-35 Problem 20.

••21 An isolated conductor has net charge $+10 \times 10^{-6}$ C and a cavity with a point charge $q = +3.0 \times 10^{-6}$ C. What is the charge on (a) the cavity wall and (b) the outer surface?

sec. 23-7 Applying Gauss' Law: Cylindrical Symmetry

•22 An electron is released 9.0 cm from a very long nonconducting rod with a uniform 6.0 μC/m. What is the magnitude of the electron's initial acceleration?

•23 (a) The drum of a photocopying machine has a length of 42 cm and a diameter of 12 cm. The electric field just above the drum's surface is 2.3×10^5 N/C. What is the total charge on the drum? (b) The manufacturer wishes to produce a desktop version of the machine. This requires reducing the drum length to 28 cm and the diameter to 8.0 cm. The electric field at the drum surface must not change. What must be the charge on this new drum?

•24 Figure 23-36 shows a section of a long, thin-walled metal tube of radius $R = 3.00$ cm, with a charge per unit length of $\lambda = 2.00 \times 10^{-8}$ C/m. What is the magnitude E of the electric field at radial distance (a) $r = R/2.00$ and (b) $r = 2.00R$? (c) Graph E versus r for the range $r = 0$ to $2.00R$.

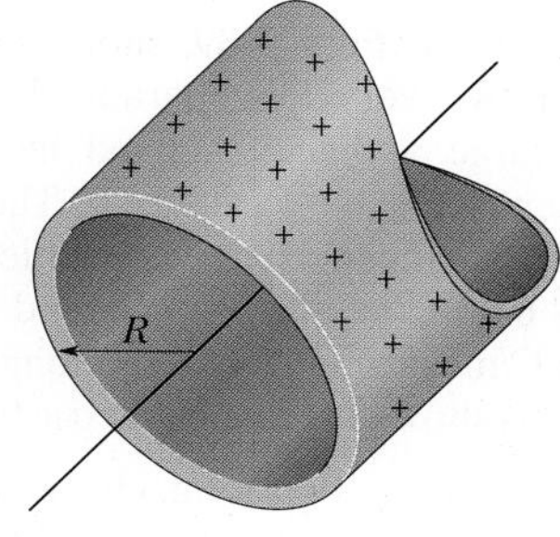

Fig. 23-36 Problem 24.

•25 SSM An infinite line of charge produces a field of magnitude 4.5×10^4 N/C at distance 2.0 m. Find the linear charge density.

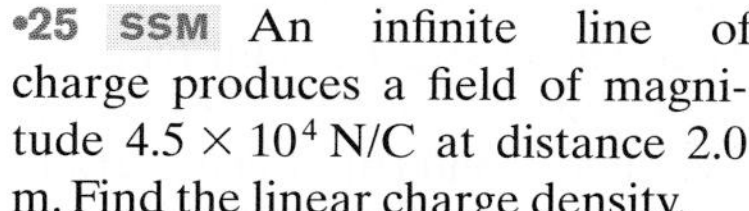

••26 Figure 23-37*a* shows a narrow charged solid cylinder that is coaxial with a larger charged cylindrical shell. Both are noncon-

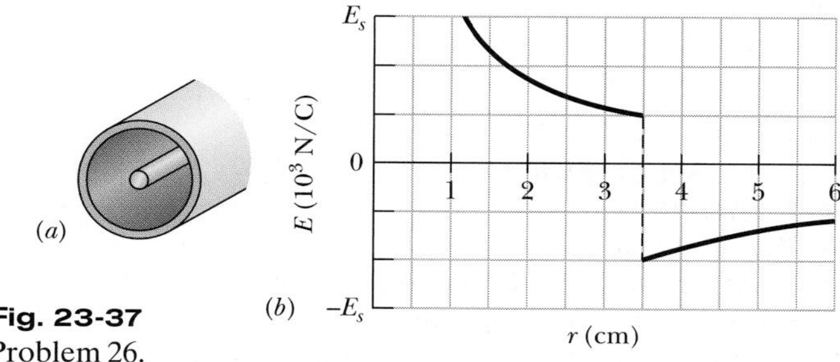

Fig. 23-37 Problem 26.

ducting and thin and have uniform surface charge densities on their outer surfaces. Figure 23-37*b* gives the radial component E of the electric field versus radial distance r from the common axis, and $E_s = 3.0 \times 10^3$ N/C. What is the shell's linear charge density?

••27 A long, straight wire has fixed negative charge with a linear charge density of magnitude 3.6 nC/m. The wire is to be enclosed by a coaxial, thin-walled nonconducting cylindrical shell of radius 1.5 cm. The shell is to have positive charge on its outside surface with a surface charge density σ that makes the net external electric field zero. Calculate σ.

••28 GO A charge of uniform linear density 2.0 nC/m is distributed along a long, thin, nonconducting rod. The rod is coaxial with a long conducting cylindrical shell (inner radius = 5.0 cm, outer radius = 10 cm). The net charge on the shell is zero. (a) What is the magnitude of the electric field 15 cm from the axis of the shell? What is the surface charge density on the (b) inner and (c) outer surface of the shell?

••29 SSM WWW Figure 23-38 is a section of a conducting rod of radius $R_1 = 1.30$ mm and length $L = 11.00$ m inside a thin-walled coaxial conducting cylindrical shell of radius $R_2 = 10.0R_1$ and the (same) length L. The net charge on the rod is $Q_1 = +3.40 \times 10^{-12}$ C; that on the shell is $Q_2 = -2.00Q_1$. What are the (a) magnitude E and (b) direction (radially inward or outward) of the electric field at radial distance $r = 2.00R_2$? What are (c) E and (d) the direction at $r = 5.00R_1$? What is the charge on the (e) interior and (f) exterior surface of the shell?

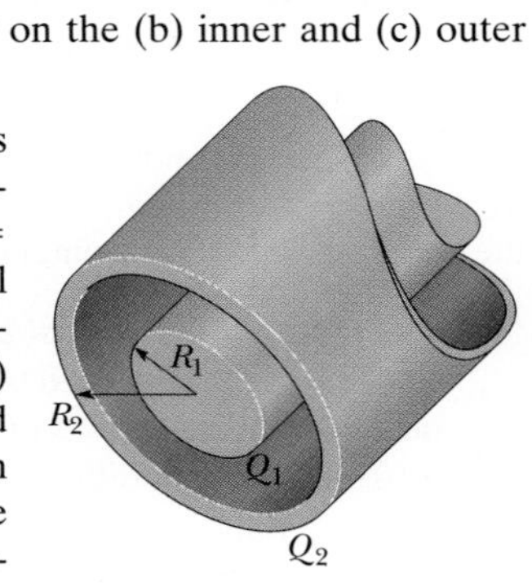

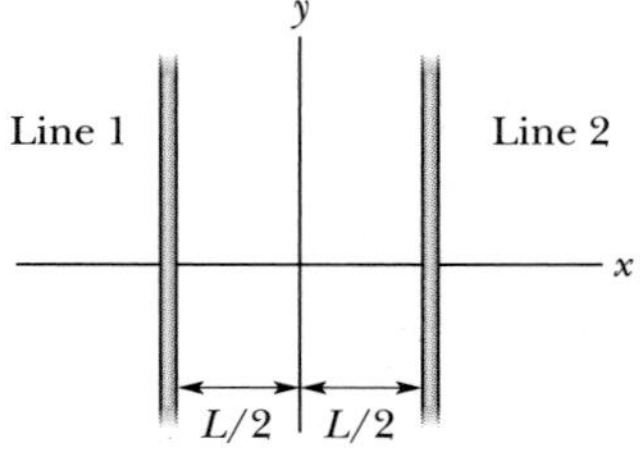

Fig. 23-38 Problem 29.

••30 In Fig. 23-39, short sections of two very long parallel lines of charge are shown, fixed in place, separated by $L = 8.0$ cm. The uniform linear charge densities are $+6.0\ \mu$C/m for line 1 and $-2.0\ \mu$C/m for line 2. Where along the x axis shown is the net electric field from the two lines zero?

Fig. 23-39 Problem 30.

••31 ILW Two long, charged, thin-walled, concentric cylindrical shells have radii of 3.0 and 6.0 cm. The charge per unit length is 5.0×10^{-6} C/m on the inner shell and -7.0×10^{-6} C/m on the outer shell. What are the (a) magnitude E and (b) direction (radially inward or outward) of the electric field at radial distance $r = 4.0$ cm? What are (c) E and (d) the direction at $r = 8.0$ cm?

•••32 A long, nonconducting, solid cylinder of radius 4.0 cm has a nonuniform volume charge density ρ that is a function of radial distance r from the cylinder axis: $\rho = Ar^2$. For $A = 2.5\ \mu$C/m^5, what is the magnitude of the electric field at (a) $r = 3.0$ cm and (b) $r = 5.0$ cm?

sec. 23-8 Applying Gauss' Law: Planar Symmetry

•33 In Fig. 23-40, two large, thin metal plates are parallel and close to each other. On their inner faces, the plates have excess surface charge densities of opposite signs and magnitude 7.00×10^{-22} C/m^2. In unit-vector notation, what is the electric field at points (a) to the left of the plates, (b) to the right of them, and (c) between them?

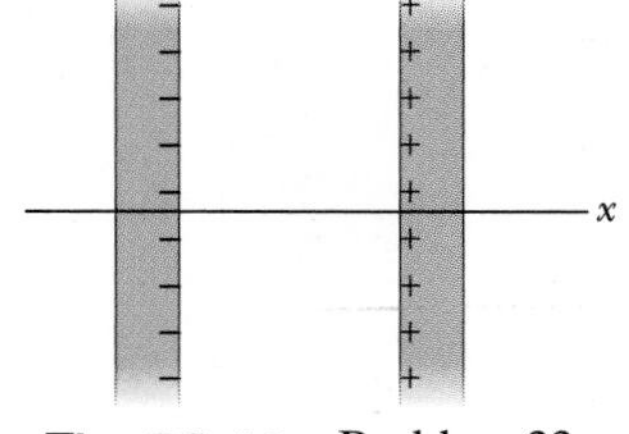

Fig. 23-40 Problem 33.

•34 In Fig. 23-41, a small circular hole of radius $R = 1.80$ cm has been cut in the middle of an infinite, flat, nonconducting surface that has uniform charge density $\sigma = 4.50$ pC/m^2. A z axis, with its origin at the hole's center, is perpendicular to the surface. In unit-vector notation, what is the electric field at point P at $z = 2.56$ cm? (*Hint:* See Eq. 22-26 and use superposition.)

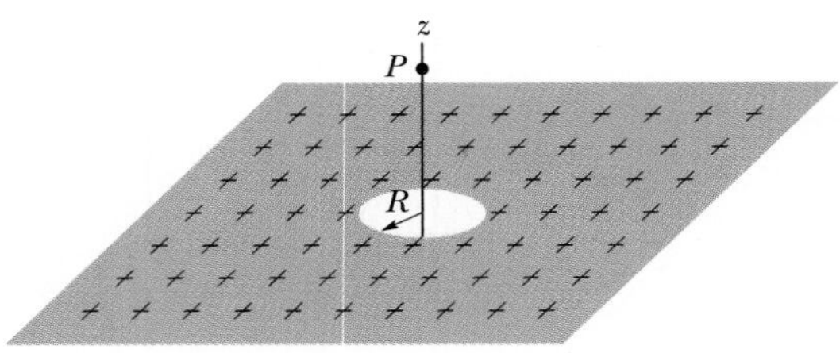

Fig. 23-41 Problem 34.

•35 GO Figure 23-42*a* shows three plastic sheets that are large, parallel, and uniformly charged. Figure 23-42*b* gives the component of the net electric field along an x axis through the sheets. The scale of the vertical axis is set by $E_s = 6.0 \times 10^5$ N/C. What is the ratio of the charge density on sheet 3 to that on sheet 2?

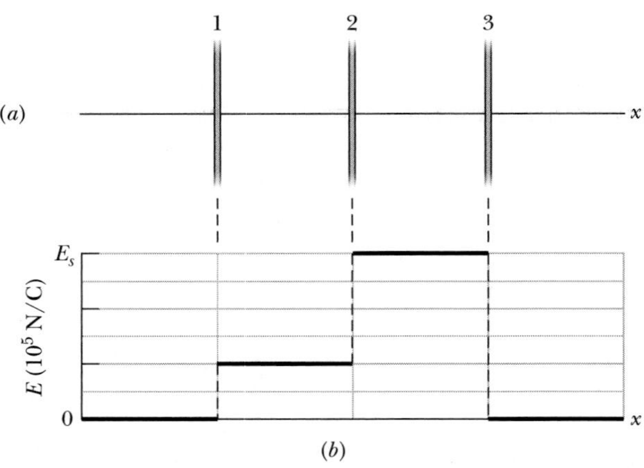

Fig. 23-42 Problem 35.

•36 Figure 23-43 shows cross sections through two large, parallel, nonconducting sheets with identical distributions of positive charge with surface charge density $\sigma = 1.77 \times 10^{-22}$ C/m^2. In unit-vector notation, what is $\vec{E}$ at points (a) above the sheets, (b) between them, and (c) below them?

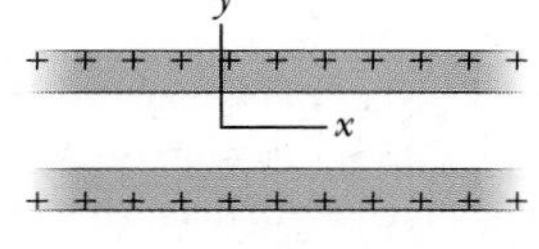

Fig. 23-43 Problem 36.

•37 SSM WWW A square metal plate of edge length 8.0 cm and negligible thickness has a total charge of 6.0×10^{-6} C. (a) Estimate the magnitude E of the electric field just off the center of the plate (at, say, a distance of 0.50 mm from the center) by assuming that the charge is spread uniformly over the two faces of the plate. (b) Estimate E at a distance of 30 m (large relative to the plate size) by assuming that the plate is a point charge.

••38 GO In Fig. 23-44*a*, an electron is shot directly away from a uniformly charged plastic sheet, at speed $v_s = 2.0 \times 10^5$ m/s. The sheet is nonconducting, flat, and very large. Figure 23-44*b* gives the electron's vertical velocity component v versus time t until the return to the launch point. What is the sheet's surface charge density?

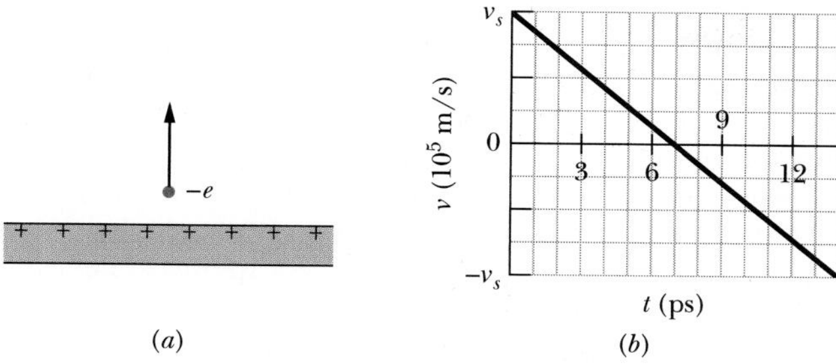

Fig. 23-44 Problem 38.

••39 SSM In Fig. 23-45, a small, nonconducting ball of mass $m = 1.0$ mg and charge $q = 2.0 \times 10^{-8}$ C (distributed uniformly through its volume) hangs from an insulating thread that makes an angle $\theta = 30°$ with a vertical, uniformly charged nonconducting sheet (shown in cross section). Considering the gravitational force on the ball and assuming the sheet extends far vertically and into and out of the page, calculate the surface charge density σ of the sheet.

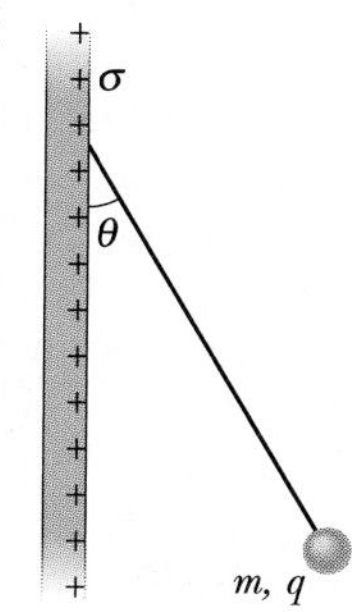

Fig. 23-45 Problem 39.

••40 Figure 23-46 shows a very large nonconducting sheet that has a uniform surface charge density of $\sigma = -2.00\ \mu\text{C/m}^2$; it also shows a particle of charge $Q = 6.00\ \mu$C, at distance d from the sheet. Both are fixed in place. If $d = 0.200$ m, at what (a) positive and (b) negative coordinate on the x axis (other than infinity) is the net electric field $\vec{E}_{net}$ of the sheet and particle zero? (c) If $d = 0.800$ m, at what coordinate on the x axis is $\vec{E}_{net} = 0$?

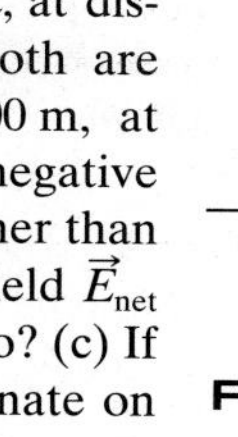

Fig. 23-46 Problem 40.

••41 GO An electron is shot directly toward the center of a large metal plate that has surface charge density -2.0×10^{-6} C/m^2. If the initial kinetic energy of the electron is 1.60×10^{-17} J and if the electron is to stop (due to electrostatic repulsion from the plate) just as it reaches the plate, how far from the plate must the launch point be?

••42 Two large metal plates of area 1.0 m^2 face each other, 5.0 cm apart, with equal charge magnitudes $|q|$ but opposite signs. The field magnitude E between them (neglect fringing) is 55 N/C. Find $|q|$.

•••43 Figure 23-47 shows a cross section through a very large nonconducting slab of thickness $d = 9.40$ mm and uniform volume charge density $\rho = 5.80$ fC/m^3. The origin of an x axis is at the slab's center. What is the magnitude of the slab's electric field at an x coordinate of (a) 0, (b) 2.00 mm, (c) 4.70 mm, and (d) 26.0 mm?

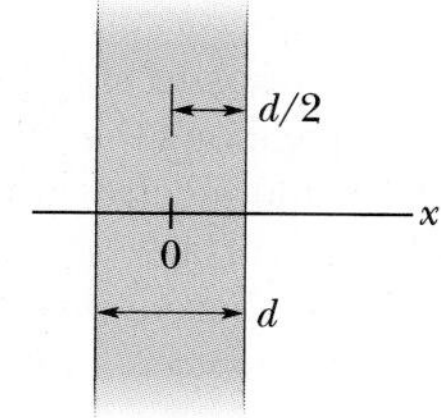

Fig. 23-47 Problem 43.

sec. 23-9 Applying Gauss' Law: Spherical Symmetry

•44 Figure 23-48 gives the magnitude of the electric field inside and outside a sphere with a positive charge distributed uniformly throughout its volume. The scale of the vertical axis is set by $E_s = 5.0 \times 10^7$ N/C. What is the charge on the sphere?

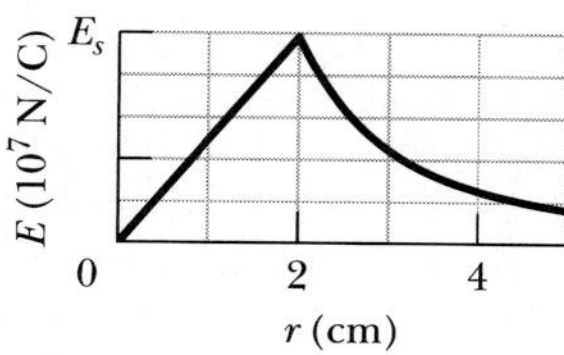

Fig. 23-48 Problem 44.

•45 Two charged concentric spherical shells have radii 10.0 cm and 15.0 cm. The charge on the inner shell is 4.00×10^{-8} C, and that on the outer shell is 2.00×10^{-8} C. Find the electric field (a) at $r = 12.0$ cm and (b) at $r = 20.0$ cm.

•46 A point charge causes an electric flux of $-750\ \text{N}\cdot\text{m}^2/\text{C}$ to pass through a spherical Gaussian surface of 10.0 cm radius centered on the charge. (a) If the radius of the Gaussian surface were doubled, how much flux would pass through the surface? (b) What is the value of the point charge?

•47 SSM An unknown charge sits on a conducting solid sphere of radius 10 cm. If the electric field 15 cm from the center of the sphere has the magnitude 3.0×10^3 N/C and is directed radially inward, what is the net charge on the sphere?

••48 A charged particle is held at the center of a spherical shell. Figure 23-49 gives the magnitude E of the electric field versus radial distance r. The scale of the vertical axis is set by $E_s = 10.0 \times 10^7$ N/C. Approximately, what is the net charge on the shell?

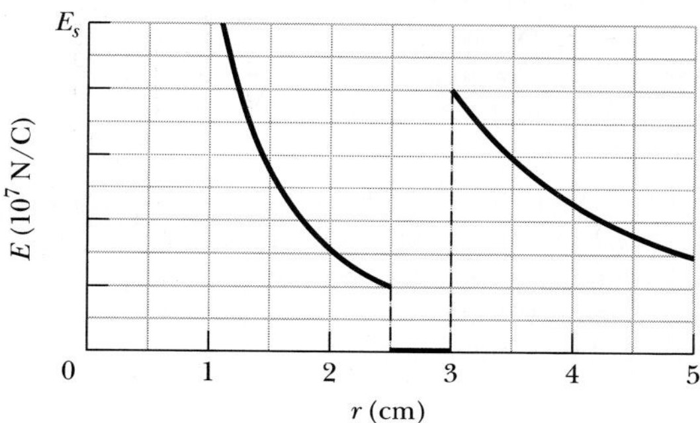

Fig. 23-49 Problem 48.

••49 In Fig. 23-50, a solid sphere of radius $a = 2.00$ cm is concentric with a spherical conducting shell of inner radius $b = 2.00a$ and outer radius $c = 2.40a$. The sphere has a net uniform charge $q_1 = +5.00$ fC; the shell has a net charge $q_2 = -q_1$. What is the magnitude of the electric field at radial distances (a) $r = 0$, (b) $r = a/2.00$, (c) $r = a$, (d) $r = 1.50a$, (e) $r = 2.30a$, and (f) $r = 3.50a$? What is the net charge on the (g) inner and (h) outer surface of the shell?

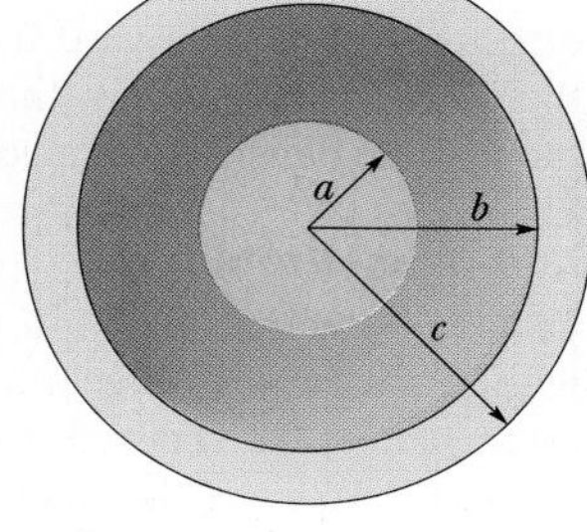

Fig. 23-50 Problem 49.

••50 GO Figure 23-51 shows two nonconducting spherical shells fixed in place on an x axis. Shell 1 has uniform surface charge density $+4.0\ \mu\text{C/m}^2$ on its outer surface and radius 0.50 cm, and shell 2 has uniform surface charge density $-2.0\ \mu\text{C/m}^2$ on its outer surface and radius 2.0 cm; the centers are separated by $L = 6.0$ cm. Other than at $x = \infty$, where on the x axis is the net electric field equal to zero?

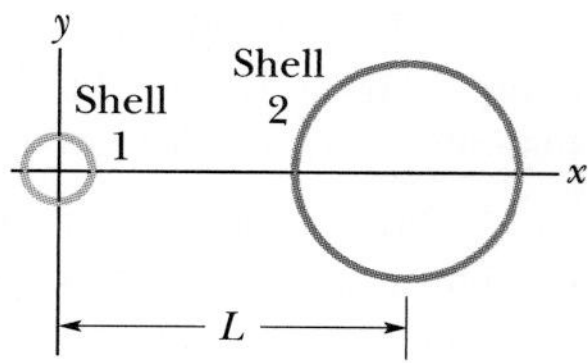

Fig. 23-51 Problem 50.

••51 SSM WWW In Fig. 23-52, a nonconducting spherical shell of inner radius $a = 2.00$ cm and outer radius $b = 2.40$ cm has (within its thickness) a positive volume charge density $\rho = A/r$, where A is a constant and r is the distance from the center of the shell. In addition, a small ball of charge $q = 45.0$ fC is located at that center. What value should A have if the electric field in the shell ($a \leq r \leq b$) is to be uniform?

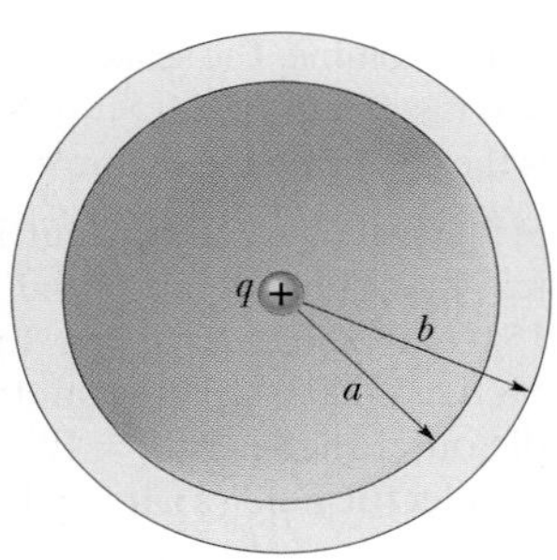

Fig. 23-52 Problem 51.

••52 Figure 23-53 shows a spherical shell with uniform volume charge density $\rho = 1.84$ nC/m^3, inner radius $a = 10.0$ cm, and outer radius $b = 2.00a$. What is the magnitude of the electric field at radial distances (a) $r = 0$; (b) $r = a/2.00$, (c) $r = a$, (d) $r = 1.50a$, (e) $r = b$, and (f) $r = 3.00b$?

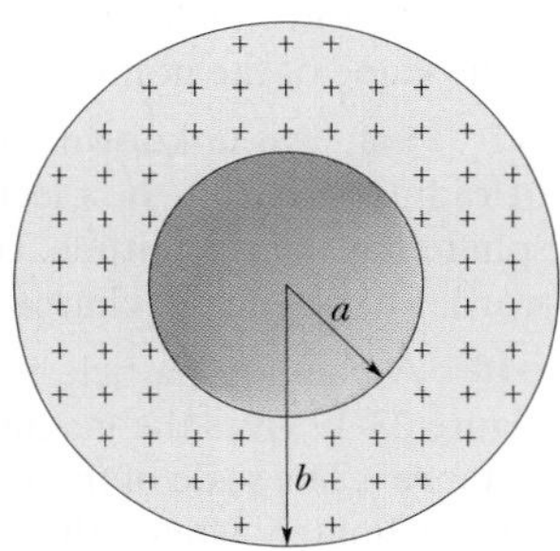

Fig. 23-53 Problem 52.

•••53 ILW The volume charge density of a solid nonconducting sphere of radius $R = 5.60$ cm varies with radial distance r as given by $\rho = (14.1 \text{ pC/m}^3)r/R$. (a) What is the sphere's total charge? What is the field magnitude E at (b) $r = 0$, (c) $r = R/2.00$, and (d) $r = R$? (e) Graph E versus r.

•••54 Figure 23-54 shows, in cross section, two solid spheres with uniformly distributed charge throughout their volumes. Each has radius R. Point P lies on a line connecting the centers of the spheres, at radial distance $R/2.00$ from the center of sphere 1. If the net electric field at point P is zero, what is the ratio q_2/q_1 of the total charges?

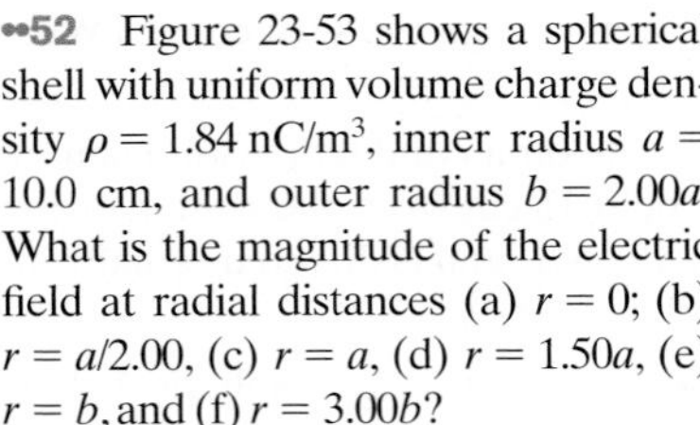

Fig. 23-54 Problem 54.

•••55 A charge distribution that is spherically symmetric but not uniform radially produces an electric field of magnitude $E = Kr^4$, directed radially outward from the center of the sphere. Here r is the radial distance from that center, and K is a constant. What is the volume density ρ of the charge distribution?

Additional Problems

56 The electric field in a particular space is $\vec{E} = (x + 2)\hat{\text{i}}$ N/C, with x in meters. Consider a cylindrical Gaussian surface of radius 20 cm that is coaxial with the x axis. One end of the cylinder is at $x = 0$. (a) What is the magnitude of the electric flux through the other end of the cylinder at $x = 2.0$ m? (b) What net charge is enclosed within the cylinder?

57 A thin-walled metal spherical shell has radius 25.0 cm and charge 2.00×10^{-7} C. Find E for a point (a) inside the shell, (b) just outside it, and (c) 3.00 m from the center.

58 A uniform surface charge of density 8.0 nC/m^2 is distributed over the entire xy plane. What is the electric flux through a spherical Gaussian surface centered on the origin and having a radius of 5.0 cm?

59 Charge of uniform volume density $\rho = 1.2$ nC/m^3 fills an infinite slab between $x = -5.0$ cm and $x = +5.0$ cm. What is the magnitude of the electric field at any point with the coordinate (a) $x = 4.0$ cm and (b) $x = 6.0$ cm?

60 *The chocolate crumb mystery.* Explosions ignited by electrostatic discharges (sparks) constitute a serious danger in facilities handling grain or powder. Such an explosion occurred in chocolate crumb powder at a biscuit factory in the 1970s. Workers usually emptied newly delivered sacks of the powder into a loading bin, from which it was blown through electrically grounded plastic pipes to a silo for storage. Somewhere along this route, two conditions for an explosion were met: (1) The magnitude of an electric field became 3.0×10^6 N/C or greater, so that electrical breakdown and thus sparking could occur. (2) The energy of a spark was 150 mJ or greater so that it could ignite the powder explosively. Let us check for the first condition in the powder flow through the plastic pipes.

Suppose a stream of *negatively* charged powder was blown through a cylindrical pipe of radius $R = 5.0$ cm. Assume that the powder and its charge were spread uniformly through the pipe with a volume charge density ρ. (a) Using Gauss' law, find an expression for the magnitude of the electric field $\vec{E}$ in the pipe as a function of radial distance r from the pipe center. (b) Does E increase or decrease with increasing r? (c) Is $\vec{E}$ directed radially inward or outward? (d) For $\rho = 1.1 \times 10^{-3}$ C/m^3 (a typical value at the factory), find the maximum E and determine where that maximum field occurs. (e) Could sparking occur, and if so, where? (The story continues with Problem 70 in Chapter 24.)

61 SSM A thin-walled metal spherical shell of radius a has a charge q_a. Concentric with it is a thin-walled metal spherical shell of radius $b > a$ and charge q_b. Find the electric field at points a distance r from the common center, where (a) $r < a$, (b) $a < r < b$, and (c) $r > b$. (d) Discuss the criterion you would use to determine how the charges are distributed on the inner and outer surfaces of the shells.

62 A point charge $q = 1.0 \times 10^{-7}$ C is at the center of a spherical cavity of radius 3.0 cm in a chunk of metal. Find the electric field (a) 1.5 cm from the cavity center and (b) anyplace in the metal.

63 A proton at speed $v = 3.00 \times 10^5$ m/s orbits at radius $r = 1.00$ cm outside a charged sphere. Find the sphere's charge.

64 Equation 23-11 ($E = \sigma/\varepsilon_0$) gives the electric field at points near a charged conducting surface. Apply this equation to a conducting sphere of radius r and charge q, and show that the electric field outside the sphere is the same as the field of a point charge located at the center of the sphere.

65 Charge Q is uniformly distributed in a sphere of radius R. (a) What fraction of the charge is contained within the radius $r = R/2.00$? (b) What is the ratio of the electric field magnitude at $r = R/2.00$ to that on the surface of the sphere?

66 Assume that a ball of charged particles has a uniformly distributed negative charge density except for a narrow radial tunnel through its center, from the surface on one side to the surface on the opposite side. Also assume that we can position a proton anywhere along the tunnel or outside the ball. Let F_R be the magnitude of the electrostatic force on the proton when it is located at the ball's surface, at radius R. As a multiple of R, how far from the surface is there a point where the force magnitude is $0.50F_R$ if we move the proton (a) away from the ball and (b) into the tunnel?

67 SSM The electric field at point P just outside the outer surface of a hollow spherical conductor of inner radius 10 cm and outer radius 20 cm has magnitude 450 N/C and is directed outward. When an unknown point charge Q is introduced into the center of the sphere, the electric field at P is still directed outward but is now 180 N/C. (a) What was the net charge enclosed by the

outer surface before Q was introduced? (b) What is charge Q? After Q is introduced, what is the charge on the (c) inner and (d) outer surface of the conductor?

68 The net electric flux through each face of a die (singular of dice) has a magnitude in units of $10^3\ \text{N}\cdot\text{m}^2/\text{C}$ that is exactly equal to the number of spots N on the face (1 through 6). The flux is inward for N odd and outward for N even. What is the net charge inside the die?

69 Figure 23-55 shows, in cross section, three infinitely large nonconducting sheets on which charge is uniformly spread. The surface charge densities are $\sigma_1 = +2.00\ \mu\text{C/m}^2$, $\sigma_2 = +4.00\ \mu\text{C/m}^2$, and $\sigma_3 = -5.00\ \mu\text{C/m}^2$, and distance $L = 1.50$ cm. In unit-vector notation, what is the net electric field at point P?

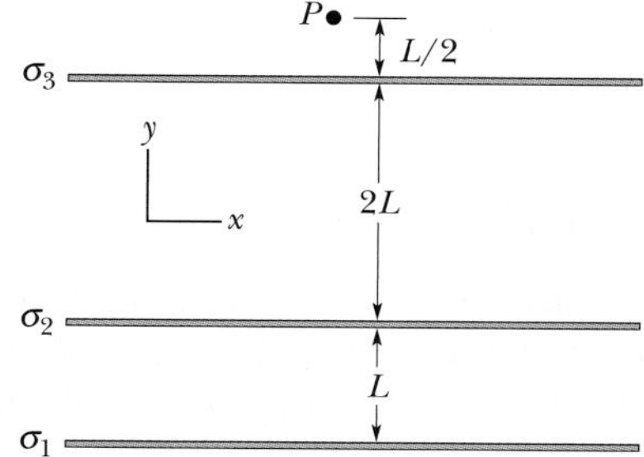

Fig. 23-55 Problem 69.

70 Charge of uniform volume density $\rho = 3.2\ \mu\text{C/m}^3$ fills a nonconducting solid sphere of radius 5.0 cm. What is the magnitude of the electric field (a) 3.5 cm and (b) 8.0 cm from the sphere's center?

71 A Gaussian surface in the form of a hemisphere of radius $R = 5.68$ cm lies in a uniform electric field of magnitude $E = 2.50$ N/C. The surface encloses no net charge. At the (flat) base of the surface, the field is perpendicular to the surface and directed into the surface. What is the flux through (a) the base and (b) the curved portion of the surface?

72 What net charge is enclosed by the Gaussian cube of Problem 2?

73 A nonconducting solid sphere has a uniform volume charge density ρ. Let $\vec{r}$ be the vector from the center of the sphere to a general point P within the sphere. (a) Show that the electric field at P is given by $\vec{E} = \rho\vec{r}/3\varepsilon_0$. (Note that the result is independent of the radius of the sphere.) (b) A spherical cavity is hollowed out of the sphere, as shown in Fig. 23-56. Using superposition concepts, show that the electric field at all points within the cavity is uniform and equal to $\vec{E} = \rho\vec{a}/3\varepsilon_0$, where $\vec{a}$ is the position vector from the center of the sphere to the center of the cavity.

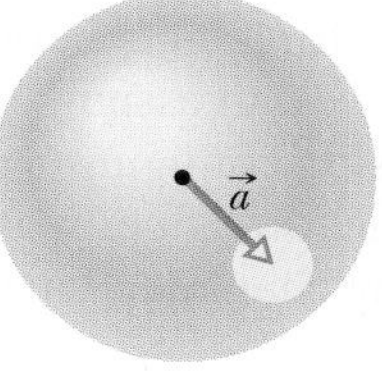

Fig. 23-56 Problem 73.

74 A uniform charge density of $500\ \text{nC/m}^3$ is distributed throughout a spherical volume of radius 6.00 cm. Consider a cubical Gaussian surface with its center at the center of the sphere. What is the electric flux through this cubical surface if its edge length is (a) 4.00 cm and (b) 14.0 cm?

75 Figure 23-57 shows a Geiger counter, a device used to detect ionizing radiation, which causes ionization of atoms. A thin, positively charged central wire is surrounded by a concentric, circular, conducting cylindrical shell with an equal negative charge, creating a strong radial electric field. The shell contains a low-pressure inert gas. A particle of radiation entering the device through the shell wall ionizes a few of the gas atoms. The resulting free electrons (e) are drawn to the positive wire. However, the electric field is so intense that, between collisions with gas atoms, the free electrons gain energy sufficient to ionize these atoms also. More free electrons are thereby created, and the process is repeated until the electrons reach the wire. The resulting "avalanche" of electrons is collected by the wire, generating a signal that is used to record the passage of the original particle of radiation. Suppose that the radius of the central wire is 25 μm, the inner radius of the shell 1.4 cm, and the length of the shell 16 cm. If the electric field at the shell's inner wall is 2.9×10^4 N/C, what is the total positive charge on the central wire?

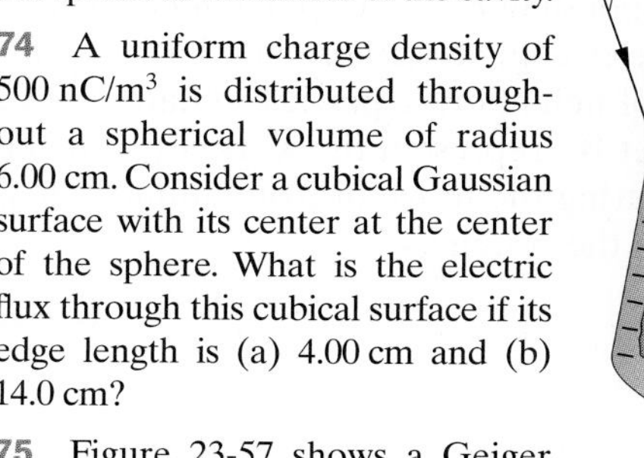
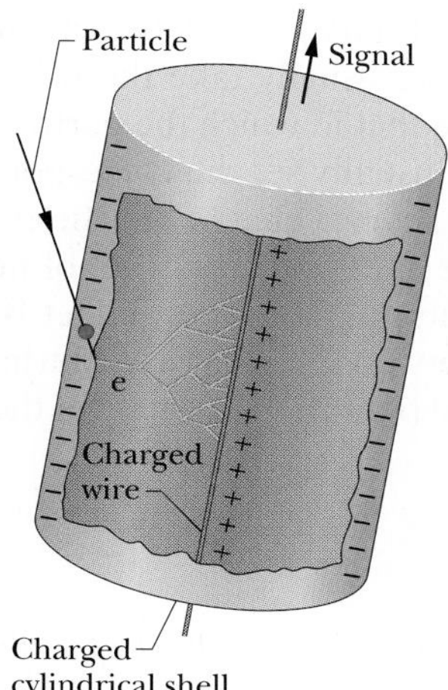

Fig. 23-57 Problem 75.

76 Charge is distributed uniformly throughout the volume of an infinitely long solid cylinder of radius R. (a) Show that, at a distance $r < R$ from the cylinder axis,

$$E = \frac{\rho r}{2\varepsilon_0},$$

where ρ is the volume charge density. (b) Write an expression for E when $r > R$.

77 SSM A spherical conducting shell has a charge of $-14\ \mu$C on its outer surface and a charged particle in its hollow. If the net charge on the shell is $-10\ \mu$C, what is the charge (a) on the inner surface of the shell and (b) of the particle?

78 A charge of 6.00 pC is spread uniformly throughout the volume of a sphere of radius $r = 4.00$ cm. What is the magnitude of the electric field at a radial distance of (a) 6.00 cm and (b) 3.00 cm?

79 Water in an irrigation ditch of width $w = 3.22$ m and depth $d = 1.04$ m flows with a speed of 0.207 m/s. The *mass flux* of the flowing water through an imaginary surface is the product of the water's density (1000 kg/m^3) and its volume flux through that surface. Find the mass flux through the following imaginary surfaces: (a) a surface of area wd, entirely in the water, perpendicular to the flow; (b) a surface with area $3wd/2$, of which wd is in the water, perpendicular to the flow; (c) a surface of area $wd/2$, entirely in the water, perpendicular to the flow; (d) a surface of area wd, half in the water and half out, perpendicular to the flow; (e) a surface of area wd, entirely in the water, with its normal 34.0° from the direction of flow.

80 Charge of uniform surface density $8.00\ \text{nC/m}^2$ is distributed over an entire xy plane; charge of uniform surface density $3.00\ \text{nC/m}^2$ is distributed over the parallel plane defined by $z = 2.00$ m. Determine the magnitude of the electric field at any point having a z coordinate of (a) 1.00 m and (b) 3.00 m.

81 A spherical ball of charged particles has a uniform charge density. In terms of the ball's radius R, at what radial distances (a) inside and (b) outside the ball is the magnitude of the ball's electric field equal to $\frac{1}{4}$ of the maximum magnitude of that field?

82 SSM A free electron is placed between two large, parallel, nonconducting plates that are horizontal and 2.3 cm apart. One plate has a uniform positive charge; the other has an equal amount of uniform negative charge. The force on the electron due to the electric field $\vec{E}$ between the plates balances the gravitational force on the electron. What are (a) the magnitude of the surface charge density on the plates and (b) the direction (up or down) of $\vec{E}$?

CHAPTER

24

ELECTRIC POTENTIAL

24-1 WHAT IS PHYSICS?

One goal of physics is to identify basic forces in our world, such as the electric force we discussed in Chapter 21. A related goal is to determine whether a force is conservative—that is, whether a potential energy can be associated with it. The motivation for associating a potential energy with a force is that we can then apply the principle of the conservation of mechanical energy to closed systems involving the force. This extremely powerful principle allows us to calculate the results of experiments for which force calculations alone would be very difficult. Experimentally, physicists and engineers discovered that the electric force is conservative and thus has an associated electric potential energy. In this chapter we first define this type of potential energy and then put it to use.

24-2 Electric Potential Energy

When an electrostatic force acts between two or more charged particles within a system of particles, we can assign an **electric potential energy** U to the system. If the system changes its configuration from an initial state i to a different final state f, the electrostatic force does work W on the particles. From Eq. 8-1, we then know that the resulting change ΔU in the potential energy of the system is

$$\Delta U = U_f - U_i = -W. \tag{24-1}$$

As with other conservative forces, the work done by the electrostatic force is *path independent.* Suppose a charged particle within the system moves from point i to point f while an electrostatic force between it and the rest of the system acts on it. Provided the rest of the system does not change, the work W done by the force on the particle is the same for *all* paths between points i and f.

For convenience, we usually take the *reference configuration* of a system of charged particles to be that in which the particles are all infinitely separated from one another. Also, we usually set the corresponding *reference potential energy* to be zero. Suppose that several charged particles come together from initially infinite separations (state i) to form a system of neighboring particles (state f). Let the initial potential energy U_i be zero, and let W_∞ represent the work done by the electrostatic forces between the particles during the move in from infinity. Then from Eq. 24-1, the final potential energy U of the system is

$$U = -W_\infty. \tag{24-2}$$

CHECKPOINT 1

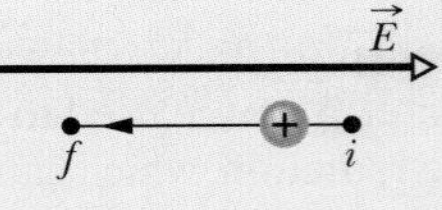

In the figure, a proton moves from point i to point f in a uniform electric field directed as shown. (a) Does the electric field do positive or negative work on the proton? (b) Does the electric potential energy of the proton increase or decrease?

Sample Problem

Work and potential energy in an electric field

Electrons are continually being knocked out of air molecules in the atmosphere by cosmic-ray particles coming in from space. Once released, each electron experiences an electrostatic force $\vec{F}$ due to the electric field $\vec{E}$ that is produced in the atmosphere by charged particles already on Earth. Near Earth's surface the electric field has the magnitude $E = 150$ N/C and is directed downward. What is the change ΔU in the electric potential energy of a released electron when the electrostatic force causes it to move vertically upward through a distance $d = 520$ m (Fig. 24-1)?

KEY IDEAS

(1) The change ΔU in the electric potential energy of the electron is related to the work W done on the electron by the electric field. Equation 24-1 ($\Delta U = -W$) gives the relation.

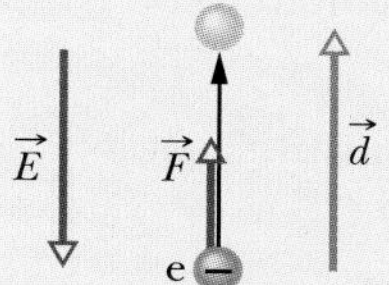

Fig. 24-1 An electron in the atmosphere is moved upward through displacement $\vec{d}$ by an electrostatic force $\vec{F}$ due to an electric field $\vec{E}$.

(2) The work done by a constant force $\vec{F}$ on a particle undergoing a displacement $\vec{d}$ is

$$W = \vec{F} \cdot \vec{d}. \qquad (24\text{-}3)$$

(3) The electrostatic force and the electric field are related by the force equation $\vec{F} = q\vec{E}$, where here q is the charge of an electron ($= -1.6 \times 10^{-19}$ C).

Calculations: Substituting for $\vec{F}$ in Eq. 24-3 and taking the dot product yield

$$W = q\vec{E} \cdot \vec{d} = qEd \cos\theta, \qquad (24\text{-}4)$$

where θ is the angle between the directions of $\vec{E}$ and $\vec{d}$. The field $\vec{E}$ is directed downward and the displacement $\vec{d}$ is directed upward; so $\theta = 180°$. Substituting this and other data into Eq. 24-4, we find

$$W = (-1.6 \times 10^{-19}\ \text{C})(150\ \text{N/C})(520\ \text{m}) \cos 180°$$
$$= 1.2 \times 10^{-14}\ \text{J}.$$

Equation 24-1 then yields

$$\Delta U = -W = -1.2 \times 10^{-14}\ \text{J}. \qquad \text{(Answer)}$$

This result tells us that during the 520 m ascent, the electric potential energy of the electron *decreases* by 1.2×10^{-14} J.

WILEY PLUS Additional examples, video, and practice available at *WileyPLUS*

24-3 Electric Potential

The potential energy of a charged particle in an electric field depends on the charge magnitude. However, the potential energy *per unit charge* has a unique value at any point in an electric field.

For an example of this, suppose we place a test particle of positive charge 1.60×10^{-19} C at a point in an electric field where the particle has an electric potential energy of 2.40×10^{-17} J. Then the potential energy per unit charge is

$$\frac{2.40 \times 10^{-17}\ \text{J}}{1.60 \times 10^{-19}\ \text{C}} = 150\ \text{J/C}.$$

Next, suppose we replace that test particle with one having twice as much positive charge, 3.20×10^{-19} C. We would find that the second particle has an electric potential energy of 4.80×10^{-17} J, twice that of the first particle. However, the potential energy per unit charge would be the same, still 150 J/C.

Thus, the potential energy per unit charge, which can be symbolized as U/q, is independent of the charge q of the particle we happen to use and is *characteristic only of the electric field* we are investigating. The potential energy per unit charge at a point in an electric field is called the **electric potential** V (or simply the **potential**) at that point. Thus,

$$V = \frac{U}{q}. \qquad (24\text{-}5)$$

Note that electric potential is a scalar, not a vector.

The *electric potential difference* ΔV between any two points i and f in an electric field is equal to the difference in potential energy per unit charge between the two points:

$$\Delta V = V_f - V_i = \frac{U_f}{q} - \frac{U_i}{q} = \frac{\Delta U}{q}. \tag{24-6}$$

Using Eq. 24-1 to substitute $-W$ for ΔU in Eq. 24-6, we can define the potential difference between points i and f as

$$\Delta V = V_f - V_i = -\frac{W}{q} \quad \text{(potential difference defined)}. \tag{24-7}$$

The potential difference between two points is thus the negative of the work done by the electrostatic force to move a unit charge from one point to the other. A potential difference can be positive, negative, or zero, depending on the signs and magnitudes of q and W.

If we set $U_i = 0$ at infinity as our reference potential energy, then by Eq. 24-5, the electric potential V must also be zero there. Then from Eq. 24-7, we can define the electric potential at any point in an electric field to be

$$V = -\frac{W_\infty}{q} \quad \text{(potential defined)}, \tag{24-8}$$

where W_∞ is the work done by the electric field on a charged particle as that particle moves in from infinity to point f. A potential V can be positive, negative, or zero, depending on the signs and magnitudes of q and W_∞.

The SI unit for potential that follows from Eq. 24-8 is the joule per coulomb. This combination occurs so often that a special unit, the *volt* (abbreviated V), is used to represent it. Thus,

$$1 \text{ volt} = 1 \text{ joule per coulomb}. \tag{24-9}$$

This new unit allows us to adopt a more conventional unit for the electric field $\vec{E}$, which we have measured up to now in newtons per coulomb. With two unit conversions, we obtain

$$\begin{aligned} 1 \text{ N/C} &= \left(1\frac{\text{N}}{\text{C}}\right)\left(\frac{1 \text{ V}\cdot\text{C}}{1 \text{ J}}\right)\left(\frac{1 \text{ J}}{1 \text{ N}\cdot\text{m}}\right) \\ &= 1 \text{ V/m}. \end{aligned} \tag{24-10}$$

The conversion factor in the second set of parentheses comes from Eq. 24-9; that in the third set of parentheses is derived from the definition of the joule. From now on, we shall express values of the electric field in volts per meter rather than in newtons per coulomb.

Finally, we can now define an energy unit that is a convenient one for energy measurements in the atomic and subatomic domain: One *electron-volt* (eV) is the energy equal to the work required to move a single elementary charge e, such as that of the electron or the proton, through a potential difference of exactly one volt. Equation 24-7 tells us that the magnitude of this work is $q\,\Delta V$; so

$$\begin{aligned} 1 \text{ eV} &= e(1 \text{ V}) \\ &= (1.60 \times 10^{-19} \text{ C})(1 \text{ J/C}) = 1.60 \times 10^{-19} \text{ J}. \end{aligned}$$

Work Done by an Applied Force

Suppose we move a particle of charge q from point i to point f in an electric field by applying a force to it. During the move, our applied force does work W_{app} on the charge while the electric field does work W on it. By the work–kinetic energy theorem of Eq. 7-10, the change ΔK in the kinetic energy of the particle is

$$\Delta K = K_f - K_i = W_{app} + W. \tag{24-11}$$

Now suppose the particle is stationary before and after the move. Then K_f and K_i are both zero, and Eq. 24-11 reduces to

$$W_{app} = -W. \tag{24-12}$$

In words, the work W_{app} done by our applied force during the move is equal to the negative of the work W done by the electric field—provided there is no change in kinetic energy.

By using Eq. 24-12 to substitute W_{app} into Eq. 24-1, we can relate the work done by our applied force to the change in the potential energy of the particle during the move. We find

$$\Delta U = U_f - U_i = W_{app}. \tag{24-13}$$

By similarly using Eq. 24-12 to substitute W_{app} into Eq. 24-7, we can relate our work W_{app} to the electric potential difference ΔV between the initial and final locations of the particle. We find

$$W_{app} = q\,\Delta V. \tag{24-14}$$

W_{app} can be positive, negative, or zero depending on the signs and magnitudes of q and ΔV.

CHECKPOINT 2

In the figure of Checkpoint 1, we move the proton from point i to point f in a uniform electric field directed as shown. (a) Does our force do positive or negative work? (b) Does the proton move to a point of higher or lower potential?

24-4 Equipotential Surfaces

Adjacent points that have the same electric potential form an **equipotential surface,** which can be either an imaginary surface or a real, physical surface. No net work W is done on a charged particle by an electric field when the particle moves between two points i and f on the same equipotential surface. This follows from Eq. 24-7, which tells us that W must be zero if $V_f = V_i$. Because of the path independence of work (and thus of potential energy and potential), $W = 0$ for *any* path connecting points i and f on a given equipotential surface regardless of whether that path lies entirely on that surface.

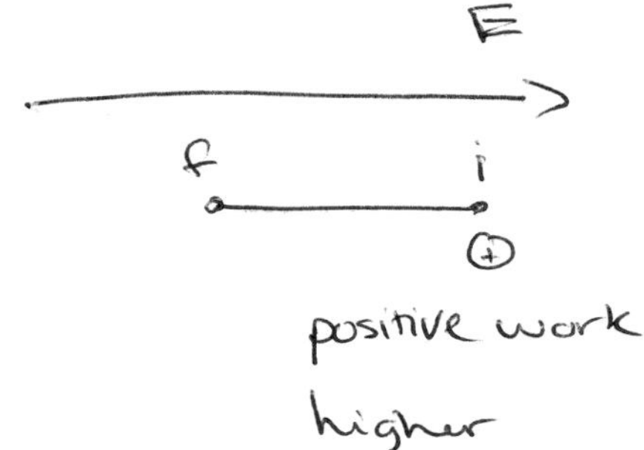

Figure 24-2 shows a *family* of equipotential surfaces associated with the electric field due to some distribution of charges. The work done by the electric field

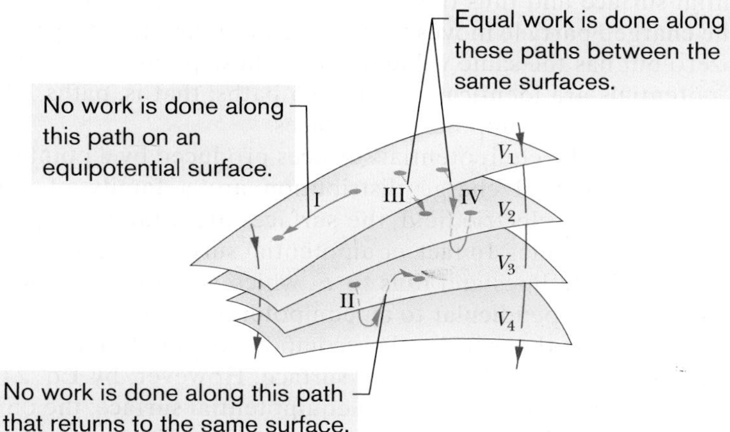

Fig. 24-2 Portions of four equipotential surfaces at electric potentials $V_1 = 100$ V, $V_2 = 80$ V, $V_3 = 60$ V, and $V_4 = 40$ V. Four paths along which a test charge may move are shown. Two electric field lines are also indicated.

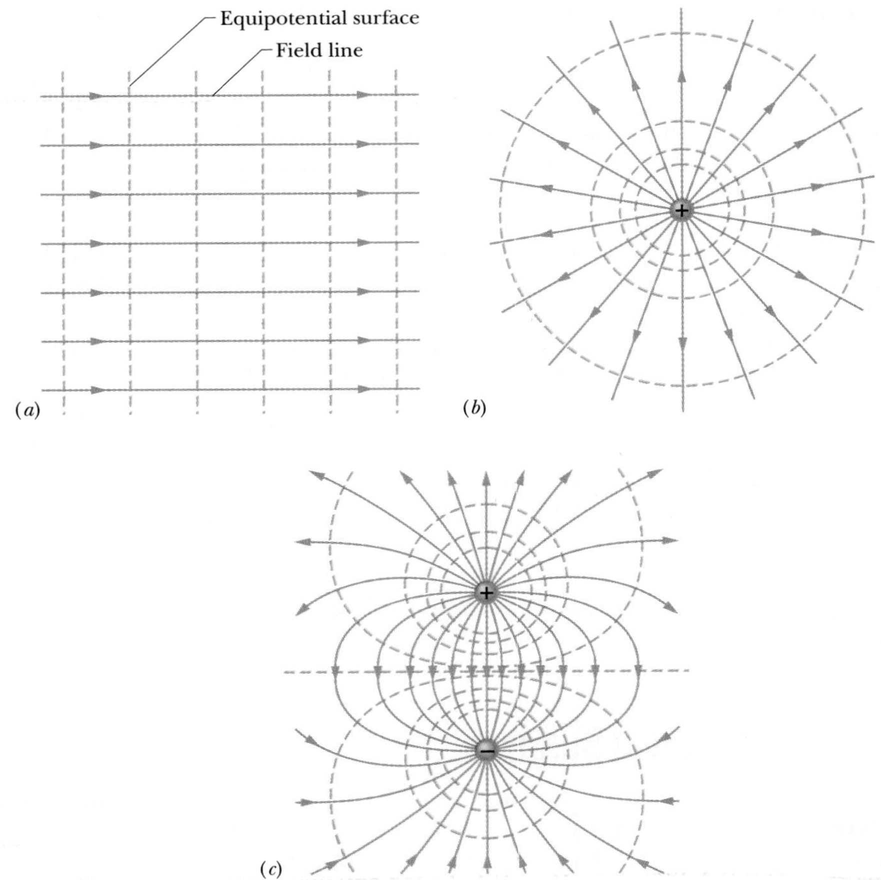

Fig. 24-3 Electric field lines (purple) and cross sections of equipotential surfaces (gold) for (*a*) a uniform electric field, (*b*) the field due to a point charge, and (*c*) the field due to an electric dipole.

on a charged particle as the particle moves from one end to the other of paths I and II is zero because each of these paths begins and ends on the same equipotential surface and thus there is no net change in potential. The work done as the charged particle moves from one end to the other of paths III and IV is not zero but has the same value for both these paths because the initial and final potentials are identical for the two paths; that is, paths III and IV connect the same pair of equipotential surfaces.

From symmetry, the equipotential surfaces produced by a point charge or a spherically symmetrical charge distribution are a family of concentric spheres. For a uniform electric field, the surfaces are a family of planes perpendicular to the field lines. In fact, equipotential surfaces are always perpendicular to electric field lines and thus to $\vec{E}$, which is always tangent to these lines. If $\vec{E}$ were *not* perpendicular to an equipotential surface, it would have a component lying along that surface. This component would then do work on a charged particle as it moved along the surface. However, by Eq. 24-7 work cannot be done if the surface is truly an equipotential surface; the only possible conclusion is that $\vec{E}$ must be everywhere perpendicular to the surface. Figure 24-3 shows electric field lines and cross sections of the equipotential surfaces for a uniform electric field and for the field associated with a point charge and with an electric dipole.

24-5 Calculating the Potential from the Field

We can calculate the potential difference between any two points i and f in an electric field if we know the electric field vector $\vec{E}$ all along any path connecting those points. To make the calculation, we find the work done on a positive test charge by the field as the charge moves from i to f, and then use Eq. 24-7.

Consider an arbitrary electric field, represented by the field lines in Fig. 24-4, and a positive test charge q_0 that moves along the path shown from point i to point f. At any point on the path, an electrostatic force $q_0\vec{E}$ acts on the charge as it moves through a differential displacement $d\vec{s}$. From Chapter 7, we know that the differential work dW done on a particle by a force $\vec{F}$ during a displacement $d\vec{s}$ is given by the dot product of the force and the displacement:

$$dW = \vec{F}\cdot d\vec{s}. \tag{24-15}$$

For the situation of Fig. 24-4, $\vec{F} = q_0\vec{E}$ and Eq. 24-15 becomes

$$dW = q_0\vec{E}\cdot d\vec{s}. \tag{24-16}$$

To find the total work W done on the particle by the field as the particle moves from point i to point f, we sum—via integration—the differential works done on the charge as it moves through all the displacements $d\vec{s}$ along the path:

$$W = q_0\int_i^f \vec{E}\cdot d\vec{s}. \tag{24-17}$$

If we substitute the total work W from Eq. 24-17 into Eq. 24-7, we find

$$V_f - V_i = -\int_i^f \vec{E}\cdot d\vec{s}. \tag{24-18}$$

Thus, the potential difference $V_f - V_i$ between any two points i and f in an electric field is equal to the negative of the *line integral* (meaning the integral along a particular path) of $\vec{E}\cdot d\vec{s}$ from i to f. However, because the electrostatic force is conservative, all paths (whether easy or difficult to use) yield the same result.

Equation 24-18 allows us to calculate the difference in potential between any two points in the field. If we set potential $V_i = 0$, then Eq. 24-18 becomes

$$V = -\int_i^f \vec{E}\cdot d\vec{s}, \tag{24-19}$$

in which we have dropped the subscript f on V_f. Equation 24-19 gives us the potential V at any point f in the electric field *relative to the zero potential* at point i. If we let point i be at infinity, then Eq. 24-19 gives us the potential V at any point f relative to the zero potential at infinity.

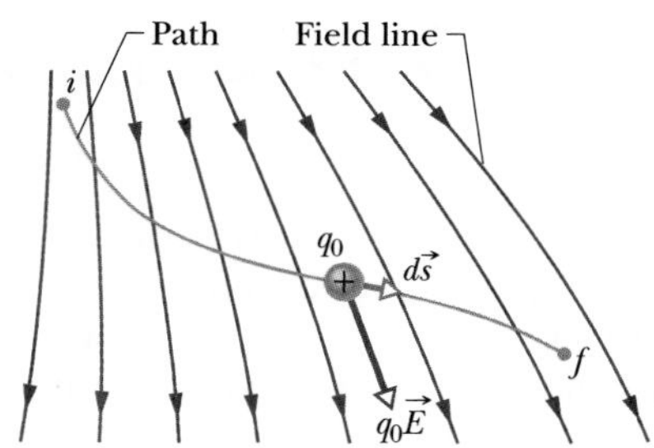

Fig. 24-4 A test charge q_0 moves from point i to point f along the path shown in a nonuniform electric field. During a displacement $d\vec{s}$, an electrostatic force $q_0\vec{E}$ acts on the test charge. This force points in the direction of the field line at the location of the test charge.

CHECKPOINT 3

The figure here shows a family of parallel equipotential surfaces (in cross section) and five paths along which we shall move an electron from one surface to another. (a) What is the direction of the electric field associated with the surfaces? (b) For each path, is the work we do positive, negative, or zero? (c) Rank the paths according to the work we do, greatest first.

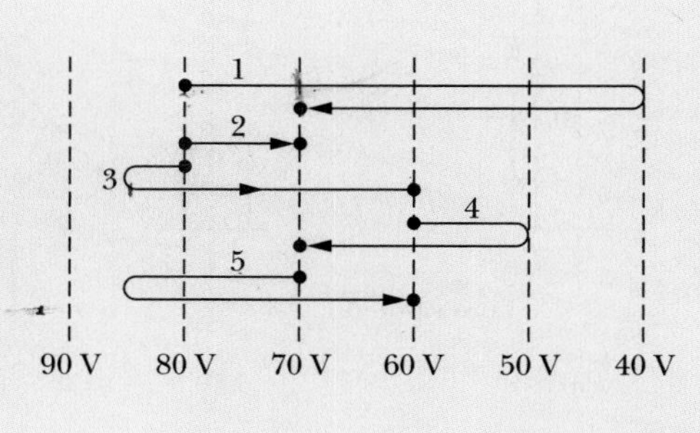

Sample Problem

Finding the potential change from the electric field

(a) Figure 24-5*a* shows two points *i* and *f* in a uniform electric field $\vec{E}$. The points lie on the same electric field line (not shown) and are separated by a distance *d*. Find the potential difference $V_f - V_i$ by moving a positive test charge q_0 from *i* to *f* along the path shown, which is parallel to the field direction.

KEY IDEA

We can find the potential difference between any two points in an electric field by integrating $\vec{E} \cdot d\vec{s}$ along a path connecting those two points according to Eq. 24-18.

Calculations: We begin by mentally moving a test charge q_0 along that path, from initial point *i* to final point *f*. As we move such a test charge along the path in Fig. 24-5*a*, its differential displacement $d\vec{s}$ always has the same direction as $\vec{E}$. Thus, the angle θ between $\vec{E}$ and $d\vec{s}$ is zero and the dot product in Eq. 24-18 is

$$\vec{E} \cdot d\vec{s} = E\, ds \cos\theta = E\, ds. \qquad (24\text{-}20)$$

Equations 24-18 and 24-20 then give us

$$V_f - V_i = -\int_i^f \vec{E} \cdot d\vec{s} = -\int_i^f E\, ds. \qquad (24\text{-}21)$$

Since the field is uniform, *E* is constant over the path and can be moved outside the integral, giving us

$$V_f - V_i = -E\int_i^f ds = -Ed, \qquad \text{(Answer)}$$

in which the integral is simply the length *d* of the path. The minus sign in the result shows that the potential at point *f* in Fig. 24-5*a* is lower than the potential at point *i*. This is a general result: The potential always decreases along a path that extends in the direction of the electric field lines.

(b) Now find the potential difference $V_f - V_i$ by moving the positive test charge q_0 from *i* to *f* along the path *icf* shown in Fig. 24-5*b*.

Calculations: The Key Idea of (a) applies here too, except now we move the test charge along a path that consists of two lines: *ic* and *cf*. At all points along line *ic*, the displacement $d\vec{s}$ of the test charge is perpendicular to $\vec{E}$. Thus, the angle θ between $\vec{E}$ and $d\vec{s}$ is 90°, and the dot product $\vec{E} \cdot d\vec{s}$ is 0. Equation 24-18 then tells us that points *i* and *c* are at the same potential: $V_c - V_i = 0$.

For line *cf* we have $\theta = 45°$ and, from Eq. 24-18,

$$V_f - V_i = -\int_c^f \vec{E} \cdot d\vec{s} = -\int_c^f E(\cos 45°)\, ds$$

$$= -E(\cos 45°)\int_c^f ds.$$

The integral in this equation is just the length of line *cf*; from Fig. 24-5*b*, that length is $d/\cos 45°$. Thus,

$$V_f - V_i = -E(\cos 45°)\frac{d}{\cos 45°} = -Ed. \qquad \text{(Answer)}$$

This is the same result we obtained in (a), as it must be; the potential difference between two points does not depend on the path connecting them. Moral: When you want to find the potential difference between two points by moving a test charge between them, you can save time and work by choosing a path that simplifies the use of Eq. 24-18.

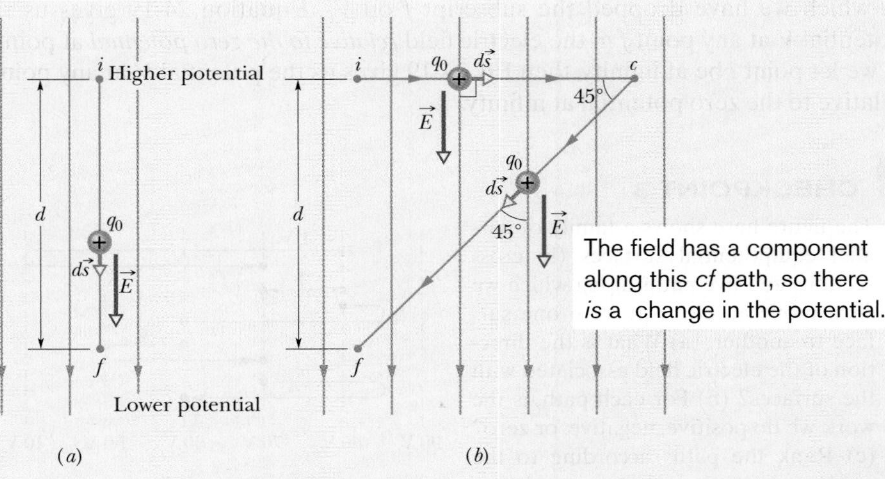

Fig. 24-5 (*a*) A test charge q_0 moves in a straight line from point *i* to point *f*, along the direction of a uniform external electric field. (*b*) Charge q_0 moves along path *icf* in the same electric field.

Additional examples, video, and practice available at *WileyPLUS*

24-6 Potential Due to a Point Charge

We now use Eq. 24-18 to derive, for the space around a charged particle, an expression for the electric potential V relative to the zero potential at infinity. Consider a point P at distance R from a fixed particle of positive charge q (Fig. 24-6). To use Eq. 24-18, we imagine that we move a positive test charge q_0 from point P to infinity. Because the path we take does not matter, let us choose the simplest one—a line that extends radially from the fixed particle through P to infinity.

To use Eq. 24-18, we must evaluate the dot product

$$\vec{E}\cdot d\vec{s} = E\cos\theta\, ds. \tag{24-22}$$

The electric field $\vec{E}$ in Fig. 24-6 is directed radially outward from the fixed particle. Thus, the differential displacement $d\vec{s}$ of the test particle along its path has the same direction as $\vec{E}$. That means that in Eq. 24-22, angle $\theta = 0$ and $\cos\theta = 1$. Because the path is radial, let us write ds as dr. Then, substituting the limits R and ∞, we can write Eq. 24-18 as

$$V_f - V_i = -\int_R^\infty E\, dr. \tag{24-23}$$

Next, we set $V_f = 0$ (at ∞) and $V_i = V$ (at R). Then, for the magnitude of the electric field at the site of the test charge, we substitute from Eq. 22-3:

$$E = \frac{1}{4\pi\varepsilon_0}\frac{q}{r^2}. \tag{24-24}$$

With these changes, Eq. 24-23 then gives us

$$0 - V = -\frac{q}{4\pi\varepsilon_0}\int_R^\infty \frac{1}{r^2}\,dr = \frac{q}{4\pi\varepsilon_0}\left[\frac{1}{r}\right]_R^\infty$$

$$= -\frac{1}{4\pi\varepsilon_0}\frac{q}{R}. \tag{24-25}$$

Solving for V and switching R to r, we then have

$$V = \frac{1}{4\pi\varepsilon_0}\frac{q}{r} \tag{24-26}$$

as the electric potential V due to a particle of charge q at any radial distance r from the particle.

Although we have derived Eq. 24-26 for a positively charged particle, the derivation holds also for a negatively charged particle, in which case, q is a negative quantity. Note that the sign of V is the same as the sign of q:

> A positively charged particle produces a positive electric potential. A negatively charged particle produces a negative electric potential.

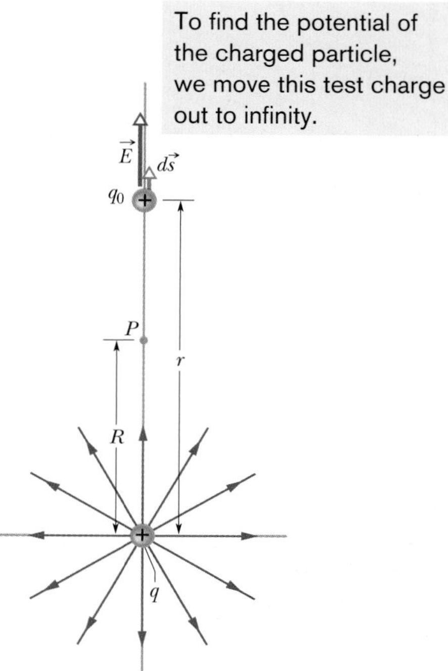

Fig. 24-6 The positive point charge q produces an electric field $\vec{E}$ and an electric potential V at point P. We find the potential by moving a test charge q_0 from P to infinity. The test charge is shown at distance r from the point charge, during differential displacement $d\vec{s}$.

Figure 24-7 shows a computer-generated plot of Eq. 24-26 for a positively charged particle; the magnitude of V is plotted vertically. Note that the magnitude increases as $r \to 0$. In fact, according to Eq. 24-26, V is infinite at $r = 0$, although Fig. 24-7 shows a finite, smoothed-off value there.

Equation 24-26 also gives the electric potential either *outside or on the external surface of* a spherically symmetric charge distribution. We can prove this by using one of the shell theorems of Sections 21-4 and 23-9 to replace the actual spherical charge distribution with an equal charge concentrated at its center. Then the derivation leading to Eq. 24-26 follows, provided we do not consider a point within the actual distribution.

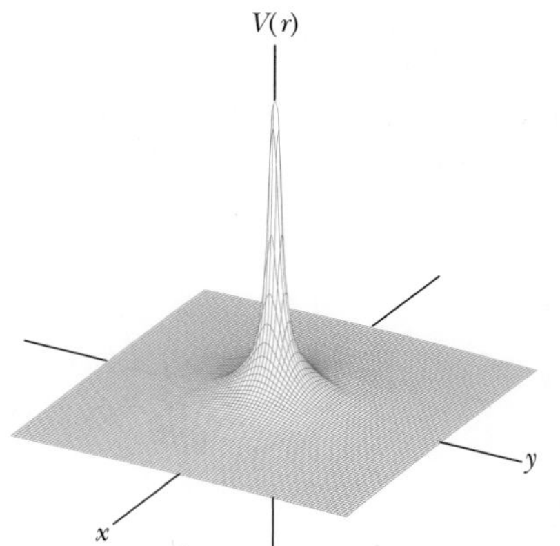

Fig. 24-7 A computer-generated plot of the electric potential $V(r)$ due to a positive point charge located at the origin of an xy plane. The potentials at points in the xy plane are plotted vertically. (Curved lines have been added to help you visualize the plot.) The infinite value of V predicted by Eq. 24-26 for $r = 0$ is not plotted.

24-7 Potential Due to a Group of Point Charges

We can find the net potential at a point due to a group of point charges with the help of the superposition principle. Using Eq. 24-26 with the sign of the charge included, we calculate separately the potential resulting from each charge at the given point. Then we sum the potentials. For n charges, the net potential is

$$V = \sum_{i=1}^{n} V_i = \frac{1}{4\pi\varepsilon_0} \sum_{i=1}^{n} \frac{q_i}{r_i} \quad (n \text{ point charges}). \tag{24-27}$$

Here q_i is the value of the ith charge and r_i is the radial distance of the given point from the ith charge. The sum in Eq. 24-27 is an *algebraic sum,* not a vector sum like the sum that would be used to calculate the electric field resulting from a group of point charges. Herein lies an important computational advantage of potential over electric field: It is a lot easier to sum several scalar quantities than to sum several vector quantities whose directions and components must be considered.

CHECKPOINT 4

The figure here shows three arrangements of two protons. Rank the arrangements according to the net electric potential produced at point P by the protons, greatest first.

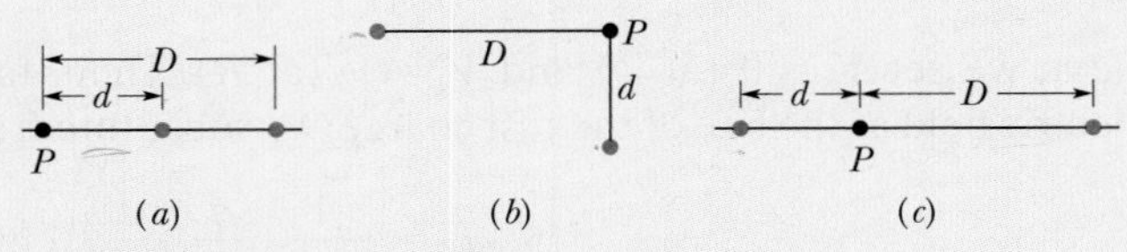

Sample Problem

Net potential of several charged particles

What is the electric potential at point P, located at the center of the square of point charges shown in Fig. 24-8a? The distance d is 1.3 m, and the charges are

$$q_1 = +12 \text{ nC}, \qquad q_3 = +31 \text{ nC},$$
$$q_2 = -24 \text{ nC}, \qquad q_4 = +17 \text{ nC}.$$

KEY IDEA

The electric potential V at point P is the algebraic sum of the electric potentials contributed by the four point charges.

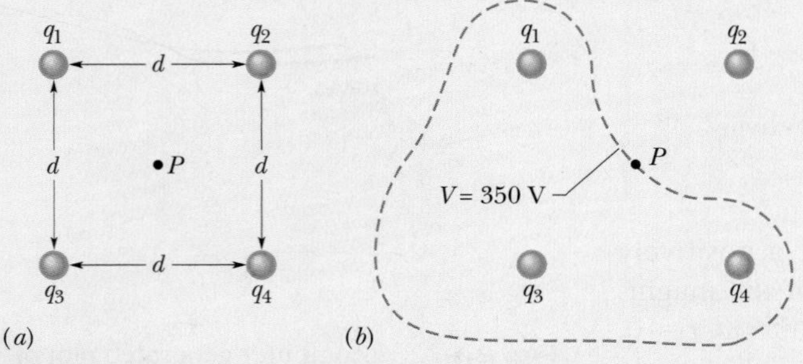

Fig. 24-8 (a) Four point charges are held fixed at the corners of a square. (b) The closed curve is a cross section, in the plane of the figure, of the equipotential surface that contains point P. (The curve is drawn only roughly.)

(Because electric potential is a scalar, the orientations of the point charges do not matter.)

Calculations: From Eq. 24-27, we have

$$V = \sum_{i=1}^{4} V_i = \frac{1}{4\pi\varepsilon_0}\left(\frac{q_1}{r} + \frac{q_2}{r} + \frac{q_3}{r} + \frac{q_4}{r}\right).$$

The distance r is $d/\sqrt{2}$, which is 0.919 m, and the sum of the charges is

$$q_1 + q_2 + q_3 + q_4 = (12 - 24 + 31 + 17) \times 10^{-9} \text{ C}$$
$$= 36 \times 10^{-9} \text{ C}.$$

Thus,
$$V = \frac{(8.99 \times 10^9 \text{ N}\cdot\text{m}^2/\text{C}^2)(36 \times 10^{-9} \text{ C})}{0.919 \text{ m}}$$
$$\approx 350 \text{ V}. \qquad \text{(Answer)}$$

Close to any of the three positive charges in Fig. 24-8a, the potential has very large positive values. Close to the single negative charge, the potential has very large negative values. Therefore, there must be points within the square that have the same intermediate potential as that at point P. The curve in Fig. 24-8b shows the intersection of the plane of the figure with the equipotential surface that contains point P. Any point along that curve has the same potential as point P.

Additional examples, video, and practice available at *WileyPLUS*

Sample Problem

Potential is not a vector, orientation is irrelevant

(a) In Fig. 24-9*a*, 12 electrons (of charge $-e$) are equally spaced and fixed around a circle of radius R. Relative to $V = 0$ at infinity, what are the electric potential and electric field at the center C of the circle due to these electrons?

KEY IDEAS

(1) The electric potential V at C is the algebraic sum of the electric potentials contributed by all the electrons. (Because electric potential is a scalar, the orientations of the electrons do not matter.) (2) The electric field at C is a vector quantity and thus the orientation of the electrons *is* important.

Calculations: Because the electrons all have the same negative charge $-e$ and are all the same distance R from C, Eq. 24-27 gives us

$$V = -12\frac{1}{4\pi\varepsilon_0}\frac{e}{R}. \quad \text{(Answer)} \quad (24\text{-}28)$$

Because of the symmetry of the arrangement in Fig. 24-9*a*, the electric field vector at C due to any given electron is canceled by the field vector due to the electron that is diametrically opposite it. Thus, at C,

$$\vec{E} = 0. \quad \text{(Answer)}$$

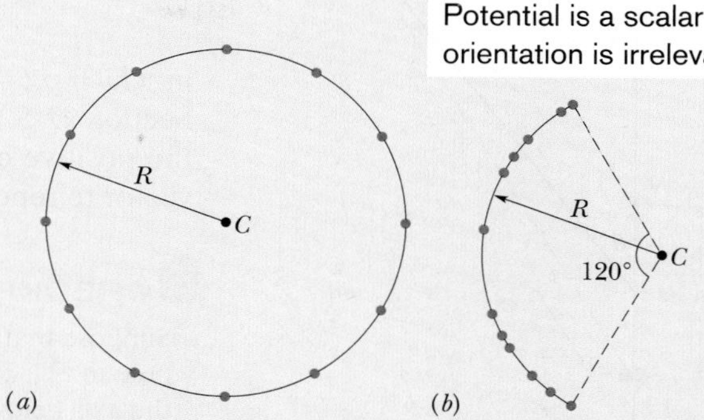

Fig. 24-9 (*a*) Twelve electrons uniformly spaced around a circle. (*b*) The electrons nonuniformly spaced along an arc of the original circle.

(b) If the electrons are moved along the circle until they are nonuniformly spaced over a 120° arc (Fig. 24-9*b*), what then is the potential at C? How does the electric field at C change (if at all)?

Reasoning: The potential is still given by Eq. 24-28, because the distance between C and each electron is unchanged and orientation is irrelevant. The electric field is no longer zero, however, because the arrangement is no longer symmetric. A net field is now directed toward the charge distribution.

Additional examples, video, and practice available at *WileyPLUS*

24-8 Potential Due to an Electric Dipole

Now let us apply Eq. 24-27 to an electric dipole to find the potential at an arbitrary point P in Fig. 24-10*a*. At P, the positive point charge (at distance $r_{(+)}$) sets up potential $V_{(+)}$ and the negative point charge (at distance $r_{(-)}$) sets up potential $V_{(-)}$. Then the net potential at P is given by Eq. 24-27 as

$$V = \sum_{i=1}^{2} V_i = V_{(+)} + V_{(-)} = \frac{1}{4\pi\varepsilon_0}\left(\frac{q}{r_{(+)}} + \frac{-q}{r_{(-)}}\right)$$
$$= \frac{q}{4\pi\varepsilon_0}\frac{r_{(-)} - r_{(+)}}{r_{(-)}r_{(+)}}. \quad (24\text{-}29)$$

Naturally occurring dipoles—such as those possessed by many molecules—are quite small; so we are usually interested only in points that are relatively far from the dipole, such that $r \gg d$, where d is the distance between the charges. Under those conditions, the approximations that follow from Fig. 24-10*b* are

$$r_{(-)} - r_{(+)} \approx d\cos\theta \quad \text{and} \quad r_{(-)}r_{(+)} \approx r^2.$$

If we substitute these quantities into Eq. 24-29, we can approximate V to be

$$V = \frac{q}{4\pi\varepsilon_0}\frac{d\cos\theta}{r^2},$$

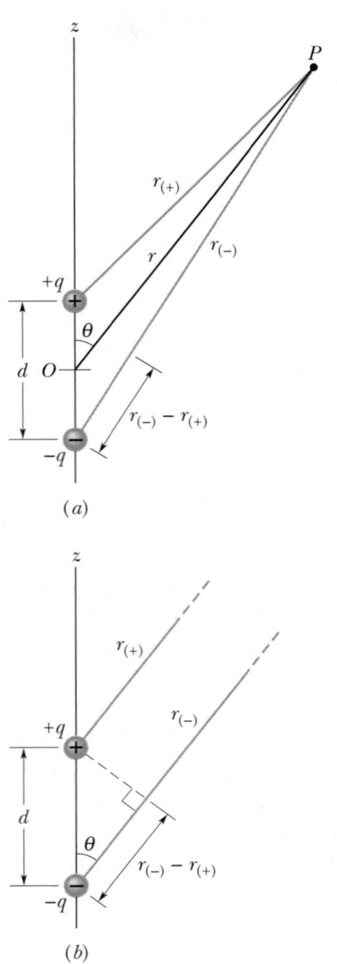

Fig. 24-10 (*a*) Point *P* is a distance *r* from the midpoint *O* of a dipole. The line *OP* makes an angle θ with the dipole axis. (*b*) If *P* is far from the dipole, the lines of lengths $r_{(+)}$ and $r_{(-)}$ are approximately parallel to the line of length *r*, and the dashed black line is approximately perpendicular to the line of length $r_{(-)}$.

where θ is measured from the dipole axis as shown in Fig. 24-10*a*. We can now write *V* as

$$V = \frac{1}{4\pi\varepsilon_0}\frac{p\cos\theta}{r^2} \quad \text{(electric dipole)}, \tag{24-30}$$

in which $p\ (= qd)$ is the magnitude of the electric dipole moment $\vec{p}$ defined in Section 22-5. The vector $\vec{p}$ is directed along the dipole axis, from the negative to the positive charge. (Thus, θ is measured from the direction of $\vec{p}$.) We use this vector to report the orientation of an electric dipole.

CHECKPOINT 5

Suppose that three points are set at equal (large) distances *r* from the center of the dipole in Fig. 24-10: Point *a* is on the dipole axis above the positive charge, point *b* is on the axis below the negative charge, and point *c* is on a perpendicular bisector through the line connecting the two charges. Rank the points according to the electric potential of the dipole there, greatest (most positive) first.

Induced Dipole Moment

Many molecules, such as water, have *permanent* electric dipole moments. In other molecules (called *nonpolar molecules*) and in every isolated atom, the centers of the positive and negative charges coincide (Fig. 24-11*a*) and thus no dipole moment is set up. However, if we place an atom or a nonpolar molecule in an external electric field, the field distorts the electron orbits and separates the centers of positive and negative charge (Fig. 24-11*b*). Because the electrons are negatively charged, they tend to be shifted in a direction opposite the field. This shift sets up a dipole moment $\vec{p}$ that points in the direction of the field. This dipole moment is said to be *induced* by the field, and the atom or molecule is then said to be *polarized* by the field (that is, it has a positive side and a negative side). When the field is removed, the induced dipole moment and the polarization disappear.

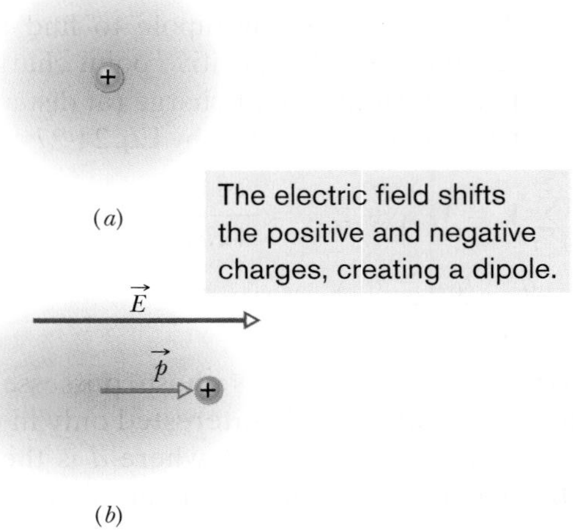

Fig. 24-11 (*a*) An atom, showing the positively charged nucleus (green) and the negatively charged electrons (gold shading). The centers of positive and negative charge coincide. (*b*) If the atom is placed in an external electric field $\vec{E}$, the electron orbits are distorted so that the centers of positive and negative charge no longer coincide. An induced dipole moment $\vec{p}$ appears. The distortion is greatly exaggerated here.

24-9 Potential Due to a Continuous Charge Distribution

When a charge distribution q is continuous (as on a uniformly charged thin rod or disk), we cannot use the summation of Eq. 24-27 to find the potential V at a point P. Instead, we must choose a differential element of charge dq, determine the potential dV at P due to dq, and then integrate over the entire charge distribution.

Let us again take the zero of potential to be at infinity. If we treat the element of charge dq as a point charge, then we can use Eq. 24-26 to express the potential dV at point P due to dq:

$$dV = \frac{1}{4\pi\varepsilon_0}\frac{dq}{r} \quad \text{(positive or negative } dq\text{).} \tag{24-31}$$

Here r is the distance between P and dq. To find the total potential V at P, we integrate to sum the potentials due to all the charge elements:

$$V = \int dV = \frac{1}{4\pi\varepsilon_0}\int \frac{dq}{r}. \tag{24-32}$$

The integral must be taken over the entire charge distribution. Note that because the electric potential is a scalar, there are *no vector components* to consider in Eq. 24-32.

We now examine two continuous charge distributions, a line and a disk.

Line of Charge

In Fig. 24-12*a*, a thin nonconducting rod of length L has a positive charge of uniform linear density λ. Let us determine the electric potential V due to the rod at point P, a perpendicular distance d from the left end of the rod.

We consider a differential element dx of the rod as shown in Fig. 24-12*b*. This (or any other) element of the rod has a differential charge of

$$dq = \lambda\, dx. \tag{24-33}$$

This element produces an electric potential dV at point P, which is a distance $r = (x^2 + d^2)^{1/2}$ from the element (Fig. 24-12*c*). Treating the element as a point

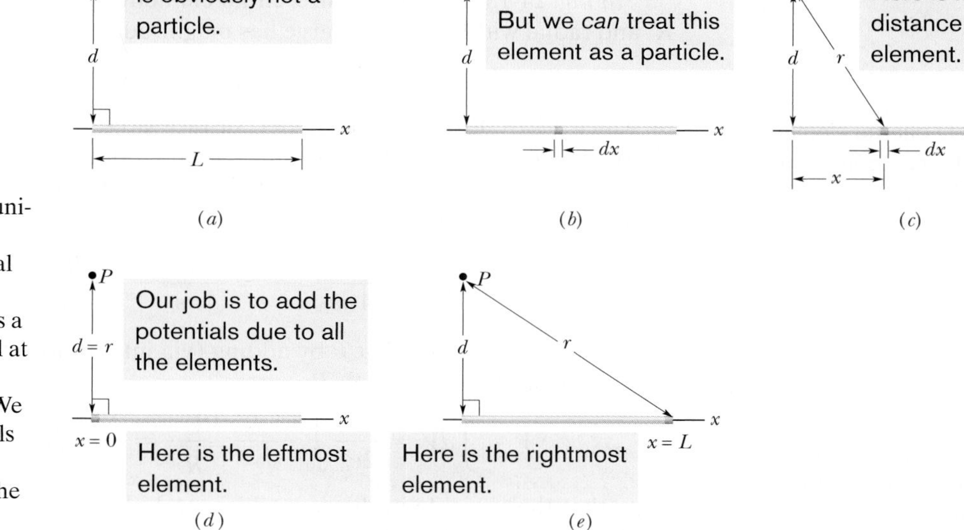

Fig. 24-12 (*a*) A thin, uniformly charged rod produces an electric potential V at point P. (*b*) An element can be treated as a particle. (*c*) The potential at P due to the element depends on the distance r. We need to sum the potentials due to all the elements, from the left side (*d*) to the right side (*e*).

charge, we can use Eq. 24-31 to write the potential dV as

$$dV = \frac{1}{4\pi\varepsilon_0}\frac{dq}{r} = \frac{1}{4\pi\varepsilon_0}\frac{\lambda\,dx}{(x^2 + d^2)^{1/2}}. \tag{24-34}$$

Since the charge on the rod is positive and we have taken $V = 0$ at infinity, we know from Section 24-6 that dV in Eq. 24-34 must be positive.

We now find the total potential V produced by the rod at point P by integrating Eq. 24-34 along the length of the rod, from $x = 0$ to $x = L$ (Figs. 24-12*d* and *e*), using integral 17 in Appendix E. We find

$$V = \int dV = \int_0^L \frac{1}{4\pi\varepsilon_0}\frac{\lambda}{(x^2 + d^2)^{1/2}}\,dx$$

$$= \frac{\lambda}{4\pi\varepsilon_0}\int_0^L \frac{dx}{(x^2 + d^2)^{1/2}}$$

$$= \frac{\lambda}{4\pi\varepsilon_0}\left[\ln\left(x + (x^2 + d^2)^{1/2}\right)\right]_0^L$$

$$= \frac{\lambda}{4\pi\varepsilon_0}\left[\ln\left(L + (L^2 + d^2)^{1/2}\right) - \ln d\right].$$

We can simplify this result by using the general relation $\ln A - \ln B = \ln(A/B)$. We then find

$$V = \frac{\lambda}{4\pi\varepsilon_0}\ln\left[\frac{L + (L^2 + d^2)^{1/2}}{d}\right]. \tag{24-35}$$

Because V is the sum of positive values of dV, it too is positive, consistent with the logarithm being positive for an argument greater than 1.

Charged Disk

In Section 22-7, we calculated the magnitude of the electric field at points on the central axis of a plastic disk of radius R that has a uniform charge density σ on one surface. Here we derive an expression for $V(z)$, the electric potential at any point on the central axis.

In Fig. 24-13, consider a differential element consisting of a flat ring of radius R' and radial width dR'. Its charge has magnitude

$$dq = \sigma(2\pi R')(dR'),$$

in which $(2\pi R')(dR')$ is the upper surface area of the ring. All parts of this charged element are the same distance r from point P on the disk's axis. With the aid of Fig. 24-13, we can use Eq. 24-31 to write the contribution of this ring to the electric potential at P as

$$dV = \frac{1}{4\pi\varepsilon_0}\frac{dq}{r} = \frac{1}{4\pi\varepsilon_0}\frac{\sigma(2\pi R')(dR')}{\sqrt{z^2 + R'^2}}. \tag{24-36}$$

We find the net potential at P by adding (via integration) the contributions of all the rings from $R' = 0$ to $R' = R$:

$$V = \int dV = \frac{\sigma}{2\varepsilon_0}\int_0^R \frac{R'\,dR'}{\sqrt{z^2 + R'^2}} = \frac{\sigma}{2\varepsilon_0}\left(\sqrt{z^2 + R^2} - z\right). \tag{24-37}$$

Note that the variable in the second integral of Eq. 24-37 is R' and not z, which remains constant while the integration over the surface of the disk is carried out. (Note also that, in evaluating the integral, we have assumed that $z \geq 0$.)

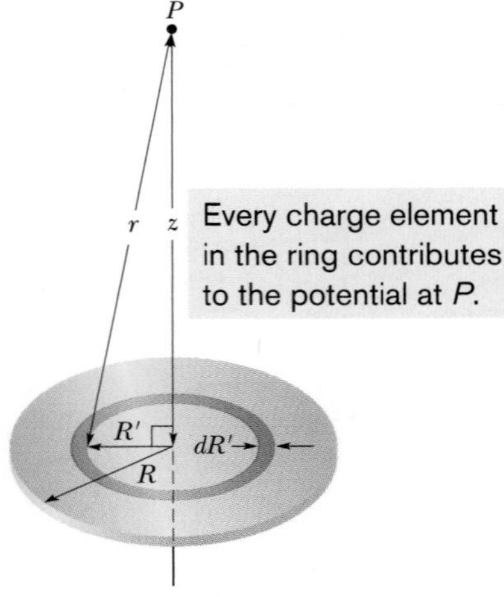

Fig. 24-13 A plastic disk of radius R, charged on its top surface to a uniform surface charge density σ. We wish to find the potential V at point P on the central axis of the disk.

24-10 Calculating the Field from the Potential

In Section 24-5, you saw how to find the potential at a point f if you know the electric field along a path from a reference point to point f. In this section, we propose to go the other way—that is, to find the electric field when we know the potential. As Fig. 24-3 shows, solving this problem graphically is easy: If we know the potential V at all points near an assembly of charges, we can draw in a family of equipotential surfaces. The electric field lines, sketched perpendicular to those surfaces, reveal the variation of $\vec{E}$. What we are seeking here is the mathematical equivalent of this graphical procedure.

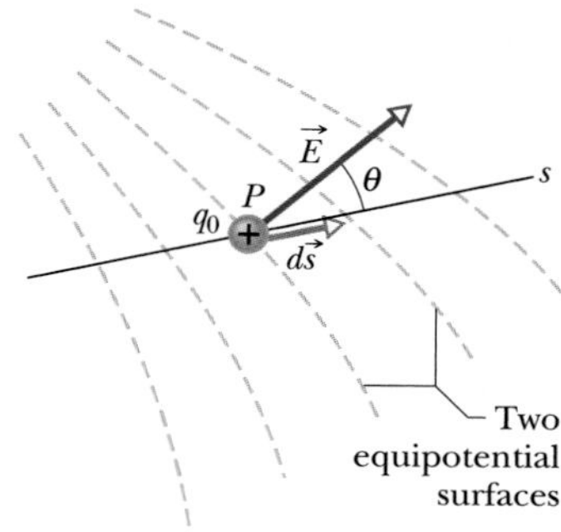

Fig. 24-14 A test charge q_0 moves a distance $d\vec{s}$ from one equipotential surface to another. (The separation between the surfaces has been exaggerated for clarity.) The displacement $d\vec{s}$ makes an angle θ with the direction of the electric field $\vec{E}$.

Figure 24-14 shows cross sections of a family of closely spaced equipotential surfaces, the potential difference between each pair of adjacent surfaces being dV. As the figure suggests, the field $\vec{E}$ at any point P is perpendicular to the equipotential surface through P.

Suppose that a positive test charge q_0 moves through a displacement $d\vec{s}$ from one equipotential surface to the adjacent surface. From Eq. 24-7, we see that the work the electric field does on the test charge during the move is $-q_0\,dV$. From Eq. 24-16 and Fig. 24-14, we see that the work done by the electric field may also be written as the scalar product $(q_0\vec{E})\cdot d\vec{s}$, or $q_0E(\cos\theta)\,ds$. Equating these two expressions for the work yields

$$-q_0\,dV = q_0E(\cos\theta)\,ds, \tag{24-38}$$

or

$$E\cos\theta = -\frac{dV}{ds}. \tag{24-39}$$

Since $E\cos\theta$ is the component of $\vec{E}$ in the direction of $d\vec{s}$, Eq. 24-39 becomes

$$E_s = -\frac{\partial V}{\partial s}. \tag{24-40}$$

We have added a subscript to E and switched to the partial derivative symbols to emphasize that Eq. 24-40 involves only the variation of V along a specified axis (here called the s axis) and only the component of $\vec{E}$ along that axis. In words, Eq. 24-40 (which is essentially the reverse operation of Eq. 24-18) states:

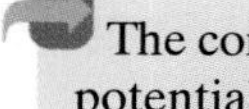

The component of $\vec{E}$ in any direction is the negative of the rate at which the electric potential changes with distance in that direction.

If we take the s axis to be, in turn, the x, y, and z axes, we find that the x, y, and z components of $\vec{E}$ at any point are

$$E_x = -\frac{\partial V}{\partial x}; \quad E_y = -\frac{\partial V}{\partial y}; \quad E_z = -\frac{\partial V}{\partial z}. \tag{24-41}$$

Thus, if we know V for all points in the region around a charge distribution—that is, if we know the function $V(x, y, z)$—we can find the components of $\vec{E}$, and thus $\vec{E}$ itself, at any point by taking partial derivatives.

For the simple situation in which the electric field $\vec{E}$ is uniform, Eq. 24-40 becomes

$$E = -\frac{\Delta V}{\Delta s}, \tag{24-42}$$

where s is perpendicular to the equipotential surfaces. The component of the electric field is zero in any direction parallel to the equipotential surfaces because there is no change in potential along the surfaces.

CHECKPOINT 6

The figure shows three pairs of parallel plates with the same separation, and the electric potential of each plate. The electric field between the plates is uniform and perpendicular to the plates. (a) Rank the pairs according to the magnitude of the electric field between the plates, greatest first. (b) For which pair is the electric field pointing rightward? (c) If an electron is released midway between the third pair of plates, does it remain there, move rightward at constant speed, move leftward at constant speed, accelerate rightward, or accelerate leftward?

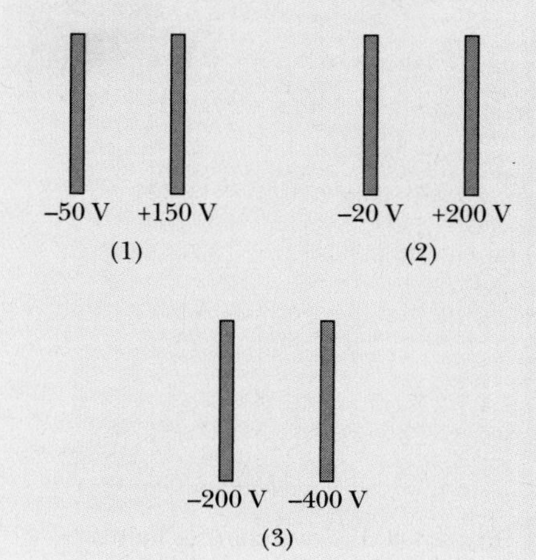

Sample Problem

Finding the field from the potential

The electric potential at any point on the central axis of a uniformly charged disk is given by Eq. 24-37,

$$V = \frac{\sigma}{2\varepsilon_0}(\sqrt{z^2 + R^2} - z).$$

Starting with this expression, derive an expression for the electric field at any point on the axis of the disk.

KEY IDEAS

We want the electric field $\vec{E}$ as a function of distance z along the axis of the disk. For any value of z, the direction of $\vec{E}$ must be along that axis because the disk has circular symmetry about that axis. Thus, we want the component E_z of $\vec{E}$ in the direction of z. This component is the negative of the rate at which the electric potential changes with distance z.

Calculation: Thus, from the last of Eqs. 24-41, we can write

$$E_z = -\frac{\partial V}{\partial z} = -\frac{\sigma}{2\varepsilon_0}\frac{d}{dz}(\sqrt{z^2 + R^2} - z)$$

$$= \frac{\sigma}{2\varepsilon_0}\left(1 - \frac{z}{\sqrt{z^2 + R^2}}\right). \quad \text{(Answer)}$$

This is the same expression that we derived in Section 22-7 by integration, using Coulomb's law.

WILEY PLUS Additional examples, video, and practice available at *WileyPLUS*

24-11 Electric Potential Energy of a System of Point Charges

In Section 24-2, we discussed the electric potential energy of a charged particle as an electrostatic force does work on it. In that section, we assumed that the charges that produced the force were fixed in place, so that neither the force nor the corresponding electric field could be influenced by the presence of the test charge. In this section we can take a broader view, to find the electric potential energy of a *system* of charges due to the electric field produced *by* those same charges.

For a simple example, suppose you push together two bodies that have charges of the same electrical sign. The work that you must do is stored as electric potential energy in the two-body system (provided the kinetic energy of the bodies does not change). If you later release the charges, you can recover this stored energy, in whole or in part, as kinetic energy of the charged bodies as they rush away from each other.

We define the electric potential energy *of a system of point charges,* held in fixed positions by forces not specified, as follows:

> The electric potential energy of a system of fixed point charges is equal to the work that must be done by an external agent to assemble the system, bringing each charge in from an infinite distance.

We assume that the charges are stationary both in their initial infinitely distant positions and in their final assembled configuration.

Figure 24-15 shows two point charges q_1 and q_2, separated by a distance r. To find the electric potential energy of this two-charge system, we must mentally build the system, starting with both charges infinitely far away and at rest. When we bring q_1 in from infinity and put it in place, we do no work because no electrostatic force acts on q_1. However, when we next bring q_2 in from infinity and put it in place, we must do work because q_1 exerts an electrostatic force on q_2 during the move.

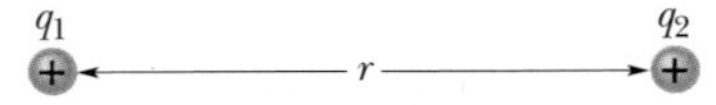

Fig. 24-15 Two charges held a fixed distance r apart.

We can calculate that work with Eq. 24-8 by dropping the minus sign (so that the equation gives the work *we* do rather than the field's work) and substituting q_2 for the general charge q. Our work is then equal to q_2V, where V is the potential that has been set up by q_1 at the point where we put q_2. From Eq. 24-26, that potential is

$$V = \frac{1}{4\pi\varepsilon_0}\frac{q_1}{r}.$$

Thus, from our definition, the electric potential energy of the pair of point charges of Fig. 24-15 is

$$U = W = q_2 V = \frac{1}{4\pi\varepsilon_0}\frac{q_1 q_2}{r}. \qquad (24\text{-}43)$$

If the charges have the same sign, we have to do positive work to push them together against their mutual repulsion. Hence, as Eq. 24-43 shows, the potential energy of the system is then positive. If the charges have opposite signs, we have to do negative work against their mutual attraction to bring them together if they are to be stationary. The potential energy of the system is then negative.

Sample Problem

Potential energy of a system of three charged particles

Figure 24-16 shows three point charges held in fixed positions by forces that are not shown. What is the electric potential energy U of this system of charges? Assume that $d = 12$ cm and that

$$q_1 = +q, \quad q_2 = -4q, \quad \text{and} \quad q_3 = +2q,$$

in which $q = 150$ nC.

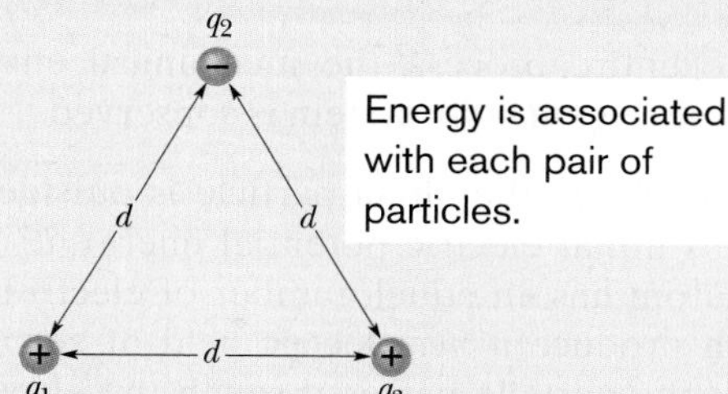

Fig. 24-16 Three charges are fixed at the vertices of an equilateral triangle. What is the electric potential energy of the system?

KEY IDEA

The potential energy U of the system is equal to the work we must do to assemble the system, bringing in each charge from an infinite distance.

Calculations: Let's mentally build the system of Fig. 24-16, starting with one of the point charges, say q_1, in place and the others at infinity. Then we bring another one, say q_2, in from infinity and put it in place. From Eq. 24-43 with d substituted for r, the potential energy U_{12} associated with the pair of point charges q_1 and q_2 is

$$U_{12} = \frac{1}{4\pi\varepsilon_0}\frac{q_1 q_2}{d}.$$

We then bring the last point charge q_3 in from infinity and put it in place. The work that we must do in this last step is equal to the sum of the work we must do to bring q_3 near q_1 and the work we must do to bring it near q_2. From Eq. 24-43, with d substituted for r, that sum is

$$W_{13} + W_{23} = U_{13} + U_{23} = \frac{1}{4\pi\varepsilon_0}\frac{q_1 q_3}{d} + \frac{1}{4\pi\varepsilon_0}\frac{q_2 q_3}{d}.$$

The total potential energy U of the three-charge system is the sum of the potential energies associated with the three pairs of charges. This sum (which is actually independent of the order in which the charges are brought together) is

$$U = U_{12} + U_{13} + U_{23}$$

$$= \frac{1}{4\pi\varepsilon_0}\left(\frac{(+q)(-4q)}{d} + \frac{(+q)(+2q)}{d} + \frac{(-4q)(+2q)}{d}\right)$$

$$= -\frac{10q^2}{4\pi\varepsilon_0 d}$$

$$= -\frac{(8.99 \times 10^9\ \text{N}\cdot\text{m}^2/\text{C}^2)(10)(150 \times 10^{-9}\ \text{C})^2}{0.12\ \text{m}}$$

$$= -1.7 \times 10^{-2}\ \text{J} = -17\ \text{mJ}. \qquad \text{(Answer)}$$

The negative potential energy means that negative work would have to be done to assemble this structure, starting with the three charges infinitely separated and at rest. Put another way, an external agent would have to do 17 mJ of work to disassemble the structure completely, ending with the three charges infinitely far apart.

Additional examples, video, and practice available at *WileyPLUS*

Sample Problem

Conservation of mechanical energy with electric potential energy

An alpha particle (two protons, two neutrons) moves into a stationary gold atom (79 protons, 118 neutrons), passing through the electron region that surrounds the gold nucleus like a shell and headed directly toward the nucleus (Fig. 24-17). The alpha particle slows until it momentarily stops when its center is at radial distance $r = 9.23$ fm from the nuclear center. Then it moves back along its incoming path. (Because the gold nucleus is much more massive than the alpha particle, we can assume the gold nucleus does not move.) What was the kinetic energy K_i of the alpha particle when it was initially far away (hence external to the gold atom)? Assume that the only force acting between the alpha particle and the gold nucleus is the (electrostatic) Coulomb force.

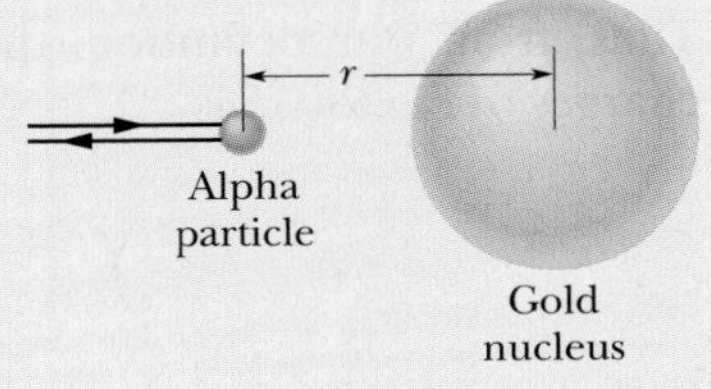

Fig. 24-17 An alpha particle, traveling head-on toward the center of a gold nucleus, comes to a momentary stop (at which time all its kinetic energy has been transferred to electric potential energy) and then reverses its path.

KEY IDEA

During the entire process, the mechanical energy of the *alpha particle + gold atom* system is conserved.

Reasoning: When the alpha particle is outside the atom, the system's initial electric potential energy U_i is zero because the atom has an equal number of electrons and protons, which produce a *net* electric field of zero. However, once the alpha particle passes through the electron region surrounding the nucleus on its way to the nucleus, the electric field due to the electrons goes to zero. The reason is that the electrons act like a closed spherical shell of uniform negative charge and, as discussed in Section 23-9, such a shell produces zero electric field in the space it encloses. The alpha particle still experiences the electric field of the protons in the nucleus, which produces a repulsive force on the protons within the alpha particle.

As the incoming alpha particle is slowed by this repulsive force, its kinetic energy is transferred to electric potential energy of the system. The transfer is complete when the alpha particle momentarily stops and the kinetic energy is $K_f = 0$.

Calculations: The principle of conservation of mechanical energy tells us that

$$K_i + U_i = K_f + U_f. \tag{24-44}$$

We know two values: $U_i = 0$ and $K_f = 0$. We also know that the potential energy U_f at the stopping point is given by the right side of Eq. 24-43, with $q_1 = 2e$, $q_2 = 79e$ (in which e is the elementary charge, 1.60×10^{-19} C), and $r = 9.23$ fm. Thus, we can rewrite Eq. 24-44 as

$$K_i = \frac{1}{4\pi\varepsilon_0}\frac{(2e)(79e)}{9.23 \text{ fm}}$$

$$= \frac{(8.99 \times 10^9 \text{ N}\cdot\text{m}^2/\text{C}^2)(158)(1.60 \times 10^{-19}\text{ C})^2}{9.23 \times 10^{-15}\text{ m}}$$

$$= 3.94 \times 10^{-12}\text{ J} = 24.6 \text{ MeV}. \qquad \text{(Answer)}$$

Additional examples, video, and practice available at *WileyPLUS*

24-12 Potential of a Charged Isolated Conductor

In Section 23-6, we concluded that $\vec{E} = 0$ for all points inside an isolated conductor. We then used Gauss' law to prove that an excess charge placed on an isolated conductor lies entirely on its surface. (This is true even if the conductor has an empty internal cavity.) Here we use the first of these facts to prove an extension of the second:

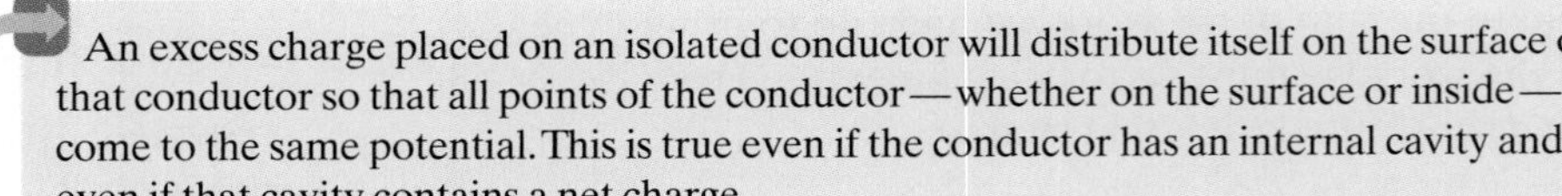
An excess charge placed on an isolated conductor will distribute itself on the surface of that conductor so that all points of the conductor—whether on the surface or inside—come to the same potential. This is true even if the conductor has an internal cavity and even if that cavity contains a net charge.

Our proof follows directly from Eq. 24-18, which is

$$V_f - V_i = -\int_i^f \vec{E}\cdot d\vec{s}.$$

Since $\vec{E} = 0$ for all points within a conductor, it follows directly that $V_f = V_i$ for all possible pairs of points i and f in the conductor.

Figure 24-18*a* is a plot of potential against radial distance r from the center for an isolated spherical conducting shell of 1.0 m radius, having a charge of 1.0 μC. For points outside the shell, we can calculate $V(r)$ from Eq. 24-26 because the charge q behaves for such external points as if it were concentrated at the center of the shell. That equation holds right up to the surface of the shell. Now let us push a small test charge through the shell—assuming a small hole exists—to its center. No extra work is needed to do this because no net electric force acts on the test charge once it is inside the shell. Thus, the potential at all points inside the shell has the same value as that on the surface, as Fig. 24-18*a* shows.

Figure 24-18*b* shows the variation of electric field with radial distance for the same shell. Note that $E = 0$ everywhere inside the shell. The curves of Fig. 24-18*b* can be derived from the curve of Fig. 24-18*a* by differentiating with respect to r, using Eq. 24-40 (recall that the derivative of any constant is zero). The curve of Fig. 24-18*a* can be derived from the curves of Fig. 24-18*b* by integrating with respect to r, using Eq. 24-19.

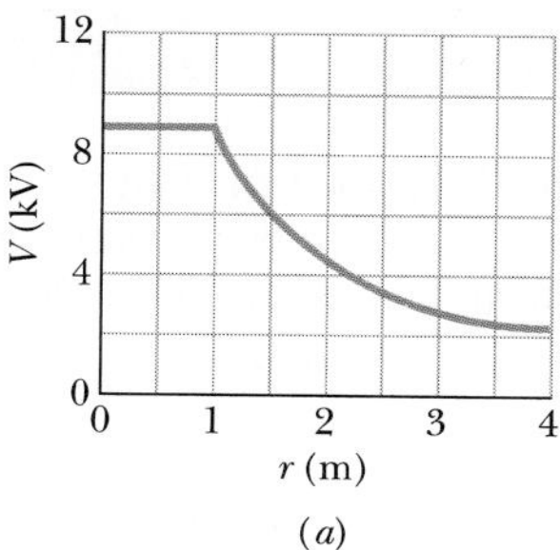

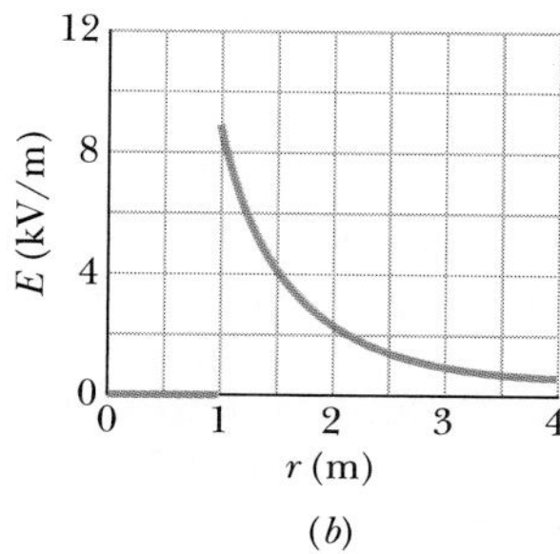

Fig. 24-18 (*a*) A plot of $V(r)$ both inside and outside a charged spherical shell of radius 1.0 m. (*b*) A plot of $E(r)$ for the same shell.

Fig. 24-19 A large spark jumps to a car's body and then exits by moving across the insulating left front tire (note the flash there), leaving the person inside unharmed. *(Courtesy Westinghouse Electric Corporation)*

Spark Discharge from a Charged Conductor

On nonspherical conductors, a surface charge does not distribute itself uniformly over the surface of the conductor. At sharp points or sharp edges, the surface charge density—and thus the external electric field, which is proportional to it—may reach very high values. The air around such sharp points or edges may become ionized, producing the corona discharge that golfers and mountaineers see on the tips of bushes, golf clubs, and rock hammers when thunderstorms threaten. Such corona discharges, like hair that stands on end, are often the precursors of lightning strikes. In such circumstances, it is wise to enclose yourself in a cavity inside a conducting shell, where the electric field is guaranteed to be zero. A car (unless it is a convertible or made with a plastic body) is almost ideal (Fig. 24-19).

Isolated Conductor in an External Electric Field

If an isolated conductor is placed in an *external electric field,* as in Fig. 24-20, all points of the conductor still come to a single potential regardless of whether the conductor has an excess charge. The free conduction electrons distribute themselves on the surface in such a way that the electric field they produce at interior points cancels the external electric field that would otherwise be there. Furthermore, the electron distribution causes the net electric field at all points on the surface to be perpendicular to the surface. If the conductor in Fig. 24-20 could be somehow removed, leaving the surface charges frozen in place, the internal and external electric field would remain absolutely unchanged.

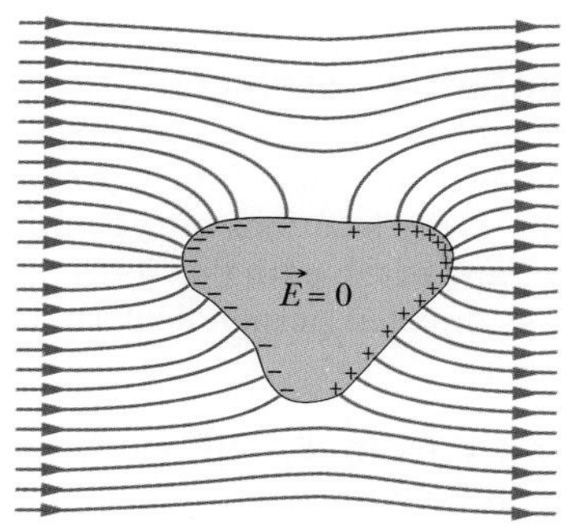

Fig. 24-20 An uncharged conductor is suspended in an external electric field. The free electrons in the conductor distribute themselves on the surface as shown, so as to reduce the net electric field inside the conductor to zero and make the net field at the surface perpendicular to the surface.

REVIEW & SUMMARY

Electric Potential Energy The change ΔU in the electric potential energy U of a point charge as the charge moves from an initial point i to a final point f in an electric field is

$$\Delta U = U_f - U_i = -W, \tag{24-1}$$

where W is the work done by the electrostatic force (due to the external electric field) on the point charge during the move from i to f. If the potential energy is defined to be zero at infinity, the **electric potential energy** U of the point charge at a particular point is

$$U = -W_\infty. \tag{24-2}$$

Here W_∞ is the work done by the electrostatic force on the point charge as the charge moves from infinity to the particular point.

Electric Potential Difference and Electric Potential We define the **potential difference** ΔV between two points i and f in an electric field as

$$\Delta V = V_f - V_i = -\frac{W}{q}, \tag{24-7}$$

where q is the charge of a particle on which work W is done by the electric field as the particle moves from point i to point f. The **potential** at a point is defined as

$$V = -\frac{W_\infty}{q}. \tag{24-8}$$

Here W_∞ is the work done on the particle by the electric field as the particle moves in from infinity to the point. The SI unit of potential is the *volt:* 1 volt = 1 joule per coulomb.

Potential and potential difference can also be written in terms of the electric potential energy U of a particle of charge q in an electric field:

$$V = \frac{U}{q}, \tag{24-5}$$

$$\Delta V = V_f - V_i = \frac{U_f}{q} - \frac{U_i}{q} = \frac{\Delta U}{q}. \tag{24-6}$$

Equipotential Surfaces The points on an **equipotential surface** all have the same electric potential. The work done on a test charge in moving it from one such surface to another is independent of the locations of the initial and final points on these surfaces and of the path that joins the points. The electric field $\vec{E}$ is always directed perpendicularly to corresponding equipotential surfaces.

Finding V from $\vec{E}$ The electric potential difference between two points i and f is

$$V_f - V_i = -\int_i^f \vec{E} \cdot d\vec{s}, \tag{24-18}$$

where the integral is taken over any path connecting the points. If the integration is difficult along any particular path, we can choose a different path along which the integration might be easier. If we choose $V_i = 0$, we have, for the potential at a particular point,

$$V = -\int_i^f \vec{E} \cdot d\vec{s}. \tag{24-19}$$

Potential Due to Point Charges The electric potential due to a single point charge at a distance r from that point charge is

$$V = \frac{1}{4\pi\varepsilon_0}\frac{q}{r}, \tag{24-26}$$

where V has the same sign as q. The potential due to a collection of point charges is

$$V = \sum_{i=1}^{n} V_i = \frac{1}{4\pi\varepsilon_0}\sum_{i=1}^{n}\frac{q_i}{r_i}. \tag{24-27}$$

Potential Due to an Electric Dipole At a distance r from an electric dipole with dipole moment magnitude $p = qd$, the electric potential of the dipole is

$$V = \frac{1}{4\pi\varepsilon_0}\frac{p\cos\theta}{r^2} \tag{24-30}$$

for $r \gg d$; the angle θ is defined in Fig. 24-10.

Potential Due to a Continuous Charge Distribution For a continuous distribution of charge, Eq. 24-27 becomes

$$V = \frac{1}{4\pi\varepsilon_0}\int\frac{dq}{r}, \tag{24-32}$$

in which the integral is taken over the entire distribution.

Calculating $\vec{E}$ from V The component of $\vec{E}$ in any direction is the negative of the rate at which the potential changes with distance in that direction:

$$E_s = -\frac{\partial V}{\partial s}. \tag{24-40}$$

The x, y, and z components of $\vec{E}$ may be found from

$$E_x = -\frac{\partial V}{\partial x}; \quad E_y = -\frac{\partial V}{\partial y}; \quad E_z = -\frac{\partial V}{\partial z}. \tag{24-41}$$

When $\vec{E}$ is uniform, Eq. 24-40 reduces to

$$E = -\frac{\Delta V}{\Delta s}, \tag{24-42}$$

where s is perpendicular to the equipotential surfaces. The electric field is zero parallel to an equipotential surface.

Electric Potential Energy of a System of Point Charges The electric potential energy of a system of point charges is equal to the work needed to assemble the system with the charges initially at rest and infinitely distant from each other. For two charges at separation r,

$$U = W = \frac{1}{4\pi\varepsilon_0}\frac{q_1 q_2}{r}. \tag{24-43}$$

Potential of a Charged Conductor An excess charge placed on a conductor will, in the equilibrium state, be located entirely on the outer surface of the conductor. The charge will distribute itself so that the following occur: (1) The entire conductor, including interior points, is at a uniform potential. (2) At every internal point, the electric field due to the charge cancels the external electric field that otherwise would have been there. (3) The net electric field at every point on the surface is perpendicular to the surface.

QUESTIONS

1 In Fig. 24-21, eight particles form a square, with distance d between adjacent particles. What is the electric potential at point P at the center of the square if the electric potential is zero at infinity?

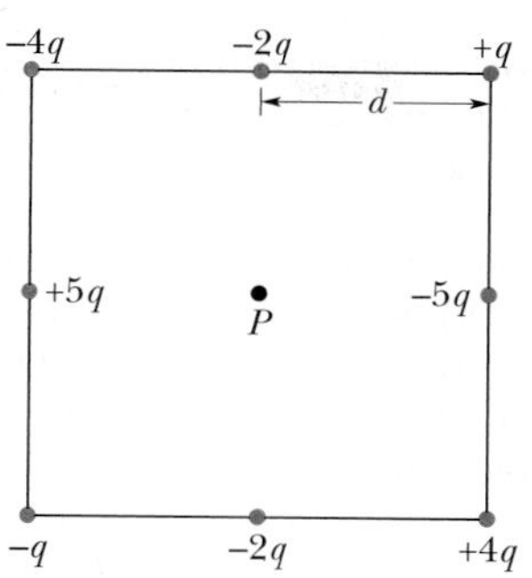

Fig. 24-21 Question 1.

2 Figure 24-22 shows three sets of cross sections of equipotential surfaces; all three cover the same size region of space. (a) Rank the arrangements according to the magnitude of the electric field present in the region, greatest first. (b) In which is the electric field directed down the page?

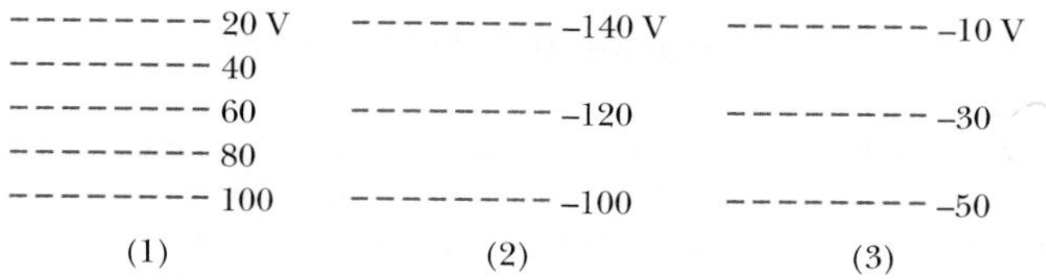

Fig. 24-22 Question 2.

3 Figure 24-23 shows four pairs of charged particles. For each pair, let $V = 0$ at infinity and consider V_{net} at points on the x axis. For which pairs is there a point at which $V_{net} = 0$ (a) between the particles and (b) to the right of the particles? (c) At such a point is $\vec{E}_{net}$ due to the particles equal to zero? (d) For each pair, are there off-axis points (other than at infinity) where $V_{net} = 0$?

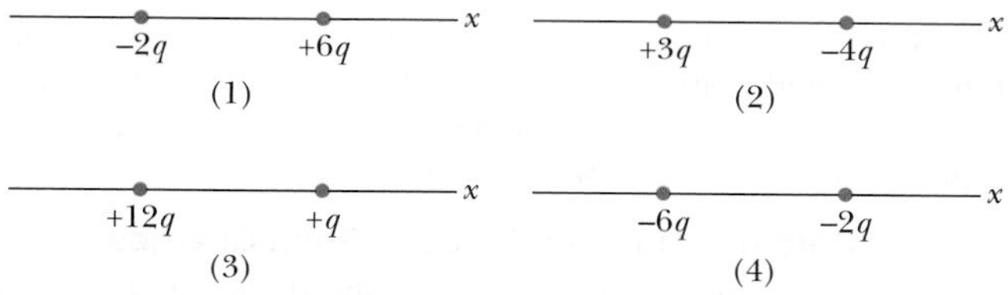

Fig. 24-23 Questions 3 and 9.

4 Figure 24-24 gives the electric potential V as a function of x. (a) Rank the five regions according to the magnitude of the x component of the electric field within them, greatest first. What is the direction of the field along the x axis in (b) region 2 and (c) region 4?

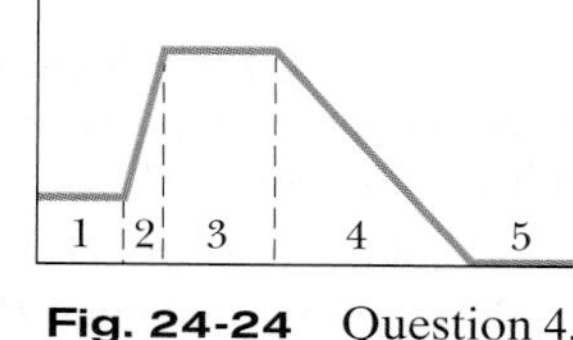

Fig. 24-24 Question 4.

5 Figure 24-25 shows three paths along which we can move the positively charged sphere A closer to positively charged sphere B, which is held fixed in place. (a) Would sphere A be moved to a higher or lower electric potential? Is the work done (b) by our force and (c) by the electric field due to B positive, negative, or zero? (d) Rank the paths according to the work our force does, greatest first.

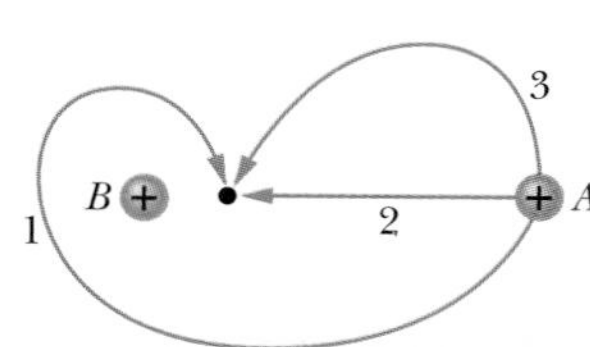

Fig. 24-25 Question 5.

6 Figure 24-26 shows four arrangements of charged particles, all the same distance from the origin. Rank the situations according to the net electric potential at the origin, most positive first. Take the potential to be zero at infinity.

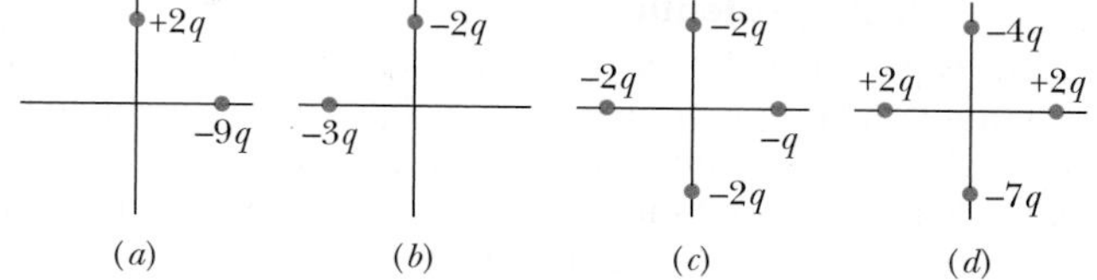

Fig. 24-26 Question 6.

7 Figure 24-27 shows a system of three charged particles. If you move the particle of charge $+q$ from point A to point D, are the following quantities positive, negative, or zero: (a) the change in the electric potential energy of the three-particle system, (b) the work done by the net electrostatic force on the particle you moved (that is, the net force due to the other two particles), and (c) the work done by your force? (d) What are the answers to (a) through (c) if, instead, the particle is moved from B to C?

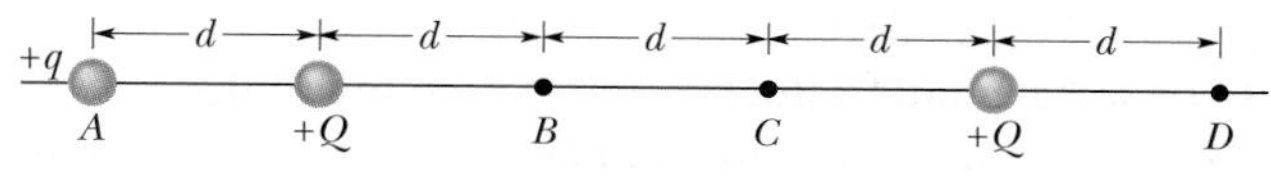

Fig. 24-27 Questions 7 and 8.

8 In the situation of Question 7, is the work done by your force positive, negative, or zero if the particle is moved (a) from A to B, (b) from A to C, and (c) from B to D? (d) Rank those moves according to the magnitude of the work done by your force, greatest first.

9 Figure 24-23 shows four pairs of charged particles with identical separations. (a) Rank the pairs according to their electric potential energy (that is, the energy of the two-particle system), greatest (most positive) first. (b) For each pair, if the separation between the particles is increased, does the potential energy of the pair increase or decrease?

10 (a) In Fig. 24-28a, what is the potential at point P due to charge Q at distance R from P? Set $V = 0$ at infinity. (b) In Fig. 24-28b, the same charge Q has been spread uniformly over a circular arc of radius R and central angle 40°. What is the potential at point P, the center of curvature of the arc? (c) In Fig. 24-28c, the same charge Q has been spread uniformly over a circle of radius R. What is the potential at point P, the center of the circle? (d) Rank the three situations according to the magnitude of the electric field that is set up at P, greatest first.

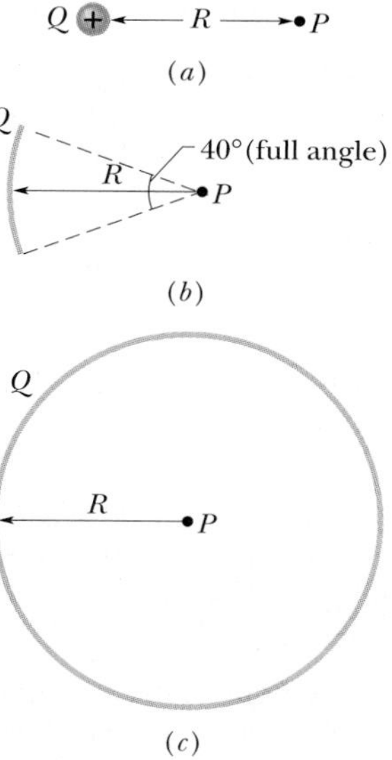

Fig. 24-28 Question 10.

PROBLEMS

Tutoring problem available (at instructor's discretion) in *WileyPLUS* and WebAssign

SSM Worked-out solution available in Student Solutions Manual

• – ••• Number of dots indicates level of problem difficulty

WWW Worked-out solution is at

ILW Interactive solution is at

http://www.wiley.com/college/halliday

Additional information available in *The Flying Circus of Physics* and at flyingcircusofphysics.com

sec. 24-3 Electric Potential

•1 SSM A particular 12 V car battery can send a total charge of 84 A · h (ampere-hours) through a circuit, from one terminal to the other. (a) How many coulombs of charge does this represent? (*Hint:* See Eq. 21-3.) (b) If this entire charge undergoes a change in electric potential of 12 V, how much energy is involved?

•2 The electric potential difference between the ground and a cloud in a particular thunderstorm is 1.2×10^9 V. In the unit electron-volts, what is the magnitude of the change in the electric potential energy of an electron that moves between the ground and the cloud?

•3 Much of the material making up Saturn's rings is in the form of tiny dust grains having radii on the order of 10^{-6} m. These grains are located in a region containing a dilute ionized gas, and they pick up excess electrons. As an approximation, suppose each grain is spherical, with radius $R = 1.0 \times 10^{-6}$ m. How many electrons would one grain have to pick up to have a potential of -400 V on its surface (taking $V = 0$ at infinity)?

sec. 24-5 Calculating the Potential from the Field

•4 Two large, parallel, conducting plates are 12 cm apart and have charges of equal magnitude and opposite sign on their facing surfaces. An electrostatic force of 3.9×10^{-15} N acts on an electron placed anywhere between the two plates. (Neglect fringing.) (a) Find the electric field at the position of the electron. (b) What is the potential difference between the plates?

•5 SSM An infinite nonconducting sheet has a surface charge density $\sigma = 0.10\ \mu\text{C/m}^2$ on one side. How far apart are equipotential surfaces whose potentials differ by 50 V?

•6 When an electron moves from A to B along an electric field line in Fig. 24-29, the electric field does 3.94×10^{-19} J of work on it. What are the electric potential differences (a) $V_B - V_A$, (b) $V_C - V_A$, and (c) $V_C - V_B$?

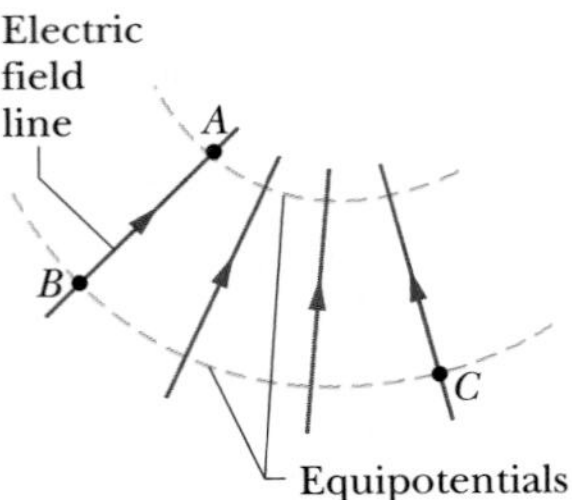

Fig. 24-29 Problem 6.

••7 The electric field in a region of space has the components $E_y = E_z = 0$ and $E_x = (4.00\ \text{N/C})x$. Point A is on the y axis at $y = 3.00$ m, and point B is on the x axis at $x = 4.00$ m. What is the potential difference $V_B - V_A$?

••8 A graph of the x component of the electric field as a function of x in a region of space is shown in Fig. 24-30. The scale of the vertical axis is set by $E_{xs} = 20.0$ N/C. The y and z components of the electric field are zero in this region. If the electric potential at the origin is 10 V, (a) what is the electric potential at $x = 2.0$ m, (b) what is the greatest positive value of the electric potential for points on the x axis for which $0 \le x \le 6.0$ m, and (c) for what value of x is the electric potential zero?

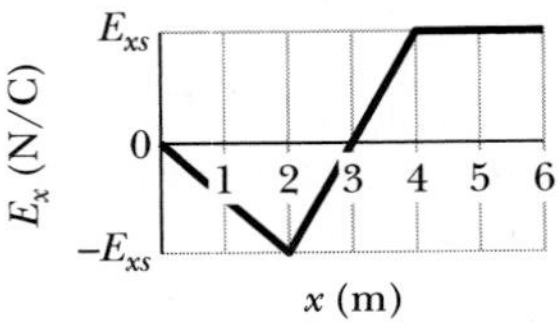

Fig. 24-30 Problem 8.

••9 An infinite nonconducting sheet has a surface charge density $\sigma = +5.80\ \text{pC/m}^2$. (a) How much work is done by the electric field due to the sheet if a particle of charge $q = +1.60 \times 10^{-19}$ C is moved from the sheet to a point P at distance $d = 3.56$ cm from the sheet? (b) If the electric potential V is defined to be zero on the sheet, what is V at P?

•••10 Two uniformly charged, infinite, nonconducting planes are parallel to a yz plane and positioned at $x = -50$ cm and $x = +50$ cm. The charge densities on the planes are -50nC/m^2 and $+25\ \text{nC/m}^2$, respectively. What is the magnitude of the potential difference between the origin and the point on the x axis at $x = +80$ cm? (*Hint:* Use Gauss' law.)

•••11 A nonconducting sphere has radius $R = 2.31$ cm and uniformly distributed charge $q = +3.50$ fC. Take the electric potential at the sphere's center to be $V_0 = 0$. What is V at radial distance (a) $r = 1.45$ cm and (b) $r = R$. (*Hint:* See Section 23-9.)

sec. 24-7 Potential Due to a Group of Point Charges

•12 As a space shuttle moves through the dilute ionized gas of Earth's ionosphere, the shuttle's potential is typically changed by -1.0 V during one revolution. Assuming the shuttle is a sphere of radius 10 m, estimate the amount of charge it collects.

•13 What are (a) the charge and (b) the charge density on the surface of a conducting sphere of radius 0.15 m whose potential is 200 V (with $V = 0$ at infinity)?

•14 Consider a point charge $q = 1.0\ \mu\text{C}$, point A at distance $d_1 = 2.0$ m from q, and point B at distance $d_2 = 1.0$ m. (a) If A and B are diametrically opposite each other, as in Fig. 24-31a, what is the elec-

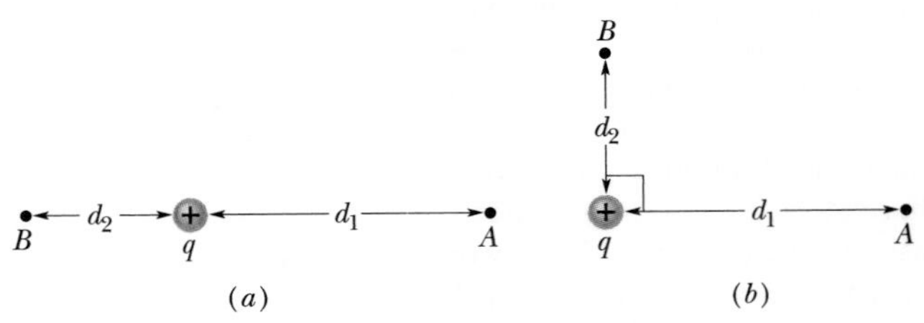

Fig. 24-31 Problem 14.

tric potential difference $V_A - V_B$? (b) What is that electric potential difference if A and B are located as in Fig. 24-31b?

••15 SSM ILW A spherical drop of water carrying a charge of 30 pC has a potential of 500 V at its surface (with $V = 0$ at infinity). (a) What is the radius of the drop? (b) If two such drops of the same charge and radius combine to form a single spherical drop, what is the potential at the surface of the new drop?

••16 GO Figure 24-32 shows a rectangular array of charged particles fixed in place, with distance $a = 39.0$ cm and the charges shown as integer multiples of $q_1 = 3.40$ pC and $q_2 = 6.00$ pC. With $V = 0$ at infinity, what is the net electric potential at the rectangle's center? (*Hint:* Thoughtful examination can reduce the calculation.)

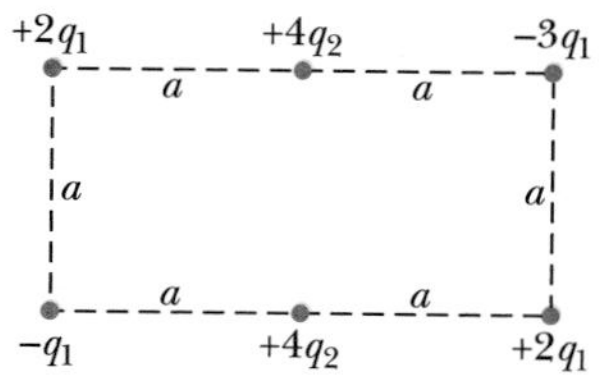

Fig. 24-32 Problem 16.

••17 GO In Fig. 24-33, what is the net electric potential at point P due to the four particles if $V = 0$ at infinity, $q = 5.00$ fC, and $d = 4.00$ cm?

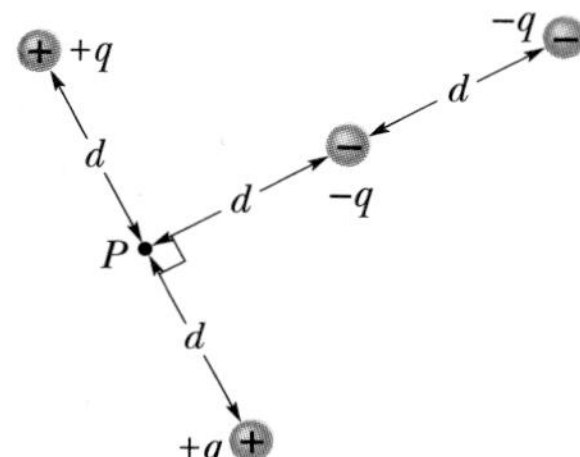

Fig. 24-33 Problem 17.

••18 GO Two charged particles are shown in Fig. 24-34a. Particle 1, with charge q_1, is fixed in place at distance d. Particle 2, with charge q_2, can be moved along the x axis. Figure 24-34b gives the net electric potential V at the origin due to the two particles as a function of the x coordinate of particle 2. The scale of the x axis is set by $x_s = 16.0$ cm. The plot has an asymptote of $V = 5.76 \times 10^{-7}$ V as $x \to \infty$. What is q_2 in terms of e?

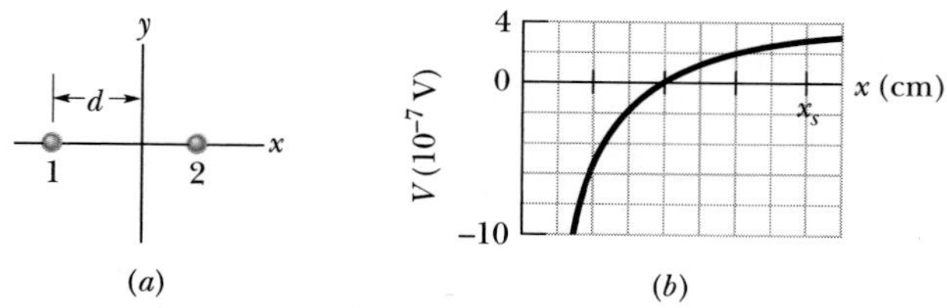

Fig. 24-34 Problem 18.

••19 In Fig. 24-35, particles with the charges $q_1 = +5e$ and $q_2 = -15e$ are fixed in place with a separation of $d = 24.0$ cm. With electric potential defined to be $V = 0$ at infinity, what are the finite (a) positive and (b) negative values of x at which the net electric potential on the x axis is zero?

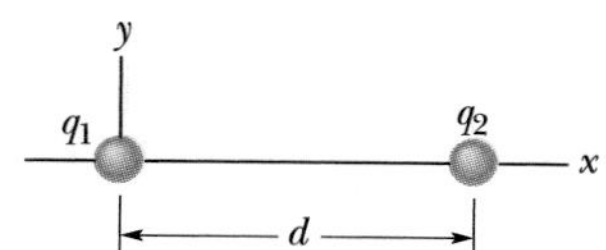

Fig. 24-35 Problems 19, 20, and 97.

••20 Two particles, of charges q_1 and q_2, are separated by distance d in Fig. 24-35. The net electric field due to the particles is zero at $x = d/4$. With $V = 0$ at infinity, locate (in terms of d) any point on the x axis (other than at infinity) at which the electric potential due to the two particles is zero.

sec. 24-8 Potential Due to an Electric Dipole

•21 ILW The ammonia molecule NH_3 has a permanent electric dipole moment equal to 1.47 D, where 1 D = 1 debye unit = 3.34×10^{-30} C·m. Calculate the electric potential due to an ammonia molecule at a point 52.0 nm away along the axis of the dipole. (Set $V = 0$ at infinity.)

••22 In Fig. 24-36a, a particle of elementary charge $+e$ is initially at coordinate $z = 20$ nm on the dipole axis (here a z axis) through an electric dipole, on the positive side of the dipole. (The origin of z is at the center of the dipole.) The particle is then moved along a circular path around the dipole center until it is at coordinate $z = -20$ nm, on the negative side of the dipole axis. Figure 24-36b gives the work W_a done by the force moving the particle versus the angle θ that locates the particle relative to the positive direction of the z axis. The scale of the vertical axis is set by $W_{as} = 4.0 \times 10^{-30}$ J. What is the magnitude of the dipole moment?

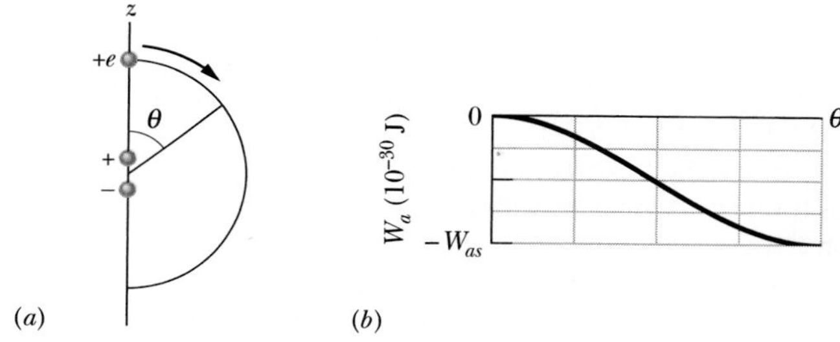

Fig. 24-36 Problem 22.

sec. 24-9 Potential Due to a Continuous Charge Distribution

•23 (a) Figure 24-37a shows a nonconducting rod of length $L = 6.00$ cm and uniform linear charge density $\lambda = +3.68$ pC/m. Assume that the electric potential is defined to be $V = 0$ at infinity. What is V at point P at distance $d = 8.00$ cm along the rod's perpendicular bisector? (b) Figure 24-37b shows an identical rod except that one half is now negatively charged. Both halves have a linear charge density of magnitude 3.68 pC/m. With $V = 0$ at infinity, what is V at P?

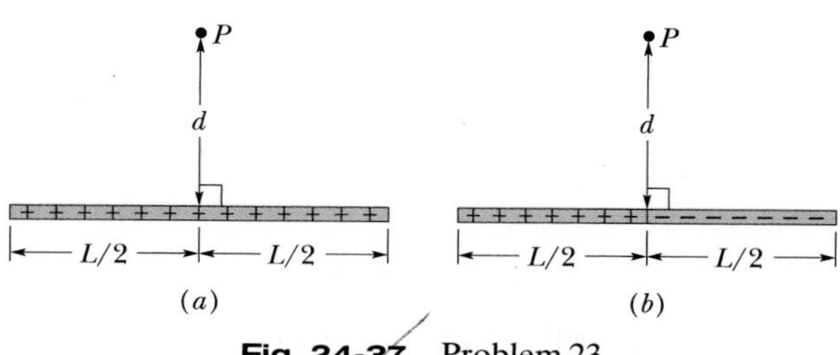

Fig. 24-37 Problem 23.

•24 In Fig. 24-38, a plastic rod having a uniformly distributed charge $Q = -25.6$ pC has been bent into a circular arc of radius $R =$ 3.71 cm and central angle $\phi = 120°$. With $V =$ 0 at infinity, what is the electric potential at P, the center of curvature of the rod?

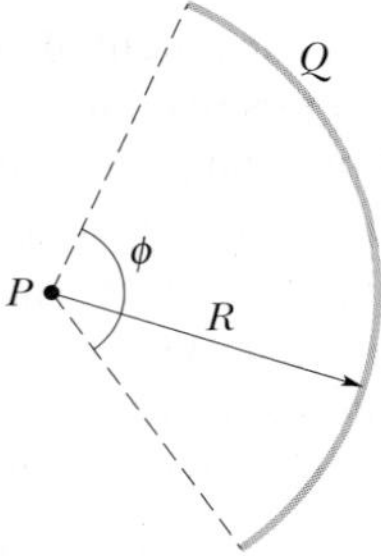

Fig. 24-38 Problem 24.

•25 A plastic rod has been bent into a circle of radius $R = 8.20$ cm. It has a charge $Q_1 =$ +4.20 pC uniformly distributed along one-quarter of its circumference and a charge $Q_2 = -6Q_1$ uniformly distributed along the rest of the circumference (Fig. 24-39). With $V = 0$ at infinity, what is the electric potential at (a) the center C of the circle and (b) point P, on the central axis of the circle at distance $D = 6.71$ cm from the center?

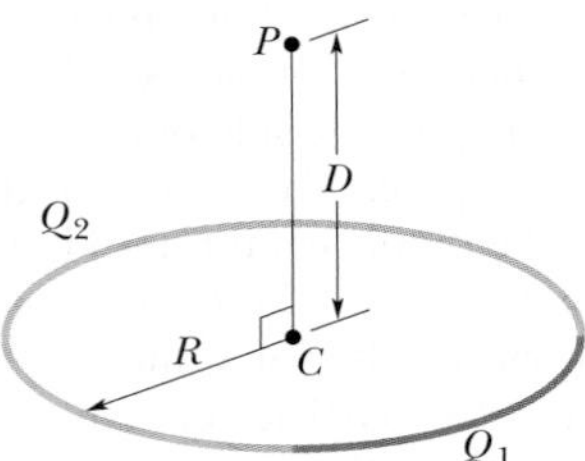

Fig. 24-39 Problem 25.

••26 GO Figure 24-40 shows a thin rod with a uniform charge density of 2.00 μC/m. Evaluate the electric potential at point P if $d = D = L/4.00$.

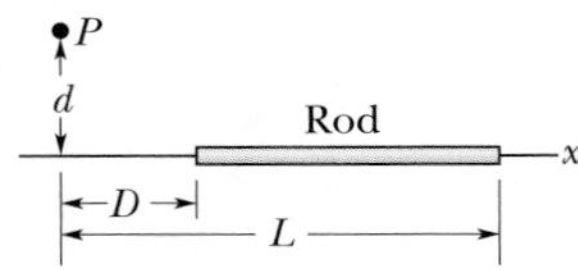

Fig. 24-40 Problem 26.

••27 In Fig. 24-41, three thin plastic rods form quarter-circles with a common center of curvature at the origin. The uniform charges on the rods are $Q_1 = +30$ nC, $Q_2 = +3.0Q_1$, and $Q_3 = -8.0Q_1$. What is the net electric potential at the origin due to the rods?

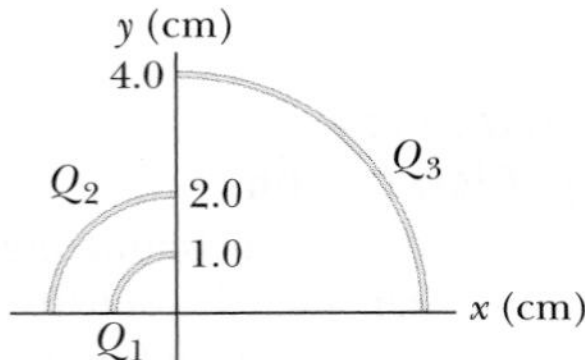

Fig. 24-41 Problem 27.

••28 GO Figure 24-42 shows a thin plastic rod of length $L = 12.0$ cm and uniform positive charge $Q = 56.1$ fC lying on an x axis. With $V = 0$ at infinity, find the electric potential at point P_1 on the axis, at distance $d = 2.50$ cm from one end of the rod.

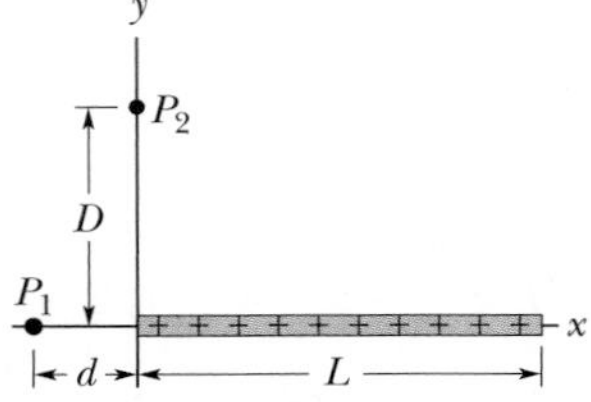

Fig. 24-42 Problems 28, 33, 38, and 40.

••29 In Fig. 24-43, what is the net electric potential at the origin due to the circular arc of charge $Q_1 = +7.21$ pC and the two particles of charges $Q_2 = 4.00Q_1$ and $Q_3 = -2.00Q_1$? The arc's center of curvature is at the origin and its radius is $R = 2.00$ m; the angle indicated is $\theta = 20.0°$.

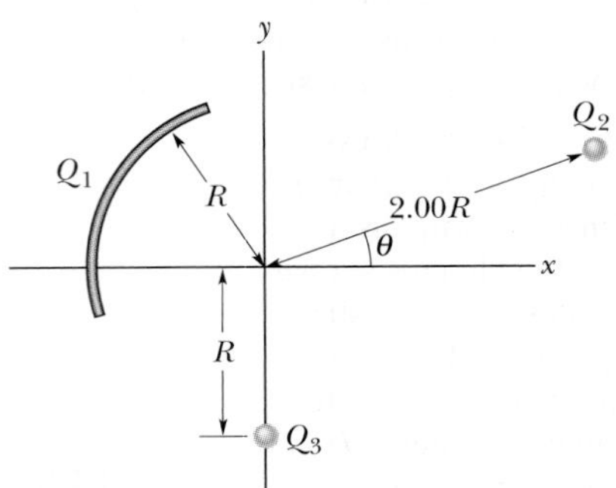

Fig. 24-43 Problem 29.

••30 GO The smiling face of Fig. 24-44 consists of three items:

1. a thin rod of charge -3.0 μC that forms a full circle of radius 6.0 cm;
2. a second thin rod of charge 2.0 μC that forms a circular arc of radius 4.0 cm, subtending an angle of 90° about the center of the full circle;
3. an electric dipole with a dipole moment that is perpendicular to a radial line and has magnitude 1.28×10^{-21} C·m.

What is the net electric potential at the center?

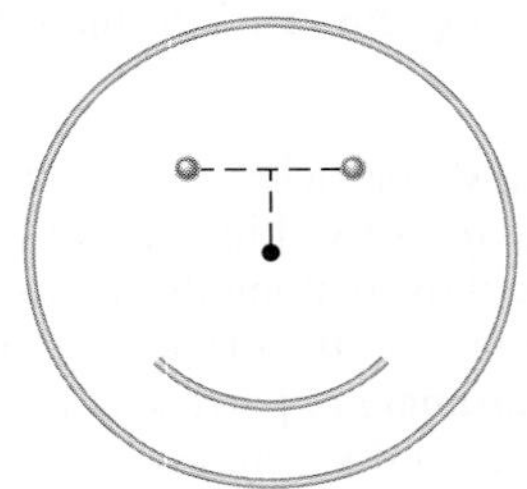

Fig. 24-44 Problem 30.

••31 SSM WWW A plastic disk of radius $R = 64.0$ cm is charged on one side with a uniform surface charge density $\sigma = 7.73$ fC/m², and then three quadrants of the disk are removed. The remaining quadrant is shown in Fig. 24-45. With $V = 0$ at infinity, what is the potential due to the remaining quadrant at point P, which is on the central axis of the original disk at distance $D = 25.9$ cm from the original center?

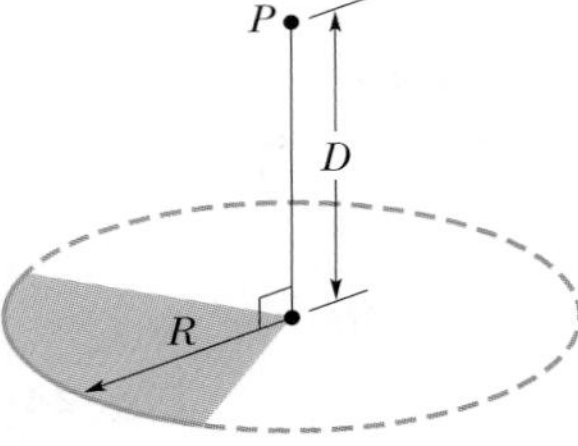

Fig. 24-45 Problem 31.

•••32 A nonuniform linear charge distribution given by $\lambda = bx$, where b is a constant, is located along an x axis from $x = 0$ to $x =$ 0.20 m. If $b = 20$ nC/m² and $V = 0$ at infinity, what is the electric potential at (a) the origin and (b) the point $y = 0.15$ m on the y axis?

•••33 The thin plastic rod shown in Fig. 24-42 has length $L = 12.0$ cm and a nonuniform linear charge density $\lambda = cx$, where $c = 28.9$

pC/m^2. With $V = 0$ at infinity, find the electric potential at point P_1 on the axis, at distance $d = 3.00$ cm from one end.

sec. 24-10 Calculating the Field from the Potential

•34 Two large parallel metal plates are 1.5 cm apart and have charges of equal magnitudes but opposite signs on their facing surfaces. Take the potential of the negative plate to be zero. If the potential halfway between the plates is then +5.0 V, what is the electric field in the region between the plates?

•35 The electric potential at points in an xy plane is given by $V = (2.0\ \text{V/m}^2)x^2 - (3.0\ \text{V/m}^2)y^2$. In unit-vector notation, what is the electric field at the point (3.0 m, 2.0 m)?

•36 The electric potential V in the space between two flat parallel plates 1 and 2 is given (in volts) by $V = 1500x^2$, where x (in meters) is the perpendicular distance from plate 1. At $x = 1.3$ cm, (a) what is the magnitude of the electric field and (b) is the field directed toward or away from plate 1?

••37 SSM What is the magnitude of the electric field at the point $(3.00\hat{i} - 2.00\hat{j} + 4.00\hat{k})$ m if the electric potential is given by $V = 2.00xyz^2$, where V is in volts and x, y, and z are in meters?

••38 Figure 24-42 shows a thin plastic rod of length $L = 13.5$ cm and uniform charge 43.6 fC. (a) In terms of distance d, find an expression for the electric potential at point P_1. (b) Next, substitute variable x for d and find an expression for the magnitude of the component E_x of the electric field at P_1. (c) What is the direction of E_x relative to the positive direction of the x axis? (d) What is the value of E_x at P_1 for $x = d = 6.20$ cm? (e) From the symmetry in Fig. 24-42, determine E_y at P_1.

••39 An electron is placed in an xy plane where the electric potential depends on x and y as shown in Fig. 24-46 (the potential does not depend on z). The scale of the vertical axis is set by $V_s = 500$ V. In unit-vector notation, what is the electric force on the electron?

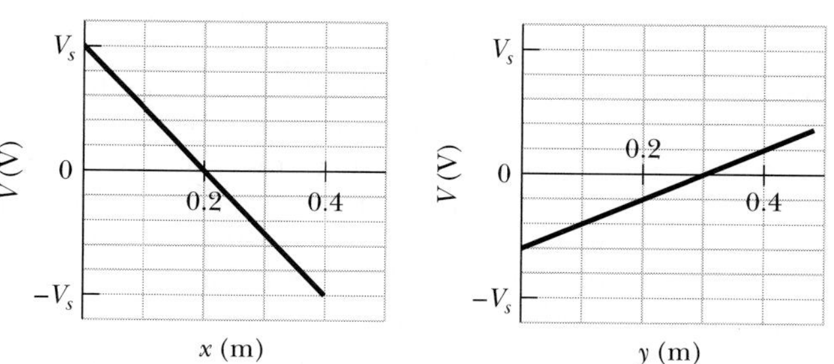

Fig. 24-46 Problem 39.

•••40 The thin plastic rod of length $L = 10.0$ cm in Fig. 24-42 has a nonuniform linear charge density $\lambda = cx$, where $c = 49.9$ pC/m^2. (a) With $V = 0$ at infinity, find the electric potential at point P_2 on the y axis at $y = D = 3.56$ cm. (b) Find the electric field component E_y at P_2. (c) Why cannot the field component E_x at P_2 be found using the result of (a)?

sec. 24-11 Electric Potential Energy of a System of Point Charges

•41 A particle of charge +7.5 μC is released from rest at the point $x = 60$ cm on an x axis. The particle begins to move due to the presence of a charge Q that remains fixed at the origin. What is the kinetic energy of the particle at the instant it has moved 40 cm if (a) $Q = +20$ μC and (b) $Q = -20$ μC?

•42 (a) What is the electric potential energy of two electrons separated by 2.00 nm? (b) If the separation increases, does the potential energy increase or decrease?

•43 SSM ILW WWW How much work is required to set up the arrangement of Fig. 24-47 if $q = 2.30$ pC, $a = 64.0$ cm, and the particles are initially infinitely far apart and at rest?

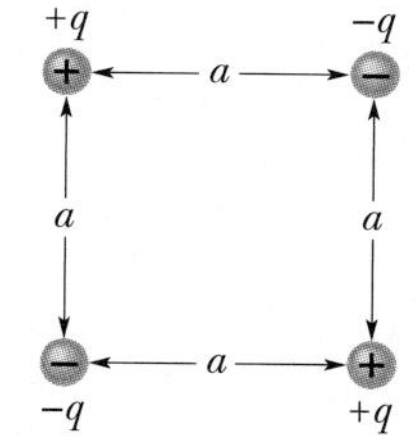

Fig. 24-47 Problem 43.

•44 In Fig. 24-48, seven charged particles are fixed in place to form a square with an edge length of 4.0 cm. How much work must we do to bring a particle of charge $+6e$ initially at rest from an infinite distance to the center of the square?

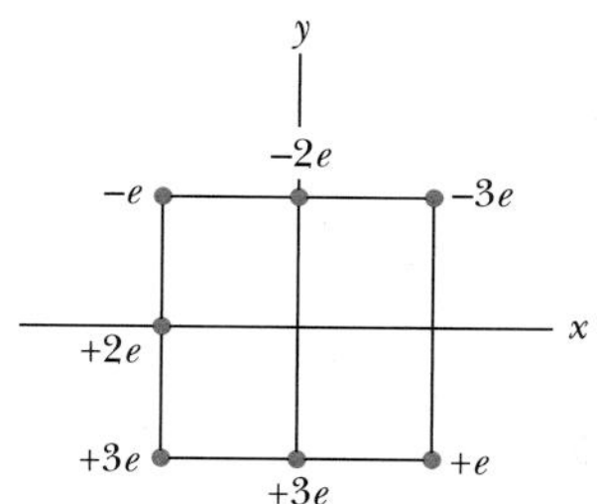

Fig. 24-48 Problem 44.

••45 ILW A particle of charge q is fixed at point P, and a second particle of mass m and the same charge q is initially held a distance r_1 from P. The second particle is then released. Determine its speed when it is a distance r_2 from P. Let $q = 3.1$ μC, $m = 20$ mg, $r_1 = 0.90$ mm, and $r_2 = 2.5$ mm.

••46 A charge of -9.0 nC is uniformly distributed around a thin plastic ring lying in a yz plane with the ring center at the origin. A -6.0 pC point charge is located on the x axis at $x = 3.0$ m. For a ring radius of 1.5 m, how much work must an external force do on the point charge to move it to the origin?

••47 GO What is the *escape speed* for an electron initially at rest on the surface of a sphere with a radius of 1.0 cm and a uniformly distributed charge of 1.6×10^{-15} C? That is, what initial speed must the electron have in order to reach an infinite distance from the sphere and have zero kinetic energy when it gets there?

••48 A thin, spherical, conducting shell of radius R is mounted on an isolating support and charged to a potential of -125 V. An electron is then fired directly toward the center of the shell, from point P at distance r from the center of the shell ($r \gg R$). What initial speed v_0 is needed for the electron to just reach the shell before reversing direction?

••49 GO Two electrons are fixed 2.0 cm apart. Another electron is shot from infinity and stops midway between the two. What is its initial speed?

••50 In Fig. 24-49, how much work must we do to bring a particle, of charge $Q = +16e$ and initially at rest, along the dashed line from infinity to

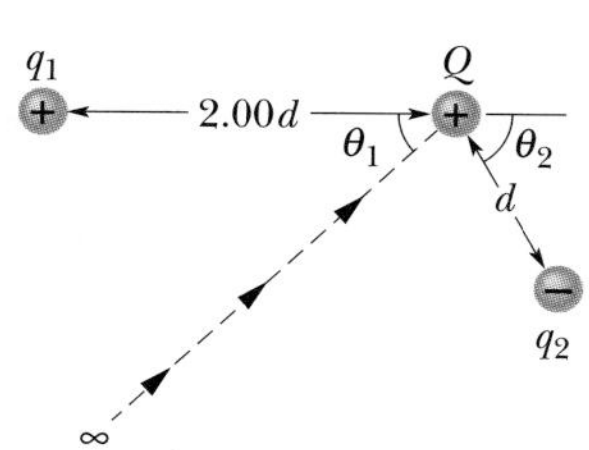

Fig. 24-49 Problem 50.

the indicated point near two fixed particles of charges $q_1 = +4e$ and $q_2 = -q_1/2$? Distance $d = 1.40$ cm, $\theta_1 = 43°$, and $\theta_2 = 60°$.

••51 GO In the rectangle of Fig. 24-50, the sides have lengths 5.0 cm and 15 cm, $q_1 = -5.0\ \mu C$, and $q_2 = +2.0\ \mu C$. With $V = 0$ at infinity, what is the electric potential at (a) corner A and (b) corner B? (c) How much work is required to move a charge $q_3 = +3.0\ \mu C$ from B to A along a diagonal of the rectangle? (d) Does this work increase or decrease the electric potential energy of the three-charge system? Is more, less, or the same work required if q_3 is moved along a path that is (e) inside the rectangle but not on a diagonal and (f) outside the rectangle?

Fig. 24-50 Problem 51.

••52 Figure 24-51a shows an electron moving along an electric dipole axis toward the negative side of the dipole. The dipole is fixed in place. The electron was initially very far from the dipole, with kinetic energy 100 eV. Figure 24-51b gives the kinetic energy K of the electron versus its distance r from the dipole center. The scale of the horizontal axis is set by $r_s = 0.10$ m. What is the magnitude of the dipole moment?

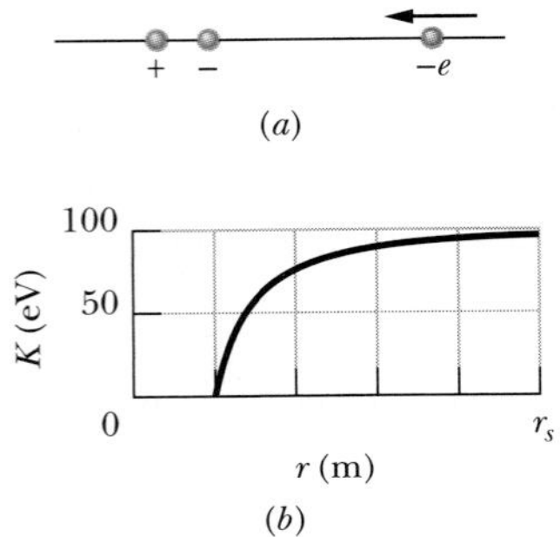

Fig. 24-51 Problem 52.

••53 Two tiny metal spheres A and B, mass $m_A = 5.00$ g and $m_B = 10.0$ g, have equal positive charge $q = 5.00\ \mu C$. The spheres are connected by a massless nonconducting string of length $d = 1.00$ m, which is much greater than the radii of the spheres. (a) What is the electric potential energy of the system? (b) Suppose you cut the string. At that instant, what is the acceleration of each sphere? (c) A long time after you cut the string, what is the speed of each sphere?

••54 A positron (charge $+e$, mass equal to the electron mass) is moving at 1.0×10^7 m/s in the positive direction of an x axis when, at $x = 0$, it encounters an electric field directed along the x axis. The electric potential V associated with the field is given in Fig. 24-52. The scale of the vertical axis is set by $V_s = 500.0$ V. (a) Does the positron emerge from the field at $x = 0$ (which means its motion is reversed) or at $x = 0.50$ m (which means its motion is not reversed)? (b) What is its speed when it emerges?

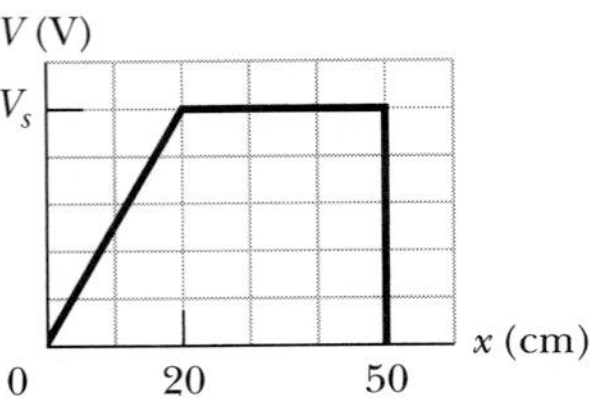

Fig. 24-52 Problem 54.

••55 An electron is projected with an initial speed of 3.2×10^5 m/s directly toward a proton that is fixed in place. If the electron is initially a great distance from the proton, at what distance from the proton is the speed of the electron instantaneously equal to twice the initial value?

••56 Figure 24-53a shows three particles on an x axis. Particle 1 (with a charge of $+5.0\ \mu C$) and particle 2 (with a charge of $+3.0\ \mu C$) are fixed in place with separation $d = 4.0$ cm. Particle 3 can be moved along the x axis to the right of particle 2. Figure 24-53b gives the electric potential energy U of the three-particle system as a function of the x coordinate of particle 3. The scale of the vertical axis is set by $U_s = 5.0$ J. What is the charge of particle 3?

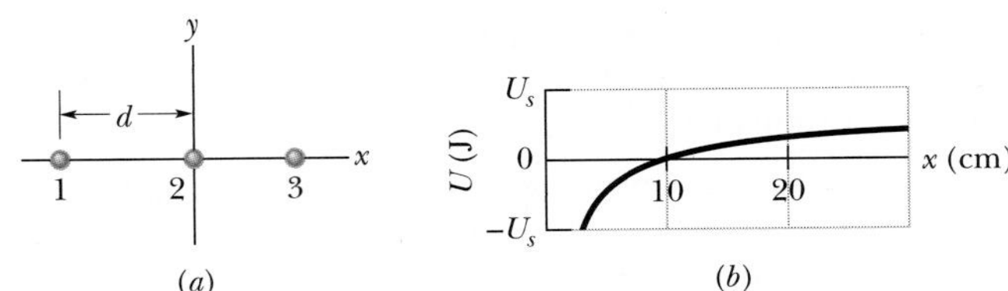

Fig. 24-53 Problem 56.

••57 SSM Identical 50 μC charges are fixed on an x axis at $x = \pm 3.0$ m. A particle of charge $q = -15\ \mu C$ is then released from rest at a point on the positive part of the y axis. Due to the symmetry of the situation, the particle moves along the y axis and has kinetic energy 1.2 J as it passes through the point $x = 0$, $y = 4.0$ m. (a) What is the kinetic energy of the particle as it passes through the origin? (b) At what negative value of y will the particle momentarily stop?

••58 GO *Proton in a well.* Figure 24-54 shows electric potential V along an x axis. The scale of the vertical axis is set by $V_s = 10.0$ V. A proton is to be released at $x = 3.5$ cm with initial kinetic energy 4.00 eV. (a) If it is initially moving in the negative direction of the axis, does it reach a turning point (if so, what is the x coordinate of that point) or does it escape from the plotted region (if so, what is its speed at $x = 0$)? (b) If it is initially moving in the positive direction of the axis, does it reach a turning point (if so, what is the x coordinate of that point) or does it escape from the plotted region (if so, what is its speed at $x = 6.0$ cm)? What are the (c) magnitude F and (d) direction (positive or negative direction of the x axis) of the electric force on the proton if the proton moves just to the left of $x = 3.0$ cm? What are (e) F and (f) the direction if the proton moves just to the right of $x = 5.0$ cm?

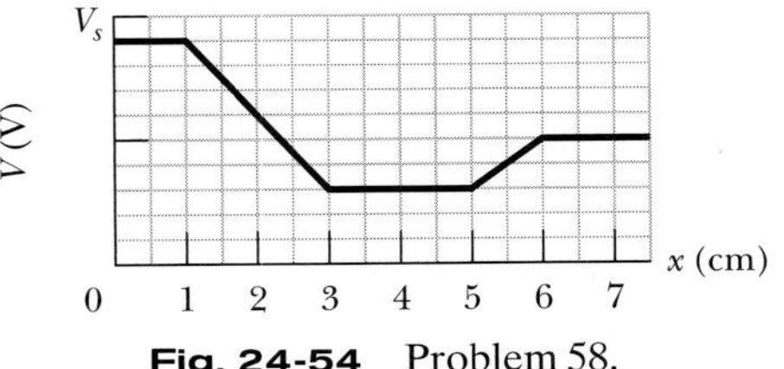

Fig. 24-54 Problem 58.

••59 In Fig. 24-55, a charged particle (either an electron or a proton) is moving rightward between two parallel charged plates separated by distance $d = 2.00$ mm. The plate potentials are $V_1 = -70.0$ V and $V_2 = -50.0$ V. The particle is slowing from an initial speed of 90.0 km/s at the left plate. (a) Is the particle an electron or a proton? (b) What is its speed just as it reaches plate 2?

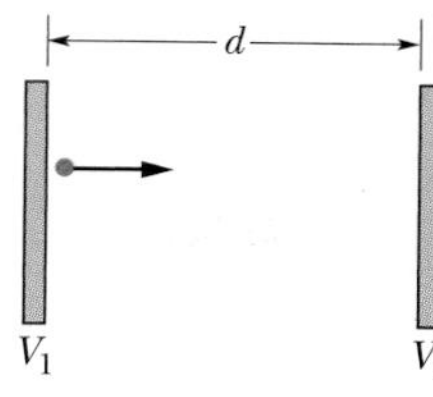

Fig. 24-55 Problem 59.

••60 In Fig. 24-56*a*, we move an electron from an infinite distance to a point at distance $R = 8.00$ cm from a tiny charged ball. The move requires work $W = 2.16 \times 10^{-13}$ J by us. (a) What is the charge Q on the ball? In Fig. 24-56*b*, the ball has been sliced up and the slices spread out so that an equal amount of charge is at the hour positions on a circular clock face of radius $R = 8.00$ cm. Now the electron is brought from an infinite distance to the center of the circle. (b) With that addition of the electron to the system of 12 charged particles, what is the change in the electric potential energy of the system?

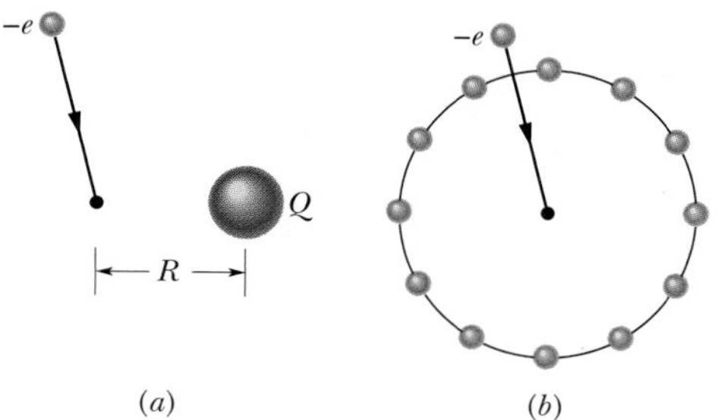

Fig. 24-56 Problem 60.

•••61 Suppose N electrons can be placed in either of two configurations. In configuration 1, they are all placed on the circumference of a narrow ring of radius R and are uniformly distributed so that the distance between adjacent electrons is the same everywhere. In configuration 2, $N - 1$ electrons are uniformly distributed on the ring and one electron is placed in the center of the ring. (a) What is the smallest value of N for which the second configuration is less energetic than the first? (b) For that value of N, consider any one circumference electron—call it e_0. How many other circumference electrons are closer to e_0 than the central electron is?

sec. 24-12 Potential of a Charged Isolated Conductor

•62 Sphere 1 with radius R_1 has positive charge q. Sphere 2 with radius $2.00R_1$ is far from sphere 1 and initially uncharged. After the separated spheres are connected with a wire thin enough to retain only negligible charge, (a) is potential V_1 of sphere 1 greater than, less than, or equal to potential V_2 of sphere 2? What fraction of q ends up on (b) sphere 1 and (c) sphere 2? (d) What is the ratio σ_1/σ_2 of the surface charge densities of the spheres?

•63 SSM WWW Two metal spheres, each of radius 3.0 cm, have a center-to-center separation of 2.0 m. Sphere 1 has charge $+1.0 \times 10^{-8}$ C; sphere 2 has charge -3.0×10^{-8} C. Assume that the separation is large enough for us to say that the charge on each sphere is uniformly distributed (the spheres do not affect each other). With $V = 0$ at infinity, calculate (a) the potential at the point halfway between the centers and the potential on the surface of (b) sphere 1 and (c) sphere 2.

•64 A hollow metal sphere has a potential of +400 V with respect to ground (defined to be at $V = 0$) and a charge of 5.0×10^{-9} C. Find the electric potential at the center of the sphere.

•65 SSM What is the excess charge on a conducting sphere of radius $r = 0.15$ m if the potential of the sphere is 1500 V and $V = 0$ at infinity?

••66 Two isolated, concentric, conducting spherical shells have radii $R_1 = 0.500$ m and $R_2 = 1.00$ m, uniform charges $q_1 = +2.00$ μC and $q_2 = +1.00$ μC, and negligible thicknesses. What is the magnitude of the electric field E at radial distance (a) $r = 4.00$ m, (b) $r = 0.700$ m, and (c) $r = 0.200$ m? With $V = 0$ at infinity, what is V at (d) $r = 4.00$ m, (e) $r = 1.00$ m, (f) $r = 0.700$ m, (g) $r = 0.500$ m, (h) $r = 0.200$ m, and (i) $r = 0$? (j) Sketch $E(r)$ and $V(r)$.

••67 A metal sphere of radius 15 cm has a net charge of 3.0×10^{-8} C. (a) What is the electric field at the sphere's surface? (b) If $V = 0$ at infinity, what is the electric potential at the sphere's surface? (c) At what distance from the sphere's surface has the electric potential decreased by 500 V?

Additional Problems

68 Here are the charges and coordinates of two point charges located in an xy plane: $q_1 = +3.00 \times 10^{-6}$ C, $x = +3.50$ cm, $y = +0.500$ cm and $q_2 = -4.00 \times 10^{-6}$ C, $x = -2.00$ cm, $y = +1.50$ cm. How much work must be done to locate these charges at their given positions, starting from infinite separation?

69 SSM A long, solid, conducting cylinder has a radius of 2.0 cm. The electric field at the surface of the cylinder is 160 N/C, directed radially outward. Let A, B, and C be points that are 1.0 cm, 2.0 cm, and 5.0 cm, respectively, from the central axis of the cylinder. What are (a) the magnitude of the electric field at C and the electric potential differences (b) $V_B - V_C$ and (c) $V_A - V_B$?

70 *The chocolate crumb mystery.* This story begins with Problem 60 in Chapter 23. (a) From the answer to part (a) of that problem, find an expression for the electric potential as a function of the radial distance r from the center of the pipe. (The electric potential is zero on the grounded pipe wall.) (b) For the typical volume charge density $\rho = -1.1 \times 10^{-3}$ C/m^3, what is the difference in the electric potential between the pipe's center and its inside wall? (The story continues with Problem 60 in Chapter 25.)

71 SSM Starting from Eq. 24-30, derive an expression for the electric field due to a dipole at a point on the dipole axis.

72 The magnitude E of an electric field depends on the radial distance r according to $E = A/r^4$, where A is a constant with the unit volt–cubic meter. As a multiple of A, what is the magnitude of the electric potential difference between $r = 2.00$ m and $r = 3.00$ m?

73 (a) If an isolated conducting sphere 10 cm in radius has a net charge of 4.0 μC and if $V = 0$ at infinity, what is the potential on the surface of the sphere? (b) Can this situation actually occur, given that the air around the sphere undergoes electrical breakdown when the field exceeds 3.0 MV/m?

74 Three particles, charge $q_1 = +10$ μC, $q_2 = -20$ μC, and $q_3 = +30\mu$C, are positioned at the vertices of an isosceles triangle as shown in Fig. 24-57. If $a = 10$ cm and $b = 6.0$ cm, how much work must an external agent do to exchange the positions of (a) q_1 and q_3 and, instead, (b) q_1 and q_2?

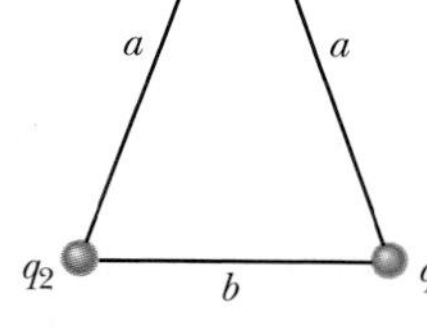

Fig. 24-57 Problem 74.

75 An electric field of approximately 100 V/m is often observed near the surface of Earth. If this were the field over the entire

surface, what would be the electric potential of a point on the surface? (Set $V = 0$ at infinity.)

76 A Gaussian sphere of radius 4.00 cm is centered on a ball that has a radius of 1.00 cm and a uniform charge distribution. The total (net) electric flux through the surface of the Gaussian sphere is $+5.60 \times 10^4\ \text{N}\cdot\text{m}^2/\text{C}$. What is the electric potential 12.0 cm from the center of the ball?

77 In a Millikan oil-drop experiment (Section 22-8), a uniform electric field of 1.92×10^5 N/C is maintained in the region between two plates separated by 1.50 cm. Find the potential difference between the plates.

78 Figure 24-58 shows three circular, nonconducting arcs of radius $R = 8.50$ cm. The charges on the arcs are $q_1 = 4.52$ pC, $q_2 = -2.00q_1$, $q_3 = +3.00q_1$. With $V = 0$ at infinity, what is the net electric potential of the arcs at the common center of curvature?

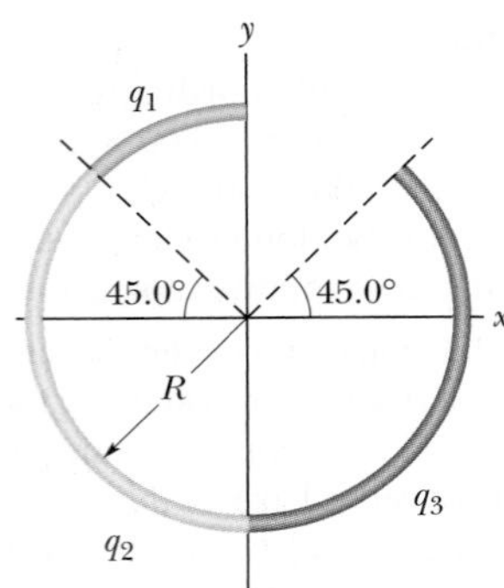

Fig. 24-58 Problem 78.

79 An electron is released from rest on the axis of an electric dipole that has charge e and charge separation $d = 20$ pm and that is fixed in place. The release point is on the positive side of the dipole, at distance $7.0d$ from the dipole center. What is the electron's speed when it reaches a point $5.0d$ from the dipole center?

80 Figure 24-59 shows a ring of outer radius $R = 13.0$ cm, inner radius $r = 0.200R$, and uniform surface charge density $\sigma = 6.20$ pC/m². With $V = 0$ at infinity, find the electric potential at point P on the central axis of the ring, at distance $z = 2.00R$ from the center of the ring.

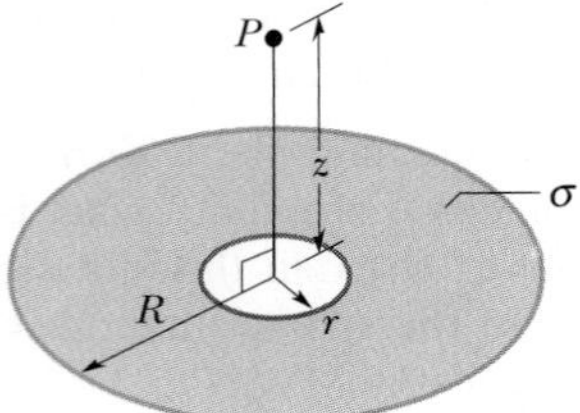

Fig. 24-59 Problem 80.

81 *Electron in a well.* Figure 24-60 shows electric potential V along an x axis. The scale of the vertical axis is set by $V_s = 8.0$ V. An electron is to be released at $x = 4.5$ cm with initial kinetic energy 3.00 eV. (a) If it is initially moving in the negative direction of the axis, does it reach a turning point (if so, what is the x coordinate of that point) or does it escape from the plotted region (if so, what is its speed at $x = 0$)? (b) If it is initially moving in the positive direction of the axis, does it reach a turning point (if so, what is the x coordinate of that point) or does it escape from the plotted region (if so, what is its speed at $x = 7.0$ cm)? What are the (c) magnitude F and (d) direction (positive or negative direction of the x axis) of the electric force on the electron if the electron moves just to the left of $x = 4.0$ cm? What are (e) F and (f) the direction if it moves just to the right of $x = 5.0$ cm?

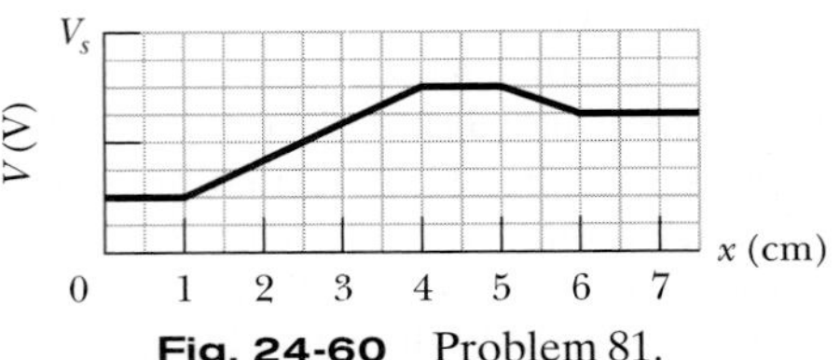

Fig. 24-60 Problem 81.

82 (a) If Earth had a uniform surface charge density of 1.0 electron/m² (a very artificial assumption), what would its potential be? (Set $V = 0$ at infinity.) What would be the (b) magnitude and (c) direction (radially inward or outward) of the electric field due to Earth just outside its surface?

83 In Fig. 24-61, point P is at distance $d_1 = 4.00$ m from particle 1 ($q_1 = -2e$) and distance $d_2 = 2.00$ m from particle 2 ($q_2 = +2e$), with both particles fixed in place. (a) With $V = 0$ at infinity, what is V at P? If we bring a particle of charge $q_3 = +2e$ from infinity to P, (b) how much work do we do and (c) what is the potential energy of the three-particle sytem?

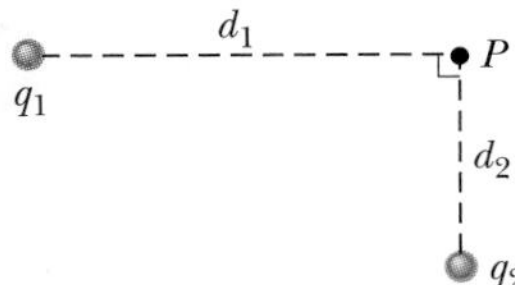

Fig. 24-61 Problem 83.

84 A solid conducting sphere of radius 3.0 cm has a charge of 30 nC distributed uniformly over its surface. Let A be a point 1.0 cm from the center of the sphere, S be a point on the surface of the sphere, and B be a point 5.0 cm from the center of the sphere. What are the electric potential differences (a) $V_S - V_B$ and (b) $V_A - V_B$?

85 In Fig. 24-62, we move a particle of charge $+2e$ in from infinity to the x axis. How much work do we do? Distance D is 4.00 m.

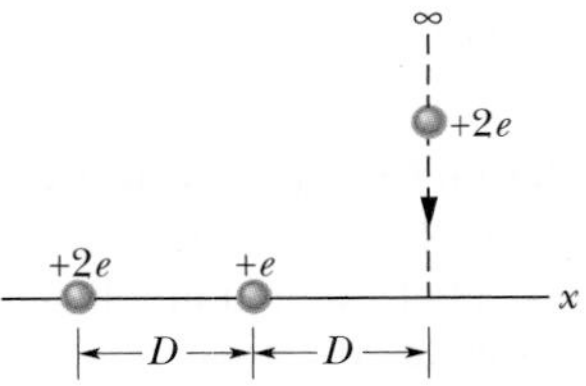

Fig. 24-62 Problem 85.

86 Figure 24-63 shows a hemisphere with a charge of 4.00 μC distributed uniformly through its volume. The hemisphere lies on an xy plane the way half a grapefruit might lie face down on a kitchen table. Point P is located on the plane, along a radial line from the hemisphere's center of curvature, at radial distance 15 cm. What is the electric potential at point P due to the hemisphere?

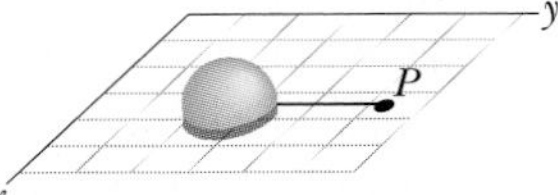

Fig. 24-63 Problem 86.

87 SSM Three +0.12 C charges form an equilateral triangle 1.7 m on a side. Using energy supplied at the rate of 0.83 kW, how many days would be required to move one of the charges to the midpoint of the line joining the other two charges?

88 Two charges $q = +2.0\ \mu C$ are fixed a distance $d = 2.0$ cm apart (Fig. 24-64). (a) With $V = 0$ at infinity, what is the electric potential at point C? (b) You bring a third charge $q = +2.0\ \mu C$ from infinity to C. How much work must you do? (c) What is the potential energy U of the three-charge configuration when the third charge is in place?

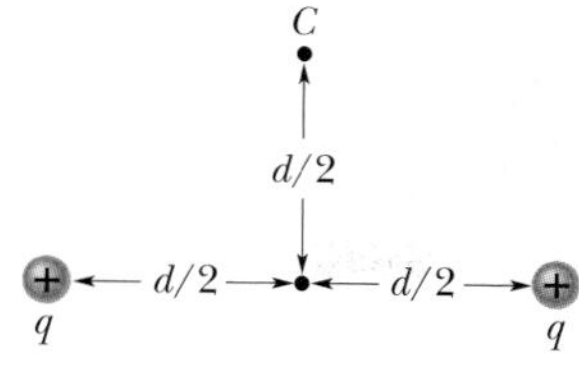

Fig. 24-64 Problem 88.

89 Initially two electrons are fixed in place with a separation of 2.00 μm. How much work must we do to bring a third electron in from infinity to complete an equilateral triangle?

90 A particle of positive charge Q is fixed at point P. A second particle of mass m and negative charge $-q$ moves at constant speed in a circle of radius r_1, centered at P. Derive an expression for the work W that must be done by an external agent on the second particle to increase the radius of the circle of motion to r_2.

91 Two charged, parallel, flat conducting surfaces are spaced $d = 1.00$ cm apart and produce a potential difference $\Delta V = 625$ V between them. An electron is projected from one surface directly toward the second. What is the initial speed of the electron if it stops just at the second surface?

92 In Fig. 24-65, point P is at the center of the rectangle. With $V = 0$ at infinity, $q_1 = 5.00$ fC, $q_2 = 2.00$ fC, $q_3 = 3.00$ fC, and $d = 2.54$ cm, what is the net electric potential at P due to the six charged particles?

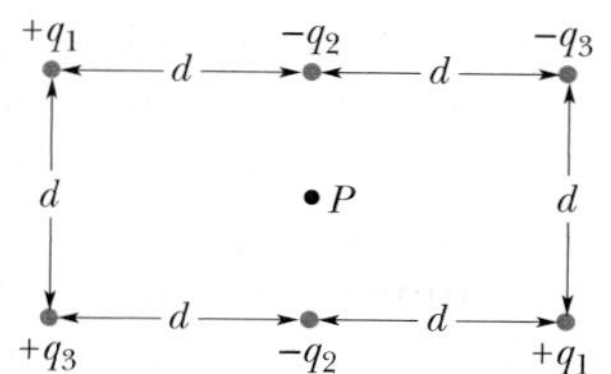

Fig. 24-65 Problem 92.

93 **SSM** A uniform charge of $+16.0\ \mu C$ is on a thin circular ring lying in an xy plane and centered on the origin. The ring's radius is 3.00 cm. If point A is at the origin and point B is on the z axis at $z = 4.00$ cm, what is $V_B - V_A$?

94 Consider a point charge $q = 1.50 \times 10^{-8}$ C, and take $V = 0$ at infinity. (a) What are the shape and dimensions of an equipotential surface having a potential of 30.0 V due to q alone? (b) Are surfaces whose potentials differ by a constant amount (1.0 V, say) evenly spaced?

95 **SSM** A thick spherical shell of charge Q and uniform volume charge density ρ is bounded by radii r_1 and $r_2 > r_1$. With $V = 0$ at infinity, find the electric potential V as a function of distance r from the center of the distribution, considering regions (a) $r > r_2$, (b) $r_2 > r > r_1$, and (c) $r < r_1$. (d) Do these solutions agree with each other at $r = r_2$ and $r = r_1$? (*Hint:* See Section 23-9.)

96 A charge q is distributed uniformly throughout a spherical volume of radius R. Let $V = 0$ at infinity. What are (a) V at radial distance $r < R$ and (b) the potential difference between points at $r = R$ and the point at $r = 0$?

97 Figure 24-35 shows two charged particles on an axis. Sketch the electric field lines and the equipotential surfaces in the plane of the page for (a) $q_1 = +q, q_2 = +2q$ and (b) $q_1 = +q, q_2 = -3q$.

98 What is the electric potential energy of the charge configuration of Fig. 24-8*a*? Use the numerical values provided in the associated sample problem.

99 (a) Using Eq. 24-32, show that the electric potential at a point on the central axis of a thin ring (of charge q and radius R) and at distance z from the ring is

$$V = \frac{1}{4\pi\varepsilon_0}\frac{q}{\sqrt{z^2 + R^2}}.$$

(b) From this result, derive an expression for the electric field magnitude E at points on the ring's axis; compare your result with the calculation of E in Section 22-6.

100 An alpha particle (which has two protons) is sent directly toward a target nucleus containing 92 protons. The alpha particle has an initial kinetic energy of 0.48 pJ. What is the least center-to-center distance the alpha particle will be from the target nucleus, assuming the nucleus does not move?

101 In the quark model of fundamental particles, a proton is composed of three quarks: two "up" quarks, each having charge $+2e/3$, and one "down" quark, having charge $-e/3$. Suppose that the three quarks are equidistant from one another. Take that separation distance to be 1.32×10^{-15} m and calculate the electric potential energy of the system of (a) only the two up quarks and (b) all three quarks.

102 (a) A proton of kinetic energy 4.80 MeV travels head-on toward a lead nucleus. Assuming that the proton does not penetrate the nucleus and that the only force between proton and nucleus is the Coulomb force, calculate the smallest center-to-center separation d_p between proton and nucleus when the proton momentarily stops. If the proton were replaced with an alpha particle (which contains two protons) of the same initial kinetic energy, the alpha particle would stop at center-to-center separation d_α. (b) What is d_α/d_p?

103 In Fig. 24-66, two particles of charges q_1 and q_2 are fixed to an x axis. If a third particle, of charge $+6.0\ \mu C$, is brought from an infinite distance to point P, the three-particle system has the same electric potential energy as the original two-particle system. What is the charge ratio q_1/q_2?

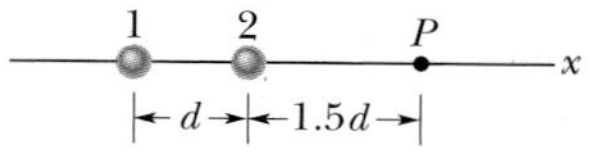

Fig. 24-66 Problem 103.

104 A charge of 1.50×10^{-8} C lies on an isolated metal sphere of radius 16.0 cm. With $V = 0$ at infinity, what is the electric potential at points on the sphere's surface?

105 **SSM** A solid copper sphere whose radius is 1.0 cm has a very thin surface coating of nickel. Some of the nickel atoms are radioactive, each atom emitting an electron as it decays. Half of these electrons enter the copper sphere, each depositing 100 keV of energy there. The other half of the electrons escape, each carrying away a charge $-e$. The nickel coating has an activity of 3.70×10^8 radioactive decays per second. The sphere is hung from a long, nonconducting string and isolated from its surroundings. (a) How long will it take for the potential of the sphere to increase by 1000 V? (b) How long will it take for the temperature of the sphere to increase by 5.0 K due to the energy deposited by the electrons? The heat capacity of the sphere is 14 J/K.

CHAPTER

25 CAPACITANCE

25-1 WHAT IS PHYSICS?

One goal of physics is to provide the basic science for practical devices designed by engineers. The focus of this chapter is on one extremely common example—the capacitor, a device in which electrical energy can be stored. For example, the batteries in a camera store energy in the photoflash unit by charging a capacitor. The batteries can supply energy at only a modest rate, too slowly for the photoflash unit to emit a flash of light. However, once the capacitor is charged, it can supply energy at a much greater rate when the photoflash unit is triggered—enough energy to allow the unit to emit a burst of bright light.

The physics of capacitors can be generalized to other devices and to any situation involving electric fields. For example, Earth's atmospheric electric field is modeled by meteorologists as being produced by a huge spherical capacitor that partially discharges via lightning. The charge that skis collect as they slide along snow can be modeled as being stored in a capacitor that frequently discharges as sparks (which can be seen by nighttime skiers on dry snow).

The first step in our discussion of capacitors is to determine how much charge can be stored. This "how much" is called capacitance.

25-2 Capacitance

Figure 25-1 shows some of the many sizes and shapes of capacitors. Figure 25-2 shows the basic elements of *any* capacitor—two isolated conductors of any

Fig. 25-1 An assortment of capacitors.

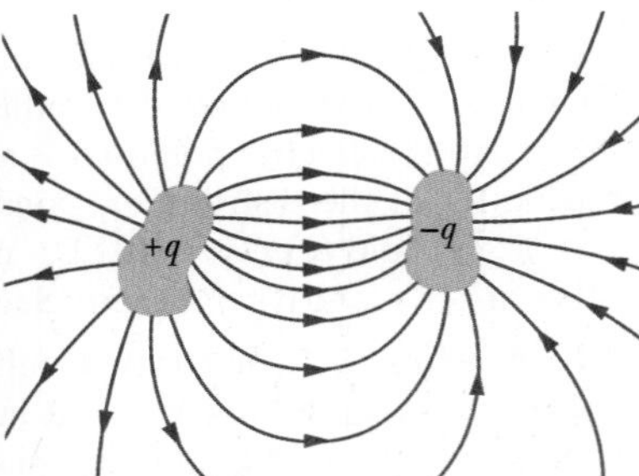

Fig. 25-2 Two conductors, isolated electrically from each other and from their surroundings, form a *capacitor.* When the capacitor is charged, the charges on the conductors, or *plates* as they are called, have the same magnitude q but opposite signs.

(Paul Silvermann/Fundamental Photographs)

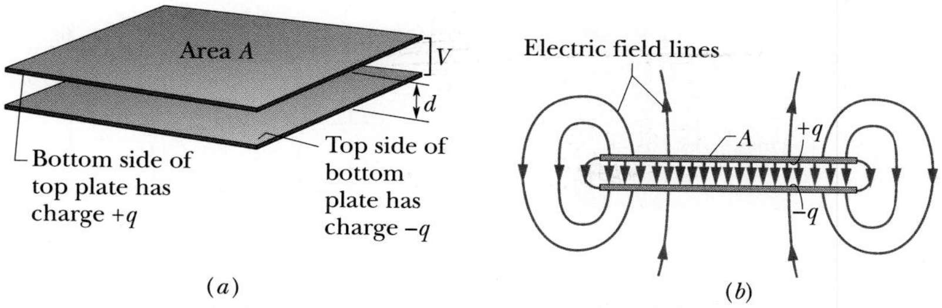

Fig. 25-3 (*a*) A parallel-plate capacitor, made up of two plates of area A separated by a distance d. The charges on the facing plate surfaces have the same magnitude q but opposite signs. (*b*) As the field lines show, the electric field due to the charged plates is uniform in the central region between the plates. The field is not uniform at the edges of the plates, as indicated by the "fringing" of the field lines there.

shape. No matter what their geometry, flat or not, we call these conductors *plates*.

Figure 25-3*a* shows a less general but more conventional arrangement, called a *parallel-plate capacitor,* consisting of two parallel conducting plates of area A separated by a distance d. The symbol we use to represent a capacitor (⊣⊢) is based on the structure of a parallel-plate capacitor but is used for capacitors of all geometries. We assume for the time being that no material medium (such as glass or plastic) is present in the region between the plates. In Section 25-6, we shall remove this restriction.

When a capacitor is *charged,* its plates have charges of equal magnitudes but opposite signs: $+q$ and $-q$. However, we refer to the *charge of a capacitor* as being q, the absolute value of these charges on the plates. (Note that q is not the net charge on the capacitor, which is zero.)

Because the plates are conductors, they are equipotential surfaces; all points on a plate are at the same electric potential. Moreover, there is a potential difference between the two plates. For historical reasons, we represent the absolute value of this potential difference with V rather than with the ΔV we used in previous notation.

The charge q and the potential difference V for a capacitor are proportional to each other; that is,

$$q = CV. \tag{25-1}$$

The proportionality constant C is called the **capacitance** of the capacitor. Its value depends only on the geometry of the plates and *not* on their charge or potential difference. The capacitance is a measure of how much charge must be put on the plates to produce a certain potential difference between them: The *greater the capacitance, the more charge is required.*

The SI unit of capacitance that follows from Eq. 25-1 is the coulomb per volt. This unit occurs so often that it is given a special name, the *farad* (F):

$$1 \text{ farad} = 1 \text{ F} = 1 \text{ coulomb per volt} = 1 \text{ C/V}. \tag{25-2}$$

As you will see, the farad is a very large unit. Submultiples of the farad, such as the microfarad ($1\ \mu\text{F} = 10^{-6}$ F) and the picofarad ($1 \text{ pF} = 10^{-12}$ F), are more convenient units in practice.

Charging a Capacitor

One way to charge a capacitor is to place it in an electric circuit with a battery. An *electric circuit* is a path through which charge can flow. A *battery* is a device

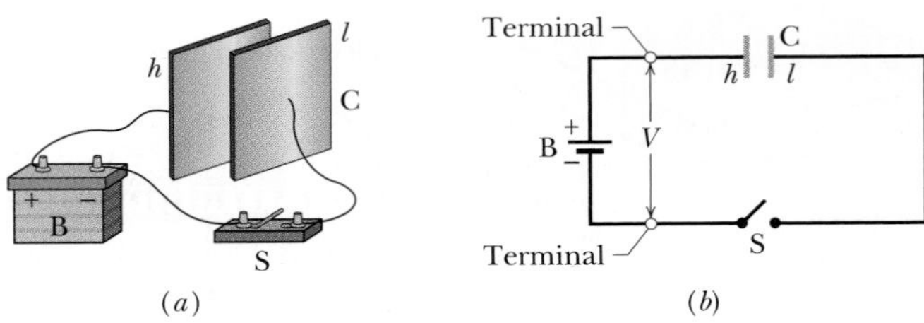

Fig. 25-4 (*a*) Battery B, switch S, and plates *h* and *l* of capacitor C, connected in a circuit. (*b*) A schematic diagram with the *circuit elements* represented by their symbols.

that maintains a certain potential difference between its *terminals* (points at which charge can enter or leave the battery) by means of internal electrochemical reactions in which electric forces can move internal charge.

In Fig. 25-4*a*, a battery B, a switch S, an uncharged capacitor C, and interconnecting wires form a circuit. The same circuit is shown in the *schematic diagram* of Fig. 25-4*b*, in which the symbols for a battery, a switch, and a capacitor represent those devices. The battery maintains potential difference V between its terminals. The terminal of higher potential is labeled $+$ and is often called the *positive* terminal; the terminal of lower potential is labeled $-$ and is often called the *negative* terminal.

The circuit shown in Figs. 25-4*a* and *b* is said to be *incomplete* because switch S is *open;* that is, the switch does not electrically connect the wires attached to it. When the switch is *closed,* electrically connecting those wires, the circuit is complete and charge can then flow through the switch and the wires. As we discussed in Chapter 21, the charge that can flow through a conductor, such as a wire, is that of electrons. When the circuit of Fig. 25-4 is completed, electrons are driven through the wires by an electric field that the battery sets up in the wires. The field drives electrons from capacitor plate h to the positive terminal of the battery; thus, plate h, losing electrons, becomes positively charged. The field drives just as many electrons from the negative terminal of the battery to capacitor plate l; thus, plate l, gaining electrons, becomes negatively charged *just as much* as plate h, losing electrons, becomes positively charged.

Initially, when the plates are uncharged, the potential difference between them is zero. As the plates become oppositely charged, that potential difference increases until it equals the potential difference V between the terminals of the battery. Then plate h and the positive terminal of the battery are at the same potential, and there is no longer an electric field in the wire between them. Similarly, plate l and the negative terminal reach the same potential, and there is then no electric field in the wire between them. Thus, with the field zero, there is no further drive of electrons. The capacitor is then said to be *fully charged,* with a potential difference V and charge q that are related by Eq. 25-1.

In this book we assume that during the charging of a capacitor and afterward, charge cannot pass from one plate to the other across the gap separating them. Also, we assume that a capacitor can retain (or *store*) charge indefinitely, until it is put into a circuit where it can be *discharged.*

CHECKPOINT 1

Does the capacitance C of a capacitor increase, decrease, or remain the same (a) when the charge q on it is doubled and (b) when the potential difference V across it is tripled?

25-3 Calculating the Capacitance

Our goal here is to calculate the capacitance of a capacitor once we know its geometry. Because we shall consider a number of different geometries, it seems wise to develop a general plan to simplify the work. In brief our plan is as follows: (1) Assume a charge q on the plates; (2) calculate the electric field $\vec{E}$ between the plates in terms of this charge, using Gauss' law; (3) knowing $\vec{E}$, calculate the potential difference V between the plates from Eq. 24-18; (4) calculate C from Eq. 25-1.

Before we start, we can simplify the calculation of both the electric field and the potential difference by making certain assumptions. We discuss each in turn.

We use Gauss' law to relate q and E. Then we integrate the E to get the potential difference.

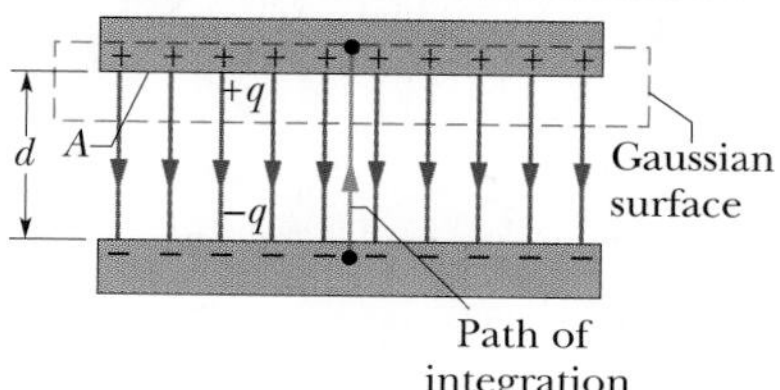

Fig. 25-5 A charged parallel-plate capacitor. A Gaussian surface encloses the charge on the positive plate. The integration of Eq. 25-6 is taken along a path extending directly from the negative plate to the positive plate.

Calculating the Electric Field

To relate the electric field $\vec{E}$ between the plates of a capacitor to the charge q on either plate, we shall use Gauss' law:

$$\varepsilon_0 \oint \vec{E} \cdot d\vec{A} = q. \tag{25-3}$$

Here q is the charge enclosed by a Gaussian surface and $\oint \vec{E} \cdot d\vec{A}$ is the net electric flux through that surface. In all cases that we shall consider, the Gaussian surface will be such that whenever there is an electric flux through it, $\vec{E}$ will have a uniform magnitude E and the vectors $\vec{E}$ and $d\vec{A}$ will be parallel. Equation 25-3 then reduces to

$$q = \varepsilon_0 EA \quad \text{(special case of Eq. 25-3)}, \tag{25-4}$$

in which A is the area of that part of the Gaussian surface through which there is a flux. For convenience, we shall always draw the Gaussian surface in such a way that it completely encloses the charge on the positive plate; see Fig. 25-5 for an example.

Calculating the Potential Difference

In the notation of Chapter 24 (Eq. 24-18), the potential difference between the plates of a capacitor is related to the field $\vec{E}$ by

$$V_f - V_i = -\int_i^f \vec{E} \cdot d\vec{s}, \tag{25-5}$$

in which the integral is to be evaluated along any path that starts on one plate and ends on the other. We shall always choose a path that follows an electric field line, from the negative plate to the positive plate. For this path, the vectors $\vec{E}$ and $d\vec{s}$ will have opposite directions; so the dot product $\vec{E} \cdot d\vec{s}$ will be equal to $-E\,ds$. Thus, the right side of Eq. 25-5 will then be positive. Letting V represent the difference $V_f - V_i$, we can then recast Eq. 25-5 as

$$V = \int_-^+ E\,ds \quad \text{(special case of Eq. 25-5)}, \tag{25-6}$$

in which the $-$ and $+$ remind us that our path of integration starts on the negative plate and ends on the positive plate.

We are now ready to apply Eqs. 25-4 and 25-6 to some particular cases.

A Parallel-Plate Capacitor

We assume, as Fig. 25-5 suggests, that the plates of our parallel-plate capacitor are so large and so close together that we can neglect the fringing of the electric field

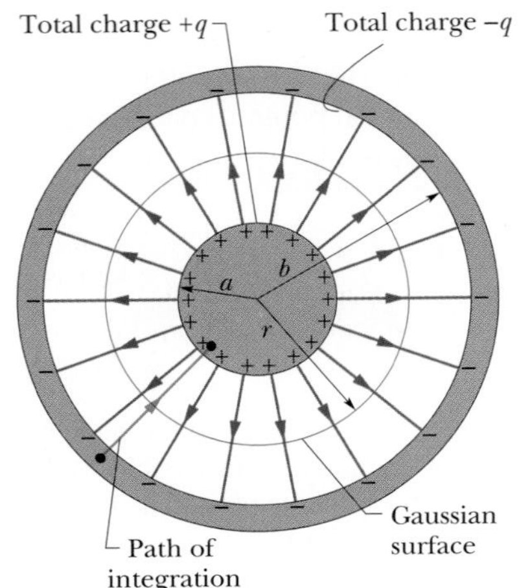

Fig. 25-6 A cross section of a long cylindrical capacitor, showing a cylindrical Gaussian surface of radius r (that encloses the positive plate) and the radial path of integration along which Eq. 25-6 is to be applied. This figure also serves to illustrate a spherical capacitor in a cross section through its center.

at the edges of the plates, taking $\vec{E}$ to be constant throughout the region between the plates.

We draw a Gaussian surface that encloses just the charge q on the positive plate, as in Fig. 25-5. From Eq. 25-4 we can then write

$$q = \varepsilon_0 EA, \tag{25-7}$$

where A is the area of the plate.

Equation 25-6 yields

$$V = \int_{-}^{+} E\,ds = E \int_0^d ds = Ed. \tag{25-8}$$

In Eq. 25-8, E can be placed outside the integral because it is a constant; the second integral then is simply the plate separation d.

If we now substitute q from Eq. 25-7 and V from Eq. 25-8 into the relation $q = CV$ (Eq. 25-1), we find

$$C = \frac{\varepsilon_0 A}{d} \quad \text{(parallel-plate capacitor).} \tag{25-9}$$

Thus, the capacitance does indeed depend only on geometrical factors—namely, the plate area A and the plate separation d. Note that C increases as we increase area A or decrease separation d.

As an aside, we point out that Eq. 25-9 suggests one of our reasons for writing the electrostatic constant in Coulomb's law in the form $1/4\pi\varepsilon_0$. If we had not done so, Eq. 25-9—which is used more often in engineering practice than Coulomb's law—would have been less simple in form. We note further that Eq. 25-9 permits us to express the permittivity constant ε_0 in a unit more appropriate for use in problems involving capacitors; namely,

$$\varepsilon_0 = 8.85 \times 10^{-12}\ \text{F/m} = 8.85\ \text{pF/m}. \tag{25-10}$$

We have previously expressed this constant as

$$\varepsilon_0 = 8.85 \times 10^{-12}\ \text{C}^2/\text{N}\cdot\text{m}^2. \tag{25-11}$$

A Cylindrical Capacitor

Figure 25-6 shows, in cross section, a cylindrical capacitor of length L formed by two coaxial cylinders of radii a and b. We assume that $L \gg b$ so that we can neglect the fringing of the electric field that occurs at the ends of the cylinders. Each plate contains a charge of magnitude q.

As a Gaussian surface, we choose a cylinder of length L and radius r, closed by end caps and placed as is shown in Fig. 25-6. It is coaxial with the cylinders and encloses the central cylinder and thus also the charge q on that cylinder. Equation 25-4 then relates that charge and the field magnitude E as

$$q = \varepsilon_0 EA = \varepsilon_0 E(2\pi rL),$$

in which $2\pi rL$ is the area of the curved part of the Gaussian surface. There is no flux through the end caps. Solving for E yields

$$E = \frac{q}{2\pi\varepsilon_0 Lr}. \tag{25-12}$$

Substitution of this result into Eq. 25-6 yields

$$V = \int_{-}^{+} E\,ds = -\frac{q}{2\pi\varepsilon_0 L}\int_b^a \frac{dr}{r} = \frac{q}{2\pi\varepsilon_0 L}\ln\left(\frac{b}{a}\right), \tag{25-13}$$

where we have used the fact that here $ds = -dr$ (we integrated radially inward). From the relation $C = q/V$, we then have

$$C = 2\pi\varepsilon_0 \frac{L}{\ln(b/a)} \quad \text{(cylindrical capacitor).} \tag{25-14}$$

We see that the capacitance of a cylindrical capacitor, like that of a parallel-plate capacitor, depends only on geometrical factors, in this case the length L and the two radii b and a.

A Spherical Capacitor

Figure 25-6 can also serve as a central cross section of a capacitor that consists of two concentric spherical shells, of radii a and b. As a Gaussian surface we draw a sphere of radius r concentric with the two shells; then Eq. 25-4 yields

$$q = \varepsilon_0 EA = \varepsilon_0 E(4\pi r^2),$$

in which $4\pi r^2$ is the area of the spherical Gaussian surface. We solve this equation for E, obtaining

$$E = \frac{1}{4\pi\varepsilon_0} \frac{q}{r^2}, \tag{25-15}$$

which we recognize as the expression for the electric field due to a uniform spherical charge distribution (Eq. 23-15).

If we substitute this expression into Eq. 25-6, we find

$$V = \int_-^+ E\,ds = -\frac{q}{4\pi\varepsilon_0} \int_b^a \frac{dr}{r^2} = \frac{q}{4\pi\varepsilon_0}\left(\frac{1}{a} - \frac{1}{b}\right) = \frac{q}{4\pi\varepsilon_0} \frac{b-a}{ab}, \tag{25-16}$$

where again we have substituted $-dr$ for ds. If we now substitute Eq. 25-16 into Eq. 25-1 and solve for C, we find

$$C = 4\pi\varepsilon_0 \frac{ab}{b-a} \quad \text{(spherical capacitor).} \tag{25-17}$$

An Isolated Sphere

We can assign a capacitance to a *single* isolated spherical conductor of radius R by assuming that the "missing plate" is a conducting sphere of infinite radius. After all, the field lines that leave the surface of a positively charged isolated conductor must end somewhere; the walls of the room in which the conductor is housed can serve effectively as our sphere of infinite radius.

To find the capacitance of the conductor, we first rewrite Eq. 25-17 as

$$C = 4\pi\varepsilon_0 \frac{a}{1 - a/b}.$$

If we then let $b \rightarrow \infty$ and substitute R for a, we find

$$C = 4\pi\varepsilon_0 R \quad \text{(isolated sphere).} \tag{25-18}$$

Note that this formula and the others we have derived for capacitance (Eqs. 25-9, 25-14, and 25-17) involve the constant ε_0 multiplied by a quantity that has the dimensions of a length.

CHECKPOINT 2

For capacitors charged by the same battery, does the charge stored by the capacitor increase, decrease, or remain the same in each of the following situations? (a) The plate separation of a parallel-plate capacitor is increased. (b) The radius of the inner cylinder of a cylindrical capacitor is increased. (c) The radius of the outer spherical shell of a spherical capacitor is increased.

Sample Problem

Charging the plates in a parallel-plate capacitor

In Fig. 25-7*a*, switch S is closed to connect the uncharged capacitor of capacitance $C = 0.25\ \mu\text{F}$ to the battery of potential difference $V = 12$ V. The lower capacitor plate has thickness $L = 0.50$ cm and face area $A = 2.0 \times 10^{-4}\ \text{m}^2$, and it consists of copper, in which the density of conduction electrons is $n = 8.49 \times 10^{28}$ electrons/m³. From what depth d within the plate (Fig. 25-7*b*) must electrons move to the plate face as the capacitor becomes charged?

KEY IDEA

The charge collected on the plate is related to the capacitance and the potential difference across the capacitor by Eq. 25-1 ($q = CV$).

Calculations: Because the lower plate is connected to the negative terminal of the battery, conduction electrons move up to the face of the plate. From Eq. 25-1, the total charge magnitude that collects there is

$$q = CV = (0.25 \times 10^{-6}\ \text{F})(12\ \text{V}) = 3.0 \times 10^{-6}\ \text{C}.$$

Dividing this result by e gives us the number N of conduction electrons that come up to the face:

$$N = \frac{q}{e} = \frac{3.0 \times 10^{-6}\ \text{C}}{1.602 \times 10^{-19}\ \text{C}} = 1.873 \times 10^{13}\ \text{electrons}.$$

These electrons come from a volume that is the product of the face area A and the depth d we seek. Thus, from the density of conduction electrons (number per volume), we can write

$$n = \frac{N}{Ad},$$

or

$$d = \frac{N}{An} = \frac{1.873 \times 10^{13}\ \text{electrons}}{(2.0 \times 10^{-4}\ \text{m}^2)(8.49 \times 10^{28}\ \text{electrons/m}^3)} = 1.1 \times 10^{-12}\ \text{m} = 1.1\ \text{pm}. \quad \text{(Answer)}$$

In common speech, we would say that the battery charges the capacitor by supplying the charged particles. But what the battery really does is set up an electric field in the wires and plate such that electrons very close to the plate face move up to the negative face.

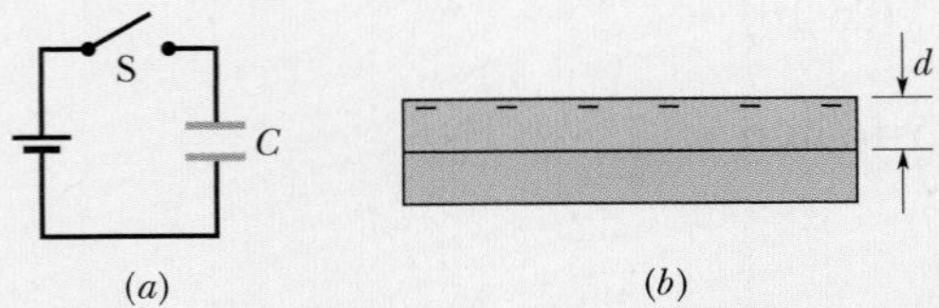

Fig. 25-7 (*a*) A battery and capacitor circuit. (*b*) The lower capacitor plate.

Additional examples, video, and practice available at *WileyPLUS*

25-4 Capacitors in Parallel and in Series

When there is a combination of capacitors in a circuit, we can sometimes replace that combination with an **equivalent capacitor**—that is, a single capacitor that has the same capacitance as the actual combination of capacitors. With such a replacement, we can simplify the circuit, affording easier solutions for unknown quantities of the circuit. Here we discuss two basic combinations of capacitors that allow such a replacement.

Capacitors in Parallel

Figure 25-8*a* shows an electric circuit in which three capacitors are connected *in parallel* to battery B. This description has little to do with how the capacitor plates are drawn. Rather, "in parallel" means that the capacitors are directly wired together at one plate and directly wired together at the other plate, and that the same potential difference V is applied across the two groups of wired-together plates. Thus, each capacitor has the same potential difference V, which produces charge on the capacitor. (In Fig. 25-8*a*, the applied potential V is maintained by the battery.) In general,

When a potential difference V is applied across several capacitors connected in parallel, that potential difference V is applied across each capacitor. The total charge q stored on the capacitors is the sum of the charges stored on all the capacitors.

When we analyze a circuit of capacitors in parallel, we can simplify it with this mental replacement:

Capacitors connected in parallel can be replaced with an equivalent capacitor that has the same *total* charge q and the same potential difference V as the actual capacitors.

(You might remember this result with the nonsense word "par-V," which is close to "party," to mean "capacitors in parallel have the same V.") Figure 25-8*b* shows the equivalent capacitor (with equivalent capacitance C_{eq}) that has replaced the three capacitors (with actual capacitances C_1, C_2, and C_3) of Fig. 25-8*a*.

To derive an expression for C_{eq} in Fig. 25-8*b*, we first use Eq. 25-1 to find the charge on each actual capacitor:

$$q_1 = C_1V, \quad q_2 = C_2V, \quad \text{and} \quad q_3 = C_3V.$$

The total charge on the parallel combination of Fig. 25-8*a* is then

$$q = q_1 + q_2 + q_3 = (C_1 + C_2 + C_3)V.$$

The equivalent capacitance, with the same total charge q and applied potential difference V as the combination, is then

$$C_{eq} = \frac{q}{V} = C_1 + C_2 + C_3,$$

a result that we can easily extend to any number n of capacitors, as

$$C_{eq} = \sum_{j=1}^{n} C_j \quad (n \text{ capacitors in parallel}). \tag{25-19}$$

Thus, to find the equivalent capacitance of a parallel combination, we simply add the individual capacitances.

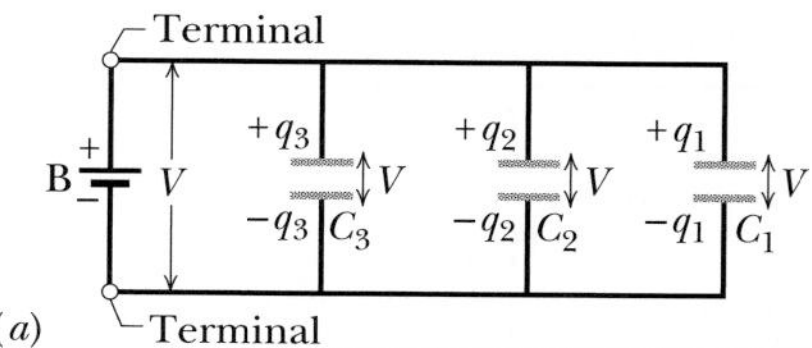

Parallel capacitors and their equivalent have the same V ("par-V").

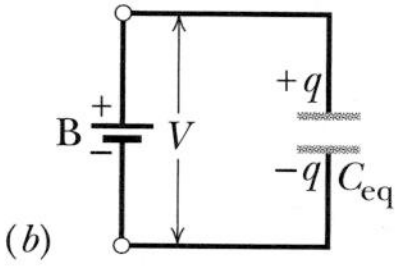

Fig. 25-8 (*a*) Three capacitors connected in parallel to battery B. The battery maintains potential difference V across its terminals and thus across *each* capacitor. (*b*) The equivalent capacitor, with capacitance C_{eq}, replaces the parallel combination.

Capacitors in Series

Figure 25-9*a* shows three capacitors connected *in series* to battery B. This description has little to do with how the capacitors are drawn. Rather, "in series" means that the capacitors are wired serially, one after the other, and that a potential difference V is applied across the two ends of the series. (In Fig. 25-9*a*, this potential difference V is maintained by battery B.) The potential differences that then exist across the capacitors in the series produce identical charges q on them.

When a potential difference V is applied across several capacitors connected in series, the capacitors have identical charge q. The sum of the potential differences across all the capacitors is equal to the applied potential difference V.

We can explain how the capacitors end up with identical charge by following a *chain reaction* of events, in which the charging of each capacitor causes the charging of the next capacitor. We start with capacitor 3 and work upward to capacitor 1. When the battery is first connected to the series of capacitors, it

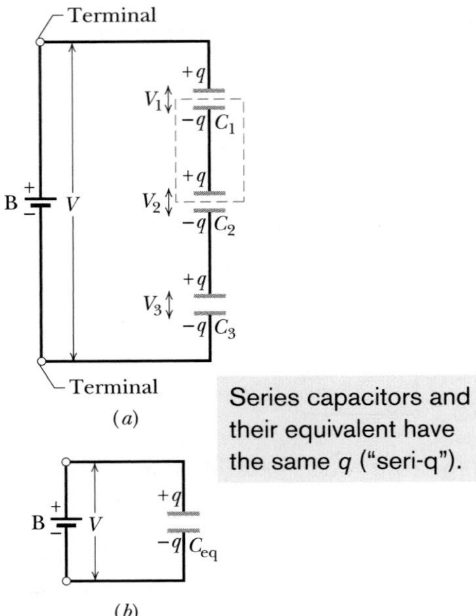

Series capacitors and their equivalent have the same q ("seri-q").

Fig. 25-9 (*a*) Three capacitors connected in series to battery B. The battery maintains potential difference V between the top and bottom plates of the series combination. (*b*) The equivalent capacitor, with capacitance C_{eq}, replaces the series combination.

produces charge $-q$ on the bottom plate of capacitor 3. That charge then repels negative charge from the top plate of capacitor 3 (leaving it with charge $+q$). The repelled negative charge moves to the bottom plate of capacitor 2 (giving it charge $-q$). That charge on the bottom plate of capacitor 2 then repels negative charge from the top plate of capacitor 2 (leaving it with charge $+q$) to the bottom plate of capacitor 1 (giving it charge $-q$). Finally, the charge on the bottom plate of capacitor 1 helps move negative charge from the top plate of capacitor 1 to the battery, leaving that top plate with charge $+q$.

Here are two important points about capacitors in series:

1. When charge is shifted from one capacitor to another in a series of capacitors, it can move along only one route, such as from capacitor 3 to capacitor 2 in Fig. 25-9*a*. If there are additional routes, the capacitors are not in series.
2. The battery directly produces charges on only the two plates to which it is connected (the bottom plate of capacitor 3 and the top plate of capacitor 1 in Fig. 25-9*a*). Charges that are produced on the other plates are due merely to the shifting of charge already there. For example, in Fig. 25-9*a*, the part of the circuit enclosed by dashed lines is electrically isolated from the rest of the circuit. Thus, the net charge of that part cannot be changed by the battery—its charge can only be redistributed.

When we analyze a circuit of capacitors in series, we can simplify it with this mental replacement:

Capacitors that are connected in series can be replaced with an equivalent capacitor that has the same charge q and the same *total* potential difference V as the actual series capacitors.

(You might remember this with the nonsense word "seri-q" to mean "capacitors in series have the same q.") Figure 25-9*b* shows the equivalent capacitor (with equivalent capacitance C_{eq}) that has replaced the three actual capacitors (with actual capacitances C_1, C_2, and C_3) of Fig. 25-9*a*.

To derive an expression for C_{eq} in Fig. 25-9*b*, we first use Eq. 25-1 to find the potential difference of each actual capacitor:

$$V_1 = \frac{q}{C_1}, \quad V_2 = \frac{q}{C_2}, \quad \text{and} \quad V_3 = \frac{q}{C_3}.$$

The total potential difference V due to the battery is the sum of these three potential differences. Thus,

$$V = V_1 + V_2 + V_3 = q\left(\frac{1}{C_1} + \frac{1}{C_2} + \frac{1}{C_3}\right).$$

The equivalent capacitance is then

$$C_{eq} = \frac{q}{V} = \frac{1}{1/C_1 + 1/C_2 + 1/C_3},$$

or

$$\frac{1}{C_{eq}} = \frac{1}{C_1} + \frac{1}{C_2} + \frac{1}{C_3}.$$

We can easily extend this to any number n of capacitors as

$$\frac{1}{C_{eq}} = \sum_{j=1}^{n} \frac{1}{C_j} \quad (n \text{ capacitors in series}). \tag{25-20}$$

Using Eq. 25-20 you can show that the equivalent capacitance of a series of capacitances is always *less* than the least capacitance in the series.

CHECKPOINT 3

A battery of potential V stores charge q on a combination of two identical capacitors. What are the potential difference across and the charge on either capacitor if the capacitors are (a) in parallel and (b) in series?

Sample Problem

Capacitors in parallel and in series

(a) Find the equivalent capacitance for the combination of capacitances shown in Fig. 25-10*a*, across which potential difference V is applied. Assume

$$C_1 = 12.0\ \mu\text{F}, \quad C_2 = 5.30\ \mu\text{F}, \quad \text{and} \quad C_3 = 4.50\ \mu\text{F}.$$

KEY IDEA

Any capacitors connected in series can be replaced with their equivalent capacitor, and any capacitors connected in parallel can be replaced with their equivalent capacitor. Therefore, we should first check whether any of the capacitors in Fig. 25-10*a* are in parallel or series.

Finding equivalent capacitance: Capacitors 1 and 3 are connected one after the other, but are they in series? No. The potential V that is applied to the capacitors produces charge on the bottom plate of capacitor 3. That charge causes charge to shift from the top plate of capacitor 3. However, note that the shifting charge can move to the bottom plates of both capacitor 1 and capacitor 2. Because there is more than one route for the shifting charge, capacitor 3 is not in series with capacitor 1 (or capacitor 2).

Are capacitor 1 and capacitor 2 in parallel? Yes. Their top plates are directly wired together and their bottom plates are directly wired together, and electric potential is applied between the top-plate pair and the bottom-plate pair. Thus, capacitor 1 and capacitor 2 are in parallel, and Eq. 25-19 tells us that their equivalent capacitance C_{12} is

$$C_{12} = C_1 + C_2 = 12.0\ \mu\text{F} + 5.30\ \mu\text{F} = 17.3\ \mu\text{F}.$$

In Fig. 25-10*b*, we have replaced capacitors 1 and 2 with their equivalent capacitor, called capacitor 12 (say "one two" and not "twelve"). (The connections at points A and B are exactly the same in Figs. 25-10*a* and *b*.)

Is capacitor 12 in series with capacitor 3? Again applying the test for series capacitances, we see that the charge that shifts from the top plate of capacitor 3 must entirely go to the bottom plate of capacitor 12. Thus, capacitor 12 and capacitor 3 are in series, and we can replace them with their equivalent C_{123} ("one two three"), as shown in Fig. 25-10*c*.

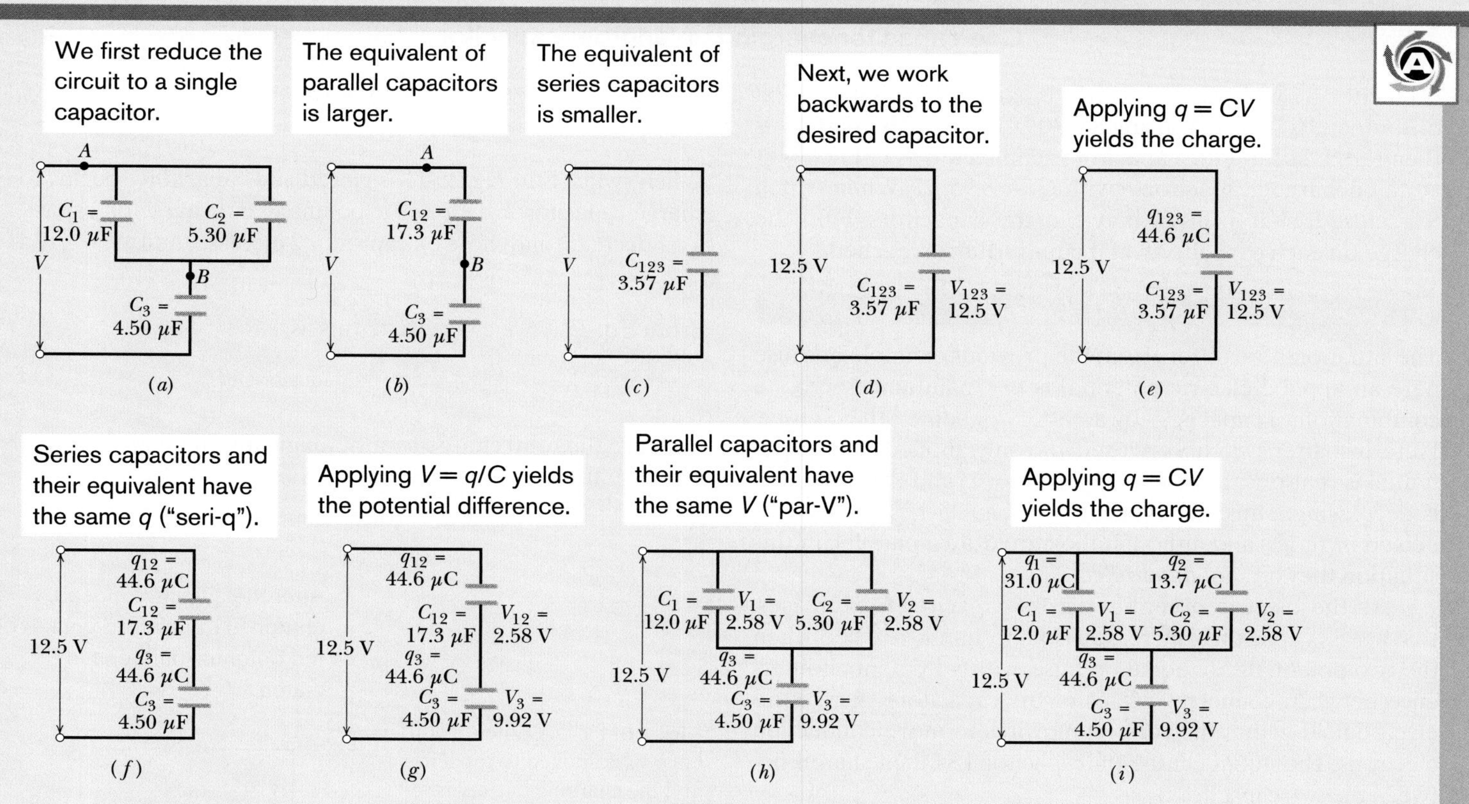

Fig. 25-10 (*a*) – (*d*) Three capacitors are reduced to one equivalent capacitor. (*e*) – (*i*) Working backwards to get the charges.

From Eq. 25-20, we have

$$\frac{1}{C_{123}} = \frac{1}{C_{12}} + \frac{1}{C_3}$$

$$= \frac{1}{17.3\ \mu\text{F}} + \frac{1}{4.50\ \mu\text{F}} = 0.280\ \mu\text{F}^{-1},$$

from which

$$C_{123} = \frac{1}{0.280\ \mu\text{F}^{-1}} = 3.57\ \mu\text{F}. \qquad \text{(Answer)}$$

(b) The potential difference applied to the input terminals in Fig. 25-10*a* is $V = 12.5$ V. What is the charge on C_1?

KEY IDEAS

We now need to work backwards from the equivalent capacitance to get the charge on a particular capacitor. We have two techniques for such "backwards work": (1) Seri-q: Series capacitors have the same charge as their equivalent capacitor. (2) Par-V: Parallel capacitors have the same potential difference as their equivalent capacitor.

Working backwards: To get the charge q_1 on capacitor 1, we work backwards to that capacitor, starting with the equivalent capacitor 123. Because the given potential difference V (= 12.5 V) is applied across the actual combination of three capacitors in Fig. 25-10*a*, it is also applied across C_{123} in Figs. 25-10*d* and *e*. Thus, Eq. 25-1 ($q = CV$) gives us

$$q_{123} = C_{123}V = (3.57\ \mu\text{F})(12.5\ \text{V}) = 44.6\ \mu\text{C}.$$

The series capacitors 12 and 3 in Fig. 25-10*b* each have the same charge as their equivalent capacitor 123 (Fig. 25-10*f*). Thus, capacitor 12 has charge $q_{12} = q_{123} = 44.6\ \mu$C. From Eq. 25-1 and Fig. 25-10*g*, the potential difference across capacitor 12 must be

$$V_{12} = \frac{q_{12}}{C_{12}} = \frac{44.6\ \mu\text{C}}{17.3\ \mu\text{F}} = 2.58\ \text{V}.$$

The parallel capacitors 1 and 2 each have the same potential difference as their equivalent capacitor 12 (Fig. 25-10*h*). Thus, capacitor 1 has potential difference $V_1 = V_{12} = 2.58$ V, and, from Eq. 25-1 and Fig. 25-10*i*, the charge on capacitor 1 must be

$$q_1 = C_1V_1 = (12.0\ \mu\text{F})(2.58\ \text{V})$$
$$= 31.0\ \mu\text{C}. \qquad \text{(Answer)}$$

Sample Problem

One capacitor charging up another capacitor

Capacitor 1, with $C_1 = 3.55\ \mu$F, is charged to a potential difference $V_0 = 6.30$ V, using a 6.30 V battery. The battery is then removed, and the capacitor is connected as in Fig. 25-11 to an uncharged capacitor 2, with $C_2 = 8.95\ \mu$F. When switch S is closed, charge flows between the capacitors. Find the charge on each capacitor when equilibrium is reached.

KEY IDEAS

The situation here differs from the previous example because here an applied electric potential is *not* maintained across a combination of capacitors by a battery or some other source. Here, just after switch S is closed, the only applied electric potential is that of capacitor 1 on capacitor 2, and that potential is decreasing. Thus, the capacitors in Fig. 25-11 are not connected *in series;* and although they are drawn parallel, in this situation they are not *in parallel.*

As the electric potential across capacitor 1 decreases, that across capacitor 2 increases. Equilibrium is reached when the two potentials are equal because, with no potential difference between connected plates of the capacitors, there is no electric field within the connecting wires to move conduction electrons. The initial charge on capacitor 1 is then shared between the two capacitors.

Calculations: Initially, when capacitor 1 is connected to the battery, the charge it acquires is, from Eq. 25-1,

$$q_0 = C_1V_0 = (3.55 \times 10^{-6}\ \text{F})(6.30\ \text{V})$$
$$= 22.365 \times 10^{-6}\ \text{C}.$$

When switch S in Fig. 25-11 is closed and capacitor 1 begins to charge capacitor 2, the electric potential and charge on capacitor 1 decrease and those on capacitor 2 increase until

$$V_1 = V_2 \qquad \text{(equilibrium).}$$

From Eq. 25-1, we can rewrite this as

$$\frac{q_1}{C_1} = \frac{q_2}{C_2} \qquad \text{(equilibrium).}$$

Because the total charge cannot magically change, the total after the transfer must be

$$q_1 + q_2 = q_0 \qquad \text{(charge conservation);}$$

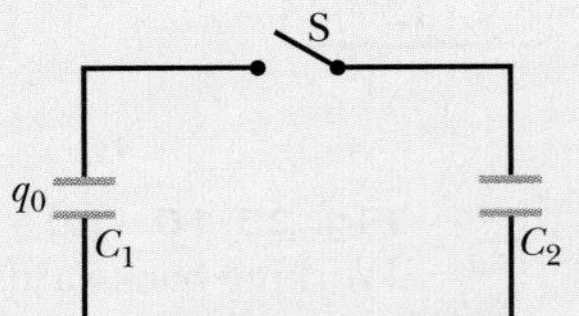

Fig. 25-11 A potential difference V_0 is applied to capacitor 1 and the charging battery is removed. Switch S is then closed so that the charge on capacitor 1 is shared with capacitor 2.

thus $q_2 = q_0 - q_1$.

We can now rewrite the second equilibrium equation as

$$\frac{q_1}{C_1} = \frac{q_0 - q_1}{C_2}.$$

Solving this for q_1 and substituting given data, we find

$$q_1 = 6.35\ \mu\text{C}. \quad \text{(Answer)}$$

The rest of the initial charge ($q_0 = 22.365\ \mu$C) must be on capacitor 2:

$$q_2 = 16.0\ \mu\text{C}. \quad \text{(Answer)}$$

Additional examples, video, and practice available at *WileyPLUS*

25-5 **Energy Stored in an Electric Field**

Work must be done by an external agent to charge a capacitor. Starting with an uncharged capacitor, for example, imagine that—using "magic tweezers"—you remove electrons from one plate and transfer them one at a time to the other plate. The electric field that builds up in the space between the plates has a direction that tends to oppose further transfer. Thus, as charge accumulates on the capacitor plates, you have to do increasingly larger amounts of work to transfer additional electrons. In practice, this work is done not by "magic tweezers" but by a battery, at the expense of its store of chemical energy.

We visualize the work required to charge a capacitor as being stored in the form of electric potential energy U in the electric field between the plates. You can recover this energy at will, by discharging the capacitor in a circuit, just as you can recover the potential energy stored in a stretched bow by releasing the bowstring to transfer the energy to the kinetic energy of an arrow.

Suppose that, at a given instant, a charge q' has been transferred from one plate of a capacitor to the other. The potential difference V' between the plates at that instant will be q'/C. If an extra increment of charge dq' is then transferred, the increment of work required will be, from Eq. 24-7,

$$dW = V'\,dq' = \frac{q'}{C}\,dq'.$$

The work required to bring the total capacitor charge up to a final value q is

$$W = \int dW = \frac{1}{C}\int_0^q q'\,dq' = \frac{q^2}{2C}.$$

This work is stored as potential energy U in the capacitor, so that

$$U = \frac{q^2}{2C} \quad \text{(potential energy).} \tag{25-21}$$

From Eq. 25-1, we can also write this as

$$U = \tfrac{1}{2}CV^2 \quad \text{(potential energy).} \tag{25-22}$$

Equations 25-21 and 25-22 hold no matter what the geometry of the capacitor is.

To gain some physical insight into energy storage, consider two parallel-plate capacitors that are identical except that capacitor 1 has twice the plate separation of capacitor 2. Then capacitor 1 has twice the volume between its plates and also, from Eq. 25-9, half the capacitance of capacitor 2. Equation 25-4 tells us that if both capacitors have the same charge q, the electric fields between their plates are identical. And Eq. 25-21 tells us that capacitor 1 has twice the stored potential energy of capacitor 2. Thus, of two otherwise identical capacitors with the same charge and same electric field, the one with twice the volume between its plates has twice the stored potential energy. Arguments like this tend to verify our earlier assumption:

The potential energy of a charged capacitor may be viewed as being stored in the electric field between its plates.

Explosions in Airborne Dust

As we discussed in Section 24-12, making contact with certain materials, such as clothing, carpets, and even playground slides, can leave you with a significant electrical potential. You might become painfully aware of that potential if a spark leaps between you and a grounded object, such as a faucet. In many industries involving the production and transport of powder, such as in the cosmetic and food industries, such a spark can be disastrous. Although the powder in bulk may not burn at all, when individual powder grains are airborne and thus surrounded by oxygen, they can burn so fiercely that a cloud of the grains burns as an explosion. Safety engineers cannot eliminate all possible sources of sparks in the powder industries. Instead, they attempt to keep the amount of energy available in the sparks below the threshold value U_t ($\approx$ 150 mJ) typically required to ignite airborne grains.

Suppose a person becomes charged by contact with various surfaces as he walks through an airborne powder. We can roughly model the person as a spherical capacitor of radius R = 1.8 m. From Eq. 25-18 ($C = 4\pi\varepsilon_0 R$) and Eq. 25-22 ($U = \frac{1}{2}CV^2$), we see that the energy of the capacitor is

$$U = \tfrac{1}{2}(4\pi\varepsilon_0 R)V^2.$$

From this we see that the threshold energy corresponds to a potential of

$$V = \sqrt{\frac{2U_t}{4\pi\varepsilon_0 R}} = \sqrt{\frac{2(150 \times 10^{-3}\ \text{J})}{4\pi(8.85 \times 10^{-12}\ \text{C}^2/\text{N}\cdot\text{m}^2)(1.8\ \text{m})}}$$
$$= 3.9 \times 10^4\ \text{V}.$$

Safety engineers attempt to keep the potential of the personnel below this level by "bleeding" off the charge through, say, a conducting floor.

Energy Density

In a parallel-plate capacitor, neglecting fringing, the electric field has the same value at all points between the plates. Thus, the **energy density** u—that is, the potential energy per unit volume between the plates—should also be uniform. We can find u by dividing the total potential energy by the volume Ad of the space between the plates. Using Eq. 25-22, we obtain

$$u = \frac{U}{Ad} = \frac{CV^2}{2Ad}. \tag{25-23}$$

With Eq. 25-9 ($C = \varepsilon_0 A/d$), this result becomes

$$u = \tfrac{1}{2}\varepsilon_0\left(\frac{V}{d}\right)^2. \tag{25-24}$$

However, from Eq. 24-42 ($E = -\Delta V/\Delta s$), V/d equals the electric field magnitude E; so

$$u = \tfrac{1}{2}\varepsilon_0 E^2 \quad \text{(energy density).} \tag{25-25}$$

Although we derived this result for the special case of an electric field of a parallel-plate capacitor, it holds generally, whatever may be the source of the electric field. If an electric field $\vec{E}$ exists at any point in space, we can think of that point as a site of electric potential energy with a density (amount per unit volume) given by Eq. 25-25.

Sample Problem

Potential energy and energy density of an electric field

An isolated conducting sphere whose radius R is 6.85 cm has a charge $q = 1.25$ nC.

(a) How much potential energy is stored in the electric field of this charged conductor?

KEY IDEAS

(1) An isolated sphere has capacitance given by Eq. 25-18 ($C = 4\pi\varepsilon_0 R$). (2) The energy U stored in a capacitor depends on the capacitor's charge q and capacitance C according to Eq. 25-21 ($U = q^2/2C$).

Calculation: Substituting $C = 4\pi\varepsilon_0 R$ into Eq. 25-21 gives us

$$U = \frac{q^2}{2C} = \frac{q^2}{8\pi\varepsilon_0 R}$$

$$= \frac{(1.25 \times 10^{-9}\text{ C})^2}{(8\pi)(8.85 \times 10^{-12}\text{ F/m})(0.0685\text{ m})}$$

$$= 1.03 \times 10^{-7}\text{ J} = 103\text{ nJ}. \qquad \text{(Answer)}$$

(b) What is the energy density at the surface of the sphere?

KEY IDEA

The density u of the energy stored in an electric field depends on the magnitude E of the field, according to Eq. 25-25 ($u = \frac{1}{2}\varepsilon_0 E^2$).

Calculations: Here we must first find E at the surface of the sphere, as given by Eq. 23-15:

$$E = \frac{1}{4\pi\varepsilon_0}\frac{q}{R^2}.$$

The energy density is then

$$u = \tfrac{1}{2}\varepsilon_0 E^2 = \frac{q^2}{32\pi^2\varepsilon_0 R^4}$$

$$= \frac{(1.25 \times 10^{-9}\text{ C})^2}{(32\pi^2)(8.85 \times 10^{-12}\text{ C}^2/\text{N}\cdot\text{m}^2)(0.0685\text{ m})^4}$$

$$= 2.54 \times 10^{-5}\text{ J/m}^3 = 25.4\ \mu\text{J/m}^3. \qquad \text{(Answer)}$$

Additional examples, video, and practice available at *WileyPLUS*

25-6 Capacitor with a Dielectric

If you fill the space between the plates of a capacitor with a *dielectric,* which is an insulating material such as mineral oil or plastic, what happens to the capacitance? Michael Faraday—to whom the whole concept of capacitance is largely due and for whom the SI unit of capacitance is named—first looked into this matter in 1837. Using simple equipment much like that shown in Fig. 25-12, he found that the capacitance *increased* by a numerical factor κ, which he called the **dielectric constant** of the insulating material. Table 25-1 shows some dielectric materials and their dielectric constants. The dielectric constant of a vacuum is unity by definition. Because air is mostly empty space, its measured dielectric constant is only slightly greater than unity. Even common paper can significantly

Fig. 25-12 The simple electrostatic apparatus used by Faraday. An assembled apparatus (second from left) forms a spherical capacitor consisting of a central brass ball and a concentric brass shell. Faraday placed dielectric materials in the space between the ball and the shell. *(The Royal Institute, England/Bridgeman Art Library/NY)*

Table 25-1

Some Properties of Dielectrics[a]

Material	Dielectric Constant κ	Dielectric Strength (kV/mm)
Air (1 atm)	1.00054	3
Polystyrene	2.6	24
Paper	3.5	16
Transformer oil	4.5	
Pyrex	4.7	14
Ruby mica	5.4	
Porcelain	6.5	
Silicon	12	
Germanium	16	
Ethanol	25	
Water (20°C)	80.4	
Water (25°C)	78.5	
Titania ceramic	130	
Strontium titanate	310	8

For a vacuum, κ = unity.

[a]Measured at room temperature, except for the water.

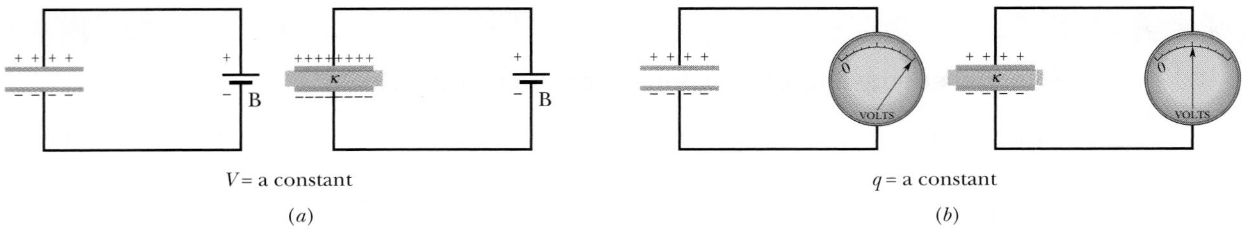

Fig. 25-13 (*a*) If the potential difference between the plates of a capacitor is maintained, as by battery B, the effect of a dielectric is to increase the charge on the plates. (*b*) If the charge on the capacitor plates is maintained, as in this case, the effect of a dielectric is to reduce the potential difference between the plates. The scale shown is that of a *potentiometer,* a device used to measure potential difference (here, between the plates). A capacitor cannot discharge through a potentiometer.

increase the capacitance of a capacitor, and some materials, such as strontium titanate, can increase the capacitance by more than two orders of magnitude.

Another effect of the introduction of a dielectric is to limit the potential difference that can be applied between the plates to a certain value V_{max}, called the *breakdown potential.* If this value is substantially exceeded, the dielectric material will break down and form a conducting path between the plates. Every dielectric material has a characteristic *dielectric strength,* which is the maximum value of the electric field that it can tolerate without breakdown. A few such values are listed in Table 25-1.

As we discussed just after Eq. 25-18, the capacitance of any capacitor can be written in the form

$$C = \varepsilon_0 \mathscr{L}, \tag{25-26}$$

in which $\mathscr{L}$ has the dimension of length. For example, $\mathscr{L} = A/d$ for a parallel-plate capacitor. Faraday's discovery was that, with a dielectric *completely* filling the space between the plates, Eq. 25-26 becomes

$$C = \kappa \varepsilon_0 \mathscr{L} = \kappa C_{air}, \tag{25-27}$$

where C_{air} is the value of the capacitance with only air between the plates. For example, if we fill a capacitor with strontium titanate, with a dielectric constant of 310, we multiply the capacitance by 310.

Figure 25-13 provides some insight into Faraday's experiments. In Fig. 25-13*a* the battery ensures that the potential difference V between the plates will remain constant. When a dielectric slab is inserted between the plates, the charge q on the plates increases by a factor of κ; the additional charge is delivered to the capacitor plates by the battery. In Fig. 25-13*b* there is no battery, and therefore the charge q must remain constant when the dielectric slab is inserted; then the potential difference V between the plates decreases by a factor of κ. Both these observations are consistent (through the relation $q = CV$) with the increase in capacitance caused by the dielectric.

Comparison of Eqs. 25-26 and 25-27 suggests that the effect of a dielectric can be summed up in more general terms:

> In a region completely filled by a dielectric material of dielectric constant κ, all electrostatic equations containing the permittivity constant ε_0 are to be modified by replacing ε_0 with $\kappa\varepsilon_0$.

Thus, the magnitude of the electric field produced by a point charge inside a dielectric is given by this modified form of Eq. 23-15:

$$E = \frac{1}{4\pi\kappa\varepsilon_0} \frac{q}{r^2}. \tag{25-28}$$

Also, the expression for the electric field just outside an isolated conductor immersed in a dielectric (see Eq. 23-11) becomes

$$E = \frac{\sigma}{\kappa\varepsilon_0}. \tag{25-29}$$

Because κ is always greater than unity, both these equations show that *for a fixed distribution of charges, the effect of a dielectric is to weaken the electric field* that would otherwise be present.

Sample Problem

Work and energy when a dielectric is inserted into a capacitor

A parallel-plate capacitor whose capacitance C is 13.5 pF is charged by a battery to a potential difference $V = 12.5$ V between its plates. The charging battery is now disconnected, and a porcelain slab ($\kappa = 6.50$) is slipped between the plates.

(a) What is the potential energy of the capacitor before the slab is inserted?

KEY IDEA

We can relate the potential energy U_i of the capacitor to the capacitance C and either the potential V (with Eq. 25-22) or the charge q (with Eq. 25-21):

$$U_i = \tfrac{1}{2}CV^2 = \frac{q^2}{2C}.$$

Calculation: Because we are given the initial potential V (= 12.5 V), we use Eq. 25-22 to find the initial stored energy:

$$U_i = \tfrac{1}{2}CV^2 = \tfrac{1}{2}(13.5 \times 10^{-12}\ \text{F})(12.5\ \text{V})^2$$
$$= 1.055 \times 10^{-9}\ \text{J} = 1055\ \text{pJ} \approx 1100\ \text{pJ}. \quad \text{(Answer)}$$

(b) What is the potential energy of the capacitor–slab device after the slab is inserted?

KEY IDEA

Because the battery has been disconnected, the charge on the capacitor cannot change when the dielectric is inserted. However, the potential *does* change.

Calculations: Thus, we must now use Eq. 25-21 to write the final potential energy U_f, but now that the slab is within the capacitor, the capacitance is κC. We then have

$$U_f = \frac{q^2}{2\kappa C} = \frac{U_i}{\kappa} = \frac{1055\ \text{pJ}}{6.50}$$
$$= 162\ \text{pJ} \approx 160\ \text{pJ}. \quad \text{(Answer)}$$

When the slab is introduced, the potential energy decreases by a factor of κ.

The "missing" energy, in principle, would be apparent to the person who introduced the slab. The capacitor would exert a tiny tug on the slab and would do work on it, in amount

$$W = U_i - U_f = (1055 - 162)\ \text{pJ} = 893\ \text{pJ}.$$

If the slab were allowed to slide between the plates with no restraint and if there were no friction, the slab would oscillate back and forth between the plates with a (constant) mechanical energy of 893 pJ, and this system energy would transfer back and forth between kinetic energy of the moving slab and potential energy stored in the electric field.

Additional examples, video, and practice available at *WileyPLUS*

25-7 Dielectrics: An Atomic View

What happens, in atomic and molecular terms, when we put a dielectric in an electric field? There are two possibilities, depending on the type of molecule:

1. *Polar dielectrics.* The molecules of some dielectrics, like water, have permanent electric dipole moments. In such materials (called *polar dielectrics*), the electric dipoles tend to line up with an external electric field as in Fig. 25-14. Because the molecules are continuously jostling each other as a result of their random thermal motion, this alignment is not complete, but it becomes more complete as the magnitude of the applied field is increased (or as the temperature, and thus the jostling, are decreased). The alignment of the electric dipoles produces an electric field that is directed opposite the applied field and is smaller in magnitude.

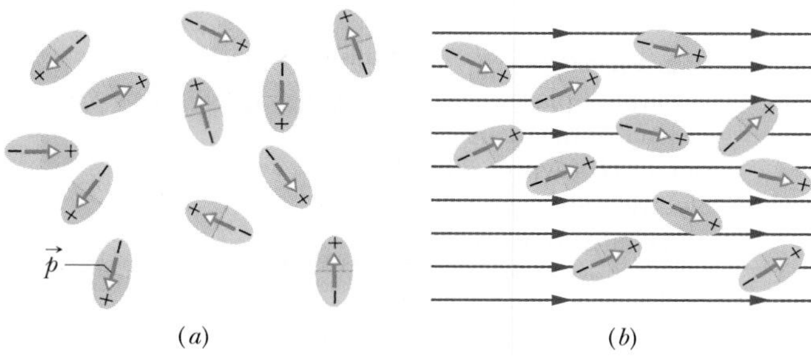

Fig. 25-14 (*a*) Molecules with a permanent electric dipole moment, showing their random orientation in the absence of an external electric field. (*b*) An electric field is applied, producing partial alignment of the dipoles. Thermal agitation prevents complete alignment.

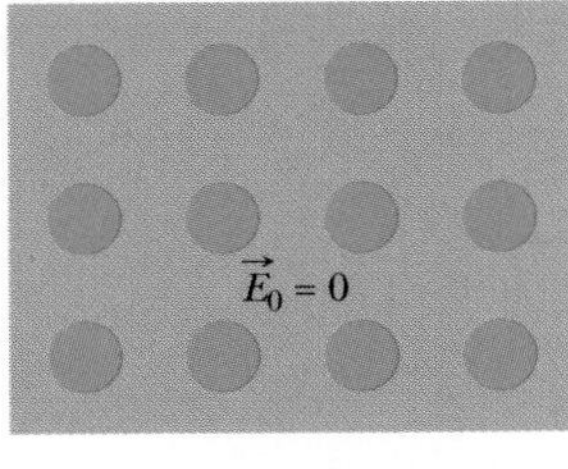

(a)

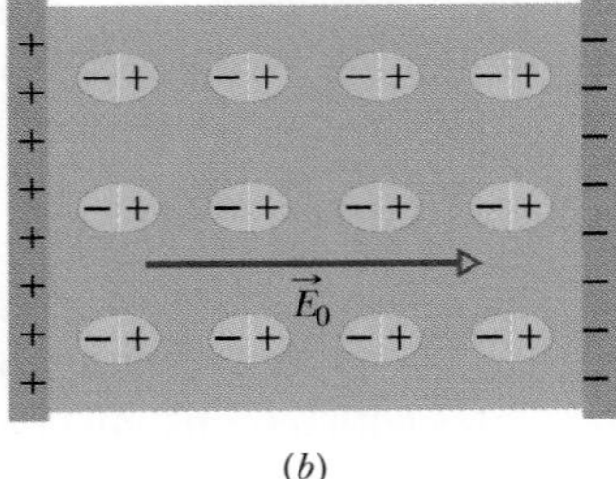

(b)

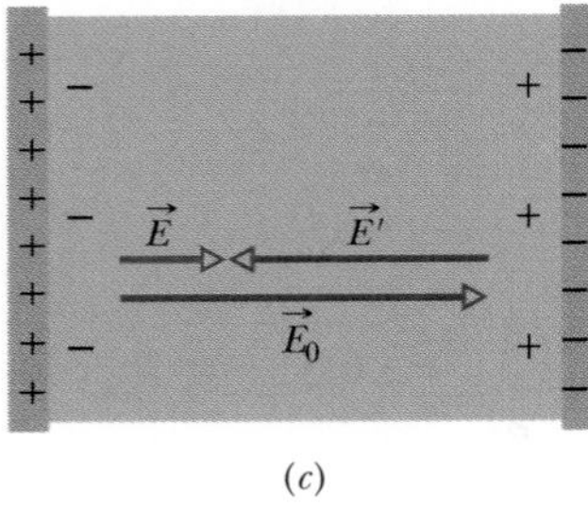

(c)

Fig. 25-15 (*a*) A nonpolar dielectric slab. The circles represent the electrically neutral atoms within the slab. (*b*) An electric field is applied via charged capacitor plates; the field slightly stretches the atoms, separating the centers of positive and negative charge. (*c*) The separation produces surface charges on the slab faces. These charges set up a field $\vec{E}'$, which opposes the applied field $\vec{E}_0$. The resultant field $\vec{E}$ inside the dielectric (the vector sum of $\vec{E}_0$ and $\vec{E}'$) has the same direction as $\vec{E}_0$ but a smaller magnitude.

2. *Nonpolar dielectrics.* Regardless of whether they have permanent electric dipole moments, molecules acquire dipole moments by induction when placed in an external electric field. In Section 24-8 (see Fig. 24-11), we saw that this occurs because the external field tends to "stretch" the molecules, slightly separating the centers of negative and positive charge.

Figure 25-15*a* shows a nonpolar dielectric slab with no external electric field applied. In Fig. 25-15*b*, an electric field $\vec{E}_0$ is applied via a capacitor, whose plates are charged as shown. The result is a slight separation of the centers of the positive and negative charge distributions within the slab, producing positive charge on one face of the slab (due to the positive ends of dipoles there) and negative charge on the opposite face (due to the negative ends of dipoles there). The slab as a whole remains electrically neutral and—within the slab—there is no excess charge in any volume element.

Figure 25-15*c* shows that the induced surface charges on the faces produce an electric field $\vec{E}'$ in the direction opposite that of the applied electric field $\vec{E}_0$. The resultant field $\vec{E}$ inside the dielectric (the vector sum of fields $\vec{E}_0$ and $\vec{E}'$) has the direction of $\vec{E}_0$ but is smaller in magnitude.

Both the field $\vec{E}'$ produced by the surface charges in Fig. 25-15*c* and the electric field produced by the permanent electric dipoles in Fig. 25-14 act in the same way—they oppose the applied field $\vec{E}$. Thus, the effect of both polar and nonpolar dielectrics is to weaken any applied field within them, as between the plates of a capacitor.

25-8 Dielectrics and Gauss' Law

In our discussion of Gauss' law in Chapter 23, we assumed that the charges existed in a vacuum. Here we shall see how to modify and generalize that law if dielectric materials, such as those listed in Table 25-1, are present. Figure 25-16 shows a parallel-plate capacitor of plate area A, both with and without a dielectric. We assume that the charge q on the plates is the same in both situations. Note that the field between the plates induces charges on the faces of the dielectric by one of the methods described in Section 25-7.

For the situation of Fig. 25-16*a*, without a dielectric, we can find the electric field $\vec{E}_0$ between the plates as we did in Fig. 25-5: We enclose the charge $+q$ on the top plate with a Gaussian surface and then apply Gauss' law. Letting E_0 represent the magnitude of the field, we find

$$\varepsilon_0 \oint \vec{E} \cdot d\vec{A} = \varepsilon_0 EA = q, \tag{25-30}$$

or

$$E_0 = \frac{q}{\varepsilon_0 A}. \tag{25-31}$$

In Fig. 25-16*b*, with the dielectric in place, we can find the electric field between the plates (and within the dielectric) by using the same Gaussian surface. However, now the surface encloses two types of charge: It still encloses

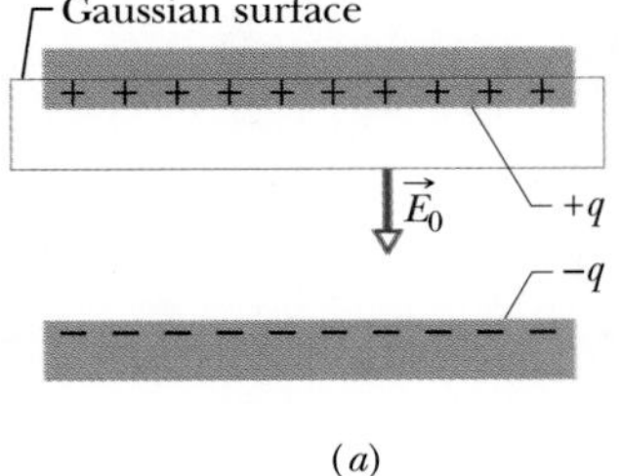

(a)

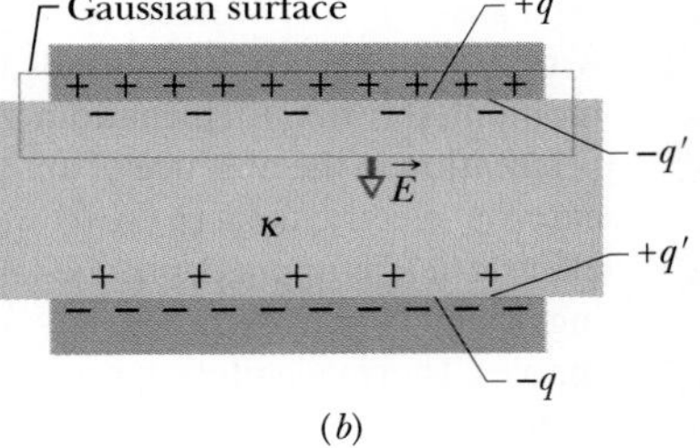

(b)

Fig. 25-16 A parallel-plate capacitor (*a*) without and (*b*) with a dielectric slab inserted. The charge q on the plates is assumed to be the same in both cases.

charge $+q$ on the top plate, but it now also encloses the induced charge $-q'$ on the top face of the dielectric. The charge on the conducting plate is said to be *free charge* because it can move if we change the electric potential of the plate; the induced charge on the surface of the dielectric is not free charge because it cannot move from that surface.

The net charge enclosed by the Gaussian surface in Fig. 25-16*b* is $q - q'$, so Gauss' law now gives

$$\varepsilon_0 \oint \vec{E} \cdot d\vec{A} = \varepsilon_0 EA = q - q', \tag{25-32}$$

or

$$E = \frac{q - q'}{\varepsilon_0 A}. \tag{25-33}$$

The effect of the dielectric is to weaken the original field E_0 by a factor of κ; so we may write

$$E = \frac{E_0}{\kappa} = \frac{q}{\kappa \varepsilon_0 A}. \tag{25-34}$$

Comparison of Eqs. 25-33 and 25-34 shows that

$$q - q' = \frac{q}{\kappa}. \tag{25-35}$$

Equation 25-35 shows correctly that the magnitude q' of the induced surface charge is less than that of the free charge q and is zero if no dielectric is present (because then $\kappa = 1$ in Eq. 25-35).

By substituting for $q - q'$ from Eq. 25-35 in Eq. 25-32, we can write Gauss' law in the form

$$\varepsilon_0 \oint \kappa \vec{E} \cdot d\vec{A} = q \quad \text{(Gauss' law with dielectric).} \tag{25-36}$$

This equation, although derived for a parallel-plate capacitor, is true generally and is the most general form in which Gauss' law can be written. Note:

1. The flux integral now involves $\kappa\vec{E}$, not just $\vec{E}$. (The vector $\varepsilon_0\kappa\vec{E}$ is sometimes called the *electric displacement* $\vec{D}$, so that Eq. 25-36 can be written in the form $\oint \vec{D} \cdot d\vec{A} = q$.)
2. The charge q enclosed by the Gaussian surface is now taken to be the *free charge only.* The induced surface charge is deliberately ignored on the right side of Eq. 25-36, having been taken fully into account by introducing the dielectric constant κ on the left side.
3. Equation 25-36 differs from Eq. 23-7, our original statement of Gauss' law, only in that ε_0 in the latter equation has been replaced by $\kappa\varepsilon_0$. We keep κ inside the integral of Eq. 25-36 to allow for cases in which κ is not constant over the entire Gaussian surface.

Sample Problem

Dielectric partially filling the gap in a capacitor

Figure 25-17 shows a parallel-plate capacitor of plate area A and plate separation d. A potential difference V_0 is applied between the plates by connecting a battery between them. The battery is then disconnected, and a dielectric slab of thickness b and dielectric constant κ is placed between the plates as shown. Assume $A = 115\ \text{cm}^2$, $d = 1.24$ cm, $V_0 = 85.5$ V, $b = 0.780$ cm, and $\kappa = 2.61$.

(a) What is the capacitance C_0 before the dielectric slab is inserted?

Calculation: From Eq. 25-9 we have

$$C_0 = \frac{\varepsilon_0 A}{d} = \frac{(8.85 \times 10^{-12}\ \text{F/m})(115 \times 10^{-4}\ \text{m}^2)}{1.24 \times 10^{-2}\ \text{m}}$$

$$= 8.21 \times 10^{-12}\ \text{F} = 8.21\ \text{pF}. \quad \text{(Answer)}$$

(b) What free charge appears on the plates?

Calculation: From Eq. 25-1,

$$q = C_0V_0 = (8.21 \times 10^{-12}\text{ F})(85.5\text{ V}) = 7.02 \times 10^{-10}\text{ C} = 702\text{ pC}. \qquad \text{(Answer)}$$

Because the battery was disconnected before the slab was inserted, the free charge is unchanged.

(c) What is the electric field E_0 in the gaps between the plates and the dielectric slab?

KEY IDEA

We need to apply Gauss' law, in the form of Eq. 25-36, to Gaussian surface I in Fig. 25-17.

Calculations: That surface passes through the gap, and so it encloses *only* the free charge on the upper capacitor plate. Electric field pierces only the bottom of the Gaussian surface. Because there the area vector $d\vec{A}$ and the field vector $\vec{E}_0$ are both directed downward, the dot product in Eq. 25-36 becomes

$$\vec{E}_0 \cdot d\vec{A} = E_0\,dA \cos 0° = E_0\,dA.$$

Equation 25-36 then becomes

$$\varepsilon_0 \kappa E_0 \oint dA = q.$$

The integration now simply gives the surface area A of the plate. Thus, we obtain

$$\varepsilon_0 \kappa E_0 A = q,$$

or

$$E_0 = \frac{q}{\varepsilon_0 \kappa A}.$$

We must put $\kappa = 1$ here because Gaussian surface I does not pass through the dielectric. Thus, we have

$$E_0 = \frac{q}{\varepsilon_0 \kappa A} = \frac{7.02 \times 10^{-10}\text{ C}}{(8.85 \times 10^{-12}\text{ F/m})(1)(115 \times 10^{-4}\text{ m}^2)} = 6900\text{ V/m} = 6.90\text{ kV/m}. \qquad \text{(Answer)}$$

Note that the value of E_0 does not change when the slab is introduced because the amount of charge enclosed by Gaussian surface I in Fig. 25-17 does not change.

(d) What is the electric field E_1 in the dielectric slab?

KEY IDEA

Now we apply Gauss' law in the form of Eq. 25-36 to Gaussian surface II in Fig. 25-17.

Calculations: That surface encloses free charge $-q$ and induced charge $+q'$, but we ignore the latter when we use Eq. 25-36. We find

$$\varepsilon_0 \oint \kappa \vec{E}_1 \cdot d\vec{A} = -\varepsilon_0 \kappa E_1 A = -q. \qquad (25\text{-}37)$$

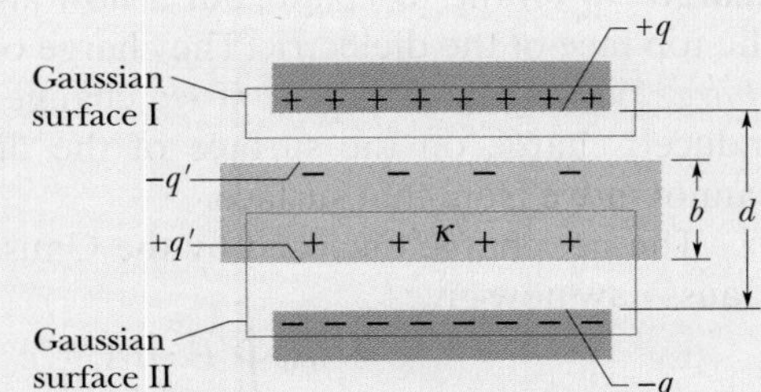

Fig. 25-17 A parallel-plate capacitor containing a dielectric slab that only partially fills the space between the plates.

The first minus sign in this equation comes from the dot product $\vec{E}_1 \cdot d\vec{A}$ along the top of the Gaussian surface because now the field vector $\vec{E}_1$ is directed downward and the area vector $d\vec{A}$ (which, as always, points outward from the interior of a closed Gaussian surface) is directed upward. With 180° between the vectors, the dot product is negative. Now $\kappa = 2.61$. Thus, Eq. 25-37 gives us

$$E_1 = \frac{q}{\varepsilon_0 \kappa A} = \frac{E_0}{\kappa} = \frac{6.90\text{ kV/m}}{2.61} = 2.64\text{ kV/m}. \qquad \text{(Answer)}$$

(e) What is the potential difference V between the plates after the slab has been introduced?

KEY IDEA

We find V by integrating along a straight line directly from the bottom plate to the top plate.

Calculation: Within the dielectric, the path length is b and the electric field is E_1. Within the two gaps above and below the dielectric, the total path length is $d - b$ and the electric field is E_0. Equation 25-6 then yields

$$V = \int_-^+ E\,ds = E_0(d - b) + E_1 b$$
$$= (6900\text{ V/m})(0.0124\text{ m} - 0.00780\text{ m}) + (2640\text{ V/m})(0.00780\text{ m})$$
$$= 52.3\text{ V}. \qquad \text{(Answer)}$$

This is less than the original potential difference of 85.5 V.

(f) What is the capacitance with the slab in place between the plates of the capacitor?

KEY IDEA

The capacitance C is related to the free charge q and the potential difference V via Eq. 25-1.

Calculation: Taking q from (b) and V from (e), we have

$$C = \frac{q}{V} = \frac{7.02 \times 10^{-10}\text{ C}}{52.3\text{ V}} = 1.34 \times 10^{-11}\text{ F} = 13.4\text{ pF}. \qquad \text{(Answer)}$$

This is greater than the original capacitance of 8.21 pF.

Additional examples, video, and practice available at *WileyPLUS*

REVIEW & SUMMARY

Capacitor; Capacitance A **capacitor** consists of two isolated conductors (the *plates*) with charges $+q$ and $-q$. Its **capacitance** C is defined from

$$q = CV, \qquad (25\text{-}1)$$

where V is the potential difference between the plates.

Determining Capacitance We generally determine the capacitance of a particular capacitor configuration by (1) assuming a charge q to have been placed on the plates, (2) finding the electric field $\vec{E}$ due to this charge, (3) evaluating the potential difference V, and (4) calculating C from Eq. 25-1. Some specific results are the following:

A *parallel-plate capacitor* with flat parallel plates of area A and spacing d has capacitance

$$C = \frac{\varepsilon_0 A}{d}. \qquad (25\text{-}9)$$

A *cylindrical capacitor* (two long coaxial cylinders) of length L and radii a and b has capacitance

$$C = 2\pi\varepsilon_0 \frac{L}{\ln(b/a)}. \qquad (25\text{-}14)$$

A *spherical capacitor* with concentric spherical plates of radii a and b has capacitance

$$C = 4\pi\varepsilon_0 \frac{ab}{b-a}. \qquad (25\text{-}17)$$

An *isolated sphere* of radius R has capacitance

$$C = 4\pi\varepsilon_0 R. \qquad (25\text{-}18)$$

Capacitors in Parallel and in Series The **equivalent capacitances** C_{eq} of combinations of individual capacitors connected in **parallel** and in **series** can be found from

$$C_{eq} = \sum_{j=1}^{n} C_j \quad (n \text{ capacitors in parallel}) \qquad (25\text{-}19)$$

and

$$\frac{1}{C_{eq}} = \sum_{j=1}^{n} \frac{1}{C_j} \quad (n \text{ capacitors in series}). \qquad (25\text{-}20)$$

Equivalent capacitances can be used to calculate the capacitances of more complicated series–parallel combinations.

Potential Energy and Energy Density The **electric potential energy** U of a charged capacitor,

$$U = \frac{q^2}{2C} = \tfrac{1}{2}CV^2, \qquad (25\text{-}21, 25\text{-}22)$$

is equal to the work required to charge the capacitor. This energy can be associated with the capacitor's electric field $\vec{E}$. By extension we can associate stored energy with any electric field. In vacuum, the **energy density** u, or potential energy per unit volume, within an electric field of magnitude E is given by

$$u = \tfrac{1}{2}\varepsilon_0 E^2. \qquad (25\text{-}25)$$

Capacitance with a Dielectric If the space between the plates of a capacitor is completely filled with a dielectric material, the capacitance C is increased by a factor κ, called the **dielectric constant,** which is characteristic of the material. In a region that is completely filled by a dielectric, all electrostatic equations containing ε_0 must be modified by replacing ε_0 with $\kappa\varepsilon_0$.

The effects of adding a dielectric can be understood physically in terms of the action of an electric field on the permanent or induced electric dipoles in the dielectric slab. The result is the formation of induced charges on the surfaces of the dielectric, which results in a weakening of the field within the dielectric for a given amount of free charge on the plates.

Gauss' Law with a Dielectric When a dielectric is present, Gauss' law may be generalized to

$$\varepsilon_0 \oint \kappa\vec{E}\cdot d\vec{A} = q. \qquad (25\text{-}36)$$

Here q is the free charge; any induced surface charge is accounted for by including the dielectric constant κ inside the integral.

QUESTIONS

1 Figure 25-18 shows plots of charge versus potential difference for three parallel-plate capacitors that have the plate areas and separations given in the table. Which plot goes with which capacitor?

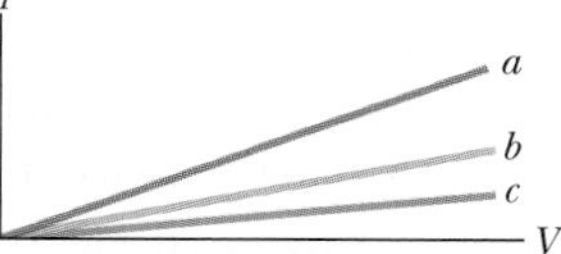

Fig. 25-18 Question 1.

Capacitor	Area	Separation
1	A	d
2	$2A$	d
3	A	$2d$

2 What is C_{eq} of three capacitors, each of capacitance C, if they are connected to a battery (a) in series with one another and (b) in parallel? (c) In which arrangement is there more charge on the equivalent capacitance?

3 (a) In Fig. 25-19*a*, are capacitors 1 and 3 in series? (b) In the same figure, are capacitors 1 and 2 in parallel? (c) Rank the equivalent capacitances of the four circuits shown in Fig. 25-19, greatest first.

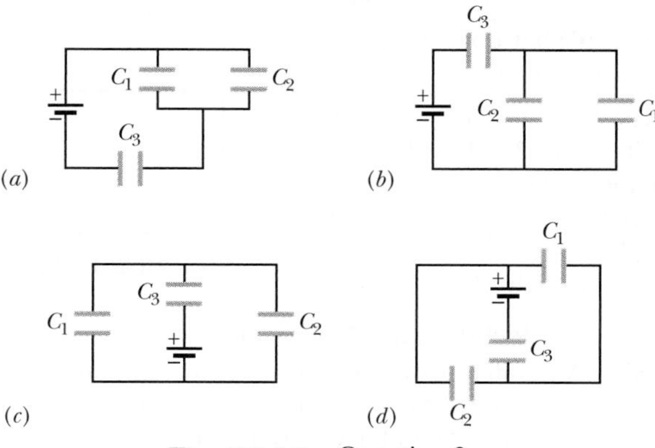

Fig. 25-19 Question 3.

4 Figure 25-20 shows three circuits, each consisting of a switch and two capacitors, initially charged as indicated (top plate positive). After the switches have been closed, in which circuit (if any) will the charge on the left-hand capacitor (a) increase, (b) decrease, and (c) remain the same?

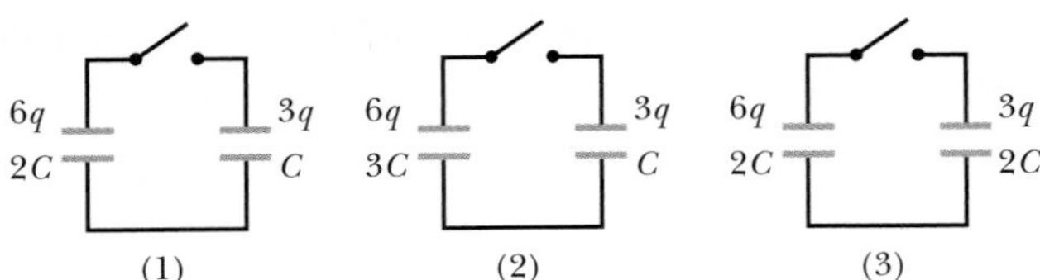

Fig. 25-20 Question 4.

5 Initially, a single capacitance C_1 is wired to a battery. Then capacitance C_2 is added in parallel. Are (a) the potential difference across C_1 and (b) the charge q_1 on C_1 now more than, less than, or the same as previously? (c) Is the equivalent capacitance C_{12} of C_1 and C_2 more than, less than, or equal to C_1? (d) Is the charge stored on C_1 and C_2 together more than, less than, or equal to the charge stored previously on C_1?

6 Repeat Question 5 for C_2 added in series rather than in parallel.

7 For each circuit in Fig. 25-21, are the capacitors connected in series, in parallel, or in neither mode?

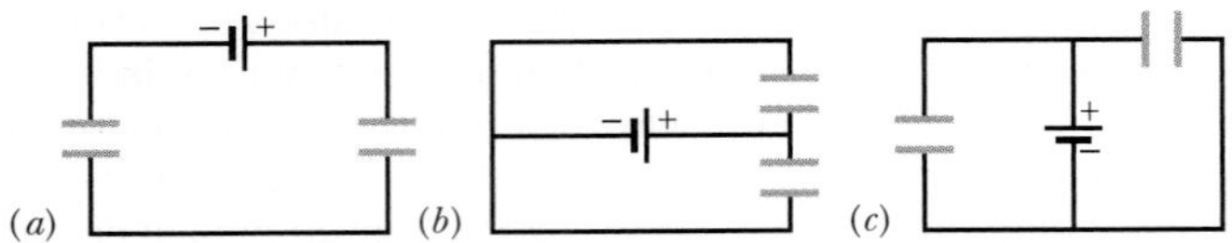

Fig. 25-21 Question 7.

8 Figure 25-22 shows an open switch, a battery of potential difference V, a current-measuring meter A, and three identical uncharged capacitors of capacitance C. When the switch is closed and the circuit reaches equilibrium, what are (a) the potential difference across each capacitor and (b) the charge on the left plate of each capacitor? (c) During charging, what net charge passes through the meter?

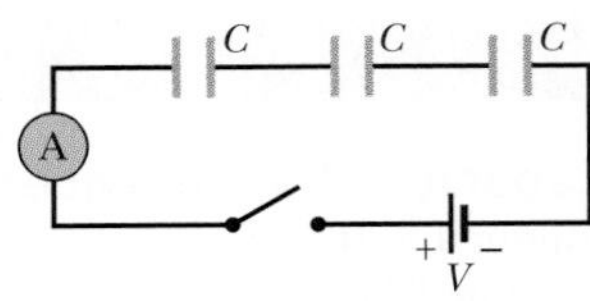

Fig. 25-22 Question 8.

9 A parallel-plate capacitor is connected to a battery of electric potential difference V. If the plate separation is decreased, do the following quantities increase, decrease, or remain the same: (a) the capacitor's capacitance, (b) the potential difference across the capacitor, (c) the charge on the capacitor, (d) the energy stored by the capacitor, (e) the magnitude of the electric field between the plates, and (f) the energy density of that electric field?

10 When a dielectric slab is inserted between the plates of one of the two identical capacitors in Fig. 25-23, do the following properties of that capacitor increase, decrease, or remain the same: (a) capacitance, (b) charge, (c) potential difference, and (d) potential energy? (e) How about the same properties of the other capacitor?

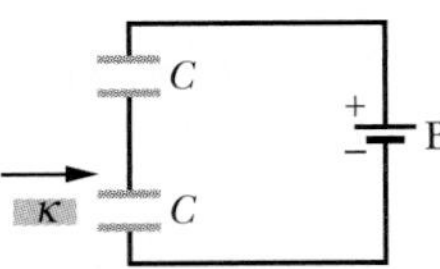

Fig. 23-19 Question 10.

11 You are to connect capacitances C_1 and C_2, with $C_1 > C_2$, to a battery, first individually, then in series, and then in parallel. Rank those arrangements according to the amount of charge stored, greatest first.

PROBLEMS

GO Tutoring problem available (at instructor's discretion) in *WileyPLUS* and WebAssign
SSM Worked-out solution available in Student Solutions Manual
• – ••• Number of dots indicates level of problem difficulty
Additional information available in *The Flying Circus of Physics* and at flyingcircusofphysics.com
WWW Worked-out solution is at
ILW Interactive solution is at
http://www.wiley.com/college/halliday

sec. 25-2 Capacitance

•1 The two metal objects in Fig. 25-24 have net charges of +70 pC and −70 pC, which result in a 20 V potential difference between them. (a) What is the capacitance of the system? (b) If the charges are changed to +200 pC and −200 pC, what does the capacitance become? (c) What does the potential difference become?

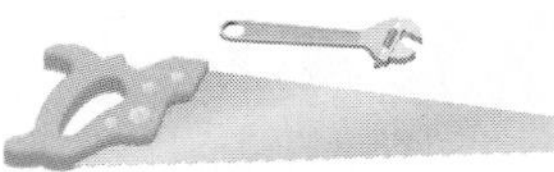

Fig. 25-24 Problem 1.

•2 The capacitor in Fig. 25-25 has a capacitance of 25 μF and is initially uncharged. The battery provides a potential difference of 120 V. After switch S is closed, how much charge will pass through it?

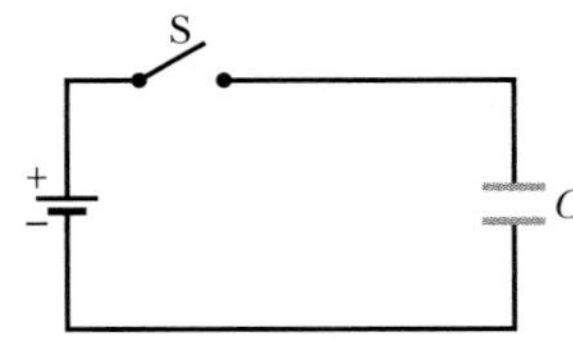

Fig. 25-25 Problem 2.

sec. 25-3 Calculating the Capacitance

•3 SSM A parallel-plate capacitor has circular plates of 8.20 cm radius and 1.30 mm separation. (a) Calculate the capacitance. (b) Find the charge for a potential difference of 120 V.

•4 The plates of a spherical capacitor have radii 38.0 mm and 40.0 mm. (a) Calculate the capacitance. (b) What must be the plate area of a parallel-plate capacitor with the same plate separation and capacitance?

•5 What is the capacitance of a drop that results when two mercury spheres, each of radius R = 2.00 mm, merge?

•6 You have two flat metal plates, each of area 1.00 m^2, with which to construct a parallel-plate capacitor. (a) If the capacitance of the device is to be 1.00 F, what must be the separation between the plates? (b) Could this capacitor actually be constructed?

•7 If an uncharged parallel-plate capacitor (capacitance C) is connected to a battery, one plate becomes negatively charged as electrons move to the plate face (area A). In Fig. 25-26, the depth d from which the electrons come in the plate in a particular capacitor is plotted against a range of values for the

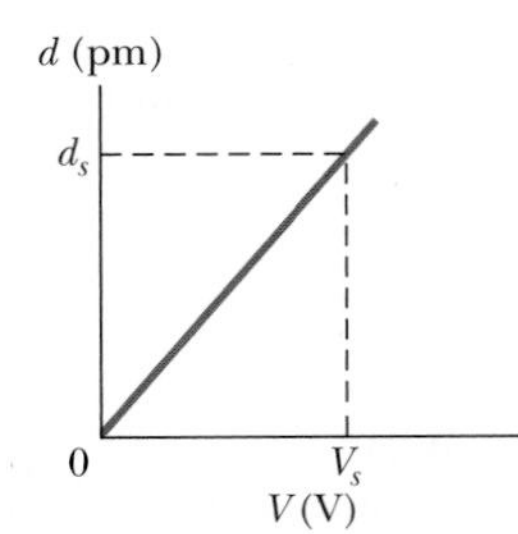

Fig. 25-26 Problem 7.

potential difference V of the battery. The density of conduction electrons in the copper plates is 8.49×10^{28} electrons/m^3. The vertical scale is set by $d_s = 1.00$ pm, and the horizontal scale is set by $V_s = 20.0$ V. What is the ratio C/A?

sec. 25-4 Capacitors in Parallel and in Series

•8 How many 1.00 μF capacitors must be connected in parallel to store a charge of 1.00 C with a potential of 110 V across the capacitors?

•9 Each of the uncharged capacitors in Fig. 25-27 has a capacitance of 25.0 μF. A potential difference of $V = 4200$ V is established when the switch is closed. How many coulombs of charge then pass through meter A?

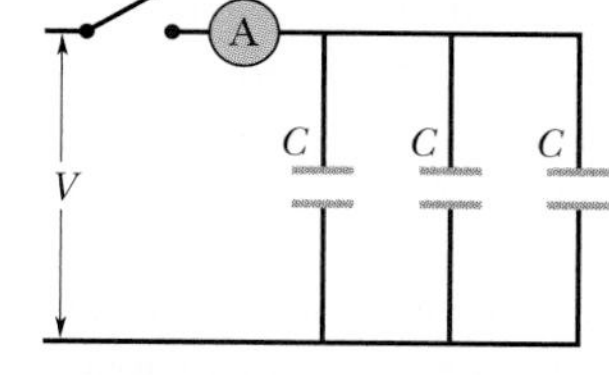

Fig. 25-27 Problem 9.

•10 In Fig. 25-28, find the equivalent capacitance of the combination. Assume that C_1 is 10.0 μF, C_2 is 5.00 μF, and C_3 is 4.00 μF.

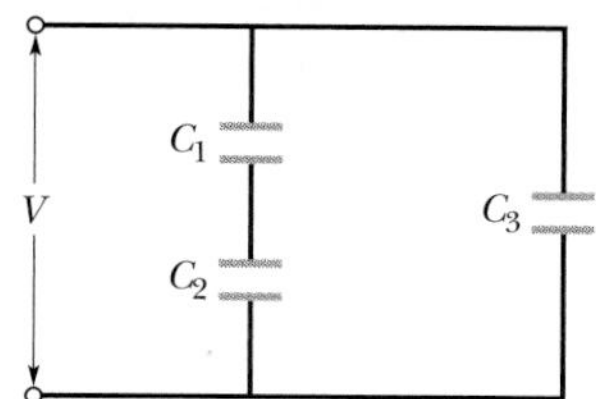

Fig. 25-28 Problems 10 and 34.

•11 ILW In Fig. 25-29, find the equivalent capacitance of the combination. Assume that $C_1 =$ 10.0 μF, $C_2 = 5.00$ μF, and C_3 = 4.00 μF.

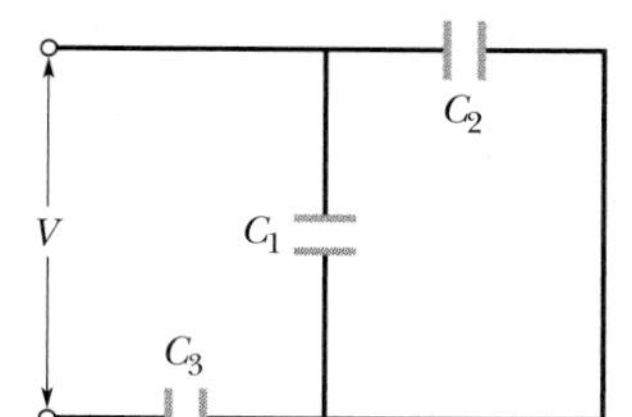

Fig. 25-29 Problems 11, 17, and 38.

••12 Two parallel-plate capacitors, 6.0 μF each, are connected in parallel to a 10 V battery. One of the capacitors is then squeezed so that its plate separation is 50.0% of its initial value. Because of the squeezing, (a) how much additional charge is transferred to the capacitors by the battery and (b) what is the increase in the total charge stored on the capacitors?

••13 SSM ILW A 100 pF capacitor is charged to a potential difference of 50 V, and the charging battery is disconnected. The capacitor is then connected in parallel with a second (initially uncharged) capacitor. If the potential difference across the first capacitor drops to 35 V, what is the capacitance of this second capacitor?

••14 In Fig. 25-30, the battery has a potential difference of $V = 10.0$ V and the five capacitors each have a capacitance of 10.0 μF. What is the charge on (a) capacitor 1 and (b) capacitor 2?

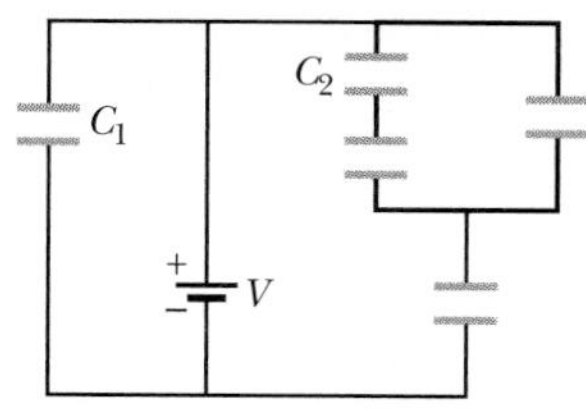

Fig. 25-30 Problem 14.

••15 GO In Fig. 25-31, a 20.0 V battery is connected across capacitors of capacitances $C_1 = C_6 = 3.00$ μF and $C_3 = C_5 = 2.00C_2 = 2.00C_4 = 4.00$ μF. What are (a) the equivalent capacitance C_{eq} of the capacitors and (b) the charge stored by C_{eq}? What are (c) V_1 and (d) q_1 of capacitor 1, (e) V_2 and (f) q_2 of capacitor 2, and (g) V_3 and (h) q_3 of capacitor 3?

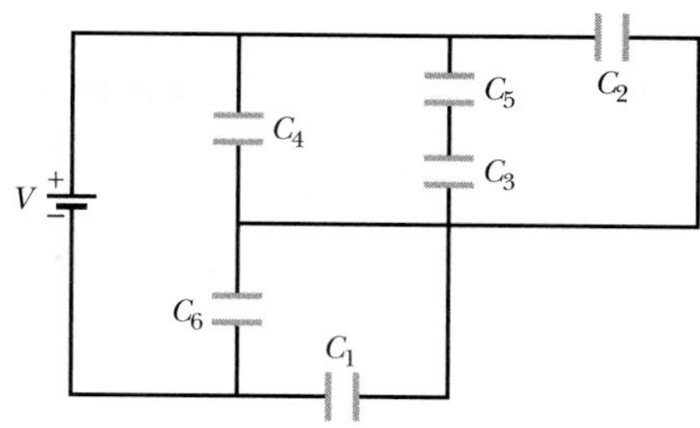

Fig. 25-31 Problem 15.

••16 Plot 1 in Fig. 25-32a gives the charge q that can be stored on capacitor 1 versus the electric potential V set up across it. The vertical scale is set by $q_s = 16.0$ μC, and the horizontal scale is set by $V_s = 2.0$ V. Plots 2 and 3 are similar plots for capacitors 2 and 3, respectively. Figure 25-32b shows a circuit with those three capacitors and a 6.0 V battery. What is the charge stored on capacitor 2 in that circuit?

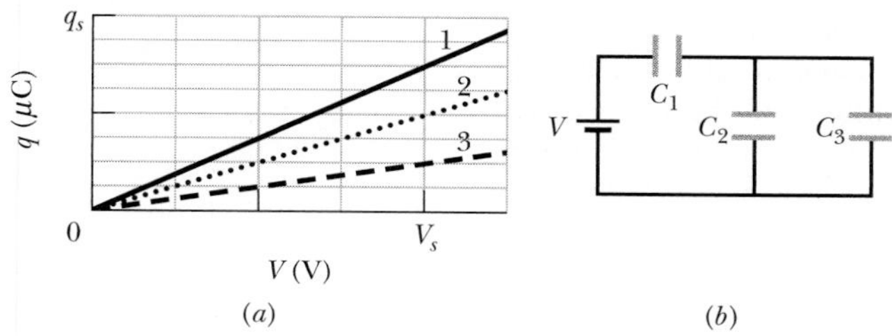

Fig. 25-32 Problem 16.

••17 GO In Fig. 25-29, a potential difference of $V = 100.0$ V is applied across a capacitor arrangement with capacitances $C_1 = 10.0$ μF, $C_2 = 5.00$ μF, and $C_3 = 4.00$ μF. If capacitor 3 undergoes electrical breakdown so that it becomes equivalent to conducting wire, what is the increase in (a) the charge on capacitor 1 and (b) the potential difference across capacitor 1?

••18 Figure 25-33 shows a circuit section of four air-filled capacitors that is connected to a larger circuit. The graph below the section shows the electric potential $V(x)$ as a function of position x

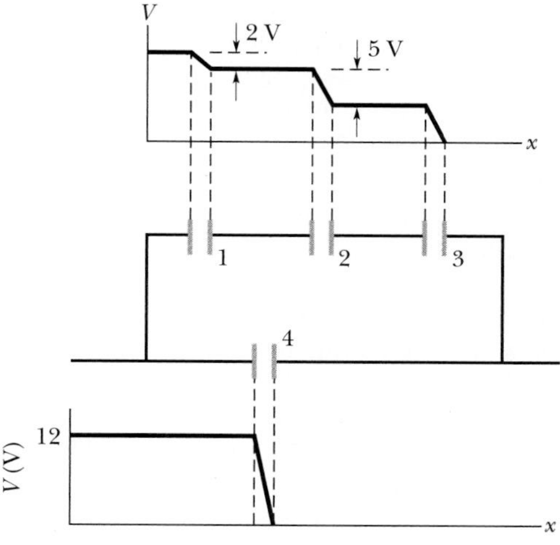

Fig. 25-33 Problem 18.

along the lower part of the section, through capacitor 4. Similarly, the graph above the section shows the electric potential $V(x)$ as a function of position x along the upper part of the section, through capacitors 1, 2, and 3. Capacitor 3 has a capacitance of 0.80 μF. What are the capacitances of (a) capacitor 1 and (b) capacitor 2?

••19 GO In Fig. 25-34, the battery has potential difference V = 9.0 V, C_2 = 3.0 μF, C_4 = 4.0 μF, and all the capacitors are initially uncharged. When switch S is closed, a total charge of 12 μC passes through point a and a total charge of 8.0 μC passes through point b. What are (a) C_1 and (b) C_3?

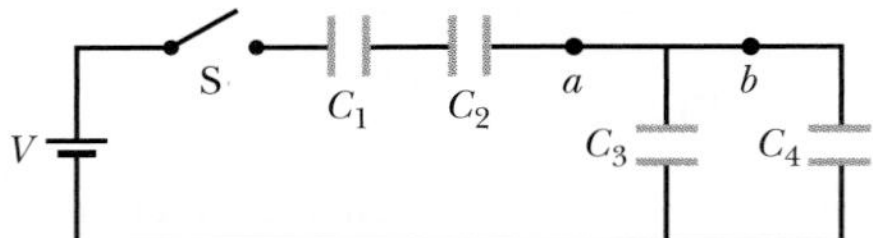

Fig. 25-34 Problem 19.

••20 Figure 25-35 shows a variable "air gap" capacitor for manual tuning. Alternate plates are connected together; one group of plates is fixed in position, and the other group is capable of rotation. Consider a capacitor of n = 8 plates of alternating polarity, each plate having area A = 1.25 cm² and separated from adjacent plates by distance d = 3.40 mm. What is the maximum capacitance of the device?

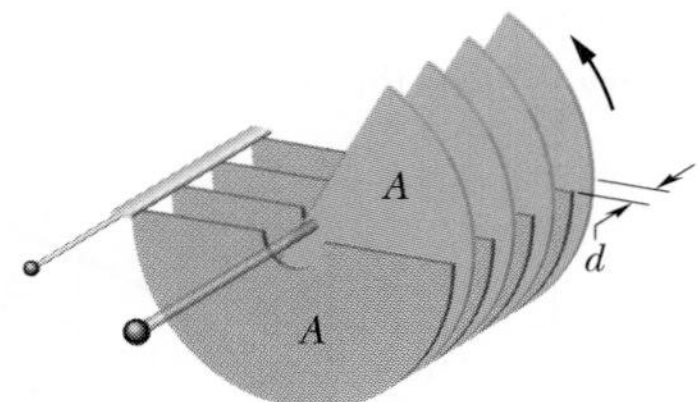

Fig. 25-35 Problem 20.

••21 SSM WWW In Fig. 25-36, the capacitances are C_1 = 1.0 μF and C_2 = 3.0 μF, and both capacitors are charged to a potential difference of V = 100 V but with opposite polarity as shown. Switches S_1 and S_2 are now closed. (a) What is now the potential difference between points a and b? What now is the charge on capacitor (b) 1 and (c) 2?

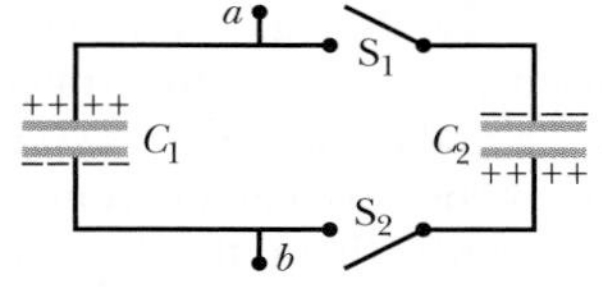

Fig. 25-36 Problem 21.

••22 In Fig. 25-37, V = 10 V, C_1 = 10 μF, and C_2 = C_3 = 20 μF. Switch S is first thrown to the left side until capacitor 1 reaches equilibrium. Then the switch is thrown to the right. When equilibrium is again reached, how much charge is on capacitor 1?

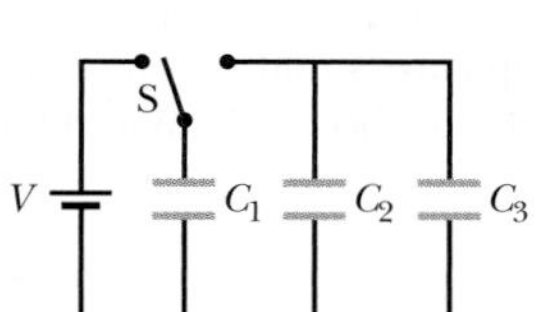

Fig. 25-37 Problem 22.

••23 The capacitors in Fig. 25-38 are initially uncharged. The capacitances are C_1 = 4.0 μF, C_2 = 8.0 μF, and C_3 = 12 μF, and the battery's potential difference is V = 12 V. When switch S is closed, how many electrons travel through (a) point a, (b) point b, (c) point c, and (d) point d? In the figure, do the electrons travel up or down through (e) point b and (f) point c?

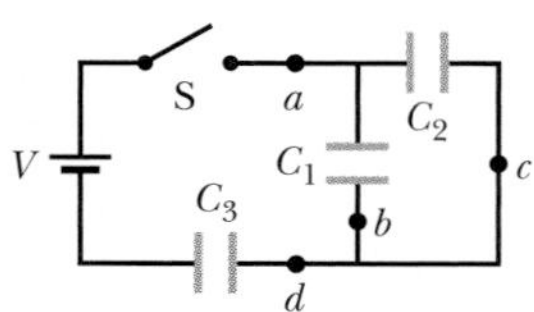

Fig. 25-38 Problem 23.

••24 Figure 25-39 represents two air-filled cylindrical capacitors connected in series across a battery with potential V = 10 V. Capacitor 1 has an inner plate radius of 5.0 mm, an outer plate radius of 1.5 cm, and a length of 5.0 cm. Capacitor 2 has an inner plate radius of 2.5 mm, an outer plate radius of 1.0 cm, and a length of 9.0 cm. The outer plate of capacitor 2 is a conducting organic membrane that can be stretched, and the capacitor can be inflated to increase the plate separation. If the outer plate radius is increased to 2.5 cm by inflation, (a) how many electrons move through point P and (b) do they move toward or away from the battery?

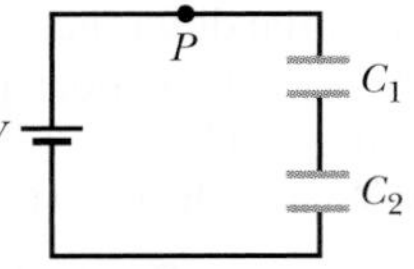

Fig. 25-39 Problem 24.

••25 In Fig. 25-40, two parallel-plate capacitors (with air between the plates) are connected to a battery. Capacitor 1 has a plate area of 1.5 cm² and an electric field (between its plates) of magnitude 2000 V/m. Capacitor 2 has a plate area of 0.70 cm² and an electric field of magnitude 1500 V/m. What is the total charge on the two capacitors?

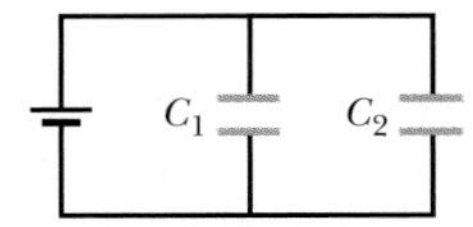

Fig. 25-40 Problem 25.

•••26 Capacitor 3 in Fig. 25-41a is a *variable capacitor* (its capacitance C_3 can be varied). Figure 25-41b gives the electric potential V_1 across capacitor 1 versus C_3. The horizontal scale is set by C_{3s} = 12.0 μF. Electric potential V_1 approaches an asymptote of 10 V as $C_3 \to \infty$. What are (a) the electric potential V across the battery, (b) C_1, and (c) C_2?

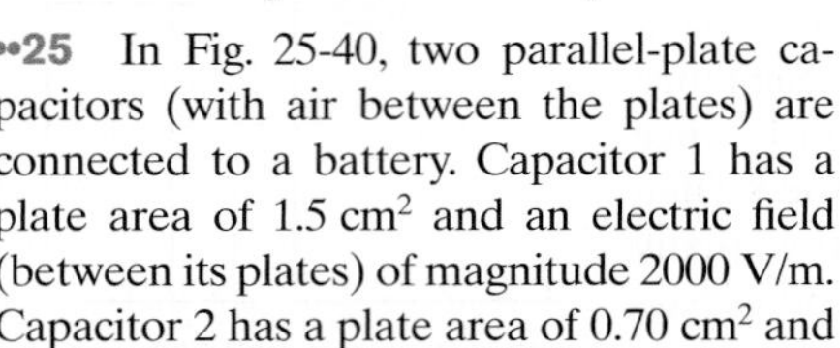

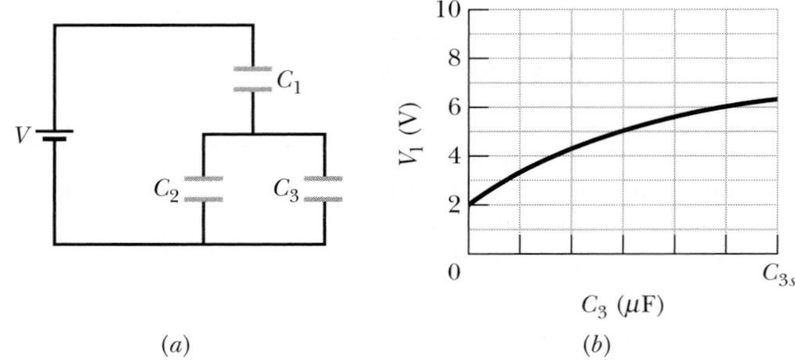

Fig. 25-41 Problem 26.

•••27 GO Figure 25-42 shows a 12.0 V battery and four uncharged capacitors of capacitances C_1 = 1.00 μF, C_2 = 2.00 μF, C_3 = 3.00 μF, and C_4 = 4.00 μF. If only switch S_1 is closed, what is the charge on (a) capacitor 1, (b) capacitor 2, (c) capacitor 3, and (d) capacitor 4? If both switches are closed, what is the charge on (e) capacitor 1, (f) capacitor 2, (g) capacitor 3, and (h) capacitor 4?

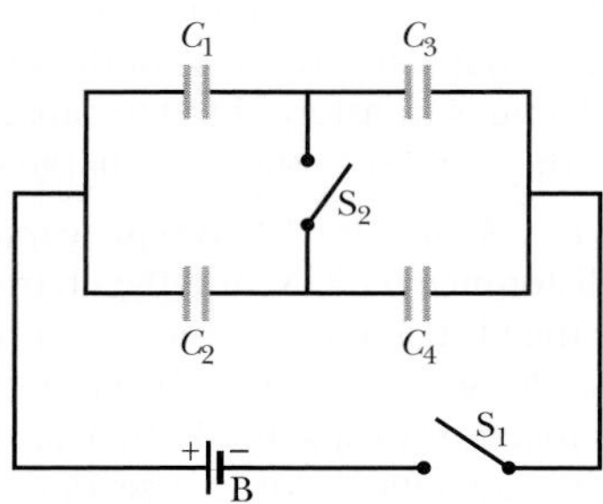

Fig. 25-42 Problem 27.

•••28 GO Figure 25-43 displays a 12.0 V battery and 3 uncharged capacitors of capacitances C_1 = 4.00 μF, C_2 = 6.00 μF, and C_3 = 3.00 μF. The switch is thrown to the left side until capacitor 1 is fully charged. Then the switch is thrown to the right. What is the final charge on (a) capacitor 1, (b) capacitor 2, and (c) capacitor 3?

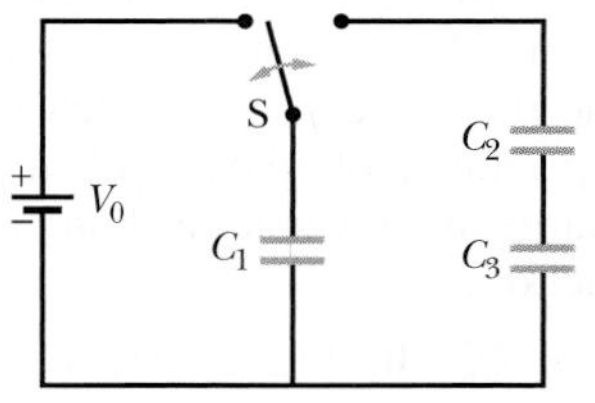

Fig. 25-43 Problem 28.

sec. 25-5 Energy Stored in an Electric Field

•29 What capacitance is required to store an energy of 10 kW · h at a potential difference of 1000 V?

•30 How much energy is stored in 1.00 m^3 of air due to the "fair weather" electric field of magnitude 150 V/m?

•31 SSM A 2.0 μF capacitor and a 4.0 μF capacitor are connected in parallel across a 300 V potential difference. Calculate the total energy stored in the capacitors.

•32 A parallel-plate air-filled capacitor having area 40 cm^2 and plate spacing 1.0 mm is charged to a potential difference of 600 V. Find (a) the capacitance, (b) the magnitude of the charge on each plate, (c) the stored energy, (d) the electric field between the plates, and (e) the energy density between the plates.

••33 A charged isolated metal sphere of diameter 10 cm has a potential of 8000 V relative to $V = 0$ at infinity. Calculate the energy density in the electric field near the surface of the sphere.

••34 In Fig. 25-28, a potential difference $V = 100$ V is applied across a capacitor arrangement with capacitances $C_1 = 10.0$ μF, $C_2 = 5.00$ μF, and $C_3 = 4.00$ μF. What are (a) charge q_3, (b) potential difference V_3, and (c) stored energy U_3 for capacitor 3, (d) q_1, (e) V_1, and (f) U_1 for capacitor 1, and (g) q_2, (h) V_2, and (i) U_2 for capacitor 2?

••35 Assume that a stationary electron is a point of charge. What is the energy density u of its electric field at radial distances (a) $r =$ 1.00 mm, (b) $r = 1.00$ μm, (c) $r = 1.00$ nm, and (d) $r = 1.00$ pm? (e) What is u in the limit as $r \to 0$?

••36 As a safety engineer, you must evaluate the practice of storing flammable conducting liquids in nonconducting containers. The company supplying a certain liquid has been using a squat, cylindrical plastic container of radius $r =$ 0.20 m and filling it to height $h = 10$ cm, which is not the container's full interior height (Fig. 25-44). Your investigation reveals that during handling at the company, the exterior surface of the container commonly acquires a negative charge density of magnitude 2.0 $\mu C/m^2$ (approximately uniform). Because the liquid is a conducting material, the charge on the container induces charge separation within the liquid. (a) How much negative charge is induced in the center of the liquid's bulk? (b) Assume the capacitance of the central portion of the liquid relative to ground is 35 pF. What is the potential energy associated with the negative charge in that effective capacitor? (c) If a spark occurs between the ground and the central portion of the liquid (through the venting port), the potential energy can be fed into the spark. The minimum spark energy needed to ignite the liquid is 10 mJ. In this situation, can a spark ignite the liquid?

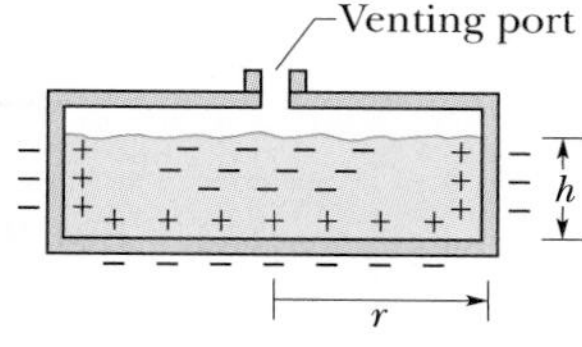

Fig. 25-44 Problem 36.

••37 SSM ILW WWW The parallel plates in a capacitor, with a plate area of 8.50 cm^2 and an air-filled separation of 3.00 mm, are charged by a 6.00 V battery. They are then disconnected from the battery and pulled apart (without discharge) to a separation of 8.00 mm. Neglecting fringing, find (a) the potential difference between the plates, (b) the initial stored energy, (c) the final stored energy, and (d) the work required to separate the plates.

••38 In Fig. 25-29, a potential difference $V = 100$ V is applied across a capacitor arrangement with capacitances $C_1 = 10.0$ μF, $C_2 = 5.00$ μF, and $C_3 = 15.0$ μF. What are (a) charge q_3, (b) potential difference V_3, and (c) stored energy U_3 for capacitor 3, (d) q_1, (e) V_1, and (f) U_1 for capacitor 1, and (g) q_2, (h) V_2, and (i) U_2 for capacitor 2?

••39 GO In Fig. 25-45, $C_1 = 10.0$ μF, $C_2 = 20.0$ μF, and $C_3 = 25.0$ μF. If no capacitor can withstand a potential difference of more than 100 V without failure, what are (a) the magnitude of the maximum potential difference that can exist between points A and B and (b) the maximum energy that can be stored in the three-capacitor arrangement?

A —C_1—C_2—C_3— B

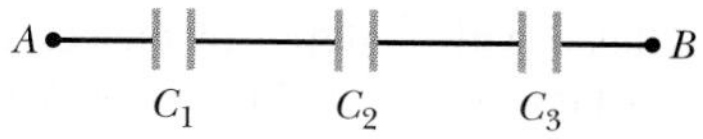

Fig. 25-45 Problem 39.

sec. 25-6 Capacitor with a Dielectric

•40 An air-filled parallel-plate capacitor has a capacitance of 1.3 pF. The separation of the plates is doubled, and wax is inserted between them. The new capacitance is 2.6 pF. Find the dielectric constant of the wax.

•41 SSM A coaxial cable used in a transmission line has an inner radius of 0.10 mm and an outer radius of 0.60 mm. Calculate the capacitance per meter for the cable. Assume that the space between the conductors is filled with polystyrene.

•42 A parallel-plate air-filled capacitor has a capacitance of 50 pF. (a) If each of its plates has an area of 0.35 m^2, what is the separation? (b) If the region between the plates is now filled with material having $\kappa = 5.6$, what is the capacitance?

•43 Given a 7.4 pF air-filled capacitor, you are asked to convert it to a capacitor that can store up to 7.4 μJ with a maximum potential difference of 652 V. Which dielectric in Table 25-1 should you use to fill the gap in the capacitor if you do not allow for a margin of error?

••44 You are asked to construct a capacitor having a capacitance near 1 nF and a breakdown potential in excess of 10 000 V. You think of using the sides of a tall Pyrex drinking glass as a dielectric, lining the inside and outside curved surfaces with aluminum foil to act as the plates. The glass is 15 cm tall with an inner radius of 3.6 cm and an outer radius of 3.8 cm. What are the (a) capacitance and (b) breakdown potential of this capacitor?

••45 A certain parallel-plate capacitor is filled with a dielectric for which $\kappa = 5.5$. The area of each plate is 0.034 m^2, and the plates are separated by 2.0 mm. The capacitor will fail (short out and burn up) if the electric field between the plates exceeds 200 kN/C. What is the maximum energy that can be stored in the capacitor?

••46 In Fig. 25-46, how much charge is stored on the parallel-plate capacitors by the 12.0 V battery? One is filled with air, and the other is filled with a dielectric for which $\kappa = 3.00$; both capacitors have a plate area of 5.00×10^{-3} m^2 and a plate separation of 2.00 mm.

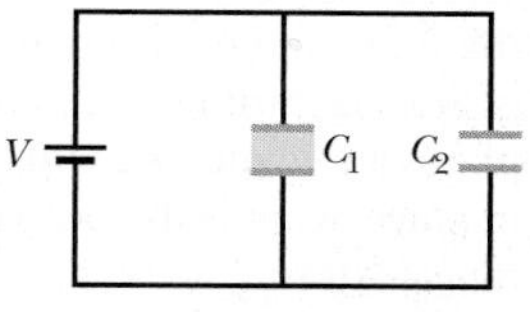

Fig. 25-46 Problem 46.

••47 SSM ILW A certain substance has a dielectric constant of 2.8 and a dielectric strength of 18 MV/m. If it is used as the dielectric material in a parallel-plate capacitor, what minimum area should the plates of the capacitor have to obtain a capacitance of 7.0×10^{-2} μF and to ensure that the capacitor will be able to withstand a potential difference of 4.0 kV?

••48 Figure 25-47 shows a parallel-plate capacitor with a plate area $A = 5.56\ \text{cm}^2$ and separation $d = 5.56$ mm. The left half of the gap is filled with material of dielectric constant $\kappa_1 = 7.00$; the right half is filled with material of dielectric constant $\kappa_2 = 12.0$. What is the capacitance?

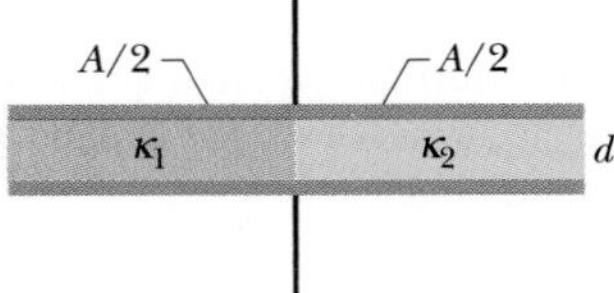

Fig. 25-47 Problem 48.

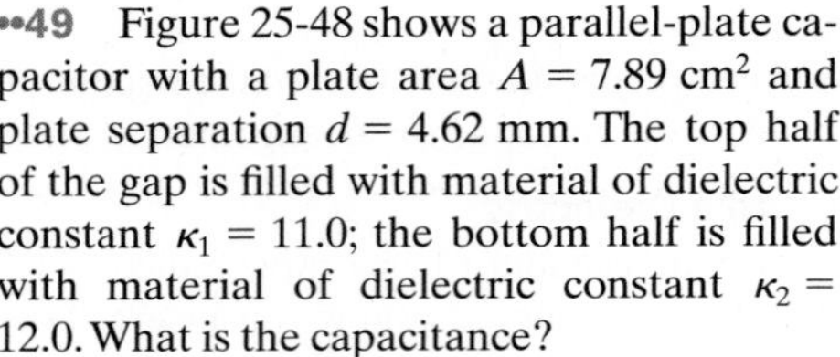

••49 Figure 25-48 shows a parallel-plate capacitor with a plate area $A = 7.89\ \text{cm}^2$ and plate separation $d = 4.62$ mm. The top half of the gap is filled with material of dielectric constant $\kappa_1 = 11.0$; the bottom half is filled with material of dielectric constant $\kappa_2 = 12.0$. What is the capacitance?

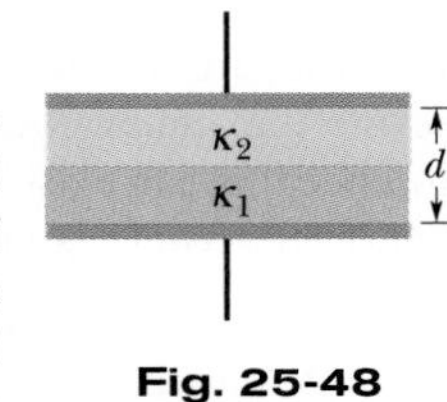

Fig. 25-48 Problem 49.

••50 GO Figure 25-49 shows a parallel-plate capacitor of plate area $A = 10.5\ \text{cm}^2$ and plate separation $2d = 7.12$ mm. The left half of the gap is filled with material of dielectric constant $\kappa_1 = 21.0$; the top of the right half is filled with material of dielectric constant $\kappa_2 = 42.0$; the bottom of the right half is filled with material of dielectric constant $\kappa_3 = 58.0$. What is the capacitance?

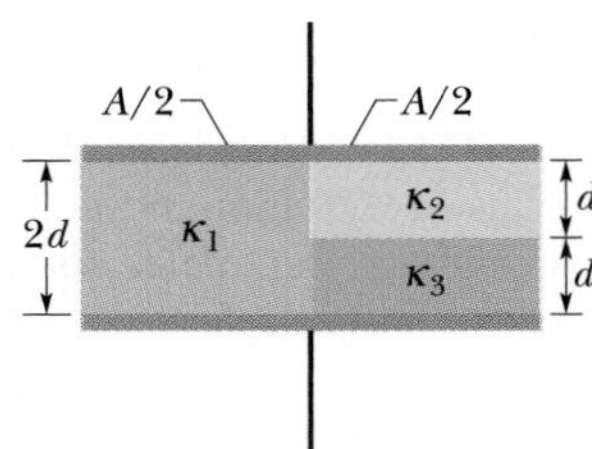

Fig. 25-49 Problem 50.

sec. 25-8 Dielectrics and Gauss' Law

•51 SSM WWW A parallel-plate capacitor has a capacitance of 100 pF, a plate area of $100\ \text{cm}^2$, and a mica dielectric ($\kappa = 5.4$) completely filling the space between the plates. At 50 V potential difference, calculate (a) the electric field magnitude E in the mica, (b) the magnitude of the free charge on the plates, and (c) the magnitude of the induced surface charge on the mica.

•52 For the arrangement of Fig. 25-17, suppose that the battery remains connected while the dielectric slab is being introduced. Calculate (a) the capacitance, (b) the charge on the capacitor plates, (c) the electric field in the gap, and (d) the electric field in the slab, after the slab is in place.

••53 A parallel-plate capacitor has plates of area $0.12\ \text{m}^2$ and a separation of 1.2 cm. A battery charges the plates to a potential difference of 120 V and is then disconnected. A dielectric slab of thickness 4.0 mm and dielectric constant 4.8 is then placed symmetrically between the plates. (a) What is the capacitance before the slab is inserted? (b) What is the capacitance with the slab in place? What is the free charge q (c) before and (d) after the slab is inserted? What is the magnitude of the electric field (e) in the space between the plates and dielectric and (f) in the dielectric itself? (g) With the slab in place, what is the potential difference across the plates? (h) How much external work is involved in inserting the slab?

••54 Two parallel plates of area $100\ \text{cm}^2$ are given charges of equal magnitudes 8.9×10^{-7} C but opposite signs. The electric field within the dielectric material filling the space between the plates is 1.4×10^6 V/m. (a) Calculate the dielectric constant of the material. (b) Determine the magnitude of the charge induced on each dielectric surface.

••55 The space between two concentric conducting spherical shells of radii $b = 1.70$ cm and $a = 1.20$ cm is filled with a substance of dielectric constant $\kappa = 23.5$. A potential difference $V = 73.0$ V is applied across the inner and outer shells. Determine (a) the capacitance of the device, (b) the free charge q on the inner shell, and (c) the charge q' induced along the surface of the inner shell.

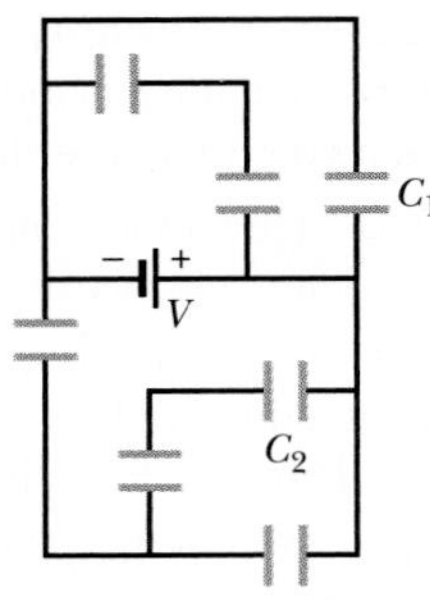

Fig. 25-50 Problem 56.

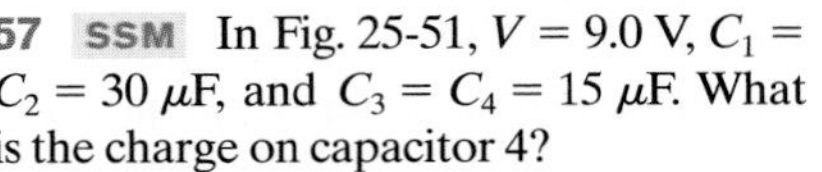

Additional Problems

56 In Fig. 25-50, the battery potential difference V is 10.0 V and each of the seven capacitors has capacitance 10.0 μF. What is the charge on (a) capacitor 1 and (b) capacitor 2?

57 SSM In Fig. 25-51, $V = 9.0$ V, $C_1 = C_2 = 30\ \mu$F, and $C_3 = C_4 = 15\ \mu$F. What is the charge on capacitor 4?

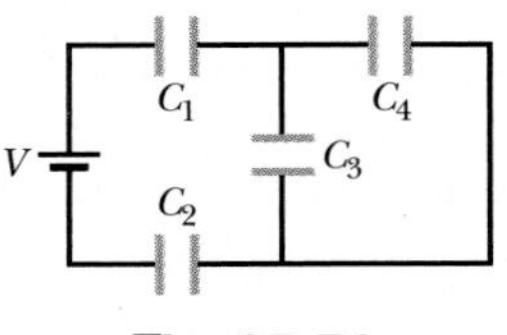

Fig. 25-51 Problem 57.

58 The capacitances of the four capacitors shown in Fig. 25-52 are given in terms of a certain quantity C. (a) If $C = 50\ \mu$F, what is the equivalent capacitance between points A and B? (*Hint:* First imagine that a battery is connected between those two points; then reduce the circuit to an equivalent capacitance.) (b) Repeat for points A and D.

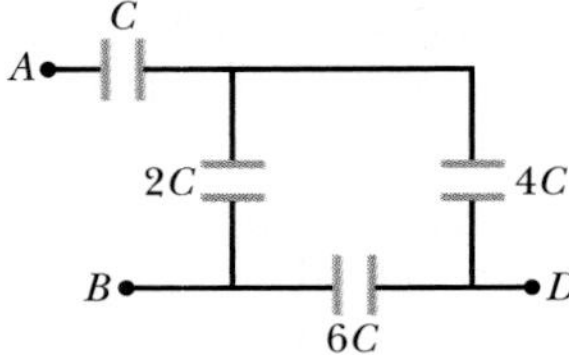

Fig. 25-52 Problem 58.

59 In Fig. 25-53, $V = 12$ V, $C_1 = C_4 = 2.0\ \mu$F, $C_2 = 4.0\ \mu$F, and $C_3 = 1.0\ \mu$F. What is the charge on capacitor 4?

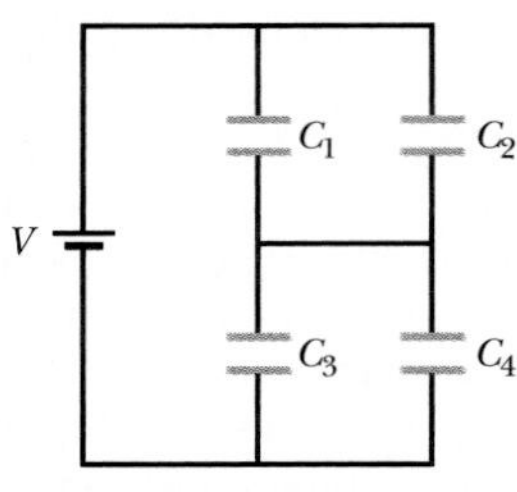

Fig. 25-53 Problem 59.

60 *The chocolate crumb mystery.* This story begins with Problem 60 in Chapter 23. As part of the investigation of the biscuit factory explosion, the electric potentials of the workers were measured as they emptied sacks of chocolate crumb powder into the loading bin, stirring up a cloud of the powder around themselves. Each worker had an electric potential of about 7.0 kV relative to the ground, which was taken as zero potential. (a) Assuming that each worker was effectively a capacitor with a typical capacitance of 200 pF, find the energy stored in that effective capacitor. If a single spark between the worker and any conducting object connected to the ground neutralized the worker, that energy would be transferred to the spark. According to measurements, a spark that could ignite a cloud of chocolate crumb powder, and thus set off an explosion, had to have an energy of at least 150 mJ. (b) Could a spark from a worker have set off an explosion in the cloud of powder in the loading bin? (The story continues with Problem 60 in Chapter 26.)

61 Figure 25-54 shows capacitor 1 ($C_1 = 8.00\ \mu$F), capacitor 2 ($C_2 = 6.00\ \mu$F), and capacitor 3 ($C_3 = 8.00\ \mu$F) connected to a 12.0 V battery. When switch S is closed so as to connect uncharged ca-

pacitor 4 ($C_4 = 6.00\ \mu F$), (a) how much charge passes through point P from the battery and (b) how much charge shows up on capacitor 4? (c) Explain the discrepancy in those two results.

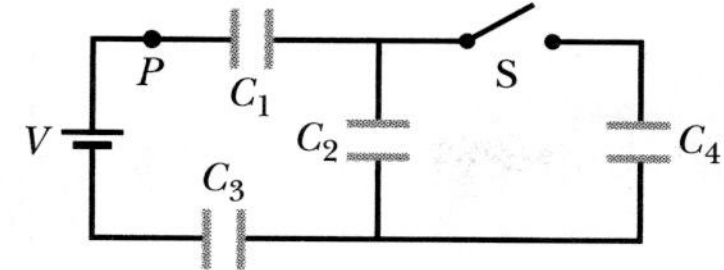

Fig. 25-54 Problem 61.

62 Two air-filled, parallel-plate capacitors are to be connected to a 10 V battery, first individually, then in series, and then in parallel. In those arrangements, the energy stored in the capacitors turns out to be, listed least to greatest: 75 μJ, 100 μJ, 300 μJ, and 400 μJ. Of the two capacitors, what is the (a) smaller and (b) greater capacitance?

63 Two parallel-plate capacitors, 6.0 μF each, are connected in series to a 10 V battery. One of the capacitors is then squeezed so that its plate separation is halved. Because of the squeezing, (a) how much additional charge is transferred to the capacitors by the battery and (b) what is the increase in the *total* charge stored on the capacitors (the charge on the positive plate of one capacitor plus the charge on the positive plate of the other capacitor)?

64 GO In Fig. 25-55, $V = 12$ V, $C_1 = C_5 = C_6 = 6.0\ \mu F$, and $C_2 = C_3 = C_4 = 4.0\ \mu F$. What are (a) the net charge stored on the capacitors and (b) the charge on capacitor 4?

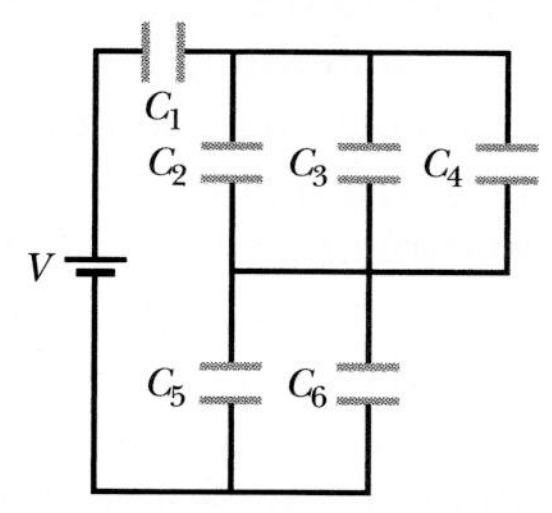

Fig. 25-55 Problem 64.

65 SSM In Fig. 25-56, the parallel-plate capacitor of plate area $2.00 \times 10^{-2}\ m^2$ is filled with two dielectric slabs, each with thickness 2.00 mm. One slab has dielectric constant 3.00, and the other, 4.00. How much charge does the 7.00 V battery store on the capacitor?

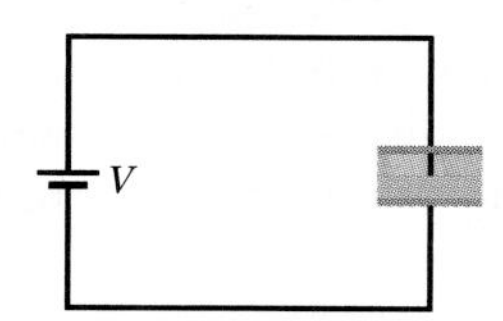

Fig. 25-56 Problem 65.

66 A cylindrical capacitor has radii a and b as in Fig. 25-6. Show that half the stored electric potential energy lies within a cylinder whose radius is $r = \sqrt{ab}$.

67 A capacitor of capacitance $C_1 = 6.00\ \mu F$ is connected in series with a capacitor of capacitance $C_2 = 4.00\ \mu F$, and a potential difference of 200 V is applied across the pair. (a) Calculate the equivalent capacitance. What are (b) charge q_1 and (c) potential difference V_1 on capacitor 1 and (d) q_2 and (e) V_2 on capacitor 2?

68 Repeat Problem 67 for the same two capacitors but with them now connected in parallel.

69 A certain capacitor is charged to a potential difference V. If you wish to increase its stored energy by 10%, by what percentage should you increase V?

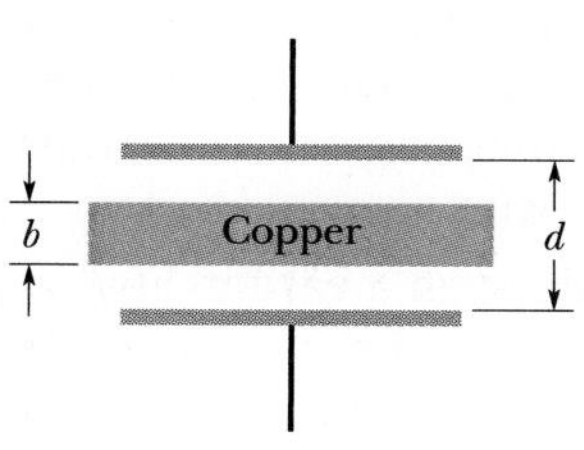

Fig. 25-57 Problems 70 and 71.

70 A slab of copper of thickness $b = 2.00$ mm is thrust into a parallel-plate capacitor of plate area $A = 2.40\ cm^2$ and plate separation $d = 5.00$ mm, as shown in Fig. 25-57; the slab is exactly halfway between the plates. (a) What is the capacitance after the slab is introduced? (b) If a charge $q = 3.40\ \mu C$ is maintained on the plates, what is the ratio of the stored energy before to that after the slab is inserted? (c) How much work is done on the slab as it is inserted? (d) Is the slab sucked in or must it be pushed in?

71 Repeat Problem 70, assuming that a potential difference $V = 85.0$ V, rather than the charge, is held constant.

72 A potential difference of 300 V is applied to a series connection of two capacitors of capacitances $C_1 = 2.00\ \mu F$ and $C_2 = 8.00\ \mu F$. What are (a) charge q_1 and (b) potential difference V_1 on capacitor 1 and (c) q_2 and (d) V_2 on capacitor 2? The *charged* capacitors are then disconnected from each other and from the battery. Then the capacitors are reconnected with plates of the *same* signs wired together (the battery is not used). What now are (e) q_1, (f) V_1, (g) q_2, and (h) V_2? Suppose, *instead,* the capacitors charged in part (a) are reconnected with plates of *opposite* signs wired together. What now are (i) q_1, (j) V_1, (k) q_2, and (l) V_2?

73 Figure 25-58 shows a four-capacitor arrangement that is connected to a larger circuit at points A and B. The capacitances are $C_1 = 10\ \mu F$ and $C_2 = C_3 = C_4 = 20\ \mu F$. The charge on capacitor 1 is 30 μC. What is the magnitude of the potential difference $V_A - V_B$?

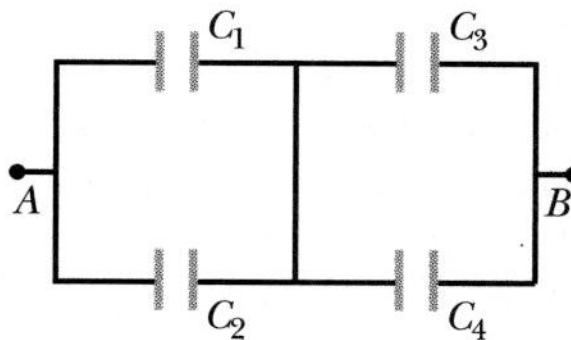

Fig. 25-58 Problem 73.

74 You have two plates of copper, a sheet of mica (thickness = 0.10 mm, $\kappa = 5.4$), a sheet of glass (thickness = 2.0 mm, $\kappa = 7.0$), and a slab of paraffin (thickness = 1.0 cm, $\kappa = 2.0$). To make a parallel-plate capacitor with the largest C, which sheet should you place between the copper plates?

75 A capacitor of unknown capacitance C is charged to 100 V and connected across an initially uncharged 60 μF capacitor. If the final potential difference across the 60 μF capacitor is 40 V, what is C?

76 A 10 V battery is connected to a series of n capacitors, each of capacitance 2.0 μF. If the total stored energy is 25 μJ, what is n?

77 SSM In Fig. 25-59, two parallel-plate capacitors A and B are connected in parallel across a 600 V battery. Each plate has area 80.0 cm^2; the plate separations are 3.00 mm. Capacitor A is filled with air; capacitor B is filled with a dielectric of dielectric constant $\kappa = 2.60$. Find the magnitude of the electric field within (a) the dielectric of capacitor B and (b) the air of capacitor A. What are the free charge densities σ on the higher-potential plate of (c) capacitor A and (d) capacitor B? (e) What is the induced charge density σ' on the top surface of the dielectric?

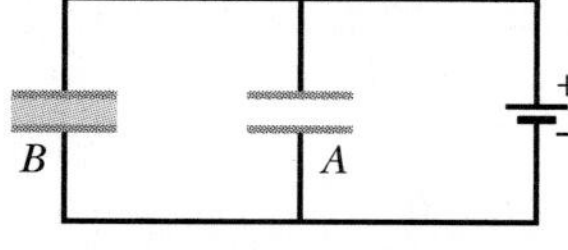

Fig. 25-59 Problem 77.

78 You have many 2.0 μF capacitors, each capable of withstanding 200 V without undergoing electrical breakdown (in which they conduct charge instead of storing it). How would you assemble a combination having an equivalent capacitance of (a) 0.40 μF and (b) 1.2 μF, each combination capable of withstanding 1000 V?

CHAPTER

26

CURRENT AND RESISTANCE

26-1 WHAT IS PHYSICS?

In the last five chapters we discussed electrostatics—the physics of stationary charges. In this and the next chapter, we discuss the physics of **electric currents**—that is, charges in motion.

Examples of electric currents abound and involve many professions. Meteorologists are concerned with lightning and with the less dramatic slow flow of charge through the atmosphere. Biologists, physiologists, and engineers working in medical technology are concerned with the nerve currents that control muscles and especially with how those currents can be reestablished after spinal cord injuries. Electrical engineers are concerned with countless electrical systems, such as power systems, lightning protection systems, information storage systems, and music systems. Space engineers monitor and study the flow of charged particles from our Sun because that flow can wipe out telecommunication systems in orbit and even power transmission systems on the ground.

In this chapter we discuss the basic physics of electric currents and why they can be established in some materials but not in others. We begin with the meaning of electric current.

26-2 Electric Current

Although an electric current is a stream of moving charges, not all moving charges constitute an electric current. If there is to be an electric current through a given surface, there must be a net flow of charge through that surface. Two examples clarify our meaning.

1. The free electrons (conduction electrons) in an isolated length of copper wire are in random motion at speeds of the order of 10^6 m/s. If you pass a hypothetical plane through such a wire, conduction electrons pass through it *in both directions* at the rate of many billions per second—but there is *no net transport* of charge and thus *no current* through the wire. However, if you connect the ends of the wire to a battery, you slightly bias the flow in one direction, with the result that there now is a net transport of charge and thus an electric current through the wire.
2. The flow of water through a garden hose represents the directed flow of positive charge (the protons in the water molecules) at a rate of perhaps several million coulombs per second. There is no net transport of charge, however, because there is a parallel flow of negative charge (the electrons in the water molecules) of exactly the same amount moving in exactly the same direction.

In this chapter we restrict ourselves largely to the study—within the framework of classical physics—of *steady* currents of *conduction electrons* moving through *metallic conductors* such as copper wires.

As Fig. 26-1*a* reminds us, any isolated conducting loop—regardless of whether it has an excess charge—is all at the same potential. No electric field can exist within it or along its surface. Although conduction electrons are available, no net electric force acts on them and thus there is no current.

If, as in Fig. 26-1*b*, we insert a battery in the loop, the conducting loop is no longer at a single potential. Electric fields act inside the material making up the loop, exerting forces on the conduction electrons, causing them to move and thus establishing a current. After a very short time, the electron flow reaches a constant value and the current is in its *steady state* (it does not vary with time).

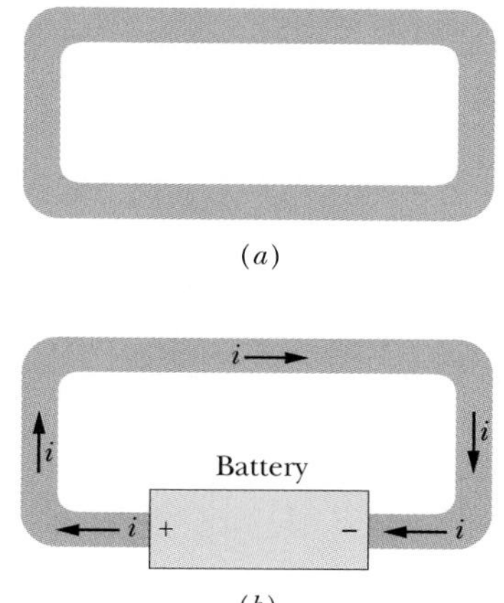

Fig. 26-1 (*a*) A loop of copper in electrostatic equilibrium. The entire loop is at a single potential, and the electric field is zero at all points inside the copper. (*b*) Adding a battery imposes an electric potential difference between the ends of the loop that are connected to the terminals of the battery. The battery thus produces an electric field within the loop, from terminal to terminal, and the field causes charges to move around the loop. This movement of charges is a current i.

Figure 26-2 shows a section of a conductor, part of a conducting loop in which current has been established. If charge dq passes through a hypothetical plane (such as aa') in time dt, then the current i through that plane is defined as

$$i = \frac{dq}{dt} \quad \text{(definition of current).} \tag{26-1}$$

We can find the charge that passes through the plane in a time interval extending from 0 to t by integration:

$$q = \int dq = \int_0^t i\, dt, \tag{26-2}$$

in which the current i may vary with time.

The current is the same in any cross section.

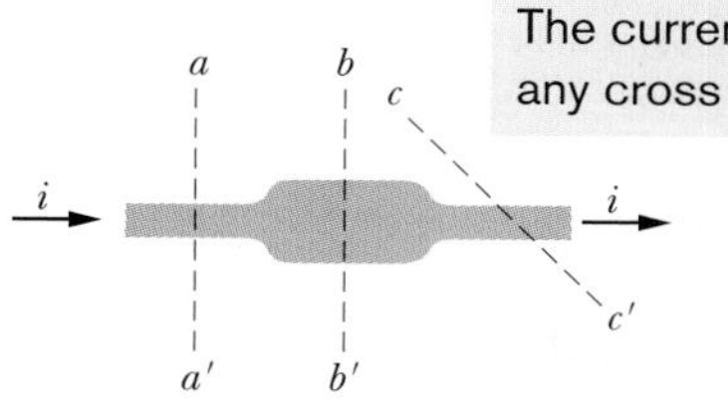

Fig. 26-2 The current i through the conductor has the same value at planes aa', bb', and cc'.

Under steady-state conditions, the current is the same for planes aa', bb', and cc' and indeed for all planes that pass completely through the conductor, no matter what their location or orientation. This follows from the fact that charge is conserved. Under the steady-state conditions assumed here, an electron must pass through plane aa' for every electron that passes through plane cc'. In the same way, if we have a steady flow of water through a garden hose, a drop of water must leave the nozzle for every drop that enters the hose at the other end. The amount of water in the hose is a conserved quantity.

The SI unit for current is the coulomb per second, or the ampere (A), which is an SI base unit:

$$1 \text{ ampere} = 1 \text{ A} = 1 \text{ coulomb per second} = 1 \text{ C/s}.$$

The formal definition of the ampere is discussed in Chapter 29.

Current, as defined by Eq. 26-1, is a scalar because both charge and time in that equation are scalars. Yet, as in Fig. 26-1*b*, we often represent a current with an arrow to indicate that charge is moving. Such arrows are not vectors, however, and they do not require vector addition. Figure 26-3*a* shows a conductor with current i_0 splitting at a junction into two branches. Because charge is conserved, the magnitudes of the currents in the branches must add to yield the magnitude of the current in the original conductor, so that

$$i_0 = i_1 + i_2. \tag{26-3}$$

As Fig. 26-3*b* suggests, bending or reorienting the wires in space does not change the validity of Eq. 26-3. Current arrows show only a direction (or sense) of flow along a conductor, not a direction in space.

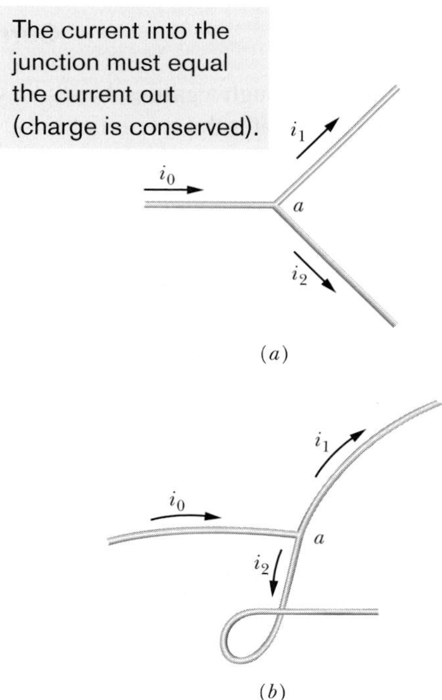

Fig. 26-3 The relation $i_0 = i_1 + i_2$ is true at junction a no matter what the orientation in space of the three wires. Currents are scalars, not vectors.

The Directions of Currents

In Fig. 26-1*b* we drew the current arrows in the direction in which positively charged particles would be forced to move through the loop by the electric field. Such positive *charge carriers,* as they are often called, would move away from the positive battery terminal and toward the negative terminal. Actually, the charge carriers in the copper loop of Fig. 26-1*b* are electrons and thus are negatively charged. The electric field forces them to move in the direction opposite the current arrows, from the negative terminal to the positive terminal. For historical reasons, however, we use the following convention:

A current arrow is drawn in the direction in which positive charge carriers would move, even if the actual charge carriers are negative and move in the opposite direction.

We can use this convention because in *most* situations, the assumed motion of positive charge carriers in one direction has the same effect as the actual motion of negative charge carriers in the opposite direction. (When the effect is not the same, we shall drop the convention and describe the actual motion.)

CHECKPOINT 1

The figure here shows a portion of a circuit. What are the magnitude and direction of the current i in the lower right-hand wire?

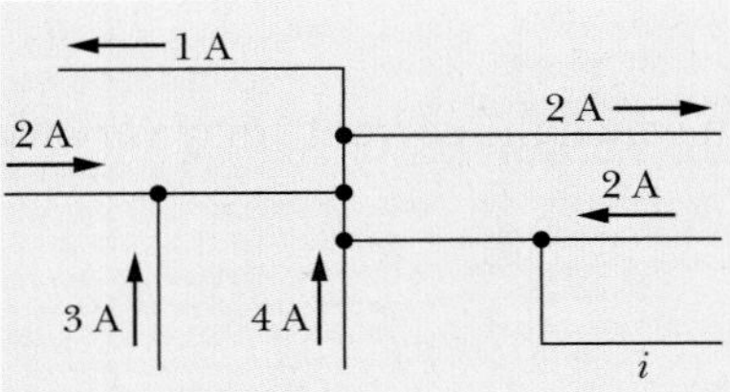

Sample Problem

Current is the rate at which charge passes a point

Water flows through a garden hose at a volume flow rate dV/dt of 450 cm^3/s. What is the current of negative charge?

KEY IDEAS

The current i of negative charge is due to the electrons in the water molecules moving through the hose. The current is the rate at which that negative charge passes through any plane that cuts completely across the hose.

Calculations: We can write the current in terms of the number of molecules that pass through such a plane per second as

$$i = \begin{pmatrix}\text{charge}\\\text{per}\\\text{electron}\end{pmatrix}\begin{pmatrix}\text{electrons}\\\text{per}\\\text{molecule}\end{pmatrix}\begin{pmatrix}\text{molecules}\\\text{per}\\\text{second}\end{pmatrix}$$

or

$$i = (e)(10)\frac{dN}{dt}.$$

We substitute 10 electrons per molecule because a water (H_2O) molecule contains 8 electrons in the single oxygen atom and 1 electron in each of the two hydrogen atoms.

We can express the rate dN/dt in terms of the given volume flow rate dV/dt by first writing

$$\begin{pmatrix}\text{molecules}\\\text{per}\\\text{second}\end{pmatrix} = \begin{pmatrix}\text{molecules}\\\text{per}\\\text{mole}\end{pmatrix}\begin{pmatrix}\text{moles}\\\text{per unit}\\\text{mass}\end{pmatrix}$$

$$\times\begin{pmatrix}\text{mass}\\\text{per unit}\\\text{volume}\end{pmatrix}\begin{pmatrix}\text{volume}\\\text{per}\\\text{second}\end{pmatrix}.$$

"Molecules per mole" is Avogadro's number N_A. "Moles per unit mass" is the inverse of the mass per mole, which is the molar mass M of water. "Mass per unit volume" is the (mass) density ρ_{mass} of water. The volume per second is the volume flow rate dV/dt. Thus, we have

$$\frac{dN}{dt} = N_A\left(\frac{1}{M}\right)\rho_{\text{mass}}\left(\frac{dV}{dt}\right) = \frac{N_A\rho_{\text{mass}}}{M}\frac{dV}{dt}.$$

Substituting this into the equation for i, we find

$$i = 10eN_AM^{-1}\rho_{\text{mass}}\frac{dV}{dt}.$$

We know that Avogadro's number N_A is 6.02×10^{23} molecules/mol, or $6.02 \times 10^{23}\ \text{mol}^{-1}$, and from Table 15-1 we know that the density of water ρ_{mass} under normal conditions is $1000\ \text{kg/m}^3$. We can get the molar mass of water from the molar masses listed in Appendix F (in grams per mole): We add the molar mass of oxygen (16 g/mol) to twice the molar mass of hydrogen (1 g/mol), obtaining 18 g/mol = 0.018 kg/mol. So, the current of negative charge due to the electrons in the water is

$$\begin{aligned} i &= (10)(1.6 \times 10^{-19}\ \text{C})(6.02 \times 10^{23}\ \text{mol}^{-1}) \\ &\quad \times (0.018\ \text{kg/mol})^{-1}(1000\ \text{kg/m}^3)(450 \times 10^{-6}\ \text{m}^3/\text{s}) \\ &= 2.41 \times 10^7\ \text{C/s} = 2.41 \times 10^7\ \text{A} \\ &= 24.1\ \text{MA}. \end{aligned} \quad \text{(Answer)}$$

This current of negative charge is exactly compensated by a current of positive charge associated with the nuclei of the three atoms that make up the water molecule. Thus, there is no net flow of charge through the hose.

Additional examples, video, and practice available at *WileyPLUS*

26-3 Current Density

Sometimes we are interested in the current i in a particular conductor. At other times we take a localized view and study the flow of charge through a cross section of the conductor at a particular point. To describe this flow, we can use the **current density** $\vec{J}$, which has the same direction as the velocity of the moving charges if they are positive and the opposite direction if they are negative. For each element of the cross section, the magnitude J is equal to the current per unit area through that element. We can write the amount of current through the element as $\vec{J} \cdot d\vec{A}$, where $d\vec{A}$ is the area vector of the element, perpendicular to the element. The total current through the surface is then

$$i = \int \vec{J} \cdot d\vec{A}. \quad (26\text{-}4)$$

If the current is uniform across the surface and parallel to $d\vec{A}$, then $\vec{J}$ is also uniform and parallel to $d\vec{A}$. Then Eq. 26-4 becomes

$$i = \int J\,dA = J \int dA = JA,$$

so

$$J = \frac{i}{A}, \quad (26\text{-}5)$$

where A is the total area of the surface. From Eq. 26-4 or 26-5 we see that the SI unit for current density is the ampere per square meter (A/m^2).

In Chapter 22 we saw that we can represent an electric field with electric field lines. Figure 26-4 shows how current density can be represented with a similar set of lines, which we can call *streamlines*. The current, which is toward the right in Fig. 26-4, makes a transition from the wider conductor at the left to the narrower conductor at the right. Because charge is conserved during the transition, the amount of charge and thus the amount of current cannot change. However, the current density does change—it is greater in the narrower conductor. The spacing of the streamlines suggests this increase in current density; streamlines that are closer together imply greater current density.

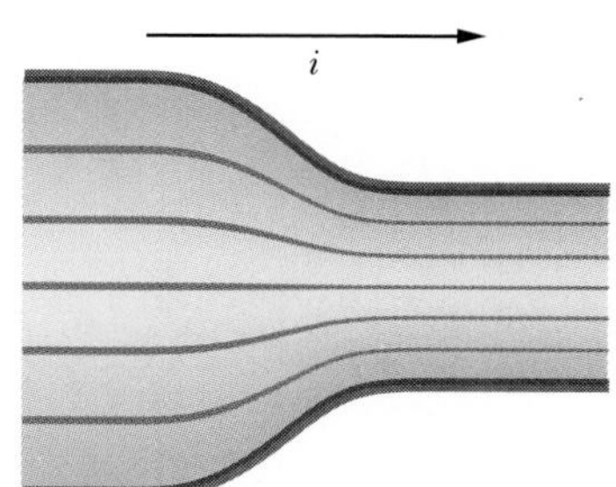

Fig. 26-4 Streamlines representing current density in the flow of charge through a constricted conductor.

Drift Speed

When a conductor does not have a current through it, its conduction electrons move randomly, with no net motion in any direction. When the conductor does have a current through it, these electrons actually still move randomly, but now

Current is said to be due to positive charges that are propelled by the electric field.

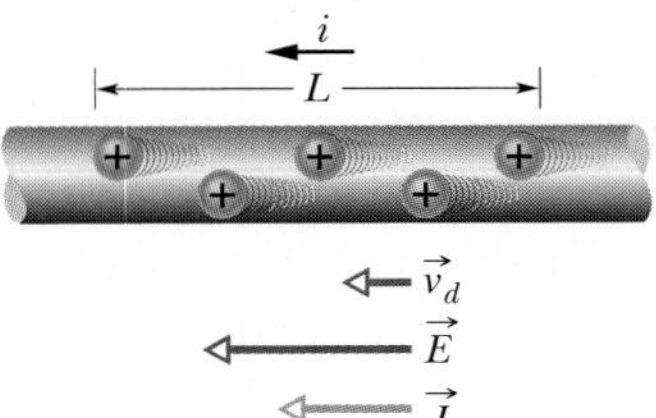

Fig. 26-5 Positive charge carriers drift at speed v_d in the direction of the applied electric field $\vec{E}$. By convention, the direction of the current density $\vec{J}$ and the sense of the current arrow are drawn in that same direction.

they tend to *drift* with a **drift speed** v_d in the direction opposite that of the applied electric field that causes the current. The drift speed is tiny compared with the speeds in the random motion. For example, in the copper conductors of household wiring, electron drift speeds are perhaps 10^{-5} or 10^{-4} m/s, whereas the random-motion speeds are around 10^6 m/s.

We can use Fig. 26-5 to relate the drift speed v_d of the conduction electrons in a current through a wire to the magnitude J of the current density in the wire. For convenience, Fig. 26-5 shows the equivalent drift of *positive* charge carriers in the direction of the applied electric field $\vec{E}$. Let us assume that these charge carriers all move with the same drift speed v_d and that the current density J is uniform across the wire's cross-sectional area A. The number of charge carriers in a length L of the wire is nAL, where n is the number of carriers per unit volume. The total charge of the carriers in the length L, each with charge e, is then

$$q = (nAL)e.$$

Because the carriers all move along the wire with speed v_d, this total charge moves through any cross section of the wire in the time interval

$$t = \frac{L}{v_d}.$$

Equation 26-1 tells us that the current i is the time rate of transfer of charge across a cross section, so here we have

$$i = \frac{q}{t} = \frac{nALe}{L/v_d} = nAev_d. \tag{26-6}$$

Solving for v_d and recalling Eq. 26-5 ($J = i/A$), we obtain

$$v_d = \frac{i}{nAe} = \frac{J}{ne}$$

or, extended to vector form,

$$\vec{J} = (ne)\vec{v}_d. \tag{26-7}$$

Here the product ne, whose SI unit is the coulomb per cubic meter (C/m^3), is the *carrier charge density.* For positive carriers, ne is positive and Eq. 26-7 predicts that $\vec{J}$ and $\vec{v}_d$ have the same direction. For negative carriers, ne is negative and $\vec{J}$ and $\vec{v}_d$ have opposite directions.

CHECKPOINT 2

The figure shows conduction electrons moving leftward in a wire. Are the following leftward or rightward: (a) the current i, (b) the current density $\vec{J}$, (c) the electric field $\vec{E}$ in the wire?

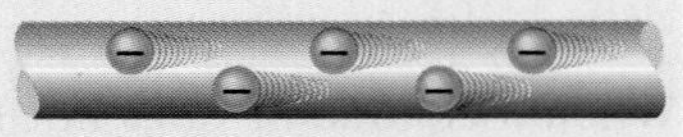

Sample Problem

Current density, uniform and nonuniform

(a) The current density in a cylindrical wire of radius $R = 2.0$ mm is uniform across a cross section of the wire and is $J = 2.0 \times 10^5$ A/m^2. What is the current through the outer portion of the wire between radial distances $R/2$ and R (Fig. 26-6*a*)?

KEY IDEA

Because the current density is uniform across the cross section, the current density J, the current i, and the cross-sectional area A are related by Eq. 26-5 ($J = i/A$).

Calculations: We want only the current through a reduced cross-sectional area A' of the wire (rather than the entire area), where

$$A' = \pi R^2 - \pi\left(\frac{R}{2}\right)^2 = \pi\left(\frac{3R^2}{4}\right)$$

$$= \frac{3\pi}{4}(0.0020\text{ m})^2 = 9.424 \times 10^{-6}\text{ m}^2.$$

So, we rewrite Eq. 26-5 as

$$i = JA'$$

and then substitute the data to find

$$i = (2.0 \times 10^5\text{ A/m}^2)(9.424 \times 10^{-6}\text{ m}^2)$$

$$= 1.9\text{ A}. \qquad \text{(Answer)}$$

(b) Suppose, instead, that the current density through a cross section varies with radial distance r as $J = ar^2$, in which $a = 3.0 \times 10^{11}$ A/m^4 and r is in meters. What now is the current through the same outer portion of the wire?

KEY IDEA

Because the current density is not uniform across a cross section of the wire, we must resort to Eq. 26-4 ($i = \int \vec{J} \cdot d\vec{A}$) and integrate the current density over the portion of the wire from $r = R/2$ to $r = R$.

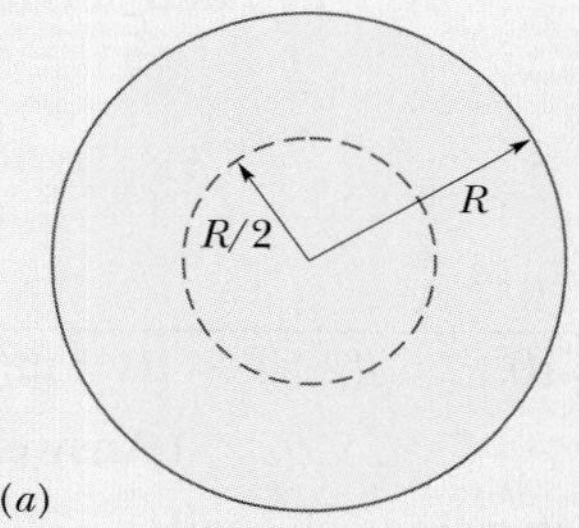

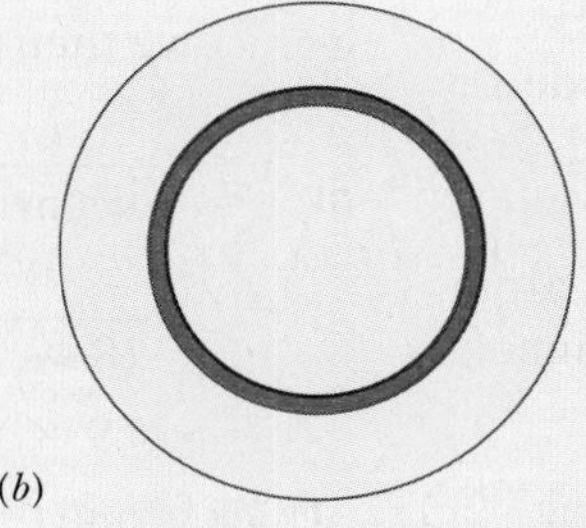

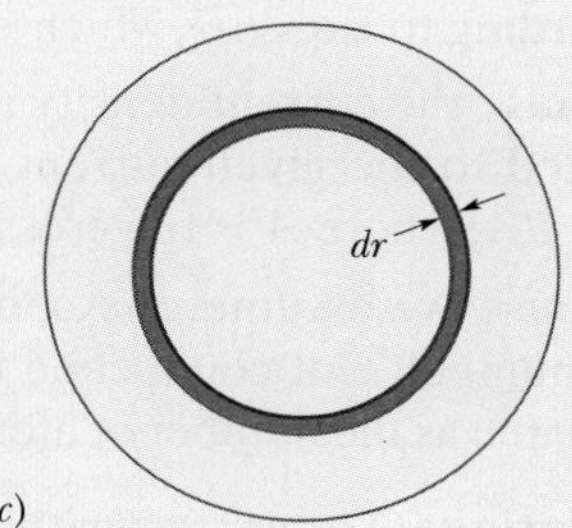

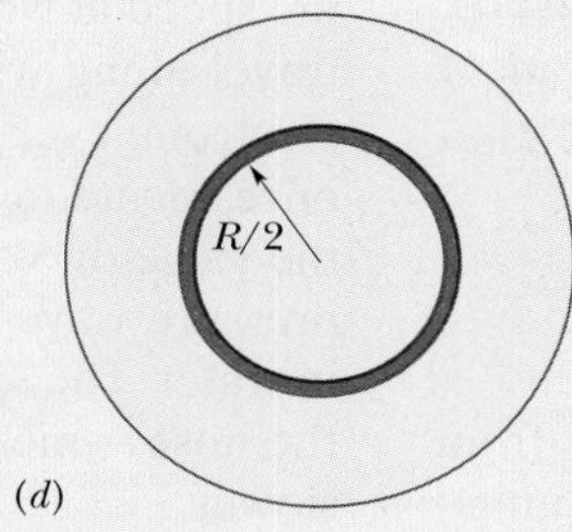

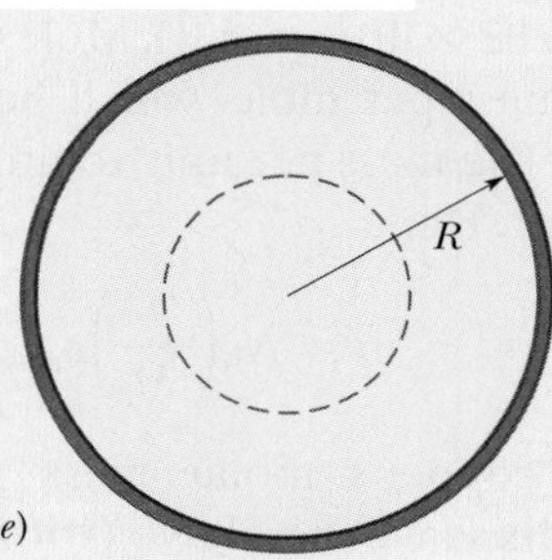

Fig. 26-6 (*a*) Cross section of a wire of radius R. If the current density is uniform, the current is just the product of the current density and the area. (*b*) – (*e*) If the current is nonuniform, we must first find the current through a thin ring and then sum (via integration) the currents in all such rings in the given area.

Calculations: The current density vector $\vec{J}$ (along the wire's length) and the differential area vector $d\vec{A}$ (perpendicular to a cross section of the wire) have the same direction. Thus,

$$\vec{J} \cdot d\vec{A} = J\,dA \cos 0 = J\,dA.$$

We need to replace the differential area dA with something we can actually integrate between the limits $r = R/2$ and $r = R$. The simplest replacement (because J is given as a function of r) is the area $2\pi r\,dr$ of a thin ring of circumference $2\pi r$ and width dr (Fig. 26-6*b*). We can then integrate with r as the variable of integration. Equation 26-4 then gives us

$$\begin{aligned} i &= \int \vec{J} \cdot d\vec{A} = \int J\,dA \\ &= \int_{R/2}^{R} ar^2\, 2\pi r\,dr = 2\pi a \int_{R/2}^{R} r^3\,dr \\ &= 2\pi a \left[\frac{r^4}{4}\right]_{R/2}^{R} = \frac{\pi a}{2}\left[R^4 - \frac{R^4}{16}\right] = \frac{15}{32}\pi a R^4 \\ &= \frac{15}{32}\pi(3.0 \times 10^{11}\ \text{A/m}^4)(0.0020\ \text{m})^4 = 7.1\ \text{A}. \end{aligned}$$

(Answer)

Sample Problem

In a current, the conduction electrons move very slowly

What is the drift speed of the conduction electrons in a copper wire with radius $r = 900\ \mu\text{m}$ when it has a uniform current $i = 17$ mA? Assume that each copper atom contributes one conduction electron to the current and that the current density is uniform across the wire's cross section.

KEY IDEAS

1. The drift speed v_d is related to the current density $\vec{J}$ and the number n of conduction electrons per unit volume according to Eq. 26-7, which we can write as $J = nev_d$.
2. Because the current density is uniform, its magnitude J is related to the given current i and wire size by Eq. 26-5 ($J = i/A$, where A is the cross-sectional area of the wire).
3. Because we assume one conduction electron per atom, the number n of conduction electrons per unit volume is the same as the number of atoms per unit volume.

Calculations: Let us start with the third idea by writing

$$n = \begin{pmatrix}\text{atoms} \\ \text{per unit} \\ \text{volume}\end{pmatrix} = \begin{pmatrix}\text{atoms} \\ \text{per} \\ \text{mole}\end{pmatrix}\begin{pmatrix}\text{moles} \\ \text{per unit} \\ \text{mass}\end{pmatrix}\begin{pmatrix}\text{mass} \\ \text{per unit} \\ \text{volume}\end{pmatrix}.$$

The number of atoms per mole is just Avogadro's number N_A ($= 6.02 \times 10^{23}\ \text{mol}^{-1}$). Moles per unit mass is the inverse of the mass per mole, which here is the molar mass M of copper. The mass per unit volume is the (mass) density ρ_{mass} of copper. Thus,

$$n = N_A\left(\frac{1}{M}\right)\rho_{\text{mass}} = \frac{N_A \rho_{\text{mass}}}{M}.$$

Taking copper's molar mass M and density ρ_{mass} from Appendix F, we then have (with some conversions of units)

$$\begin{aligned} n &= \frac{(6.02 \times 10^{23}\ \text{mol}^{-1})(8.96 \times 10^3\ \text{kg/m}^3)}{63.54 \times 10^{-3}\ \text{kg/mol}} \\ &= 8.49 \times 10^{28}\ \text{electrons/m}^3 \end{aligned}$$

or

$$n = 8.49 \times 10^{28}\ \text{m}^{-3}.$$

Next let us combine the first two key ideas by writing

$$\frac{i}{A} = nev_d.$$

Substituting for A with πr^2 ($= 2.54 \times 10^{-6}\ \text{m}^2$) and solving for v_d, we then find

$$\begin{aligned} v_d &= \frac{i}{ne(\pi r^2)} \\ &= \frac{17 \times 10^{-3}\ \text{A}}{(8.49 \times 10^{28}\ \text{m}^{-3})(1.6 \times 10^{-19}\ \text{C})(2.54 \times 10^{-6}\ \text{m}^2)} \\ &= 4.9 \times 10^{-7}\ \text{m/s}, \end{aligned}$$

(Answer)

which is only 1.8 mm/h, slower than a sluggish snail.

Lights are fast: You may well ask: "If the electrons drift so slowly, why do the room lights turn on so quickly when I throw the switch?" Confusion on this point results from not distinguishing between the drift speed of the electrons and the speed at which *changes* in the electric field configuration travel along wires. This latter speed is nearly that of light; electrons everywhere in the wire begin drifting almost at once, including into the lightbulbs. Similarly, when you open the valve on your garden hose with the hose full of water, a pressure wave travels along the hose at the speed of sound in water. The speed at which the water itself moves through the hose—measured perhaps with a dye marker—is much slower.

Additional examples, video, and practice available at *WileyPLUS*

26-4 Resistance and Resistivity

If we apply the same potential difference between the ends of geometrically similar rods of copper and of glass, very different currents result. The characteristic of the conductor that enters here is its electrical **resistance.** We determine the resistance between any two points of a conductor by applying a potential difference V between those points and measuring the current i that results. The resistance R is then

$$R = \frac{V}{i} \quad \text{(definition of } R\text{)}. \tag{26-8}$$

The SI unit for resistance that follows from Eq. 26-8 is the volt per ampere. This combination occurs so often that we give it a special name, the **ohm** (symbol Ω); that is,

$$1 \text{ ohm} = 1\ \Omega = 1 \text{ volt per ampere} = 1 \text{ V/A}. \tag{26-9}$$

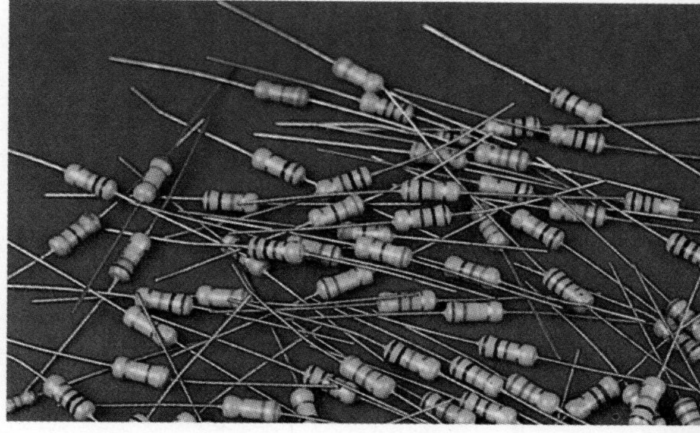

Fig. 26-7 An assortment of resistors. The circular bands are color-coding marks that identify the value of the resistance. *(The Image Works)*

A conductor whose function in a circuit is to provide a specified resistance is called a **resistor** (see Fig. 26-7). In a circuit diagram, we represent a resistor and a resistance with the symbol -WW-. If we write Eq. 26-8 as

$$i = \frac{V}{R},$$

we see that, for a given V, the greater the resistance, the smaller the current.

The resistance of a conductor depends on the manner in which the potential difference is applied to it. Figure 26-8, for example, shows a given potential difference applied in two different ways to the same conductor. As the current density streamlines suggest, the currents in the two cases—hence the measured resistances—will be different. Unless otherwise stated, we shall assume that any given potential difference is applied as in Fig. 26-8*b*.

Fig. 26-8 Two ways of applying a potential difference to a conducting rod. The gray connectors are assumed to have negligible resistance. When they are arranged as in (*a*) in a small region at each rod end, the measured resistance is larger than when they are arranged as in (*b*) to cover the entire rod end.

As we have done several times in other connections, we often wish to take a general view and deal not with particular objects but with materials. Here we do so by focusing not on the potential difference V across a particular resistor but on the electric field $\vec{E}$ at a point in a resistive material. Instead of dealing with the current i through the resistor, we deal with the current density $\vec{J}$ at the point in question. Instead of the resistance R of an object, we deal with the **resistivity** ρ of the *material:*

$$\rho = \frac{E}{J} \quad \text{(definition of } \rho\text{)}. \tag{26-10}$$

(Compare this equation with Eq. 26-8.)

If we combine the SI units of E and J according to Eq. 26-10, we get, for the unit of ρ, the ohm-meter ($\Omega \cdot$m):

$$\frac{\text{unit } (E)}{\text{unit } (J)} = \frac{\text{V/m}}{\text{A/m}^2} = \frac{\text{V}}{\text{A}}\text{m} = \Omega \cdot \text{m}.$$

(Do not confuse the *ohm-meter,* the unit of resistivity, with the *ohmmeter,* which is an instrument that measures resistance.) Table 26-1 lists the resistivities of some materials.

Table 26-1

Resistivities of Some Materials at Room Temperature (20°C)

Material	Resistivity, ρ ($\Omega \cdot$m)	Temperature Coefficient of Resistivity, α (K^{-1})
	Typical Metals	
Silver	1.62×10^{-8}	4.1×10^{-3}
Copper	1.69×10^{-8}	4.3×10^{-3}
Gold	2.35×10^{-8}	4.0×10^{-3}
Aluminum	2.75×10^{-8}	4.4×10^{-3}
Manganin[a]	4.82×10^{-8}	0.002×10^{-3}
Tungsten	5.25×10^{-8}	4.5×10^{-3}
Iron	9.68×10^{-8}	6.5×10^{-3}
Platinum	10.6×10^{-8}	3.9×10^{-3}
	Typical Semiconductors	
Silicon, pure	2.5×10^{3}	-70×10^{-3}
Silicon, *n*-type[b]	8.7×10^{-4}	
Silicon, *p*-type[c]	2.8×10^{-3}	
	Typical Insulators	
Glass	$10^{10} - 10^{14}$	
Fused quartz	$\sim 10^{16}$	

[a]An alloy specifically designed to have a small value of α.
[b]Pure silicon doped with phosphorus impurities to a charge carrier density of 10^{23} m^{-3}.
[c]Pure silicon doped with aluminum impurities to a charge carrier density of 10^{23} m^{-3}.

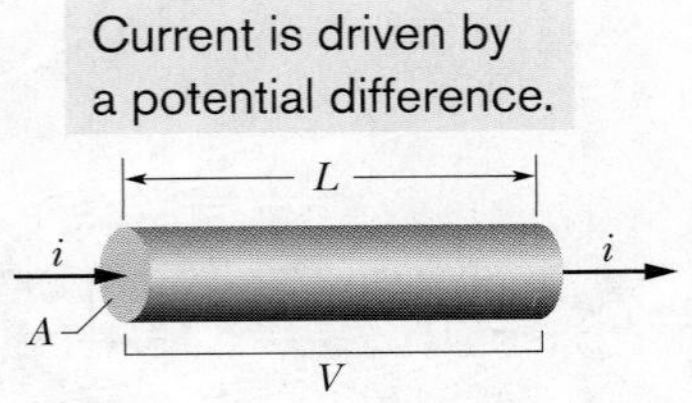

Fig. 26-9 A potential difference V is applied between the ends of a wire of length L and cross section A, establishing a current i.

We can write Eq. 26-10 in vector form as

$$\vec{E} = \rho\vec{J}. \tag{26-11}$$

Equations 26-10 and 26-11 hold only for *isotropic* materials—materials whose electrical properties are the same in all directions.

We often speak of the **conductivity** σ of a material. This is simply the reciprocal of its resistivity, so

$$\sigma = \frac{1}{\rho} \quad \text{(definition of } \sigma\text{)}. \tag{26-12}$$

The SI unit of conductivity is the reciprocal ohm-meter, $(\Omega \cdot \text{m})^{-1}$. The unit name mhos per meter is sometimes used (mho is ohm backwards). The definition of σ allows us to write Eq. 26-11 in the alternative form

$$\vec{J} = \sigma\vec{E}. \tag{26-13}$$

Calculating Resistance from Resistivity

We have just made an important distinction:

Resistance is a property of an object. Resistivity is a property of a material.

If we know the resistivity of a substance such as copper, we can calculate the resistance of a length of wire made of that substance. Let A be the cross-sectional area of the wire, let L be its length, and let a potential difference V exist between its ends (Fig. 26-9). If the streamlines representing the current density are uniform throughout the wire, the electric field and the current density will be constant for all points within the wire and, from Eqs. 24-42 and 26-5, will have the values

$$E = V/L \quad \text{and} \quad J = i/A. \tag{26-14}$$

We can then combine Eqs. 26-10 and 26-14 to write

$$\rho = \frac{E}{J} = \frac{V/L}{i/A}. \tag{26-15}$$

However, V/i is the resistance R, which allows us to recast Eq. 26-15 as

$$R = \rho\frac{L}{A}. \tag{26-16}$$

Equation 26-16 can be applied only to a homogeneous isotropic conductor of uniform cross section, with the potential difference applied as in Fig. 26-8*b*.

The macroscopic quantities V, i, and R are of greatest interest when we are making electrical measurements on specific conductors. They are the quantities that we read directly on meters. We turn to the microscopic quantities E, J, and ρ when we are interested in the fundamental electrical properties of materials.

CHECKPOINT 3

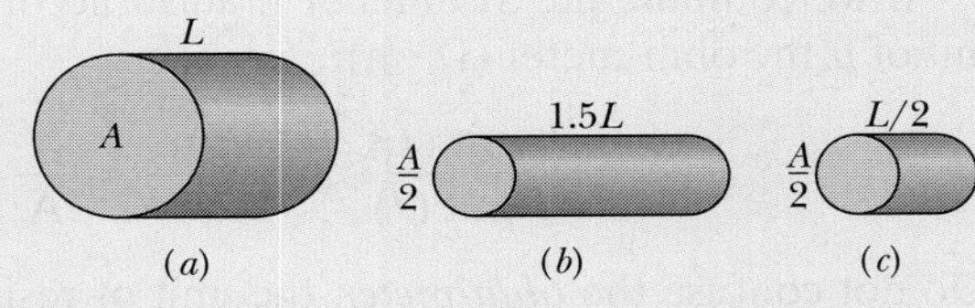

The figure here shows three cylindrical copper conductors along with their face areas and lengths. Rank them according to the current through them, greatest first, when the same potential difference V is placed across their lengths.

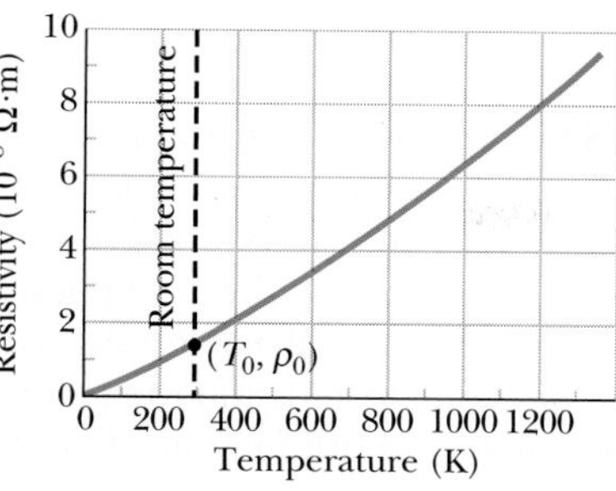

Fig. 26-10 The resistivity of copper as a function of temperature. The dot on the curve marks a convenient reference point at temperature $T_0 = 293$ K and resistivity $\rho_0 = 1.69 \times 10^{-8}\ \Omega \cdot \mathrm{m}$.

Resistivity can depend on temperature.

Variation with Temperature

The values of most physical properties vary with temperature, and resistivity is no exception. Figure 26-10, for example, shows the variation of this property for copper over a wide temperature range. The relation between temperature and resistivity for copper—and for metals in general—is fairly linear over a rather broad temperature range. For such linear relations we can write an empirical approximation that is good enough for most engineering purposes:

$$\rho - \rho_0 = \rho_0\alpha(T - T_0). \tag{26-17}$$

Here T_0 is a selected reference temperature and ρ_0 is the resistivity at that temperature. Usually $T_0 = 293$ K (room temperature), for which $\rho_0 = 1.69 \times 10^{-8}$ $\Omega \cdot$m for copper.

Because temperature enters Eq. 26-17 only as a difference, it does not matter whether you use the Celsius or Kelvin scale in that equation because the sizes of degrees on these scales are identical. The quantity α in Eq. 26-17, called the *temperature coefficient of resistivity,* is chosen so that the equation gives good agreement with experiment for temperatures in the chosen range. Some values of α for metals are listed in Table 26-1.

Sample Problem

A material has resistivity, a block of the material has resistance

A rectangular block of iron has dimensions 1.2 cm × 1.2 cm × 15 cm. A potential difference is to be applied to the block between parallel sides and in such a way that those sides are equipotential surfaces (as in Fig. 26-8*b*). What is the resistance of the block if the two parallel sides are (1) the square ends (with dimensions 1.2 cm × 1.2 cm) and (2) two rectangular sides (with dimensions 1.2 cm × 15 cm)?

KEY IDEA

The resistance R of an object depends on how the electric potential is applied to the object. In particular, it depends on the ratio L/A, according to Eq. 26-16 ($R = \rho L/A$), where A is the area of the surfaces to which the potential difference is applied and L is the distance between those surfaces.

Calculations: For arrangement 1, we have $L = 15$ cm = 0.15 m and

$$A = (1.2\ \mathrm{cm})^2 = 1.44 \times 10^{-4}\ \mathrm{m}^2.$$

Substituting into Eq. 26-16 with the resistivity ρ from Table 26-1, we then find that for arrangement 1,

$$R = \frac{\rho L}{A} = \frac{(9.68 \times 10^{-8}\ \Omega \cdot \mathrm{m})(0.15\ \mathrm{m})}{1.44 \times 10^{-4}\ \mathrm{m}^2}$$
$$= 1.0 \times 10^{-4}\ \Omega = 100\ \mu\Omega. \quad \text{(Answer)}$$

Similarly, for arrangement 2, with distance $L = 1.2$ cm and area $A = (1.2\ \mathrm{cm})(15\ \mathrm{cm})$, we obtain

$$R = \frac{\rho L}{A} = \frac{(9.68 \times 10^{-8}\ \Omega \cdot \mathrm{m})(1.2 \times 10^{-2}\ \mathrm{m})}{1.80 \times 10^{-3}\ \mathrm{m}^2}$$
$$= 6.5 \times 10^{-7}\ \Omega = 0.65\ \mu\Omega. \quad \text{(Answer)}$$

Additional examples, video, and practice available at *WileyPLUS*

26-5 Ohm's Law

As we just discussed in Section 26-4, a resistor is a conductor with a specified resistance. It has that same resistance no matter what the magnitude and direction (*polarity*) of the applied potential difference are. Other conducting devices, however, might have resistances that change with the applied potential difference.

Figure 26-11*a* shows how to distinguish such devices. A potential difference V is applied across the device being tested, and the resulting current i through the device is measured as V is varied in both magnitude and polarity. The polarity of V is arbitrarily taken to be positive when the left terminal of the device is at a higher potential than the right terminal. The direction of the resulting current (from left to right) is arbitrarily assigned a plus sign. The reverse polarity of V (with the right terminal at a higher potential) is then negative; the current it causes is assigned a minus sign.

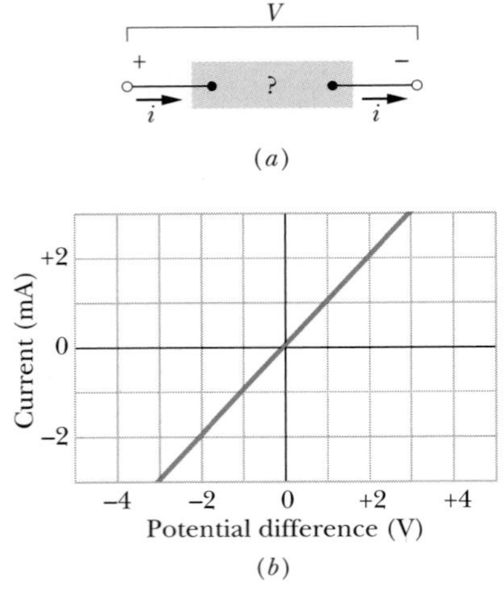

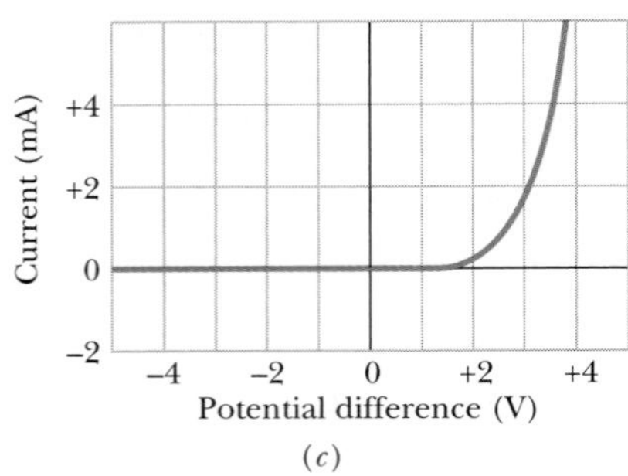

Fig. 26-11 (*a*) A potential difference V is applied to the terminals of a device, establishing a current i. (*b*) A plot of current i versus applied potential difference V when the device is a 1000 Ω resistor. (*c*) A plot when the device is a semiconducting *pn* junction diode.

Figure 26-11*b* is a plot of i versus V for one device. This plot is a straight line passing through the origin, so the ratio i/V (which is the slope of the straight line) is the same for all values of V. This means that the resistance $R = V/i$ of the device is independent of the magnitude and polarity of the applied potential difference V.

Figure 26-11*c* is a plot for another conducting device. Current can exist in this device only when the polarity of V is positive and the applied potential difference is more than about 1.5 V. When current does exist, the relation between i and V is not linear; it depends on the value of the applied potential difference V.

We distinguish between the two types of device by saying that one obeys Ohm's law and the other does not.

> **Ohm's law** is an assertion that the current through a device is *always* directly proportional to the potential difference applied to the device.

(This assertion is correct only in certain situations; still, for historical reasons, the term "law" is used.) The device of Fig. 26-11*b*—which turns out to be a 1000 Ω resistor—obeys Ohm's law. The device of Fig. 26-11*c*—which is called a *pn* junction diode—does not.

> A conducting device obeys Ohm's law when the resistance of the device is independent of the magnitude and polarity of the applied potential difference.

It is often contended that $V = iR$ is a statement of Ohm's law. That is not true! This equation is the defining equation for resistance, and it applies to all conducting devices, whether they obey Ohm's law or not. If we measure the potential difference V across, and the current i through, any device, even a *pn* junction diode, we can find its resistance *at that value of* V as $R = V/i$. The essence of Ohm's law, however, is that a plot of i versus V is linear; that is, R is independent of V.

We can express Ohm's law in a more general way if we focus on conducting *materials* rather than on conducting *devices*. The relevant relation is then Eq. 26-11 ($\vec{E} = \rho\vec{J}$), which corresponds to $V = iR$.

> A conducting material obeys Ohm's law when the resistivity of the material is independent of the magnitude and direction of the applied electric field.

All homogeneous materials, whether they are conductors like copper or semiconductors like pure silicon or silicon containing special impurities, obey Ohm's law within some range of values of the electric field. If the field is too strong, however, there are departures from Ohm's law in all cases.

CHECKPOINT 4

The following table gives the current i (in amperes) through two devices for several values of potential difference V (in volts). From these data, determine which device does not obey Ohm's law.

Device 1		Device 2	
V	i	V	i
2.00	4.50	2.00	1.50
3.00	6.75	3.00	2.20
4.00	9.00	4.00	2.80

26-6 A Microscopic View of Ohm's Law

To find out *why* particular materials obey Ohm's law, we must look into the details of the conduction process at the atomic level. Here we consider only conduction in metals, such as copper. We base our analysis on the *free-electron model,* in which we assume that the conduction electrons in the metal are free to move throughout the volume of a sample, like the molecules of a gas in a closed container. We also assume that the electrons collide not with one another but only with atoms of the metal.

According to classical physics, the electrons should have a Maxwellian speed distribution somewhat like that of the molecules in a gas (Section 19-7), and thus the average electron speed should depend on the temperature. The motions of electrons are, however, governed not by the laws of classical physics but by those of quantum physics. As it turns out, an assumption that is much closer to the quantum reality is that conduction electrons in a metal move with a single effective speed v_{eff}, and this speed is essentially independent of the temperature. For copper, $v_{eff} \approx 1.6 \times 10^6$ m/s.

When we apply an electric field to a metal sample, the electrons modify their random motions slightly and drift very slowly—in a direction opposite that of the field—with an average drift speed v_d. The drift speed in a typical metallic conductor is about 5×10^{-7} m/s, less than the effective speed (1.6×10^6 m/s) by many orders of magnitude. Figure 26-12 suggests the relation between these two speeds. The gray lines show a possible random path for an electron in the absence of an applied field; the electron proceeds from A to B, making six collisions along the way. The green lines show how the same events *might* occur when an electric field $\vec{E}$ is applied. We see that the electron drifts steadily to the right, ending at B' rather than at B. Figure 26-12 was drawn with the assumption that $v_d \approx 0.02v_{eff}$. However, because the actual value is more like $v_d \approx (10^{-13})v_{eff}$, the drift displayed in the figure is greatly exaggerated.

The motion of conduction electrons in an electric field $\vec{E}$ is thus a combination of the motion due to random collisions and that due to $\vec{E}$. When we consider all the free electrons, their random motions average to zero and make no contribution to the drift speed. Thus, the drift speed is due only to the effect of the electric field on the electrons.

If an electron of mass m is placed in an electric field of magnitude E, the electron will experience an acceleration given by Newton's second law:

$$a = \frac{F}{m} = \frac{eE}{m}. \tag{26-18}$$

The nature of the collisions experienced by conduction electrons is such that, after a typical collision, each electron will—so to speak—completely lose its memory of its previous drift velocity. Each electron will then start off fresh after every encounter, moving off in a random direction. In the average time τ between collisions, the average electron will acquire a drift speed of $v_d = a\tau$. Moreover, if we measure the drift speeds of all the electrons at any instant, we will find that their average drift speed is also $a\tau$. Thus, at any instant, on average, the electrons will have drift speed $v_d = a\tau$. Then Eq. 26-18 gives us

$$v_d = a\tau = \frac{eE\tau}{m}. \tag{26-19}$$

Combining this result with Eq. 26-7 ($\vec{J} = ne\vec{v}_d$), in magnitude form, yields

$$v_d = \frac{J}{ne} = \frac{eE\tau}{m}, \tag{26-20}$$

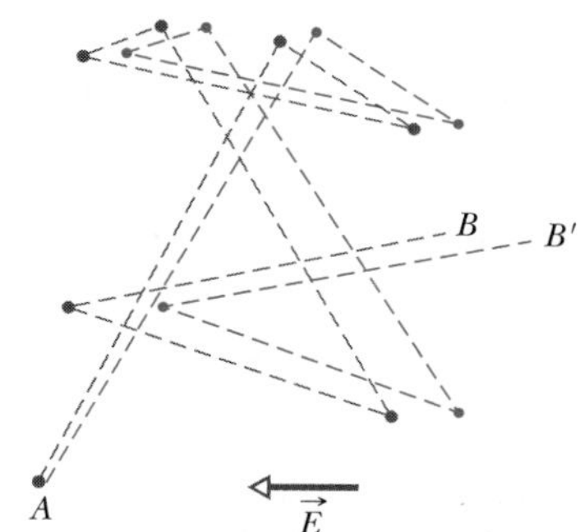

Fig. 26-12 The gray lines show an electron moving from A to B, making six collisions en route. The green lines show what the electron's path might be in the presence of an applied electric field $\vec{E}$. Note the steady drift in the direction of $-\vec{E}$. (Actually, the green lines should be slightly curved, to represent the parabolic paths followed by the electrons between collisions, under the influence of an electric field.)

which we can write as

$$E = \left(\frac{m}{e^2 n\tau}\right) J. \tag{26-21}$$

Comparing this with Eq. 26-11 ($\vec{E} = \rho\vec{J}$), in magnitude form, leads to

$$\rho = \frac{m}{e^2 n\tau}. \tag{26-22}$$

Equation 26-22 may be taken as a statement that metals obey Ohm's law if we can show that, for metals, their resistivity ρ is a constant, independent of the strength of the applied electric field $\vec{E}$. Let's consider the quantities in Eq. 26-22. We can reasonably assume that n, the number of conduction electrons per volume, is independent of the field, and m and e are constants. Thus, we only need to convince ourselves that τ, the average time (or *mean free time*) between collisions, is a constant, independent of the strength of the applied electric field. Indeed, τ can be considered to be a constant because the drift speed v_d caused by the field is so much smaller than the effective speed v_{eff} that the electron speed—and thus τ—is hardly affected by the field.

Sample Problem

Mean free time and mean free distance

(a) What is the mean free time τ between collisions for the conduction electrons in copper?

KEY IDEAS

The mean free time τ of copper is approximately constant, and in particular does not depend on any electric field that might be applied to a sample of the copper. Thus, we need not consider any particular value of applied electric field. However, because the resistivity ρ displayed by copper under an electric field depends on τ, we can find the mean free time τ from Eq. 26-22 ($\rho = m/e^2 n\tau$).

Calculations: That equation gives us

$$\tau = \frac{m}{ne^2\rho}. \tag{26-23}$$

The number of conduction electrons per unit volume in copper is $8.49 \times 10^{28}\ \text{m}^{-3}$. We take the value of ρ from Table 26-1. The denominator then becomes

$$(8.49 \times 10^{28}\ \text{m}^{-3})(1.6 \times 10^{-19}\ \text{C})^2(1.69 \times 10^{-8}\ \Omega\cdot\text{m})$$
$$= 3.67 \times 10^{-17}\ \text{C}^2\cdot\Omega/\text{m}^2 = 3.67 \times 10^{-17}\ \text{kg/s},$$

where we converted units as

$$\frac{\text{C}^2\cdot\Omega}{\text{m}^2} = \frac{\text{C}^2\cdot\text{V}}{\text{m}^2\cdot\text{A}} = \frac{\text{C}^2\cdot\text{J/C}}{\text{m}^2\cdot\text{C/s}} = \frac{\text{kg}\cdot\text{m}^2/\text{s}^2}{\text{m}^2/\text{s}} = \frac{\text{kg}}{\text{s}}.$$

Using these results and substituting for the electron mass m, we then have

$$\tau = \frac{9.1 \times 10^{-31}\ \text{kg}}{3.67 \times 10^{-17}\ \text{kg/s}} = 2.5 \times 10^{-14}\ \text{s}. \quad \text{(Answer)}$$

(b) The mean free path λ of the conduction electrons in a conductor is the average distance traveled by an electron between collisions. (This definition parallels that in Section 19-6 for the mean free path of molecules in a gas.) What is λ for the conduction electrons in copper, assuming that their effective speed v_{eff} is 1.6×10^6 m/s?

KEY IDEA

The distance d any particle travels in a certain time t at a constant speed v is $d = vt$.

Calculation: For the electrons in copper, this gives us

$$\lambda = v_{\text{eff}}\tau \tag{26-24}$$
$$= (1.6 \times 10^6\ \text{m/s})(2.5 \times 10^{-14}\ \text{s})$$
$$= 4.0 \times 10^{-8}\ \text{m} = 40\ \text{nm}. \quad \text{(Answer)}$$

This is about 150 times the distance between nearest-neighbor atoms in a copper lattice. Thus, on the average, each conduction electron passes many copper atoms before finally hitting one.

Additional examples, video, and practice available at *WileyPLUS*

26-7 Power in Electric Circuits

Figure 26-13 shows a circuit consisting of a battery B that is connected by wires, which we assume have negligible resistance, to an unspecified conducting device. The device might be a resistor, a storage battery (a rechargeable battery), a motor, or some other electrical device. The battery maintains a potential difference of magnitude V across its own terminals and thus (because of the wires) across the terminals of the unspecified device, with a greater potential at terminal a of the device than at terminal b.

Because there is an external conducting path between the two terminals of the battery, and because the potential differences set up by the battery are maintained, a steady current i is produced in the circuit, directed from terminal a to terminal b. The amount of charge dq that moves between those terminals in time interval dt is equal to $i\,dt$. This charge dq moves through a decrease in potential of magnitude V, and thus its electric potential energy decreases in magnitude by the amount

$$dU = dq\,V = i\,dt\,V. \tag{26-25}$$

The principle of conservation of energy tells us that the decrease in electric potential energy from a to b is accompanied by a transfer of energy to some other form. The power P associated with that transfer is the rate of transfer dU/dt, which is given by Eq. 26-25 as

$$P = iV \quad \text{(rate of electrical energy transfer).} \tag{26-26}$$

Moreover, this power P is also the rate at which energy is transferred from the battery to the unspecified device. If that device is a motor connected to a mechanical load, the energy is transferred as work done on the load. If the device is a storage battery that is being charged, the energy is transferred to stored chemical energy in the storage battery. If the device is a resistor, the energy is transferred to internal thermal energy, tending to increase the resistor's temperature.

The unit of power that follows from Eq. 26-26 is the volt-ampere (V·A). We can write it as

$$1\,\text{V}\cdot\text{A} = \left(1\,\frac{\text{J}}{\text{C}}\right)\left(1\,\frac{\text{C}}{\text{s}}\right) = 1\,\frac{\text{J}}{\text{s}} = 1\,\text{W}.$$

As an electron moves through a resistor at constant drift speed, its average kinetic energy remains constant and its lost electric potential energy appears as thermal energy in the resistor and the surroundings. On a microscopic scale this energy transfer is due to collisions between the electron and the molecules of the resistor, which leads to an increase in the temperature of the resistor lattice. The mechanical energy thus transferred to thermal energy is *dissipated* (lost) because the transfer cannot be reversed.

For a resistor or some other device with resistance R, we can combine Eqs. 26-8 ($R = V/i$) and 26-26 to obtain, for the rate of electrical energy dissipation due to a resistance, either

$$P = i^2R \quad \text{(resistive dissipation)} \tag{26-27}$$

or

$$P = \frac{V^2}{R} \quad \text{(resistive dissipation).} \tag{26-28}$$

Caution: We must be careful to distinguish these two equations from Eq. 26-26: $P = iV$ applies to electrical energy transfers of all kinds; $P = i^2R$ and $P = V^2/R$ apply only to the transfer of electric potential energy to thermal energy in a device with resistance.

The battery at the left supplies energy to the conduction electrons that form the current.

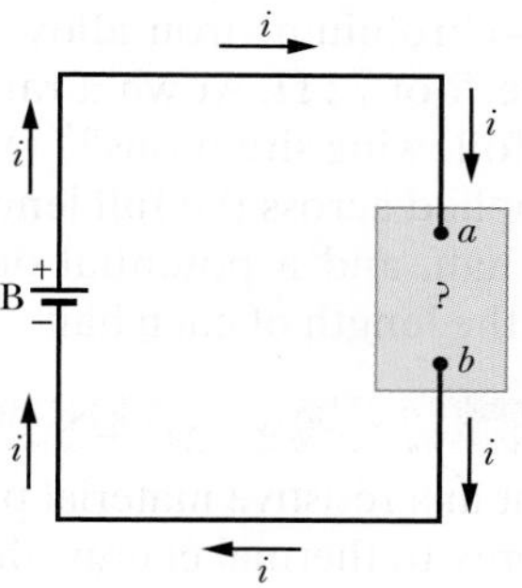

Fig. 26-13 A battery B sets up a current i in a circuit containing an unspecified conducting device.

CHECKPOINT 5

A potential difference V is connected across a device with resistance R, causing current i through the device. Rank the following variations according to the change in the rate at which electrical energy is converted to thermal energy due to the resistance, greatest change first: (a) V is doubled with R unchanged, (b) i is doubled with R unchanged, (c) R is doubled with V unchanged, (d) R is doubled with i unchanged.

Sample Problem

Rate of energy dissipation in a wire carrying current

You are given a length of uniform heating wire made of a nickel–chromium–iron alloy called Nichrome; it has a resistance R of 72 Ω. At what rate is energy dissipated in each of the following situations? (1) A potential difference of 120 V is applied across the full length of the wire. (2) The wire is cut in half, and a potential difference of 120 V is applied across the length of each half.

KEY IDEA

Current in a resistive material produces a transfer of mechanical energy to thermal energy; the rate of transfer (dissipation) is given by Eqs. 26-26 to 26-28.

Calculations: Because we know the potential V and resistance R, we use Eq. 26-28, which yields, for situation 1,

$$P = \frac{V^2}{R} = \frac{(120\ \text{V})^2}{72\ \Omega} = 200\ \text{W}. \qquad \text{(Answer)}$$

In situation 2, the resistance of each half of the wire is (72 Ω)/2, or 36 Ω. Thus, the dissipation rate for each half is

$$P' = \frac{(120\ \text{V})^2}{36\ \Omega} = 400\ \text{W},$$

and that for the two halves is

$$P = 2P' = 800\ \text{W}. \qquad \text{(Answer)}$$

This is four times the dissipation rate of the full length of wire. Thus, you might conclude that you could buy a heating coil, cut it in half, and reconnect it to obtain four times the heat output. Why is this unwise? (What would happen to the amount of current in the coil?)

Additional examples, video, and practice available at *WileyPLUS*

26-8 Semiconductors

Semiconducting devices are at the heart of the microelectronic revolution that ushered in the information age. Table 26-2 compares the properties of silicon—a typical semiconductor—and copper—a typical metallic conductor. We see that silicon has many fewer charge carriers, a much higher resistivity, and a temperature coefficient of resistivity that is both large and negative. Thus, although the resistivity of copper increases with increasing temperature, that of pure silicon decreases.

Pure silicon has such a high resistivity that it is effectively an insulator and thus not of much direct use in microelectronic circuits. However, its resistivity can be greatly reduced in a controlled way by adding minute amounts of specific "impurity" atoms in a process called *doping.* Table 26-1 gives typical values of resistivity for silicon before and after doping with two different impurities.

We can roughly explain the differences in resistivity (and thus in conductivity) between semiconductors, insulators, and metallic conductors in terms of the energies of their electrons. (We need quantum physics to explain in more detail.) In a metallic conductor such as copper wire, most of the electrons are firmly locked in place within the atoms; much energy would be required to free them so they could move and participate in an electric current. However, there are also some electrons that, roughly speaking, are only loosely held in place and that require only little energy to become free. Thermal energy can supply that energy,

Table 26-2

Some Electrical Properties of Copper and Silicon

Property	Copper	Silicon
Type of material	Metal	Semiconductor
Charge carrier density, m^{-3}	8.49×10^{28}	1×10^{16}
Resistivity, Ω · m	1.69×10^{-8}	2.5×10^{3}
Temperature coefficient of resistivity, K^{-1}	$+4.3 \times 10^{-3}$	-70×10^{-3}

as can an electric field applied across the conductor. The field would not only free these loosely held electrons but would also propel them along the wire; thus, the field would drive a current through the conductor.

In an insulator, significantly greater energy is required to free electrons so they can move through the material. Thermal energy cannot supply enough energy, and neither can any reasonable electric field applied to the insulator. Thus, no electrons are available to move through the insulator, and hence no current occurs even with an applied electric field.

A semiconductor is like an insulator *except* that the energy required to free some electrons is not quite so great. More important, doping can supply electrons or positive charge carriers that are very loosely held within the material and thus are easy to get moving. Moreover, by controlling the doping of a semiconductor, we can control the density of charge carriers that can participate in a current and thereby can control some of its electrical properties. Most semiconducting devices, such as transistors and junction diodes, are fabricated by the selective doping of different regions of the silicon with impurity atoms of different kinds.

Let us now look again at Eq. 26-25 for the resistivity of a conductor:

$$\rho = \frac{m}{e^2 n \tau}, \tag{26-29}$$

where n is the number of charge carriers per unit volume and τ is the mean time between collisions of the charge carriers. (We derived this equation for conductors, but it also applies to semiconductors.) Let us consider how the variables n and τ change as the temperature is increased.

In a conductor, n is large but very nearly constant with any change in temperature. The increase of resistivity with temperature for metals (Fig. 26-10) is due to an increase in the collision rate of the charge carriers, which shows up in Eq. 26-29 as a decrease in τ, the mean time between collisions.

In a semiconductor, n is small but increases very rapidly with temperature as the increased thermal agitation makes more charge carriers available. This causes a *decrease* of resistivity with increasing temperature, as indicated by the negative temperature coefficient of resistivity for silicon in Table 26-2. The same increase in collision rate that we noted for metals also occurs for semiconductors, but its effect is swamped by the rapid increase in the number of charge carriers.

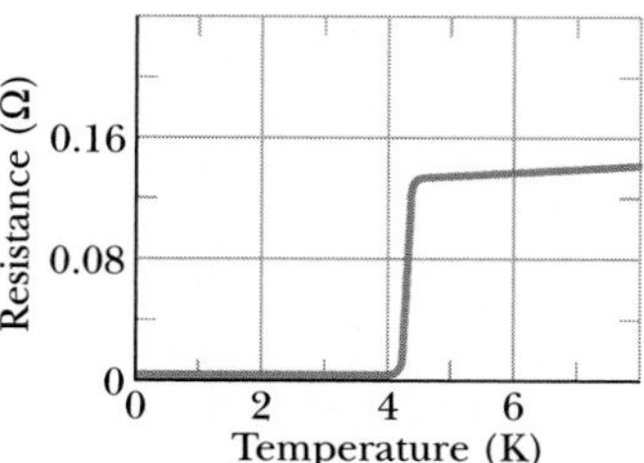

Fig. 26-14 The resistance of mercury drops to zero at a temperature of about 4 K.

26-9 Superconductors

In 1911, Dutch physicist Kamerlingh Onnes discovered that the resistivity of mercury absolutely disappears at temperatures below about 4 K (Fig. 26-14). This phenomenon of **superconductivity** is of vast potential importance in technology because it means that charge can flow through a superconducting conductor without losing its energy to thermal energy. Currents created in a superconducting ring, for example, have persisted for several years without loss; the electrons making up the current require a force and a source of energy at start-up time but not thereafter.

Prior to 1986, the technological development of superconductivity was throttled by the cost of producing the extremely low temperatures required to achieve the effect. In 1986, however, new ceramic materials were discovered that become superconducting at considerably higher (and thus cheaper to produce) temperatures. Practical application of superconducting devices at room temperature may eventually become commonplace.

Superconductivity is a phenomenon much different from conductivity. In fact, the best of the normal conductors, such as silver and copper, cannot become superconducting at any temperature, and the new ceramic superconductors are actually good insulators when they are not at low enough temperatures to be in a superconducting state.

A disk-shaped magnet is levitated above a superconducting material that has been cooled by liquid nitrogen. The goldfish is along for the ride. *(Courtesy Shoji Tonaka/International Superconductivity Technology Center, Tokyo, Japan)*

One explanation for superconductivity is that the electrons that make up the current move in coordinated pairs. One of the electrons in a pair may electrically distort the molecular structure of the superconducting material as it moves through, creating nearby a short-lived concentration of positive charge. The other electron in the pair may then be attracted toward this positive charge. According to the theory, such coordination between electrons would prevent them from colliding with the molecules of the material and thus would eliminate electrical resistance. The theory worked well to explain the pre-1986, lower temperature superconductors, but new theories appear to be needed for the newer, higher temperature superconductors.

REVIEW & SUMMARY

Current An **electric current** i in a conductor is defined by

$$i = \frac{dq}{dt}. \tag{26-1}$$

Here dq is the amount of (positive) charge that passes in time dt through a hypothetical surface that cuts across the conductor. By convention, the direction of electric current is taken as the direction in which positive charge carriers would move. The SI unit of electric current is the **ampere** (A): 1 A = 1 C/s.

Current Density Current (a scalar) is related to **current density** $\vec{J}$ (a vector) by

$$i = \int \vec{J} \cdot d\vec{A}, \tag{26-4}$$

where $d\vec{A}$ is a vector perpendicular to a surface element of area dA and the integral is taken over any surface cutting across the conductor. $\vec{J}$ has the same direction as the velocity of the moving charges if they are positive and the opposite direction if they are negative.

Drift Speed of the Charge Carriers When an electric field $\vec{E}$ is established in a conductor, the charge carriers (assumed positive) acquire a **drift speed** v_d in the direction of $\vec{E}$; the velocity $\vec{v}_d$ is related to the current density by

$$\vec{J} = (ne)\vec{v}_d, \tag{26-7}$$

where ne is the *carrier charge density*.

Resistance of a Conductor The **resistance** R of a conductor is defined as

$$R = \frac{V}{i} \quad \text{(definition of } R\text{)}, \tag{26-8}$$

where V is the potential difference across the conductor and i is the current. The SI unit of resistance is the **ohm** (Ω): 1 Ω = 1 V/A. Similar equations define the **resistivity** ρ and **conductivity** σ of a material:

$$\rho = \frac{1}{\sigma} = \frac{E}{J} \quad \text{(definitions of } \rho \text{ and } \sigma\text{)}, \tag{26-12, 26-10}$$

where E is the magnitude of the applied electric field. The SI unit of resistivity is the ohm-meter ($\Omega \cdot$m). Equation 26-10 corresponds to the vector equation

$$\vec{E} = \rho\vec{J}. \tag{26-11}$$

The resistance R of a conducting wire of length L and uniform cross section is

$$R = \rho\frac{L}{A}, \tag{26-16}$$

where A is the cross-sectional area.

Change of ρ with Temperature The resistivity ρ for most materials changes with temperature. For many materials, including metals, the relation between ρ and temperature T is approximated by the equation

$$\rho - \rho_0 = \rho_0\alpha(T - T_0). \tag{26-17}$$

Here T_0 is a reference temperature, ρ_0 is the resistivity at T_0, and α is the temperature coefficient of resistivity for the material.

Ohm's Law A given device (conductor, resistor, or any other electrical device) obeys *Ohm's law* if its resistance R, defined by Eq. 26-8 as V/i, is independent of the applied potential difference V. A given *material* obeys Ohm's law if its resistivity, defined by Eq. 26-10, is independent of the magnitude and direction of the applied electric field $\vec{E}$.

Resistivity of a Metal By assuming that the conduction electrons in a metal are free to move like the molecules of a gas, it is possible to derive an expression for the resistivity of a metal:

$$\rho = \frac{m}{e^2 n\tau}. \tag{26-22}$$

Here n is the number of free electrons per unit volume and τ is the mean time between the collisions of an electron with the atoms of the metal. We can explain why metals obey Ohm's law by pointing out that τ is essentially independent of the magnitude E of any electric field applied to a metal.

Power The power P, or rate of energy transfer, in an electrical device across which a potential difference V is maintained is

$$P = iV \quad \text{(rate of electrical energy transfer)}. \tag{26-26}$$

Resistive Dissipation If the device is a resistor, we can write Eq. 26-26 as

$$P = i^2R = \frac{V^2}{R} \quad \text{(resistive dissipation)}. \tag{26-27, 26-28}$$

In a resistor, electric potential energy is converted to internal thermal energy via collisions between charge carriers and atoms.

Semiconductors *Semiconductors* are materials that have few conduction electrons but can become conductors when they are *doped* with other atoms that contribute free electrons.

Superconductors *Superconductors* are materials that lose all electrical resistance at low temperatures. Recent research has discovered materials that are superconducting at surprisingly high temperatures.

QUESTIONS

1 Figure 26-15 shows cross sections through three long conductors of the same length and material, with square cross sections of edge lengths as shown. Conductor B fits snugly within conductor A, and conductor C fits snugly within conductor B. Rank the following according to their end-to-end resistances, greatest first: the individual conductors and the combinations of $A + B$ (B inside A), $B + C$ (C inside B), and $A + B + C$ (B inside A inside C).

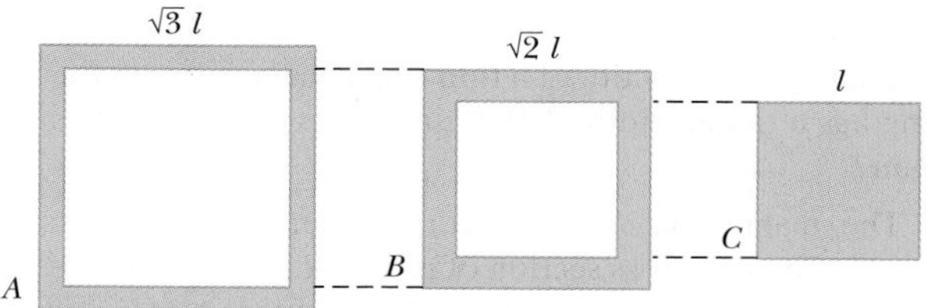

Fig. 26-15 Question 1.

2 Figure 26-16 shows cross sections through three wires of identical length and material; the sides are given in millimeters. Rank the wires according to their resistance (measured end to end along each wire's length), greatest first.

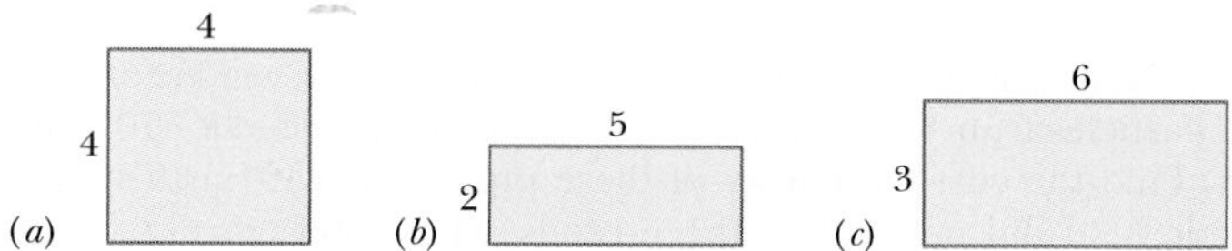

Fig. 26-16 Question 2.

3 Figure 26-17 shows a rectangular solid conductor of edge lengths L, $2L$, and $3L$. A potential difference V is to be applied uniformly between pairs of opposite faces of the conductor as in Fig. 26-8*b*. First V is applied between the left–right faces, then between the top–bottom faces, and then between the front–back faces. Rank those pairs, greatest first, according to the following (within the conductor): (a) the magnitude of the electric field, (b) the current density, (c) the current, and (d) the drift speed of the electrons.

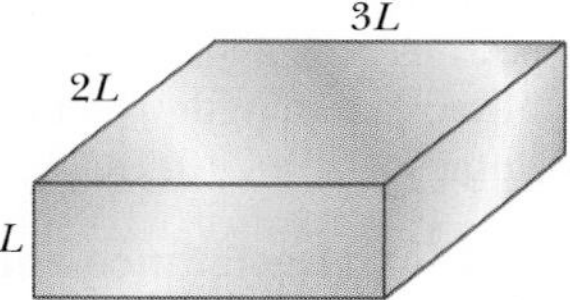

Fig. 26-17 Question 3.

4 Figure 26-18 shows plots of the current i through a certain cross section of a wire over four different time periods. Rank the periods according to the net charge that passes through the cross section during the period, greatest first.

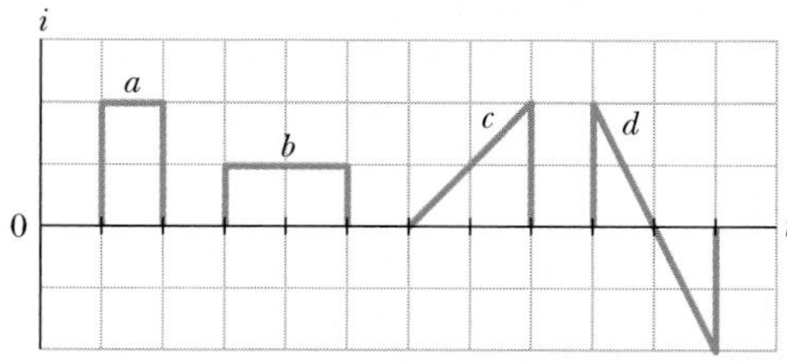

Fig. 26-18 Question 4.

5 Figure 26-19 shows four situations in which positive and negative charges move horizontally and gives the rate at which each charge moves. Rank the situations according to the effective current through the regions, greatest first.

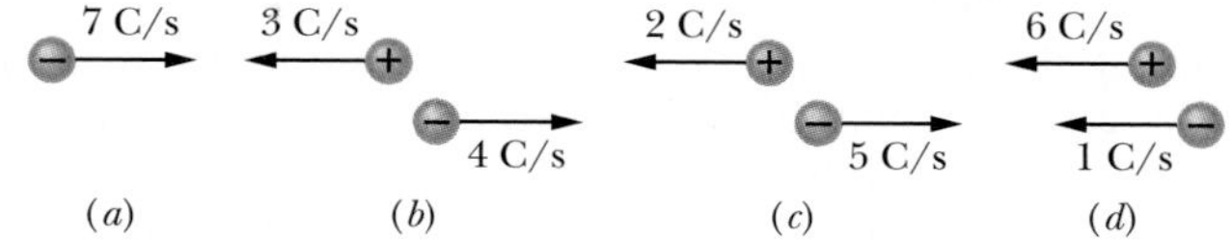

Fig. 26-19 Question 5.

6 In Fig. 26-20, a wire that carries a current consists of three sections with different radii. Rank the sections according to the following quantities, greatest first: (a) current, (b) magnitude of current density, and (c) magnitude of electric field.

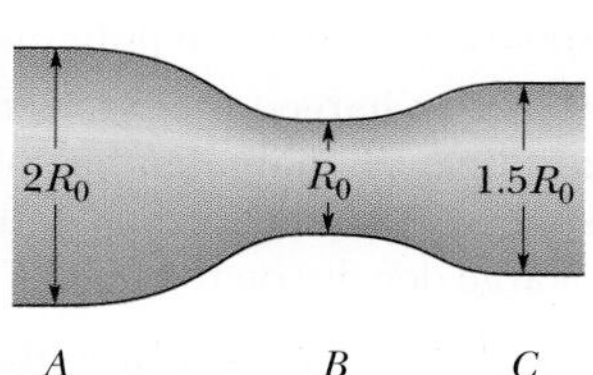

Fig. 26-20 Question 6.

7 Figure 26-21 gives the electric potential $V(x)$ versus position x along a copper wire carrying current. The wire consists of three sections that differ in radius. Rank the three sections according to the magnitude of the (a) electric field and (b) current density, greatest first.

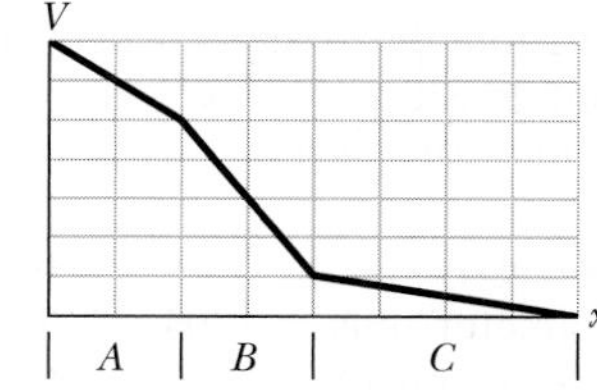

Fig. 26-21 Question 7.

8 The following table gives the lengths of three copper rods, their diameters, and the potential differences between their ends. Rank the rods according to (a) the magnitude of the electric field within them, (b) the current density within them, and (c) the drift speed of electrons through them, greatest first.

Rod	Length	Diameter	Potential Difference
1	L	$3d$	V
2	$2L$	d	$2V$
3	$3L$	$2d$	$2V$

9 Figure 26-22 gives the drift speed v_d of conduction electrons in a copper wire versus position x along the wire. The wire consists of three sections that differ in radius. Rank the three sections according to the following quantities, greatest first: (a) radius, (b) number of conduction electrons per cubic meter, (c) magnitude of electric field, (d) conductivity.

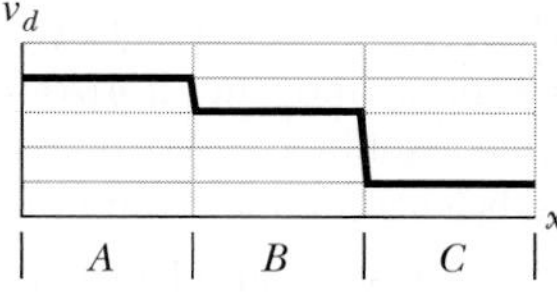

Fig. 26-22 Question 9.

10 Three wires, of the same diameter, are connected in turn between two points maintained at a constant potential difference. Their resistivities and lengths are ρ and L (wire A), 1.2ρ and $1.2L$ (wire B), and 0.9ρ and L (wire C). Rank the wires according to the rate at which energy is transferred to thermal energy, greatest first.

PROBLEMS

 Tutoring problem available (at instructor's discretion) in *WileyPLUS* and WebAssign

SSM Worked-out solution available in Student Solutions Manual

• – ••• Number of dots indicates level of problem difficulty

WWW Worked-out solution is at / ILW Interactive solution is at http://www.wiley.com/college/halliday

Additional information available in *The Flying Circus of Physics* and at flyingcircusofphysics.com

sec. 26-2 Electric Current

•1 During the 4.0 min a 5.0 A current is set up in a wire, how many (a) coulombs and (b) electrons pass through any cross section across the wire's width?

••2 An isolated conducting sphere has a 10 cm radius. One wire carries a current of 1.000 002 0 A into it. Another wire carries a current of 1.000 000 0 A out of it. How long would it take for the sphere to increase in potential by 1000 V?

••3 A charged belt, 50 cm wide, travels at 30 m/s between a source of charge and a sphere. The belt carries charge into the sphere at a rate corresponding to 100 μA. Compute the surface charge density on the belt.

sec. 26-3 Current Density

•4 The (United States) National Electric Code, which sets maximum safe currents for insulated copper wires of various diameters, is given (in part) in the table. Plot the safe current density as a function of diameter. Which wire gauge has the maximum safe current density? ("Gauge" is a way of identifying wire diameters, and 1 mil = 10^{-3} in.)

Gauge	4	6	8	10	12	14	16	18
Diameter, mils	204	162	129	102	81	64	51	40
Safe current, A	70	50	35	25	20	15	6	3

•5 SSM WWW A beam contains 2.0×10^8 doubly charged positive ions per cubic centimeter, all of which are moving north with a speed of 1.0×10^5 m/s. What are the (a) magnitude and (b) direction of the current density $\vec{J}$? (c) What additional quantity do you need to calculate the total current i in this ion beam?

•6 A certain cylindrical wire carries current. We draw a circle of radius r around its central axis in Fig. 26-23a to determine the current i within the circle. Figure 26-23b shows current i as a function of r^2. The vertical scale is set by i_s = 4.0 mA, and the horizontal scale is set by r_s^2 = 4.0 mm². (a) Is the current density uniform? (b) If so, what is its magnitude?

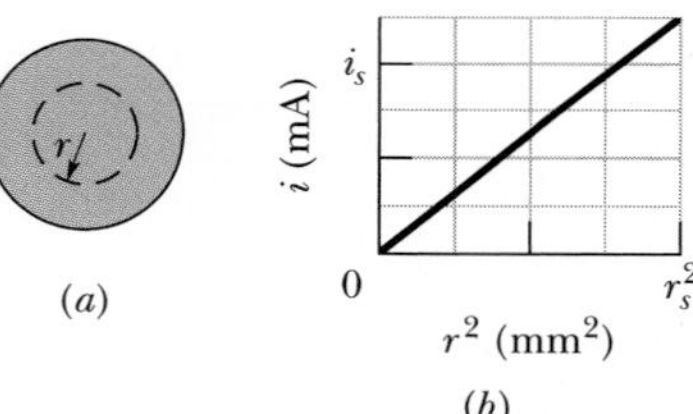

Fig. 26-23 Problem 6.

•7 A fuse in an electric circuit is a wire that is designed to melt, and thereby open the circuit, if the current exceeds a predetermined value. Suppose that the material to be used in a fuse melts when the current density rises to 440 A/cm². What diameter of cylindrical wire should be used to make a fuse that will limit the current to 0.50 A?

•8 A small but measurable current of 1.2×10^{-10} A exists in a copper wire whose diameter is 2.5 mm. The number of charge carriers per unit volume is 8.49×10^{28} m^{-3}. Assuming the current is uniform, calculate the (a) current density and (b) electron drift speed.

••9 The magnitude $J(r)$ of the current density in a certain cylindrical wire is given as a function of radial distance from the center of the wire's cross section as $J(r) = Br$, where r is in meters, J is in amperes per square meter, and $B = 2.00 \times 10^5$ A/m³. This function applies out to the wire's radius of 2.00 mm. How much current is contained within the width of a thin ring concentric with the wire if the ring has a radial width of 10.0 μm and is at a radial distance of 1.20 mm?

••10 The magnitude J of the current density in a certain lab wire with a circular cross section of radius R = 2.00 mm is given by $J = (3.00 \times 10^8)r^2$, with J in amperes per square meter and radial distance r in meters. What is the current through the outer section bounded by $r = 0.900R$ and $r = R$?

••11 What is the current in a wire of radius R = 3.40 mm if the magnitude of the current density is given by (a) $J_a = J_0 r/R$ and (b) $J_b = J_0(1 - r/R)$, in which r is the radial distance and $J_0 = 5.50 \times 10^4$ A/m²? (c) Which function maximizes the current density near the wire's surface?

••12 Near Earth, the density of protons in the solar wind (a stream of particles from the Sun) is 8.70 cm^{-3}, and their speed is 470 km/s. (a) Find the current density of these protons. (b) If Earth's magnetic field did not deflect the protons, what total current would Earth receive?

••13 ILW GO How long does it take electrons to get from a car battery to the starting motor? Assume the current is 300 A and the electrons travel through a copper wire with cross-sectional area 0.21 cm² and length 0.85 m. The number of charge carriers per unit volume is 8.49×10^{28} m^{-3}.

sec. 26-4 Resistance and Resistivity

•14 A human being can be electrocuted if a current as small as 50 mA passes near the heart. An electrician working with sweaty hands makes good contact with the two conductors he is holding, one in each hand. If his resistance is 2000 Ω, what might the fatal voltage be?

•15 SSM A coil is formed by winding 250 turns of insulated 16-gauge copper wire (diameter = 1.3 mm) in a single layer on a cylindrical form of radius 12 cm. What is the resistance of the coil? Neglect the thickness of the insulation. (Use Table 26-1.)

•16 Copper and aluminum are being considered for a high-voltage transmission line that must carry a current of 60.0 A. The resistance per unit length is to be 0.150 Ω/km. The densities of copper and aluminum are 8960 and 2600 kg/m³, respectively. Compute (a) the magnitude J of the current density and (b) the mass per unit length λ for a copper cable and (c) J and (d) λ for an aluminum cable.

•17 A wire of Nichrome (a nickel–chromium–iron alloy commonly used in heating elements) is 1.0 m long and 1.0 mm² in cross-sectional area. It carries a current of 4.0 A when a 2.0 V potential difference is applied between its ends. Calculate the conductivity σ of Nichrome.

•18 A wire 4.00 m long and 6.00 mm in diameter has a resistance of 15.0 mΩ. A potential difference of 23.0 V is applied between the ends. (a) What is the current in the wire? (b) What is the magnitude of the current density? (c) Calculate the resistivity of the wire material. (d) Using Table 26-1, identify the material.

•19 SSM What is the resistivity of a wire of 1.0 mm diameter, 2.0 m length, and 50 mΩ resistance?

•20 A certain wire has a resistance R. What is the resistance of a second wire, made of the same material, that is half as long and has half the diameter?

••21 ILW A common flashlight bulb is rated at 0.30 A and 2.9 V (the values of the current and voltage under operating conditions). If the resistance of the tungsten bulb filament at room temperature (20°C) is 1.1 Ω, what is the temperature of the filament when the bulb is on?

••22 *Kiting during a storm.* The legend that Benjamin Franklin flew a kite as a storm approached is only a legend—he was neither stupid nor suicidal. Suppose a kite string of radius 2.00 mm extends directly upward by 0.800 km and is coated with a 0.500 mm layer of water having resistivity 150 Ω · m. If the potential difference between the two ends of the string is 160 MV, what is the current through the water layer? The danger is not this current but the chance that the string draws a lightning strike, which can have a current as large as 500 000 A (way beyond just being lethal).

••23 When 115 V is applied across a wire that is 10 m long and has a 0.30 mm radius, the magnitude of the current density is 1.4×10^4 A/m^2. Find the resistivity of the wire.

••24 Figure 26-24a gives the magnitude $E(x)$ of the electric fields that have been set up by a battery along a resistive rod of length 9.00 mm (Fig. 26-24b). The vertical scale is set by $E_s = 4.00 \times 10^3$ V/m. The rod consists of three sections of the same material but with different radii. (The schematic diagram of Fig. 26-24b does not indicate the different radii.) The radius of section 3 is 2.00 mm. What is the radius of (a) section 1 and (b) section 2?

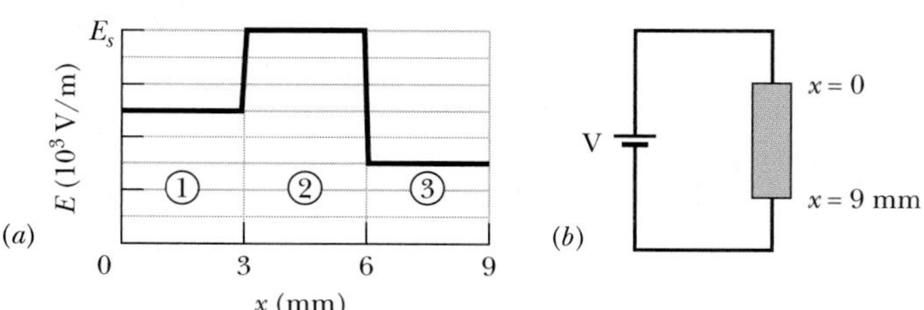

Fig. 26-24 Problem 24.

••25 SSM ILW A wire with a resistance of 6.0 Ω is drawn out through a die so that its new length is three times its original length. Find the resistance of the longer wire, assuming that the resistivity and density of the material are unchanged.

••26 In Fig. 26-25a, a 9.00 V battery is connected to a resistive strip that consists of three sections with the same cross-sectional areas but different conductivities. Figure 26-25b gives the electric potential $V(x)$ versus position x along the strip. The horizontal scale is set by $x_s = 8.00$ mm. Section 3 has conductivity 3.00×10^7 $(\Omega \cdot \text{m})^{-1}$. What is the conductivity of section (a) 1 and (b) 2?

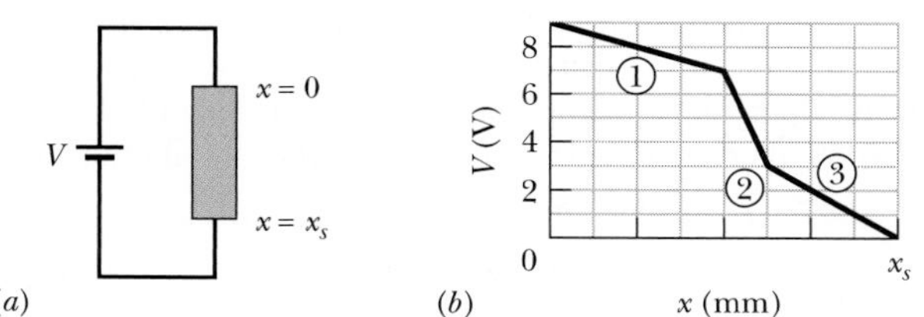

Fig. 26-25 Problem 26.

••27 SSM WWW Two conductors are made of the same material and have the same length. Conductor A is a solid wire of diameter 1.0 mm. Conductor B is a hollow tube of outside diameter 2.0 mm and inside diameter 1.0 mm. What is the resistance ratio R_A/R_B, measured between their ends?

••28 GO Figure 26-26 gives the electric potential $V(x)$ along a copper wire carrying uniform current, from a point of higher potential $V_s = 12.0$ μV at $x = 0$ to a point of zero potential at $x_s = 3.00$ m. The wire has a radius of 2.00 mm. What is the current in the wire?

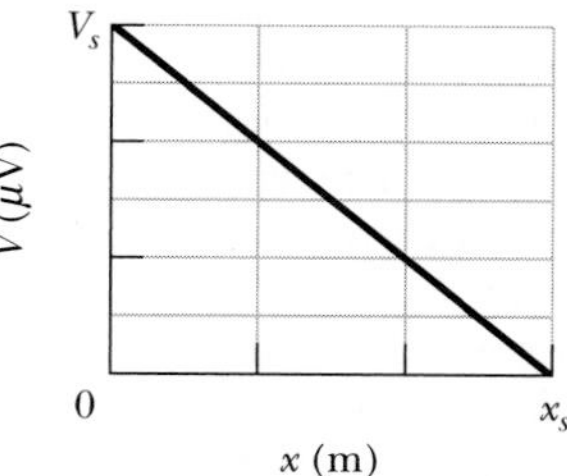

Fig. 26-26 Problem 28.

••29 A potential difference of 3.00 nV is set up across a 2.00 cm length of copper wire that has a radius of 2.00 mm. How much charge drifts through a cross section in 3.00 ms?

••30 If the gauge number of a wire is increased by 6, the diameter is halved; if a gauge number is increased by 1, the diameter decreases by the factor $2^{1/6}$ (see the table in Problem 4). Knowing this, and knowing that 1000 ft of 10-gauge copper wire has a resistance of approximately 1.00 Ω, estimate the resistance of 25 ft of 22-gauge copper wire.

••31 An electrical cable consists of 125 strands of fine wire, each having 2.65 μΩ resistance. The same potential difference is applied between the ends of all the strands and results in a total current of 0.750 A. (a) What is the current in each strand? (b) What is the applied potential difference? (c) What is the resistance of the cable?

••32 Earth's lower atmosphere contains negative and positive ions that are produced by radioactive elements in the soil and cosmic rays from space. In a certain region, the atmospheric electric field strength is 120 V/m and the field is directed vertically down. This field causes singly charged positive ions, at a density of 620 cm^{-3}, to drift downward and singly charged negative ions, at a density of 550 cm^{-3}, to drift upward (Fig. 26-27). The measured conductivity of the air in that region is 2.70×10^{-14} $(\Omega \cdot \text{m})^{-1}$. Calculate (a) the magnitude of the current density and (b) the ion drift speed, assumed to be the same for positive and negative ions.

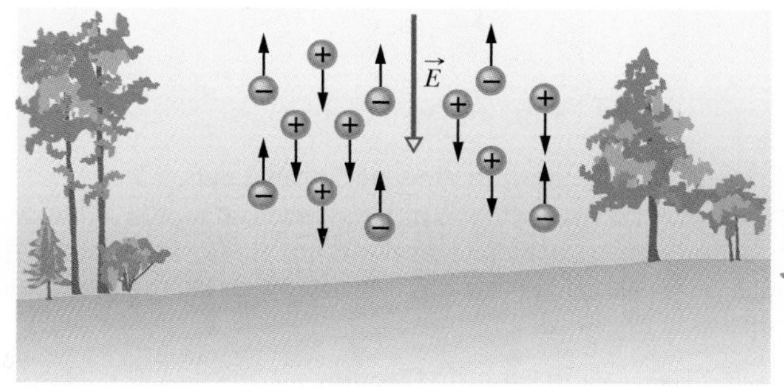

Fig. 26-27 Problem 32.

••33 A block in the shape of a rectangular solid has a cross-sectional area of 3.50 cm^2 across its width, a front-to-rear length of 15.8 cm, and a resistance of 935 Ω. The block's material contains 5.33×10^{22} conduction electrons/m^3. A potential difference of 35.8 V is maintained between its front and rear faces. (a) What is the current in the block? (b) If the current density is uniform, what is its magnitude? What are (c) the drift velocity of the conduction electrons and (d) the magnitude of the electric field in the block?

•••34 GO Figure 26-28 shows wire section 1 of diameter $D_1 = 4.00R$ and wire section 2 of diameter $D_2 = 2.00R$, connected by a tapered section. The wire is copper and carries a current. Assume that the current is uniformly distributed across any cross-sectional area through the wire's width. The electric potential change V along the length $L = 2.00$ m shown in section 2 is 10.0 μV. The number of charge carriers per unit volume is 8.49×10^{28} m^{-3}. What is the drift speed of the conduction electrons in section 1?

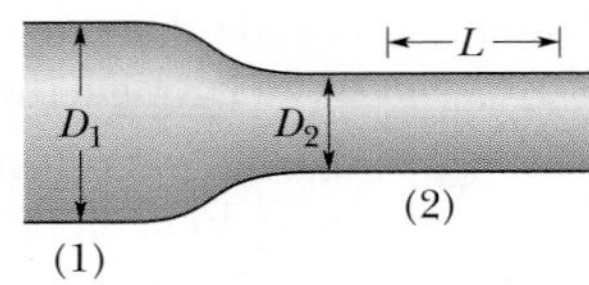

Fig. 26-28 Problem 34.

•••35 In Fig. 26-29, current is set up through a truncated right circular cone of resistivity 731 Ω · m, left radius $a = 2.00$ mm, right radius $b = 2.30$ mm, and length $L = 1.94$ cm. Assume that the current density is uniform across any cross section taken perpendicular to the length. What is the resistance of the cone?

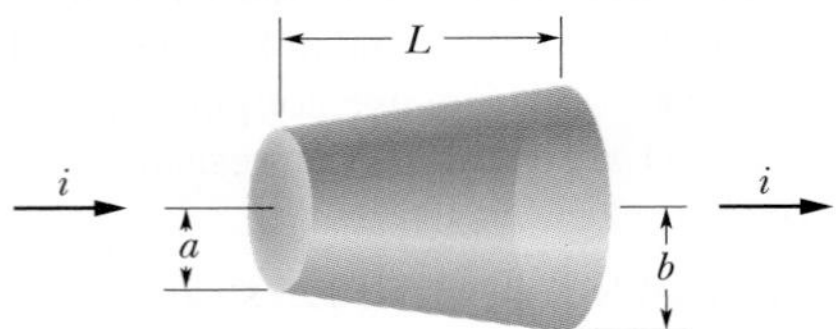

Fig. 26-29 Problem 35.

•••36 GO *Swimming during a storm.* Figure 26-30 shows a swimmer at distance $D = 35.0$ m from a lightning strike to the water, with current $I = 78$ kA. The water has resistivity 30 Ω · m, the width of the swimmer along a radial line from the strike is 0.70 m, and his resistance across that width is 4.00 kΩ. Assume that the current spreads through the water over a hemisphere centered on the strike point. What is the current through the swimmer?

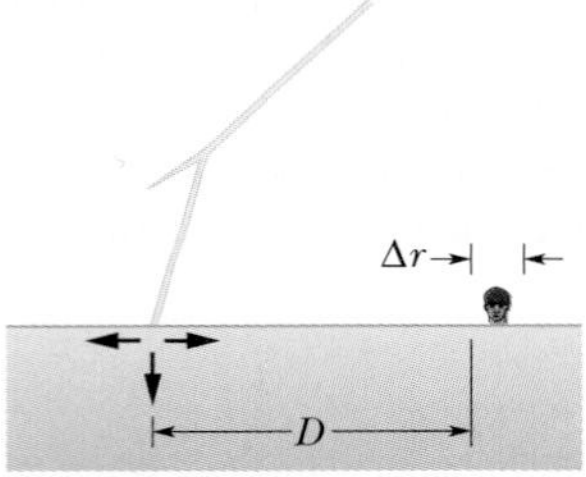

Fig. 26-30 Problem 36.

sec. 26-6 A Microscopic View of Ohm's Law

••37 Show that, according to the free-electron model of electrical conduction in metals and classical physics, the resistivity of metals should be proportional to $\sqrt{T}$, where T is the temperature in kelvins. (See Eq. 19-31.)

sec. 26-7 Power in Electric Circuits

•38 In Fig. 26-31a, a 20 Ω resistor is connected to a battery. Figure 26-31b shows the increase of thermal energy E_{th} in the resistor as a function of time t. The vertical scale is set by $E_{th,s} = 2.50$ mJ, and the horizontal scale is set by $t_s = 4.0$ s. What is the electric potential across the battery?

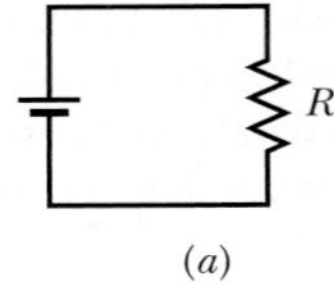

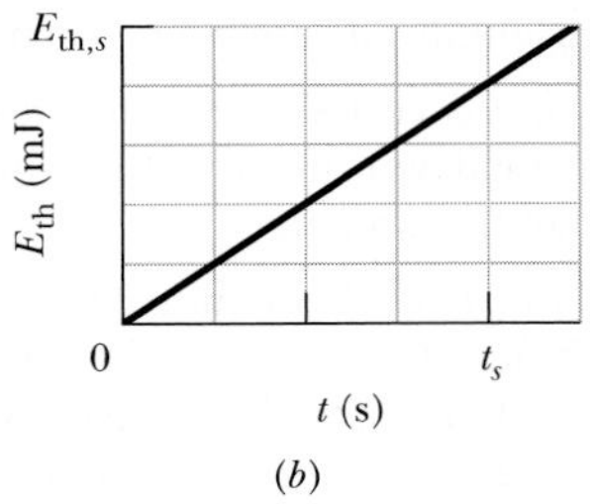

Fig. 26-31 Problem 38.

•39 A certain brand of hot-dog cooker works by applying a potential difference of 120 V across opposite ends of a hot dog and allowing it to cook by means of the thermal energy produced. The current is 10.0 A, and the energy required to cook one hot dog is 60.0 kJ. If the rate at which energy is supplied is unchanged, how long will it take to cook three hot dogs simultaneously?

•40 Thermal energy is produced in a resistor at a rate of 100 W when the current is 3.00 A. What is the resistance?

•41 SSM A 120 V potential difference is applied to a space heater whose resistance is 14 Ω when hot. (a) At what rate is electrical energy transferred to thermal energy? (b) What is the cost for 5.0 h at US$0.05/kW · h?

•42 In Fig. 26-32, a battery of potential difference $V = 12$ V is connected to a resistive strip of resistance $R = 6.0$ Ω. When an electron moves through the strip from one end to the other, (a) in which direction in the figure does the electron move, (b) how much work is done on the electron by the electric field in the strip, and (c) how much energy is transferred to the thermal energy of the strip by the electron?

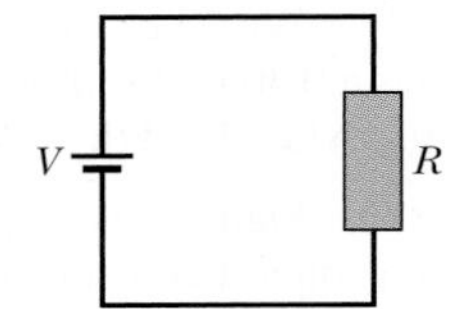

Fig. 26-32 Problem 42.

•43 ILW An unknown resistor is connected between the terminals of a 3.00 V battery. Energy is dissipated in the resistor at the rate of 0.540 W. The same resistor is then connected between the terminals of a 1.50 V battery. At what rate is energy now dissipated?

•44 A student kept his 9.0 V, 7.0 W radio turned on at full volume from 9:00 P.M. until 2:00 A.M. How much charge went through it?

•45 SSM ILW A 1250 W radiant heater is constructed to operate at 115 V. (a) What is the current in the heater when the unit is operating? (b) What is the resistance of the heating coil? (c) How much thermal energy is produced in 1.0 h?

••46 GO A copper wire of cross-sectional area 2.00×10^{-6} m^2 and length 4.00 m has a current of 2.00 A uniformly distributed across that area. (a) What is the magnitude of the electric field along the wire? (b) How much electrical energy is transferred to thermal energy in 30 min?

••47 A heating element is made by maintaining a potential difference of 75.0 V across the length of a Nichrome wire that has a 2.60×10^{-6} m^2 cross section. Nichrome has a resistivity of 5.00×10^{-7} Ω · m. (a) If the element dissipates 5000 W, what is its length? (b) If 100 V is used to obtain the same dissipation rate, what should the length be?

••48 *Exploding shoes.* The rain-soaked shoes of a person may explode if ground current from nearby lightning vaporizes the water. The sudden conversion of water to water vapor causes a dramatic expansion that can rip apart shoes. Water has density 1000 kg/m^3 and requires 2256 kJ/kg to be vaporized. If horizontal current lasts 2.00 ms and encounters water with resistivity 150 $\Omega \cdot$ m, length 12.0 cm, and vertical cross-sectional area 15×10^{-5} m^2, what average current is required to vaporize the water?

••49 A 100 W lightbulb is plugged into a standard 120 V outlet. (a) How much does it cost per 31-day month to leave the light turned on continuously? Assume electrical energy costs US\$0.06/kW $\cdot$ h. (b) What is the resistance of the bulb? (c) What is the current in the bulb?

••50 GO The current through the battery and resistors 1 and 2 in Fig. 26-33*a* is 2.00 A. Energy is transferred from the current to thermal energy E_{th} in both resistors. Curves 1 and 2 in Fig. 26-33*b* give that thermal energy E_{th} for resistors 1 and 2, respectively, as a function of time *t*. The vertical scale is set by $E_{th,s} = 40.0$ mJ, and the horizontal scale is set by $t_s = 5.00$ s. What is the power of the battery?

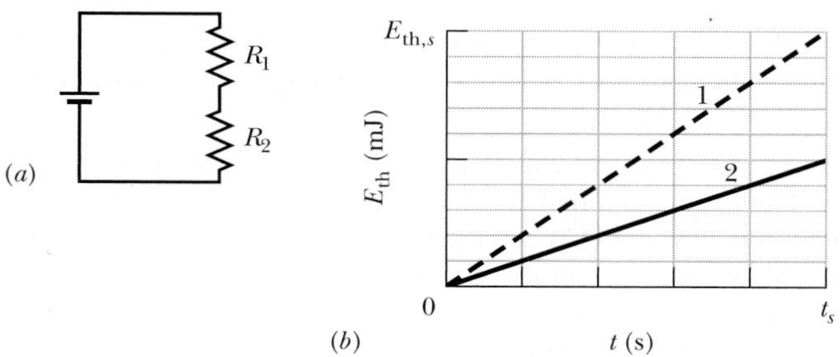

Fig. 26-33 Problem 50.

••51 SSM WWW Wire *C* and wire *D* are made from different materials and have length $L_C = L_D = 1.0$ m. The resistivity and diameter of wire *C* are 2.0×10^{-6} $\Omega \cdot$ m and 1.00 mm, and those of wire *D* are 1.0×10^{-6} $\Omega \cdot$ m and 0.50 mm. The wires are joined as shown in Fig. 26-34, and a current of 2.0 A is set up in them. What is the electric potential difference between (a) points 1 and 2 and (b) points 2 and 3? What is the rate at which energy is dissipated between (c) points 1 and 2 and (d) points 2 and 3?

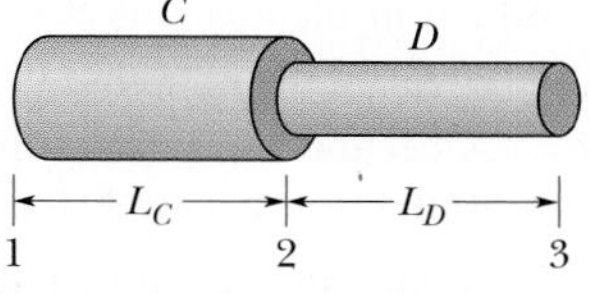

Fig. 26-34 Problem 51.

••52 GO The current-density magnitude in a certain circular wire is $J = (2.75 \times 10^{10}$ A/m$^4)r^2$, where *r* is the radial distance out to the wire's radius of 3.00 mm. The potential applied to the wire (end to end) is 60.0 V. How much energy is converted to thermal energy in 1.00 h?

••53 A 120 V potential difference is applied to a space heater that dissipates 500 W during operation. (a) What is its resistance during operation? (b) At what rate do electrons flow through any cross section of the heater element?

•••54 Figure 26-35*a* shows a rod of resistive material. The resistance per unit length of the rod increases in the positive direction of the *x* axis. At any position *x* along the rod, the resistance *dR* of a narrow (differential) section of width *dx* is given by $dR = 5.00x\,dx$, where *dR* is in ohms and *x* is in meters. Figure 26-35*b* shows such a narrow section. You are to slice off a length of the rod between $x = 0$ and some position $x = L$ and then connect that length to a battery with potential difference $V = 5.0$ V (Fig. 26-35*c*). You want the current in the length to transfer energy to thermal energy at the rate of 200 W. At what position $x = L$ should you cut the rod?

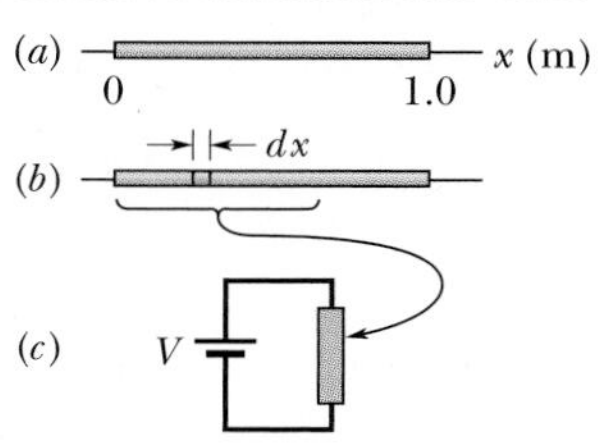

Fig. 26-35 Problem 54.

Additional Problems

55 SSM A Nichrome heater dissipates 500 W when the applied potential difference is 110 V and the wire temperature is 800°C. What would be the dissipation rate if the wire temperature were held at 200°C by immersing the wire in a bath of cooling oil? The applied potential difference remains the same, and α for Nichrome at 800°C is 4.0×10^{-4} K^{-1}.

56 A potential difference of 1.20 V will be applied to a 33.0 m length of 18-gauge copper wire (diameter = 0.0400 in.). Calculate (a) the current, (b) the magnitude of the current density, (c) the magnitude of the electric field within the wire, and (d) the rate at which thermal energy will appear in the wire.

57 An 18.0 W device has 9.00 V across it. How much charge goes through the device in 4.00 h?

58 An aluminum rod with a square cross section is 1.3 m long and 5.2 mm on edge. (a) What is the resistance between its ends? (b) What must be the diameter of a cylindrical copper rod of length 1.3 m if its resistance is to be the same as that of the aluminum rod?

59 A cylindrical metal rod is 1.60 m long and 5.50 mm in diameter. The resistance between its two ends (at 20°C) is 1.09×10^{-3} Ω. (a) What is the material? (b) A round disk, 2.00 cm in diameter and 1.00 mm thick, is formed of the same material. What is the resistance between the round faces, assuming that each face is an equipotential surface?

60 *The chocolate crumb mystery.* This story begins with Problem 60 in Chapter 23 and continues through Chapters 24 and 25. The chocolate crumb powder moved to the silo through a pipe of radius *R* with uniform speed *v* and uniform charge density ρ. (a) Find an expression for the current *i* (the rate at which charge on the powder moved) through a perpendicular cross section of the pipe. (b) Evaluate *i* for the conditions at the factory: pipe radius $R = 5.0$ cm, speed $v = 2.0$ m/s, and charge density $\rho = 1.1 \times 10^{-3}$ C/m^3.

If the powder were to flow through a change *V* in electric potential, its energy could be transferred to a spark at the rate $P = iV$. (c) Could there be such a transfer within the pipe due to the radial potential difference discussed in Problem 70 of Chapter 24?

As the powder flowed from the pipe into the silo, the electric potential of the powder changed. The magnitude of that change was at least equal to the radial potential difference within the pipe (as evaluated in Problem 70 of Chapter 24). (d) Assuming that value for the potential difference and using the current found in (b) above, find the rate at which energy could have been transferred from the powder to a spark as the powder exited the pipe. (e) If a spark did occur at the exit and lasted for 0.20 s (a reasonable expectation), how much energy would have been transferred to the spark?

Recall from Problem 60 in Chapter 23 that a minimum energy transfer of 150 mJ is needed to cause an explosion. (f) Where did the powder explosion most likely occur: in the powder cloud at the unloading bin (Problem 60 of Chapter 25), within the pipe, or at the exit of the pipe into the silo?

61 SSM A steady beam of alpha particles ($q = +2e$) traveling with constant kinetic energy 20 MeV carries a current of 0.25 μA. (a) If the beam is directed perpendicular to a flat surface, how many alpha particles strike the surface in 3.0 s? (b) At any instant, how many alpha particles are there in a given 20 cm length of the beam? (c) Through what potential difference is it necessary to accelerate each alpha particle from rest to bring it to an energy of 20 MeV?

62 A resistor with a potential difference of 200 V across it transfers electrical energy to thermal energy at the rate of 3000 W. What is the resistance of the resistor?

63 A 2.0 kW heater element from a dryer has a length of 80 cm. If a 10 cm section is removed, what power is used by the now shortened element at 120 V?

64 A cylindrical resistor of radius 5.0 mm and length 2.0 cm is made of material that has a resistivity of $3.5 \times 10^{-5}\ \Omega \cdot \text{m}$. What are (a) the magnitude of the current density and (b) the potential difference when the energy dissipation rate in the resistor is 1.0 W?

65 A potential difference V is applied to a wire of cross-sectional area A, length L, and resistivity ρ. You want to change the applied potential difference and stretch the wire so that the energy dissipation rate is multiplied by 30.0 and the current is multiplied by 4.00. Assuming the wire's density does not change, what are (a) the ratio of the new length to L and (b) the ratio of the new cross-sectional area to A?

66 The headlights of a moving car require about 10 A from the 12 V alternator, which is driven by the engine. Assume the alternator is 80% efficient (its output electrical power is 80% of its input mechanical power), and calculate the horsepower the engine must supply to run the lights.

67 A 500 W heating unit is designed to operate with an applied potential difference of 115 V. (a) By what percentage will its heat output drop if the applied potential difference drops to 110 V? Assume no change in resistance. (b) If you took the variation of resistance with temperature into account, would the actual drop in heat output be larger or smaller than that calculated in (a)?

68 The copper windings of a motor have a resistance of 50 Ω at 20°C when the motor is idle. After the motor has run for several hours, the resistance rises to 58 Ω. What is the temperature of the windings now? Ignore changes in the dimensions of the windings. (Use Table 26-1.)

69 How much electrical energy is transferred to thermal energy in 2.00 h by an electrical resistance of 400 Ω when the potential applied across it is 90.0 V?

70 A caterpillar of length 4.0 cm crawls in the direction of electron drift along a 5.2-mm-diameter bare copper wire that carries a uniform current of 12 A. (a) What is the potential difference between the two ends of the caterpillar? (b) Is its tail positive or negative relative to its head? (c) How much time does the caterpillar take to crawl 1.0 cm if it crawls at the drift speed of the electrons in the wire? (The number of charge carriers per unit volume is $8.49 \times 10^{28}\ \text{m}^{-3}$.)

71 SSM (a) At what temperature would the resistance of a copper conductor be double its resistance at 20.0°C? (Use 20.0°C as the reference point in Eq. 26-17; compare your answer with Fig. 26-10.) (b) Does this same "doubling temperature" hold for all copper conductors, regardless of shape or size?

72 A steel trolley-car rail has a cross-sectional area of 56.0 cm^2. What is the resistance of 10.0 km of rail? The resistivity of the steel is $3.00 \times 10^{-7}\ \Omega \cdot \text{m}$.

73 A coil of current-carrying Nichrome wire is immersed in a liquid. (Nichrome is a nickel–chromium–iron alloy commonly used in heating elements.) When the potential difference across the coil is 12 V and the current through the coil is 5.2 A, the liquid evaporates at the steady rate of 21 mg/s. Calculate the heat of vaporization of the liquid (see Section 18-8).

74 GO The current density in a wire is uniform and has magnitude $2.0 \times 10^6\ \text{A/m}^2$, the wire's length is 5.0 m, and the density of conduction electrons is $8.49 \times 10^{28}\ \text{m}^{-3}$. How long does an electron take (on the average) to travel the length of the wire?

75 A certain x-ray tube operates at a current of 7.00 mA and a potential difference of 80.0 kV. What is its power in watts?

76 A current is established in a gas discharge tube when a sufficiently high potential difference is applied across the two electrodes in the tube. The gas ionizes; electrons move toward the positive terminal and singly charged positive ions toward the negative terminal. (a) What is the current in a hydrogen discharge tube in which 3.1×10^{18} electrons and 1.1×10^{18} protons move past a cross-sectional area of the tube each second? (b) Is the direction of the current density $\vec{J}$ toward or away from the negative terminal?

Wet wiz

CHAPTER 27

CIRCUITS

27-1 WHAT IS PHYSICS?

You are surrounded by electric circuits. You might take pride in the number of electrical devices you own and might even carry a mental list of the devices you wish you owned. Every one of those devices, as well as the electrical grid that powers your home, depends on modern electrical engineering. We cannot easily estimate the current financial worth of electrical engineering and its products, but we can be certain that the financial worth continues to grow yearly as more and more tasks are handled electrically. Radios are now tuned electronically instead of manually. Messages are now sent by email instead of through the postal system. Research journals are now read on a computer instead of in a library building, and research papers are now copied and filed electronically instead of photocopied and tucked into a filing cabinet.

The basic science of electrical engineering is physics. In this chapter we cover the physics of electric circuits that are combinations of resistors and batteries (and, in Section 27-9, capacitors). We restrict our discussion to circuits through which charge flows in one direction, which are called either *direct-current circuits* or *DC circuits.* We begin with the question: How can you get charges to flow?

27-2 "Pumping" Charges

If you want to make charge carriers flow through a resistor, you must establish a potential difference between the ends of the device. One way to do this is to connect each end of the resistor to one plate of a charged capacitor. The trouble with this scheme is that the flow of charge acts to discharge the capacitor, quickly bringing the plates to the same potential. When that happens, there is no longer an electric field in the resistor, and thus the flow of charge stops.

To produce a steady flow of charge, you need a "charge pump," a device that—by doing work on the charge carriers—maintains a potential difference between a pair of terminals. We call such a device an **emf device,** and the device is said to provide an **emf** $\mathscr{E}$, which means that it does work on charge carriers. An emf device is sometimes called a *seat of emf.* The term *emf* comes from the outdated phrase *electromotive force,* which was adopted before scientists clearly understood the function of an emf device.

In Chapter 26, we discussed the motion of charge carriers through a circuit in terms of the electric field set up in the circuit—the field produces forces that move the charge carriers. In this chapter we take a different approach: We discuss the motion of the charge carriers in terms of the required energy—an emf device supplies the energy for the motion via the work it does.

A common emf device is the *battery,* used to power a wide variety of machines from wristwatches to submarines. The emf device that most influences our daily lives, however, is the *electric generator,* which, by means of electrical connections (wires) from a generating plant, creates a potential difference in our

The world's largest battery energy storage plant (dismantled in 1996) connected over 8000 large lead-acid batteries in 8 strings at 1000 V each with a capability of 10 MW of power for 4 hours. Charged up at night, the batteries were then put to use during peak power demands on the electrical system. *(Courtesy Southern California Edison Company)*

homes and workplaces. The emf devices known as *solar cells,* long familiar as the wing-like panels on spacecraft, also dot the countryside for domestic applications. Less familiar emf devices are the *fuel cells* that power the space shuttles and the *thermopiles* that provide onboard electrical power for some spacecraft and for remote stations in Antarctica and elsewhere. An emf device does not have to be an instrument—living systems, ranging from electric eels and human beings to plants, have physiological emf devices.

Although the devices we have listed differ widely in their modes of operation, they all perform the same basic function—they do work on charge carriers and thus maintain a potential difference between their terminals.

27-3 Work, Energy, and Emf

Figure 27-1 shows an emf device (consider it to be a battery) that is part of a simple circuit containing a single resistance R (the symbol for resistance and a resistor is -WW-). The emf device keeps one of its terminals (called the positive terminal and often labeled +) at a higher electric potential than the other terminal (called the negative terminal and labeled −). We can represent the emf of the device with an arrow that points from the negative terminal toward the positive terminal as in Fig. 27-1. A small circle on the tail of the emf arrow distinguishes it from the arrows that indicate current direction.

When an emf device is not connected to a circuit, the internal chemistry of the device does not cause any net flow of charge carriers within it. However, when it is connected to a circuit as in Fig. 27-1, its internal chemistry causes a net flow of positive charge carriers from the negative terminal to the positive terminal, in the direction of the emf arrow. This flow is part of the current that is set up around the circuit in that same direction (clockwise in Fig. 27-1).

Within the emf device, positive charge carriers move from a region of low electric potential and thus low electric potential energy (at the negative terminal) to a region of higher electric potential and higher electric potential energy (at the positive terminal). This motion is just the opposite of what the electric field between the terminals (which is directed from the positive terminal toward the negative terminal) would cause the charge carriers to do.

Thus, there must be some source of energy within the device, enabling it to do work on the charges by forcing them to move as they do. The energy source may be chemical, as in a battery or a fuel cell. It may involve mechanical forces, as in an electric generator. Temperature differences may supply the energy, as in a thermopile; or the Sun may supply it, as in a solar cell.

Let us now analyze the circuit of Fig. 27-1 from the point of view of work and energy transfers. In any time interval dt, a charge dq passes through any cross section of this circuit, such as aa'. This same amount of charge must enter the emf device at its low-potential end and leave at its high-potential end. The device must do an amount of work dW on the charge dq to force it to move in this way. We define the emf of the emf device in terms of this work:

$$\mathscr{E} = \frac{dW}{dq} \quad \text{(definition of } \mathscr{E}\text{).} \tag{27-1}$$

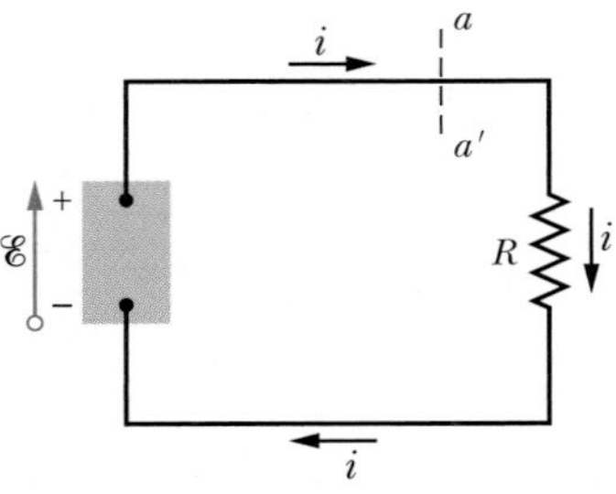

Fig. 27-1 A simple electric circuit, in which a device of emf $\mathscr{E}$ does work on the charge carriers and maintains a steady current i in a resistor of resistance R.

In words, the emf of an emf device is the work per unit charge that the device does in moving charge from its low-potential terminal to its high-potential terminal. The SI unit for emf is the joule per coulomb; in Chapter 24 we defined that unit as the *volt.*

An **ideal emf device** is one that lacks any internal resistance to the internal movement of charge from terminal to terminal. The potential difference between the terminals of an ideal emf device is equal to the emf of the device. For example, an ideal battery with an emf of 12.0 V always has a potential difference of 12.0 V between its terminals.

A **real emf device,** such as any real battery, has internal resistance to the internal movement of charge. When a real emf device is not connected to a circuit, and thus does not have current through it, the potential difference between its terminals is equal to its emf. However, when that device has current through it, the potential difference between its terminals differs from its emf. We shall discuss such real batteries in Section 27-5.

When an emf device is connected to a circuit, the device transfers energy to the charge carriers passing through it. This energy can then be transferred from the charge carriers to other devices in the circuit, for example, to light a bulb. Figure 27-2*a* shows a circuit containing two ideal rechargeable (*storage*) batteries A and B, a resistance R, and an electric motor M that can lift an object by using energy it obtains from charge carriers in the circuit. Note that the batteries are connected so that they tend to send charges around the circuit in opposite directions. The actual direction of the current in the circuit is determined by the battery with the larger emf, which happens to be battery B, so the chemical energy within battery B is decreasing as energy is transferred to the charge carriers passing through it. However, the chemical energy within battery A is increasing because the current in it is directed from the positive terminal to the negative terminal. Thus, battery B is charging battery A. Battery B is also providing energy to motor M and energy that is being dissipated by resistance R. Figure 27-2*b* shows all three energy transfers from battery B; each decreases that battery's chemical energy.

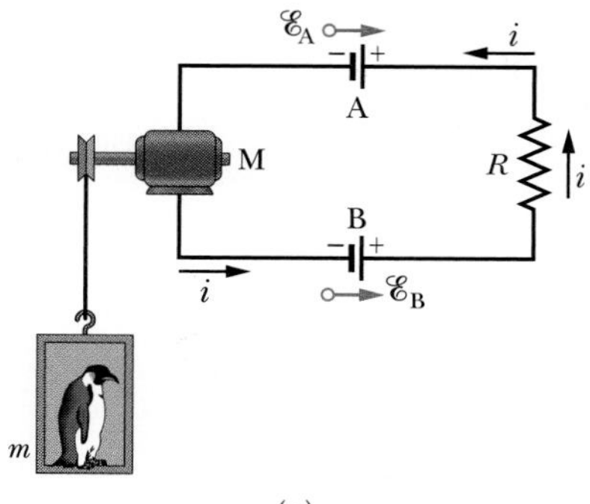

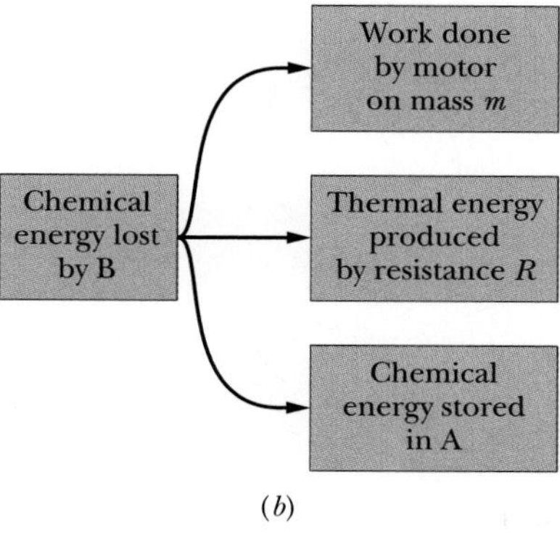

Fig. 27-2 (*a*) In the circuit, $\mathscr{E}_B > \mathscr{E}_A$; so battery B determines the direction of the current. (*b*) The energy transfers in the circuit.

27-4 Calculating the Current in a Single-Loop Circuit

We discuss here two equivalent ways to calculate the current in the simple *single-loop* circuit of Fig. 27-3; one method is based on energy conservation considerations, and the other on the concept of potential. The circuit consists of an ideal battery B with emf $\mathscr{E}$, a resistor of resistance R, and two connecting wires. (Unless otherwise indicated, we assume that wires in circuits have negligible resistance. Their function, then, is merely to provide pathways along which charge carriers can move.)

Energy Method

Equation 26-27 ($P = i^2R$) tells us that in a time interval dt an amount of energy given by $i^2R\,dt$ will appear in the resistor of Fig. 27-3 as thermal energy. As noted in Section 26-7, this energy is said to be *dissipated.* (Because we assume the wires to have negligible resistance, no thermal energy will appear in them.) During the same interval, a charge $dq = i\,dt$ will have moved through battery B, and the work that the battery will have done on this charge, according to Eq. 27-1, is

$$dW = \mathscr{E}\,dq = \mathscr{E}i\,dt.$$

From the principle of conservation of energy, the work done by the (ideal) battery must equal the thermal energy that appears in the resistor:

$$\mathscr{E}i\,dt = i^2R\,dt.$$

This gives us

$$\mathscr{E} = iR.$$

The emf $\mathscr{E}$ is the energy per unit charge transferred to the moving charges by the battery. The quantity iR is the energy per unit charge transferred *from* the moving charges to thermal energy within the resistor. Therefore, this equation means that the energy per unit charge transferred to the moving charges is equal to the

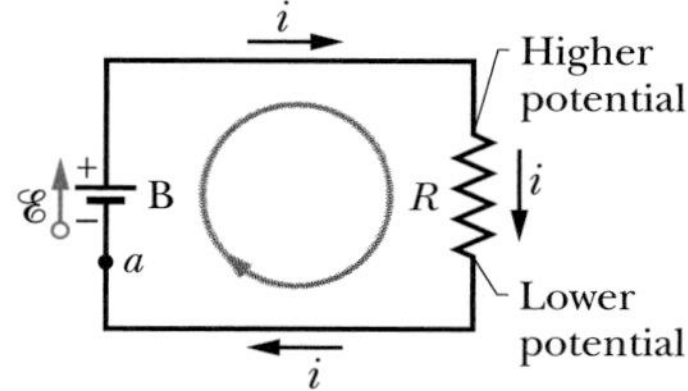

Fig. 27-3 A single-loop circuit in which a resistance R is connected across an ideal battery B with emf $\mathscr{E}$. The resulting current i is the same throughout the circuit.

energy per unit charge transferred from them. Solving for i, we find

$$i = \frac{\mathscr{E}}{R}. \tag{27-2}$$

Potential Method

Suppose we start at any point in the circuit of Fig. 27-3 and mentally proceed around the circuit in either direction, adding algebraically the potential differences that we encounter. Then when we return to our starting point, we must also have returned to our starting potential. Before actually doing so, we shall formalize this idea in a statement that holds not only for single-loop circuits such as that of Fig. 27-3 but also for any complete loop in a *multiloop* circuit, as we shall discuss in Section 27-7:

LOOP RULE: The algebraic sum of the changes in potential encountered in a complete traversal of any loop of a circuit must be zero.

This is often referred to as *Kirchhoff's loop rule* (or *Kirchhoff's voltage law*), after German physicist Gustav Robert Kirchhoff. This rule is equivalent to saying that each point on a mountain has only one elevation above sea level. If you start from any point and return to it after walking around the mountain, the algebraic sum of the changes in elevation that you encounter must be zero.

In Fig. 27-3, let us start at point a, whose potential is V_a, and mentally walk clockwise around the circuit until we are back at a, keeping track of potential changes as we move. Our starting point is at the low-potential terminal of the battery. Because the battery is ideal, the potential difference between its terminals is equal to $\mathscr{E}$. When we pass through the battery to the high-potential terminal, the change in potential is $+\mathscr{E}$.

As we walk along the top wire to the top end of the resistor, there is no potential change because the wire has negligible resistance; it is at the same potential as the high-potential terminal of the battery. So too is the top end of the resistor. When we pass through the resistor, however, the potential changes according to Eq. 26-8 (which we can rewrite as $V = iR$). Moreover, the potential must decrease because we are moving from the higher potential side of the resistor. Thus, the change in potential is $-iR$.

We return to point a by moving along the bottom wire. Because this wire also has negligible resistance, we again find no potential change. Back at point a, the potential is again V_a. Because we traversed a complete loop, our initial potential, as modified for potential changes along the way, must be equal to our final potential; that is,

$$V_a + \mathscr{E} - iR = V_a.$$

The value of V_a cancels from this equation, which becomes

$$\mathscr{E} - iR = 0.$$

Solving this equation for i gives us the same result, $i = \mathscr{E}/R$, as the energy method (Eq. 27-2).

If we apply the loop rule to a complete *counterclockwise* walk around the circuit, the rule gives us

$$-\mathscr{E} + iR = 0$$

and we again find that $i = \mathscr{E}/R$. Thus, you may mentally circle a loop in either direction to apply the loop rule.

To prepare for circuits more complex than that of Fig. 27-3, let us set down two rules for finding potential differences as we move around a loop:

RESISTANCE RULE: For a move through a resistance in the direction of the current, the change in potential is $-iR$; in the opposite direction it is $+iR$.

EMF RULE: For a move through an ideal emf device in the direction of the emf arrow, the change in potential is $+\mathscr{E}$; in the opposite direction it is $-\mathscr{E}$.

CHECKPOINT 1

The figure shows the current i in a single-loop circuit with a battery B and a resistance R (and wires of negligible resistance). (a) Should the emf arrow at B be drawn pointing leftward or rightward? At points a, b, and c, rank (b) the magnitude of the current, (c) the electric potential, and (d) the electric potential energy of the charge carriers, greatest first.

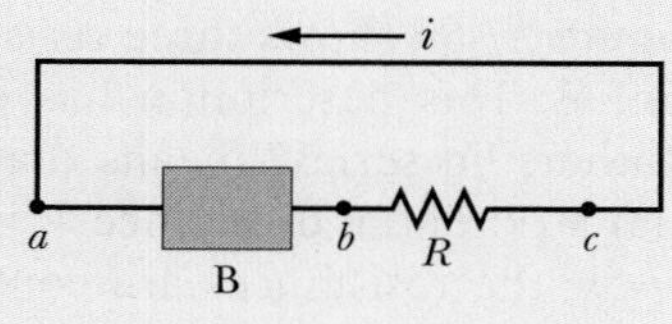

27-5 Other Single-Loop Circuits

In this section we extend the simple circuit of Fig. 27-3 in two ways.

Internal Resistance

Figure 27-4a shows a real battery, with internal resistance r, wired to an external resistor of resistance R. The internal resistance of the battery is the electrical resistance of the conducting materials of the battery and thus is an unremovable feature of the battery. In Fig. 27-4a, however, the battery is drawn as if it could be separated into an ideal battery with emf $\mathscr{E}$ and a resistor of resistance r. The order in which the symbols for these separated parts are drawn does not matter.

If we apply the loop rule clockwise beginning at point a, the *changes* in potential give us

$$\mathscr{E} - ir - iR = 0. \tag{27-3}$$

Solving for the current, we find

$$i = \frac{\mathscr{E}}{R + r}. \tag{27-4}$$

Note that this equation reduces to Eq. 27-2 if the battery is ideal—that is, if $r = 0$.

Figure 27-4b shows graphically the changes in electric potential around the circuit. (To better link Fig. 27-4b with the *closed circuit* in Fig. 27-4a, imagine curling the graph into a cylinder with point a at the left overlapping point a at

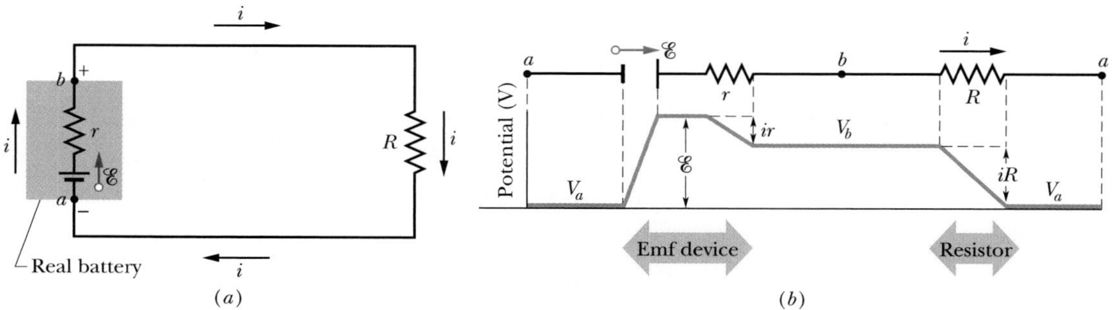

Fig. 27-4 (a) A single-loop circuit containing a real battery having internal resistance r and emf $\mathscr{E}$. (b) The same circuit, now spread out in a line. The potentials encountered in traversing the circuit clockwise from a are also shown. The potential V_a is arbitrarily assigned a value of zero, and other potentials in the circuit are graphed relative to V_a.

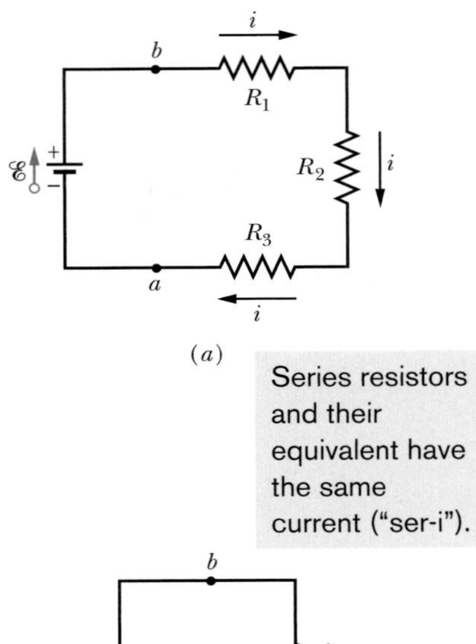

Series resistors and their equivalent have the same current ("ser-i").

(b)

Fig. 27-5 (*a*) Three resistors are connected in series between points *a* and *b*. (*b*) An equivalent circuit, with the three resistors replaced with their equivalent resistance R_{eq}.

the right.) Note how traversing the circuit is like walking around a (potential) mountain back to your starting point—you return to the starting elevation.

In this book, when a battery is not described as real or if no internal resistance is indicated, you can generally assume that it is ideal—but, of course, in the real world batteries are always real and have internal resistance.

Resistances in Series

Figure 27-5*a* shows three resistances connected **in series** to an ideal battery with emf $\mathscr{E}$. This description has little to do with how the resistances are drawn. Rather, "in series" means that the resistances are wired one after another and that a potential difference V is applied across the two ends of the series. In Fig. 27-5*a*, the resistances are connected one after another between a and b, and a potential difference is maintained across a and b by the battery. The potential differences that then exist across the resistances in the series produce identical currents i in them. In general,

> When a potential difference V is applied across resistances connected in series, the resistances have identical currents i. The sum of the potential differences across the resistances is equal to the applied potential difference V.

Note that charge moving through the series resistances can move along only a single route. If there are additional routes, so that the currents in different resistances are different, the resistances are not connected in series.

> Resistances connected in series can be replaced with an equivalent resistance R_{eq} that has the same current i and the same *total* potential difference V as the actual resistances.

You might remember that R_{eq} and all the actual series resistances have the same current i with the nonsense word "ser-i." Figure 27-5*b* shows the equivalent resistance R_{eq} that can replace the three resistances of Fig. 27-5*a*.

To derive an expression for R_{eq} in Fig. 27-5*b*, we apply the loop rule to both circuits. For Fig. 27-5*a*, starting at a and going clockwise around the circuit, we find

$$\mathscr{E} - iR_1 - iR_2 - iR_3 = 0,$$

or

$$i = \frac{\mathscr{E}}{R_1 + R_2 + R_3}. \tag{27-5}$$

For Fig. 27-5*b*, with the three resistances replaced with a single equivalent resistance R_{eq}, we find

$$\mathscr{E} - iR_{eq} = 0,$$

or

$$i = \frac{\mathscr{E}}{R_{eq}}. \tag{27-6}$$

Comparison of Eqs. 27-5 and 27-6 shows that

$$R_{eq} = R_1 + R_2 + R_3.$$

The extension to n resistances is straightforward and is

$$R_{eq} = \sum_{j=1}^{n} R_j \qquad (n \text{ resistances in series}). \tag{27-7}$$

Note that when resistances are in series, their equivalent resistance is greater than any of the individual resistances.

CHECKPOINT 2

In Fig. 27-5*a*, if $R_1 > R_2 > R_3$, rank the three resistances according to (a) the current through them and (b) the potential difference across them, greatest first.

27-6 Potential Difference Between Two Points

We often want to find the potential difference between two points in a circuit. For example, in Fig. 27-6, what is the potential difference $V_b - V_a$ between points a and b? To find out, let's start at point a (at potential V_a) and move through the battery to point b (at potential V_b) while keeping track of the potential changes we encounter. When we pass through the battery's emf, the potential increases by $\mathscr{E}$. When we pass through the battery's internal resistance r, we move in the direction of the current and thus the potential decreases by ir. We are then at the potential of point b and we have

$$V_a + \mathscr{E} - ir = V_b,$$

or

$$V_b - V_a = \mathscr{E} - ir. \quad (27\text{-}8)$$

The internal resistance reduces the potential difference between the terminals.

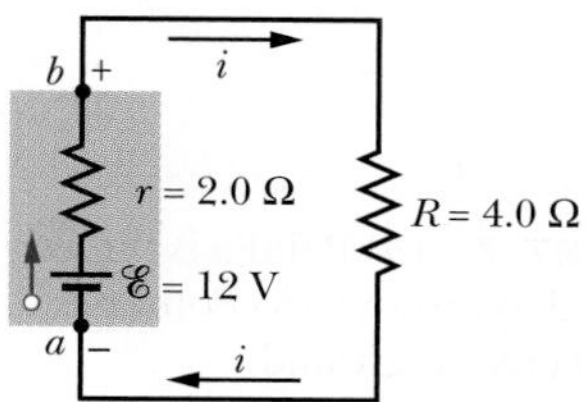

Fig. 27-6 Points a and b, which are at the terminals of a real battery, differ in potential.

To evaluate this expression, we need the current i. Note that the circuit is the same as in Fig. 27-4a, for which Eq. 27-4 gives the current as

$$i = \frac{\mathscr{E}}{R + r}. \quad (27\text{-}9)$$

Substituting this equation into Eq. 27-8 gives us

$$V_b - V_a = \mathscr{E} - \frac{\mathscr{E}}{R + r} r$$

$$= \frac{\mathscr{E}}{R + r} R. \quad (27\text{-}10)$$

Now substituting the data given in Fig. 27-6, we have

$$V_b - V_a = \frac{12\text{ V}}{4.0\ \Omega + 2.0\ \Omega} 4.0\ \Omega = 8.0\text{ V}. \quad (27\text{-}11)$$

Suppose, instead, we move from a to b counterclockwise, passing through resistor R rather than through the battery. Because we move opposite the current, the potential increases by iR. Thus,

$$V_a + iR = V_b$$

or

$$V_b - V_a = iR. \quad (27\text{-}12)$$

Substituting for i from Eq. 27-9, we again find Eq. 27-10. Hence, substitution of the data in Fig. 27-6 yields the same result, $V_b - V_a = 8.0$ V. In general,

> To find the potential between any two points in a circuit, start at one point and traverse the circuit to the other point, following any path, and add algebraically the changes in potential you encounter.

Potential Difference Across a Real Battery

In Fig. 27-6, points a and b are located at the terminals of the battery. Thus, the potential difference $V_b - V_a$ is the terminal-to-terminal potential difference V across the battery. From Eq. 27-8, we see that

$$V = \mathscr{E} - ir. \quad (27\text{-}13)$$

If the internal resistance r of the battery in Fig. 27-6 were zero, Eq. 27-13 tells us that V would be equal to the emf $\mathscr{E}$ of the battery—namely, 12 V. However, because $r = 2.0\ \Omega$, Eq. 27-13 tells us that V is less than $\mathscr{E}$. From Eq. 27-11, we know that V is only 8.0 V. Note that the result depends on the value of the current through the battery. If the same battery were in a different circuit and had a different current through it, V would have some other value.

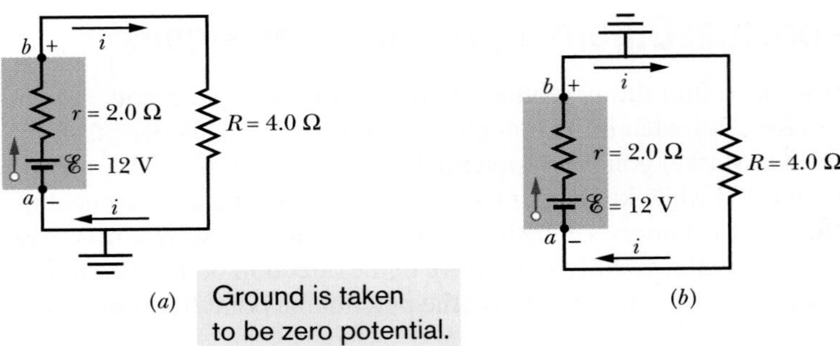

Fig. 27-7 (*a*) Point *a* is directly connected to ground. (*b*) Point *b* is directly connected to ground.

Grounding a Circuit

Figure 27-7*a* shows the same circuit as Fig. 27-6 except that here point *a* is directly connected to *ground,* as indicated by the common symbol ⏚. *Grounding a circuit* usually means connecting the circuit to a conducting path to Earth's surface (actually to the electrically conducting moist dirt and rock below ground). Here, such a connection means only that the potential is defined to be zero at the grounding point in the circuit. Thus in Fig. 27-7*a*, the potential at *a* is defined to be $V_a = 0$. Equation 27-11 then tells us that the potential at *b* is $V_b = 8.0$ V.

Figure 27-7*b* is the same circuit except that point *b* is now directly connected to ground. Thus, the potential there is defined to be $V_b = 0$. Equation 27-11 now tells us that the potential at *a* is $V_a = -8.0$ V.

Power, Potential, and Emf

When a battery or some other type of emf device does work on the charge carriers to establish a current *i*, the device transfers energy from its source of energy (such as the chemical source in a battery) to the charge carriers. Because a real emf device has an internal resistance *r*, it also transfers energy to internal thermal energy via resistive dissipation (Section 26-7). Let us relate these transfers.

The net rate *P* of energy transfer from the emf device to the charge carriers is given by Eq. 26-26:

$$P = iV, \tag{27-14}$$

where *V* is the potential across the terminals of the emf device. From Eq. 27-13, we can substitute $V = \mathscr{E} - ir$ into Eq. 27-14 to find

$$P = i(\mathscr{E} - ir) = i\mathscr{E} - i^2 r. \tag{27-15}$$

From Eq. 26-27, we recognize the term $i^2 r$ in Eq. 27-15 as the rate P_r of energy transfer to thermal energy within the emf device:

$$P_r = i^2 r \quad \text{(internal dissipation rate).} \tag{27-16}$$

Then the term $i\mathscr{E}$ in Eq. 27-15 must be the rate P_{emf} at which the emf device transfers energy *both* to the charge carriers and to internal thermal energy. Thus,

$$P_{emf} = i\mathscr{E} \quad \text{(power of emf device).} \tag{27-17}$$

If a battery is being *recharged,* with a "wrong way" current through it, the energy transfer is then *from* the charge carriers *to* the battery—both to the battery's chemical energy and to the energy dissipated in the internal resistance *r*. The rate of change of the chemical energy is given by Eq. 27-17, the rate of dissipation is given by Eq. 27-16, and the rate at which the carriers supply energy is given by Eq. 27-14.

CHECKPOINT 3

A battery has an emf of 12 V and an internal resistance of 2 Ω. Is the terminal-to-terminal potential difference greater than, less than, or equal to 12 V if the current in the battery is (a) from the negative to the positive terminal, (b) from the positive to the negative terminal, and (c) zero?

Sample Problem

Single-loop circuit with two real batteries

The emfs and resistances in the circuit of Fig. 27-8a have the following values:

$$\mathscr{E}_1 = 4.4\text{ V},\quad \mathscr{E}_2 = 2.1\text{ V},$$
$$r_1 = 2.3\ \Omega,\quad r_2 = 1.8\ \Omega,\quad R = 5.5\ \Omega.$$

(a) What is the current i in the circuit?

KEY IDEA

We can get an expression involving the current i in this single-loop circuit by applying the loop rule.

Calculations: Although knowing the direction of i is not necessary, we can easily determine it from the emfs of the two batteries. Because $\mathscr{E}_1$ is greater than $\mathscr{E}_2$, battery 1 controls the direction of i, so the direction is clockwise. (These decisions about where to start and which way you go are arbitrary but, once made, you must be consistent with decisions about the plus and minus signs.) Let us then apply the loop rule by going counterclockwise—against the current—and starting at point a. We find

$$-\mathscr{E}_1 + ir_1 + iR + ir_2 + \mathscr{E}_2 = 0.$$

Check that this equation also results if we apply the loop rule clockwise or start at some point other than a. Also, take the time to compare this equation term by term with Fig. 27-8b, which shows the potential changes graphically (with the potential at point a arbitrarily taken to be zero).

Solving the above loop equation for the current i, we obtain

$$i = \frac{\mathscr{E}_1 - \mathscr{E}_2}{R + r_1 + r_2} = \frac{4.4\text{ V} - 2.1\text{ V}}{5.5\ \Omega + 2.3\ \Omega + 1.8\ \Omega}$$
$$= 0.2396\text{ A} \approx 240\text{ mA}. \qquad \text{(Answer)}$$

(b) What is the potential difference between the terminals of battery 1 in Fig. 27-8a?

KEY IDEA

We need to sum the potential differences between points a and b.

Calculations: Let us start at point b (effectively the negative terminal of battery 1) and travel clockwise through battery 1 to point a (effectively the positive terminal), keeping track of potential changes. We find that

$$V_b - ir_1 + \mathscr{E}_1 = V_a,$$

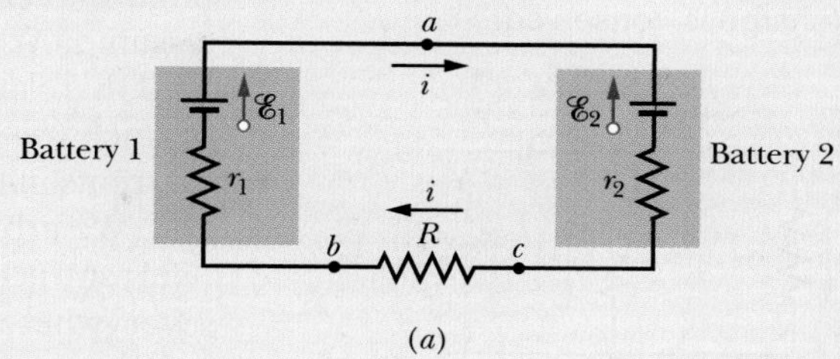

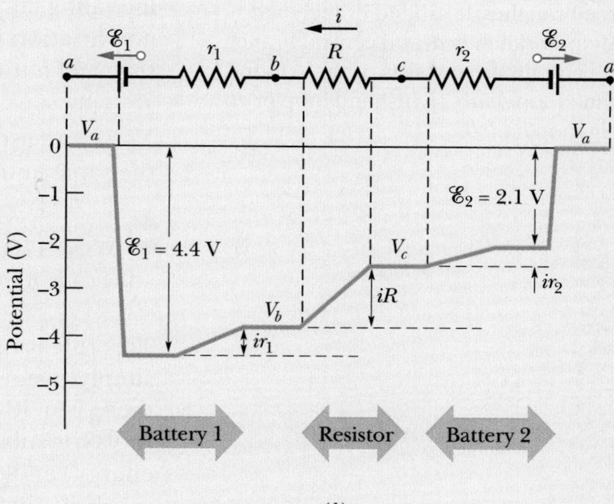

Fig. 27-8 (a) A single-loop circuit containing two real batteries and a resistor. The batteries oppose each other; that is, they tend to send current in opposite directions through the resistor. (b) A graph of the potentials, counterclockwise from point a, with the potential at a arbitrarily taken to be zero. (To better link the circuit with the graph, mentally cut the circuit at a and then unfold the left side of the circuit toward the left and the right side of the circuit toward the right.)

which gives us

$$V_a - V_b = -ir_1 + \mathscr{E}_1$$
$$= -(0.2396\text{ A})(2.3\ \Omega) + 4.4\text{ V}$$
$$= +3.84\text{ V} \approx 3.8\text{ V}, \qquad \text{(Answer)}$$

which is less than the emf of the battery. You can verify this result by starting at point b in Fig. 27-8a and traversing the circuit counterclockwise to point a. We learn two points here. (1) The potential difference between two points in a circuit is independent of the path we choose to go from one to the other. (2) When the current in the battery is in the "proper" direction, the terminal-to-terminal potential difference is low.

Additional examples, video, and practice available at *WileyPLUS*

The current into the junction must equal the current out (charge is conserved).

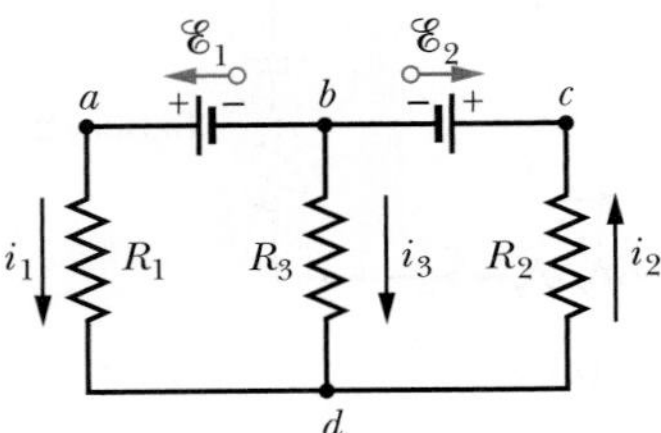

Fig. 27-9 A multiloop circuit consisting of three branches: left-hand branch *bad*, right-hand branch *bcd*, and central branch *bd*. The circuit also consists of three loops: left-hand loop *badb*, right-hand loop *bcdb*, and big loop *badcb*.

27-7 Multiloop Circuits

Figure 27-9 shows a circuit containing more than one loop. For simplicity, we assume the batteries are ideal. There are two *junctions* in this circuit, at b and d, and there are three *branches* connecting these junctions. The branches are the left branch (*bad*), the right branch (*bcd*), and the central branch (*bd*). What are the currents in the three branches?

We arbitrarily label the currents, using a different subscript for each branch. Current i_1 has the same value everywhere in branch *bad*, i_2 has the same value everywhere in branch *bcd*, and i_3 is the current through branch *bd*. The directions of the currents are assumed arbitrarily.

Consider junction d for a moment: Charge comes into that junction via incoming currents i_1 and i_3, and it leaves via outgoing current i_2. Because there is no variation in the charge at the junction, the total incoming current must equal the total outgoing current:

$$i_1 + i_3 = i_2. \tag{27-18}$$

You can easily check that applying this condition to junction b leads to exactly the same equation. Equation 27-18 thus suggests a general principle:

JUNCTION RULE: The sum of the currents entering any junction must be equal to the sum of the currents leaving that junction.

This rule is often called *Kirchhoff's junction rule* (or *Kirchhoff's current law*). It is simply a statement of the conservation of charge for a steady flow of charge—there is neither a buildup nor a depletion of charge at a junction. Thus, our basic tools for solving complex circuits are the *loop rule* (based on the conservation of energy) and the *junction rule* (based on the conservation of charge).

Equation 27-18 is a single equation involving three unknowns. To solve the circuit completely (that is, to find all three currents), we need two more equations involving those same unknowns. We obtain them by applying the loop rule twice. In the circuit of Fig. 27-9, we have three loops from which to choose: the left-hand loop (*badb*), the right-hand loop (*bcdb*), and the big loop (*badcb*). Which two loops we choose does not matter—let's choose the left-hand loop and the right-hand loop.

If we traverse the left-hand loop in a counterclockwise direction from point b, the loop rule gives us

$$\mathscr{E}_1 - i_1R_1 + i_3R_3 = 0. \tag{27-19}$$

If we traverse the right-hand loop in a counterclockwise direction from point b, the loop rule gives us

$$-i_3R_3 - i_2R_2 - \mathscr{E}_2 = 0. \tag{27-20}$$

We now have three equations (Eqs. 27-18, 27-19, and 27-20) in the three unknown currents, and they can be solved by a variety of techniques.

If we had applied the loop rule to the big loop, we would have obtained (moving counterclockwise from b) the equation

$$\mathscr{E}_1 - i_1R_1 - i_2R_2 - \mathscr{E}_2 = 0.$$

However, this is merely the sum of Eqs. 27-19 and 27-20.

Parallel resistors and their equivalent have the same potential difference ("par-V").

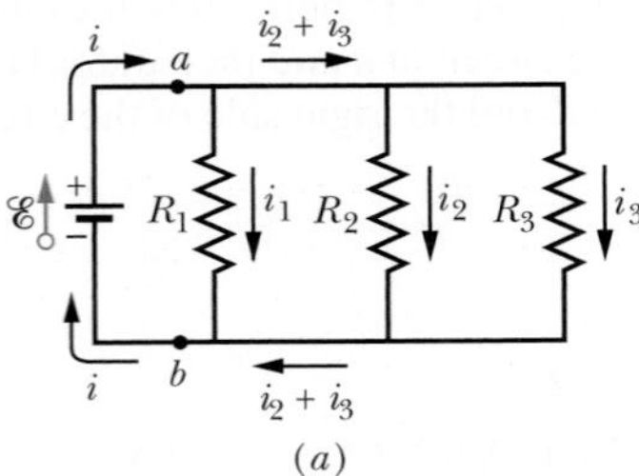

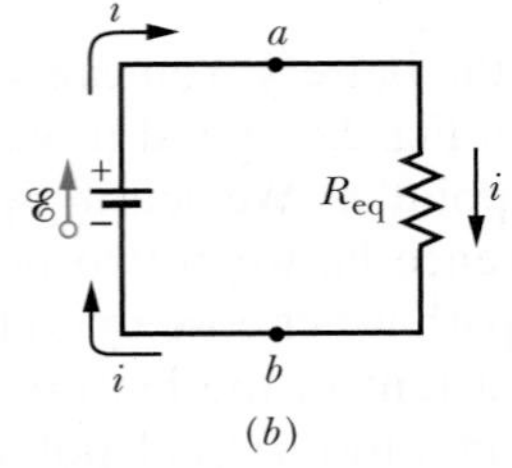

Fig. 27-10 (*a*) Three resistors connected in parallel across points a and b. (*b*) An equivalent circuit, with the three resistors replaced with their equivalent resistance R_{eq}.

Resistances in Parallel

Figure 27-10*a* shows three resistances connected *in parallel* to an ideal battery of emf $\mathscr{E}$. The term "in parallel" means that the resistances are directly wired together on one side and directly wired together on the other side, and that a potential difference V is applied across the pair of connected sides. Thus, all three resistances have the same potential difference V across them, producing a current through each. In general,

When a potential difference V is applied across resistances connected in parallel, the resistances all have that same potential difference V.

In Fig. 27-10*a*, the applied potential difference V is maintained by the battery. In Fig. 27-10*b*, the three parallel resistances have been replaced with an equivalent resistance R_{eq}.

Resistances connected in parallel can be replaced with an equivalent resistance R_{eq} that has the same potential difference V and the same *total* current i as the actual resistances.

You might remember that R_{eq} and all the actual parallel resistances have the same potential difference V with the nonsense word "par-V."

To derive an expression for R_{eq} in Fig. 27-10*b*, we first write the current in each actual resistance in Fig. 27-10*a* as

$$i_1 = \frac{V}{R_1}, \quad i_2 = \frac{V}{R_2}, \quad \text{and} \quad i_3 = \frac{V}{R_3},$$

where V is the potential difference between a and b. If we apply the junction rule at point a in Fig. 27-10*a* and then substitute these values, we find

$$i = i_1 + i_2 + i_3 = V\left(\frac{1}{R_1} + \frac{1}{R_2} + \frac{1}{R_3}\right). \tag{27-21}$$

If we replaced the parallel combination with the equivalent resistance R_{eq} (Fig. 27-10*b*), we would have

$$i = \frac{V}{R_{eq}}. \tag{27-22}$$

Comparing Eqs. 27-21 and 27-22 leads to

$$\frac{1}{R_{eq}} = \frac{1}{R_1} + \frac{1}{R_2} + \frac{1}{R_3}. \tag{27-23}$$

Extending this result to the case of n resistances, we have

$$\frac{1}{R_{eq}} = \sum_{j=1}^{n} \frac{1}{R_j} \quad (n \text{ resistances in parallel}). \tag{27-24}$$

For the case of two resistances, the equivalent resistance is their product divided by their sum; that is,

$$R_{eq} = \frac{R_1 R_2}{R_1 + R_2}. \tag{27-25}$$

Note that when two or more resistances are connected in parallel, the equivalent resistance is smaller than any of the combining resistances. Table 27-1 summarizes the equivalence relations for resistors and capacitors in series and in parallel.

CHECKPOINT 4

A battery, with potential V across it, is connected to a combination of two identical resistors and then has current i through it. What are the potential difference across and the current through either resistor if the resistors are (a) in series and (b) in parallel?

Table 27-1

Series and Parallel Resistors and Capacitors

Series	Parallel	Series	Parallel
Resistors		Capacitors	
$R_{eq} = \sum_{j=1}^{n} R_j$ Eq. 27-7	$\frac{1}{R_{eq}} = \sum_{j=1}^{n} \frac{1}{R_j}$ Eq. 27-24	$\frac{1}{C_{eq}} = \sum_{j=1}^{n} \frac{1}{C_j}$ Eq. 25-20	$C_{eq} = \sum_{j=1}^{n} C_j$ Eq. 25-19
Same current through all resistors	Same potential difference across all resistors	Same charge on all capacitors	Same potential difference across all capacitors

Sample Problem

Resistors in parallel and in series

Figure 27-11*a* shows a multiloop circuit containing one ideal battery and four resistances with the following values:

$$R_1 = 20\ \Omega, \quad R_2 = 20\ \Omega, \quad \mathscr{E} = 12\ \text{V},$$

$$R_3 = 30\ \Omega, \quad R_4 = 8.0\ \Omega.$$

(a) What is the current through the battery?

KEY IDEA

Noting that the current through the battery must also be the current through R_1, we see that we might find the current by applying the loop rule to a loop that includes R_1 because the current would be included in the potential difference across R_1.

Incorrect method: Either the left-hand loop or the big loop should do. Noting that the emf arrow of the battery points upward, so the current the battery supplies is clockwise, we might apply the loop rule to the left-hand loop, clockwise from point *a*. With *i* being the current through the battery, we would get

$$+\mathscr{E} - iR_1 - iR_2 - iR_4 = 0 \qquad \text{(incorrect).}$$

However, this equation is incorrect because it assumes that R_1, R_2, and R_4 all have the same current *i*. Resistances R_1 and R_4 do have the same current, because the current passing through R_4 must pass through the battery and then through R_1 with no change in value. However, that current splits at junction point *b*—only part passes through R_2, the rest through R_3.

Dead-end method: To distinguish the several currents in the circuit, we must label them individually as in Fig. 27-11*b*. Then, circling clockwise from *a*, we can write the loop rule for the left-hand loop as

$$+\mathscr{E} - i_1R_1 - i_2R_2 - i_1R_4 = 0.$$

Unfortunately, this equation contains two unknowns, i_1 and i_2; we would need at least one more equation to find them.

Successful method: A much easier option is to simplify the circuit of Fig. 27-11*b* by finding equivalent resistances. Note carefully that R_1 and R_2 are *not* in series and thus cannot be replaced with an equivalent resistance. However, R_2 and R_3 are in parallel, so we can use either Eq. 27-24 or Eq. 27-25 to find their equivalent resistance R_{23}. From the latter,

$$R_{23} = \frac{R_2R_3}{R_2 + R_3} = \frac{(20\ \Omega)(30\ \Omega)}{50\ \Omega} = 12\ \Omega.$$

We can now redraw the circuit as in Fig. 27-11*c*; note that the current through R_{23} must be i_1 because charge that moves through R_1 and R_4 must also move through R_{23}. For this simple one-loop circuit, the loop rule (applied clockwise from point *a* as in Fig. 27-11*d*) yields

$$+\mathscr{E} - i_1R_1 - i_1R_{23} - i_1R_4 = 0.$$

Substituting the given data, we find

$$12\ \text{V} - i_1(20\ \Omega) - i_1(12\ \Omega) - i_1(8.0\ \Omega) = 0,$$

which gives us

$$i_1 = \frac{12\ \text{V}}{40\ \Omega} = 0.30\ \text{A}. \qquad \text{(Answer)}$$

(b) What is the current i_2 through R_2?

KEY IDEAS

(1) We must now work backward from the equivalent circuit of Fig. 27-11*d*, where R_{23} has replaced R_2 and R_3. (2) Because R_2 and R_3 are in parallel, they both have the same potential difference across them as R_{23}.

Working backward: We know that the current through R_{23} is $i_1 = 0.30$ A. Thus, we can use Eq. 26-8 ($R = V/i$) and Fig. 27-11*e* to find the potential difference V_{23} across R_{23}. Setting $R_{23} = 12\ \Omega$ from (a), we write Eq. 26-8 as

$$V_{23} = i_1R_{23} = (0.30\ \text{A})(12\ \Omega) = 3.6\ \text{V}.$$

The potential difference across R_2 is thus also 3.6 V (Fig. 27-11*f*), so the current i_2 in R_2 must be, by Eq. 26-8 and Fig. 27-11*g*,

$$i_2 = \frac{V_2}{R_2} = \frac{3.6\ \text{V}}{20\ \Omega} = 0.18\ \text{A}. \qquad \text{(Answer)}$$

(c) What is the current i_3 through R_3?

KEY IDEAS

We can answer by using either of two techniques: (1) Apply Eq. 26-8 as we just did. (2) Use the junction rule, which tells us that at point *b* in Fig. 27-11*b*, the incoming current i_1 and the outgoing currents i_2 and i_3 are related by

$$i_1 = i_2 + i_3.$$

Calculation: Rearranging this junction-rule result yields the result displayed in Fig. 27-11*g*:

$$i_3 = i_1 - i_2 = 0.30\ \text{A} - 0.18\ \text{A}$$
$$= 0.12\ \text{A}. \qquad \text{(Answer)}$$

Additional examples, video, and practice available at *WileyPLUS*

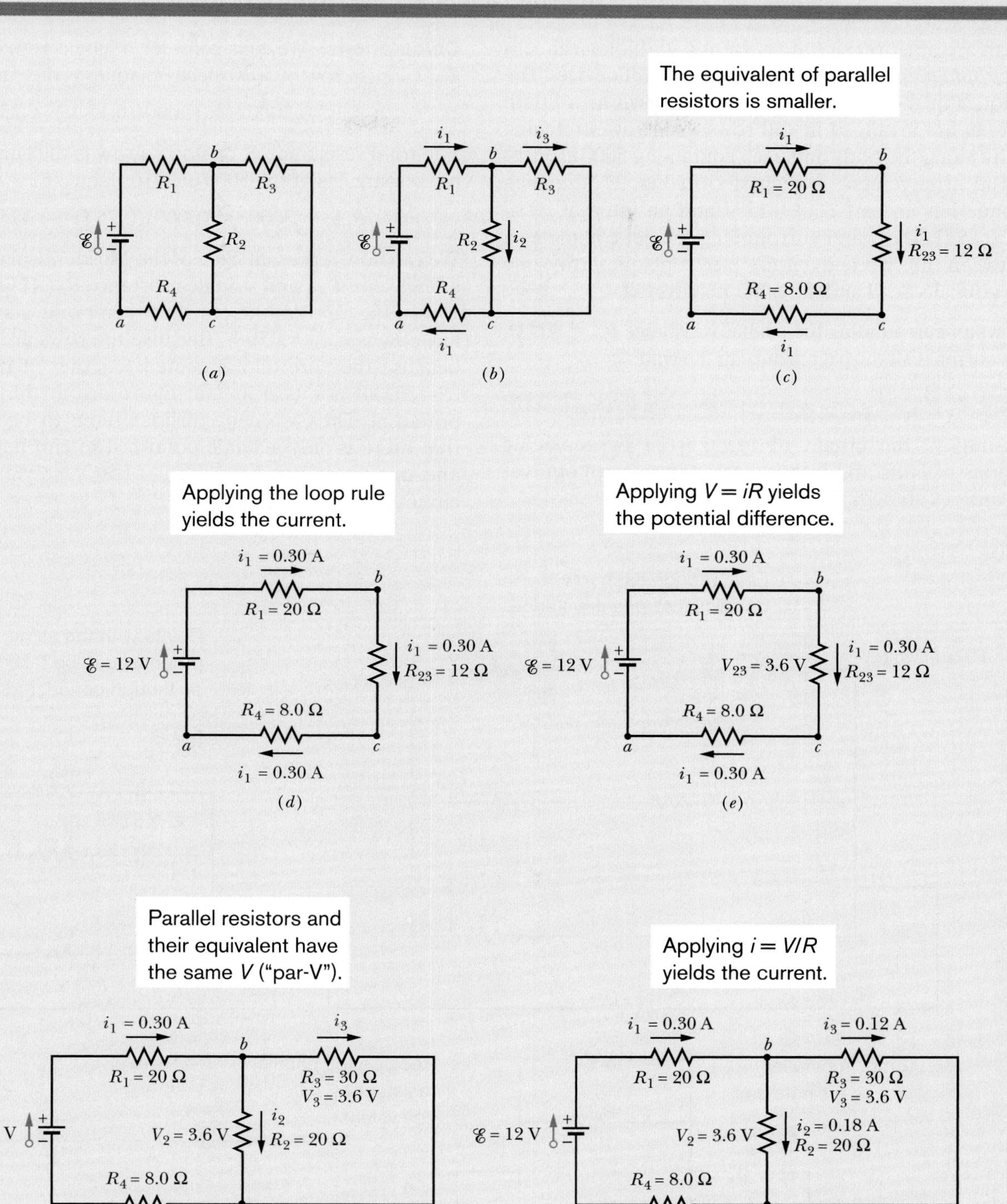

Fig. 27-11 (*a*) A circuit with an ideal battery. (*b*) Label the currents. (*c*) Replacing the parallel resistors with their equivalent. (*d*) – (*g*) Working backward to find the currents through the parallel resistors.

Sample Problem

Many real batteries in series and in parallel in an electric fish

Electric fish are able to generate current with biological cells called *electroplaques*, which are physiological emf devices. The electroplaques in the type of electric fish known as a South American eel are arranged in 140 rows, each row stretching horizontally along the body and each containing 5000 electroplaques. The arrangement is suggested in Fig. 27-12*a*; each electroplaque has an emf $\mathscr{E}$ of 0.15 V and an internal resistance r of 0.25 Ω. The water surrounding the eel completes a circuit between the two ends of the electroplaque array, one end at the animal's head and the other near its tail.

(a) If the water surrounding the eel has resistance $R_w = 800\ \Omega$, how much current can the eel produce in the water?

KEY IDEA

We can simplify the circuit of Fig. 27-12*a* by replacing combinations of emfs and internal resistances with equivalent emfs and resistances.

Calculations: We first consider a single row. The total emf $\mathscr{E}_{\text{row}}$ along a row of 5000 electroplaques is the sum of the emfs:

$$\mathscr{E}_{\text{row}} = 5000\mathscr{E} = (5000)(0.15\text{ V}) = 750\text{ V}.$$

The total resistance R_{row} along a row is the sum of the internal resistances of the 5000 electroplaques:

$$R_{\text{row}} = 5000r = (5000)(0.25\ \Omega) = 1250\ \Omega.$$

We can now represent each of the 140 identical rows as having a single emf $\mathscr{E}_{\text{row}}$ and a single resistance R_{row} (Fig. 27-12*b*).

In Fig. 27-12*b*, the emf between point *a* and point *b* on any row is $\mathscr{E}_{\text{row}} = 750$ V. Because the rows are identical and because they are all connected together at the left in Fig. 27-12*b*, all points *b* in that figure are at the same electric potential. Thus, we can consider them to be connected so that there is only a single point *b*. The emf between point *a* and this single point *b* is $\mathscr{E}_{\text{row}} = 750$ V, so we can draw the circuit as shown in Fig. 27-12*c*.

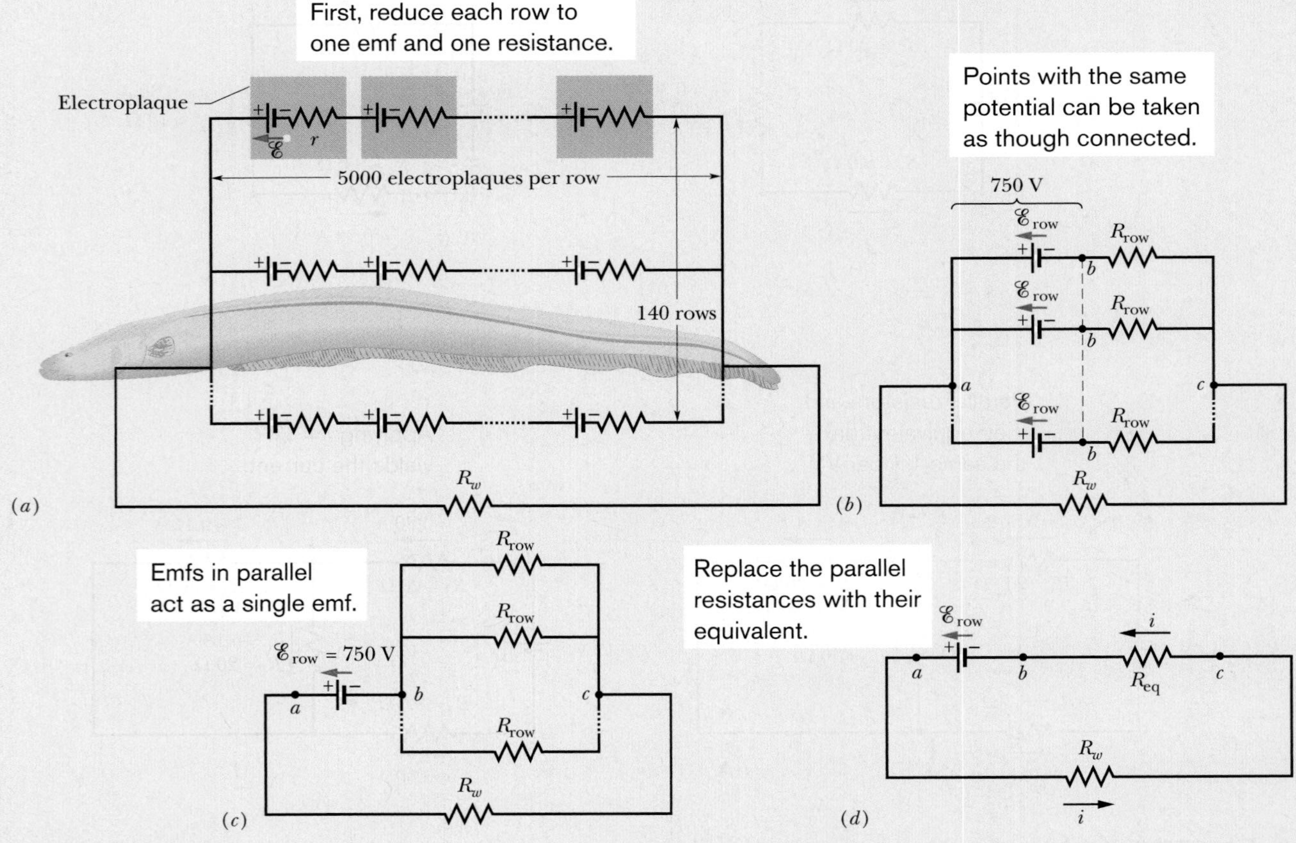

Fig. 27-12 (*a*) A model of the electric circuit of an eel in water. Each electroplaque of the eel has an emf $\mathscr{E}$ and internal resistance r. Along each of 140 rows extending from the head to the tail of the eel, there are 5000 electroplaques. The surrounding water has resistance R_w. (*b*) The emf $\mathscr{E}_{\text{row}}$ and resistance R_{row} of each row. (*c*) The emf between points *a* and *b* is $\mathscr{E}_{\text{row}}$. Between points *b* and *c* are 140 parallel resistances R_{row}. (*d*) The simplified circuit, with R_{eq} replacing the parallel combination.

Between points b and c in Fig. 27-12c are 140 resistances $R_{row} = 1250\ \Omega$, all in parallel. The equivalent resistance R_{eq} of this combination is given by Eq. 27-24 as

$$\frac{1}{R_{eq}} = \sum_{j=1}^{140} \frac{1}{R_j} = 140 \frac{1}{R_{row}},$$

or

$$R_{eq} = \frac{R_{row}}{140} = \frac{1250\ \Omega}{140} = 8.93\ \Omega.$$

Replacing the parallel combination with R_{eq}, we obtain the simplified circuit of Fig. 27-12d. Applying the loop rule to this circuit counterclockwise from point b, we have

$$\mathscr{E}_{row} - iR_w - iR_{eq} = 0.$$

Solving for i and substituting the known data, we find

$$i = \frac{\mathscr{E}_{row}}{R_w + R_{eq}} = \frac{750\ \text{V}}{800\ \Omega + 8.93\ \Omega}$$
$$= 0.927\ \text{A} \approx 0.93\ \text{A}. \qquad \text{(Answer)}$$

If the head or tail of the eel is near a fish, some of this current could pass along a narrow path through the fish, stunning or killing it.

(b) How much current i_{row} travels through each row of Fig. 27-12a?

KEY IDEA

Because the rows are identical, the current into and out of the eel is evenly divided among them.

Calculation: Thus, we write

$$i_{row} = \frac{i}{140} = \frac{0.927\ \text{A}}{140} = 6.6 \times 10^{-3}\ \text{A}. \qquad \text{(Answer)}$$

Thus, the current through each row is small, about two orders of magnitude smaller than the current through the water. This tends to spread the current through the eel's body, so that the eel need not stun or kill itself when it stuns or kills a fish.

Sample Problem

Multiloop circuit and simultaneous loop equations

Figure 27-13 shows a circuit whose elements have the following values:

$$\mathscr{E}_1 = 3.0\ \text{V}, \quad \mathscr{E}_2 = 6.0\ \text{V},$$
$$R_1 = 2.0\ \Omega, \quad R_2 = 4.0\ \Omega.$$

The three batteries are ideal batteries. Find the magnitude and direction of the current in each of the three branches.

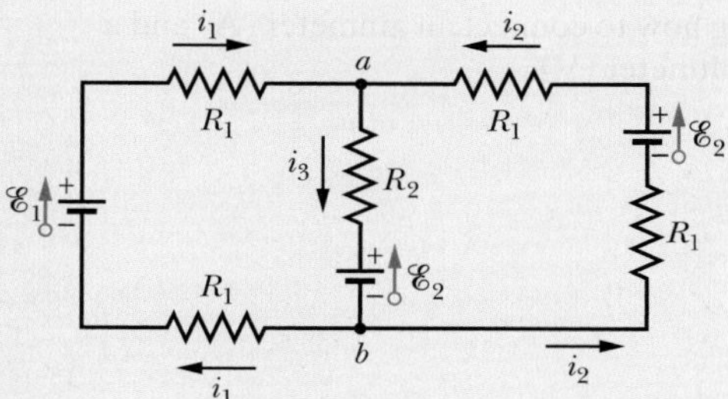

Fig. 27-13 A multiloop circuit with three ideal batteries and five resistances.

KEY IDEAS

It is not worthwhile to try to simplify this circuit, because no two resistors are in parallel, and the resistors that are in series (those in the right branch or those in the left branch) present no problem. So, our plan is to apply the junction and loop rules.

Junction rule: Using arbitrarily chosen directions for the currents as shown in Fig. 27-13, we apply the junction rule at point a by writing

$$i_3 = i_1 + i_2. \qquad (27\text{-}26)$$

An application of the junction rule at junction b gives only the same equation, so we next apply the loop rule to any two of the three loops of the circuit.

Left-hand loop: We first arbitrarily choose the left-hand loop, arbitrarily start at point b, and arbitrarily traverse the loop in the clockwise direction, obtaining

$$-i_1R_1 + \mathscr{E}_1 - i_1R_1 - (i_1 + i_2)R_2 - \mathscr{E}_2 = 0,$$

where we have used $(i_1 + i_2)$ instead of i_3 in the middle branch. Substituting the given data and simplifying yield

$$i_1(8.0\ \Omega) + i_2(4.0\ \Omega) = -3.0\ \text{V}. \qquad (27\text{-}27)$$

Right-hand loop: For our second application of the loop rule, we arbitrarily choose to traverse the right-hand loop counterclockwise from point b, finding

$$-i_2R_1 + \mathscr{E}_2 - i_2R_1 - (i_1 + i_2)R_2 - \mathscr{E}_2 = 0.$$

Substituting the given data and simplifying yield

$$i_1(4.0\ \Omega) + i_2(8.0\ \Omega) = 0. \qquad (27\text{-}28)$$

Combining equations: We now have a system of two equations (Eqs. 27-27 and 27-28) in two unknowns (i_1 and i_2) to solve either "by hand" (which is easy enough here) or with a "math package." (One solution technique is Cramer's rule, given in Appendix E.) We find

$$i_1 = -0.50\ \text{A}. \qquad (27\text{-}29)$$

(The minus sign signals that our arbitrary choice of direction for i_1 in Fig. 27-13 is wrong, but we must wait to correct it.) Substituting $i_1 = -0.50$ A into Eq. 27-28 and solving for i_2 then give us

$$i_2 = 0.25\ \text{A}. \qquad \text{(Answer)}$$

With Eq. 27-26 we then find that

$$i_3 = i_1 + i_2 = -0.50\text{ A} + 0.25\text{ A}$$
$$= -0.25\text{ A}.$$

The positive answer we obtained for i_2 signals that our choice of direction for that current is correct. However, the negative answers for i_1 and i_3 indicate that our choices for those currents are wrong. Thus, as a *last step* here, we correct the answers by reversing the arrows for i_1 and i_3 in Fig. 27-13 and then writing

$$i_1 = 0.50\text{ A} \quad \text{and} \quad i_3 = 0.25\text{ A}. \quad \text{(Answer)}$$

Caution: Always make any such correction as the last step and not before calculating *all* the currents.

Additional examples, video, and practice available at *WileyPLUS*

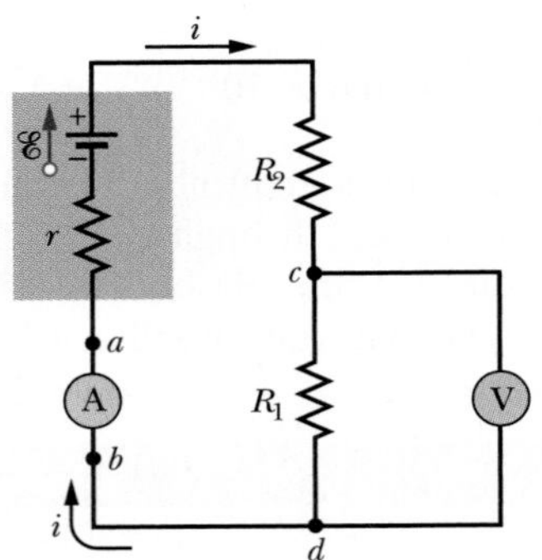

Fig. 27-14 A single-loop circuit, showing how to connect an ammeter (A) and a voltmeter (V).

27-8 The Ammeter and the Voltmeter

An instrument used to measure currents is called an *ammeter.* To measure the current in a wire, you usually have to break or cut the wire and insert the ammeter so that the current to be measured passes through the meter. (In Fig. 27-14, ammeter A is set up to measure current i.)

It is essential that the resistance R_A of the ammeter be very much smaller than other resistances in the circuit. Otherwise, the very presence of the meter will change the current to be measured.

A meter used to measure potential differences is called a *voltmeter.* To find the potential difference between any two points in the circuit, the voltmeter terminals are connected between those points without breaking or cutting the wire. (In Fig. 27-14, voltmeter V is set up to measure the voltage across R_1.)

It is essential that the resistance R_V of a voltmeter be very much larger than the resistance of any circuit element across which the voltmeter is connected. Otherwise, the meter itself becomes an important circuit element and alters the potential difference that is to be measured.

Often a single meter is packaged so that, by means of a switch, it can be made to serve as either an ammeter or a voltmeter—and usually also as an *ohmmeter,* designed to measure the resistance of any element connected between its terminals. Such a versatile unit is called a *multimeter.*

27-9 *RC* Circuits

In preceding sections we dealt only with circuits in which the currents did not vary with time. Here we begin a discussion of time-varying currents.

Charging a Capacitor

The capacitor of capacitance C in Fig. 27-15 is initially uncharged. To charge it, we close switch S on point a. This completes an *RC series circuit* consisting of the capacitor, an ideal battery of emf ℰ, and a resistance R.

From Section 25-2, we already know that as soon as the circuit is complete, charge begins to flow (current exists) between a capacitor plate and a battery terminal on each side of the capacitor. This current increases the charge q on the plates and the potential difference V_C (= q/C) across the capacitor. When that potential difference equals the potential difference across the battery (which here is equal to the emf ℰ), the current is zero. From Eq. 25-1 ($q = CV$), the *equilibrium* (final) *charge* on the then fully charged capacitor is equal to Cℰ.

Here we want to examine the charging process. In particular we want to know how the charge $q(t)$ on the capacitor plates, the potential difference $V_C(t)$ across the capacitor, and the current $i(t)$ in the circuit vary with time during the charging process. We begin by applying the loop rule to the circuit, traversing it

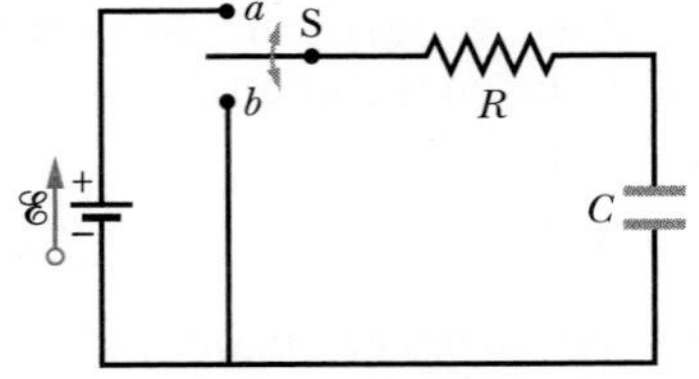

Fig. 27-15 When switch S is closed on a, the capacitor is *charged* through the resistor. When the switch is afterward closed on b, the capacitor *discharges* through the resistor.

clockwise from the negative terminal of the battery. We find

$$\mathscr{E} - iR - \frac{q}{C} = 0. \tag{27-30}$$

The last term on the left side represents the potential difference across the capacitor. The term is negative because the capacitor's top plate, which is connected to the battery's positive terminal, is at a higher potential than the lower plate. Thus, there is a drop in potential as we move down through the capacitor.

We cannot immediately solve Eq. 27-30 because it contains two variables, i and q. However, those variables are not independent but are related by

$$i = \frac{dq}{dt}. \tag{27-31}$$

Substituting this for i in Eq. 27-30 and rearranging, we find

$$R\frac{dq}{dt} + \frac{q}{C} = \mathscr{E} \quad \text{(charging equation).} \tag{27-32}$$

This differential equation describes the time variation of the charge q on the capacitor in Fig. 27-15. To solve it, we need to find the function $q(t)$ that satisfies this equation and also satisfies the condition that the capacitor be initially uncharged; that is, $q = 0$ at $t = 0$.

We shall soon show that the solution to Eq. 27-32 is

$$q = C\mathscr{E}(1 - e^{-t/RC}) \quad \text{(charging a capacitor).} \tag{27-33}$$

(Here e is the exponential base, 2.718 . . . , and not the elementary charge.) Note that Eq. 27-33 does indeed satisfy our required initial condition, because at $t = 0$ the term $e^{-t/RC}$ is unity; so the equation gives $q = 0$. Note also that as t goes to infinity (that is, a long time later), the term $e^{-t/RC}$ goes to zero; so the equation gives the proper value for the full (equilibrium) charge on the capacitor—namely, $q = C\mathscr{E}$. A plot of $q(t)$ for the charging process is given in Fig. 27-16a.

The derivative of $q(t)$ is the current $i(t)$ charging the capacitor:

$$i = \frac{dq}{dt} = \left(\frac{\mathscr{E}}{R}\right)e^{-t/RC} \quad \text{(charging a capacitor).} \tag{27-34}$$

A plot of $i(t)$ for the charging process is given in Fig. 27-16b. Note that the current has the initial value $\mathscr{E}/R$ and that it decreases to zero as the capacitor becomes fully charged.

A capacitor that is being charged initially acts like ordinary connecting wire relative to the charging current. A long time later, it acts like a broken wire.

By combining Eq. 25-1 ($q = CV$) and Eq. 27-33, we find that the potential difference $V_C(t)$ across the capacitor during the charging process is

$$V_C = \frac{q}{C} = \mathscr{E}(1 - e^{-t/RC}) \quad \text{(charging a capacitor).} \tag{27-35}$$

This tells us that $V_C = 0$ at $t = 0$ and that $V_C = \mathscr{E}$ when the capacitor becomes fully charged as $t \to \infty$.

The capacitor's charge grows as the resistor's current dies out.

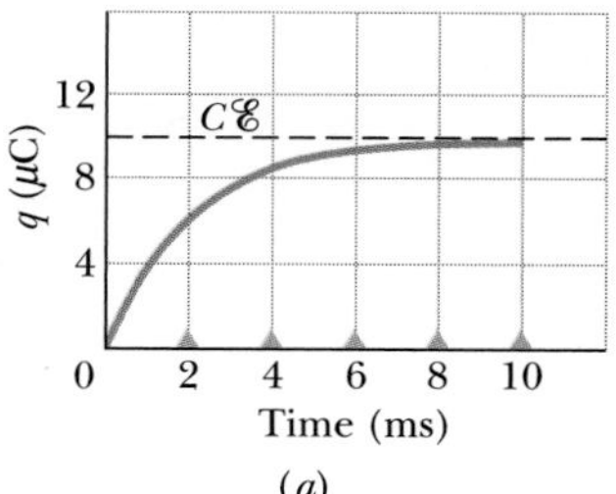

(a)

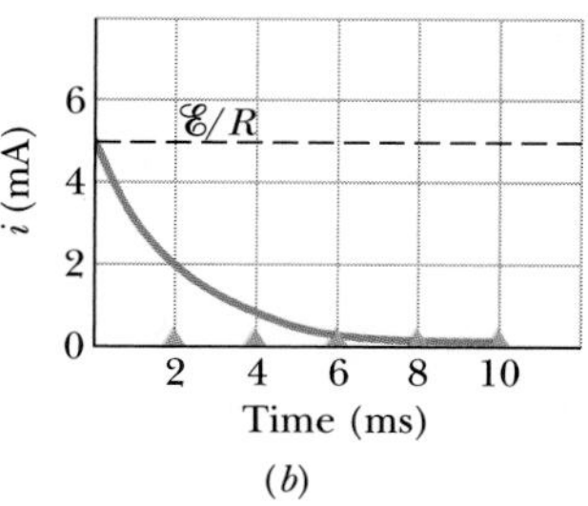

(b)

Fig. 27-16 (a) A plot of Eq. 27-33, which shows the buildup of charge on the capacitor of Fig. 27-15. (b) A plot of Eq. 27-34, which shows the decline of the charging current in the circuit of Fig. 27-15. The curves are plotted for $R = 2000\ \Omega$, $C = 1\ \mu$F, and $\mathscr{E} = 10$ V; the small triangles represent successive intervals of one time constant τ.

The Time Constant

The product RC that appears in Eqs. 27-33, 27-34, and 27-35 has the dimensions of time (both because the argument of an exponential must be dimensionless and because, in fact, 1.0 Ω × 1.0 F = 1.0 s). The product RC is called the **capacitive time constant** of the circuit and is represented with the symbol τ:

$$\tau = RC \quad \text{(time constant).} \tag{27-36}$$

From Eq. 27-33, we can now see that at time $t = \tau (= RC)$, the charge on the initially uncharged capacitor of Fig. 27-15 has increased from zero to

$$q = C\mathscr{E}(1 - e^{-1}) = 0.63C\mathscr{E}. \tag{27-37}$$

In words, during the first time constant τ the charge has increased from zero to 63% of its final value $C\mathscr{E}$. In Fig. 27-16, the small triangles along the time axes mark successive intervals of one time constant during the charging of the capacitor. The charging times for RC circuits are often stated in terms of τ.

Discharging a Capacitor

Assume now that the capacitor of Fig. 27-15 is fully charged to a potential V_0 equal to the emf $\mathscr{E}$ of the battery. At a new time $t = 0$, switch S is thrown from a to b so that the capacitor can *discharge* through resistance R. How do the charge $q(t)$ on the capacitor and the current $i(t)$ through the discharge loop of capacitor and resistance now vary with time?

The differential equation describing $q(t)$ is like Eq. 27-32 except that now, with no battery in the discharge loop, $\mathscr{E} = 0$. Thus,

$$R\frac{dq}{dt} + \frac{q}{C} = 0 \quad \text{(discharging equation).} \tag{27-38}$$

The solution to this differential equation is

$$q = q_0 e^{-t/RC} \quad \text{(discharging a capacitor),} \tag{27-39}$$

where $q_0 (= CV_0)$ is the initial charge on the capacitor. You can verify by substitution that Eq. 27-39 is indeed a solution of Eq. 27-38.

Equation 27-39 tells us that q decreases exponentially with time, at a rate that is set by the capacitive time constant $\tau = RC$. At time $t = \tau$, the capacitor's charge has been reduced to $q_0 e^{-1}$, or about 37% of the initial value. Note that a greater τ means a greater discharge time.

Differentiating Eq. 27-39 gives us the current $i(t)$:

$$i = \frac{dq}{dt} = -\left(\frac{q_0}{RC}\right)e^{-t/RC} \quad \text{(discharging a capacitor).} \tag{27-40}$$

This tells us that the current also decreases exponentially with time, at a rate set by τ. The initial current i_0 is equal to q_0/RC. Note that you can find i_0 by simply applying the loop rule to the circuit at $t = 0$; just then the capacitor's initial potential V_0 is connected across the resistance R, so the current must be $i_0 = V_0/R = (q_0/C)/R = q_0/RC$. The minus sign in Eq. 27-40 can be ignored; it merely means that the capacitor's charge q is decreasing.

Derivation of Eq. 27-33

To solve Eq. 27-32, we first rewrite it as

$$\frac{dq}{dt} + \frac{q}{RC} = \frac{\mathscr{E}}{R}. \tag{27-41}$$

The general solution to this differential equation is of the form

$$q = q_p + Ke^{-at}, \tag{27-42}$$

where q_p is a *particular solution* of the differential equation, K is a constant to be evaluated from the initial conditions, and $a = 1/RC$ is the coefficient of q in Eq. 27-41. To find q_p, we set $dq/dt = 0$ in Eq. 27-41 (corresponding to the final condition of no further charging), let $q = q_p$, and solve, obtaining

$$q_p = C\mathscr{E}. \tag{27-43}$$

To evaluate K, we first substitute this into Eq. 27-42 to get

$$q = C\mathscr{E} + Ke^{-at}.$$

Then substituting the initial conditions $q = 0$ and $t = 0$ yields

$$0 = C\mathscr{E} + K,$$

or $K = -C\mathscr{E}$. Finally, with the values of q_p, a, and K inserted, Eq. 27-42 becomes

$$q = C\mathscr{E} - C\mathscr{E}e^{-t/RC},$$

which, with a slight modification, is Eq. 27-33.

CHECKPOINT 5

The table gives four sets of values for the circuit elements in Fig. 27-15. Rank the sets according to (a) the initial current (as the switch is closed on a) and (b) the time required for the current to decrease to half its initial value, greatest first.

	1	2	3	4
$\mathscr{E}$ (V)	12	12	10	10
R (Ω)	2	3	10	5
C (μF)	3	2	0.5	2

Sample Problem

Discharging an *RC* circuit to avoid a fire in a race car pit stop

As a car rolls along pavement, electrons move from the pavement first onto the tires and then onto the car body. The car stores this excess charge and the associated electric potential energy as if the car body were one plate of a capacitor and the pavement were the other plate (Fig. 27-17a). When the car stops, it discharges its excess charge and energy through the tires, just as a capacitor can discharge through a resistor. If a conducting object comes within a few centimeters of the car before the car is discharged, the remaining energy can be suddenly transferred to a spark between the car and the object. Suppose the conducting object is a fuel dispenser. The spark will not ignite the fuel and cause a fire if the spark energy is less than the critical value $U_{fire} = 50$ mJ.

When the car of Fig. 27-17a stops at time $t = 0$, the car–ground potential difference is $V_0 = 30$ kV. The car–ground capacitance is $C = 500$ pF, and the resistance of *each* tire is $R_{tire} = 100$ GΩ. How much time does the car take to discharge through the tires to drop below the critical value U_{fire}?

KEY IDEAS

(1) At any time t, a capacitor's stored electric potential energy U is related to its stored charge q according to Eq. 25-21 ($U = q^2/2C$). (2) While a capacitor is discharging, the charge decreases with time according to Eq. 27-39 ($q = q_0e^{-t/RC}$).

Calculations: We can treat the tires as resistors that are connected to one another at their tops via the car body and at

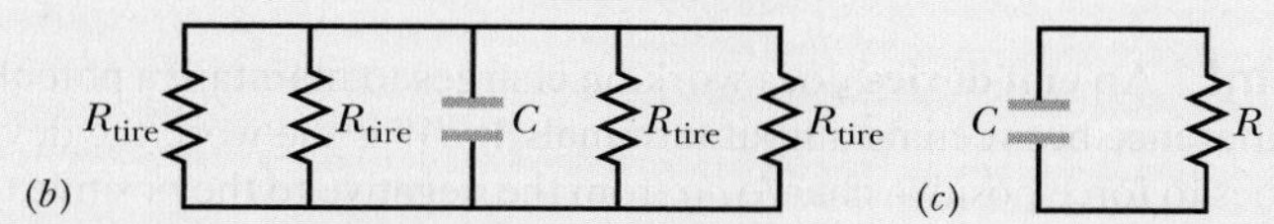

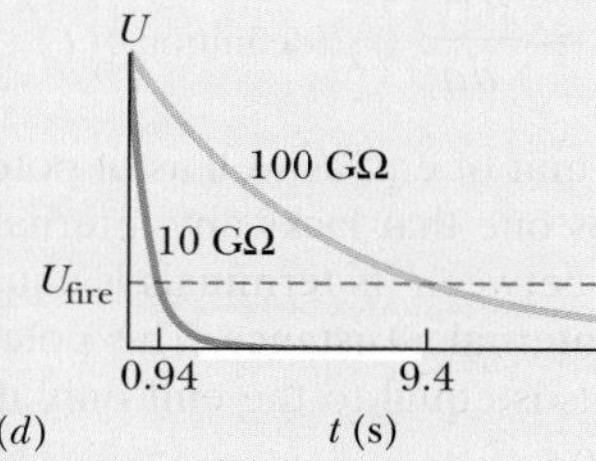

Fig. 27-17 (a) A charged car and the pavement acts like a capacitor that can discharge through the tires. (b) The effective circuit of the car–pavement capacitor, with four tire resistances R_{tire} connected in parallel. (c) The equivalent resistance R of the tires. (d) The electric potential energy U in the car–pavement capacitor decreases during discharge.

their bottoms via the pavement. Figure 27-17*b* shows how the four resistors are connected in parallel across the car's capacitance, and Fig. 27-17*c* shows their equivalent resistance R. From Eq. 27-24, R is given by

$$\frac{1}{R} = \frac{1}{R_{\text{tire}}} + \frac{1}{R_{\text{tire}}} + \frac{1}{R_{\text{tire}}} + \frac{1}{R_{\text{tire}}},$$

or
$$R = \frac{R_{\text{tire}}}{4} = \frac{100 \times 10^9\,\Omega}{4} = 25 \times 10^9\,\Omega. \tag{27-44}$$

When the car stops, it discharges its excess charge and energy through R.

We now use our two Key Ideas to analyze the discharge. Substituting Eq. 27-39 into Eq. 25-21 gives

$$U = \frac{q^2}{2C} = \frac{(q_0 e^{-t/RC})^2}{2C}$$

$$= \frac{q_0^2}{2C} e^{-2t/RC}. \tag{27-45}$$

From Eq. 25-1 ($q = CV$), we can relate the initial charge q_0 on the car to the given initial potential difference V_0: $q_0 = CV_0$. Substituting this equation into Eq. 27-45 brings us to

$$U = \frac{(CV_0)^2}{2C} e^{-2t/RC} = \frac{CV_0^2}{2} e^{-2t/RC},$$

or
$$e^{-2t/RC} = \frac{2U}{CV_0^2}. \tag{27-46}$$

Taking the natural logarithms of both sides, we obtain

$$-\frac{2t}{RC} = \ln\left(\frac{2U}{CV_0^2}\right),$$

or
$$t = -\frac{RC}{2} \ln\left(\frac{2U}{CV_0^2}\right). \tag{27-47}$$

Substituting the given data, we find that the time the car takes to discharge to the energy level $U_{\text{fire}} = 50$ mJ is

$$t = -\frac{(25 \times 10^9\,\Omega)(500 \times 10^{-12}\,\text{F})}{2}$$

$$\times \ln\left(\frac{2(50 \times 10^{-3}\,\text{J})}{(500 \times 10^{-12}\,\text{F})(30 \times 10^3\,\text{V})^2}\right)$$

$$= 9.4\ \text{s}. \qquad \text{(Answer)}$$

Fire or no fire: This car requires at least 9.4 s before fuel or a fuel dispenser can be brought safely near it. During a race, a pit crew cannot wait that long. Instead, tires for race cars include some type of conducting material (such as carbon black) to lower the tire resistance and thus increase the car's discharge rate. Figure 27-17*d* shows the stored energy U versus time t for tire resistances of $R = 100$ GΩ (the value we used in our calculations here) and $R = 10$ GΩ. Note how much more rapidly a car discharges to level U_{fire} with the lower R value.

Additional examples, video, and practice available at *WileyPLUS*

REVIEW & SUMMARY

Emf An **emf device** does work on charges to maintain a potential difference between its output terminals. If dW is the work the device does to force positive charge dq from the negative to the positive terminal, then the **emf** (work per unit charge) of the device is

$$\mathscr{E} = \frac{dW}{dq} \quad \text{(definition of } \mathscr{E}\text{)}. \tag{27-1}$$

The volt is the SI unit of emf as well as of potential difference. An **ideal emf device** is one that lacks any internal resistance. The potential difference between its terminals is equal to the emf. A **real emf device** has internal resistance. The potential difference between its terminals is equal to the emf only if there is no current through the device.

Analyzing Circuits The change in potential in traversing a resistance R in the direction of the current is $-iR$; in the opposite direction it is $+iR$ (resistance rule). The change in potential in traversing an ideal emf device in the direction of the emf arrow is $+\mathscr{E}$; in the opposite direction it is $-\mathscr{E}$ (emf rule). Conservation of energy leads to the loop rule:

Loop Rule. *The algebraic sum of the changes in potential encountered in a complete traversal of any loop of a circuit must be zero.*

Conservation of charge gives us the junction rule:

Junction Rule. *The sum of the currents entering any junction must be equal to the sum of the currents leaving that junction.*

Single-Loop Circuits The current in a single-loop circuit containing a single resistance R and an emf device with emf $\mathscr{E}$ and internal resistance r is

$$i = \frac{\mathscr{E}}{R + r}, \tag{27-4}$$

which reduces to $i = \mathscr{E}/R$ for an ideal emf device with $r = 0$.

Power When a real battery of emf $\mathscr{E}$ and internal resistance r does work on the charge carriers in a current i through the battery, the rate P of energy transfer to the charge carriers is

$$P = iV, \tag{27-14}$$

where V is the potential across the terminals of the battery. The rate

P_r at which energy is dissipated as thermal energy in the battery is

$$P_r = i^2 r. \tag{27-16}$$

The rate P_{emf} at which the chemical energy in the battery changes is

$$P_{emf} = i\mathscr{E}. \tag{27-17}$$

Series Resistances When resistances are in **series,** they have the same current. The equivalent resistance that can replace a series combination of resistances is

$$R_{eq} = \sum_{j=1}^{n} R_j \quad (n \text{ resistances in series}). \tag{27-7}$$

Parallel Resistances When resistances are in **parallel,** they have the same potential difference. The equivalent resistance that can replace a parallel combination of resistances is given by

$$\frac{1}{R_{eq}} = \sum_{j=1}^{n} \frac{1}{R_j} \quad (n \text{ resistances in parallel}). \tag{27-24}$$

RC Circuits When an emf $\mathscr{E}$ is applied to a resistance R and capacitance C in series, as in Fig. 27-15 with the switch at a, the charge on the capacitor increases according to

$$q = C\mathscr{E}(1 - e^{-t/RC}) \quad \text{(charging a capacitor)}, \tag{27-33}$$

in which $C\mathscr{E} = q_0$ is the equilibrium (final) charge and $RC = \tau$ is the **capacitive time constant** of the circuit. During the charging, the current is

$$i = \frac{dq}{dt} = \left(\frac{\mathscr{E}}{R}\right)e^{-t/RC} \quad \text{(charging a capacitor)}. \tag{27-34}$$

When a capacitor discharges through a resistance R, the charge on the capacitor decays according to

$$q = q_0 e^{-t/RC} \quad \text{(discharging a capacitor)}. \tag{27-39}$$

During the discharging, the current is

$$i = \frac{dq}{dt} = -\left(\frac{q_0}{RC}\right)e^{-t/RC} \quad \text{(discharging a capacitor)}. \tag{27-40}$$

QUESTIONS

1 (a) In Fig. 27-18a, with $R_1 > R_2$, is the potential difference across R_2 more than, less than, or equal to that across R_1? (b) Is the current through resistor R_2 more than, less than, or equal to that through resistor R_1?

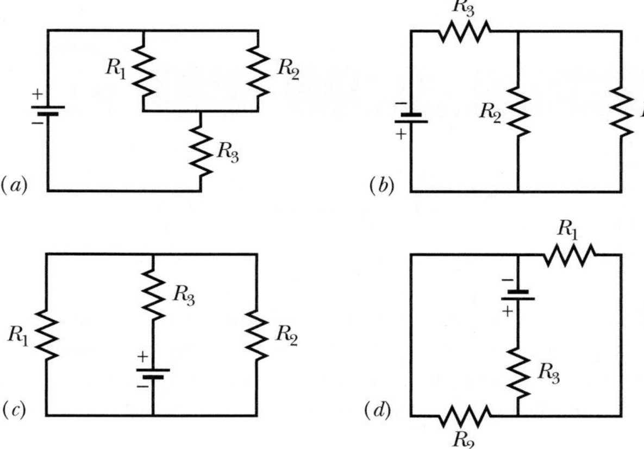

Fig. 27-18 Questions 1 and 2.

2 (a) In Fig. 27-18a, are resistors R_1 and R_3 in series? (b) Are resistors R_1 and R_2 in parallel? (c) Rank the equivalent resistances of the four circuits shown in Fig. 27-18, greatest first.

3 You are to connect resistors R_1 and R_2, with $R_1 > R_2$, to a battery, first individually, then in series, and then in parallel. Rank those arrangements according to the amount of current through the battery, greatest first.

4 In Fig. 27-19, a circuit consists of a battery and two uniform resistors, and the section lying along an x axis is divided into five segments of equal lengths. (a) Assume that $R_1 = R_2$ and rank the segments according to the magnitude of the average electric field in them, greatest first. (b) Now assume that $R_1 > R_2$ and then again rank the segments. (c) What is the direction of the electric field along the x axis?

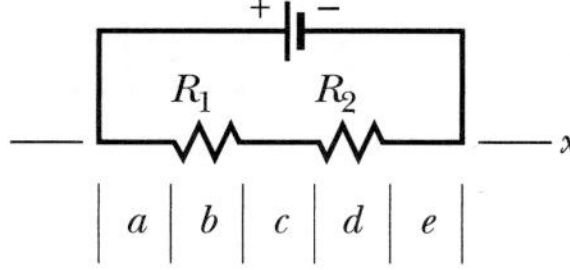

Fig. 27-19 Question 4.

5 For each circuit in Fig. 27-20, are the resistors connected in series, in parallel, or neither?

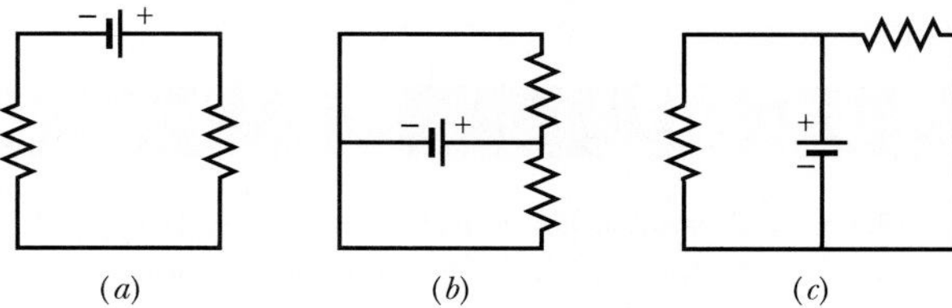

Fig. 27-20 Question 5.

6 *Res-monster maze.* In Fig. 27-21, all the resistors have a resistance of 4.0 Ω and all the (ideal) batteries have an emf of 4.0 V. What is the current through resistor R? (If you can find the proper loop through this maze, you can answer the question with a few seconds of mental calculation.)

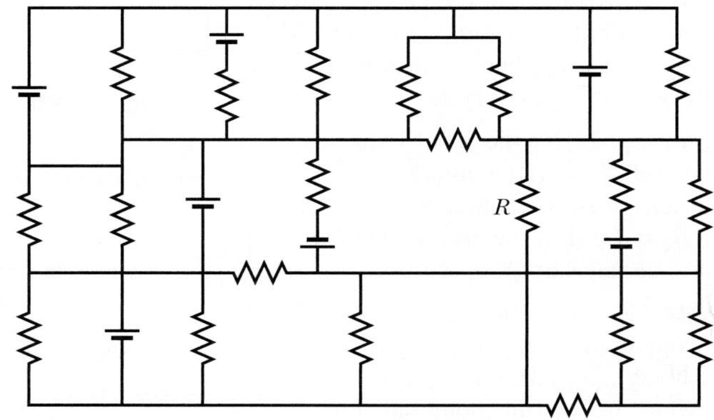

Fig. 27-21 Question 6.

7 A resistor R_1 is wired to a battery, then resistor R_2 is added in series. Are (a) the potential difference across R_1 and (b) the current i_1 through R_1 now more than, less than, or the same as previously? (c) Is the equivalent resistance R_{12} of R_1 and R_2 more than, less than, or equal to R_1?

8 *Cap-monster maze.* In Fig. 27-22, all the capacitors have a capacitance of 6.0 μF, and all the batteries have an emf of 10 V. What is the charge on capacitor C? (If you can find the proper loop through this maze, you can answer the question with a few seconds of mental calculation.)

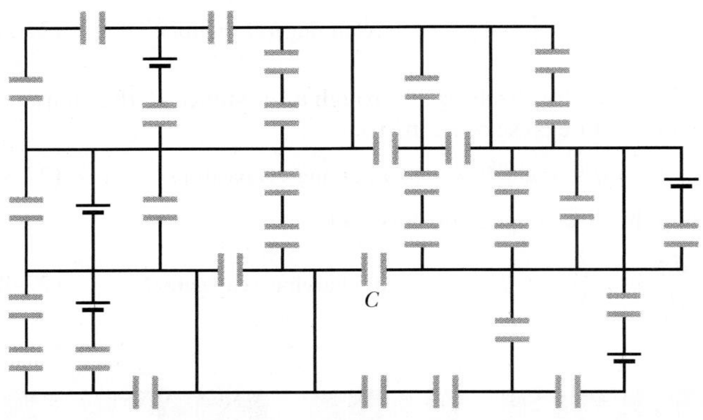

Fig. 27-22 Question 8.

9 Initially, a single resistor R_1 is wired to a battery. Then resistor R_2 is added in parallel. Are (a) the potential difference across R_1 and (b) the current i_1 through R_1 now more than, less than, or the same as previously? (c) Is the equivalent resistance R_{12} of R_1 and R_2 more than, less than, or equal to R_1? (d) Is the total current through R_1 and R_2 together more than, less than, or equal to the current through R_1 previously?

10 After the switch in Fig. 27-15 is closed on point a, there is current i through resistance R. Figure 27-23 gives that current for four sets of values of R and capacitance C: (1) R_0 and C_0, (2) $2R_0$ and C_0, (3) R_0 and $2C_0$, (4) $2R_0$ and $2C_0$. Which set goes with which curve?

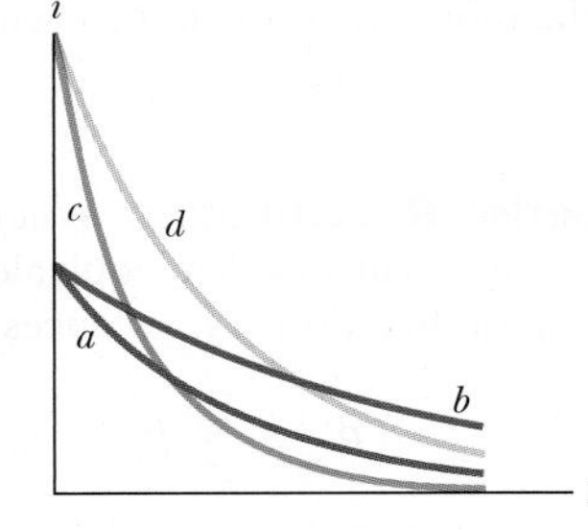

Fig. 27-23 Question 10.

11 Figure 27-24 shows three sections of circuit that are to be connected in turn to the same battery via a switch as in Fig. 27-15. The resistors are all identical, as are the capacitors. Rank the sections according to (a) the final (equilibrium) charge on the capacitor and (b) the time required for the capacitor to reach 50% of its final charge, greatest first.

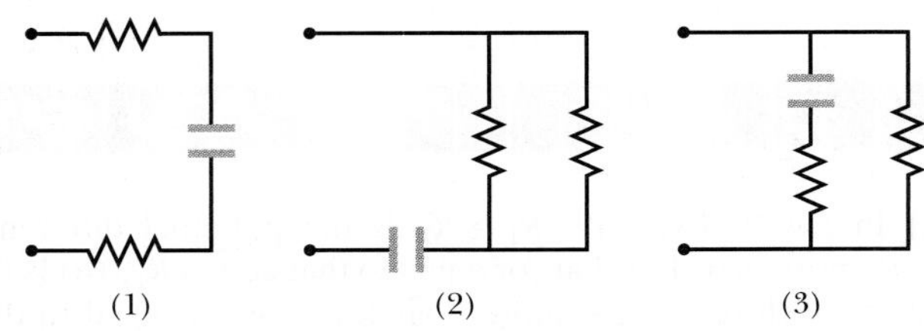

Fig. 27-24 Question 11.

PROBLEMS

GO Tutoring problem available (at instructor's discretion) in *WileyPLUS* and WebAssign
SSM Worked-out solution available in Student Solutions Manual
• – ••• Number of dots indicates level of problem difficulty
Additional information available in *The Flying Circus of Physics* and at flyingcircusofphysics.com
WWW Worked-out solution is at
ILW Interactive solution is at
http://www.wiley.com/college/halliday

sec. 27-6 Potential Difference Between Two Points

•1 SSM WWW In Fig. 27-25, the ideal batteries have emfs $\mathscr{E}_1 = 12$ V and $\mathscr{E}_2 = 6.0$ V. What are (a) the current, the dissipation rate in (b) resistor 1 (4.0 Ω) and (c) resistor 2 (8.0 Ω), and the energy transfer rate in (d) battery 1 and (e) battery 2? Is energy being supplied or absorbed by (f) battery 1 and (g) battery 2?

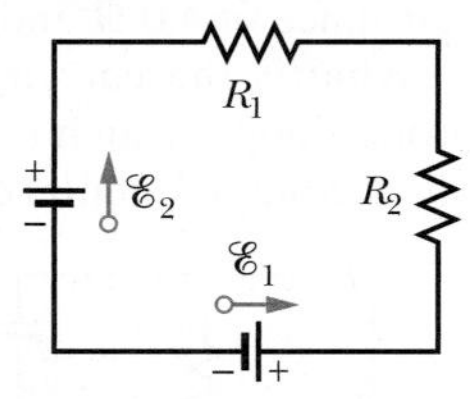

Fig. 27-25 Problem 1.

•2 In Fig. 27-26, the ideal batteries have emfs $\mathscr{E}_1 = 150$ V and $\mathscr{E}_2 = 50$ V and the resistances are $R_1 = 3.0\ \Omega$ and $R_2 = 2.0\ \Omega$. If the potential at P is 100 V, what is it at Q?

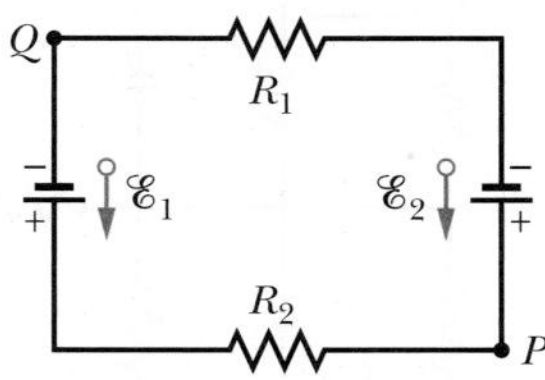

Fig. 27-26 Problem 2.

•3 ILW A car battery with a 12 V emf and an internal resistance of 0.040 Ω is being charged with a current of 50 A. What are (a) the potential difference V across the terminals, (b) the rate P_r of energy dissipation inside the battery, and (c) the rate P_{emf} of energy conversion to chemical form? When the battery is used to supply 50 A to the starter motor, what are (d) V and (e) P_r?

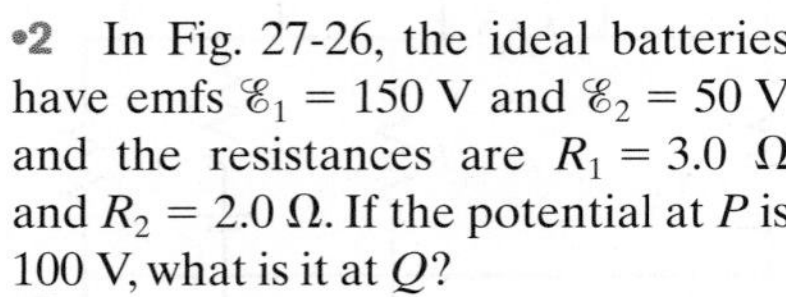

•4 GO Figure 27-27 shows a circuit of four resistors that are connected to a larger circuit. The graph below the circuit shows the electric potential $V(x)$ as a function of position x along the lower branch of the circuit, through resistor 4; the potential V_A is 12.0 V. The graph

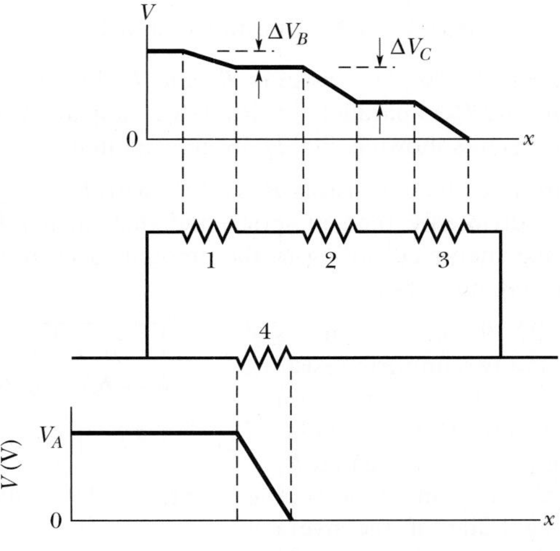

Fig. 27-27 Problem 4.

above the circuit shows the electric potential $V(x)$ versus position x along the upper branch of the circuit, through resistors 1, 2, and 3; the potential differences are $\Delta V_B = 2.00$ V and $\Delta V_C = 5.00$ V. Resistor 3 has a resistance of 200 Ω. What is the resistance of (a) resistor 1 and (b) resistor 2?

•5 A 5.0 A current is set up in a circuit for 6.0 min by a rechargeable battery with a 6.0 V emf. By how much is the chemical energy of the battery reduced?

•6 A standard flashlight battery can deliver about 2.0 W · h of energy before it runs down. (a) If a battery costs US$0.80, what is the cost of operating a 100 W lamp for 8.0 h using batteries? (b) What is the cost if energy is provided at the rate of US$0.06 per kilowatt-hour?

•7 A wire of resistance 5.0 Ω is connected to a battery whose emf $\mathscr{E}$ is 2.0 V and whose internal resistance is 1.0 Ω. In 2.0 min, how much energy is (a) transferred from chemical form in the battery, (b) dissipated as thermal energy in the wire, and (c) dissipated as thermal energy in the battery?

•8 A certain car battery with a 12.0 V emf has an initial charge of 120 A · h. Assuming that the potential across the terminals stays constant until the battery is completely discharged, for how many hours can it deliver energy at the rate of 100 W?

•9 (a) In electron-volts, how much work does an ideal battery with a 12.0 V emf do on an electron that passes through the battery from the positive to the negative terminal? (b) If 3.40×10^{18} electrons pass through each second, what is the power of the battery in watts?

••10 (a) In Fig. 27-28, what value must R have if the current in the circuit is to be 1.0 mA? Take $\mathscr{E}_1 = 2.0$ V, $\mathscr{E}_2 = 3.0$ V, and $r_1 = r_2 = 3.0$ Ω. (b) What is the rate at which thermal energy appears in R?

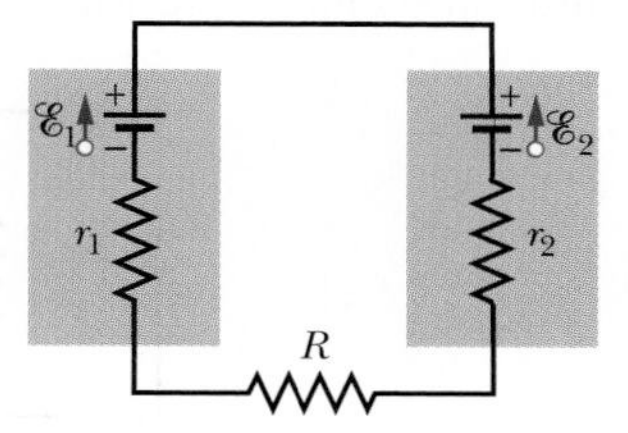

Fig. 27-28 Problem 10.

••11 SSM In Fig. 27-29, circuit section AB absorbs energy at a rate of 50 W when current $i = 1.0$ A through it is in the indicated direction. Resistance $R = 2.0$ Ω. (a) What is the potential difference between A and B? Emf device X lacks internal resistance. (b) What is its emf? (c) Is point B connected to the positive terminal of X or to the negative terminal?

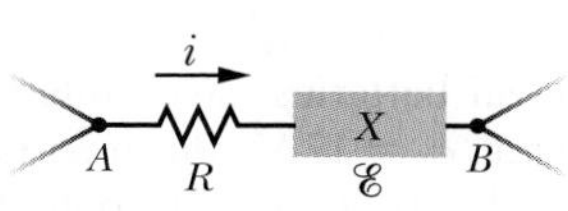

Fig. 27-29 Problem 11.

••12 Figure 27-30 shows a resistor of resistance $R = 6.00$ Ω connected to an ideal battery of emf $\mathscr{E} = 12.0$ V by means of two copper wires. Each wire has length 20.0 cm and radius 1.00 mm. In dealing with such circuits in this chapter, we generally neglect the potential differences along the wires and the transfer of energy to thermal energy in them. Check the validity of this neglect for the circuit of Fig. 27-30: What is the potential difference across (a) the resistor and (b) each of the two sections of wire? At what rate is energy lost to thermal energy in (c) the resistor and (d) each section of wire?

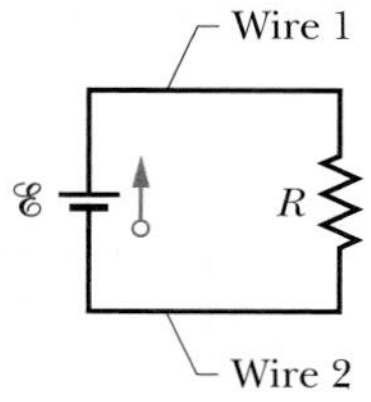

Fig. 27-30 Problem 12.

••13 A 10-km-long underground cable extends east to west and consists of two parallel wires, each of which has resistance 13 Ω/km. An electrical short develops at distance x from the west end when a conducting path of resistance R connects the wires (Fig. 27-31). The resistance of the wires and the short is then 100 Ω when measured from the east end and 200 Ω when measured from the west end. What are (a) x and (b) R?

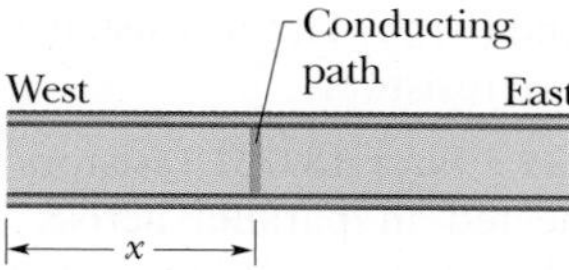

Fig. 27-31 Problem 13.

••14 GO In Fig. 27-32a, both batteries have emf $\mathscr{E} = 1.20$ V and the external resistance R is a variable resistor. Figure 27-32b gives the electric potentials V between the terminals of each battery as functions of R: Curve 1 corresponds to battery 1, and curve 2 corresponds to battery 2. The horizontal scale is set by $R_s = 0.20$ Ω. What is the internal resistance of (a) battery 1 and (b) battery 2?

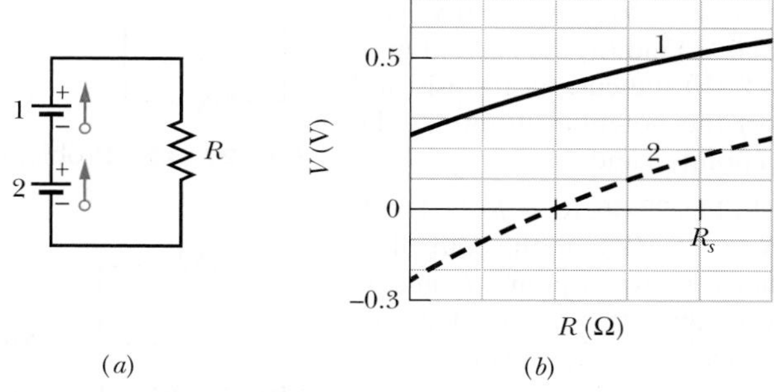

Fig. 27-32 Problem 14.

••15 ILW The current in a single-loop circuit with one resistance R is 5.0 A. When an additional resistance of 2.0 Ω is inserted in series with R, the current drops to 4.0 A. What is R?

•••16 A solar cell generates a potential difference of 0.10 V when a 500 Ω resistor is connected across it, and a potential difference of 0.15 V when a 1000 Ω resistor is substituted. What are the (a) internal resistance and (b) emf of the solar cell? (c) The area of the cell is 5.0 cm², and the rate per unit area at which it receives energy from light is 2.0 mW/cm². What is the efficiency of the cell for converting light energy to thermal energy in the 1000 Ω external resistor?

•••17 SSM In Fig. 27-33, battery 1 has emf $\mathscr{E}_1 = 12.0$ V and internal resistance $r_1 = 0.016$ Ω and battery 2 has emf $\mathscr{E}_2 = 12.0$ V and internal resistance $r_2 = 0.012$ Ω. The batteries are connected in series with an external resistance R. (a) What R value makes the terminal-to-terminal potential difference of one of the batteries zero? (b) Which battery is that?

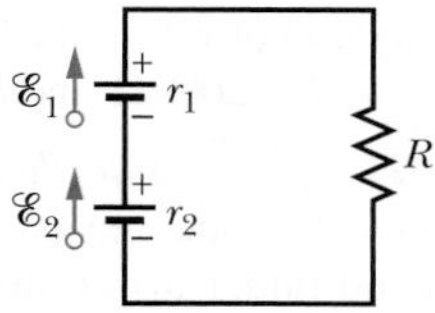

Fig. 27-33 Problem 17.

sec. 27-7 Multiloop Circuits

•18 In Fig. 27-9, what is the potential difference $V_d - V_c$ between points d and c if $\mathscr{E}_1 = 4.0$ V, $\mathscr{E}_2 = 1.0$ V, $R_1 = R_2 = 10$ Ω, and $R_3 = 5.0$ Ω, and the battery is ideal?

•19 A total resistance of 3.00 Ω is to be produced by connecting an unknown resistance to a 12.0 Ω resistance. (a) What must be the value of the unknown resistance, and (b) should it be connected in series or in parallel?

•20 When resistors 1 and 2 are connected in series, the equivalent resistance is 16.0 Ω. When they are connected in parallel, the equivalent resistance is 3.0 Ω. What are (a) the smaller resistance and (b) the larger resistance of these two resistors?

•21 Four 18.0 Ω resistors are connected in parallel across a 25.0 V ideal battery. What is the current through the battery?

•22 Figure 27-34 shows five 5.00 Ω resistors. Find the equivalent resistance between points (a) F and H and (b) F and G. (*Hint:* For each pair of points, imagine that a battery is connected across the pair.)

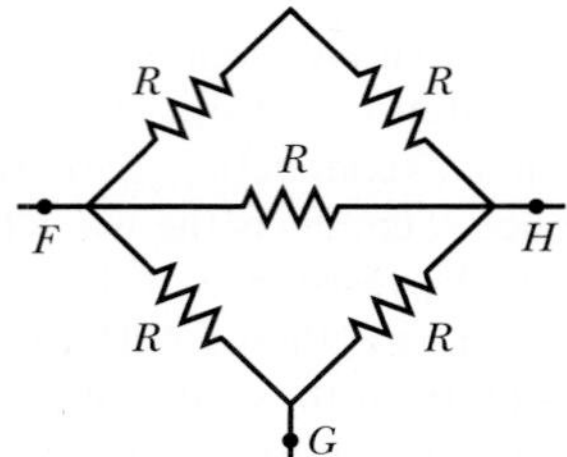

Fig. 27-34 Problem 22.

•23 In Fig. 27-35, $R_1 = 100\ \Omega$, $R_2 = 50\ \Omega$, and the ideal batteries have emfs $\mathscr{E}_1 = 6.0$ V, $\mathscr{E}_2 = 5.0$ V, and $\mathscr{E}_3 = 4.0$ V. Find (a) the current in resistor 1, (b) the current in resistor 2, and (c) the potential difference between points a and b.

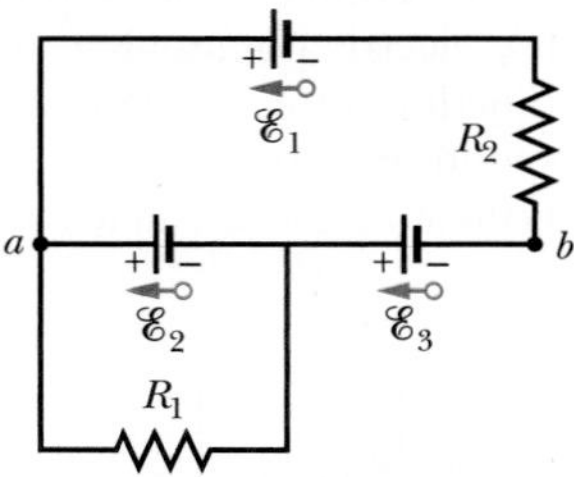

Fig. 27-35 Problem 23.

•24 In Fig. 27-36, $R_1 = R_2 = 4.00\ \Omega$ and $R_3 = 2.50\ \Omega$. Find the equivalent resistance between points D and E. (*Hint:* Imagine that a battery is connected across those points.)

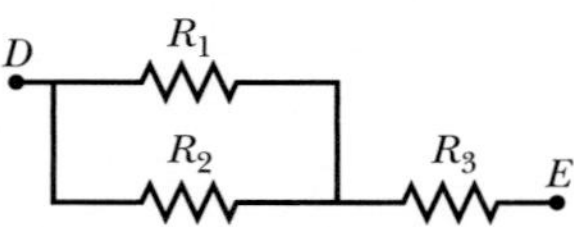

Fig. 27-36 Problem 24.

•25 SSM Nine copper wires of length l and diameter d are connected in parallel to form a single composite conductor of resistance R. What must be the diameter D of a single copper wire of length l if it is to have the same resistance?

••26 Figure 27-37 shows a battery connected across a uniform resistor R_0. A sliding contact can move across the resistor from $x = 0$ at the left to $x = 10$ cm at the right. Moving the contact changes how much resistance is to the left of the contact and how much is to the right. Find the rate at which energy is dissipated in resistor R as a function of x. Plot the function for $\mathscr{E} = 50$ V, $R = 2000\ \Omega$, and $R_0 = 100\ \Omega$.

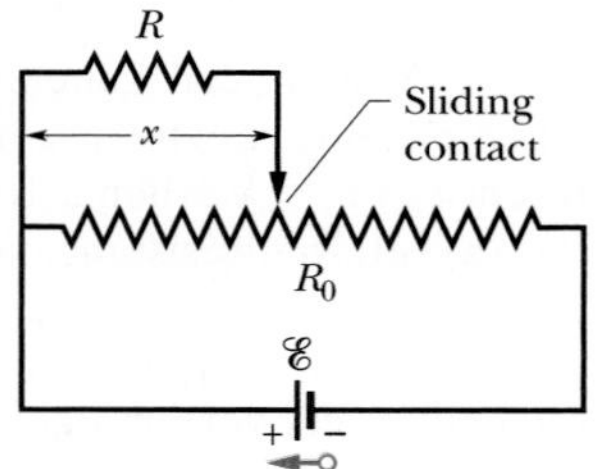

Fig. 27-37 Problem 26.

••27 *Side flash.* Figure 27-38 indicates one reason no one should stand under a tree during a lightning storm. If lightning comes down the side of the tree, a portion can jump over to the person, especially if the current on the tree reaches a dry region on the bark and thereafter must travel through air to reach the ground. In the figure, part of the lightning jumps through distance d in air and then travels through the person (who has negligible resistance relative to that of air). The rest of the current travels through air alongside the tree, for a distance h. If $d/h = 0.400$ and the total current is $I = 5000$ A, what is the current through the person?

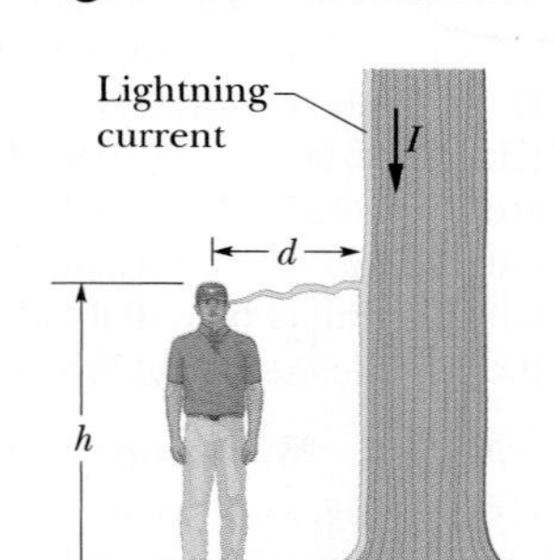

Fig. 27-38 Problem 27.

••28 The ideal battery in Fig. 27-39a has emf $\mathscr{E} = 6.0$ V. Plot 1 in Fig. 27-39b gives the electric potential difference V that can appear across resistor 1 of the circuit versus the current i in that resistor. The scale of the V axis is set by $V_s = 18.0$ V, and the scale of the i axis is set by $i_s = 3.00$ mA. Plots 2 and 3 are similar plots for resistors 2 and 3, respectively. What is the current in resistor 2?

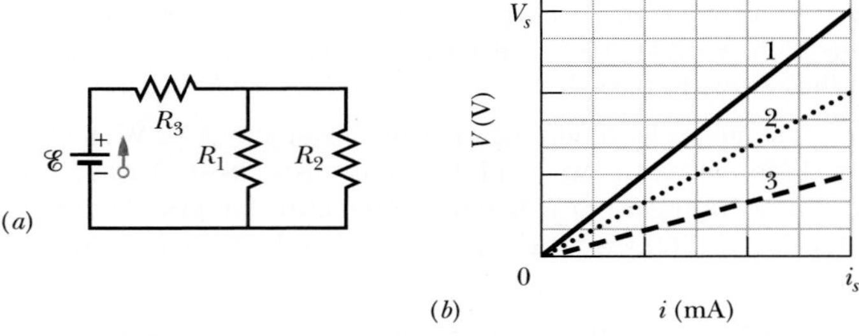

Fig. 27-39 Problem 28.

••29 In Fig. 27-40, $R_1 = 6.00\ \Omega$, $R_2 = 18.0\ \Omega$, and the ideal battery has emf $\mathscr{E} = 12.0$ V. What are the (a) size and (b) direction (left or right) of current i_1? (c) How much energy is dissipated by all four resistors in 1.00 min?

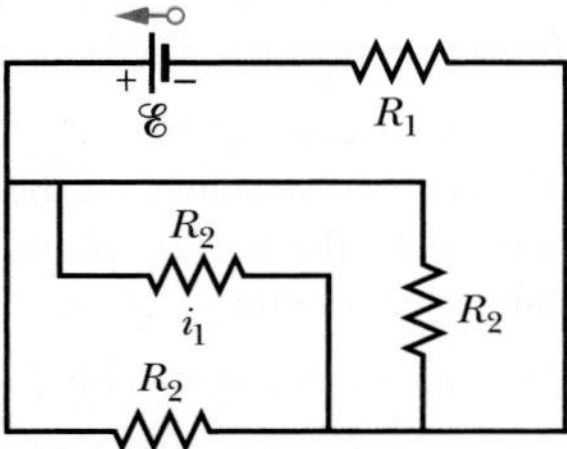

Fig. 27-40 Problem 29.

••30 GO In Fig. 27-41, the ideal batteries have emfs $\mathscr{E}_1 = 10.0$ V and $\mathscr{E}_2 = 0.500\mathscr{E}_1$, and the resistances are each 4.00 Ω. What is the current in (a) resistance 2 and (b) resistance 3?

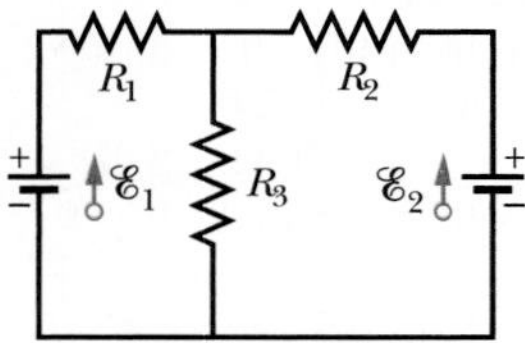

Fig. 27-41 Problems 30, 41, and 88.

••31 SSM GO In Fig. 27-42, the ideal batteries have emfs $\mathscr{E}_1 = 5.0$ V and $\mathscr{E}_2 = 12$ V, the resistances are each 2.0 Ω, and the potential is defined to be zero at the grounded point of the circuit. What are potentials (a) V_1 and (b) V_2 at the indicated points?

Fig. 27-42 Problem 31.

••32 Both batteries in Fig. 27-43a are ideal. Emf $\mathscr{E}_1$ of battery 1 has a fixed value, but emf $\mathscr{E}_2$ of battery 2 can be varied between 1.0 V

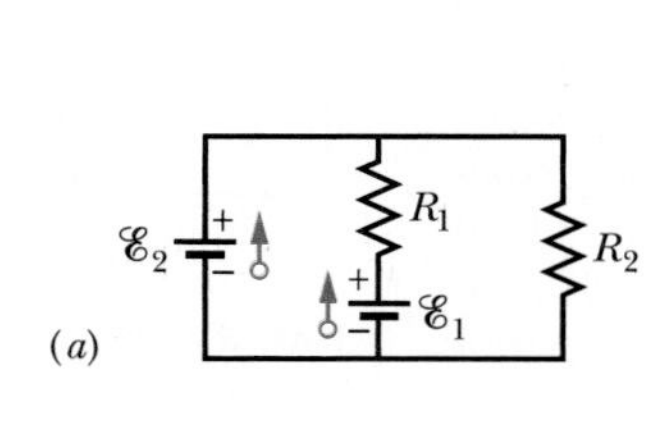

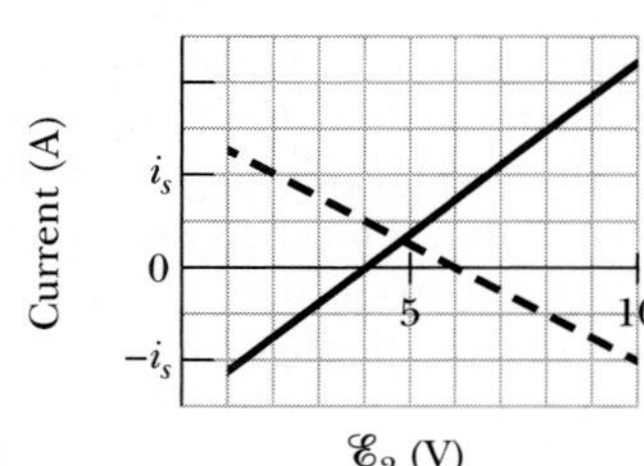

Fig. 27-43 Problem 32.

and 10 V. The plots in Fig. 27-43*b* give the currents through the two batteries as a function of $\mathscr{E}_2$. The vertical scale is set by i_s = 0.20 A. You must decide which plot corresponds to which battery, but for both plots, a negative current occurs when the direction of the current through the battery is opposite the direction of that battery's emf. What are (a) emf $\mathscr{E}_1$, (b) resistance R_1, and (c) resistance R_2?

••33 GO In Fig. 27-44, the current in resistance 6 is i_6 = 1.40 A and the resistances are $R_1 = R_2 = R_3$ = 2.00 Ω, R_4 = 16.0 Ω, R_5 = 8.00 Ω, and R_6 = 4.00 Ω. What is the emf of the ideal battery?

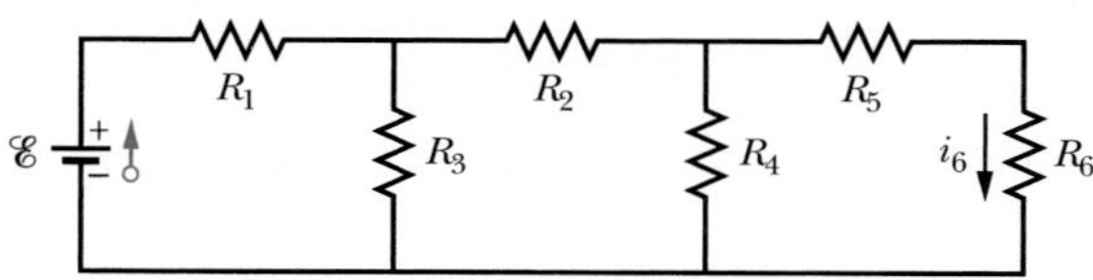

Fig. 27-44 Problem 33.

••34 The resistances in Figs. 27-45*a* and *b* are all 6.0 Ω, and the batteries are ideal 12 V batteries. (a) When switch S in Fig. 27-45*a* is closed, what is the change in the electric potential V_1 across resistor 1, or does V_1 remain the same? (b) When switch S in Fig. 27-45*b* is closed, what is the change in V_1 across resistor 1, or does V_1 remain the same?

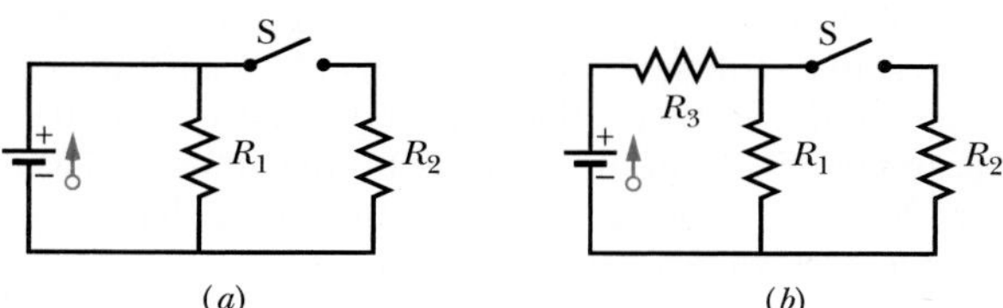

Fig. 27-45 Problem 34.

••35 In Fig. 27-46, $\mathscr{E}$ = 12.0 V, R_1 = 2000 Ω, R_2 = 3000 Ω, and R_3 = 4000 Ω. What are the potential differences (a) $V_A - V_B$, (b) $V_B - V_C$, (c) $V_C - V_D$, and (d) $V_A - V_C$?

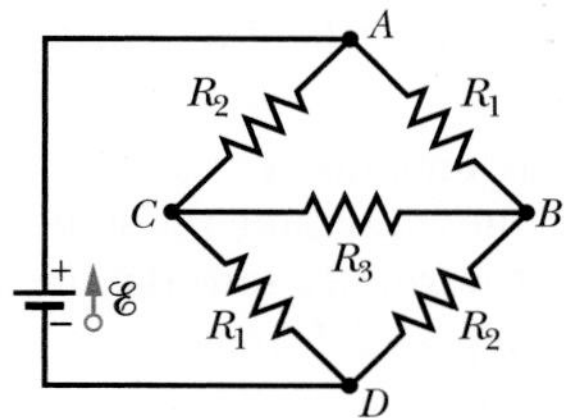

Fig. 27-46 Problem 35.

••36 GO In Fig. 27-47, $\mathscr{E}_1$ = 6.00 V, $\mathscr{E}_2$ = 12.0 V, R_1 = 100 Ω, R_2 = 200 Ω, and R_3 = 300 Ω. One point of the circuit is grounded (V = 0). What are the (a) size and (b) direction (up or down) of the current through resistance 1, the (c) size and (d) direction (left or right) of the current through resistance 2, and the (e) size and (f) direction of the current through resistance 3? (g) What is the electric potential at point *A*?

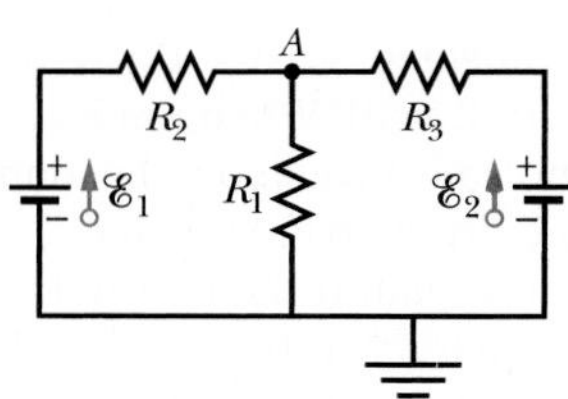

Fig. 27-47 Problem 36.

••37 In Fig. 27-48, the resistances are R_1 = 2.00 Ω, R_2 = 5.00 Ω, and the battery is ideal. What value of R_3 maximizes the dissipation rate in resistance 3?

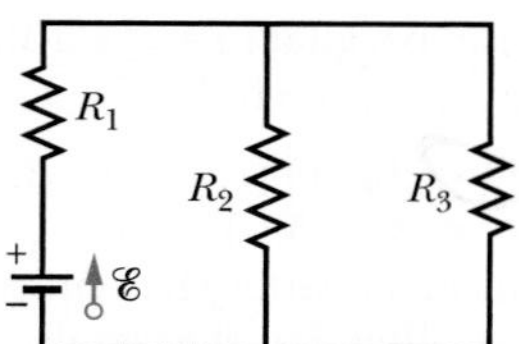

Fig. 27-48 Problems 37 and 98.

••38 Figure 27-49 shows a section of a circuit. The resistances are R_1 = 2.0 Ω, R_2 = 4.0 Ω, and R_3 = 6.0 Ω, and the indicated current is i = 6.0 A. The electric potential difference between points *A* and *B* that connect the section to the rest of the circuit is $V_A - V_B$ = 78 V. (a) Is the device represented by "Box" absorbing or providing energy to the circuit, and (b) at what rate?

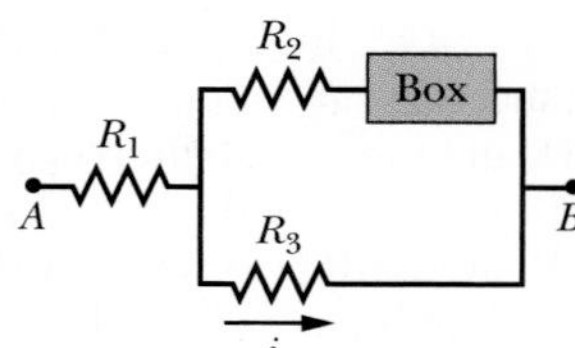

Fig. 27-49 Problem 38.

••39 GO In Fig. 27-50, two batteries of emf $\mathscr{E}$ = 12.0 V and internal resistance r = 0.300 Ω are connected in parallel across a resistance R. (a) For what value of R is the dissipation rate in the resistor a maximum? (b) What is that maximum?

••40 Two identical batteries of emf $\mathscr{E}$ = 12.0 V and internal resistance r = 0.200 Ω are to be connected to an external resistance R, either in parallel (Fig. 27-50) or in series (Fig. 27-51). If $R = 2.00r$, what is the current i in the external resistance in the (a) parallel and (b) series arrangements? (c) For which arrangement is i greater? If $R = r/2.00$, what is i in the external resistance in the (d) parallel and (e) series arrangements? (f) For which arrangement is i greater now?

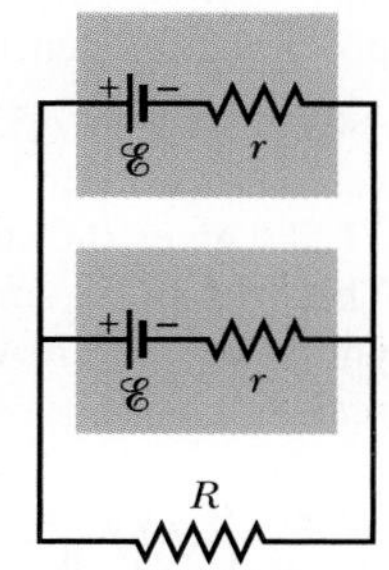

Fig. 27-50 Problems 39 and 40.

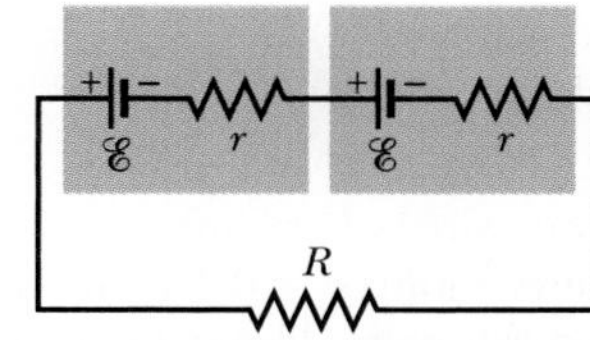

Fig. 27-51 Problem 40.

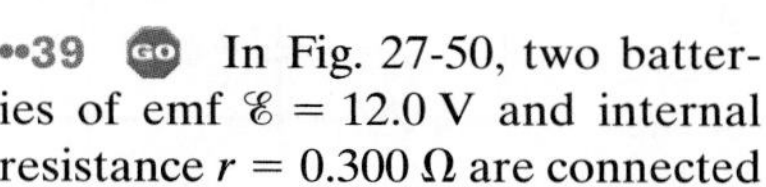

••41 In Fig. 27-41, $\mathscr{E}_1$ = 3.00 V, $\mathscr{E}_2$ = 1.00 V, R_1 = 4.00 Ω, R_2 = 2.00 Ω, R_3 = 5.00 Ω, and both batteries are ideal. What is the rate at which energy is dissipated in (a) R_1, (b) R_2, and (c) R_3? What is the power of (d) battery 1 and (e) battery 2?

••42 In Fig. 27-52, an array of n parallel resistors is connected in series to a resistor and an ideal battery. All the resistors have the same resistance. If an identical resistor were added in parallel to the parallel array, the current through the battery would change by 1.25%. What is the value of n?

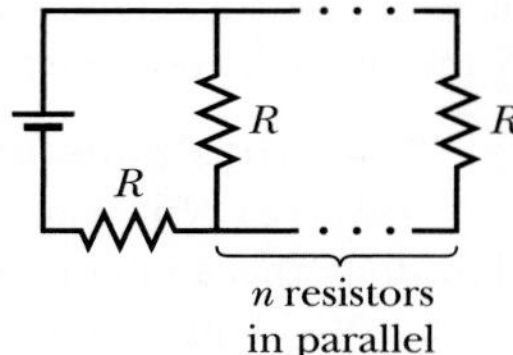

Fig. 27-52 Problem 42.

••43 You are given a number of 10 Ω resistors, each capable of dissipating only 1.0 W without being destroyed. What is the minimum number of such resistors that you need to combine in series or in parallel to make a 10 Ω resistance that is capable of dissipating at least 5.0 W?

••44 In Fig. 27-53, $R_1 = 100\ \Omega$, $R_2 = R_3 = 50.0\ \Omega$, $R_4 = 75.0\ \Omega$, and the ideal battery has emf $\mathscr{E} = 6.00$ V. (a) What is the equivalent resistance? What is i in (b) resistance 1, (c) resistance 2, (d) resistance 3, and (e) resistance 4?

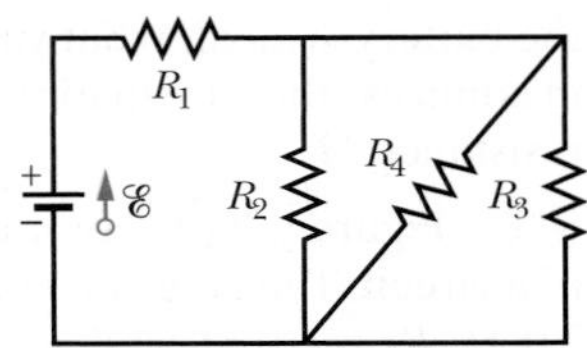

Fig. 27-53 Problems 44 and 48.

••45 ILW In Fig. 27-54, the resistances are $R_1 = 1.0\ \Omega$ and $R_2 = 2.0\ \Omega$, and the ideal batteries have emfs $\mathscr{E}_1 = 2.0$ V and $\mathscr{E}_2 = \mathscr{E}_3 = 4.0$ V. What are the (a) size and (b) direction (up or down) of the current in battery 1, the (c) size and (d) direction of the current in battery 2, and the (e) size and (f) direction of the current in battery 3? (g) What is the potential difference $V_a - V_b$?

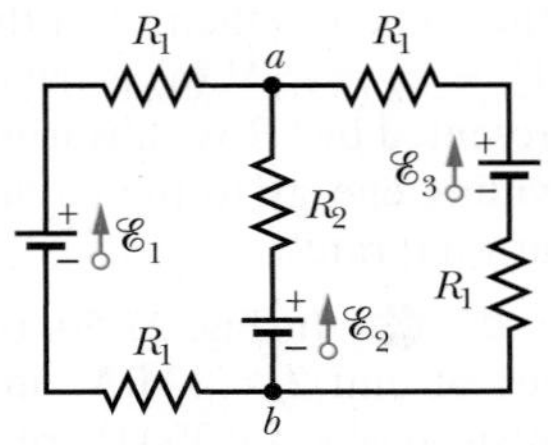

Fig. 27-54 Problem 45.

••46 In Fig. 27-55a, resistor 3 is a variable resistor and the ideal battery has emf $\mathscr{E} = 12$ V. Figure 27-55b gives the current i through the battery as a function of R_3. The horizontal scale is set by $R_{3s} = 20\ \Omega$. The curve has an asymptote of 2.0 mA as $R_3 \to \infty$. What are (a) resistance R_1 and (b) resistance R_2?

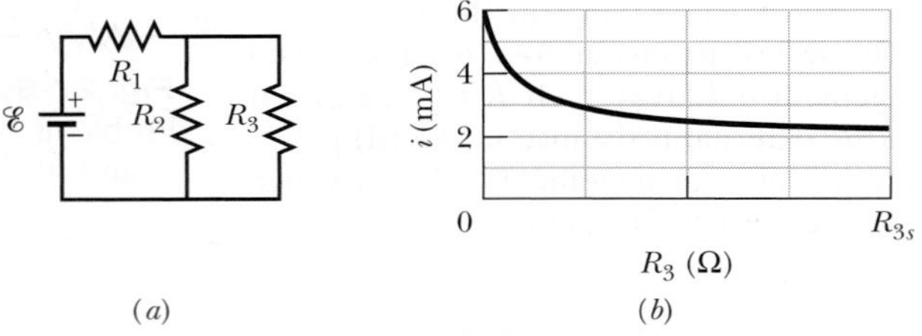

Fig. 27-55 Problem 46.

•••47 SSM A copper wire of radius $a = 0.250$ mm has an aluminum jacket of outer radius $b = 0.380$ mm. There is a current $i = 2.00$ A in the composite wire. Using Table 26-1, calculate the current in (a) the copper and (b) the aluminum. (c) If a potential difference $V = 12.0$ V between the ends maintains the current, what is the length of the composite wire?

•••48 In Fig. 27-53, the resistors have the values $R_1 = 7.00\ \Omega$, $R_2 = 12.0\ \Omega$, and $R_3 = 4.00\ \Omega$, and the ideal battery's emf is $\mathscr{E} = 24.0$ V. For what value of R_4 will the rate at which the battery transfers energy to the resistors equal (a) 60.0 W, (b) the maximum possible rate P_{max}, and (c) the minimum possible rate P_{min}? What are (d) P_{max} and (e) P_{min}?

sec. 27-8 The Ammeter and the Voltmeter

••49 ILW (a) In Fig. 27-56, what does the ammeter read if $\mathscr{E} = 5.0$ V (ideal battery), $R_1 = 2.0\ \Omega$, $R_2 = 4.0\ \Omega$, and $R_3 = 6.0\ \Omega$? (b) The ammeter and battery are now interchanged. Show that the ammeter reading is unchanged.

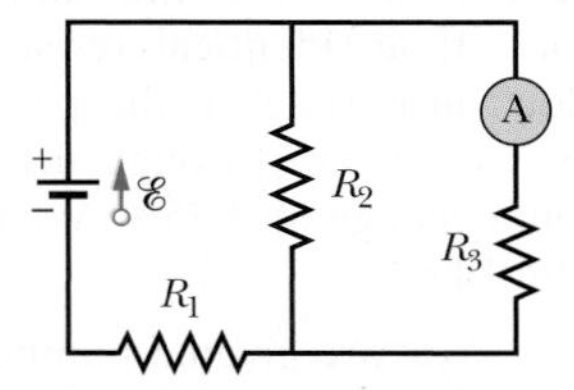

Fig. 27-56 Problem 49.

••50 In Fig. 27-57, $R_1 = 2.00R$, the ammeter resistance is zero, and the battery is ideal. What multiple of $\mathscr{E}/R$ gives the current in the ammeter?

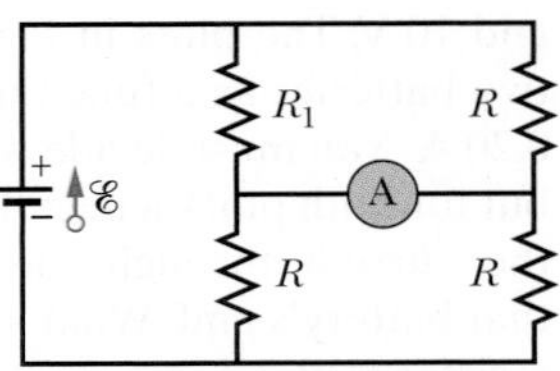

Fig. 27-57 Problem 50.

••51 In Fig. 27-58, a voltmeter of resistance $R_V = 300\ \Omega$ and an ammeter of resistance $R_A = 3.00\ \Omega$ are being used to measure a resistance R in a circuit that also contains a resistance $R_0 = 100\ \Omega$ and an ideal battery of emf $\mathscr{E} = 12.0$ V. Resistance R is given by $R = V/i$, where V is the potential across R and i is the ammeter reading. The voltmeter reading is V', which is V plus the potential difference across the ammeter. Thus, the ratio of the two meter readings is not R but only an *apparent* resistance $R' = V'/i$. If $R = 85.0\ \Omega$, what are (a) the ammeter reading, (b) the voltmeter reading, and (c) R'? (d) If R_A is decreased, does the difference between R' and R increase, decrease, or remain the same?

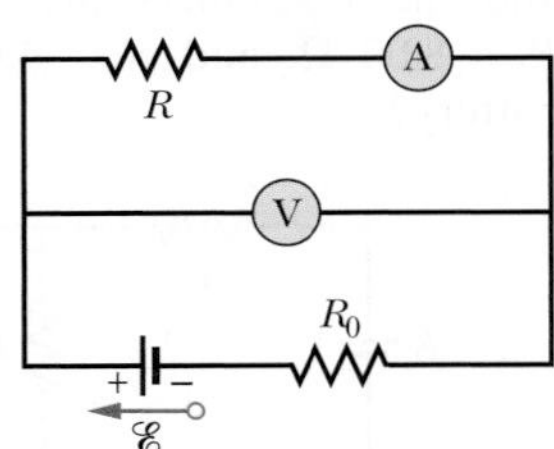

Fig. 27-58 Problem 51.

••52 A simple ohmmeter is made by connecting a 1.50 V flashlight battery in series with a resistance R and an ammeter that reads from 0 to 1.00 mA, as shown in Fig. 27-59. Resistance R is adjusted so that when the clip leads are shorted together, the meter deflects to its full-scale value of 1.00 mA. What external resistance across the leads results in a deflection of (a) 10.0%, (b) 50.0%, and (c) 90.0% of full scale? (d) If the ammeter has a resistance of 20.0 Ω and the internal resistance of the battery is negligible, what is the value of R?

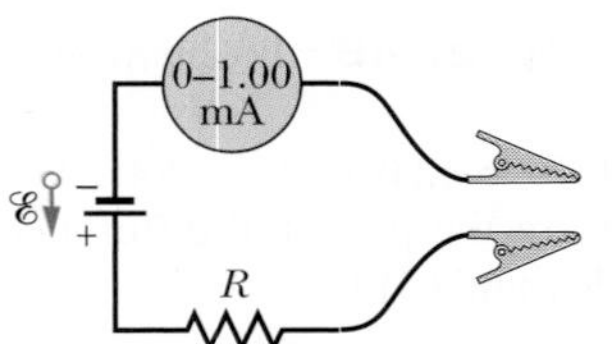

Fig. 27-59 Problem 52.

••53 In Fig. 27-14, assume that $\mathscr{E} = 3.0$ V, $r = 100\ \Omega$, $R_1 = 250\ \Omega$, and $R_2 = 300\ \Omega$. If the voltmeter resistance R_V is 5.0 kΩ, what percent error does it introduce into the measurement of the potential difference across R_1? Ignore the presence of the ammeter.

••54 When the lights of a car are switched on, an ammeter in series with them reads 10.0 A and a voltmeter connected across them reads 12.0 V (Fig. 27-60). When the electric starting motor is turned on, the ammeter reading drops to 8.00 A and the lights dim somewhat. If the internal resistance of the battery is 0.0500 Ω and that of the ammeter is negligible, what are (a) the emf of the battery and (b) the current through the starting motor when the lights are on?

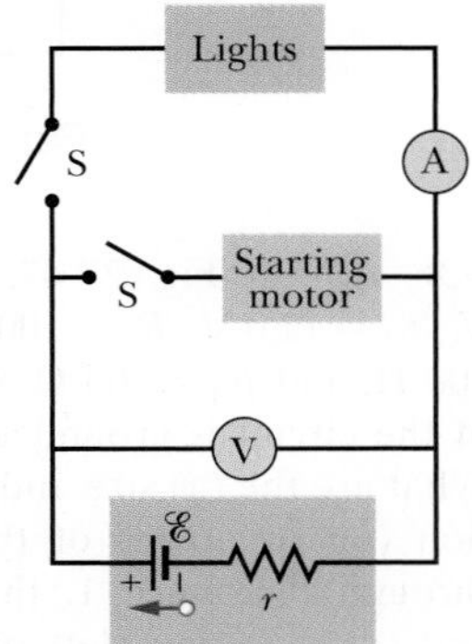

Fig. 27-60 Problem 54.

••55 In Fig. 27-61, R_s is to be adjusted in value by moving the sliding contact across it until points a and b are brought to the same potential. (One tests for this condition by momentarily connecting a sensitive ammeter between a and b; if these points are at the same potential, the ammeter will not deflect.) Show that when this adjustment is made, the following relation holds: $R_x = R_sR_2/R_1$. An unknown resistance (R_x) can be measured in terms of a standard (R_s) using this device, which is called a Wheatstone bridge.

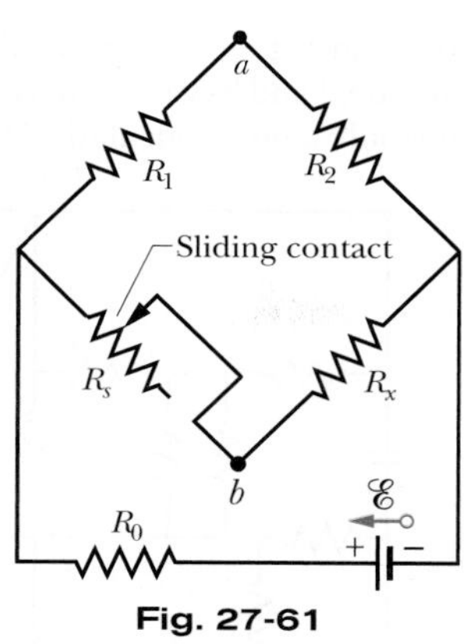

Fig. 27-61 Problem 55.

••56 In Fig. 27-62, a voltmeter of resistance $R_V = 300\ \Omega$ and an ammeter of resistance $R_A = 3.00\ \Omega$ are being used to measure a resistance R in a circuit that also contains a resistance $R_0 = 100\ \Omega$ and an ideal battery of emf $\mathscr{E} = 12.0$ V. Resistance R is given by $R = V/i$, where V is the voltmeter reading and i is the current in resistance R. However, the ammeter reading is not i but rather i', which is i plus the current through the voltmeter. Thus, the ratio of the two meter readings is not R but only an *apparent* resistance $R' = V/i'$. If $R = 85.0\ \Omega$, what are (a) the ammeter reading, (b) the voltmeter reading, and (c) R'? (d) If R_V is increased, does the difference between R' and R increase, decrease, or remain the same?

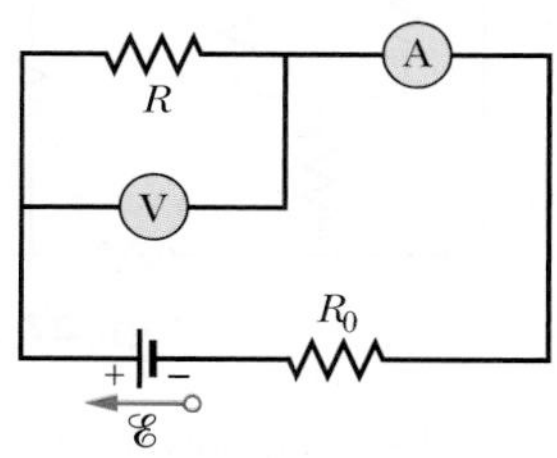

Fig. 27-62 Problem 56.

sec. 27-9 *RC* Circuits

•57 Switch S in Fig. 27-63 is closed at time $t = 0$, to begin charging an initially uncharged capacitor of capacitance $C = 15.0\ \mu$F through a resistor of resistance $R = 20.0\ \Omega$. At what time is the potential across the capacitor equal to that across the resistor?

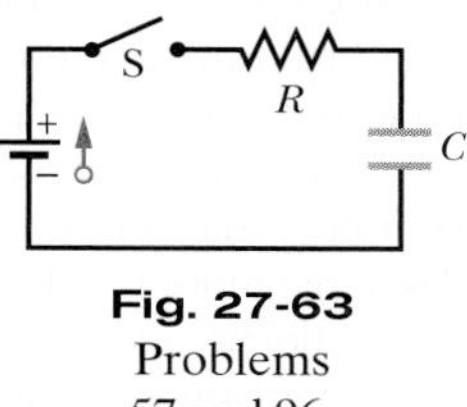

Fig. 27-63 Problems 57 and 96.

•58 In an *RC* series circuit, emf $\mathscr{E} = 12.0$ V, resistance $R = 1.40$ MΩ, and capacitance $C = 1.80\ \mu$F. (a) Calculate the time constant. (b) Find the maximum charge that will appear on the capacitor during charging. (c) How long does it take for the charge to build up to 16.0 μC?

•59 SSM What multiple of the time constant τ gives the time taken by an initially uncharged capacitor in an *RC* series circuit to be charged to 99.0% of its final charge?

•60 A capacitor with initial charge q_0 is discharged through a resistor. What multiple of the time constant τ gives the time the capacitor takes to lose (a) the first one-third of its charge and (b) two-thirds of its charge?

•61 ILW A 15.0 kΩ resistor and a capacitor are connected in series, and then a 12.0 V potential difference is suddenly applied across them. The potential difference across the capacitor rises to 5.00 V in 1.30 μs. (a) Calculate the time constant of the circuit. (b) Find the capacitance of the capacitor.

••62 Figure 27-64 shows the circuit of a flashing lamp, like those attached to barrels at highway construction sites. The fluorescent lamp L (of negligible capacitance) is connected in parallel across the capacitor C of an *RC* circuit. There is a current through the lamp only when the potential difference across it reaches the breakdown voltage V_L; then the capacitor discharges completely through the lamp and the lamp flashes briefly. For a lamp with breakdown voltage $V_L = 72.0$ V, wired to a 95.0 V ideal battery and a 0.150 μF capacitor, what resistance R is needed for two flashes per second?

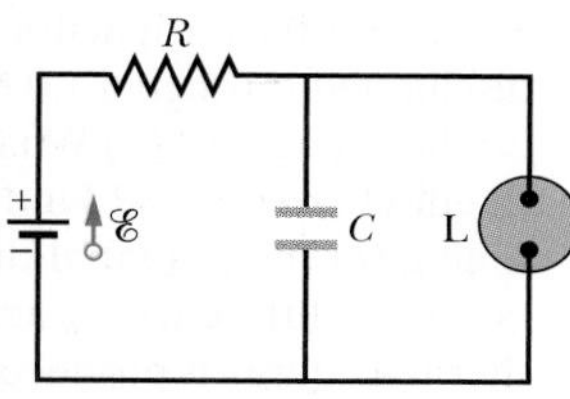

Fig. 27-64 Problem 62.

••63 SSM WWW In the circuit of Fig. 27-65, $\mathscr{E} = 1.2$ kV, $C = 6.5\ \mu$F, $R_1 = R_2 = R_3 = 0.73$ MΩ. With C completely uncharged, switch S is suddenly closed (at $t = 0$). At $t = 0$, what are (a) current i_1 in resistor 1, (b) current i_2 in resistor 2, and (c) current i_3 in resistor 3? At $t = \infty$ (that is, after many time constants), what are (d) i_1, (e) i_2, and (f) i_3? What is the potential difference V_2 across resistor 2 at (g) $t = 0$ and (h) $t = \infty$? (i) Sketch V_2 versus t between these two extreme times.

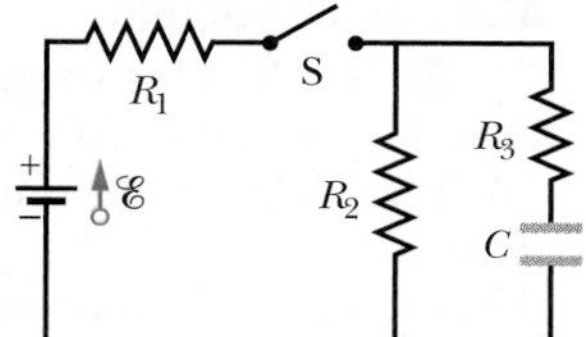

Fig. 27-65 Problem 63.

••64 A capacitor with an initial potential difference of 100 V is discharged through a resistor when a switch between them is closed at $t = 0$. At $t = 10.0$ s, the potential difference across the capacitor is 1.00 V. (a) What is the time constant of the circuit? (b) What is the potential difference across the capacitor at $t = 17.0$ s?

••65 GO In Fig. 27-66, $R_1 = 10.0$ kΩ, $R_2 = 15.0$ kΩ, $C = 0.400\ \mu$F, and the ideal battery has emf $\mathscr{E} = 20.0$ V. First, the switch is closed a long time so that the steady state is reached. Then the switch is opened at time $t = 0$. What is the current in resistor 2 at $t = 4.00$ ms?

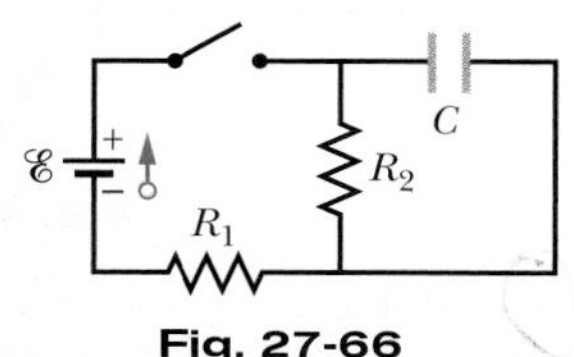

Fig. 27-66 Problems 65 and 99.

••66 Figure 27-67 displays two circuits with a charged capacitor that is to be discharged through a resistor when a switch is closed. In Fig. 27-67*a*, $R_1 = 20.0\ \Omega$ and $C_1 = 5.00\ \mu$F. In Fig. 27-67*b*, $R_2 = 10.0\ \Omega$ and $C_2 = 8.00\ \mu$F. The ratio of the initial charges on the two capacitors is $q_{02}/q_{01} = 1.50$. At time $t = 0$, both switches are closed. At what time t do the two capacitors have the same charge?

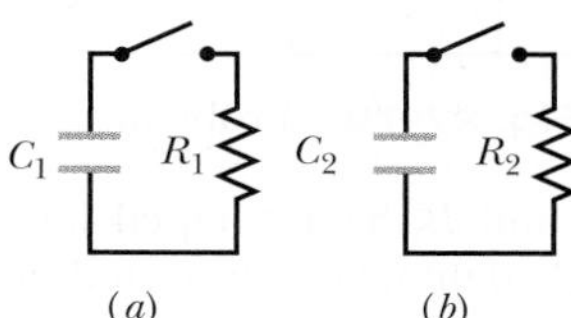

Fig. 27-67 Problem 66.

••67 The potential difference between the plates of a leaky (meaning that charge leaks from one plate to the other) 2.0 μF capacitor drops to one-fourth its initial value in 2.0 s. What is the equivalent resistance between the capacitor plates?

••68 A 1.0 μF capacitor with an initial stored energy of 0.50 J is discharged through a 1.0 MΩ resistor. (a) What is the initial charge on the capacitor? (b) What is the current through the resistor when the discharge starts? Find an expression that gives, as a function of time t, (c) the potential difference V_C across the capacitor, (d) the potential difference V_R across the resistor, and (e) the rate at which thermal energy is produced in the resistor.

•••69 A 3.00 MΩ resistor and a 1.00 μF capacitor are connected in series with an ideal battery of emf $\mathscr{E}$ = 4.00 V. At 1.00 s after the connection is made, what is the rate at which (a) the charge of the capacitor is increasing, (b) energy is being stored in the capacitor, (c) thermal energy is appearing in the resistor, and (d) energy is being delivered by the battery?

Additional Problems

70 GO Each of the six real batteries in Fig. 27-68 has an emf of 20 V and a resistance of 4.0 Ω. (a) What is the current through the (external) resistance R = 4.0 Ω? (b) What is the potential difference across each battery? (c) What is the power of each battery? (d) At what rate does each battery transfer energy to internal thermal energy?

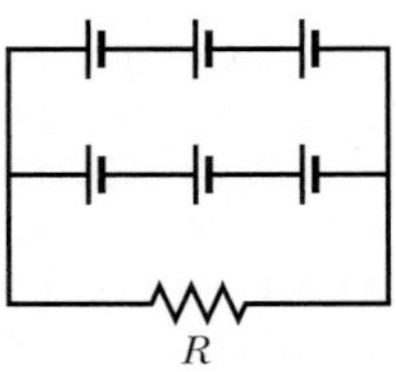

Fig. 27-68 Problem 70.

71 In Fig. 27-69, R_1 = 20.0 Ω, R_2 = 10.0 Ω, and the ideal battery has emf $\mathscr{E}$ = 120 V. What is the current at point a if we close (a) only switch S_1, (b) only switches S_1 and S_2, and (c) all three switches?

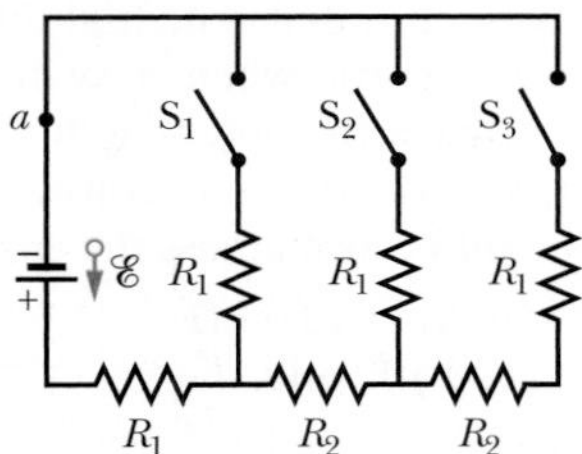

Fig. 27-69 Problem 71.

72 In Fig. 27-70, the ideal battery has emf $\mathscr{E}$ = 30.0 V, and the resistances are $R_1 = R_2$ = 14 Ω, $R_3 = R_4 = R_5$ = 6.0 Ω, R_6 = 2.0 Ω, and R_7 = 1.5 Ω. What are currents (a) i_2, (b) i_4, (c) i_1, (d) i_3, and (e) i_5?

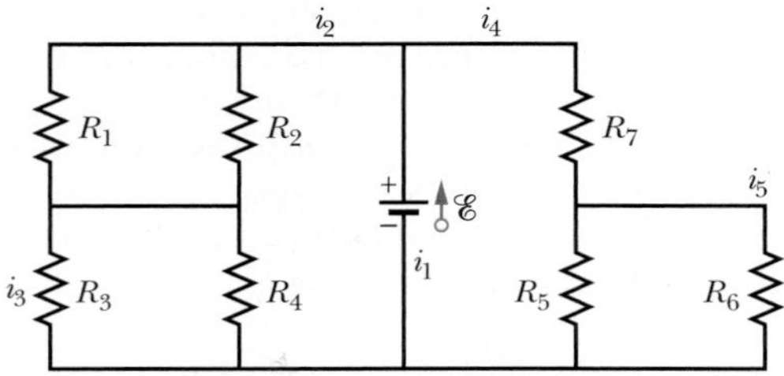

Fig. 27-70 Problem 72.

73 SSM Wires A and B, having equal lengths of 40.0 m and equal diameters of 2.60 mm, are connected in series. A potential difference of 60.0 V is applied between the ends of the composite wire. The resistances are R_A = 0.127 Ω and R_B = 0.729 Ω. For wire A, what are (a) magnitude J of the current density and (b) potential difference V? (c) Of what type material is wire A made (see Table 26-1)? For wire B, what are (d) J and (e) V? (f) Of what type material is B made?

74 What are the (a) size and (b) direction (up or down) of current i in Fig. 27-71, where all resistances are 4.0 Ω and all batteries are ideal and have an emf of 10 V? (*Hint:* This can be answered using only mental calculation.)

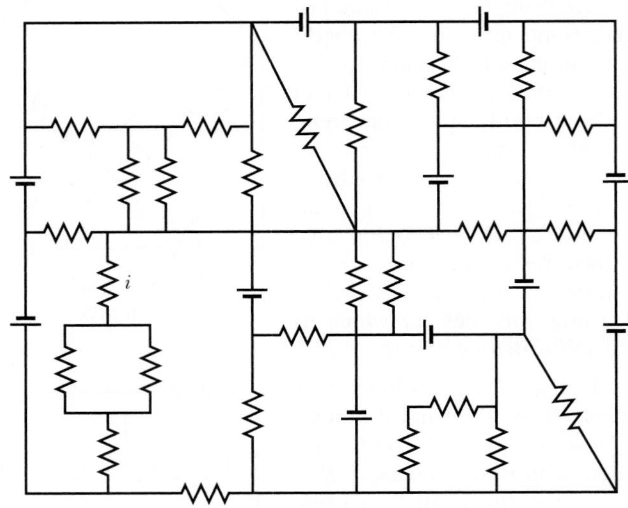

Fig. 27-71 Problem 74.

75 Suppose that, while you are sitting in a chair, charge separation between your clothing and the chair puts you at a potential of 200 V, with the capacitance between you and the chair at 150 pF. When you stand up, the increased separation between your body and the chair decreases the capacitance to 10 pF. (a) What then is the potential of your body? That potential is reduced over time, as the charge on you drains through your body and shoes (you are a capacitor discharging through a resistance). Assume that the resistance along that route is 300 GΩ. If you touch an electrical component while your potential is greater than 100 V, you could ruin the component. (b) How long must you wait until your potential reaches the safe level of 100 V?

If you wear a conducting wrist strap that is connected to ground, your potential does not increase as much when you stand up; you also discharge more rapidly because the resistance through the grounding connection is much less than through your body and shoes. (c) Suppose that when you stand up, your potential is 1400 V and the chair-to-you capacitance is 10 pF. What resistance in that wrist-strap grounding connection will allow you to discharge to 100 V in 0.30 s, which is less time than you would need to reach for, say, your computer?

76 GO In Fig. 27-72, the ideal batteries have emfs $\mathscr{E}_1$ = 20.0 V,

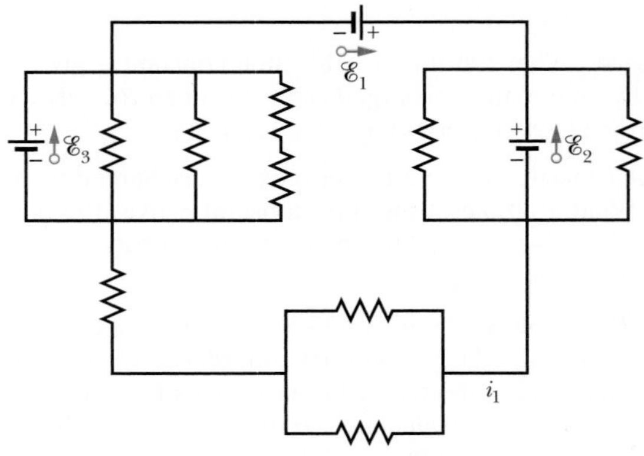

Fig. 27-71 Problem 76.

$\mathscr{E}_2 = 10.0$ V, and $\mathscr{E}_3 = 5.00$ V, and the resistances are each 2.00 Ω. What are the (a) size and (b) direction (left or right) of current i_1? (c) Does battery 1 supply or absorb energy, and (d) what is its power? (e) Does battery 2 supply or absorb energy, and (f) what is its power? (g) Does battery 3 supply or absorb energy, and (h) what is its power?

77 SSM A temperature-stable resistor is made by connecting a resistor made of silicon in series with one made of iron. If the required total resistance is 1000 Ω in a wide temperature range around 20°C, what should be the resistance of the (a) silicon resistor and (b) iron resistor? (See Table 26-1.)

78 In Fig. 27-14, assume that $\mathscr{E} = 5.0$ V, $r = 2.0$ Ω, $R_1 = 5.0$ Ω, and $R_2 = 4.0$ Ω. If the ammeter resistance R_A is 0.10 Ω, what percent error does it introduce into the measurement of the current? Assume that the voltmeter is not present.

79 SSM An initially uncharged capacitor C is fully charged by a device of constant emf $\mathscr{E}$ connected in series with a resistor R. (a) Show that the final energy stored in the capacitor is half the energy supplied by the emf device. (b) By direct integration of i^2R over the charging time, show that the thermal energy dissipated by the resistor is also half the energy supplied by the emf device.

80 In Fig. 27-73, $R_1 = 5.00$ Ω, $R_2 = 10.0$ Ω, $R_3 = 15.0$ Ω, $C_1 = 5.00$ μF, $C_2 = 10.0$ μF, and the ideal battery has emf $\mathscr{E} = 20.0$ V. Assuming that the circuit is in the steady state, what is the total energy stored in the two capacitors?

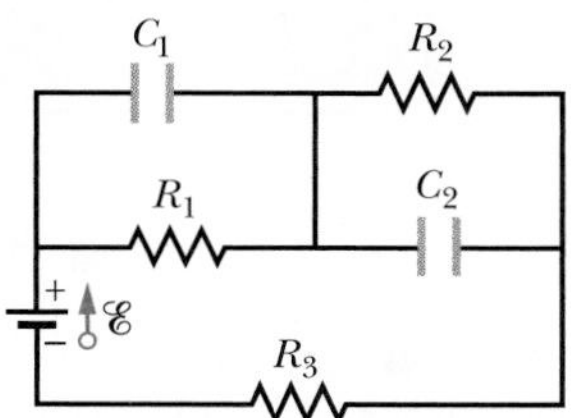

Fig. 27-73 Problem 80.

81 In Fig. 27-5*a*, find the potential difference across R_2 if $\mathscr{E} = 12$ V, $R_1 = 3.0$ Ω, $R_2 = 4.0$ Ω, and $R_3 = 5.0$ Ω.

82 In Fig. 27-8*a*, calculate the potential difference between a and c by considering a path that contains R, r_1, and $\mathscr{E}_1$.

83 SSM A controller on an electronic arcade game consists of a variable resistor connected across the plates of a 0.220 μF capacitor. The capacitor is charged to 5.00 V, then discharged through the resistor. The time for the potential difference across the plates to decrease to 0.800 V is measured by a clock inside the game. If the range of discharge times that can be handled effectively is from 10.0 μs to 6.00 ms, what should be the (a) lower value and (b) higher value of the resistance range of the resistor?

84 An automobile gasoline gauge is shown schematically in Fig. 27-74. The indicator (on the dashboard) has a resistance of 10 Ω. The tank unit is a float connected to a variable resistor whose resistance varies linearly with the volume of gasoline. The resistance is 140 Ω when the tank is empty and 20 Ω when the tank is full. Find the current in the circuit when the tank is (a) empty, (b) half-full, and (c) full. Treat the battery as ideal.

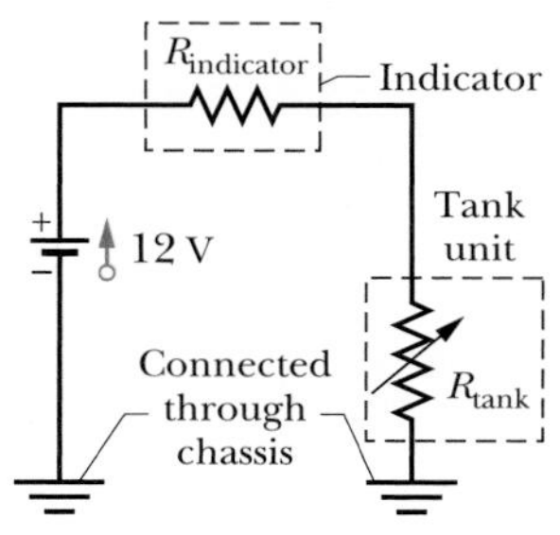

Fig. 27-74 Problem 84.

85 SSM The starting motor of a car is turning too slowly, and the mechanic has to decide whether to replace the motor, the cable, or the battery. The car's manual says that the 12 V battery should have no more than 0.020 Ω internal resistance, the motor no more than 0.200 Ω resistance, and the cable no more than 0.040 Ω resistance. The mechanic turns on the motor and measures 11.4 V across the battery, 3.0 V across the cable, and a current of 50 A. Which part is defective?

86 Two resistors R_1 and R_2 may be connected either in series or in parallel across an ideal battery with emf $\mathscr{E}$. We desire the rate of energy dissipation of the parallel combination to be five times that of the series combination. If $R_1 = 100$ Ω, what are the (a) smaller and (b) larger of the two values of R_2 that result in that dissipation rate?

87 The circuit of Fig. 27-75 shows a capacitor, two ideal batteries, two resistors, and a switch S. Initially S has been open for a long time. If it is then closed for a long time, what is the change in the charge on the capacitor? Assume $C = 10$ μF, $\mathscr{E}_1 = 1.0$ V, $\mathscr{E}_2 = 3.0$ V, $R_1 = 0.20$ Ω, and $R_2 = 0.40$ Ω.

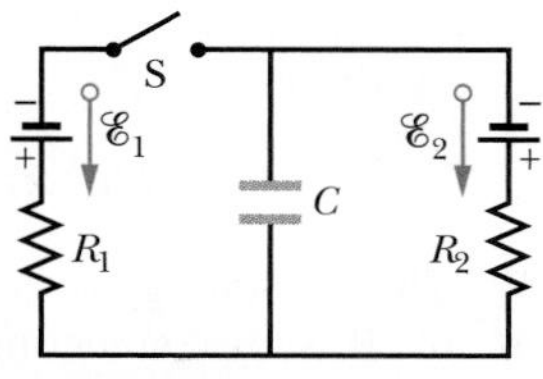

Fig. 27-75 Problem 87.

88 In Fig. 27-41, $R_1 = 10.0$ Ω, $R_2 = 20.0$ Ω, and the ideal batteries have emfs $\mathscr{E}_1 = 20.0$ V and $\mathscr{E}_2 = 50.0$ V. What value of R_3 results in no current through battery 1?

89 In Fig. 27-76, $R = 10$ Ω. What is the equivalent resistance between points A and B? (*Hint:* This circuit section might look simpler if you first assume that points A and B are connected to a battery.)

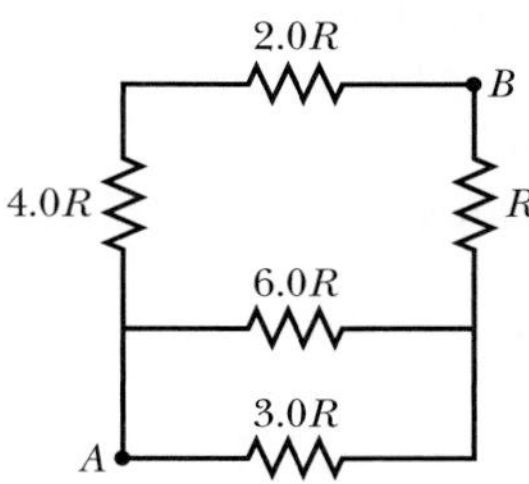

Fig. 27-76 Problem 89.

90 (a) In Fig. 27-4*a*, show that the rate at which energy is dissipated in R as thermal energy is a maximum when $R = r$. (b) Show that this maximum power is $P = \mathscr{E}^2/4r$.

91 In Fig. 27-77, the ideal batteries have emfs $\mathscr{E}_1 = 12.0$ V and

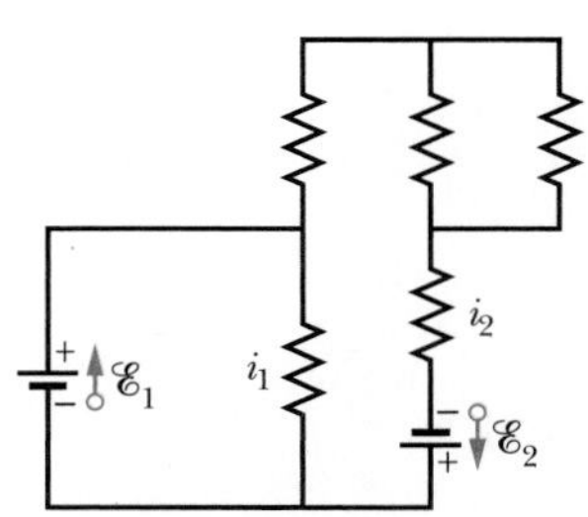

Fig. 27-77 Problem 91.

$\mathscr{E}_2 = 4.00$ V, and the resistances are each 4.00 Ω. What are the (a) size and (b) direction (up or down) of i_1 and the (c) size and (d) direction of i_2? (e) Does battery 1 supply or absorb energy, and (f) what is its energy transfer rate? (g) Does battery 2 supply or absorb energy, and (h) what is its energy transfer rate?

92 Figure 27-78 shows a portion of a circuit through which there is a current $I = 6.00$ A. The resistances are $R_1 = R_2 = 2.00R_3 = 2.00R_4 = 4.00$ Ω. What is the current i_1 through resistor 1?

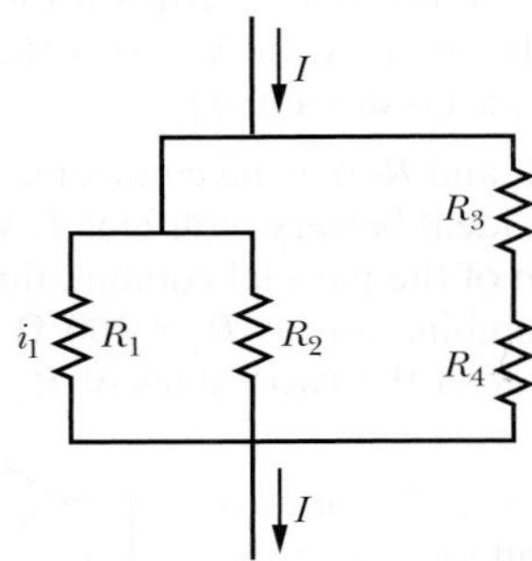

Fig. 27-78 Problem 92.

93 Thermal energy is to be generated in a 0.10 Ω resistor at the rate of 10 W by connecting the resistor to a battery whose emf is 1.5 V. (a) What potential difference must exist across the resistor? (b) What must be the internal resistance of the battery?

94 Figure 27-79 shows three 20.0 Ω resistors. Find the equivalent resistance between points (a) A and B, (b) A and C, and (c) B and C. (*Hint:* Imagine that a battery is connected between a given pair of points.)

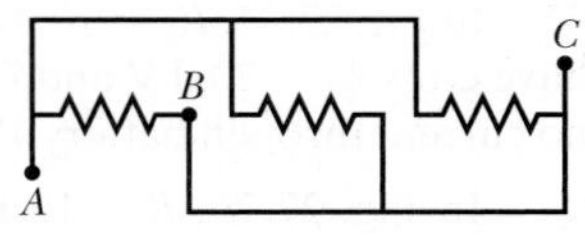

Fig. 27-79 Problem 94.

95 A 120 V power line is protected by a 15 A fuse. What is the maximum number of 500 W lamps that can be simultaneously operated in parallel on this line without "blowing" the fuse because of an excess of current?

96 Figure 27-63 shows an ideal battery of emf $\mathscr{E} = 12$ V, a resistor of resistance $R = 4.0$ Ω, and an uncharged capacitor of capacitance $C = 4.0$ μF. After switch S is closed, what is the current through the resistor when the charge on the capacitor is 8.0 μC?

97 SSM A group of N identical batteries of emf $\mathscr{E}$ and internal resistance r may be connected all in series (Fig. 27-80a) or all in parallel (Fig. 27-80b) and then across a resistor R. Show that both arrangements give the same current in R if $R = r$.

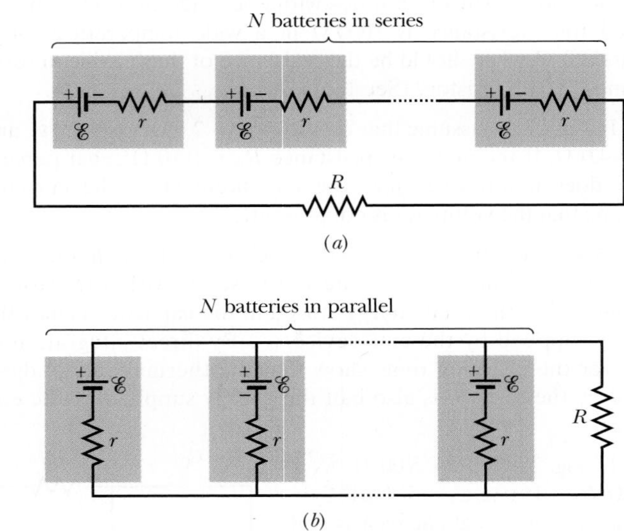

Fig. 27-80 Problem 97.

98 SSM In Fig. 27-48, $R_1 = R_2 = 10.0$ Ω, and the ideal battery has emf $\mathscr{E} = 12.0$ V. (a) What value of R_3 maximizes the rate at which the battery supplies energy and (b) what is that maximum rate?

99 SSM In Fig. 27-66, the ideal battery has emf $\mathscr{E} = 30$ V, the resistances are $R_1 = 20$ kΩ and $R_2 = 10$ kΩ, and the capacitor is uncharged. When the switch is closed at time $t = 0$, what is the current in (a) resistance 1 and (b) resistance 2? (c) A long time later, what is the current in resistance 2?

CHAPTER 28

MAGNETIC FIELDS

28-1 WHAT IS PHYSICS?

As we have discussed, one major goal of physics is the study of how an *electric field* can produce an *electric force* on a charged object. A closely related goal is the study of how a *magnetic field* can produce a *magnetic force* on a (moving) charged particle or on a magnetic object such as a magnet. You may already have a hint of what a magnetic field is if you have ever attached a note to a refrigerator door with a small magnet or accidentally erased a credit card by moving it near a magnet. The magnet acts on the door or credit card via its magnetic field.

The applications of magnetic fields and magnetic forces are countless and changing rapidly every year. Here are just a few examples. For decades, the entertainment industry depended on the magnetic recording of music and images on audiotape and videotape. Although digital technology has largely replaced magnetic recording, the industry still depends on the magnets that control CD and DVD players and computer hard drives; magnets also drive the speaker cones in headphones, TVs, computers, and telephones. A modern car comes equipped with dozens of magnets because they are required in the motors for engine ignition, automatic window control, sunroof control, and windshield wiper control. Most security alarm systems, doorbells, and automatic door latches employ magnets. In short, you are surrounded by magnets.

The science of magnetic fields is physics; the application of magnetic fields is engineering. Both the science and the application begin with the question "What produces a magnetic field?"

28-2 What Produces a Magnetic Field?

Because an electric field $\vec{E}$ is produced by an electric charge, we might reasonably expect that a magnetic field $\vec{B}$ is produced by a magnetic charge. Although individual magnetic charges (called *magnetic monopoles*) are predicted by certain theories, their existence has not been confirmed. How then are magnetic fields produced? There are two ways.

One way is to use moving electrically charged particles, such as a current in a wire, to make an **electromagnet.** The current produces a magnetic field that can be used, for example, to control a computer hard drive or to sort scrap metal (Fig. 28-1). In Chapter 29, we discuss the magnetic field due to a current.

The other way to produce a magnetic field is by means of elementary particles such as electrons because these particles have an *intrinsic* magnetic field around them. That is, the magnetic field is a basic characteristic of each particle

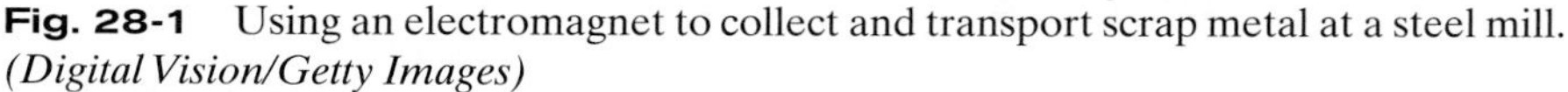

Fig. 28-1 Using an electromagnet to collect and transport scrap metal at a steel mill. *(Digital Vision/Getty Images)*

just as mass and electric charge (or lack of charge) are basic characteristics. As we discuss in Chapter 32, the magnetic fields of the electrons in certain materials add together to give a net magnetic field around the material. Such addition is the reason why a **permanent magnet**, the type used to hang refrigerator notes, has a permanent magnetic field. In other materials, the magnetic fields of the electrons cancel out, giving no net magnetic field surrounding the material. Such cancellation is the reason you do not have a permanent field around your body, which is good because otherwise you might be slammed up against a refrigerator door every time you passed one.

Our first job in this chapter is to define the magnetic field $\vec{B}$. We do so by using the experimental fact that when a charged particle moves through a magnetic field, a magnetic force $\vec{F}_B$ acts on the particle.

28-3 The Definition of $\vec{B}$

We determined the electric field $\vec{E}$ at a point by putting a test particle of charge q at rest at that point and measuring the electric force $\vec{F}_E$ acting on the particle. We then defined $\vec{E}$ as

$$\vec{E} = \frac{\vec{F}_E}{q}. \tag{28-1}$$

If a magnetic monopole were available, we could define $\vec{B}$ in a similar way. Because such particles have not been found, we must define $\vec{B}$ in another way, in terms of the magnetic force $\vec{F}_B$ exerted on a moving electrically charged test particle.

In principle, we do this by firing a charged particle through the point at which $\vec{B}$ is to be defined, using various directions and speeds for the particle and determining the force $\vec{F}_B$ that acts on the particle at that point. After many such trials we would find that when the particle's velocity $\vec{v}$ is along a particular axis through the point, force $\vec{F}_B$ is zero. For all other directions of $\vec{v}$, the magnitude of $\vec{F}_B$ is always proportional to $v \sin \phi$, where ϕ is the angle between the zero-force axis and the direction of $\vec{v}$. Furthermore, the direction of $\vec{F}_B$ is always perpendicular to the direction of $\vec{v}$. (These results suggest that a cross product is involved.)

We can then define a **magnetic field** $\vec{B}$ to be a vector quantity that is directed along the zero-force axis. We can next measure the magnitude of $\vec{F}_B$ when $\vec{v}$ is directed perpendicular to that axis and then define the magnitude of $\vec{B}$ in terms of that force magnitude:

$$B = \frac{F_B}{|q|v},$$

where q is the charge of the particle.

We can summarize all these results with the following vector equation:

$$\vec{F}_B = q\vec{v} \times \vec{B}; \tag{28-2}$$

that is, the force $\vec{F}_B$ on the particle is equal to the charge q times the cross product of its velocity $\vec{v}$ and the field $\vec{B}$ (all measured in the same reference frame). Using Eq. 3-27 for the cross product, we can write the magnitude of $\vec{F}_B$ as

$$F_B = |q|vB \sin \phi, \tag{28-3}$$

where ϕ is the angle between the directions of velocity $\vec{v}$ and magnetic field $\vec{B}$.

Finding the Magnetic Force on a Particle

Equation 28-3 tells us that the magnitude of the force $\vec{F}_B$ acting on a particle in a magnetic field is proportional to the charge q and speed v of the particle. Thus,

Cross $\vec{v}$ into $\vec{B}$ to get the new vector $\vec{v} \times \vec{B}$.

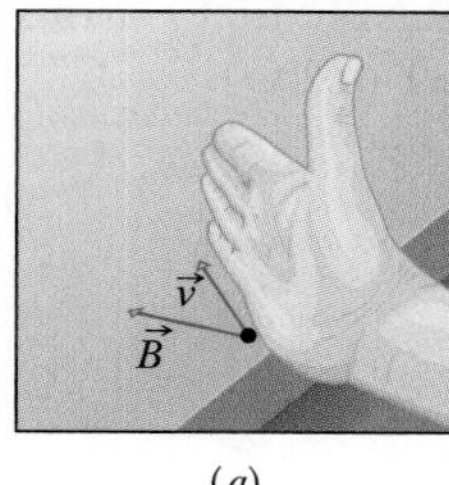

(a)

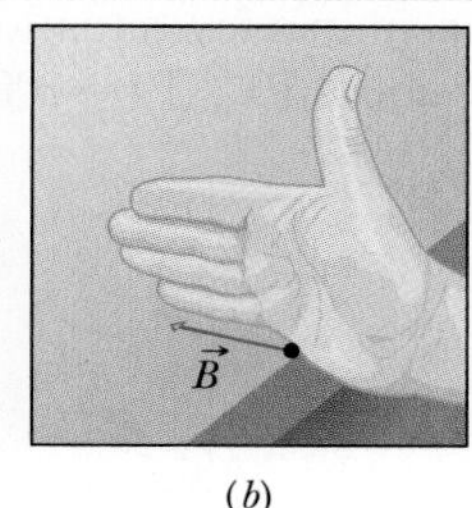

(b)

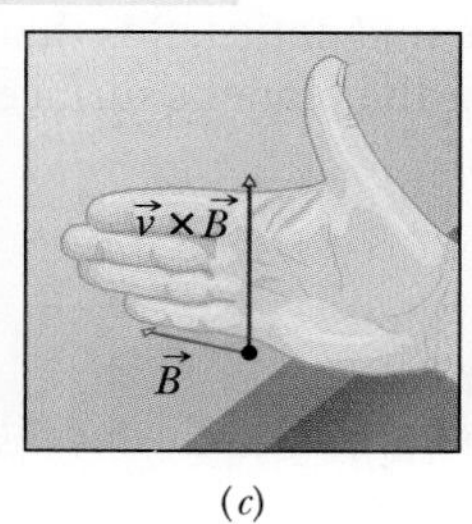

(c)

Force on positive particle

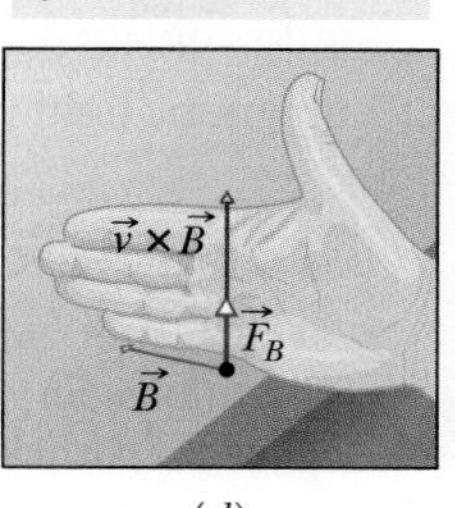

(d)

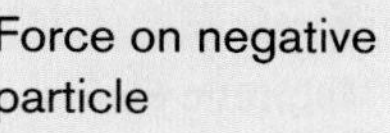

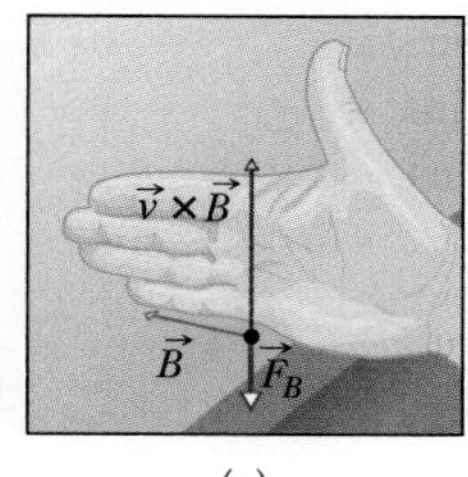

(e)

Fig. 28-2 (a) – (c) The right-hand rule (in which $\vec{v}$ is swept into $\vec{B}$ through the smaller angle ϕ between them) gives the direction of $\vec{v} \times \vec{B}$ as the direction of the thumb. (d) If q is positive, then the direction of $\vec{F}_B = q\vec{v} \times \vec{B}$ is in the direction of $\vec{v} \times \vec{B}$. (e) If q is negative, then the direction of $\vec{F}_B$ is opposite that of $\vec{v} \times \vec{B}$.

the force is equal to zero if the charge is zero or if the particle is stationary. Equation 28-3 also tells us that the magnitude of the force is zero if $\vec{v}$ and $\vec{B}$ are either parallel ($\phi = 0°$) or antiparallel ($\phi = 180°$), and the force is at its maximum when $\vec{v}$ and $\vec{B}$ are perpendicular to each other.

Equation 28-2 tells us all this plus the direction of $\vec{F}_B$. From Section 3-8, we know that the cross product $\vec{v} \times \vec{B}$ in Eq. 28-2 is a vector that is perpendicular to the two vectors $\vec{v}$ and $\vec{B}$. The right-hand rule (Figs. 28-2a through c) tells us that the thumb of the right hand points in the direction of $\vec{v} \times \vec{B}$ when the fingers sweep $\vec{v}$ into $\vec{B}$. If q is positive, then (by Eq. 28-2) the force $\vec{F}_B$ has the same sign as $\vec{v} \times \vec{B}$ and thus must be in the same direction; that is, for positive q, $\vec{F}_B$ is directed along the thumb (Fig. 28-2d). If q is negative, then the force $\vec{F}_B$ and cross product $\vec{v} \times \vec{B}$ have opposite signs and thus must be in opposite directions. For negative q, $\vec{F}_B$ is directed opposite the thumb (Fig. 28-2e).

Regardless of the sign of the charge, however,

The force $\vec{F}_B$ acting on a charged particle moving with velocity $\vec{v}$ through a magnetic field $\vec{B}$ is *always* perpendicular to $\vec{v}$ and $\vec{B}$.

Thus, $\vec{F}_B$ *never* has a component parallel to $\vec{v}$. This means that $\vec{F}_B$ cannot change the particle's speed v (and thus it cannot change the particle's kinetic energy). The force can change only the direction of $\vec{v}$ (and thus the direction of travel); only in this sense can $\vec{F}_B$ accelerate the particle.

To develop a feeling for Eq. 28-2, consider Fig. 28-3, which shows some tracks left by charged particles moving rapidly through a *bubble chamber.* The chamber, which is filled with liquid hydrogen, is immersed in a strong uniform magnetic field that is directed out of the plane of the figure. An incoming gamma ray particle—which leaves no track because it is uncharged—transforms into an electron (spiral track marked e^-) and a positron (track marked e^+) while it knocks an electron out of a hydrogen atom (long track marked e^-). Check with Eq. 28-2 and Fig. 28-2 that the three tracks made by these two negative particles and one positive particle curve in the proper directions.

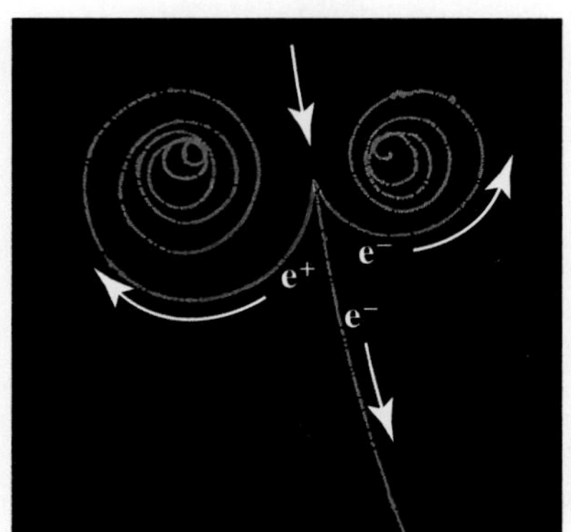

Fig. 28-3 The tracks of two electrons (e^-) and a positron (e^+) in a bubble chamber that is immersed in a uniform magnetic field that is directed out of the plane of the page. *(Lawrence Berkeley Laboratory/Photo Researchers)*

The SI unit for $\vec{B}$ that follows from Eqs. 28-2 and 28-3 is the newton per coulomb-meter per second. For convenience, this is called the **tesla** (T):

$$1 \text{ tesla} = 1 \text{ T} = 1 \frac{\text{newton}}{(\text{coulomb})(\text{meter/second})}.$$

Recalling that a coulomb per second is an ampere, we have

$$1 \text{ T} = 1 \frac{\text{newton}}{(\text{coulomb/second})(\text{meter})} = 1 \frac{\text{N}}{\text{A} \cdot \text{m}}. \quad (28\text{-}4)$$

Table 28-1

Some Approximate Magnetic Fields

At surface of neutron star	10^8 T
Near big electromagnet	1.5 T
Near small bar magnet	10^{-2} T
At Earth's surface	10^{-4} T
In interstellar space	10^{-10} T
Smallest value in magnetically shielded room	10^{-14} T

An earlier (non-SI) unit for $\vec{B}$, still in common use, is the *gauss* (G), and

$$1 \text{ tesla} = 10^4 \text{ gauss}. \tag{28-5}$$

Table 28-1 lists the magnetic fields that occur in a few situations. Note that Earth's magnetic field near the planet's surface is about 10^{-4} T ($= 100\ \mu$T or 1 G).

CHECKPOINT 1

The figure shows three situations in which a charged particle with velocity $\vec{v}$ travels through a uniform magnetic field $\vec{B}$. In each situation, what is the direction of the magnetic force $\vec{F}_B$ on the particle?

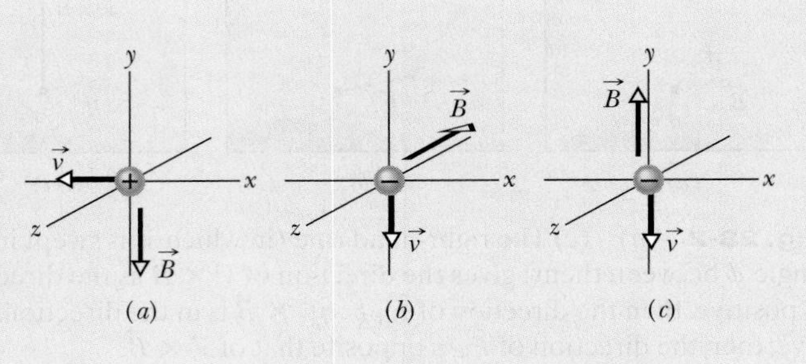

Magnetic Field Lines

We can represent magnetic fields with field lines, as we did for electric fields. Similar rules apply: (1) the direction of the tangent to a magnetic field line at any point gives the direction of $\vec{B}$ at that point, and (2) the spacing of the lines represents the magnitude of $\vec{B}$—the magnetic field is stronger where the lines are closer together, and conversely.

Figure 28-4*a* shows how the magnetic field near a *bar magnet* (a permanent magnet in the shape of a bar) can be represented by magnetic field lines. The lines all pass through the magnet, and they all form closed loops (even those that are not shown closed in the figure). The external magnetic effects of a bar magnet are strongest near its ends, where the field lines are most closely spaced. Thus, the magnetic field of the bar magnet in Fig. 28-4*b* collects the iron filings mainly near the two ends of the magnet.

The (closed) field lines enter one end of a magnet and exit the other end. The end of a magnet from which the field lines emerge is called the *north pole* of the magnet; the other end, where field lines enter the magnet, is called the *south pole*. Because a magnet has two poles, it is said to be a **magnetic dipole.** The magnets we use to fix notes on refrigerators are short bar magnets. Figure 28-5 shows two other common shapes for magnets: a *horseshoe magnet* and a magnet that has been bent around into the shape of a **C** so that the *pole faces* are facing each other. (The magnetic field between the pole faces can then be approximately uniform.) Regardless of the shape of the magnets, if we place two of them near each other we find:

Opposite magnetic poles attract each other, and like magnetic poles repel each other.

Earth has a magnetic field that is produced in its core by still unknown mechanisms. On Earth's surface, we can detect this magnetic field with a compass, which is essentially a slender bar magnet on a low-friction pivot. This bar magnet, or this needle, turns because its north-pole end is attracted toward the Arctic region of Earth. Thus, the *south* pole of Earth's magnetic field must be located toward the Arctic. Logically, we then should call the pole there a south pole. However, because we call that direction north, we are trapped into the statement that Earth has a *geomagnetic north pole* in that direction.

With more careful measurement we would find that in the Northern Hemisphere, the magnetic field lines of Earth generally point down into Earth and toward the Arctic. In the Southern Hemisphere, they generally point up out of Earth and away from the Antarctic—that is, away from Earth's *geomagnetic south pole.*

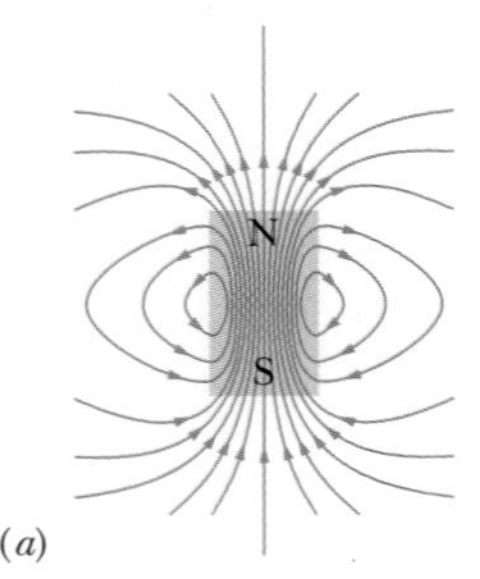

(*a*)

(*b*)

Fig. 28-4 (*a*) The magnetic field lines for a bar magnet. (*b*) A "cow magnet"—a bar magnet that is intended to be slipped down into the rumen of a cow to prevent accidentally ingested bits of scrap iron from reaching the cow's intestines. The iron filings at its ends reveal the magnetic field lines. *(Courtesy Dr. Richard Cannon, Southeast Missouri State University, Cape Girardeau)*

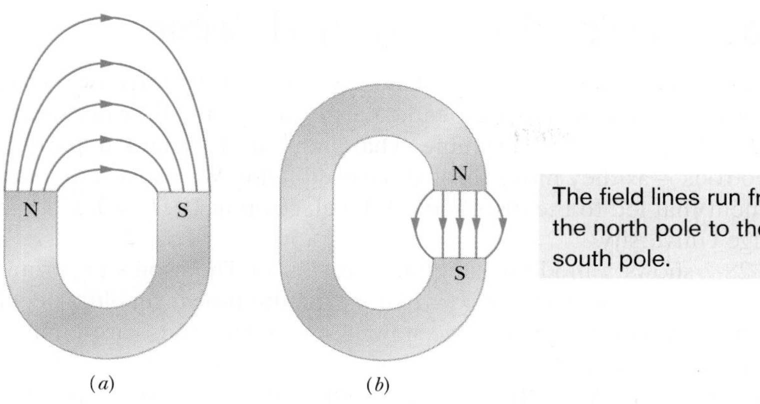

Fig. 28-5 (*a*) A horseshoe magnet and (*b*) a C-shaped magnet. (Only some of the external field lines are shown.)

Sample Problem

Magnetic force on a moving charged particle

A uniform magnetic field $\vec{B}$, with magnitude 1.2 mT, is directed vertically upward throughout the volume of a laboratory chamber. A proton with kinetic energy 5.3 MeV enters the chamber, moving horizontally from south to north. What magnetic deflecting force acts on the proton as it enters the chamber? The proton mass is 1.67×10^{-27} kg. (Neglect Earth's magnetic field.)

KEY IDEAS

Because the proton is charged and moving through a magnetic field, a magnetic force $\vec{F}_B$ can act on it. Because the initial direction of the proton's velocity is not along a magnetic field line, $\vec{F}_B$ is not simply zero.

Magnitude: To find the magnitude of $\vec{F}_B$, we can use Eq. 28-3 ($F_B = |q|vB \sin\phi$) provided we first find the proton's speed v. We can find v from the given kinetic energy because $K = \frac{1}{2}mv^2$. Solving for v, we obtain

$$v = \sqrt{\frac{2K}{m}} = \sqrt{\frac{(2)(5.3\text{ MeV})(1.60 \times 10^{-13}\text{ J/MeV})}{1.67 \times 10^{-27}\text{ kg}}}$$
$$= 3.2 \times 10^7\text{ m/s}.$$

Equation 28-3 then yields

$$F_B = |q|vB \sin\phi$$
$$= (1.60 \times 10^{-19}\text{ C})(3.2 \times 10^7\text{ m/s}) \times (1.2 \times 10^{-3}\text{ T})(\sin 90^\circ)$$
$$= 6.1 \times 10^{-15}\text{ N}. \quad \text{(Answer)}$$

This may seem like a small force, but it acts on a particle of small mass, producing a large acceleration; namely,

$$a = \frac{F_B}{m} = \frac{6.1 \times 10^{-15}\text{ N}}{1.67 \times 10^{-27}\text{ kg}} = 3.7 \times 10^{12}\text{ m/s}^2.$$

Direction: To find the direction of $\vec{F}_B$, we use the fact that $\vec{F}_B$ has the direction of the cross product $q\vec{v} \times \vec{B}$. Because the charge q is positive, $\vec{F}_B$ must have the same direction as $\vec{v} \times \vec{B}$, which can be determined with the right-hand rule for cross products (as in Fig. 28-2*d*). We know that $\vec{v}$ is directed horizontally from south to north and $\vec{B}$ is directed vertically up. The right-hand rule shows us that the deflecting force $\vec{F}_B$ must be directed horizontally from west to east, as Fig. 28-6 shows. (The array of dots in the figure represents a magnetic field directed out of the plane of the figure. An array of Xs would have represented a magnetic field directed into that plane.)

If the charge of the particle were negative, the magnetic deflecting force would be directed in the opposite direction—that is, horizontally from east to west. This is predicted automatically by Eq. 28-2 if we substitute a negative value for q.

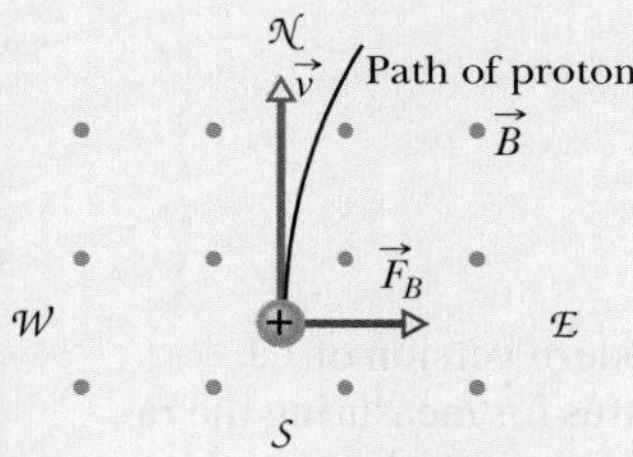

Fig. 28-6 An overhead view of a proton moving from south to north with velocity $\vec{v}$ in a chamber. A magnetic field is directed vertically upward in the chamber, as represented by the array of dots (which resemble the tips of arrows). The proton is deflected toward the east.

Additional examples, video, and practice available at *WileyPLUS*

28-4 Crossed Fields: Discovery of the Electron

Both an electric field $\vec{E}$ and a magnetic field $\vec{B}$ can produce a force on a charged particle. When the two fields are perpendicular to each other, they are said to be *crossed fields.* Here we shall examine what happens to charged particles—namely, electrons—as they move through crossed fields. We use as our example the experiment that led to the discovery of the electron in 1897 by J. J. Thomson at Cambridge University.

Figure 28-7 shows a modern, simplified version of Thomson's experimental apparatus—a *cathode ray tube* (which is like the picture tube in an old type television set). Charged particles (which we now know as electrons) are emitted by a hot filament at the rear of the evacuated tube and are accelerated by an applied potential difference V. After they pass through a slit in screen C, they form a narrow beam. They then pass through a region of crossed $\vec{E}$ and $\vec{B}$ fields, headed toward a fluorescent screen S, where they produce a spot of light (on a television screen the spot is part of the picture). The forces on the charged particles in the crossed-fields region can deflect them from the center of the screen. By controlling the magnitudes and directions of the fields, Thomson could thus control where the spot of light appeared on the screen. Recall that the force on a negatively charged particle due to an electric field is directed opposite the field. Thus, for the arrangement of Fig. 28-7, electrons are forced up the page by electric field $\vec{E}$ and down the page by magnetic field $\vec{B}$; that is, the forces are *in opposition.* Thomson's procedure was equivalent to the following series of steps.

1. Set $E = 0$ and $B = 0$ and note the position of the spot on screen S due to the undeflected beam.
2. Turn on $\vec{E}$ and measure the resulting beam deflection.
3. Maintaining $\vec{E}$, now turn on $\vec{B}$ and adjust its value until the beam returns to the undeflected position. (With the forces in opposition, they can be made to cancel.)

We discussed the deflection of a charged particle moving through an electric field $\vec{E}$ between two plates (step 2 here) in the sample problem in the preceding section. We found that the deflection of the particle at the far end of the plates is

$$y = \frac{|q|EL^2}{2mv^2}, \tag{28-6}$$

where v is the particle's speed, m its mass, and q its charge, and L is the length of the plates. We can apply this same equation to the beam of electrons in Fig. 28-7; if need be, we can calculate the deflection by measuring the deflection of the beam on screen S and then working back to calculate the deflection y at the end of the plates. (Because the direction of the deflection is set by the sign of the

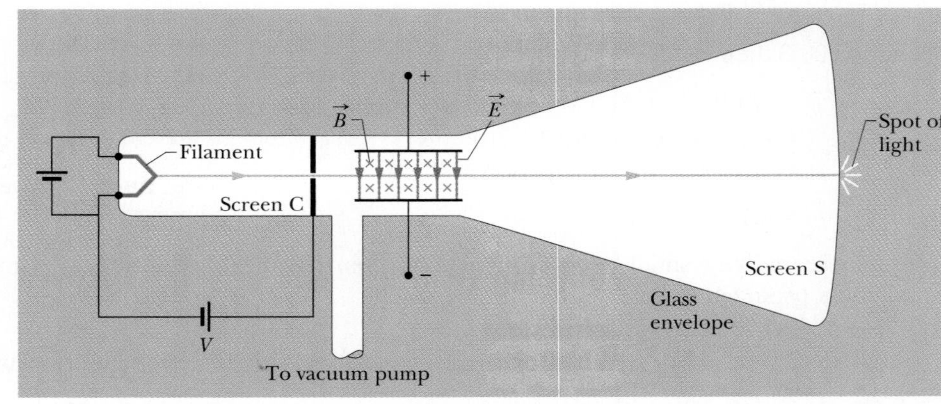

Fig. 28-7 A modern version of J. J. Thomson's apparatus for measuring the ratio of mass to charge for the electron. An electric field $\vec{E}$ is established by connecting a battery across the deflecting-plate terminals. The magnetic field $\vec{B}$ is set up by means of a current in a system of coils (not shown). The magnetic field shown is into the plane of the figure, as represented by the array of **X**s (which resemble the feathered ends of arrows).

particle's charge, Thomson was able to show that the particles that were lighting up his screen were negatively charged.)

When the two fields in Fig. 28-7 are adjusted so that the two deflecting forces cancel (step 3), we have from Eqs. 28-1 and 28-3

$$|q|E = |q|vB \sin(90°) = |q|vB$$

or

$$v = \frac{E}{B}. \qquad (28\text{-}7)$$

Thus, the crossed fields allow us to measure the speed of the charged particles passing through them. Substituting Eq. 28-7 for v in Eq. 28-6 and rearranging yield

$$\frac{m}{|q|} = \frac{B^2L^2}{2yE}, \qquad (28\text{-}8)$$

in which all quantities on the right can be measured. Thus, the crossed fields allow us to measure the ratio $m/|q|$ of the particles moving through Thomson's apparatus.

Thomson claimed that these particles are found in all matter. He also claimed that they are lighter than the lightest known atom (hydrogen) by a factor of more than 1000. (The exact ratio proved later to be 1836.15.) His $m/|q|$ measurement, coupled with the boldness of his two claims, is considered to be the "discovery of the electron."

CHECKPOINT 2

The figure shows four directions for the velocity vector $\vec{v}$ of a positively charged particle moving through a uniform electric field $\vec{E}$ (directed out of the page and represented with an encircled dot) and a uniform magnetic field $\vec{B}$. (a) Rank directions 1, 2, and 3 according to the magnitude of the net force on the particle, greatest first. (b) Of all four directions, which might result in a net force of zero?

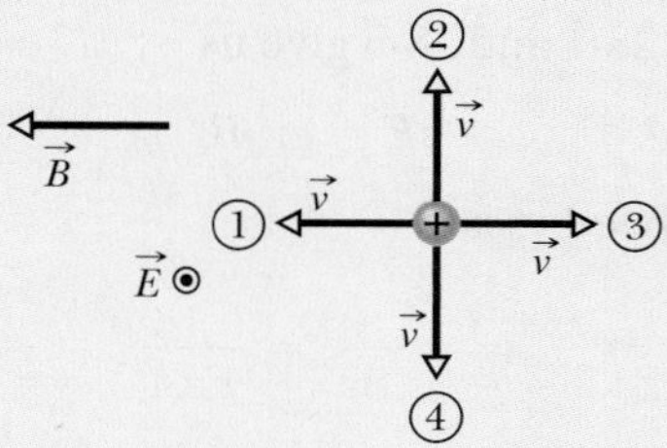

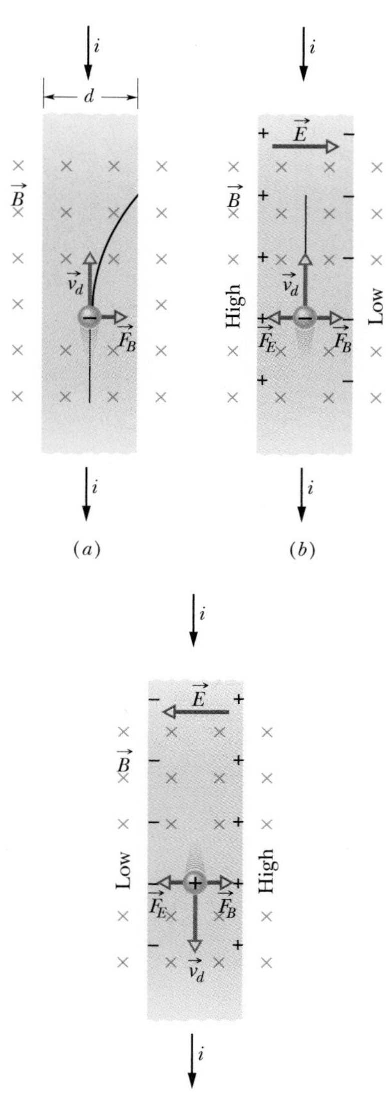

Fig. 28-8 A strip of copper carrying a current i is immersed in a magnetic field $\vec{B}$. (*a*) The situation immediately after the magnetic field is turned on. The curved path that will then be taken by an electron is shown. (*b*) The situation at equilibrium, which quickly follows. Note that negative charges pile up on the right side of the strip, leaving uncompensated positive charges on the left. Thus, the left side is at a higher potential than the right side. (*c*) For the same current direction, if the charge carriers were positively charged, *they* would pile up on the right side, and the right side would be at the higher potential.

28-5 Crossed Fields: The Hall Effect

As we just discussed, a beam of electrons in a vacuum can be deflected by a magnetic field. Can the drifting conduction electrons in a copper wire also be deflected by a magnetic field? In 1879, Edwin H. Hall, then a 24-year-old graduate student at the Johns Hopkins University, showed that they can. This **Hall effect** allows us to find out whether the charge carriers in a conductor are positively or negatively charged. Beyond that, we can measure the number of such carriers per unit volume of the conductor.

Figure 28-8*a* shows a copper strip of width d, carrying a current i whose conventional direction is from the top of the figure to the bottom. The charge carriers are electrons and, as we know, they drift (with drift speed v_d) in the opposite direction, from bottom to top. At the instant shown in Fig. 28-8*a*, an external magnetic field $\vec{B}$, pointing into the plane of the figure, has just been turned on. From Eq. 28-2 we see that a magnetic deflecting force $\vec{F}_B$ will act on each drifting electron, pushing it toward the right edge of the strip.

As time goes on, electrons move to the right, mostly piling up on the right edge of the strip, leaving uncompensated positive charges in fixed positions at the left edge. The separation of positive charges on the left edge and negative charges on the right edge produces an electric field $\vec{E}$ within the strip, pointing from left to right in Fig. 28-8*b*. This field exerts an electric force $\vec{F}_E$ on each electron, tending to push it to the left. Thus, this electric force on the electrons, which opposes the magnetic force on them, begins to build up.

An equilibrium quickly develops in which the electric force on each electron has increased enough to match the magnetic force. When this happens, as Fig. 28-8*b* shows, the force due to $\vec{B}$ and the force due to $\vec{E}$ are in balance. The drifting electrons then move along the strip toward the top of the page at velocity $\vec{v}_d$ with no further collection of electrons on the right edge of the strip and thus no further increase in the electric field $\vec{E}$.

A *Hall potential difference V* is associated with the electric field across strip width d. From Eq. 24-42, the magnitude of that potential difference is

$$V = Ed. \tag{28-9}$$

By connecting a voltmeter across the width, we can measure the potential difference between the two edges of the strip. Moreover, the voltmeter can tell us which edge is at higher potential. For the situation of Fig. 28-8*b*, we would find that the left edge is at higher potential, which is consistent with our assumption that the charge carriers are negatively charged.

For a moment, let us make the opposite assumption, that the charge carriers in current i are positively charged (Fig. 28-8*c*). Convince yourself that as these charge carriers move from top to bottom in the strip, they are pushed to the right edge by $\vec{F}_B$ and thus that the *right* edge is at higher potential. Because that last statement is contradicted by our voltmeter reading, the charge carriers must be negatively charged.

Now for the quantitative part. When the electric and magnetic forces are in balance (Fig. 28-8*b*), Eqs. 28-1 and 28-3 give us

$$eE = ev_dB. \tag{28-10}$$

From Eq. 26-7, the drift speed v_d is

$$v_d = \frac{J}{ne} = \frac{i}{neA}, \tag{28-11}$$

in which J (= i/A) is the current density in the strip, A is the cross-sectional area of the strip, and n is the *number density* of charge carriers (their number per unit volume).

In Eq. 28-10, substituting for E with Eq. 28-9 and substituting for v_d with Eq. 28-11, we obtain

$$n = \frac{Bi}{Vle}, \tag{28-12}$$

in which l (= A/d) is the thickness of the strip. With this equation we can find n from measurable quantities.

It is also possible to use the Hall effect to measure directly the drift speed v_d of the charge carriers, which you may recall is of the order of centimeters per hour. In this clever experiment, the metal strip is moved mechanically through the magnetic field in a direction opposite that of the drift velocity of the charge carriers. The speed of the moving strip is then adjusted until the Hall potential difference vanishes. At this condition, with no Hall effect, the velocity of the charge carriers *with respect to the laboratory frame* must be zero, so the velocity of the strip must be equal in magnitude but opposite the direction of the velocity of the negative charge carriers.

Sample Problem

Potential difference set up across a moving conductor

Figure 28-9*a* shows a solid metal cube, of edge length $d = 1.5$ cm, moving in the positive y direction at a constant velocity $\vec{v}$ of magnitude 4.0 m/s. The cube moves through a uniform magnetic field $\vec{B}$ of magnitude 0.050 T in the positive z direction.

(a) Which cube face is at a lower electric potential and which is at a higher electric potential because of the motion through the field?

KEY IDEA

Because the cube is moving through a magnetic field $\vec{B}$, a magnetic force $\vec{F}_B$ acts on its charged particles, including its conduction electrons.

Reasoning: When the cube first begins to move through the magnetic field, its electrons do also. Because each electron has charge q and is moving through a magnetic field with velocity $\vec{v}$, the magnetic force $\vec{F}_B$ acting on the electron is given by Eq. 28-2. Because q is negative, the direction of $\vec{F}_B$ is opposite the cross product $\vec{v} \times \vec{B}$, which is in the positive direction of the x axis (Fig. 28-9*b*). Thus, $\vec{F}_B$ acts in the negative direction of the x axis, toward the left face of the cube (Fig. 28-9*c*).

Most of the electrons are fixed in place in the atoms of the cube. However, because the cube is a metal, it contains conduction electrons that are free to move. Some of those conduction electrons are deflected by $\vec{F}_B$ to the left cube face, making that face negatively charged and leaving the right face positively charged (Fig. 28-9*d*). This charge separation produces an electric field $\vec{E}$ directed from the positively charged right face to the negatively charged left face (Fig. 28-9*e*). Thus, the left face is at a lower electric potential, and the right face is at a higher electric potential.

(b) What is the potential difference between the faces of higher and lower electric potential?

KEY IDEAS

1. The electric field $\vec{E}$ created by the charge separation produces an electric force $\vec{F}_E = q\vec{E}$ on each electron

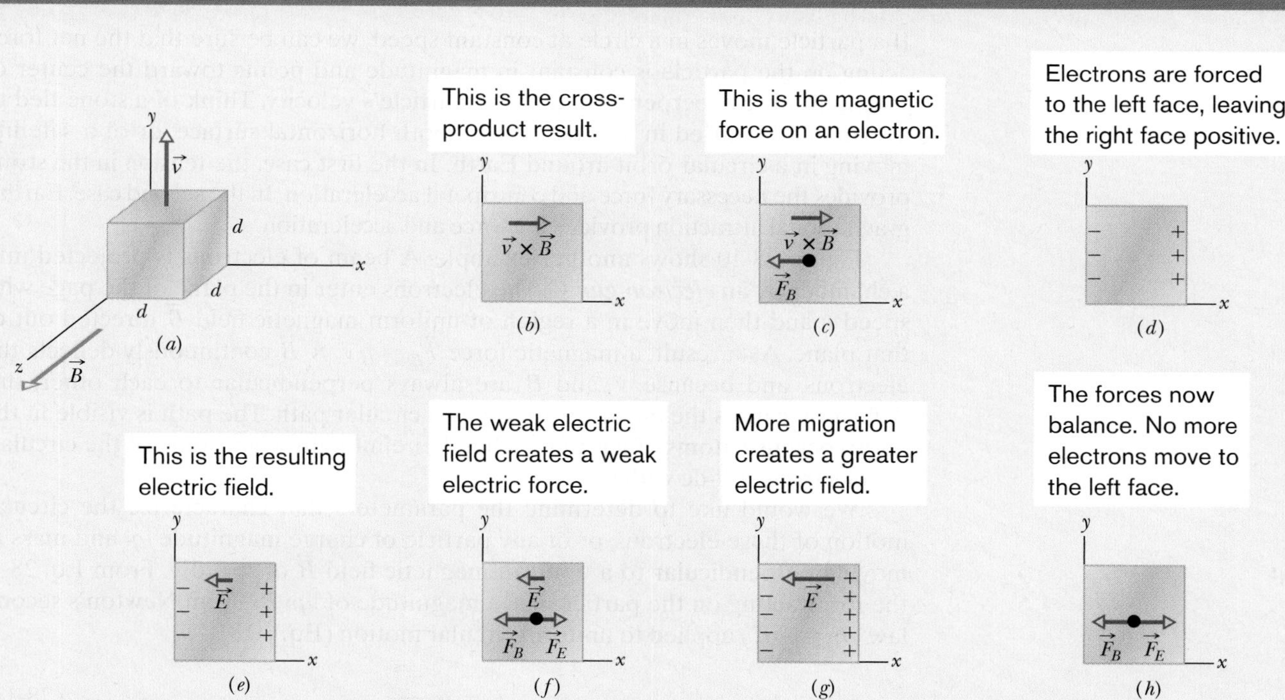

Fig. 28-9 (*a*) A solid metal cube moves at constant velocity through a uniform magnetic field. (*b*) – (*d*) In these front views, the magnetic force acting on an electron forces the electron to the left face, making that face negative and leaving the opposite face positive. (*e*) – (*f*) The resulting weak electric field creates a weak electric force on the next electron, but it too is forced to the left face. Now (*g*) the electric field is stronger and (*h*) the electric force matches the magnetic force.

(Fig. 28-9*f*). Because q is negative, this force is directed opposite the field $\vec{E}$—that is, rightward. Thus on each electron, $\vec{F}_E$ acts toward the right and $\vec{F}_B$ acts toward the left.

2. When the cube had just begun to move through the magnetic field and the charge separation had just begun, the magnitude of $\vec{E}$ began to increase from zero. Thus, the magnitude of $\vec{F}_E$ also began to increase from zero and was initially smaller than the magnitude $\vec{F}_B$. During this early stage, the net force on any electron was dominated by $\vec{F}_B$, which continuously moved additional electrons to the left cube face, increasing the charge separation (Fig. 28-9*g*).
3. However, as the charge separation increased, eventually magnitude F_E became equal to magnitude F_B (Fig. 28-9*h*). The net force on any electron was then zero, and no additional electrons were moved to the left cube face. Thus, the magnitude of $\vec{F}_E$ could not increase further, and the electrons were then in equilibrium.

Calculations: We seek the potential difference V between the left and right cube faces after equilibrium was reached (which occurred quickly). We can obtain V with Eq. 28-9 ($V = Ed$) provided we first find the magnitude E of the electric field at equilibrium. We can do so with the equation for the balance of forces ($F_E = F_B$).

For F_E, we substitute $|q|E$, and then for F_B, we substitute $|q|vB \sin \phi$ from Eq. 28-3. From Fig. 28-9*a*, we see that the angle ϕ between velocity vector $\vec{v}$ and magnetic field vector $\vec{B}$ is 90°; thus $\sin \phi = 1$ and $F_E = F_B$ yields

$$|q|E = |q|vB \sin 90° = |q|vB.$$

This gives us $E = vB$; so $V = Ed$ becomes

$$V = vBd. \tag{28-13}$$

Substituting known values gives us

$$V = (4.0 \text{ m/s})(0.050 \text{ T})(0.015 \text{ m})$$
$$= 0.0030 \text{ V} = 3.0 \text{ mV}. \quad \text{(Answer)}$$

Additional examples, video, and practice available at *WileyPLUS*

28-6 A Circulating Charged Particle

If a particle moves in a circle at constant speed, we can be sure that the net force acting on the particle is constant in magnitude and points toward the center of the circle, always perpendicular to the particle's velocity. Think of a stone tied to a string and whirled in a circle on a smooth horizontal surface, or of a satellite moving in a circular orbit around Earth. In the first case, the tension in the string provides the necessary force and centripetal acceleration. In the second case, Earth's gravitational attraction provides the force and acceleration.

Figure 28-10 shows another example: A beam of electrons is projected into a chamber by an *electron gun* G. The electrons enter in the plane of the page with speed v and then move in a region of uniform magnetic field $\vec{B}$ directed out of that plane. As a result, a magnetic force $\vec{F}_B = q\vec{v} \times \vec{B}$ continuously deflects the electrons, and because $\vec{v}$ and $\vec{B}$ are always perpendicular to each other, this deflection causes the electrons to follow a circular path. The path is visible in the photo because atoms of gas in the chamber emit light when some of the circulating electrons collide with them.

We would like to determine the parameters that characterize the circular motion of these electrons, or of any particle of charge magnitude $|q|$ and mass m moving perpendicular to a uniform magnetic field $\vec{B}$ at speed v. From Eq. 28-3, the force acting on the particle has a magnitude of $|q|vB$. From Newton's second law ($\vec{F} = m\vec{a}$) applied to uniform circular motion (Eq. 6-18),

$$F = m\frac{v^2}{r}, \tag{28-14}$$

we have

$$|q|vB = \frac{mv^2}{r}. \tag{28-15}$$

Solving for r, we find the radius of the circular path as

Fig. 28-10 Electrons circulating in a chamber containing gas at low pressure (their path is the glowing circle). A uniform magnetic field $\vec{B}$, pointing directly out of the plane of the page, fills the chamber. Note the radially directed magnetic force $\vec{F}_B$; for circular motion to occur, $\vec{F}_B$ *must* point toward the center of the circle. Use the right-hand rule for cross products to confirm that $\vec{F}_B = q\vec{v} \times \vec{B}$ gives $\vec{F}_B$ the proper direction. (Don't forget the sign of q.) *(Courtesy John Le P. Webb, Sussex University, England)*

$$r = \frac{mv}{|q|B} \quad \text{(radius).} \tag{28-16}$$

The period T (the time for one full revolution) is equal to the circumference divided by the speed:

$$T = \frac{2\pi r}{v} = \frac{2\pi}{v}\frac{mv}{|q|B} = \frac{2\pi m}{|q|B} \quad \text{(period).} \tag{28-17}$$

The frequency f (the number of revolutions per unit time) is

$$f = \frac{1}{T} = \frac{|q|B}{2\pi m} \quad \text{(frequency).} \tag{28-18}$$

The angular frequency ω of the motion is then

$$\omega = 2\pi f = \frac{|q|B}{m} \quad \text{(angular frequency).} \tag{28-19}$$

The quantities T, f, and ω do not depend on the speed of the particle (provided the speed is much less than the speed of light). Fast particles move in large circles and slow ones in small circles, but all particles with the same charge-to-mass ratio $|q|/m$ take the same time T (the period) to complete one round trip. Using Eq. 28-2, you can show that if you are looking in the direction of $\vec{B}$, the direction of rotation for a positive particle is always counterclockwise, and the direction for a negative particle is always clockwise.

Helical Paths

If the velocity of a charged particle has a component parallel to the (uniform) magnetic field, the particle will move in a helical path about the direction of the field

The velocity component perpendicular to the field causes circling, which is stretched upward by the parallel component.

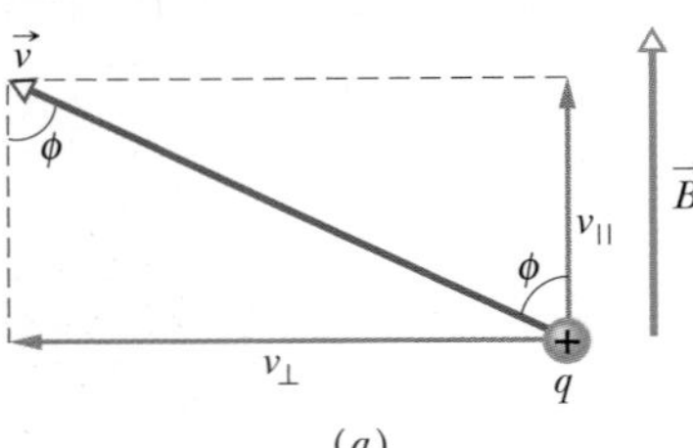

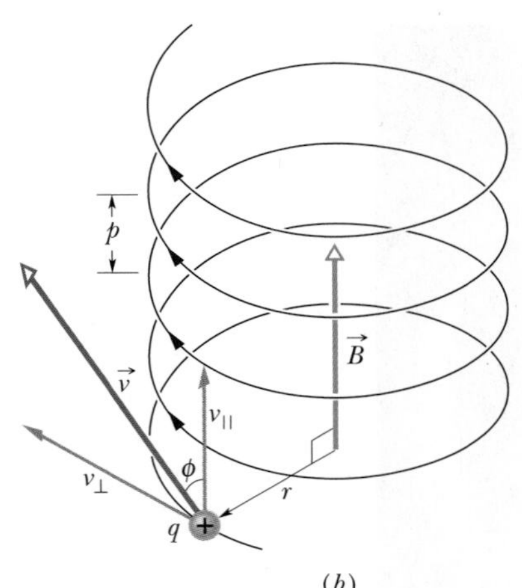

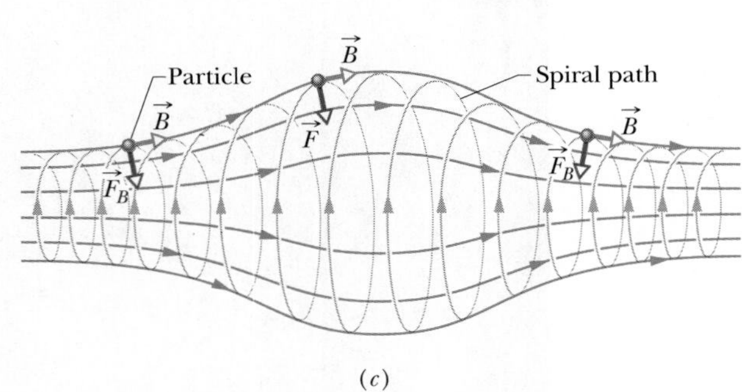

Fig. 28-11 (*a*) A charged particle moves in a uniform magnetic field $\vec{B}$, the particle's velocity $\vec{v}$ making an angle ϕ with the field direction. (*b*) The particle follows a helical path of radius r and pitch p. (*c*) A charged particle spiraling in a nonuniform magnetic field. (The particle can become trapped, spiraling back and forth between the strong field regions at either end.) Note that the magnetic force vectors at the left and right sides have a component pointing toward the center of the figure.

vector. Figure 28-11*a*, for example, shows the velocity vector $\vec{v}$ of such a particle resolved into two components, one parallel to $\vec{B}$ and one perpendicular to it:

$$v_{\parallel} = v \cos \phi \quad \text{and} \quad v_{\perp} = v \sin \phi. \tag{28-20}$$

The parallel component determines the *pitch p* of the helix—that is, the distance between adjacent turns (Fig. 28-11*b*). The perpendicular component determines the radius of the helix and is the quantity to be substituted for v in Eq. 28-16.

Figure 28-11*c* shows a charged particle spiraling in a nonuniform magnetic field. The more closely spaced field lines at the left and right sides indicate that the magnetic field is stronger there. When the field at an end is strong enough, the particle "reflects" from that end. If the particle reflects from both ends, it is said to be trapped in a *magnetic bottle*.

CHECKPOINT 3

The figure here shows the circular paths of two particles that travel at the same speed in a uniform magnetic field $\vec{B}$, which is directed into the page. One particle is a proton; the other is an electron (which is less massive). (a) Which particle follows the smaller circle, and (b) does that particle travel clockwise or counterclockwise?

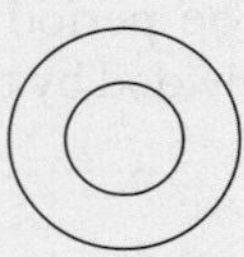

Sample Problem

Helical motion of a charged particle in a magnetic field

An electron with a kinetic energy of 22.5 eV moves into a region of uniform magnetic field $\vec{B}$ of magnitude 4.55×10^{-4} T. The angle between the directions of $\vec{B}$ and the electron's velocity $\vec{v}$ is 65.5°. What is the pitch of the helical path taken by the electron?

KEY IDEAS

(1) The pitch p is the distance the electron travels parallel to the magnetic field $\vec{B}$ during one period T of circulation. (2) The period T is given by Eq. 28-17 regardless of the angle between the directions of $\vec{v}$ and $\vec{B}$ (provided the angle is not zero, for which there is no circulation of the electron).

Calculations: Using Eqs. 28-20 and 28-17, we find

$$p = v_{\parallel} T = (v \cos \phi) \frac{2\pi m}{|q|B}. \tag{28-21}$$

Calculating the electron's speed v from its kinetic energy, find that $v = 2.81 \times 10^6$ m/s. Substituting this and known data in Eq. 28-21 gives us

$$p = (2.81 \times 10^6 \text{ m/s})(\cos 65.5°) \times \frac{2\pi(9.11 \times 10^{-31} \text{ kg})}{(1.60 \times 10^{-19} \text{ C})(4.55 \times 10^{-4} \text{ T})}$$
$$= 9.16 \text{ cm.} \qquad \text{(Answer)}$$

Additional examples, video, and practice available at *WileyPLUS*

Sample Problem

Uniform circular motion of a charged particle in a magnetic field

Figure 28-12 shows the essentials of a *mass spectrometer,* which can be used to measure the mass of an ion; an ion of mass m (to be measured) and charge q is produced in source S. The initially stationary ion is accelerated by the electric field due to a potential difference V. The ion leaves S and enters a separator chamber in which a uniform magnetic field $\vec{B}$ is perpendicular to the path of the ion. A wide detector lines the bottom wall of the chamber, and the $\vec{B}$ causes the ion to move in a semicircle and thus strike the detector. Suppose that $B = 80.000$ mT, $V = 1000.0$ V, and ions of charge $q = +1.6022 \times 10^{-19}$ C strike the detector at a point that lies at $x = 1.6254$ m. What is the mass m of the individual ions, in atomic mass units (Eq. 1-7: 1 u = 1.6605×10^{-27} kg)?

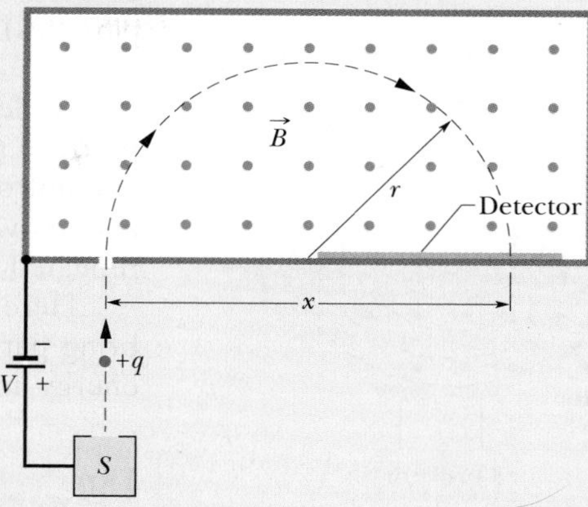

Fig. 28-12 Essentials of a mass spectrometer. A positive ion, after being accelerated from its source S by a potential difference V, enters a chamber of uniform magnetic field $\vec{B}$. There it travels through a semicircle of radius r and strikes a detector at a distance x from where it entered the chamber.

KEY IDEAS

(1) Because the (uniform) magnetic field causes the (charged) ion to follow a circular path, we can relate the ion's mass m to the path's radius r with Eq. 28-16 ($r = mv/|q|B$). From Fig. 28-12 we see that $r = x/2$ (the radius is half the diameter). From the problem statement, we know the magnitude B of the magnetic field. However, we lack the ion's speed v in the magnetic field after the ion has been accelerated due to the potential difference V. (2) To relate v and V, we use the fact that mechanical energy ($E_{mec} = K + U$) is conserved during the acceleration.

Finding speed: When the ion emerges from the source, its kinetic energy is approximately zero. At the end of the acceleration, its kinetic energy is $\frac{1}{2}mv^2$. Also, during the acceleration, the positive ion moves through a change in potential of $-V$. Thus, because the ion has positive charge q, its potential energy changes by $-qV$. If we now write the conservation of mechanical energy as

$$\Delta K + \Delta U = 0,$$

we get

$$\tfrac{1}{2}mv^2 - qV = 0$$

or

$$v = \sqrt{\frac{2qV}{m}}. \qquad (28\text{-}22)$$

Finding mass: Substituting this value for v into Eq. 28-16 gives us

$$r = \frac{mv}{qB} = \frac{m}{qB}\sqrt{\frac{2qV}{m}} = \frac{1}{B}\sqrt{\frac{2mV}{q}}.$$

Thus,

$$x = 2r = \frac{2}{B}\sqrt{\frac{2mV}{q}}.$$

Solving this for m and substituting the given data yield

$$m = \frac{B^2qx^2}{8V} = \frac{(0.080000\ \text{T})^2(1.6022 \times 10^{-19}\ \text{C})(1.6254\ \text{m})^2}{8(1000.0\ \text{V})}$$
$$= 3.3863 \times 10^{-25}\ \text{kg} = 203.93\ \text{u}. \qquad \text{(Answer)}$$

Additional examples, video, and practice available at *WileyPLUS*

28-7 Cyclotrons and Synchrotrons

Beams of high-energy particles, such as high-energy electrons and protons, have been enormously useful in probing atoms and nuclei to reveal the fundamental structure of matter. Such beams were instrumental in the discovery that atomic nuclei consist of protons and neutrons and in the discovery that protons and neutrons consist of quarks and gluons. The challenge of such beams is how to

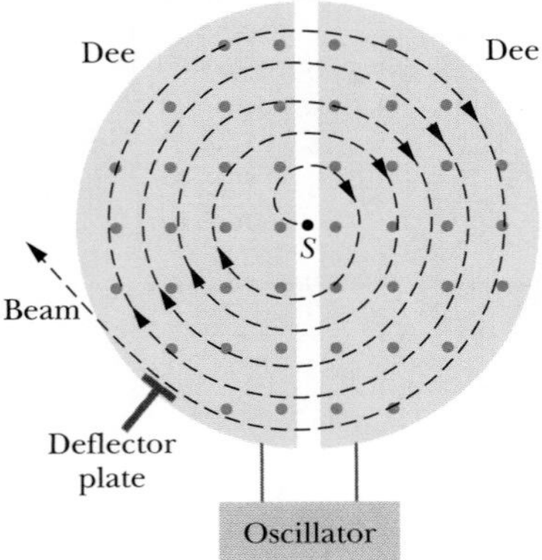

Fig. 28-13 The elements of a cyclotron, showing the particle source S and the dees. A uniform magnetic field is directed up from the plane of the page. Circulating protons spiral outward within the hollow dees, gaining energy every time they cross the gap between the dees.

make and control them. Because electrons and protons are charged, they can be accelerated to the required high energy if they move through large potential differences. Because electrons have low mass, accelerating them in this way can be done in a reasonable distance. However, because protons (and other charged particles) have greater mass, the distance required for the acceleration is too long.

A clever solution to this problem is first to let protons and other massive particles move through a modest potential difference (so that they gain a modest amount of energy) and then use a magnetic field to cause them to circle back and move through a modest potential difference again. If this procedure is repeated thousands of times, the particles end up with a very large energy.

Here we discuss two *accelerators* that employ a magnetic field to repeatedly bring particles back to an accelerating region, where they gain more and more energy until they finally emerge as a high-energy beam.

The Cyclotron

Figure 28-13 is a top view of the region of a *cyclotron* in which the particles (protons, say) circulate. The two hollow **D**-shaped objects (each open on its straight edge) are made of sheet copper. These *dees,* as they are called, are part of an electrical oscillator that alternates the electric potential difference across the gap between the dees. The electrical signs of the dees are alternated so that the electric field in the gap alternates in direction, first toward one dee and then toward the other dee, back and forth. The dees are immersed in a large magnetic field directed out of the plane of the page. The magnitude B of this field is set via a control on the electromagnet producing the field.

Suppose that a proton, injected by source S at the center of the cyclotron in Fig. 28-13, initially moves toward a negatively charged dee. It will accelerate toward this dee and enter it. Once inside, it is shielded from electric fields by the copper walls of the dee; that is, the electric field does not enter the dee. The magnetic field, however, is not screened by the (nonmagnetic) copper dee, so the proton moves in a circular path whose radius, which depends on its speed, is given by Eq. 28-16 ($r = mv/|q|B$).

Let us assume that at the instant the proton emerges into the center gap from the first dee, the potential difference between the dees is reversed. Thus, the proton *again* faces a negatively charged dee and is *again* accelerated. This process continues, the circulating proton always being in step with the oscillations of the dee potential, until the proton has spiraled out to the edge of the dee system. There a deflector plate sends it out through a portal.

The key to the operation of the cyclotron is that the frequency f at which the proton circulates in the magnetic field (and that does *not* depend on its speed) must be equal to the fixed frequency f_{osc} of the electrical oscillator, or

$$f = f_{osc} \qquad \text{(resonance condition).} \tag{28-23}$$

This *resonance condition* says that, if the energy of the circulating proton is to increase, energy must be fed to it at a frequency f_{osc} that is equal to the natural frequency f at which the proton circulates in the magnetic field.

Combining Eqs. 28-18 ($f = |q|B/2\pi m$) and 28-23 allows us to write the resonance condition as

$$|q|B = 2\pi m f_{osc}. \tag{28-24}$$

For the proton, q and m are fixed. The oscillator (we assume) is designed to work at a single fixed frequency f_{osc}. We then "tune" the cyclotron by varying B until Eq. 28-24 is satisfied, and then many protons circulate through the magnetic field, to emerge as a beam.

The Proton Synchrotron

At proton energies above 50 MeV, the conventional cyclotron begins to fail because one of the assumptions of its design—that the frequency of revolution of a charged particle circulating in a magnetic field is independent of the particle's speed—is true only for speeds that are much less than the speed of light. At greater proton speeds (above about 10% of the speed of light), we must treat the problem relativistically. According to relativity theory, as the speed of a circulating proton approaches that of light, the proton's frequency of revolution decreases steadily. Thus, the proton gets out of step with the cyclotron's oscillator—whose frequency remains fixed at f_{osc}—and eventually the energy of the still circulating proton stops increasing.

There is another problem. For a 500 GeV proton in a magnetic field of 1.5 T, the path radius is 1.1 km. The corresponding magnet for a conventional cyclotron of the proper size would be impossibly expensive, the area of its pole faces being about 4×10^6 m^2.

The *proton synchrotron* is designed to meet these two difficulties. The magnetic field B and the oscillator frequency f_{osc}, instead of having fixed values as in the conventional cyclotron, are made to vary with time during the accelerating cycle. When this is done properly, (1) the frequency of the circulating protons remains in step with the oscillator at all times, and (2) the protons follow a circular—not a spiral—path. Thus, the magnet need extend only along that circular path, not over some 4×10^6 m^2. The circular path, however, still must be large if high energies are to be achieved. The proton synchrotron at the Fermi National Accelerator Laboratory (Fermilab) in Illinois has a circumference of 6.3 km and can produce protons with energies of about 1 TeV ($= 10^{12}$ eV).

Sample Problem

Accelerating a charged particle in a cyclotron

Suppose a cyclotron is operated at an oscillator frequency of 12 MHz and has a dee radius $R = 53$ cm.

(a) What is the magnitude of the magnetic field needed for deuterons to be accelerated in the cyclotron? The deuteron mass is $m = 3.34 \times 10^{-27}$ kg (twice the proton mass).

KEY IDEA

For a given oscillator frequency f_{osc}, the magnetic field magnitude B required to accelerate any particle in a cyclotron depends on the ratio $m/|q|$ of mass to charge for the particle, according to Eq. 28-24 ($|q|B = 2\pi m f_{osc}$).

Calculation: For deuterons and the oscillator frequency $f_{osc} =$ 12 MHz, we find

$$B = \frac{2\pi m f_{osc}}{|q|} = \frac{(2\pi)(3.34 \times 10^{-27}\,\text{kg})(12 \times 10^6\,\text{s}^{-1})}{1.60 \times 10^{-19}\,\text{C}}$$

$$= 1.57\,\text{T} \approx 1.6\,\text{T}. \qquad \text{(Answer)}$$

Note that, to accelerate protons, B would have to be reduced by a factor of 2, provided the oscillator frequency remained fixed at 12 MHz.

(b) What is the resulting kinetic energy of the deuterons?

KEY IDEAS

(1) The kinetic energy ($\frac{1}{2}mv^2$) of a deuteron exiting the cyclotron is equal to the kinetic energy it had just before exiting, when it was traveling in a circular path with a radius approximately equal to the radius R of the cyclotron dees. (2) We can find the speed v of the deuteron in that circular path with Eq. 28-16 ($r = mv/|q|B$).

Calculations: Solving that equation for v, substituting R for r, and then substituting known data, we find

$$v = \frac{R|q|B}{m} = \frac{(0.53\,\text{m})(1.60 \times 10^{-19}\,\text{C})(1.57\,\text{T})}{3.34 \times 10^{-27}\,\text{kg}}$$

$$= 3.99 \times 10^7\,\text{m/s}.$$

This speed corresponds to a kinetic energy of

$$K = \tfrac{1}{2}mv^2$$

$$= \tfrac{1}{2}(3.34 \times 10^{-27}\,\text{kg})(3.99 \times 10^7\,\text{m/s})^2$$

$$= 2.7 \times 10^{-12}\,\text{J}, \qquad \text{(Answer)}$$

or about 17 MeV.

Additional examples, video, and practice available at *WileyPLUS*

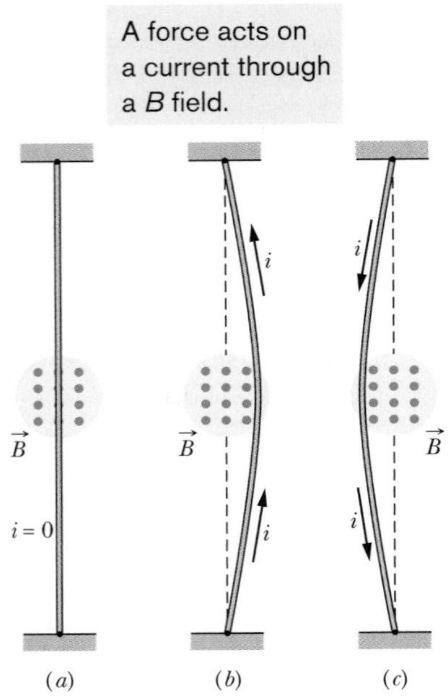

A force acts on a current through a B field.

Fig. 28-14 A flexible wire passes between the pole faces of a magnet (only the farther pole face is shown). (*a*) Without current in the wire, the wire is straight. (*b*) With upward current, the wire is deflected rightward. (*c*) With downward current, the deflection is leftward. The connections for getting the current into the wire at one end and out of it at the other end are not shown.

28-8 Magnetic Force on a Current-Carrying Wire

We have already seen (in connection with the Hall effect) that a magnetic field exerts a sideways force on electrons moving in a wire. This force must then be transmitted to the wire itself, because the conduction electrons cannot escape sideways out of the wire.

In Fig. 28-14*a*, a vertical wire, carrying no current and fixed in place at both ends, extends through the gap between the vertical pole faces of a magnet. The magnetic field between the faces is directed outward from the page. In Fig. 28-14*b*, a current is sent upward through the wire; the wire deflects to the right. In Fig. 28-14*c*, we reverse the direction of the current and the wire deflects to the left.

Figure 28-15 shows what happens inside the wire of Fig. 28-14*b*. We see one of the conduction electrons, drifting downward with an assumed drift speed v_d. Equation 28-3, in which we must put $\phi = 90°$, tells us that a force $\vec{F}_B$ of magnitude ev_dB must act on each such electron. From Eq. 28-2 we see that this force must be directed to the right. We expect then that the wire as a whole will experience a force to the right, in agreement with Fig. 28-14*b*.

If, in Fig. 28-15, we were to reverse *either* the direction of the magnetic field *or* the direction of the current, the force on the wire would reverse, being directed now to the left. Note too that it does not matter whether we consider negative charges drifting downward in the wire (the actual case) or positive charges drifting upward. The direction of the deflecting force on the wire is the same. We are safe then in dealing with a current of positive charge, as we usually do in dealing with circuits.

Consider a length L of the wire in Fig. 28-15. All the conduction electrons in this section of wire will drift past plane xx in Fig. 28-15 in a time $t = L/v_d$. Thus, in that time a charge given by

$$q = it = i\frac{L}{v_d}$$

will pass through that plane. Substituting this into Eq. 28-3 yields

$$F_B = qv_dB\sin\phi = \frac{iL}{v_d}v_dB\sin 90°$$

or

$$F_B = iLB. \tag{28-25}$$

Note that this equation gives the magnetic force that acts on a length L of straight wire carrying a current i and immersed in a uniform magnetic field $\vec{B}$ that is *perpendicular* to the wire.

If the magnetic field is *not* perpendicular to the wire, as in Fig. 28-16, the magnetic force is given by a generalization of Eq. 28-25:

$$\vec{F}_B = i\vec{L} \times \vec{B} \qquad \text{(force on a current).} \tag{28-26}$$

Here $\vec{L}$ is a *length vector* that has magnitude L and is directed along the wire segment in the direction of the (conventional) current. The force magnitude F_B is

$$F_B = iLB\sin\phi, \tag{28-27}$$

where ϕ is the angle between the directions of $\vec{L}$ and $\vec{B}$. The direction of $\vec{F}_B$ is that of the cross product $\vec{L} \times \vec{B}$ because we take current i to be a positive quantity. Equation 28-26 tells us that $\vec{F}_B$ is always perpendicular to the plane defined by vectors $\vec{L}$ and $\vec{B}$, as indicated in Fig. 28-16.

Equation 28-26 is equivalent to Eq. 28-2 in that either can be taken as the defining equation for $\vec{B}$. In practice, we define $\vec{B}$ from Eq. 28-26 because it is much easier to measure the magnetic force acting on a wire than that on a single moving charge.

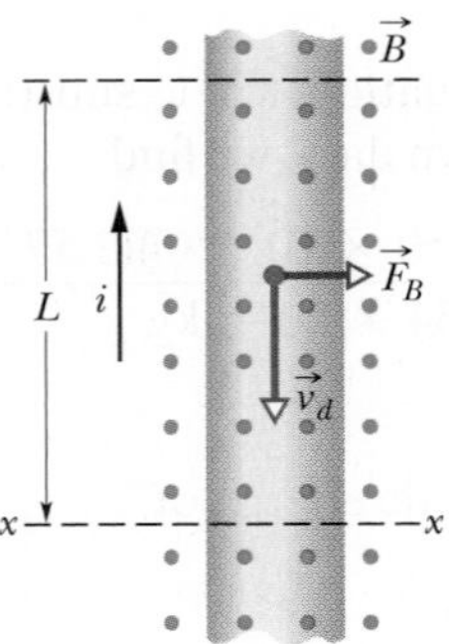

Fig. 28-15 A close-up view of a section of the wire of Fig. 28-14*b*. The current direction is upward, which means that electrons drift downward. A magnetic field that emerges from the plane of the page causes the electrons and the wire to be deflected to the right.

If a wire is not straight or the field is not uniform, we can imagine the wire broken up into small straight segments and apply Eq. 28-26 to each segment. The force on the wire as a whole is then the vector sum of all the forces on the segments that make it up. In the differential limit, we can write

$$d\vec{F}_B = i\,d\vec{L} \times \vec{B}, \qquad (28\text{-}28)$$

and we can find the resultant force on any given arrangement of currents by integrating Eq. 28-28 over that arrangement.

In using Eq. 28-28, bear in mind that there is no such thing as an isolated current-carrying wire segment of length dL. There must always be a way to introduce the current into the segment at one end and take it out at the other end.

The force is perpendicular to both the field and the length.

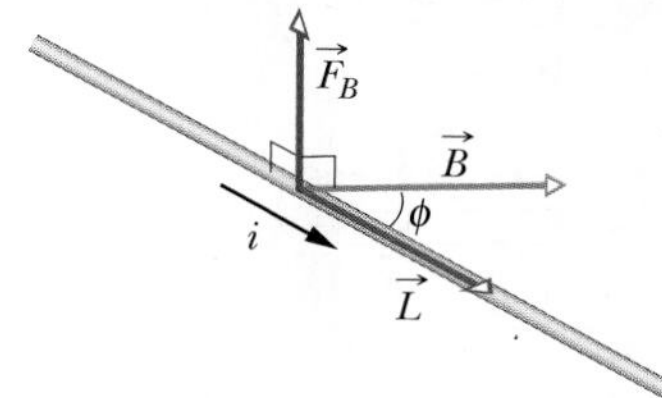

Fig. 28-16 A wire carrying current i makes an angle ϕ with magnetic field $\vec{B}$. The wire has length L in the field and length vector $\vec{L}$ (in the direction of the current). A magnetic force $\vec{F}_B = i\vec{L} \times \vec{B}$ acts on the wire.

CHECKPOINT 4

The figure shows a current i through a wire in a uniform magnetic field $\vec{B}$, as well as the magnetic force $\vec{F}_B$ acting on the wire. The field is oriented so that the force is maximum. In what direction is the field?

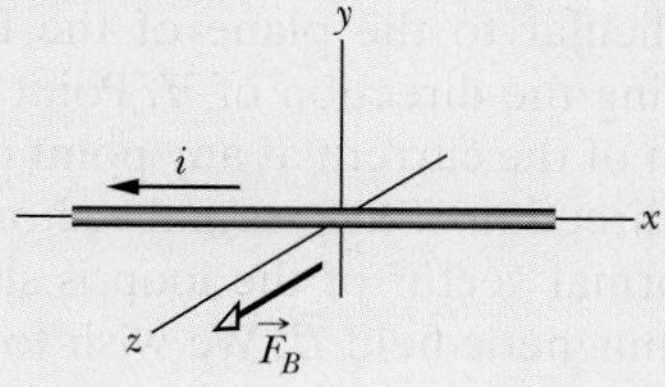

Sample Problem

Magnetic force on a wire carrying current

A straight, horizontal length of copper wire has a current $i = 28$ A through it. What are the magnitude and direction of the minimum magnetic field $\vec{B}$ needed to suspend the wire—that is, to balance the gravitational force on it? The linear density (mass per unit length) of the wire is 46.6 g/m.

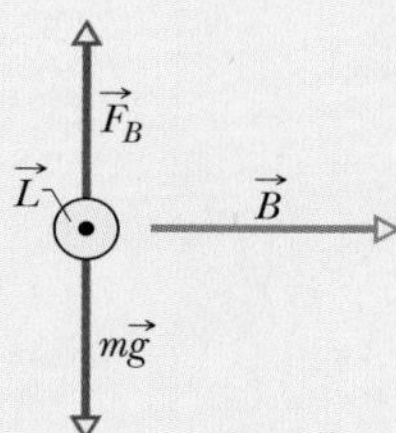

Fig. 28-17 A wire (shown in cross section) carrying current out of the page.

KEY IDEAS

(1) Because the wire carries a current, a magnetic force $\vec{F}_B$ can act on the wire if we place it in a magnetic field $\vec{B}$. To balance the downward gravitational force $\vec{F}_g$ on the wire, we want $\vec{F}_B$ to be directed upward (Fig. 28-17). (2) The direction of $\vec{F}_B$ is related to the directions of $\vec{B}$ and the wire's length vector $\vec{L}$ by Eq. 28-26 ($\vec{F}_B = i\vec{L} \times \vec{B}$).

Calculations: Because $\vec{L}$ is directed horizontally (and the current is taken to be positive), Eq. 28-26 and the right-hand rule for cross products tell us that $\vec{B}$ must be horizontal and rightward (in Fig. 28-17) to give the required upward $\vec{F}_B$.

The magnitude of $\vec{F}_B$ is $F_B = iLB \sin \phi$ (Eq. 28-27). Because we want $\vec{F}_B$ to balance $\vec{F}_g$, we want

$$iLB \sin \phi = mg, \qquad (28\text{-}29)$$

where mg is the magnitude of $\vec{F}_g$ and m is the mass of the wire.

We also want the minimal field magnitude B for $\vec{F}_B$ to balance $\vec{F}_g$. Thus, we need to maximize $\sin \phi$ in Eq. 28-29. To do so, we set $\phi = 90°$, thereby arranging for $\vec{B}$ to be perpendicular to the wire. We then have $\sin \phi = 1$, so Eq. 28-29 yields

$$B = \frac{mg}{iL \sin \phi} = \frac{(m/L)g}{i}. \qquad (28\text{-}30)$$

We write the result this way because we know m/L, the linear density of the wire. Substituting known data then gives us

$$B = \frac{(46.6 \times 10^{-3}\ \text{kg/m})(9.8\ \text{m/s}^2)}{28\ \text{A}}$$
$$= 1.6 \times 10^{-2}\ \text{T}. \qquad \text{(Answer)}$$

This is about 160 times the strength of Earth's magnetic field.

Additional examples, video, and practice available at *WileyPLUS*

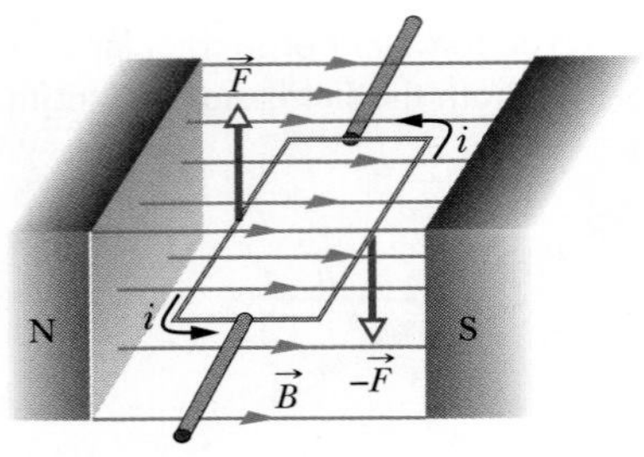

Fig. 28-18 The elements of an electric motor. A rectangular loop of wire, carrying a current and free to rotate about a fixed axis, is placed in a magnetic field. Magnetic forces on the wire produce a torque that rotates it. A commutator (not shown) reverses the direction of the current every half-revolution so that the torque always acts in the same direction.

28-9 Torque on a Current Loop

Much of the world's work is done by electric motors. The forces behind this work are the magnetic forces that we studied in the preceding section—that is, the forces that a magnetic field exerts on a wire that carries a current.

Figure 28-18 shows a simple motor, consisting of a single current-carrying loop immersed in a magnetic field $\vec{B}$. The two magnetic forces $\vec{F}$ and $-\vec{F}$ produce a torque on the loop, tending to rotate it about its central axis. Although many essential details have been omitted, the figure does suggest how the action of a magnetic field on a current loop produces rotary motion. Let us analyze that action.

Figure 28-19a shows a rectangular loop of sides a and b, carrying current i through uniform magnetic field $\vec{B}$. We place the loop in the field so that its long sides, labeled 1 and 3, are perpendicular to the field direction (which is into the page), but its short sides, labeled 2 and 4, are not. Wires to lead the current into and out of the loop are needed but, for simplicity, are not shown.

To define the orientation of the loop in the magnetic field, we use a normal vector $\vec{n}$ that is perpendicular to the plane of the loop. Figure 28-19b shows a right-hand rule for finding the direction of $\vec{n}$. Point or curl the fingers of your right hand in the direction of the current at any point on the loop. Your extended thumb then points in the direction of the normal vector $\vec{n}$.

In Fig. 28-19c, the normal vector of the loop is shown at an arbitrary angle θ to the direction of the magnetic field $\vec{B}$. We wish to find the net force and net torque acting on the loop in this orientation.

The net force on the loop is the vector sum of the forces acting on its four sides. For side 2 the vector $\vec{L}$ in Eq. 28-26 points in the direction of the current and has magnitude b. The angle between $\vec{L}$ and $\vec{B}$ for side 2 (see Fig. 28-19c) is $90° - \theta$. Thus, the magnitude of the force acting on this side is

$$F_2 = ibB \sin(90° - \theta) = ibB \cos \theta. \tag{28-31}$$

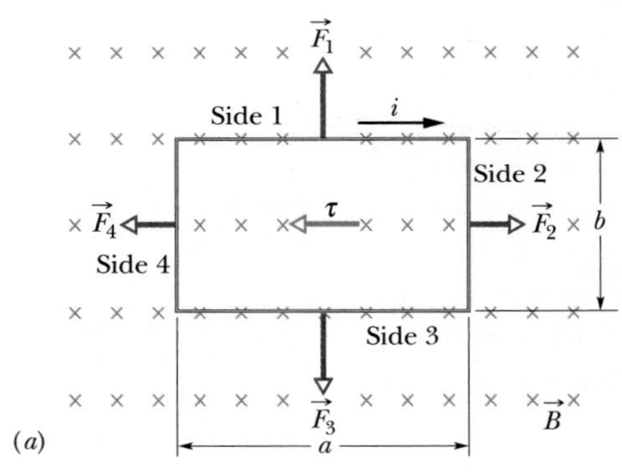

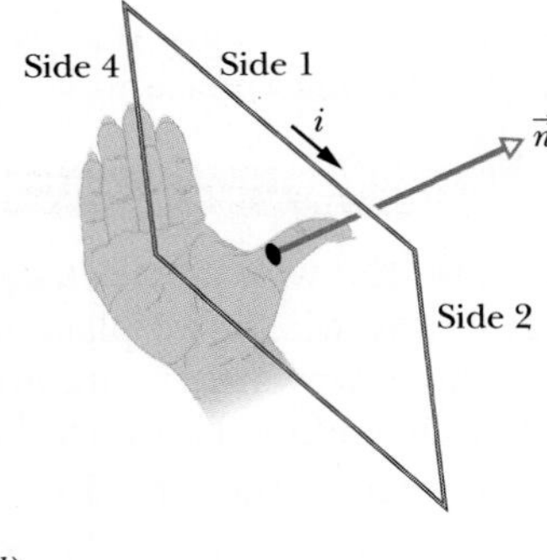

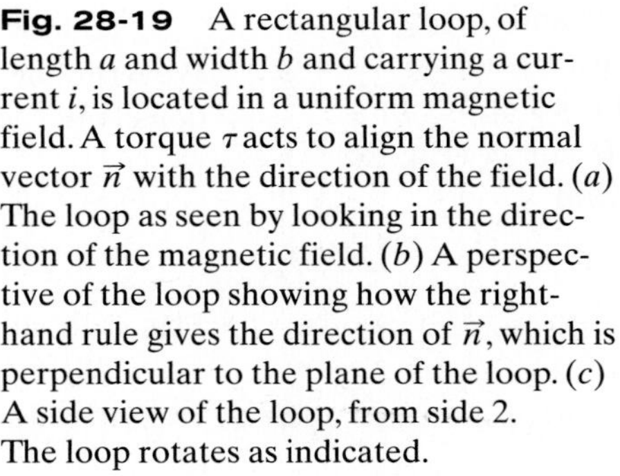

Fig. 28-19 A rectangular loop, of length a and width b and carrying a current i, is located in a uniform magnetic field. A torque τ acts to align the normal vector $\vec{n}$ with the direction of the field. (a) The loop as seen by looking in the direction of the magnetic field. (b) A perspective of the loop showing how the right-hand rule gives the direction of $\vec{n}$, which is perpendicular to the plane of the loop. (c) A side view of the loop, from side 2. The loop rotates as indicated.

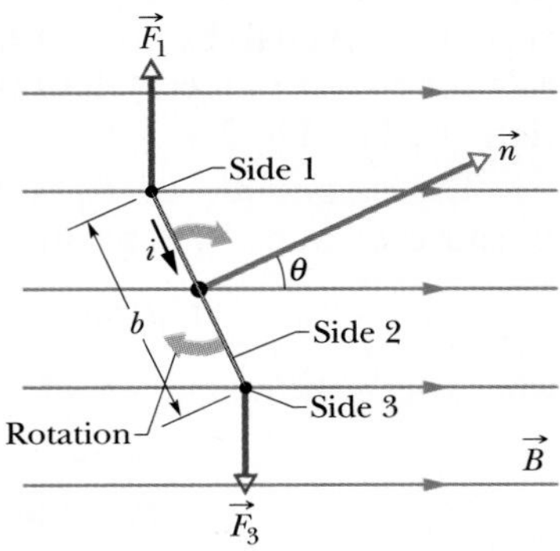

You can show that the force $\vec{F}_4$ acting on side 4 has the same magnitude as $\vec{F}_2$ but the opposite direction. Thus, $\vec{F}_2$ and $\vec{F}_4$ cancel out exactly. Their net force is zero and, because their common line of action is through the center of the loop, their net torque is also zero.

The situation is different for sides 1 and 3. For them, $\vec{L}$ is perpendicular to $\vec{B}$, so the forces $\vec{F}_1$ and $\vec{F}_3$ have the common magnitude iaB. Because these two forces have opposite directions, they do not tend to move the loop up or down. However, as Fig. 28-19*c* shows, these two forces do *not* share the same line of action; so they *do* produce a net torque. The torque tends to rotate the loop so as to align its normal vector $\vec{n}$ with the direction of the magnetic field $\vec{B}$. That torque has moment arm $(b/2)\sin\theta$ about the central axis of the loop. The magnitude τ' of the torque due to forces $\vec{F}_1$ and $\vec{F}_3$ is then (see Fig. 28-19*c*)

$$\tau' = \left(iaB\,\frac{b}{2}\sin\theta\right) + \left(iaB\,\frac{b}{2}\sin\theta\right) = iabB\sin\theta. \qquad (28\text{-}32)$$

Suppose we replace the single loop of current with a *coil* of N loops, or *turns*. Further, suppose that the turns are wound tightly enough that they can be approximated as all having the same dimensions and lying in a plane. Then the turns form a *flat coil*, and a torque τ' with the magnitude given in Eq. 28-32 acts on each of them. The total torque on the coil then has magnitude

$$\tau = N\tau' = NiabB\sin\theta = (NiA)B\sin\theta, \qquad (28\text{-}33)$$

in which A $(= ab)$ is the area enclosed by the coil. The quantities in parentheses (NiA) are grouped together because they are all properties of the coil: its number of turns, its area, and the current it carries. Equation 28-33 holds for all flat coils, no matter what their shape, provided the magnetic field is uniform. For example, for the common circular coil, with radius r, we have

$$\tau = (Ni\pi r^2)B\sin\theta. \qquad (28\text{-}34)$$

Instead of focusing on the motion of the coil, it is simpler to keep track of the vector $\vec{n}$, which is normal to the plane of the coil. Equation 28-33 tells us that a current-carrying flat coil placed in a magnetic field will tend to rotate so that $\vec{n}$ has the same direction as the field. In a motor, the current in the coil is reversed as $\vec{n}$ begins to line up with the field direction, so that a torque continues to rotate the coil. This automatic reversal of the current is done via a commutator that electrically connects the rotating coil with the stationary contacts on the wires that supply the current from some source.

28-10 The Magnetic Dipole Moment

As we have just discussed, a torque acts to rotate a current-carrying coil placed in a magnetic field. In that sense, the coil behaves like a bar magnet placed in the magnetic field. Thus, like a bar magnet, a current-carrying coil is said to be a *magnetic dipole.* Moreover, to account for the torque on the coil due to the magnetic field, we assign a **magnetic dipole moment** $\vec{\mu}$ to the coil. The direction of $\vec{\mu}$ is that of the normal vector $\vec{n}$ to the plane of the coil and thus is given by the same right-hand rule shown in Fig. 28-19. That is, grasp the coil with the fingers of your right hand in the direction of current i; the outstretched thumb of that hand gives the direction of $\vec{\mu}$. The magnitude of $\vec{\mu}$ is given by

$$\mu = NiA \qquad \text{(magnetic moment)}, \qquad (28\text{-}35)$$

in which N is the number of turns in the coil, i is the current through the coil, and A is the area enclosed by each turn of the coil. From this equation, with i in amperes and A in square meters, we see that the unit of $\vec{\mu}$ is the ampere–square meter ($A \cdot m^2$).

The magnetic moment vector attempts to align with the magnetic field.

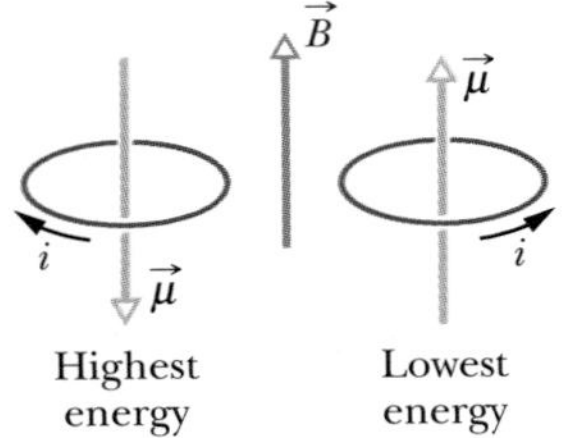

Fig. 28-20 The orientations of highest and lowest energy of a magnetic dipole (here a coil carrying current) in an external magnetic field $\vec{B}$. The direction of the current i gives the direction of the magnetic dipole moment $\vec{\mu}$ via the right-hand rule shown for $\vec{n}$ in Fig. 28-19*b*.

Using $\vec{\mu}$, we can rewrite Eq. 28-33 for the torque on the coil due to a magnetic field as

$$\tau = \mu B \sin \theta, \tag{28-36}$$

in which θ is the angle between the vectors $\vec{\mu}$ and $\vec{B}$.

We can generalize this to the vector relation

$$\vec{\tau} = \vec{\mu} \times \vec{B}, \tag{28-37}$$

which reminds us very much of the corresponding equation for the torque exerted by an *electric* field on an *electric* dipole—namely, Eq. 22-34:

$$\vec{\tau} = \vec{p} \times \vec{E}.$$

In each case the torque due to the field—either magnetic or electric—is equal to the vector product of the corresponding dipole moment and the field vector.

A magnetic dipole in an external magnetic field has an energy that depends on the dipole's orientation in the field. For electric dipoles we have shown (Eq. 22-38) that

$$U(\theta) = -\vec{p} \cdot \vec{E}.$$

In strict analogy, we can write for the magnetic case

$$U(\theta) = -\vec{\mu} \cdot \vec{B}. \tag{28-38}$$

In each case the energy due to the field is equal to the negative of the scalar product of the corresponding dipole moment and the field vector.

A magnetic dipole has its lowest energy ($= -\mu B \cos 0 = -\mu B$) when its dipole moment $\vec{\mu}$ is lined up with the magnetic field (Fig. 28-20). It has its highest energy ($= -\mu B \cos 180° = +\mu B$) when $\vec{\mu}$ is directed opposite the field. From Eq. 28-38, with U in joules and $\vec{B}$ in teslas, we see that the unit of $\vec{\mu}$ can be the joule per tesla (J/T) instead of the ampere–square meter as suggested by Eq. 28-35.

If an applied torque (due to "an external agent") rotates a magnetic dipole from an initial orientation θ_i to another orientation θ_f, then work W_a is done on the dipole by the applied torque. *If the dipole is stationary* before and after the change in its orientation, then work W_a is

$$W_a = U_f - U_i, \tag{28-39}$$

where U_f and U_i are calculated with Eq. 28-38.

So far, we have identified only a current-carrying coil as a magnetic dipole. However, a simple bar magnet is also a magnetic dipole, as is a rotating sphere of charge. Earth itself is (approximately) a magnetic dipole. Finally, most subatomic particles, including the electron, the proton, and the neutron, have magnetic dipole moments. As you will see in Chapter 32, all these quantities can be viewed as current loops. For comparison, some approximate magnetic dipole moments are shown in Table 28-2.

Table 28-2

Some Magnetic Dipole Moments

Small bar magnet	5 J/T
Earth	8.0×10^{22} J/T
Proton	1.4×10^{-26} J/T
Electron	9.3×10^{-24} J/T

CHECKPOINT 5

The figure shows four orientations, at angle θ, of a magnetic dipole moment $\vec{\mu}$ in a magnetic field. Rank the orientations according to (a) the magnitude of the torque on the dipole and (b) the orientation energy of the dipole, greatest first.

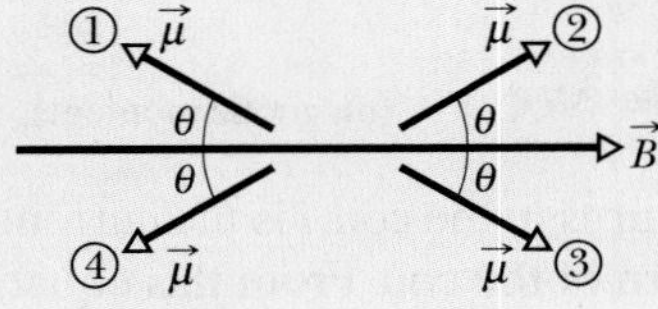

Sample Problem

Rotating a magnetic dipole in a magnetic field

Figure 28-21 shows a circular coil with 250 turns, an area A of 2.52×10^{-4} m^2, and a current of 100 μA. The coil is at rest in a uniform magnetic field of magnitude $B = 0.85$ T, with its magnetic dipole moment $\vec{\mu}$ initially aligned with $\vec{B}$.

(a) In Fig. 28-21, what is the direction of the current in the coil?

Right-hand rule: Imagine cupping the coil with your right hand so that your right thumb is outstretched in the direction of $\vec{\mu}$. The direction in which your fingers curl around the coil is the direction of the current in the coil. Thus, in the wires on the near side of the coil—those we see in Fig. 28-21—the current is from top to bottom.

(b) How much work would the torque applied by an external agent have to do on the coil to rotate it 90° from its initial orientation, so that $\vec{\mu}$ is perpendicular to $\vec{B}$ and the coil is again at rest?

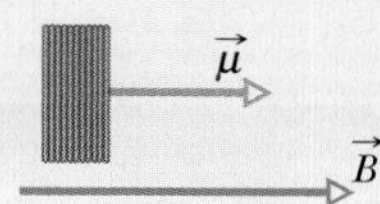

Fig. 28-21 A side view of a circular coil carrying a current and oriented so that its magnetic dipole moment is aligned with magnetic field $\vec{B}$.

KEY IDEA

The work W_a done by the applied torque would be equal to the change in the coil's orientation energy due to its change in orientation.

Calculations: From Eq. 28-39 ($W_a = U_f - U_i$), we find

$$W_a = U(90°) - U(0°)$$
$$= -\mu B \cos 90° - (-\mu B \cos 0°) = 0 + \mu B$$
$$= \mu B.$$

Substituting for μ from Eq. 28-35 ($\mu = NiA$), we find that

$$W_a = (NiA)B$$
$$= (250)(100 \times 10^{-6}\ \text{A})(2.52 \times 10^{-4}\ \text{m}^2)(0.85\ \text{T})$$
$$= 5.355 \times 10^{-6}\ \text{J} \approx 5.4\ \mu\text{J}. \qquad \text{(Answer)}$$

Additional examples, video, and practice available at *WileyPLUS*

REVIEW & SUMMARY

Magnetic Field $\vec{B}$ A **magnetic field** $\vec{B}$ is defined in terms of the force $\vec{F}_B$ acting on a test particle with charge q moving through the field with velocity $\vec{v}$:

$$\vec{F}_B = q\vec{v} \times \vec{B}. \qquad (28\text{-}2)$$

The SI unit for $\vec{B}$ is the **tesla** (T): 1 T = 1 N/(A · m) = 10^4 gauss.

The Hall Effect When a conducting strip carrying a current i is placed in a uniform magnetic field $\vec{B}$, some charge carriers (with charge e) build up on one side of the conductor, creating a potential difference V across the strip. The polarities of the sides indicate the sign of the charge carriers.

A Charged Particle Circulating in a Magnetic Field A charged particle with mass m and charge magnitude $|q|$ moving with velocity $\vec{v}$ perpendicular to a uniform magnetic field $\vec{B}$ will travel in a circle. Applying Newton's second law to the circular motion yields

$$|q|vB = \frac{mv^2}{r}, \qquad (28\text{-}15)$$

from which we find the radius r of the circle to be

$$r = \frac{mv}{|q|B}. \qquad (28\text{-}16)$$

The frequency of revolution f, the angular frequency ω, and the period of the motion T are given by

$$f = \frac{\omega}{2\pi} = \frac{1}{T} = \frac{|q|B}{2\pi m}. \qquad (28\text{-}19, 28\text{-}18, 28\text{-}17)$$

Magnetic Force on a Current-Carrying Wire A straight wire carrying a current i in a uniform magnetic field experiences a sideways force

$$\vec{F}_B = i\vec{L} \times \vec{B}. \qquad (28\text{-}26)$$

The force acting on a current element $i\,d\vec{L}$ in a magnetic field is

$$d\vec{F}_B = i\,d\vec{L} \times \vec{B}. \qquad (28\text{-}28)$$

The direction of the length vector $\vec{L}$ or $d\vec{L}$ is that of the current i.

Torque on a Current-Carrying Coil A coil (of area A and N turns, carrying current i) in a uniform magnetic field $\vec{B}$ will experience a torque $\vec{\tau}$ given by

$$\vec{\tau} = \vec{\mu} \times \vec{B}. \qquad (28\text{-}37)$$

Here $\vec{\mu}$ is the **magnetic dipole moment** of the coil, with magnitude $\mu = NiA$ and direction given by the right-hand rule.

Orientation Energy of a Magnetic Dipole The orientaion energy of a magnetic dipole in a magnetic field is

$$U(\theta) = -\vec{\mu} \cdot \vec{B}. \qquad (28\text{-}38)$$

If an external agent rotates a magnetic dipole from an initial orientation θ_i to some other orientation θ_f and the dipole is stationary both initially and finally, the work W_a done on the dipole by the agent is

$$W_a = \Delta U = U_f - U_i. \qquad (28\text{-}39)$$

QUESTIONS

1 Figure 28-22 shows three situations in which a positively charged particle moves at velocity $\vec{v}$ through a uniform magnetic field $\vec{B}$ and experiences a magnetic force $\vec{F}_B$. In each situation, determine whether the orientations of the vectors are physically reasonable.

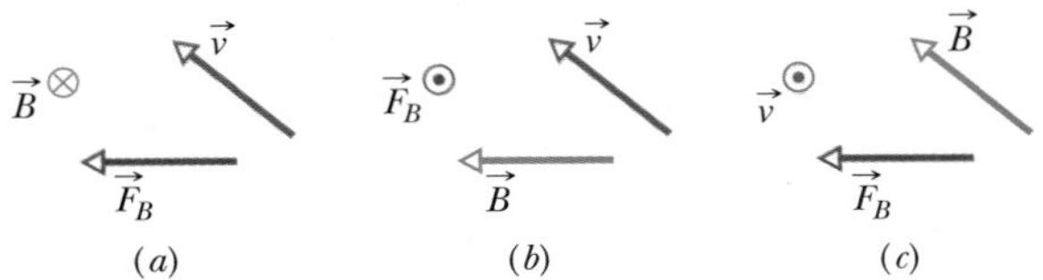

Fig. 28-22 Question 1.

2 Figure 28-23 shows a wire that carries current to the right through a uniform magnetic field. It also shows four choices for the direction of that field. (a) Rank the choices according to the magnitude of the electric potential difference that would be set up across the width of the wire, greatest first. (b) For which choice is the top side of the wire at higher potential than the bottom side of the wire?

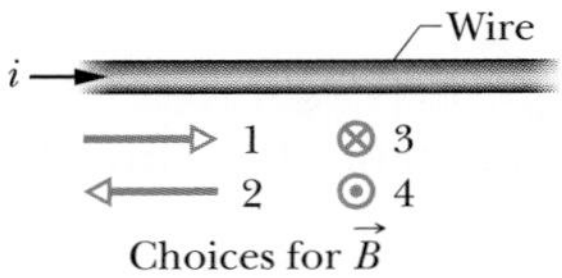

Fig. 28-23 Question 2.

3 Figure 28-24 shows a metallic, rectangular solid that is to move at a certain speed v through the uniform magnetic field $\vec{B}$. The dimensions of the solid are multiples of d, as shown. You have six choices for the direction of the velocity: parallel to x, y, or z in either the positive or negative direction. (a) Rank the six choices according to the potential difference set up across the solid, greatest first. (b) For which choice is the front face at lower potential?

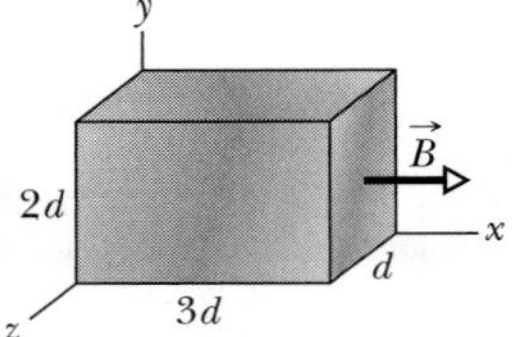

Fig. 28-24 Question 3.

4 Figure 28-25 shows the path of a particle through six regions of uniform magnetic field, where the path is either a half-circle or a quarter-circle. Upon leaving the last region, the particle travels between two charged, parallel plates and is deflected toward the plate of higher potential. What is the direction of the magnetic field in each of the six regions?

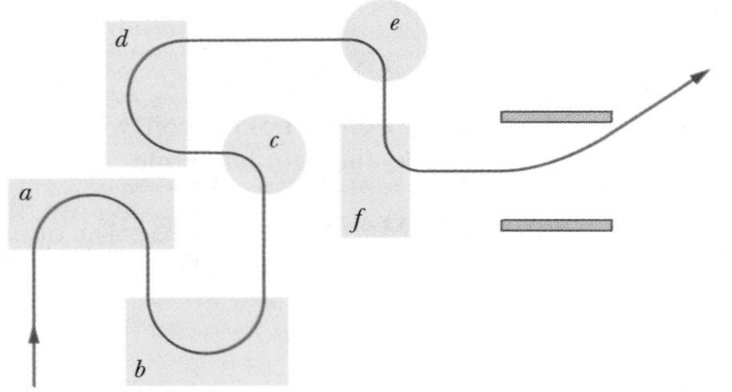

Fig. 28-25 Question 4.

5 In Section 28-4, we discussed a charged particle moving through crossed fields with the forces $\vec{F}_E$ and $\vec{F}_B$ in opposition. We found that the particle moves in a straight line (that is, neither force dominates the motion) if its speed is given by Eq. 28-7 ($v = E/B$). Which of the two forces dominates if the speed of the particle is (a) $v < E/B$ and (b) $v > E/B$?

6 Figure 28-26 shows crossed uniform electric and magnetic fields $\vec{E}$ and $\vec{B}$ and, at a certain instant, the velocity vectors of the 10 charged particles listed in Table 28-3. (The vectors are not drawn to scale.) The speeds given in the table are either less than or greater than E/B (see Question 5). Which particles will move out of the page toward you after the instant shown in Fig. 28-26?

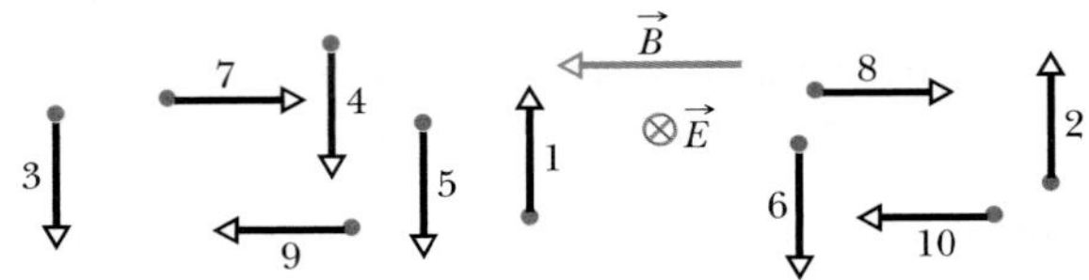

Fig. 28-26 Question 6.

Table 28-3

Question 6

Particle	Charge	Speed	Particle	Charge	Speed
1	+	Less	6	−	Greater
2	+	Greater	7	+	Less
3	+	Less	8	+	Greater
4	+	Greater	9	−	Less
5	−	Less	10	−	Greater

7 Figure 28-27 shows the path of an electron that passes through two regions containing uniform magnetic fields of magnitudes B_1 and B_2. Its path in each region is a half-circle. (a) Which field is stronger? (b) What is the direction of each field? (c) Is the time spent by the electron in the $\vec{B}_1$ region greater than, less than, or the same as the time spent in the $\vec{B}_2$ region?

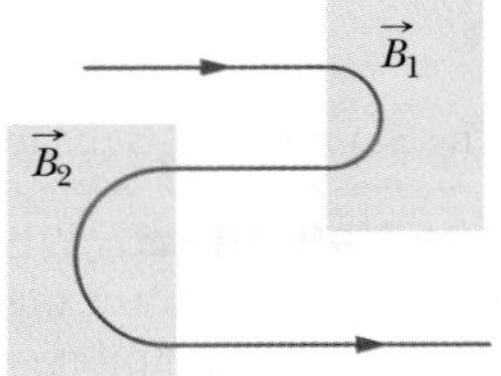

Fig. 28-27 Question 7.

8 Figure 28-28 shows the path of an electron in a region of uniform magnetic field. The path consists of two straight sections, each between a pair of uniformly charged plates, and two half-circles. Which plate is at the higher electric potential in (a) the top pair of plates and (b) the bottom pair? (c) What is the direction of the magnetic field?

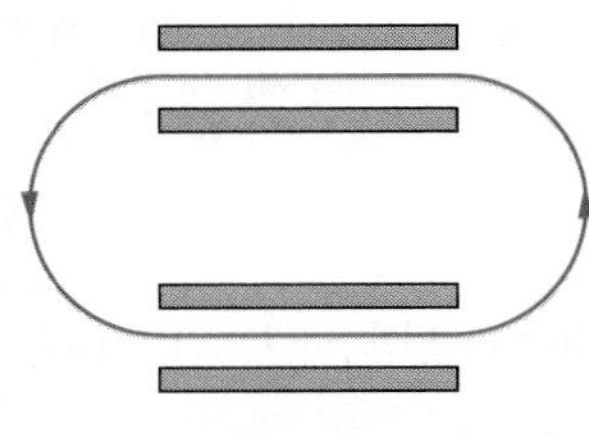

Fig. 28-28 Question 8.

9 (a) In Checkpoint 5, if the dipole moment $\vec{\mu}$ is rotated from orientation 2 to orientation 1 by an external agent, is the work done on the dipole by the agent positive, negative, or zero? (b) Rank the work done on the dipole by the agent for these three rotations, greatest first: $2 \rightarrow 1$, $2 \rightarrow 4$, $2 \rightarrow 3$.

10 *Particle roundabout.* Figure 28-29 shows 11 paths through a region of uniform magnetic field. One path is a straight line; the rest are half-circles. Table 28-4 gives the masses, charges, and speeds of 11 particles that take these paths through the field in the directions shown. Which path in the figure corresponds to which particle in the table? (The direction of the magnetic field can be determined by means of one of the paths, which is unique.)

Table 28-4

Question 10

Particle	Mass	Charge	Speed
1	$2m$	q	v
2	m	$2q$	v
3	$m/2$	q	$2v$
4	$3m$	$3q$	$3v$
5	$2m$	q	$2v$
6	m	$-q$	$2v$
7	m	$-4q$	v
8	m	$-q$	v
9	$2m$	$-2q$	$3v$
10	m	$-2q$	$8v$
11	$3m$	0	$3v$

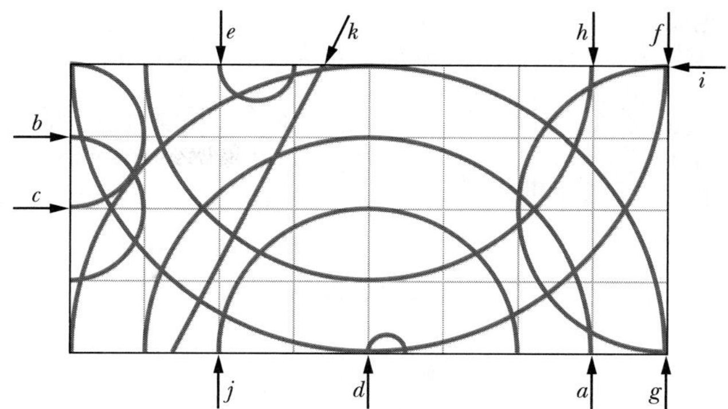

Fig. 28-29 Question 10.

11 In Fig. 28-30, a charged particle enters a uniform magnetic field $\vec{B}$ with speed v_0, moves through a half-circle in time T_0, and then leaves the field. (a) Is the charge positive or negative? (b) Is the final speed of the particle greater than, less than, or equal to v_0? (c) If the initial speed had been $0.5v_0$, would the time spent in field $\vec{B}$ have been greater than, less than, or equal to T_0? (d) Would the path have been a half-circle, more than a half-circle, or less than a half-circle?

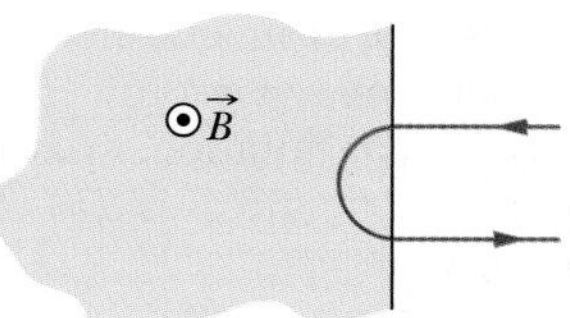

Fig. 28-30 Question 11.

PROBLEMS

 Tutoring problem available (at instructor's discretion) in *WileyPLUS* and WebAssign

SSM Worked-out solution available in Student Solutions Manual

 Number of dots indicates level of problem difficulty

Additional information available in *The Flying Circus of Physics* and at flyingcircusofphysics.com

WWW Worked-out solution is at

ILW Interactive solution is at

http://www.wiley.com/college/halliday

sec. 28-3 The Definition of $\vec{B}$

•**1** SSM ILW A proton traveling at 23.0° with respect to the direction of a magnetic field of strength 2.60 mT experiences a magnetic force of 6.50×10^{-17} N. Calculate (a) the proton's speed and (b) its kinetic energy in electron-volts.

•**2** A particle of mass 10 g and charge 80 μC moves through a uniform magnetic field, in a region where the free-fall acceleration is $-9.8\hat{j}$ m/s^2. The velocity of the particle is a constant $20\hat{i}$ km/s, which is perpendicular to the magnetic field. What, then, is the magnetic field?

•**3** An electron that has velocity

$$\vec{v} = (2.0 \times 10^6 \text{ m/s})\hat{i} + (3.0 \times 10^6 \text{ m/s})\hat{j}$$

moves through the uniform magnetic field $\vec{B} = (0.030 \text{ T})\hat{i} - (0.15 \text{ T})\hat{j}$. (a) Find the force on the electron due to the magnetic field. (b) Repeat your calculation for a proton having the same velocity.

•**4** An alpha particle travels at a velocity $\vec{v}$ of magnitude 550 m/s through a uniform magnetic field $\vec{B}$ of magnitude 0.045 T. (An alpha particle has a charge of $+3.2 \times 10^{-19}$ C and a mass of 6.6×10^{-27} kg.) The angle between $\vec{v}$ and $\vec{B}$ is 52°. What is the magnitude of (a) the force $\vec{F}_B$ acting on the particle due to the field and (b) the acceleration of the particle due to $\vec{F}_B$? (c) Does the speed of the particle increase, decrease, or remain the same?

••**5** An electron moves through a uniform magnetic field given by $\vec{B} = B_x\hat{i} + (3.0B_x)\hat{j}$. At a particular instant, the electron has velocity $\vec{v} = (2.0\hat{i} + 4.0\hat{j})$ m/s and the magnetic force acting on it is $(6.4 \times 10^{-19} \text{ N})\hat{k}$. Find B_x.

••**6** GO A proton moves through a uniform magnetic field given by $\vec{B} = (10\hat{i} - 20\hat{j} + 30\hat{k})$ mT. At time t_1, the proton has a velocity given by $\vec{v} = v_x\hat{i} + v_y\hat{j} + (2.0 \text{ km/s})\hat{k}$ and the magnetic force on the proton is $\vec{F}_B = (4.0 \times 10^{-17} \text{ N})\hat{i} + (2.0 \times 10^{-17} \text{ N})\hat{j}$. At that instant, what are (a) v_x and (b) v_y?

sec. 28-4 Crossed Fields: Discovery of the Electron

•**7** An electron has an initial velocity of $(12.0\hat{j} + 15.0\hat{k})$ km/s and a constant acceleration of $(2.00 \times 10^{12} \text{ m/s}^2)\hat{i}$ in a region in which uniform electric and magnetic fields are present. If $\vec{B} = (400\ \mu\text{T})\hat{i}$, find the electric field $\vec{E}$.

•**8** An electric field of 1.50 kV/m and a perpendicular magnetic field of 0.400 T act on a moving electron to produce no net force. What is the electron's speed?

•**9** ILW In Fig. 28-31, an electron accelerated from rest through potential difference $V_1 = 1.00$ kV enters the gap between two parallel plates having separation $d = 20.0$ mm and potential difference $V_2 = 100$ V. The lower plate is at the lower potential. Neglect fringing and assume that the electron's velocity vector is perpendicular to the electric field vector between the plates. In unit-vector notation, what uniform magnetic field allows the electron to travel in a straight line in the gap?

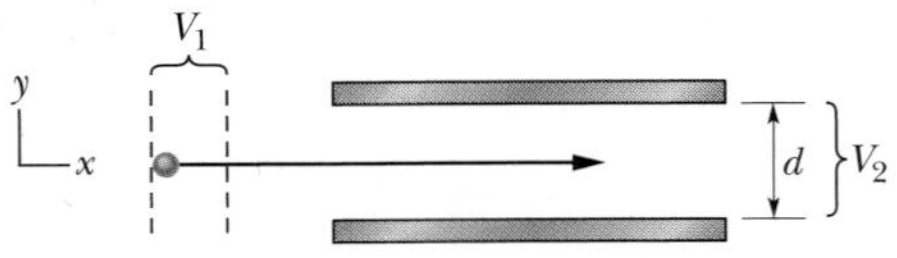

Fig. 28-31 Problem 9.

••10 A proton travels through uniform magnetic and electric fields. The magnetic field is $\vec{B} = -2.50\hat{i}$ mT. At one instant the velocity of the proton is $\vec{v} = 2000\hat{j}$ m/s. At that instant and in unit-vector notation, what is the net force acting on the proton if the electric field is (a) $4.00\hat{k}$ V/m, (b) $-4.00\hat{k}$ V/m, and (c) $4.00\hat{i}$ V/m?

••11 An ion source is producing ^{6}Li ions, which have charge $+e$ and mass 9.99×10^{-27} kg. The ions are accelerated by a potential difference of 10 kV and pass horizontally into a region in which there is a uniform vertical magnetic field of magnitude $B = 1.2$ T. Calculate the strength of the smallest electric field, to be set up over the same region, that will allow the ^{6}Li ions to pass through undeflected.

•••12 GO At time t_1, an electron is sent along the positive direction of an x axis, through both an electric field $\vec{E}$ and a magnetic field $\vec{B}$, with $\vec{E}$ directed parallel to the y axis. Figure 28-32 gives the y component $F_{net,y}$ of the net force on the electron due to the two fields, as a function of the electron's speed v at time t_1. The scale of the velocity axis is set by $v_s = 100.0$ m/s. The x and z components of the net force are zero at t_1. Assuming $B_x = 0$, find (a) the magnitude E and (b) $\vec{B}$ in unit-vector notation.

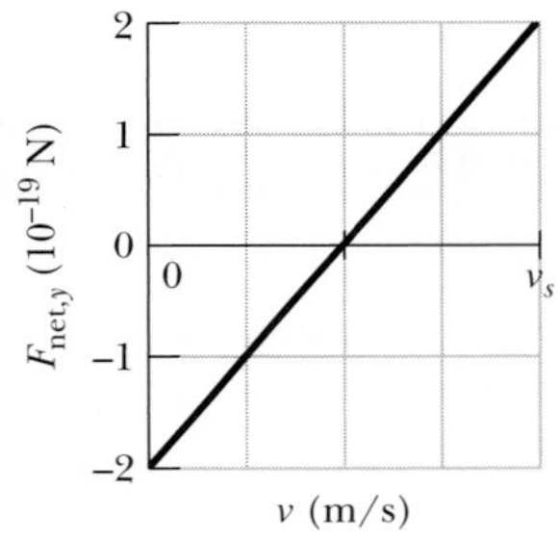

Fig. 28-32 Problem 12.

sec. 28-5 Crossed Fields: The Hall Effect

•13 A strip of copper 150 μm thick and 4.5 mm wide is placed in a uniform magnetic field $\vec{B}$ of magnitude 0.65 T, with $\vec{B}$ perpendicular to the strip. A current $i = 23$ A is then sent through the strip such that a Hall potential difference V appears across the width of the strip. Calculate V. (The number of charge carriers per unit volume for copper is 8.47×10^{28} electrons/m^3.)

•14 A metal strip 6.50 cm long, 0.850 cm wide, and 0.760 mm thick moves with constant velocity $\vec{v}$ through a uniform magnetic field $B = 1.20$ mT directed perpendicular to the strip, as shown in Fig. 28-33. A potential difference of 3.90 μV is measured between points x and y across the strip. Calculate the speed v.

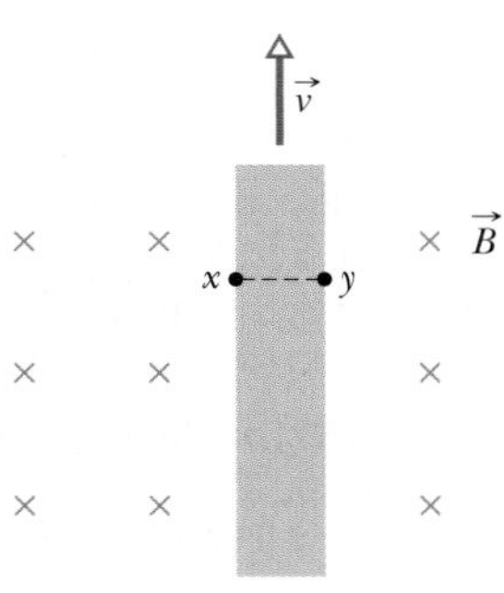

Fig. 28-33 Problem 14.

••15 In Fig. 28-34, a conducting rectangular solid of dimensions $d_x = 5.00$ m, $d_y = 3.00$ m, and $d_z = 2.00$ m moves at constant velocity $\vec{v} = (20.0 \text{ m/s})\hat{i}$ through a uniform magnetic field $\vec{B} = (30.0 \text{ mT})\hat{j}$. What are the resulting (a) electric field within the solid, in unit-vector notation, and (b) potential difference across the solid?

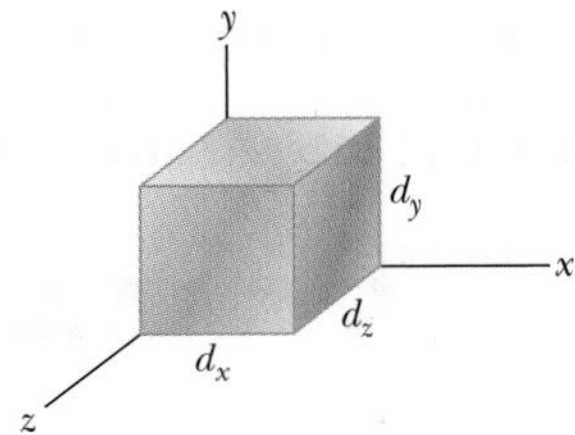

Fig. 28-34 Problems 15 and 16.

•••16 GO Figure 28-34 shows a metallic block, with its faces parallel to coordinate axes. The block is in a uniform magnetic field of magnitude 0.020 T. One edge length of the block is 25 cm; the block is *not* drawn to scale. The block is moved at 3.0 m/s parallel to each axis, in turn, and the resulting potential difference V that appears across the block is measured. With the motion parallel to the y axis, $V = 12$ mV; with the motion parallel to the z axis, $V = 18$ mV; with the motion parallel to the x axis, $V = 0$. What are the block lengths (a) d_x, (b) d_y, and (c) d_z?

sec. 28-6 A Circulating Charged Particle

•17 An alpha particle can be produced in certain radioactive decays of nuclei and consists of two protons and two neutrons. The particle has a charge of $q = +2e$ and a mass of 4.00 u, where u is the atomic mass unit, with $1 \text{ u} = 1.661 \times 10^{-27}$ kg. Suppose an alpha particle travels in a circular path of radius 4.50 cm in a uniform magnetic field with $B = 1.20$ T. Calculate (a) its speed, (b) its period of revolution, (c) its kinetic energy, and (d) the potential difference through which it would have to be accelerated to achieve this energy.

•18 GO In Fig. 28-35, a particle moves along a circle in a region of uniform magnetic field of magnitude $B = 4.00$ mT. The particle is either a proton or an electron (you must decide which). It experiences a magnetic force of magnitude 3.20×10^{-15} N. What are (a) the particle's speed, (b) the radius of the circle, and (c) the period of the motion?

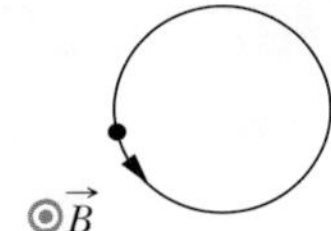

Fig. 28-35 Problem 18.

•19 A certain particle is sent into a uniform magnetic field, with the particle's velocity vector perpendicular to the direction of the field. Figure 28-36 gives the period T of the particle's motion versus the *inverse* of the field magnitude B. The vertical axis scale is set by $T_s = 40.0$ ns, and the horizontal axis scale is set by $B_s^{-1} = 5.0\ \text{T}^{-1}$. What is the ratio m/q of the particle's mass to the magnitude of its charge?

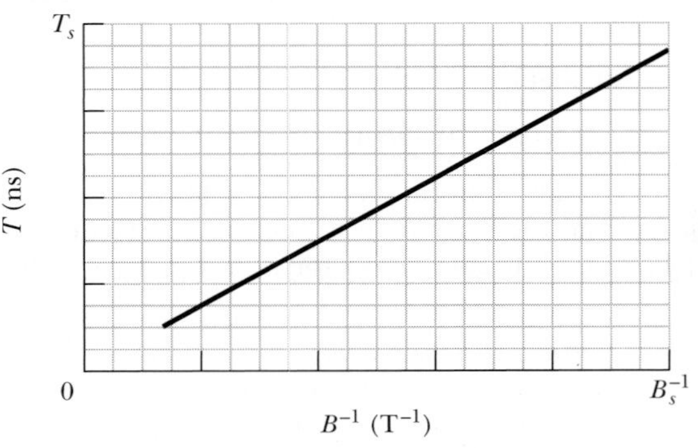

Fig. 28-36 Problem 19.

•20 An electron is accelerated from rest through potential difference V and then enters a region of uniform magnetic field, where it undergoes uniform circular motion. Figure 28-37 gives the radius r of that motion versus $V^{1/2}$. The vertical axis scale is set by $r_s = 3.0$ mm, and the horizontal axis scale is set by $V_s^{1/2} = 40.0\ V^{1/2}$. What is the magnitude of the magnetic field?

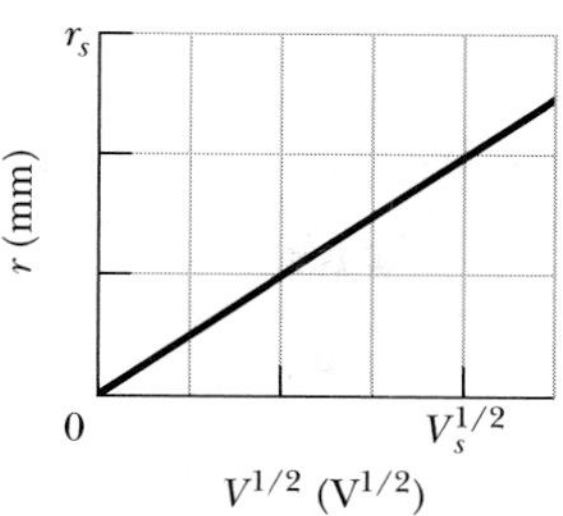

Fig. 28-37 Problem 20.

•21 SSM An electron of kinetic energy 1.20 keV circles in a plane perpendicular to a uniform magnetic field. The orbit radius is 25.0 cm. Find (a) the electron's speed, (b) the magnetic field magnitude, (c) the circling frequency, and (d) the period of the motion.

•22 In a nuclear experiment a proton with kinetic energy 1.0 MeV moves in a circular path in a uniform magnetic field. What energy must (a) an alpha particle ($q = +2e$, $m = 4.0$ u) and (b) a deuteron ($q = +e$, $m = 2.0$ u) have if they are to circulate in the same circular path?

•23 What uniform magnetic field, applied perpendicular to a beam of electrons moving at 1.30×10^6 m/s, is required to make the electrons travel in a circular arc of radius 0.350 m?

•24 An electron is accelerated from rest by a potential difference of 350 V. It then enters a uniform magnetic field of magnitude 200 mT with its velocity perpendicular to the field. Calculate (a) the speed of the electron and (b) the radius of its path in the magnetic field.

•25 (a) Find the frequency of revolution of an electron with an energy of 100 eV in a uniform magnetic field of magnitude 35.0 μT. (b) Calculate the radius of the path of this electron if its velocity is perpendicular to the magnetic field.

••26 In Fig. 28-38, a charged particle moves into a region of uniform magnetic field $\vec{B}$, goes through half a circle, and then exits that region. The particle is either a proton or an electron (you must decide which). It spends 130 ns in the region. (a) What is the magnitude of $\vec{B}$? (b) If the particle is sent back through the magnetic field (along the same initial path) but with 2.00 times its previous kinetic energy, how much time does it spend in the field during this trip?

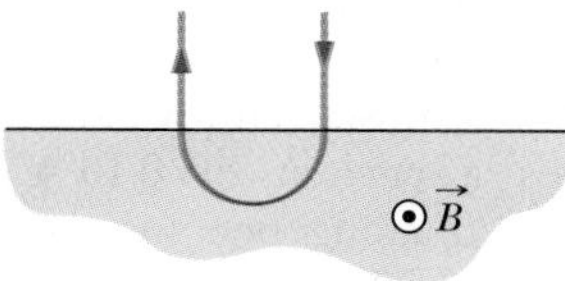

Fig. 28-38 Problem 26.

••27 A mass spectrometer (Fig. 28-12) is used to separate uranium ions of mass 3.92×10^{-25} kg and charge 3.20×10^{-19} C from related species. The ions are accelerated through a potential difference of 100 kV and then pass into a uniform magnetic field, where they are bent in a path of radius 1.00 m. After traveling through 180° and passing through a slit of width 1.00 mm and height 1.00 cm, they are collected in a cup. (a) What is the magnitude of the (perpendicular) magnetic field in the separator? If the machine is used to separate out 100 mg of material per hour, calculate (b) the current of the desired ions in the machine and (c) the thermal energy produced in the cup in 1.00 h.

••28 A particle undergoes uniform circular motion of radius 26.1 μm in a uniform magnetic field. The magnetic force on the particle has a magnitude of 1.60×10^{-17} N. What is the kinetic energy of the particle?

••29 An electron follows a helical path in a uniform magnetic field of magnitude 0.300 T. The pitch of the path is 6.00 μm, and the magnitude of the magnetic force on the electron is 2.00×10^{-15} N. What is the electron's speed?

••30 GO In Fig. 28-39, an electron with an initial kinetic energy of 4.0 keV enters region 1 at time $t = 0$. That region contains a uniform magnetic field directed into the page, with magnitude 0.010 T. The electron goes through a half-circle and then exits region 1, headed toward region 2 across a gap of 25.0 cm. There is an electric potential difference $\Delta V = 2000$ V across the gap, with a polarity such that the electron's speed increases uniformly as it traverses the gap. Region 2 contains a uniform magnetic field directed out of the page, with magnitude 0.020 T. The electron goes through a half-circle and then leaves region 2. At what time t does it leave?

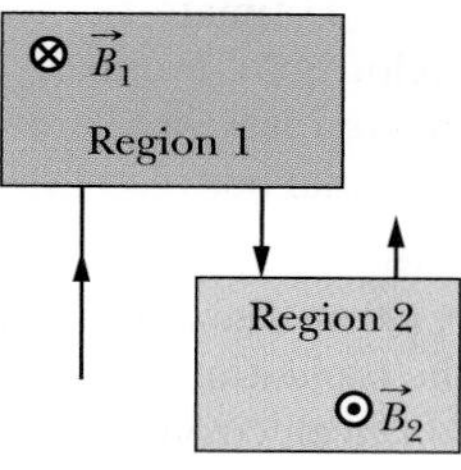

Fig. 28-39 Problem 30.

••31 A particular type of fundamental particle decays by transforming into an electron e^- and a positron e^+. Suppose the decaying particle is at rest in a uniform magnetic field $\vec{B}$ of magnitude 3.53 mT and the e^- and e^+ move away from the decay point in paths lying in a plane perpendicular to $\vec{B}$. How long after the decay do the e^- and e^+ collide?

••32 A source injects an electron of speed $v = 1.5 \times 10^7$ m/s into a uniform magnetic field of magnitude $B = 1.0 \times 10^{-3}$ T. The velocity of the electron makes an angle $\theta = 10°$ with the direction of the magnetic field. Find the distance d from the point of injection at which the electron next crosses the field line that passes through the injection point.

••33 SSM WWW A positron with kinetic energy 2.00 keV is projected into a uniform magnetic field $\vec{B}$ of magnitude 0.100 T, with its velocity vector making an angle of 89.0° with $\vec{B}$. Find (a) the period, (b) the pitch p, and (c) the radius r of its helical path.

••34 An electron follows a helical path in a uniform magnetic field given by $\vec{B} = (20\hat{i} - 50\hat{j} - 30\hat{k})$ mT. At time $t = 0$, the electron's velocity is given by $\vec{v} = (20\hat{i} - 30\hat{j} + 50\hat{k})$ m/s. (a) What is the angle ϕ between $\vec{v}$ and $\vec{B}$? The electron's velocity changes with time. Do (b) its speed and (c) the angle ϕ change with time? (d) What is the radius of the helical path?

sec. 28-7 Cyclotrons and Synchrotrons

••35 A proton circulates in a cyclotron, beginning approximately at rest at the center. Whenever it passes through the gap between dees, the electric potential difference between the dees is 200 V. (a) By how much does its kinetic energy increase with each passage through the gap? (b) What is its kinetic energy as it completes 100 passes through the gap? Let r_{100} be the radius of the proton's circular path as it completes those 100 passes and enters a dee, and let r_{101} be its next radius, as it enters a dee the next time. (c) By what percentage does the radius increase when it changes from r_{100} to r_{101}? That is, what is

$$\text{percentage increase} = \frac{r_{101} - r_{100}}{r_{100}}\,100\%?$$

••36 A cyclotron with dee radius 53.0 cm is operated at an oscillator frequency of 12.0 MHz to accelerate protons. (a) What magnitude B of magnetic field is required to achieve resonance? (b) At that field magnitude, what is the kinetic energy of a proton emerging from the cyclotron? Suppose, instead, that $B = 1.57$ T. (c) What oscillator frequency is required to achieve resonance now? (d) At that frequency, what is the kinetic energy of an emerging proton?

••37 Estimate the total path length traveled by a deuteron in a cyclotron of radius 53 cm and operating frequency 12 MHz during the (entire) acceleration process. Assume that the accelerating potential between the dees is 80 kV.

••38 In a certain cyclotron a proton moves in a circle of radius 0.500 m. The magnitude of the magnetic field is 1.20 T. (a) What is the oscillator frequency? (b) What is the kinetic energy of the proton, in electron-volts?

sec. 28-8 Magnetic Force on a Current-Carrying Wire

•39 SSM A horizontal power line carries a current of 5000 A from south to north. Earth's magnetic field (60.0 μT) is directed toward the north and inclined downward at 70.0° to the horizontal. Find the (a) magnitude and (b) direction of the magnetic force on 100 m of the line due to Earth's field.

•40 A wire 1.80 m long carries a current of 13.0 A and makes an angle of 35.0° with a uniform magnetic field of magnitude $B = 1.50$ T. Calculate the magnetic force on the wire.

•41 ILW A 13.0 g wire of length $L = 62.0$ cm is suspended by a pair of flexible leads in a uniform magnetic field of magnitude 0.440 T (Fig. 28-40). What are the (a) magnitude and (b) direction (left or right) of the current required to remove the tension in the supporting leads?

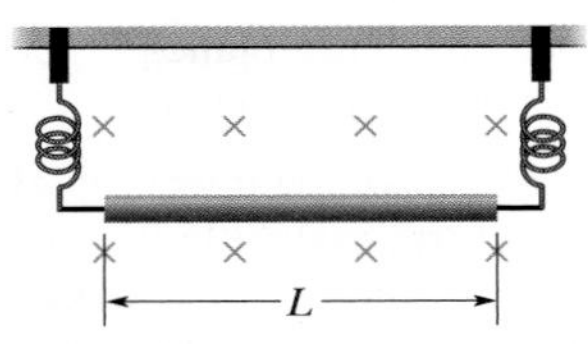

Fig. 28-40 Problem 41.

•42 The bent wire shown in Fig. 28-41 lies in a uniform magnetic field. Each straight section is 2.0 m long and makes an angle of $\theta = 60°$ with the x axis, and the wire carries a current of 2.0 A. What is the net magnetic force on the wire in unit-vector notation if the magnetic field is given by (a) $4.0\hat{k}$ T and (b) $4.0\hat{i}$ T?

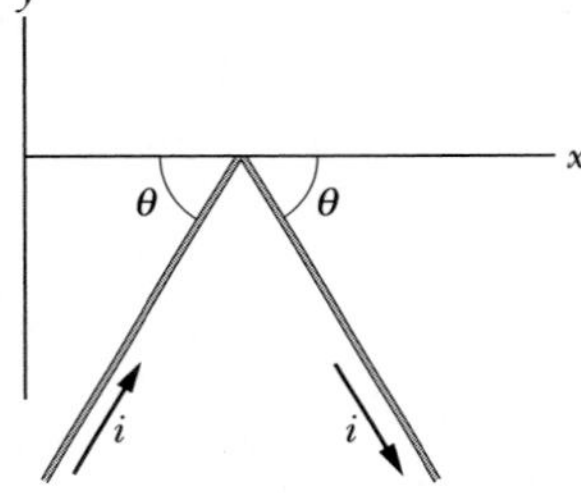

Fig. 28-41 Problem 42.

•43 A single-turn current loop, carrying a current of 4.00 A, is in the shape of a right triangle with sides 50.0, 120, and 130 cm. The loop is in a uniform magnetic field of magnitude 75.0 mT whose direction is parallel to the current in the 130 cm side of the loop. What is the magnitude of the magnetic force on (a) the 130 cm side, (b) the 50.0 cm side, and (c) the 120 cm side? (d) What is the magnitude of the net force on the loop?

••44 Figure 28-42 shows a wire ring of radius $a = 1.8$ cm that is perpendicular to the general direction of a radially symmetric, diverging magnetic field. The magnetic field at the ring is everywhere of the same magnitude $B = 3.4$ mT, and its direction at the ring everywhere makes an angle $\theta = 20°$ with a normal to the plane of the ring. The twisted lead wires have no effect on the problem. Find the magnitude of the force the field exerts on the ring if the ring carries a current $i = 4.6$ mA.

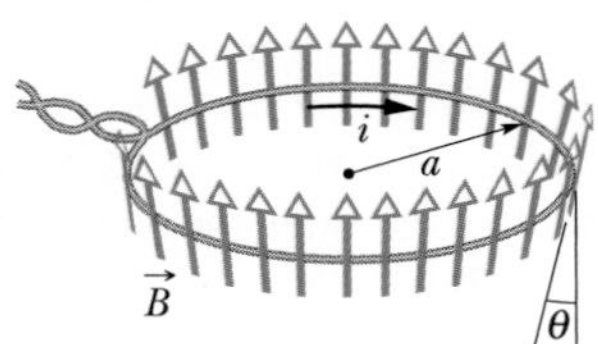

Fig. 28-42 Problem 44.

••45 A wire 50.0 cm long carries a 0.500 A current in the positive direction of an x axis through a magnetic field $\vec{B} = (3.00 \text{ mT})\hat{j} + (10.0 \text{ mT})\hat{k}$. In unit-vector notation, what is the magnetic force on the wire?

••46 In Fig. 28-43, a metal wire of mass $m = 24.1$ mg can slide with negligible friction on two horizontal parallel rails separated by distance $d = 2.56$ cm. The track lies in a vertical uniform magnetic field of magnitude 56.3 mT. At time $t = 0$, device G is connected to the rails, producing a constant current $i = 9.13$ mA in the wire and rails (even as the wire moves). At $t = 61.1$ ms, what are the wire's (a) speed and (b) direction of motion (left or right)?

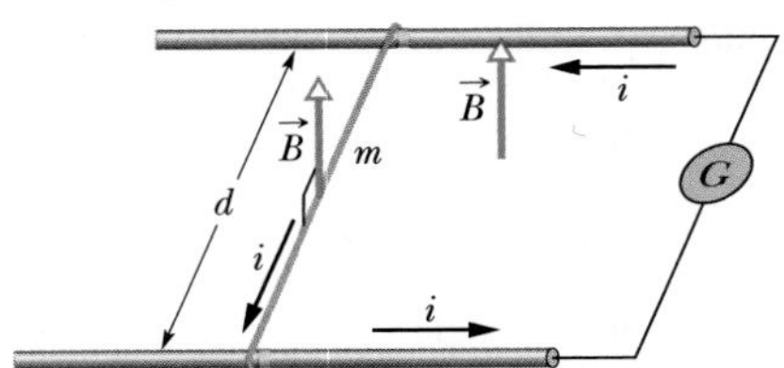

Fig. 28-43 Problem 46.

•••47 GO A 1.0 kg copper rod rests on two horizontal rails 1.0 m apart and carries a current of 50 A from one rail to the other. The coefficient of static friction between rod and rails is 0.60. What are the (a) magnitude and (b) angle (relative to the vertical) of the smallest magnetic field that puts the rod on the verge of sliding?

•••48 A long, rigid conductor, lying along an x axis, carries a current of 5.0 A in the negative x direction. A magnetic field $\vec{B}$ is present, given by $\vec{B} = 3.0\hat{i} + 8.0x^2\hat{j}$, with x in meters and $\vec{B}$ in milliteslas. Find, in unit-vector notation, the force on the 2.0 m segment of the conductor that lies between $x = 1.0$ m and $x = 3.0$ m.

sec. 28-9 Torque on a Current Loop

•49 SSM Figure 28-44 shows a rectangular 20-turn coil of wire, of dimensions 10 cm by 5.0 cm. It carries a current of 0.10 A and is hinged along one long side. It is mounted in the xy plane, at angle $\theta = 30°$ to the direction of a uniform magnetic field of magnitude 0.50 T. In unit-vector notation, what is the torque acting on the coil about the hinge line?

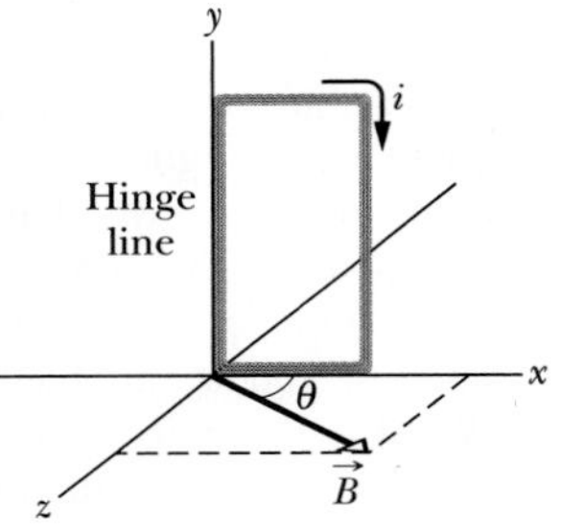

Fig. 28-44 Problem 49.

••50 An electron moves in a circle of radius $r = 5.29 \times 10^{-11}$ m with speed 2.19×10^6 m/s. Treat the circular path as a current loop with a constant current equal to the ratio of the electron's charge magnitude to the period of the motion. If the circle lies in a uniform magnetic field of magnitude $B = 7.10$ mT, what is the maximum possible magnitude of the torque produced on the loop by the field?

••51 Figure 28-45 shows a wood cylinder of mass $m = 0.250$ kg and length $L = 0.100$ m, with $N = 10.0$ turns of wire wrapped around it longitudinally, so that the plane of the wire coil contains the long central

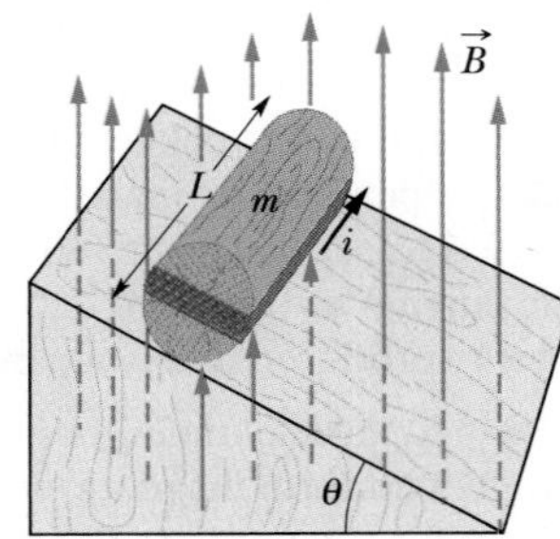

Fig. 28-45 Problem 51.

axis of the cylinder. The cylinder is released on a plane inclined at an angle θ to the horizontal, with the plane of the coil parallel to the incline plane. If there is a vertical uniform magnetic field of magnitude 0.500 T, what is the least current i through the coil that keeps the cylinder from rolling down the plane?

••52 In Fig. 28-46, a rectangular loop carrying current lies in the plane of a uniform magnetic field of magnitude 0.040 T. The loop consists of a single turn of flexible conducting wire that is wrapped around a flexible mount such that the dimensions of the rectangle can be changed. (The total length of the wire is not changed.) As edge length x is varied from approximately zero to its maximum value of approximately 4.0 cm, the magnitude τ of the torque on the loop changes. The maximum value of τ is 4.80×10^{-8} N·m. What is the current in the loop?

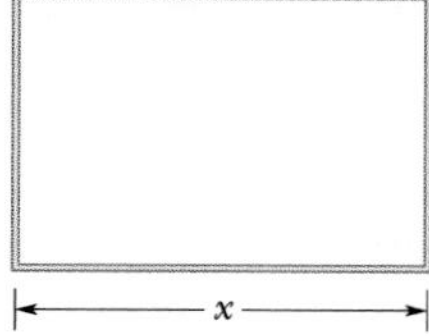

Fig. 28-46 Problem 52.

••53 Prove that the relation $\tau = NiAB \sin\theta$ holds not only for the rectangular loop of Fig. 28-19 but also for a closed loop of any shape. (*Hint:* Replace the loop of arbitrary shape with an assembly of adjacent long, thin, approximately rectangular loops that are nearly equivalent to the loop of arbitrary shape as far as the distribution of current is concerned.)

sec. 28-10 The Magnetic Dipole Moment

•54 A magnetic dipole with a dipole moment of magnitude 0.020 J/T is released from rest in a uniform magnetic field of magnitude 52 mT. The rotation of the dipole due to the magnetic force on it is unimpeded. When the dipole rotates through the orientation where its dipole moment is aligned with the magnetic field, its kinetic energy is 0.80 mJ. (a) What is the initial angle between the dipole moment and the magnetic field? (b) What is the angle when the dipole is next (momentarily) at rest?

•55 SSM Two concentric, circular wire loops, of radii $r_1 = 20.0$ cm and $r_2 = 30.0$ cm, are located in an xy plane; each carries a clockwise current of 7.00 A (Fig. 28-47). (a) Find the magnitude of the net magnetic dipole moment of the system. (b) Repeat for reversed current in the inner loop.

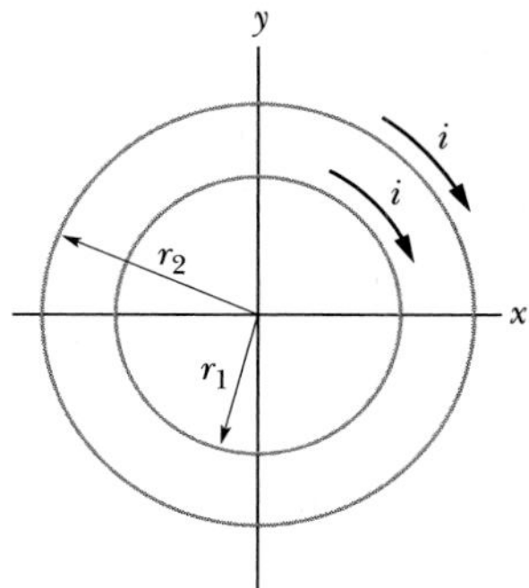

Fig. 28-47 Problem 55.

•56 A circular wire loop of radius 15.0 cm carries a current of 2.60 A. It is placed so that the normal to its plane makes an angle of 41.0° with a uniform magnetic field of magnitude 12.0 T. (a) Calculate the magnitude of the magnetic dipole moment of the loop. (b) What is the magnitude of the torque acting on the loop?

•57 SSM A circular coil of 160 turns has a radius of 1.90 cm. (a) Calculate the current that results in a magnetic dipole moment of magnitude 2.30 A·m². (b) Find the maximum magnitude of the torque that the coil, carrying this current, can experience in a uniform 35.0 mT magnetic field.

•58 The magnetic dipole moment of Earth has magnitude 8.00×10^{22} J/T. Assume that this is produced by charges flowing in Earth's molten outer core. If the radius of their circular path is 3500 km, calculate the current they produce.

•59 A current loop, carrying a current of 5.0 A, is in the shape of a right triangle with sides 30, 40, and 50 cm. The loop is in a uniform magnetic field of magnitude 80 mT whose direction is parallel to the current in the 50 cm side of the loop. Find the magnitude of (a) the magnetic dipole moment of the loop and (b) the torque on the loop.

••60 Figure 28-48 shows a current loop $ABCDEFA$ carrying a current $i = 5.00$ A. The sides of the loop are parallel to the coordinate axes shown, with $AB = 20.0$ cm, $BC = 30.0$ cm, and $FA = 10.0$ cm. In unit-vector notation, what is the magnetic dipole moment of this loop? (*Hint:* Imagine equal and opposite currents i in the line segment AD; then treat the two rectangular loops $ABCDA$ and $ADEFA$.)

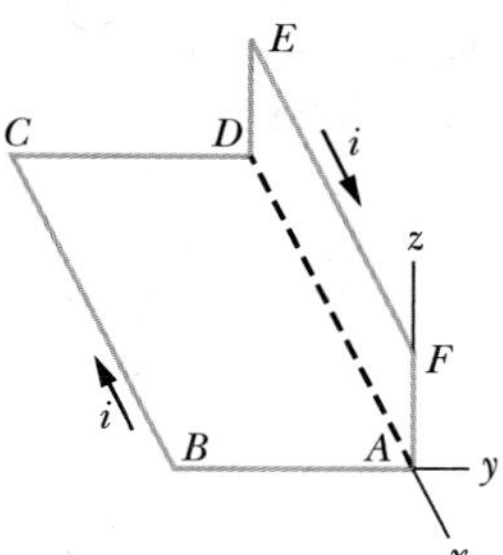

Fig. 28-48 Problem 60.

••61 SSM The coil in Fig. 28-49 carries current $i = 2.00$ A in the direction indicated, is parallel to an xz plane, has 3.00 turns and an area of 4.00×10^{-3} m², and lies in a uniform magnetic field $\vec{B} = (2.00\hat{i} - 3.00\hat{j} - 4.00\hat{k})$ mT. What are (a) the orientation energy of the coil in the magnetic field and (b) the torque (in unit-vector notation) on the coil due to the magnetic field?

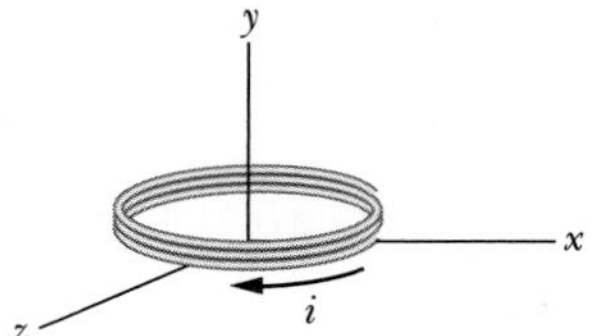

Fig. 28-49 Problem 61.

••62 GO In Fig. 28-50a, two concentric coils, lying in the same plane, carry currents in opposite directions. The current in the larger coil 1 is fixed. Current i_2 in coil 2 can be varied. Figure 28-50b gives the net magnetic moment of the two-coil system as a function of i_2. The vertical axis scale is set by $\mu_{net,s} = 2.0 \times 10^{-5}$ A·m², and the horizontal axis scale is set by $i_{2s} = 10.0$ mA. If the current in coil 2 is then reversed, what is the magnitude of the net magnetic moment of the two-coil system when $i_2 = 7.0$ mA?

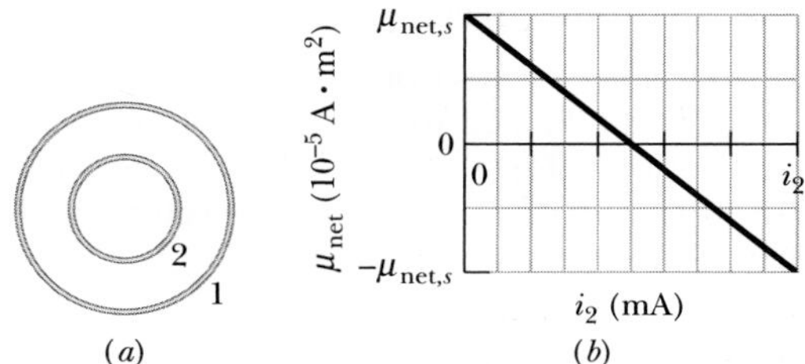

Fig. 28-50 Problem 62.

••63 A circular loop of wire having a radius of 8.0 cm carries a current of 0.20 A. A vector of unit length and parallel to the dipole moment $\vec{\mu}$ of the loop is given by $0.60\hat{i} - 0.80\hat{j}$. (This unit vector gives the orientation of the magnetic dipole moment vector.) If the loop is located in a uniform magnetic field given by $\vec{B} = (0.25\text{ T})\hat{i} + (0.30\text{ T})\hat{k}$, find (a) the torque on the loop (in unit-vector notation) and (b) the orientation energy of the loop.

••64 GO Figure 28-51 gives the orientation energy U of a magnetic dipole in an external magnetic field $\vec{B}$, as a function of angle ϕ between the directions of $\vec{B}$ and the dipole moment. The vertical axis scale is set by $U_s = 2.0 \times 10^{-4}$ J. The dipole can be rotated about an axle with negligible friction in order that to change ϕ. Counterclockwise rotation from $\phi = 0$ yields positive values of ϕ, and clockwise rotations yield negative values. The dipole is to be released at angle $\phi = 0$ with a rotational kinetic energy of 6.7×10^{-4} J, so that it rotates counterclockwise. To what maximum value of ϕ will it rotate? (In the language of Section 8-6, what value ϕ is the turning point in the potential well of Fig. 28-51?)

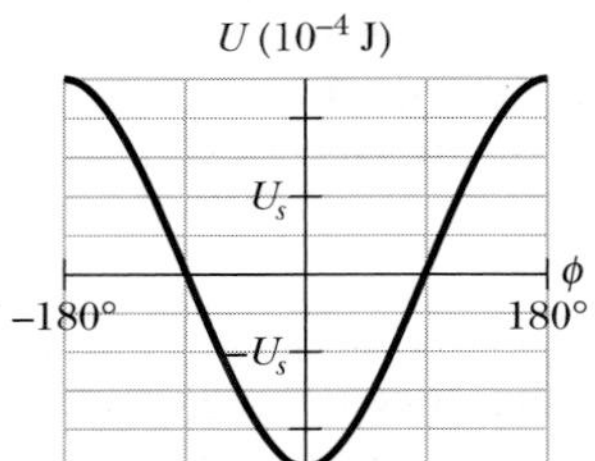

Fig. 28-51 Problem 64.

••65 SSM ILW A wire of length 25.0 cm carrying a current of 4.51 mA is to be formed into a circular coil and placed in a uniform magnetic field $\vec{B}$ of magnitude 5.71 mT. If the torque on the coil from the field is maximized, what are (a) the angle between $\vec{B}$ and the coil's magnetic dipole moment and (b) the number of turns in the coil? (c) What is the magnitude of that maximum torque?

Additional Problems

66 A proton of charge $+e$ and mass m enters a uniform magnetic field $\vec{B} = B\hat{i}$ with an initial velocity $\vec{v} = v_{0x}\hat{i} + v_{0y}\hat{j}$. Find an expression in unit-vector notation for its velocity $\vec{v}$ at any later time t.

67 A stationary circular wall clock has a face with a radius of 15 cm. Six turns of wire are wound around its perimeter; the wire carries a current of 2.0 A in the clockwise direction. The clock is located where there is a constant, uniform external magnetic field of magnitude 70 mT (but the clock still keeps perfect time). At exactly 1:00 P.M., the hour hand of the clock points in the direction of the external magnetic field. (a) After how many minutes will the minute hand point in the direction of the torque on the winding due to the magnetic field? (b) Find the torque magnitude.

68 A wire lying along a y axis from $y = 0$ to $y = 0.250$ m carries a current of 2.00 mA in the negative direction of the axis. The wire fully lies in a nonuniform magnetic field that is given by $\vec{B} = (0.300 \text{ T/m})y\hat{i} + (0.400 \text{ T/m})y\hat{j}$. In unit-vector notation, what is the magnetic force on the wire?

69 Atom 1 of mass 35 u and atom 2 of mass 37 u are both singly ionized with a charge of $+e$. After being introduced into a mass spectrometer (Fig. 28-12) and accelerated from rest through a potential difference $V = 7.3$ kV, each ion follows a circular path in a uniform magnetic field of magnitude $B = 0.50$ T. What is the distance Δx between the points where the ions strike the detector?

70 An electron with kinetic energy 2.5 keV moving along the positive direction of an x axis enters a region in which a uniform electric field of magnitude 10 kV/m is in the negative direction of the y axis. A uniform magnetic field $\vec{B}$ is to be set up to keep the electron moving along the x axis, and the direction of $\vec{B}$ is to be chosen to minimize the required magnitude of $\vec{B}$. In unit-vector notation, what $\vec{B}$ should be set up?

71 Physicist S. A. Goudsmit devised a method for measuring the mass of heavy ions by timing their period of revolution in a known magnetic field. A singly charged ion of iodine makes 7.00 rev in a 45.0 mT field in 1.29 ms. Calculate its mass in atomic mass units.

72 A beam of electrons whose kinetic energy is K emerges from a thin-foil "window" at the end of an accelerator tube. A metal plate at distance d from this window is perpendicular to the direction of the emerging beam (Fig. 28-52). (a) Show that we can prevent the beam from hitting the plate if we apply a uniform magnetic field such that

$$B \geq \sqrt{\frac{2mK}{e^2d^2}},$$

in which m and e are the electron mass and charge. (b) How should $\vec{B}$ be oriented?

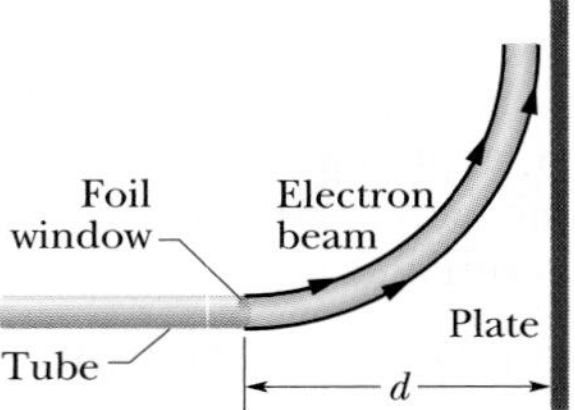

Fig. 28-52 Problem 72.

73 SSM At time $t = 0$, an electron with kinetic energy 12 keV moves through $x = 0$ in the positive direction of an x axis that is parallel to the horizontal component of Earth's magnetic field $\vec{B}$. The field's vertical component is downward and has magnitude 55.0 μT. (a) What is the magnitude of the electron's acceleration due to $\vec{B}$? (b) What is the electron's distance from the x axis when the electron reaches coordinate $x = 20$ cm?

74 GO A particle with charge 2.0 C moves through a uniform magnetic field. At one instant the velocity of the particle is $(2.0\hat{i} + 4.0\hat{j} + 6.0\hat{k})$ m/s and the magnetic force on the particle is $(4.0\hat{i} - 20\hat{j} + 12\hat{k})$ N. The x and y components of the magnetic field are equal. What is $\vec{B}$?

75 A proton, a deuteron ($q = +e, m = 2.0$ u), and an alpha particle ($q = +2e, m = 4.0$ u) all having the same kinetic energy enter a region of uniform magnetic field $\vec{B}$, moving perpendicular to $\vec{B}$. What is the ratio of (a) the radius r_d of the deuteron path to the radius r_p of the proton path and (b) the radius r_α of the alpha particle path to r_p?

76 Bainbridge's mass spectrometer, shown in Fig. 28-53, separates ions having the same velocity. The ions, after entering through slits, S_1 and S_2, pass through a velocity selector composed of an electric field produced by the charged plates P and P′, and a magnetic field $\vec{B}$ perpendicular to the electric field and the ion path. The ions that then pass undeviated through the crossed $\vec{E}$ and $\vec{B}$ fields enter into a region where a second magnetic field $\vec{B}'$ exists, where they are made to follow circular

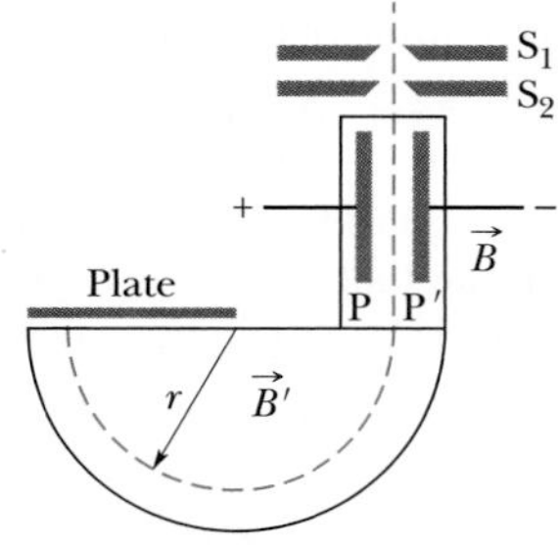

Fig. 28-53 Problem 76.

paths. A photographic plate (or a modern detector) registers their arrival. Show that, for the ions, $q/m = E/rBB'$, where r is the radius of the circular orbit.

77 SSM In Fig. 28-54, an electron moves at speed $v = 100$ m/s along an x axis through uniform electric and magnetic fields. The magnetic field $\vec{B}$ is directed into the page and has magnitude 5.00 T. In unit-vector notation, what is the electric field?

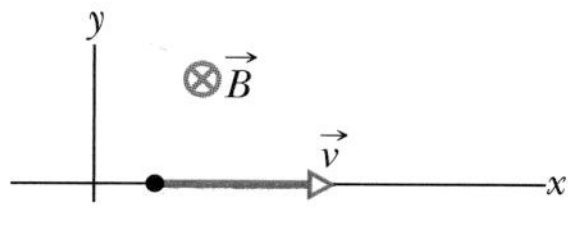

Fig. 28-54 Problem 77.

78 (a) In Fig. 28-8, show that the ratio of the Hall electric field magnitude E to the magnitude E_C of the electric field responsible for moving charge (the current) along the length of the strip is

$$\frac{E}{E_C} = \frac{B}{ne\rho},$$

where ρ is the resistivity of the material and n is the number density of the charge carriers. (b) Compute this ratio numerically for Problem 13. (See Table 26-1.)

79 SSM A proton, a deuteron ($q = +e$, $m = 2.0$ u), and an alpha particle ($q = +2e$, $m = 4.0$ u) are accelerated through the same potential difference and then enter the same region of uniform magnetic field $\vec{B}$, moving perpendicular to $\vec{B}$. What is the ratio of (a) the proton's kinetic energy K_p to the alpha particle's kinetic energy K_α and (b) the deuteron's kinetic energy K_d to K_α? If the radius of the proton's circular path is 10 cm, what is the radius of (c) the deuteron's path and (d) the alpha particle's path?

80 An electron in an old-fashioned TV camera tube is moving at 7.20×10^6 m/s in a magnetic field of strength 83.0 mT. What is the (a) maximum and (b) minimum magnitude of the force acting on the electron due to the field? (c) At one point the electron has an acceleration of magnitude 4.90×10^{14} m/s^2. What is the angle between the electron's velocity and the magnetic field?

81 A 5.0 μC particle moves through a region containing the uniform magnetic field $-20\hat{\mathrm{i}}$ mT and the uniform electric field $300\hat{\mathrm{j}}$ V/m. At a certain instant the velocity of the particle is $(17\hat{\mathrm{i}} - 11\hat{\mathrm{j}} + 7.0\hat{\mathrm{k}})$ km/s. At that instant and in unit-vector notation, what is the net electromagnetic force (the sum of the electric and magnetic forces) on the particle?

82 In a Hall-effect experiment, a current of 3.0 A sent lengthwise through a conductor 1.0 cm wide, 4.0 cm long, and 10 μm thick produces a transverse (across the width) Hall potential difference of 10 μV when a magnetic field of 1.5 T is passed perpendicularly through the thickness of the conductor. From these data, find (a) the drift velocity of the charge carriers and (b) the number density of charge carriers. (c) Show on a diagram the polarity of the Hall potential difference with assumed current and magnetic field directions, assuming also that the charge carriers are electrons.

83 SSM A particle of mass 6.0 g moves at 4.0 km/s in an xy plane, in a region with a uniform magnetic field given by $5.0\hat{\mathrm{i}}$ mT. At one instant, when the particle's velocity is directed 37° counterclockwise from the positive direction of the x axis, the magnetic force on the particle is $0.48\hat{\mathrm{k}}$ N. What is the particle's charge?

84 A wire lying along an x axis from $x = 0$ to $x = 1.00$ m carries a current of 3.00 A in the positive x direction. The wire is immersed in a nonuniform magnetic field that is given by $\vec{B} = (4.00\ \mathrm{T/m^2})x^2\hat{\mathrm{i}} - (0.600\ \mathrm{T/m^2})x^2\hat{\mathrm{j}}$. In unit-vector notation, what is the magnetic force on the wire?

85 At one instant, $\vec{v} = (-2.00\hat{\mathrm{i}} + 4.00\hat{\mathrm{j}} - 6.00\hat{\mathrm{k}})$ m/s is the velocity of a proton in a uniform magnetic field $\vec{B} = (2.00\hat{\mathrm{i}} - 4.00\hat{\mathrm{j}} + 8.00\hat{\mathrm{k}})$ mT. At that instant, what are (a) the magnetic force $\vec{F}$ acting on the proton, in unit-vector notation, (b) the angle between $\vec{v}$ and $\vec{F}$, and (c) the angle between $\vec{v}$ and $\vec{B}$?

86 An electron has velocity $\vec{v} = (32\hat{\mathrm{i}} + 40\hat{\mathrm{j}})$ km/s as it enters a uniform magnetic field $\vec{B} = 60\hat{\mathrm{i}}\ \mu$T. What are (a) the radius of the helical path taken by the electron and (b) the pitch of that path? (c) To an observer looking into the magnetic field region from the entrance point of the electron, does the electron spiral clockwise or counterclockwise as it moves?

CHAPTER

29

MAGNETIC FIELDS DUE TO CURRENTS

29-1 WHAT IS PHYSICS?

One basic observation of physics is that a moving charged particle produces a magnetic field around itself. Thus a current of moving charged particles produces a magnetic field around the current. This feature of *electromagnetism,* which is the combined study of electric and magnetic effects, came as a surprise to the people who discovered it. Surprise or not, this feature has become enormously important in everyday life because it is the basis of countless electromagnetic devices. For example, a magnetic field is produced in maglev trains and other devices used to lift heavy loads.

Our first step in this chapter is to find the magnetic field due to the current in a very small section of current-carrying wire. Then we shall find the magnetic field due to the entire wire for several different arrangements of the wire.

29-2 Calculating the Magnetic Field Due to a Current

Figure 29-1 shows a wire of arbitrary shape carrying a current i. We want to find the magnetic field $\vec{B}$ at a nearby point P. We first mentally divide the wire into differential elements ds and then define for each element a length vector $d\vec{s}$ that has length ds and whose direction is the direction of the current in ds. We can then define a differential *current-length element* to be $i\,d\vec{s}$; we wish to calculate the field $d\vec{B}$ produced at P by a typical current-length element. From experiment we find that magnetic fields, like electric fields, can be superimposed to find a net field. Thus, we can calculate the net field $\vec{B}$ at P by summing, via integration, the

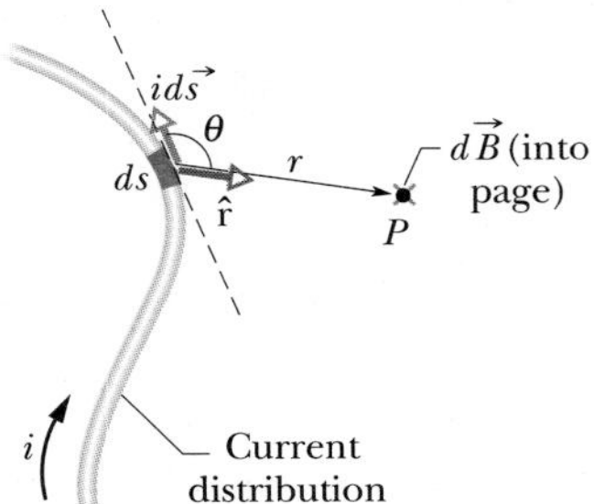

Fig. 29-1 A current-length element $i\,d\vec{s}$ produces a differential magnetic field $d\vec{B}$ at point P. The green × (the tail of an arrow) at the dot for point P indicates that $d\vec{B}$ is directed *into* the page there.

contributions $d\vec{B}$ from all the current-length elements. However, this summation is more challenging than the process associated with electric fields because of a complexity; whereas a charge element dq producing an electric field is a scalar, a current-length element $i\,d\vec{s}$ producing a magnetic field is a vector, being the product of a scalar and a vector.

The magnitude of the field $d\vec{B}$ produced at point P at distance r by a current-length element $i\,d\vec{s}$ turns out to be

$$dB = \frac{\mu_0}{4\pi}\frac{i\,ds\sin\theta}{r^2}, \qquad (29\text{-}1)$$

where θ is the angle between the directions of $d\vec{s}$ and $\hat{r}$, a unit vector that points from ds toward P. Symbol μ_0 is a constant, called the *permeability constant,* whose value is defined to be exactly

$$\mu_0 = 4\pi \times 10^{-7}\ \mathrm{T\cdot m/A} \approx 1.26 \times 10^{-6}\ \mathrm{T\cdot m/A}. \qquad (29\text{-}2)$$

The direction of $d\vec{B}$, shown as being into the page in Fig. 29-1, is that of the cross product $d\vec{s} \times \hat{r}$. We can therefore write Eq. 29-1 in vector form as

$$d\vec{B} = \frac{\mu_0}{4\pi}\frac{i\,d\vec{s} \times \hat{r}}{r^2} \quad \text{(Biot–Savart law).} \qquad (29\text{-}3)$$

This vector equation and its scalar form, Eq. 29-1, are known as the **law of Biot and Savart** (rhymes with "Leo and bazaar"). The law, which is experimentally deduced, is an inverse-square law. We shall use this law to calculate the net magnetic field $\vec{B}$ produced at a point by various distributions of current.

The magnetic field vector at any point is tangent to a circle.

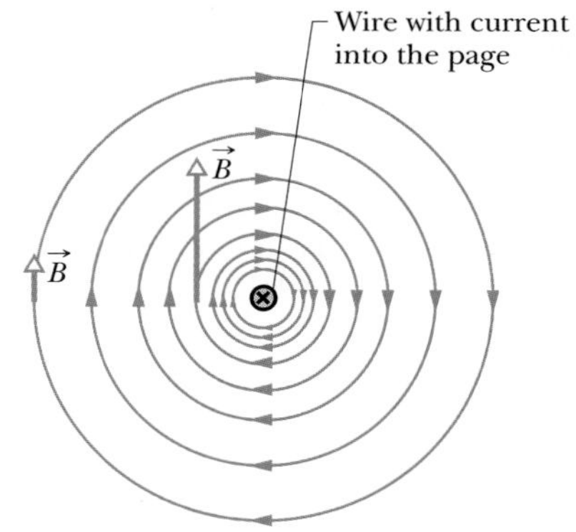

Fig. 29-2 The magnetic field lines produced by a current in a long straight wire form concentric circles around the wire. Here the current is into the page, as indicated by the $\times$.

Magnetic Field Due to a Current in a Long Straight Wire

Shortly we shall use the law of Biot and Savart to prove that the magnitude of the magnetic field at a perpendicular distance R from a long (infinite) straight wire carrying a current i is given by

$$B = \frac{\mu_0 i}{2\pi R} \quad \text{(long straight wire).} \qquad (29\text{-}4)$$

The field magnitude B in Eq. 29-4 depends only on the current and the perpendicular distance R of the point from the wire. We shall show in our derivation that the field lines of $\vec{B}$ form concentric circles around the wire, as Fig. 29-2 shows and as the iron filings in Fig. 29-3 suggest. The increase in the spacing of the lines in Fig. 29-2 with increasing distance from the wire represents the $1/R$ decrease in the magnitude of $\vec{B}$ predicted by Eq. 29-4. The lengths of the two vectors $\vec{B}$ in the figure also show the $1/R$ decrease.

Fig. 29-3 Iron filings that have been sprinkled onto cardboard collect in concentric circles when current is sent through the central wire. The alignment, which is along magnetic field lines, is caused by the magnetic field produced by the current. *(Courtesy Education Development Center)*

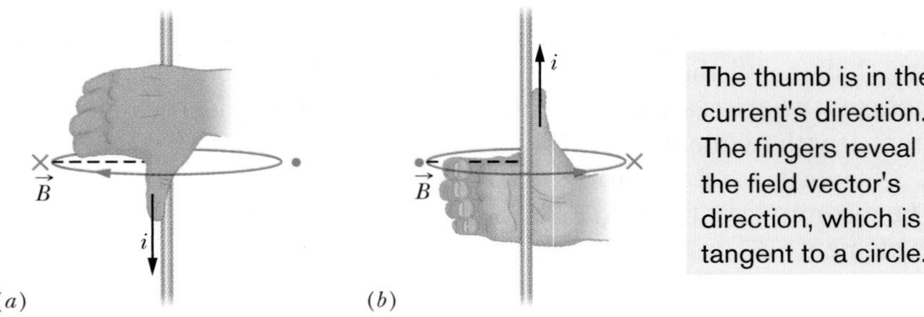

Fig. 29-4 A right-hand rule gives the direction of the magnetic field due to a current in a wire. (*a*) The situation of Fig. 29-2, seen from the side. The magnetic field $\vec{B}$ at any point to the left of the wire is perpendicular to the dashed radial line and directed into the page, in the direction of the fingertips, as indicated by the ×. (*b*) If the current is reversed, $\vec{B}$ at any point to the left is still perpendicular to the dashed radial line but now is directed out of the page, as indicated by the dot.

Here is a simple right-hand rule for finding the direction of the magnetic field set up by a current-length element, such as a section of a long wire:

Right-hand rule: Grasp the element in your right hand with your extended thumb pointing in the direction of the current. Your fingers will then naturally curl around in the direction of the magnetic field lines due to that element.

The result of applying this right-hand rule to the current in the straight wire of Fig. 29-2 is shown in a side view in Fig. 29-4*a*. To determine the direction of the magnetic field $\vec{B}$ set up at any particular point by this current, mentally wrap your right hand around the wire with your thumb in the direction of the current. Let your fingertips pass through the point; their direction is then the direction of the magnetic field at that point. In the view of Fig. 29-2, $\vec{B}$ at any point is *tangent to a magnetic field line;* in the view of Fig. 29-4, it is *perpendicular to a dashed radial line connecting the point and the current.*

Proof of Equation 29-4

Figure 29-5, which is just like Fig. 29-1 except that now the wire is straight and of infinite length, illustrates the task at hand. We seek the field $\vec{B}$ at point P, a perpendicular distance R from the wire. The magnitude of the differential magnetic field produced at P by the current-length element $i\,d\vec{s}$ located a distance r from P is given by Eq. 29-1:

$$dB = \frac{\mu_0}{4\pi}\frac{i\,ds\sin\theta}{r^2}.$$

The direction of $d\vec{B}$ in Fig. 29-5 is that of the vector $d\vec{s} \times \hat{r}$—namely, directly into the page.

Note that $d\vec{B}$ at point P has this same direction for all the current-length elements into which the wire can be divided. Thus, we can find the magnitude of the magnetic field produced at P by the current-length elements in the upper half of the infinitely long wire by integrating dB in Eq. 29-1 from 0 to ∞.

Now consider a current-length element in the lower half of the wire, one that is as far below P as $d\vec{s}$ is above P. By Eq. 29-3, the magnetic field produced at P by this current-length element has the same magnitude and direction as that from element $i\,d\vec{s}$ in Fig. 29-5. Further, the magnetic field produced by the lower half of the wire is exactly the same as that produced by the upper half. To find the magnitude of the *total* magnetic field $\vec{B}$ at P, we need only multiply the result of our integration by 2. We get

$$B = 2\int_0^\infty dB = \frac{\mu_0 i}{2\pi}\int_0^\infty \frac{\sin\theta\,ds}{r^2}. \qquad (29\text{-}5)$$

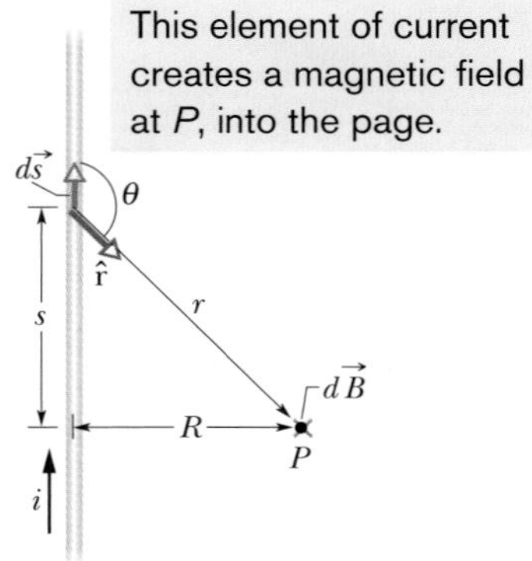

Fig. 29-5 Calculating the magnetic field produced by a current i in a long straight wire. The field $d\vec{B}$ at P associated with the current-length element $i\,d\vec{s}$ is directed into the page, as shown.

The variables θ, s, and r in this equation are not independent; Fig. 29-5 shows that they are related by

$$r = \sqrt{s^2 + R^2}$$

and $$\sin\theta = \sin(\pi - \theta) = \frac{R}{\sqrt{s^2 + R^2}}.$$

With these substitutions and integral 19 in Appendix E, Eq. 29-5 becomes

$$B = \frac{\mu_0 i}{2\pi}\int_0^\infty \frac{R\,ds}{(s^2 + R^2)^{3/2}}$$
$$= \frac{\mu_0 i}{2\pi R}\left[\frac{s}{(s^2 + R^2)^{1/2}}\right]_0^\infty = \frac{\mu_0 i}{2\pi R}, \qquad (29\text{-}6)$$

as we wanted. Note that the magnetic field at P due to either the lower half or the upper half of the infinite wire in Fig. 29-5 is half this value; that is,

$$B = \frac{\mu_0 i}{4\pi R} \quad \text{(semi-infinite straight wire).} \qquad (29\text{-}7)$$

Magnetic Field Due to a Current in a Circular Arc of Wire

To find the magnetic field produced at a point by a current in a curved wire, we would again use Eq. 29-1 to write the magnitude of the field produced by a single current-length element, and we would again integrate to find the net field produced by all the current-length elements. That integration can be difficult, depending on the shape of the wire; it is fairly straightforward, however, when the wire is a circular arc and the point is the center of curvature.

Figure 29-6*a* shows such an arc-shaped wire with central angle ϕ, radius R, and center C, carrying current i. At C, each current-length element $i\,d\vec{s}$ of the wire produces a magnetic field of magnitude dB given by Eq. 29-1. Moreover, as Fig. 29-6*b* shows, no matter where the element is located on the wire, the angle θ between the vectors $d\vec{s}$ and $\hat{r}$ is 90°; also, $r = R$. Thus, by substituting R for r and 90° for θ in Eq. 29-1, we obtain

$$dB = \frac{\mu_0}{4\pi}\frac{i\,ds\sin 90°}{R^2} = \frac{\mu_0}{4\pi}\frac{i\,ds}{R^2}. \qquad (29\text{-}8)$$

The field at C due to each current-length element in the arc has this magnitude.

An application of the right-hand rule anywhere along the wire (as in Fig. 29-6*c*) will show that all the differential fields $d\vec{B}$ have the same direction at C—directly out of the page. Thus, the total field at C is simply the sum (via integration) of all the differential fields $d\vec{B}$. We use the identity $ds = R\,d\phi$ to change the variable of integration from ds to $d\phi$ and obtain, from Eq. 29-8,

$$B = \int dB = \int_0^\phi \frac{\mu_0}{4\pi}\frac{iR\,d\phi}{R^2} = \frac{\mu_0 i}{4\pi R}\int_0^\phi d\phi.$$

Integrating, we find that

$$B = \frac{\mu_0 i\phi}{4\pi R} \quad \text{(at center of circular arc).} \qquad (29\text{-}9)$$

Note that this equation gives us the magnetic field *only* at the center of curvature of a circular arc of current. When you insert data into the equation, you must be careful to express ϕ in radians rather than degrees. For example, to find the magnitude of the magnetic field at the center of a full circle of current, you would substitute 2π rad for ϕ in Eq. 29-9, finding

$$B = \frac{\mu_0 i(2\pi)}{4\pi R} = \frac{\mu_0 i}{2R} \quad \text{(at center of full circle).} \qquad (29\text{-}10)$$

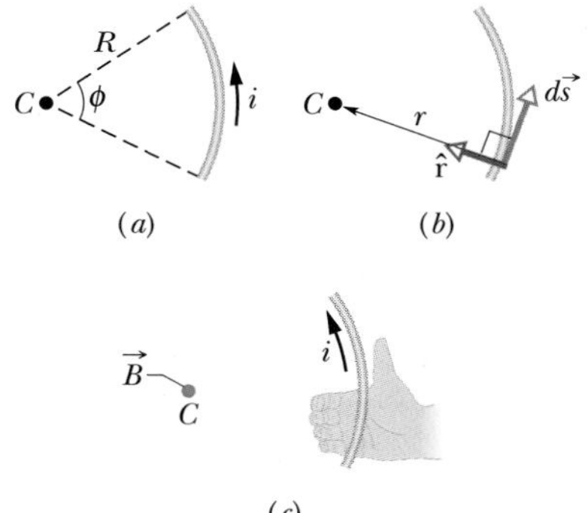

Fig. 29-6 (*a*) A wire in the shape of a circular arc with center C carries current i. (*b*) For any element of wire along the arc, the angle between the directions of $d\vec{s}$ and $\hat{r}$ is 90°. (*c*) Determining the direction of the magnetic field at the center C due to the current in the wire; the field is out of the page, in the direction of the fingertips, as indicated by the colored dot at C.

Sample Problem

Magnetic field at the center of a circular arc of current

The wire in Fig. 29-7*a* carries a current i and consists of a circular arc of radius R and central angle $\pi/2$ rad, and two straight sections whose extensions intersect the center C of the arc. What magnetic field $\vec{B}$ (magnitude and direction) does the current produce at C?

KEY IDEAS

We can find the magnetic field $\vec{B}$ at point C by applying the Biot–Savart law of Eq. 29-3 to the wire, point by point along the full length of the wire. However, the application of Eq. 29-3 can be simplified by evaluating $\vec{B}$ separately for the three distinguishable sections of the wire—namely, (1) the straight section at the left, (2) the straight section at the right, and (3) the circular arc.

Straight sections: For any current-length element in section 1, the angle θ between $d\vec{s}$ and $\hat{r}$ is zero (Fig. 29-7*b*); so Eq. 29-1 gives us

$$dB_1 = \frac{\mu_0}{4\pi}\frac{i\,ds\sin\theta}{r^2} = \frac{\mu_0}{4\pi}\frac{i\,ds\sin 0}{r^2} = 0.$$

Thus, the current along the entire length of straight section 1 contributes no magnetic field at C:

$$B_1 = 0.$$

The same situation prevails in straight section 2, where the angle θ between $d\vec{s}$ and $\hat{r}$ for any current-length element is 180°. Thus,

$$B_2 = 0.$$

Circular arc: Application of the Biot–Savart law to evaluate the magnetic field at the center of a circular arc leads to Eq. 29-9 ($B = \mu_0 i\phi/4\pi R$). Here the central angle ϕ of the arc is $\pi/2$ rad. Thus from Eq. 29-9, the magnitude of the magnetic field $\vec{B}_3$ at the arc's center C is

$$B_3 = \frac{\mu_0 i(\pi/2)}{4\pi R} = \frac{\mu_0 i}{8R}.$$

To find the direction of $\vec{B}_3$, we apply the right-hand rule displayed in Fig. 29-4. Mentally grasp the circular arc with your right hand as in Fig. 29-7*c*, with your thumb in the direction of the current. The direction in which your fingers curl around the wire indicates the direction of the magnetic field lines around the wire. They form circles around the wire, coming out of the page above the arc and going into the page inside the arc. In the region of point C (inside the arc), your fingertips point *into the plane* of the page. Thus, $\vec{B}_3$ is directed into that plane.

Net field: Generally, when we must combine two or more magnetic fields to find the net magnetic field, we must combine the fields as vectors and not simply add their magnitudes. Here, however, only the circular arc produces a magnetic field at point C. Thus, we can write the magnitude of the net field $\vec{B}$ as

$$B = B_1 + B_2 + B_3 = 0 + 0 + \frac{\mu_0 i}{8R} = \frac{\mu_0 i}{8R}. \quad \text{(Answer)}$$

The direction of $\vec{B}$ is the direction of $\vec{B}_3$—namely, into the plane of Fig. 29-7.

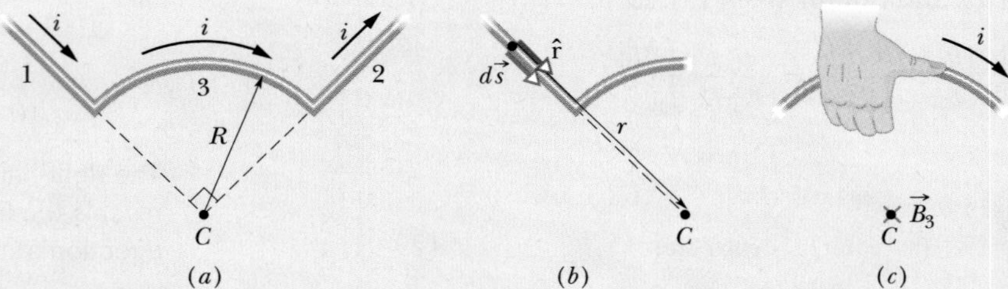

Fig. 29-7 (*a*) A wire consists of two straight sections (1 and 2) and a circular arc (3), and carries current i. (*b*) For a current-length element in section 1, the angle between $d\vec{s}$ and $\hat{r}$ is zero. (*c*) Determining the direction of magnetic field $\vec{B}_3$ at C due to the current in the circular arc; the field is into the page there.

Additional examples, video, and practice available at *WileyPLUS*

Sample Problem

Magnetic field off to the side of two long straight currents

Figure 29-8a shows two long parallel wires carrying currents i_1 and i_2 in opposite directions. What are the magnitude and direction of the net magnetic field at point P? Assume the following values: $i_1 = 15$ A, $i_2 = 32$ A, and $d = 5.3$ cm.

KEY IDEAS

(1) The net magnetic field $\vec{B}$ at point P is the vector sum of the magnetic fields due to the currents in the two wires. (2) We can find the magnetic field due to any current by applying the Biot–Savart law to the current. For points near the current in a long straight wire, that law leads to Eq. 29-4.

Finding the vectors: In Fig. 29-8a, point P is distance R from both currents i_1 and i_2. Thus, Eq. 29-4 tells us that at point P those currents produce magnetic fields $\vec{B}_1$ and $\vec{B}_2$ with magnitudes

$$B_1 = \frac{\mu_0 i_1}{2\pi R} \quad \text{and} \quad B_2 = \frac{\mu_0 i_2}{2\pi R}.$$

In the right triangle of Fig. 29-8a, note that the base angles (between sides R and d) are both 45°. This allows us to write $\cos 45° = R/d$ and replace R with $d \cos 45°$. Then the field magnitudes B_1 and B_2 become

$$B_1 = \frac{\mu_0 i_1}{2\pi d \cos 45°} \quad \text{and} \quad B_2 = \frac{\mu_0 i_2}{2\pi d \cos 45°}.$$

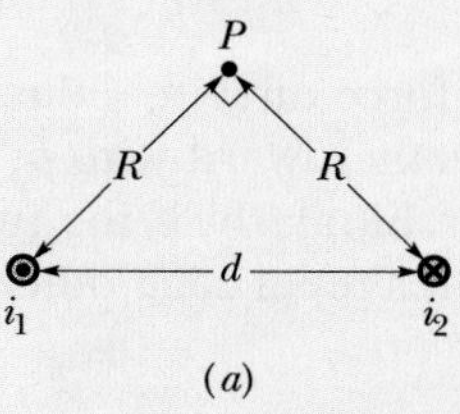

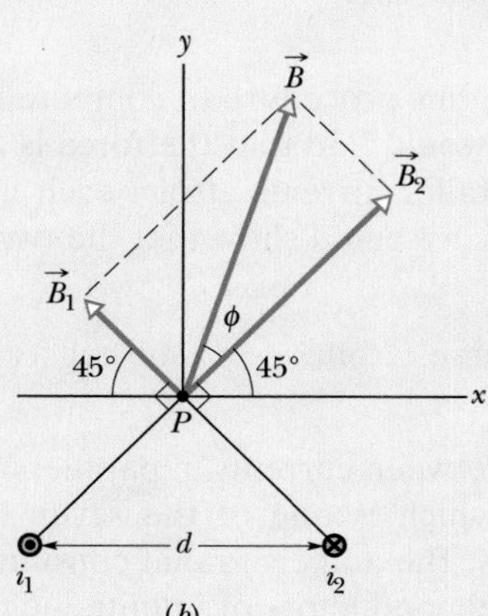

The two currents create magnetic fields that must be added as vectors to get the net field.

Fig. 29-8 (a) Two wires carry currents i_1 and i_2 in opposite directions (out of and into the page). Note the right angle at P. (b) The separate fields $\vec{B}_1$ and $\vec{B}_2$ are combined vectorially to yield the net field $\vec{B}$.

We want to combine $\vec{B}_1$ and $\vec{B}_2$ to find their vector sum, which is the net field $\vec{B}$ at P. To find the directions of $\vec{B}_1$ and $\vec{B}_2$, we apply the right-hand rule of Fig. 29-4 to each current in Fig. 29-8a. For wire 1, with current out of the page, we mentally grasp the wire with the right hand, with the thumb pointing out of the page. Then the curled fingers indicate that the field lines run counterclockwise. In particular, in the region of point P, they are directed upward to the left. Recall that the magnetic field at a point near a long, straight current-carrying wire must be directed perpendicular to a radial line between the point and the current. Thus, $\vec{B}_1$ must be directed upward to the left as drawn in Fig. 29-8b. (Note carefully the perpendicular symbol between vector $\vec{B}_1$ and the line connecting point P and wire 1.)

Repeating this analysis for the current in wire 2, we find that $\vec{B}_2$ is directed upward to the right as drawn in Fig. 29-8b. (Note the perpendicular symbol between vector $\vec{B}_2$ and the line connecting point P and wire 2.)

Adding the vectors: We can now vectorially add $\vec{B}_1$ and $\vec{B}_2$ to find the net magnetic field $\vec{B}$ at point P, either by using a vector-capable calculator or by resolving the vectors into components and then combining the components of $\vec{B}$. However, in Fig. 29-8b, there is a third method: Because $\vec{B}_1$ and $\vec{B}_2$ are perpendicular to each other, they form the legs of a right triangle, with $\vec{B}$ as the hypotenuse. The Pythagorean theorem then gives us

$$B = \sqrt{B_1^2 + B_2^2} = \frac{\mu_0}{2\pi d(\cos 45°)} \sqrt{i_1^2 + i_2^2}$$

$$= \frac{(4\pi \times 10^{-7}\ \text{T}\cdot\text{m/A})\sqrt{(15\ \text{A})^2 + (32\ \text{A})^2}}{(2\pi)(5.3 \times 10^{-2}\ \text{m})(\cos 45°)}$$

$$= 1.89 \times 10^{-4}\ \text{T} \approx 190\ \mu\text{T}. \qquad \text{(Answer)}$$

The angle ϕ between the directions of $\vec{B}$ and $\vec{B}_2$ in Fig. 29-8b follows from

$$\phi = \tan^{-1} \frac{B_1}{B_2},$$

which, with B_1 and B_2 as given above, yields

$$\phi = \tan^{-1} \frac{i_1}{i_2} = \tan^{-1} \frac{15\ \text{A}}{32\ \text{A}} = 25°.$$

The angle between the direction of $\vec{B}$ and the x axis shown in Fig. 29-8b is then

$$\phi + 45° = 25° + 45° = 70°. \qquad \text{(Answer)}$$

Additional examples, video, and practice available at *WileyPLUS*

29-3 Force Between Two Parallel Currents

Two long parallel wires carrying currents exert forces on each other. Figure 29-9 shows two such wires, separated by a distance d and carrying currents i_a and i_b. Let us analyze the forces on these wires due to each other.

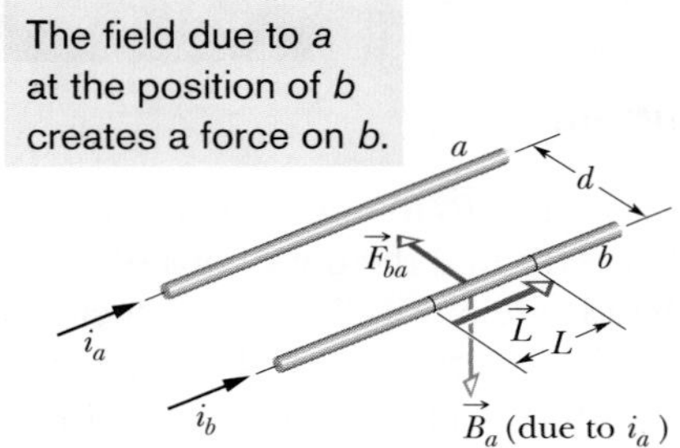

Fig. 29-9 Two parallel wires carrying currents in the same direction attract each other. $\vec{B}_a$ is the magnetic field at wire b produced by the current in wire a. $\vec{F}_{ba}$ is the resulting force acting on wire b because it carries current in $\vec{B}_a$.

We seek first the force on wire b in Fig. 29-9 due to the current in wire a. That current produces a magnetic field $\vec{B}_a$, and it is this magnetic field that actually causes the force we seek. To find the force, then, we need the magnitude and direction of the field $\vec{B}_a$ *at the site of wire b*. The magnitude of $\vec{B}_a$ at every point of wire b is, from Eq. 29-4,

$$B_a = \frac{\mu_0 i_a}{2\pi d}. \tag{29-11}$$

The curled–straight right-hand rule tells us that the direction of $\vec{B}_a$ at wire b is down, as Fig. 29-9 shows.

Now that we have the field, we can find the force it produces on wire b. Equation 28-26 tells us that the force $\vec{F}_{ba}$ on a length L of wire b due to the external magnetic field $\vec{B}_a$ is

$$\vec{F}_{ba} = i_b \vec{L} \times \vec{B}_a, \tag{29-12}$$

where $\vec{L}$ is the length vector of the wire. In Fig. 29-9, vectors $\vec{L}$ and $\vec{B}_a$ are perpendicular to each other, and so with Eq. 29-11, we can write

$$F_{ba} = i_b L B_a \sin 90° = \frac{\mu_0 L i_a i_b}{2\pi d}. \tag{29-13}$$

The direction of $\vec{F}_{ba}$ is the direction of the cross product $\vec{L} \times \vec{B}_a$. Applying the right-hand rule for cross products to $\vec{L}$ and $\vec{B}_a$ in Fig. 29-9, we see that $\vec{F}_{ba}$ is directly toward wire a, as shown.

The general procedure for finding the force on a current-carrying wire is this:

> To find the force on a current-carrying wire due to a second current-carrying wire, first find the field due to the second wire at the site of the first wire. Then find the force on the first wire due to that field.

We could now use this procedure to compute the force on wire a due to the current in wire b. We would find that the force is directly toward wire b; hence, the two wires with parallel currents attract each other. Similarly, if the two currents were antiparallel, we could show that the two wires repel each other. Thus,

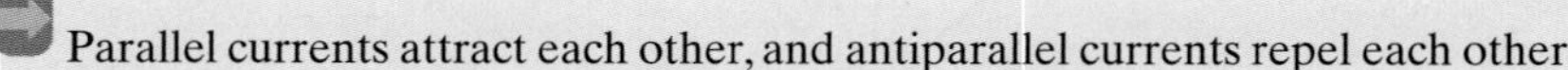

> Parallel currents attract each other, and antiparallel currents repel each other.

The force acting between currents in parallel wires is the basis for the definition of the ampere, which is one of the seven SI base units. The definition, adopted in 1946, is this: The ampere is that constant current which, if maintained in two straight, parallel conductors of infinite length, of negligible circular cross section, and placed 1 m apart in vacuum, would produce on each of these conductors a force of magnitude 2×10^{-7} newton per meter of wire length.

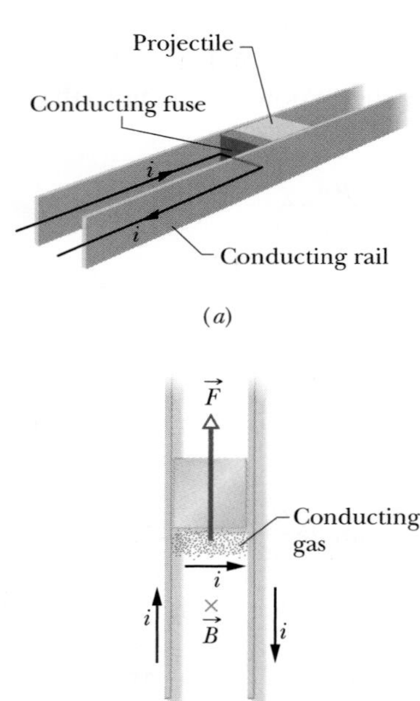

Fig. 29-10 (*a*) A rail gun, as a current i is set up in it. The current rapidly causes the conducting fuse to vaporize. (*b*) The current produces a magnetic field $\vec{B}$ between the rails, and the field causes a force $\vec{F}$ to act on the conducting gas, which is part of the current path. The gas propels the projectile along the rails, launching it.

Rail Gun

One application of the physics of Eq. 29-13 is a rail gun. In this device, a magnetic force accelerates a projectile to a high speed in a short time. The basics of a rail gun are shown in Fig. 29-10*a*. A large current is sent out along one of two parallel conducting rails, across a conducting "fuse" (such as a narrow piece of copper) between the rails, and then back to the current source along the second rail. The projectile to be fired lies on the far side of the fuse and fits loosely between the rails. Immediately after the current begins, the fuse element melts and vaporizes, creating a conducting gas between the rails where the fuse had been.

The curled–straight right-hand rule of Fig. 29-4 reveals that the currents in the rails of Fig. 29-10*a* produce magnetic fields that are directed downward between the rails. The net magnetic field $\vec{B}$ exerts a force $\vec{F}$ on the gas due to the current i through the gas (Fig. 29-10*b*). With Eq. 29-12 and the right-hand rule for cross products, we find that $\vec{F}$ points outward along the rails. As the gas is forced outward along the rails, it pushes the projectile, accelerating it by as much as $5 \times 10^6 g$, and then launches it with a speed of 10 km/s, all within 1 ms. Someday rail guns may be used to launch materials into space from mining operations on the Moon or an asteroid.

CHECKPOINT 1

The figure here shows three long, straight, parallel, equally spaced wires with identical currents either into or out of the page. Rank the wires according to the magnitude of the force on each due to the currents in the other two wires, greatest first.

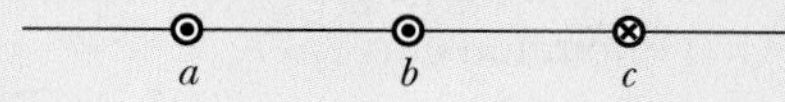

29-4 Ampere's Law

We can find the net electric field due to *any* distribution of charges by first writing the differential electric field $d\vec{E}$ due to a charge element and then summing the contributions of $d\vec{E}$ from all the elements. However, if the distribution is complicated, we may have to use a computer. Recall, however, that if the distribution has planar, cylindrical, or spherical symmetry, we can apply Gauss' law to find the net electric field with considerably less effort.

Similarly, we can find the net magnetic field due to *any* distribution of currents by first writing the differential magnetic field $d\vec{B}$ (Eq. 29-3) due to a current-length element and then summing the contributions of $d\vec{B}$ from all the elements. Again we may have to use a computer for a complicated distribution. However, if the distribution has some symmetry, we may be able to apply **Ampere's law** to find the magnetic field with considerably less effort. This law, which can be derived from the Biot–Savart law, has traditionally been credited to André-Marie Ampère (1775–1836), for whom the SI unit of current is named. However, the law actually was advanced by English physicist James Clerk Maxwell.

Ampere's law is

$$\oint \vec{B} \cdot d\vec{s} = \mu_0 i_{enc} \quad \text{(Ampere's law).} \tag{29-14}$$

The loop on the integral sign means that the scalar (dot) product $\vec{B} \cdot d\vec{s}$ is to be integrated around a *closed* loop, called an *Amperian loop.* The current i_{enc} is the *net* current encircled by that closed loop.

To see the meaning of the scalar product $\vec{B} \cdot d\vec{s}$ and its integral, let us first apply Ampere's law to the general situation of Fig. 29-11. The figure shows cross sections of three long straight wires that carry currents i_1, i_2, and i_3 either directly into or directly out of the page. An arbitrary Amperian loop lying in the plane of the page encircles two of the currents but not the third. The counterclockwise direction marked on the loop indicates the arbitrarily chosen direction of integration for Eq. 29-14.

To apply Ampere's law, we mentally divide the loop into differential vector elements $d\vec{s}$ that are everywhere directed along the tangent to the loop in the

Only the currents encircled by the loop are used in Ampere's law.

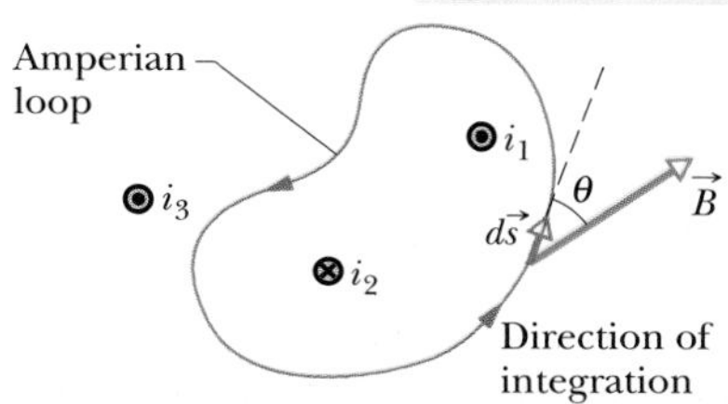

Fig. 29-11 Ampere's law applied to an arbitrary Amperian loop that encircles two long straight wires but excludes a third wire. Note the directions of the currents.

direction of integration. Assume that at the location of the element $d\vec{s}$ shown in Fig. 29-11, the net magnetic field due to the three currents is $\vec{B}$. Because the wires are perpendicular to the page, we know that the magnetic field at $d\vec{s}$ due to each current is in the plane of Fig. 29-11; thus, their net magnetic field $\vec{B}$ at $d\vec{s}$ must also be in that plane. However, we do not know the orientation of $\vec{B}$ within the plane. In Fig. 29-11, $\vec{B}$ is arbitrarily drawn at an angle θ to the direction of $d\vec{s}$.

The scalar product $\vec{B} \cdot d\vec{s}$ on the left side of Eq. 29-14 is equal to $B \cos \theta \, ds$. Thus, Ampere's law can be written as

$$\oint \vec{B} \cdot d\vec{s} = \oint B \cos \theta \, ds = \mu_0 i_{\text{enc}}. \tag{29-15}$$

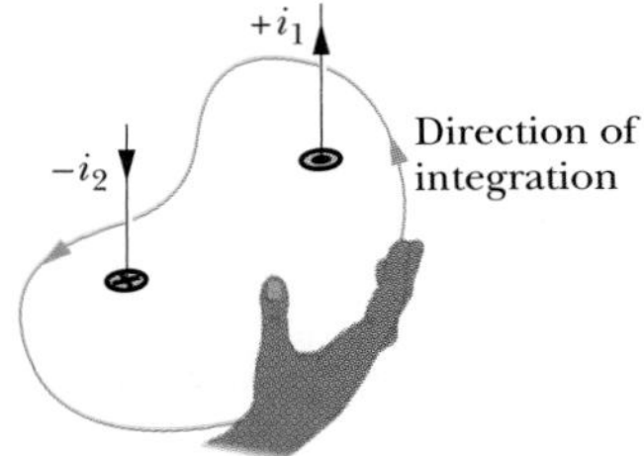

Fig. 29-12 A right-hand rule for Ampere's law, to determine the signs for currents encircled by an Amperian loop. The situation is that of Fig. 29-11.

We can now interpret the scalar product $\vec{B} \cdot d\vec{s}$ as being the product of a length ds of the Amperian loop and the field component $B \cos \theta$ tangent to the loop. Then we can interpret the integration as being the summation of all such products around the entire loop.

When we can actually perform this integration, we do not need to know the direction of $\vec{B}$ before integrating. Instead, we arbitrarily assume $\vec{B}$ to be generally in the direction of integration (as in Fig. 29-11). Then we use the following curled–straight right-hand rule to assign a plus sign or a minus sign to each of the currents that make up the net encircled current i_{enc}:

> Curl your right hand around the Amperian loop, with the fingers pointing in the direction of integration. A current through the loop in the general direction of your outstretched thumb is assigned a plus sign, and a current generally in the opposite direction is assigned a minus sign.

Finally, we solve Eq. 29-15 for the magnitude of $\vec{B}$. If B turns out positive, then the direction we assumed for $\vec{B}$ is correct. If it turns out negative, we neglect the minus sign and redraw $\vec{B}$ in the opposite direction.

In Fig. 29-12 we apply the curled–straight right-hand rule for Ampere's law to the situation of Fig. 29-11. With the indicated counterclockwise direction of integration, the net current encircled by the loop is

$$i_{\text{enc}} = i_1 - i_2.$$

(Current i_3 is not encircled by the loop.) We can then rewrite Eq. 29-15 as

$$\oint B \cos \theta \, ds = \mu_0(i_1 - i_2). \tag{29-16}$$

You might wonder why, since current i_3 contributes to the magnetic-field magnitude B on the left side of Eq. 29-16, it is not needed on the right side. The answer is that the contributions of current i_3 to the magnetic field cancel out because the integration in Eq. 29-16 is made around the full loop. In contrast, the contributions of an encircled current to the magnetic field do not cancel out.

We cannot solve Eq. 29-16 for the magnitude B of the magnetic field because for the situation of Fig. 29-11 we do not have enough information to simplify and solve the integral. However, we do know the outcome of the integration; it must be equal to $\mu_0(i_1 - i_2)$, the value of which is set by the net current passing through the loop.

We shall now apply Ampere's law to two situations in which symmetry does allow us to simplify and solve the integral, hence to find the magnetic field.

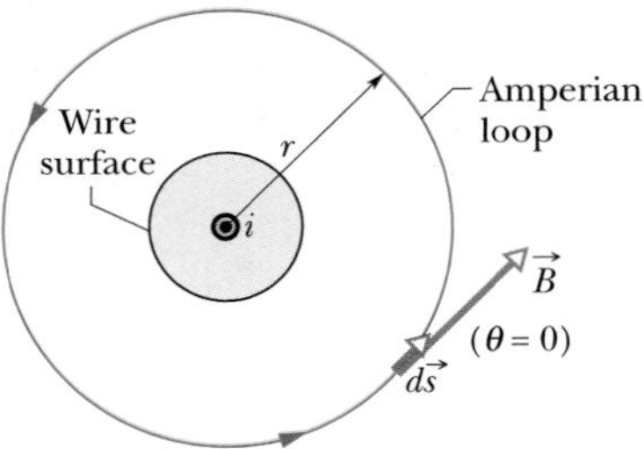

Fig. 29-13 Using Ampere's law to find the magnetic field that a current i produces outside a long straight wire of circular cross section. The Amperian loop is a concentric circle that lies outside the wire.

Magnetic Field Outside a Long Straight Wire with Current

Figure 29-13 shows a long straight wire that carries current i directly out of the page. Equation 29-4 tells us that the magnetic field $\vec{B}$ produced by the current has the same magnitude at all points that are the same distance r from the wire;

that is, the field $\vec{B}$ has cylindrical symmetry about the wire. We can take advantage of that symmetry to simplify the integral in Ampere's law (Eqs. 29-14 and 29-15) if we encircle the wire with a concentric circular Amperian loop of radius r, as in Fig. 29-13. The magnetic field $\vec{B}$ then has the same magnitude B at every point on the loop. We shall integrate counterclockwise, so that $d\vec{s}$ has the direction shown in Fig. 29-13.

We can further simplify the quantity $B \cos \theta$ in Eq. 29-15 by noting that $\vec{B}$ is tangent to the loop at every point along the loop, as is $d\vec{s}$. Thus, $\vec{B}$ and $d\vec{s}$ are either parallel or antiparallel at each point of the loop, and we shall arbitrarily assume the former. Then at every point the angle θ between $d\vec{s}$ and $\vec{B}$ is 0°, so $\cos \theta = \cos 0° = 1$. The integral in Eq. 29-15 then becomes

$$\oint \vec{B} \cdot d\vec{s} = \oint B \cos \theta \, ds = B \oint ds = B(2\pi r).$$

Note that $\oint ds$ is the summation of all the line segment lengths ds around the circular loop; that is, it simply gives the circumference $2\pi r$ of the loop.

Our right-hand rule gives us a plus sign for the current of Fig. 29-13. The right side of Ampere's law becomes $+\mu_0 i$, and we then have

$$B(2\pi r) = \mu_0 i$$

or

$$B = \frac{\mu_0 i}{2\pi r} \quad \text{(outside straight wire).} \tag{29-17}$$

With a slight change in notation, this is Eq. 29-4, which we derived earlier—with considerably more effort—using the law of Biot and Savart. In addition, because the magnitude B turned out positive, we know that the correct direction of $\vec{B}$ must be the one shown in Fig. 29-13.

Magnetic Field Inside a Long Straight Wire with Current

Figure 29-14 shows the cross section of a long straight wire of radius R that carries a uniformly distributed current i directly out of the page. Because the current is uniformly distributed over a cross section of the wire, the magnetic field $\vec{B}$ produced by the current must be cylindrically symmetrical. Thus, to find the magnetic field at points inside the wire, we can again use an Amperian loop of radius r, as shown in Fig. 29-14, where now $r < R$. Symmetry again suggests that $\vec{B}$ is tangent to the loop, as shown; so the left side of Ampere's law again yields

$$\oint \vec{B} \cdot d\vec{s} = B \oint ds = B(2\pi r). \tag{29-18}$$

To find the right side of Ampere's law, we note that because the current is uniformly distributed, the current i_{enc} encircled by the loop is proportional to the area encircled by the loop; that is,

$$i_{enc} = i \frac{\pi r^2}{\pi R^2}. \tag{29-19}$$

Our right-hand rule tells us that i_{enc} gets a plus sign. Then Ampere's law gives us

$$B(2\pi r) = \mu_0 i \frac{\pi r^2}{\pi R^2}$$

or

$$B = \left(\frac{\mu_0 i}{2\pi R^2}\right) r \quad \text{(inside straight wire).} \tag{29-20}$$

Thus, inside the wire, the magnitude B of the magnetic field is proportional to r, is zero at the center, and is maximum at $r = R$ (the surface). Note that Eqs. 29-17 and 29-20 give the same value for B at the surface.

Only the current encircled by the loop is used in Ampere's law.

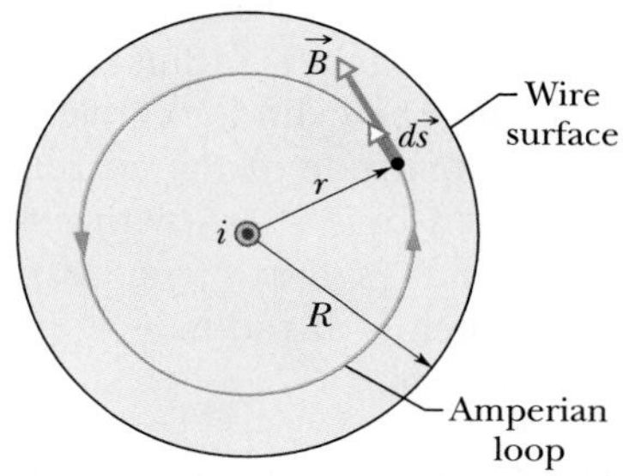

Fig. 29-14 Using Ampere's law to find the magnetic field that a current i produces inside a long straight wire of circular cross section. The current is uniformly distributed over the cross section of the wire and emerges from the page. An Amperian loop is drawn inside the wire.

CHECKPOINT 2

The figure here shows three equal currents i (two parallel and one antiparallel) and four Amperian loops. Rank the loops according to the magnitude of $\oint \vec{B} \cdot d\vec{s}$ along each, greatest first.

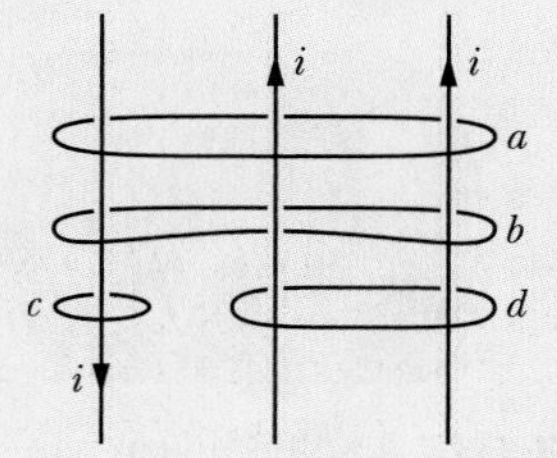

Sample Problem

Ampere's law to find the field inside a long cylinder of current

Figure 29-15*a* shows the cross section of a long conducting cylinder with inner radius $a = 2.0$ cm and outer radius $b = 4.0$ cm. The cylinder carries a current out of the page, and the magnitude of the current density in the cross section is given by $J = cr^2$, with $c = 3.0 \times 10^6$ A/m^4 and r in meters. What is the magnetic field $\vec{B}$ at the dot in Fig. 29-15*a*, which is at radius $r = 3.0$ cm from the central axis of the cylinder?

KEY IDEAS

The point at which we want to evaluate $\vec{B}$ is inside the material of the conducting cylinder, between its inner and outer radii. We note that the current distribution has cylindrical symmetry (it is the same all around the cross section for any given radius). Thus, the symmetry allows us to use Ampere's law to find $\vec{B}$ at the point. We first draw the Amperian loop shown in Fig. 29-15*b*. The loop is concentric with the cylinder and has radius $r = 3.0$ cm because we want to evaluate $\vec{B}$ at that distance from the cylinder's central axis.

Next, we must compute the current i_{enc} that is encircled by the Amperian loop. However, we *cannot* set up a proportionality as in Eq. 29-19, because here the current is not uniformly distributed. Instead, we must integrate the current density magnitude from the cylinder's inner radius a to the loop radius r, using the steps shown in Figs. 29-15*c* through *h*.

Calculations: We write the integral as

$$i_{enc} = \int J\,dA = \int_a^r cr^2(2\pi r\,dr)$$

$$= 2\pi c \int_a^r r^3\,dr = 2\pi c\left[\frac{r^4}{4}\right]_a^r$$

$$= \frac{\pi c(r^4 - a^4)}{2}.$$

Note that in these steps we took the differential area dA to be the area of the thin ring in Figs. 29-15*d–f* and then replaced it with its equivalent, the product of the ring's circumference $2\pi r$ and its thickness dr.

For the Amperian loop, the direction of integration indicated in Fig. 29-15*b* is (arbitrarily) clockwise. Applying the right-hand rule for Ampere's law to that loop, we find that we should take i_{enc} as negative because the current is directed out of the page but our thumb is directed into the page.

We next evaluate the left side of Ampere's law exactly as we did in Fig. 29-14, and we again obtain Eq. 29-18. Then Ampere's law,

$$\oint \vec{B}\cdot d\vec{s} = \mu_0 i_{enc},$$

gives us

$$B(2\pi r) = -\frac{\mu_0 \pi c}{2}(r^4 - a^4).$$

Solving for B and substituting known data yield

$$B = -\frac{\mu_0 c}{4r}(r^4 - a^4)$$

$$= -\frac{(4\pi \times 10^{-7}\ \text{T}\cdot\text{m/A})(3.0 \times 10^6\ \text{A/m}^4)}{4(0.030\ \text{m})} \times [(0.030\ \text{m})^4 - (0.020\ \text{m})^4]$$

$$= -2.0 \times 10^{-5}\ \text{T}.$$

Thus, the magnetic field $\vec{B}$ at a point 3.0 cm from the central axis has magnitude

$$B = 2.0 \times 10^{-5}\ \text{T} \qquad \text{(Answer)}$$

and forms magnetic field lines that are directed opposite our direction of integration, hence counterclockwise in Fig. 29-15*b*.

WILEY PLUS Additional examples, video, and practice available at *WileyPLUS*

29-5 Solenoids and Toroids

Magnetic Field of a Solenoid

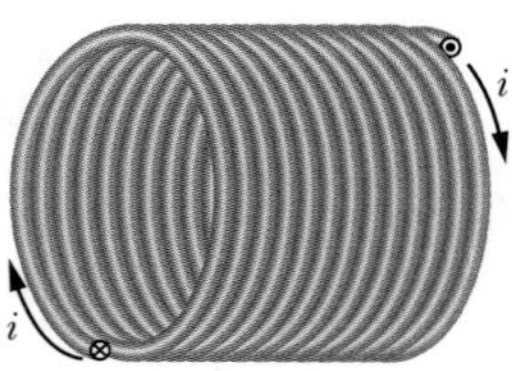

Fig. 29-16 A solenoid carrying current i.

We now turn our attention to another situation in which Ampere's law proves useful. It concerns the magnetic field produced by the current in a long, tightly wound helical coil of wire. Such a coil is called a **solenoid** (Fig. 29-16). We assume that the length of the solenoid is much greater than the diameter.

Figure 29-17 shows a section through a portion of a "stretched-out" solenoid. The solenoid's magnetic field is the vector sum of the fields produced by the indi-

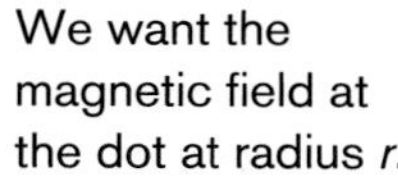

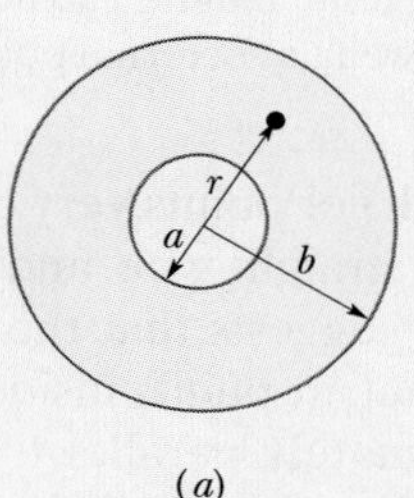

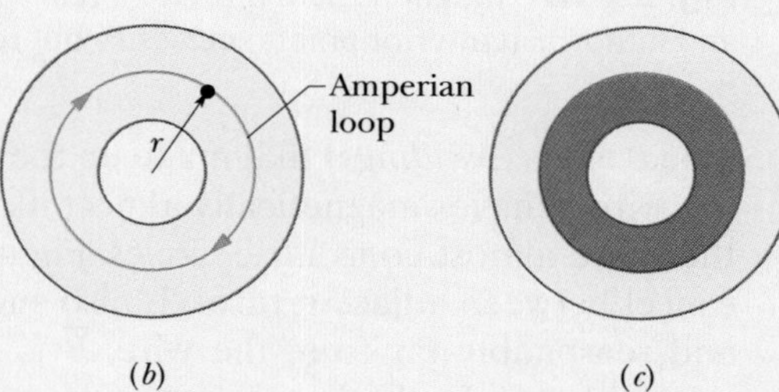

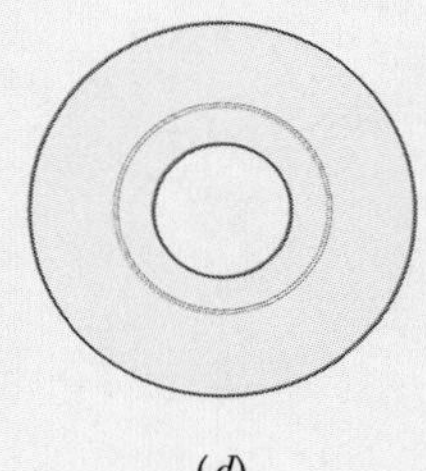

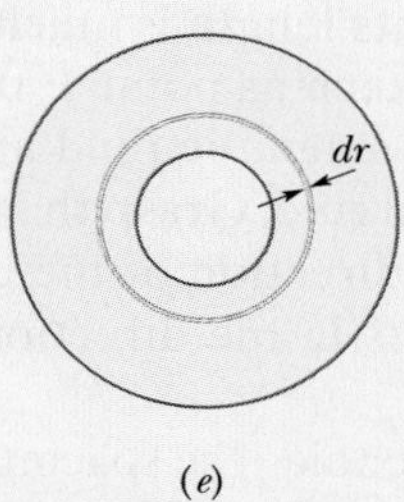

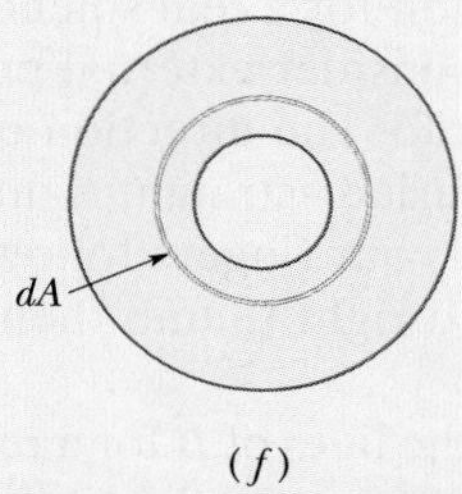

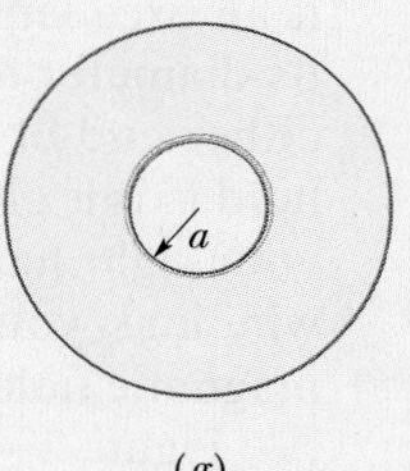

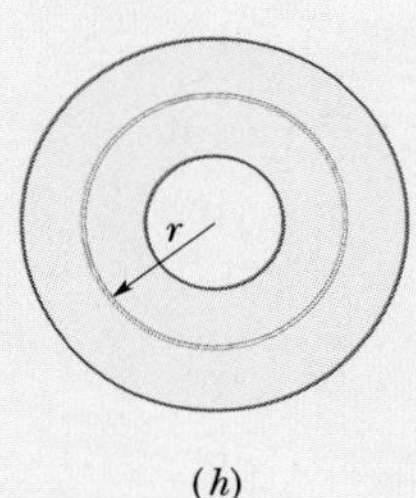

Fig. 29-15 (*a*) – (*b*) To find the magnetic field at a point within this conducting cylinder, we use a concentric Amperian loop through the point. We then need the current encircled by the loop. (*c*) – (*h*) Because the current density is nonuniform, we start with a thin ring and then sum (via integration) the currents in all such rings in the encircled area.

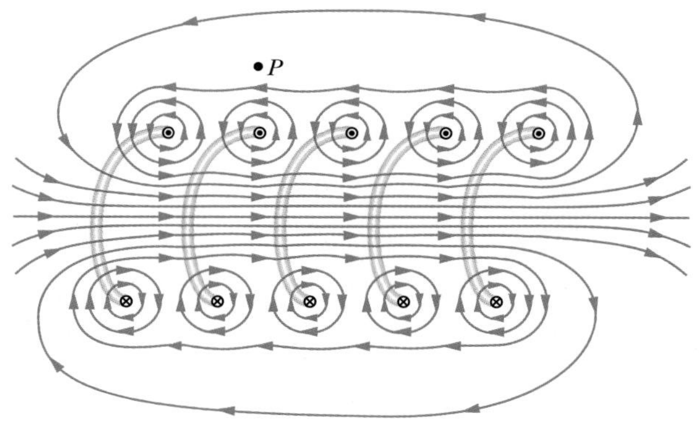

Fig. 29-17 A vertical cross section through the central axis of a "stretched-out" solenoid. The back portions of five turns are shown, as are the magnetic field lines due to a current through the solenoid. Each turn produces circular magnetic field lines near itself. Near the solenoid's axis, the field lines combine into a net magnetic field that is directed along the axis. The closely spaced field lines there indicate a strong magnetic field. Outside the solenoid the field lines are widely spaced; the field there is very weak.

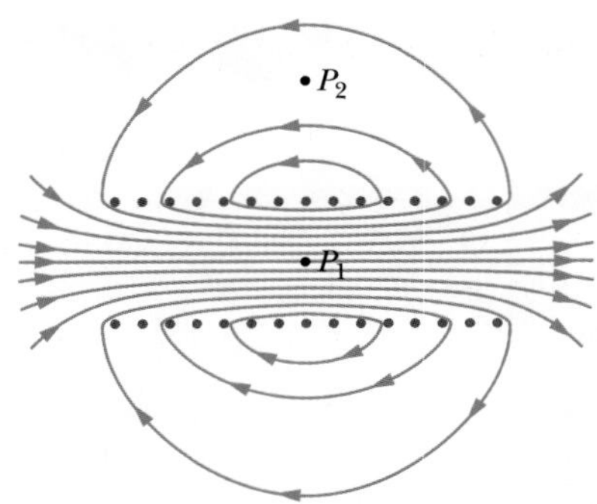

Fig. 29-18 Magnetic field lines for a real solenoid of finite length. The field is strong and uniform at interior points such as P_1 but relatively weak at external points such as P_2.

vidual turns (*windings*) that make up the solenoid. For points very close to a turn, the wire behaves magnetically almost like a long straight wire, and the lines of $\vec{B}$ there are almost concentric circles. Figure 29-17 suggests that the field tends to cancel between adjacent turns. It also suggests that, at points inside the solenoid and reasonably far from the wire, $\vec{B}$ is approximately parallel to the (central) solenoid axis. In the limiting case of an *ideal solenoid,* which is infinitely long and consists of tightly packed (*close-packed*) turns of square wire, the field inside the coil is uniform and parallel to the solenoid axis.

At points above the solenoid, such as P in Fig. 29-17, the magnetic field set up by the upper parts of the solenoid turns (these upper turns are marked ⊙) is directed to the left (as drawn near P) and tends to cancel the field set up at P by the lower parts of the turns (these lower turns are marked ⊗), which is directed to the right (not drawn). In the limiting case of an ideal solenoid, the magnetic field outside the solenoid is zero. Taking the external field to be zero is an excellent assumption for a real solenoid if its length is much greater than its diameter and if we consider external points such as point P that are not at either end of the solenoid. The direction of the magnetic field along the solenoid axis is given by a curled–straight right-hand rule: Grasp the solenoid with your right hand so that your fingers follow the direction of the current in the windings; your extended right thumb then points in the direction of the axial magnetic field.

Figure 29-18 shows the lines of $\vec{B}$ for a real solenoid. The spacing of these lines in the central region shows that the field inside the coil is fairly strong and uniform over the cross section of the coil. The external field, however, is relatively weak.

Let us now apply Ampere's law,

$$\oint \vec{B} \cdot d\vec{s} = \mu_0 i_{\text{enc}}, \tag{29-21}$$

to the ideal solenoid of Fig. 29-19, where $\vec{B}$ is uniform within the solenoid and zero outside it, using the rectangular Amperian loop *abcda*. We write $\oint \vec{B} \cdot d\vec{s}$ as

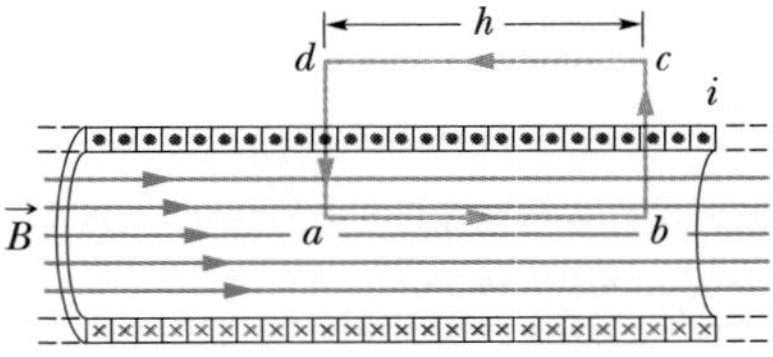

Fig. 29-19 Application of Ampere's law to a section of a long ideal solenoid carrying a current i. The Amperian loop is the rectangle *abcda*.

the sum of four integrals, one for each loop segment:

$$\oint \vec{B} \cdot d\vec{s} = \int_a^b \vec{B} \cdot d\vec{s} + \int_b^c \vec{B} \cdot d\vec{s} + \int_c^d \vec{B} \cdot d\vec{s} + \int_d^a \vec{B} \cdot d\vec{s}. \qquad (29\text{-}22)$$

The first integral on the right of Eq. 29-22 is Bh, where B is the magnitude of the uniform field $\vec{B}$ inside the solenoid and h is the (arbitrary) length of the segment from a to b. The second and fourth integrals are zero because for every element ds of these segments, $\vec{B}$ either is perpendicular to ds or is zero, and thus the product $\vec{B} \cdot d\vec{s}$ is zero. The third integral, which is taken along a segment that lies outside the solenoid, is zero because $B = 0$ at all external points. Thus, $\oint \vec{B} \cdot d\vec{s}$ for the entire rectangular loop has the value Bh.

The net current i_{enc} encircled by the rectangular Amperian loop in Fig. 29-19 is not the same as the current i in the solenoid windings because the windings pass more than once through this loop. Let n be the number of turns per unit length of the solenoid; then the loop encloses nh turns and

$$i_{enc} = i(nh).$$

Ampere's law then gives us

$$Bh = \mu_0 inh$$

or

$$B = \mu_0 in \quad \text{(ideal solenoid).} \qquad (29\text{-}23)$$

Although we derived Eq. 29-23 for an infinitely long ideal solenoid, it holds quite well for actual solenoids if we apply it only at interior points and well away from the solenoid ends. Equation 29-23 is consistent with the experimental fact that the magnetic field magnitude B within a solenoid does not depend on the diameter or the length of the solenoid and that B is uniform over the solenoidal cross section. A solenoid thus provides a practical way to set up a known uniform magnetic field for experimentation, just as a parallel-plate capacitor provides a practical way to set up a known uniform electric field.

Magnetic Field of a Toroid

Figure 29-20a shows a **toroid,** which we may describe as a (hollow) solenoid that has been curved until its two ends meet, forming a sort of hollow bracelet. What magnetic field $\vec{B}$ is set up inside the toroid (inside the hollow of the bracelet)? We can find out from Ampere's law and the symmetry of the bracelet.

From the symmetry, we see that the lines of $\vec{B}$ form concentric circles inside the toroid, directed as shown in Fig. 29-20b. Let us choose a concentric circle of

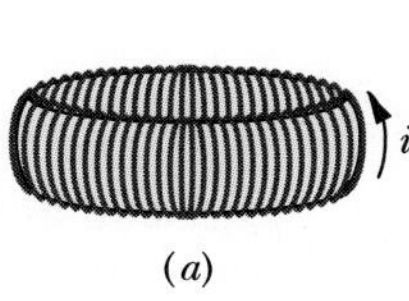

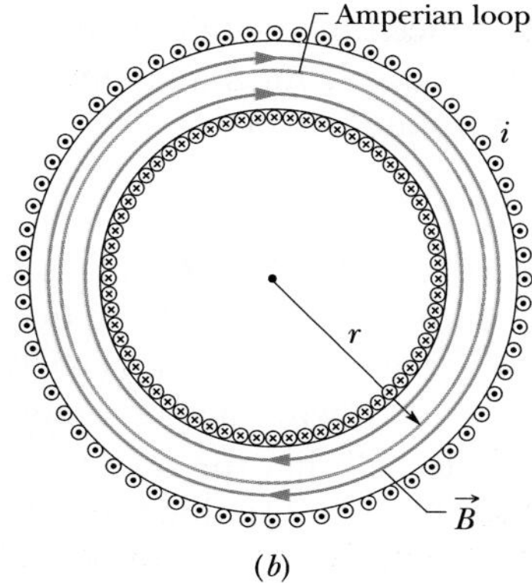

Fig. 29-20 (a) A toroid carrying a current i. (b) A horizontal cross section of the toroid. The interior magnetic field (inside the bracelet-shaped tube) can be found by applying Ampere's law with the Amperian loop shown.

radius r as an Amperian loop and traverse it in the clockwise direction. Ampere's law (Eq. 29-14) yields

$$(B)(2\pi r) = \mu_0 iN,$$

where i is the current in the toroid windings (and is positive for those windings enclosed by the Amperian loop) and N is the total number of turns. This gives

$$B = \frac{\mu_0 iN}{2\pi}\frac{1}{r} \quad \text{(toroid).} \qquad (29\text{-}24)$$

In contrast to the situation for a solenoid, B is not constant over the cross section of a toroid.

It is easy to show, with Ampere's law, that $B = 0$ for points outside an ideal toroid (as if the toroid were made from an ideal solenoid). The direction of the magnetic field within a toroid follows from our curled–straight right-hand rule: Grasp the toroid with the fingers of your right hand curled in the direction of the current in the windings; your extended right thumb points in the direction of the magnetic field.

Sample Problem

The field inside a solenoid (a long coil of current)

A solenoid has length $L = 1.23$ m and inner diameter $d = 3.55$ cm, and it carries a current $i = 5.57$ A. It consists of five close-packed layers, each with 850 turns along length L. What is B at its center?

KEY IDEA

The magnitude B of the magnetic field along the solenoid's central axis is related to the solenoid's current i and number of turns per unit length n by Eq. 29-23 ($B = \mu_0 in$).

Calculation: Because B does not depend on the diameter of the windings, the value of n for five identical layers is simply five times the value for each layer. Equation 29-23 then tells us

$$B = \mu_0 in = (4\pi \times 10^{-7}\ \text{T}\cdot\text{m/A})(5.57\ \text{A})\frac{5 \times 850\ \text{turns}}{1.23\ \text{m}}$$

$$= 2.42 \times 10^{-2}\ \text{T} = 24.2\ \text{mT}. \qquad \text{(Answer)}$$

To a good approximation, this is the field magnitude throughout most of the solenoid.

Additional examples, video, and practice available at *WileyPLUS*

29-6 A Current-Carrying Coil as a Magnetic Dipole

So far we have examined the magnetic fields produced by current in a long straight wire, a solenoid, and a toroid. We turn our attention here to the field produced by a coil carrying a current. You saw in Section 28-10 that such a coil behaves as a magnetic dipole in that, if we place it in an external magnetic field $\vec{B}$, a torque $\vec{\tau}$ given by

$$\vec{\tau} = \vec{\mu} \times \vec{B} \qquad (29\text{-}25)$$

acts on it. Here $\vec{\mu}$ is the magnetic dipole moment of the coil and has the magnitude NiA, where N is the number of turns, i is the current in each turn, and A is the area enclosed by each turn. (*Caution:* Don't confuse the magnetic dipole moment $\vec{\mu}$ with the permeability constant μ_0.)

Recall that the direction of $\vec{\mu}$ is given by a curled–straight right-hand rule: Grasp the coil so that the fingers of your right hand curl around it in the direction of the current; your extended thumb then points in the direction of the dipole moment $\vec{\mu}$.

Magnetic Field of a Coil

We turn now to the other aspect of a current-carrying coil as a magnetic dipole. What magnetic field does *it* produce at a point in the surrounding space? The problem does not have enough symmetry to make Ampere's law useful; so we must turn to the law of Biot and Savart. For simplicity, we first consider only a coil with a single circular loop and only points on its perpendicular central axis, which we take to be a z axis. We shall show that the magnitude of the magnetic field at such points is

$$B(z) = \frac{\mu_0 i R^2}{2(R^2 + z^2)^{3/2}}, \tag{29-26}$$

in which R is the radius of the circular loop and z is the distance of the point in question from the center of the loop. Furthermore, the direction of the magnetic field $\vec{B}$ is the same as the direction of the magnetic dipole moment $\vec{\mu}$ of the loop.

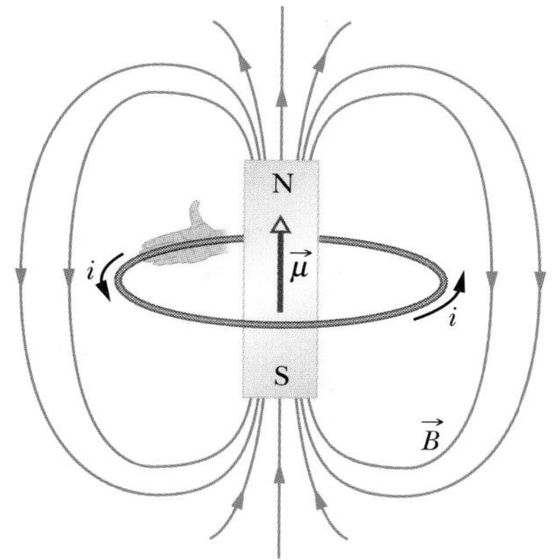

Fig. 29-21 A current loop produces a magnetic field like that of a bar magnet and thus has associated north and south poles. The magnetic dipole moment $\vec{\mu}$ of the loop, its direction given by a curled–straight right-hand rule, points from the south pole to the north pole, in the direction of the field $\vec{B}$ within the loop.

For axial points far from the loop, we have $z \gg R$ in Eq. 29-26. With that approximation, the equation reduces to

$$B(z) \approx \frac{\mu_0 i R^2}{2z^3}.$$

Recalling that πR^2 is the area A of the loop and extending our result to include a coil of N turns, we can write this equation as

$$B(z) = \frac{\mu_0}{2\pi}\frac{NiA}{z^3}.$$

Further, because $\vec{B}$ and $\vec{\mu}$ have the same direction, we can write the equation in vector form, substituting from the identity $\mu = NiA$:

$$\vec{B}(z) = \frac{\mu_0}{2\pi}\frac{\vec{\mu}}{z^3} \quad \text{(current-carrying coil).} \tag{29-27}$$

Thus, we have two ways in which we can regard a current-carrying coil as a magnetic dipole: (1) it experiences a torque when we place it in an external magnetic field; (2) it generates its own intrinsic magnetic field, given, for distant points along its axis, by Eq. 29-27. Figure 29-21 shows the magnetic field of a current loop; one side of the loop acts as a north pole (in the direction of $\vec{\mu}$) and the other side as a south pole, as suggested by the lightly drawn magnet in the figure. If we were to place a current-carrying coil in an external magnetic field, it would tend to rotate just like a bar magnet would.

CHECKPOINT 3

The figure here shows four arrangements of circular loops of radius r or $2r$, centered on vertical axes (perpendicular to the loops) and carrying identical currents in the directions indicated. Rank the arrangements according to the magnitude of the net magnetic field at the dot, midway between the loops on the central axis, greatest first.

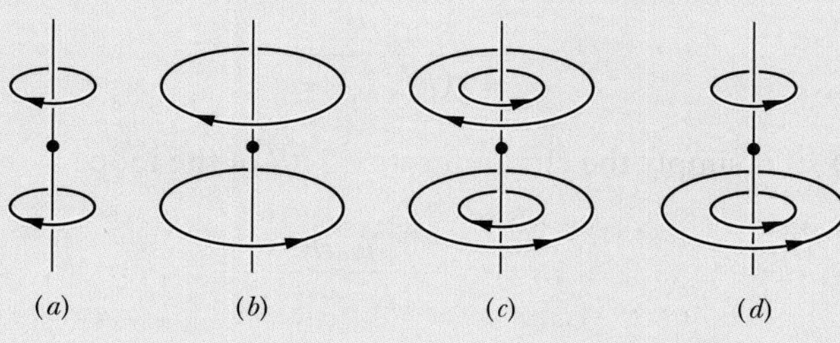

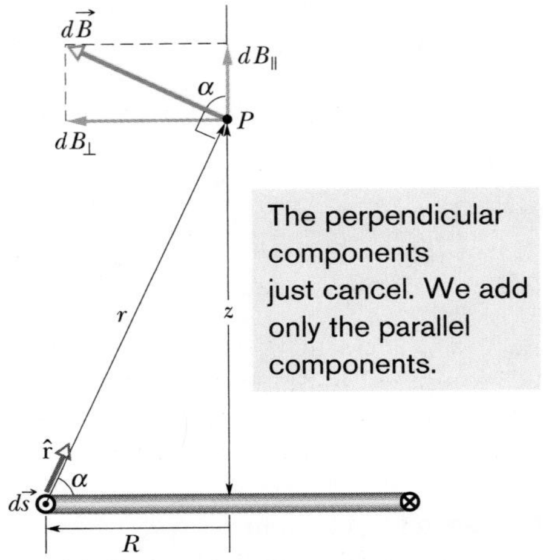

The perpendicular components just cancel. We add only the parallel components.

Fig. 29-22 Cross section through a current loop of radius R. The plane of the loop is perpendicular to the page, and only the back half of the loop is shown. We use the law of Biot and Savart to find the magnetic field at point P on the central perpendicular axis of the loop.

Proof of Equation 29-26

Figure 29-22 shows the back half of a circular loop of radius R carrying a current i. Consider a point P on the central axis of the loop, a distance z from its plane. Let us apply the law of Biot and Savart to a differential element ds of the loop, located at the left side of the loop. The length vector $d\vec{s}$ for this element points perpendicularly out of the page. The angle θ between $d\vec{s}$ and $\hat{r}$ in Fig. 29-22 is 90°; the plane formed by these two vectors is perpendicular to the plane of the page and contains both $\hat{r}$ and $d\vec{s}$. From the law of Biot and Savart (and the right-hand rule), the differential field $d\vec{B}$ produced at point P by the current in this element is perpendicular to this plane and thus is directed in the plane of the figure, perpendicular to $\hat{r}$, as indicated in Fig. 29-22.

Let us resolve $d\vec{B}$ into two components: $dB_{\parallel}$ along the axis of the loop and $dB_{\perp}$ perpendicular to this axis. From the symmetry, the vector sum of all the perpendicular components $dB_{\perp}$ due to all the loop elements ds is zero. This leaves only the axial (parallel) components $dB_{\parallel}$ and we have

$$B = \int dB_{\parallel}.$$

For the element $d\vec{s}$ in Fig. 29-22, the law of Biot and Savart (Eq. 29-1) tells us that the magnetic field at distance r is

$$dB = \frac{\mu_0}{4\pi} \frac{i\, ds \sin 90^\circ}{r^2}.$$

We also have

$$dB_{\parallel} = dB \cos \alpha.$$

Combining these two relations, we obtain

$$dB_{\parallel} = \frac{\mu_0 i \cos \alpha\, ds}{4\pi r^2}. \tag{29-28}$$

Figure 29-22 shows that r and α are related to each other. Let us express each in terms of the variable z, the distance between point P and the center of the loop. The relations are

$$r = \sqrt{R^2 + z^2} \tag{29-29}$$

and

$$\cos \alpha = \frac{R}{r} = \frac{R}{\sqrt{R^2 + z^2}}. \tag{29-30}$$

Substituting Eqs. 29-29 and 29-30 into Eq. 29-28, we find

$$dB_{\parallel} = \frac{\mu_0 i R}{4\pi (R^2 + z^2)^{3/2}}\, ds.$$

Note that i, R, and z have the same values for all elements ds around the loop; so when we integrate this equation, we find that

$$B = \int dB_{\parallel} = \frac{\mu_0 i R}{4\pi (R^2 + z^2)^{3/2}} \int ds$$

or, because $\int ds$ is simply the circumference $2\pi R$ of the loop,

$$B(z) = \frac{\mu_0 i R^2}{2(R^2 + z^2)^{3/2}}.$$

This is Eq. 29-26, the relation we sought to prove.

REVIEW & SUMMARY

The Biot–Savart Law The magnetic field set up by a current-carrying conductor can be found from the *Biot–Savart law.* This law asserts that the contribution $d\vec{B}$ to the field produced by a current-length element $i\,d\vec{s}$ at a point P located a distance r from the current element is

$$d\vec{B} = \frac{\mu_0}{4\pi}\frac{i\,d\vec{s} \times \hat{r}}{r^2} \quad \text{(Biot–Savart law).} \quad (29\text{-}3)$$

Here $\hat{r}$ is a unit vector that points from the element toward P. The quantity μ_0, called the permeability constant, has the value

$$4\pi \times 10^{-7}\ \text{T}\cdot\text{m/A} \approx 1.26 \times 10^{-6}\ \text{T}\cdot\text{m/A}.$$

Magnetic Field of a Long Straight Wire For a long straight wire carrying a current i, the Biot–Savart law gives, for the magnitude of the magnetic field at a perpendicular distance R from the wire,

$$B = \frac{\mu_0 i}{2\pi R} \quad \text{(long straight wire).} \quad (29\text{-}4)$$

Magnetic Field of a Circular Arc The magnitude of the magnetic field at the center of a circular arc, of radius R and central angle ϕ (in radians), carrying current i, is

$$B = \frac{\mu_0 i \phi}{4\pi R} \quad \text{(at center of circular arc).} \quad (29\text{-}9)$$

Force Between Parallel Currents Parallel wires carrying currents in the same direction attract each other, whereas parallel wires carrying currents in opposite directions repel each other. The magnitude of the force on a length L of either wire is

$$F_{ba} = i_b L B_a \sin 90° = \frac{\mu_0 L i_a i_b}{2\pi d}, \quad (29\text{-}13)$$

where d is the wire separation, and i_a and i_b are the currents in the wires.

Ampere's Law **Ampere's law** states that

$$\oint \vec{B}\cdot d\vec{s} = \mu_0 i_{\text{enc}} \quad \text{(Ampere's law).} \quad (29\text{-}14)$$

The line integral in this equation is evaluated around a closed loop called an *Amperian loop.* The current i on the right side is the *net* current encircled by the loop. For some current distributions, Eq. 29-14 is easier to use than Eq. 29-3 to calculate the magnetic field due to the currents.

Fields of a Solenoid and a Toroid Inside a *long solenoid* carrying current i, at points not near its ends, the magnitude B of the magnetic field is

$$B = \mu_0 i n \quad \text{(ideal solenoid),} \quad (29\text{-}23)$$

where n is the number of turns per unit length. At a point inside a *toroid*, the magnitude B of the magnetic field is

$$B = \frac{\mu_0 i N}{2\pi}\frac{1}{r} \quad \text{(toroid),} \quad (29\text{-}24)$$

where r is the distance from the center of the toroid to the point.

Field of a Magnetic Dipole The magnetic field produced by a current-carrying coil, which is a *magnetic dipole,* at a point P located a distance z along the coil's perpendicular central axis is parallel to the axis and is given by

$$\vec{B}(z) = \frac{\mu_0}{2\pi}\frac{\vec{\mu}}{z^3}, \quad (29\text{-}27)$$

where $\vec{\mu}$ is the dipole moment of the coil. This equation applies only when z is much greater than the dimensions of the coil.

QUESTIONS

1 Figure 29-23 shows three circuits, each consisting of two radial lengths and two concentric circular arcs, one of radius r and the other of radius $R > r$. The circuits have the same current through them and the same angle between the two radial lengths. Rank the circuits according to the magnitude of the net magnetic field at the center, greatest first.

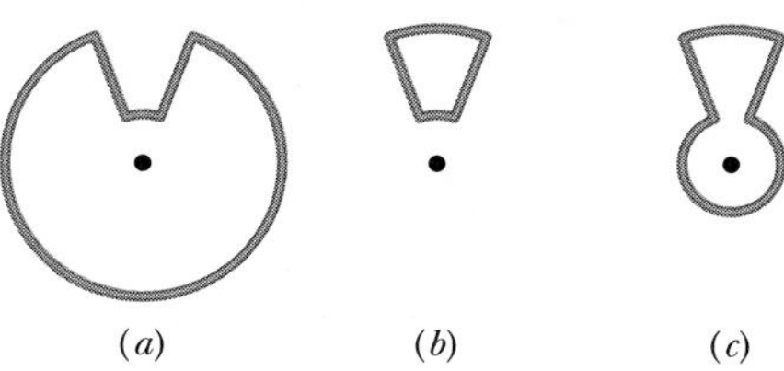

Fig. 29-23 Question 1.

2 Figure 29-24 represents a snapshot of the velocity vectors of four electrons near a wire carrying current i. The four velocities have the same magnitude; velocity $\vec{v}_2$ is directed into the page. Electrons 1 and 2 are at the same distance from the wire, as are electrons 3 and 4. Rank the electrons according to the magnitudes of the magnetic forces on them due to current i, greatest first.

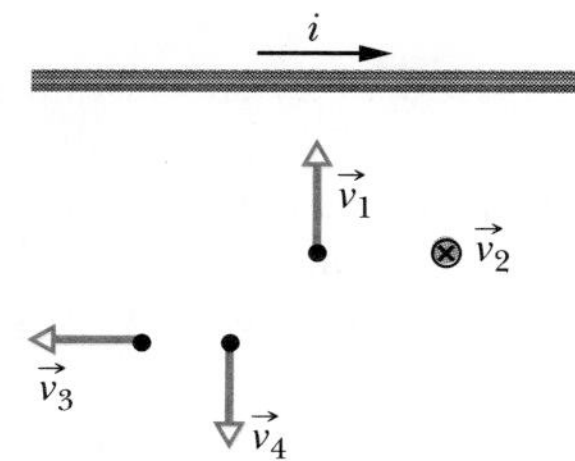

Fig. 29-24 Question 2.

3 Figure 29-25 shows four arrangements in which long parallel wires

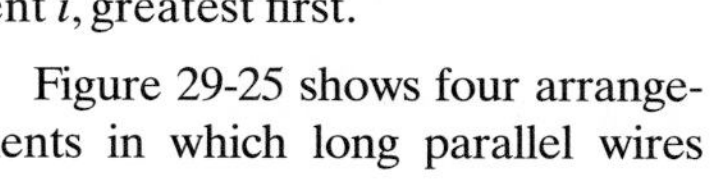

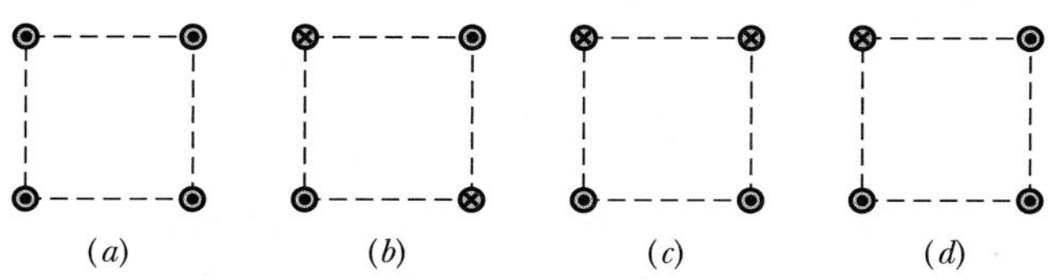

Fig. 29-25 Question 3.

carry equal currents directly into or out of the page at the corners of identical squares. Rank the arrangements according to the magnitude of the net magnetic field at the center of the square, greatest first.

4 Figure 29-26 shows cross sections of two long straight wires; the left-hand wire carries current i_1 directly out of the page. If the net magnetic field due to the two currents is to be zero at point P, (a) should the direction of current i_2 in the right-hand wire be directly into or out of the page and (b) should i_2 be greater than, less than, or equal to i_1?

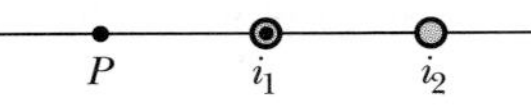

Fig. 29-26 Question 4.

5 Figure 29-27 shows three circuits consisting of straight radial lengths and concentric circular arcs (either half- or quarter-circles of radii r, $2r$, and $3r$). The circuits carry the same current. Rank them according to the magnitude of the magnetic field produced at the center of curvature (the dot), greatest first.

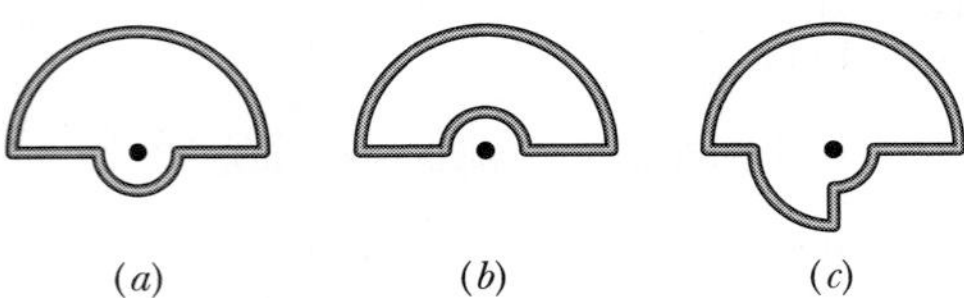

Fig. 29-27 Question 5.

6 Figure 29-28 gives, as a function of radial distance r, the magnitude B of the magnetic field inside and outside four wires (a, b, c, and d), each of which carries a current that is uniformly distributed across the wire's cross section. Overlapping portions of the plots are indicated by double labels. Rank the wires according to (a) radius, (b) the magnitude of the magnetic field on the surface, and (c) the value of the current, greatest first. (d) Is the magnitude of the current density in wire a greater than, less than, or equal to that in wire c?

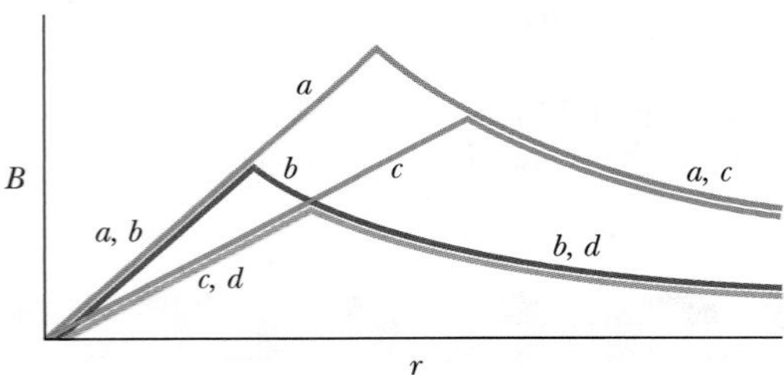

Fig. 29-28 Question 6.

7 Figure 29-29 shows four circular Amperian loops (a, b, c, d) concentric with a wire whose current is directed out of the page. The current is uniform across the wire's circular cross section (the shaded region). Rank the loops according to the magnitude of $\oint \vec{B} \cdot d\vec{s}$ around each, greatest first.

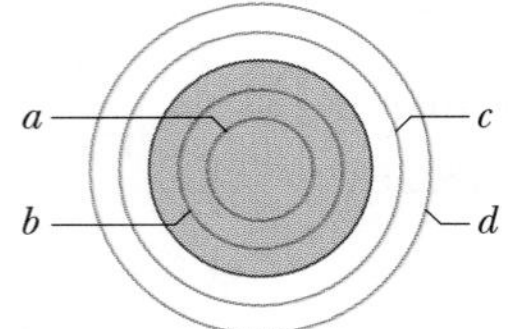

Fig. 29-29 Question 7.

8 Figure 29-30 shows four arrangements in which long, parallel, equally spaced wires carry equal currents directly into or out of the page. Rank the arrangements according to the magnitude of the net force on the central wire due to the currents in the other wires, greatest first.

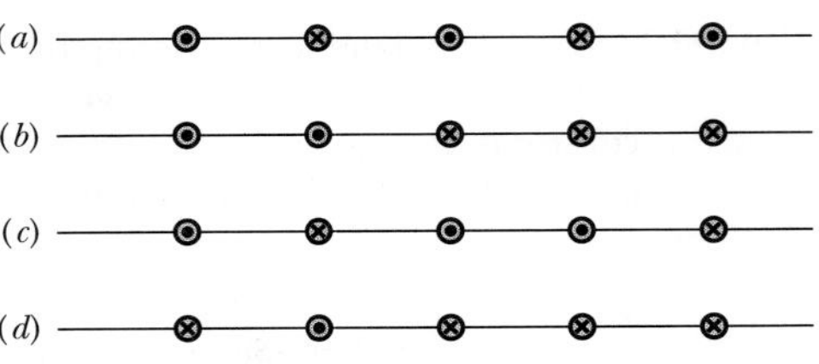

Fig. 29-30 Question 8.

9 Figure 29-31 shows four circular Amperian loops (a, b, c, d) and, in cross section, four long circular conductors (the shaded regions), all of which are concentric. Three of the conductors are hollow cylinders; the central conductor is a solid cylinder. The currents in the conductors are, from smallest radius to largest radius, 4 A out of the page, 9 A into the page, 5 A out of the page, and 3 A into the page. Rank the Amperian loops according to the magnitude of $\oint \vec{B} \cdot d\vec{s}$ around each, greatest first.

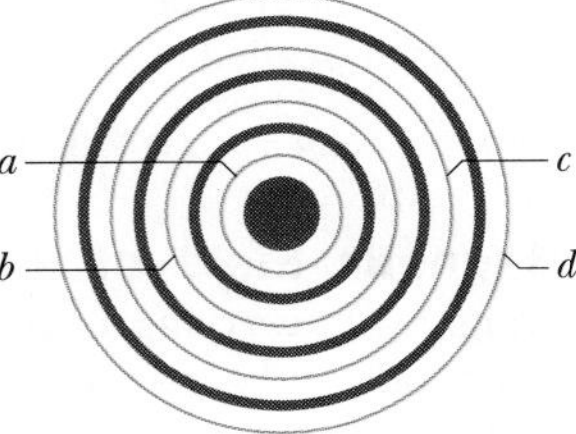

Fig. 29-31 Question 9.

10 Figure 29-32 shows four identical currents i and five Amperian paths (a through e) encircling them. Rank the paths according to the value of $\oint \vec{B} \cdot d\vec{s}$ taken in the directions shown, most positive first.

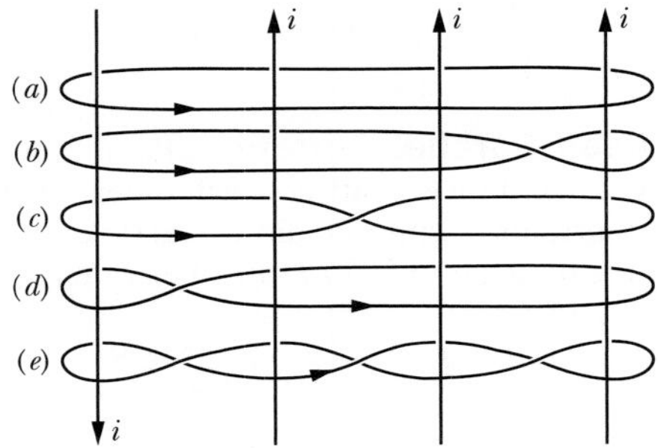

Fig. 29-32 Question 10.

11 Figure 29-33 shows three arrangements of three long straight wires carrying equal currents directly into or out of the page. (a) Rank the arrangements according to the magnitude of the net force on wire A due to the currents in the other wires, greatest first. (b) In arrangement 3, is the angle between the net force on wire A and the dashed line equal to, less than, or more than 45°?

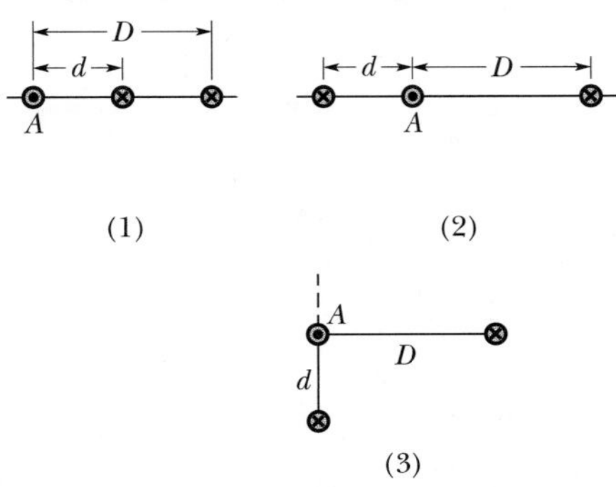

Fig. 29-33 Question 11.

PROBLEMS

GO Tutoring problem available (at instructor's discretion) in *WileyPLUS* and WebAssign
SSM Worked-out solution available in Student Solutions Manual
WWW Worked-out solution is at
ILW Interactive solution is at
http://www.wiley.com/college/halliday
• – ••• Number of dots indicates level of problem difficulty
Additional information available in *The Flying Circus of Physics* and at flyingcircusofphysics.com

sec. 29-2 Calculating the Magnetic Field Due to a Current

•1 A surveyor is using a magnetic compass 6.1 m below a power line in which there is a steady current of 100 A. (a) What is the magnetic field at the site of the compass due to the power line? (b) Will this field interfere seriously with the compass reading? The horizontal component of Earth's magnetic field at the site is 20 μT.

•2 Figure 29-34*a* shows an element of length ds = 1.00 μm in a very long straight wire carrying current. The current in that element sets up a differential magnetic field $d\vec{B}$ at points in the surrounding space. Figure 29-34*b* gives the magnitude dB of the field for points 2.5 cm from the element, as a function of angle θ between the wire and a straight line to the point. The vertical scale is set by dB_s = 60.0 pT. What is the magnitude of the magnetic field set up by the entire wire at perpendicular distance 2.5 cm from the wire?

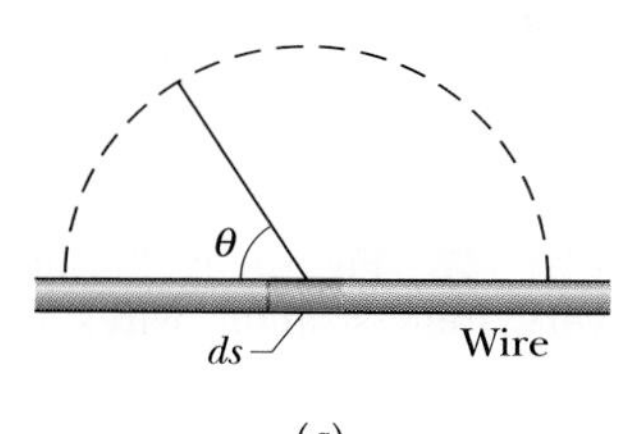

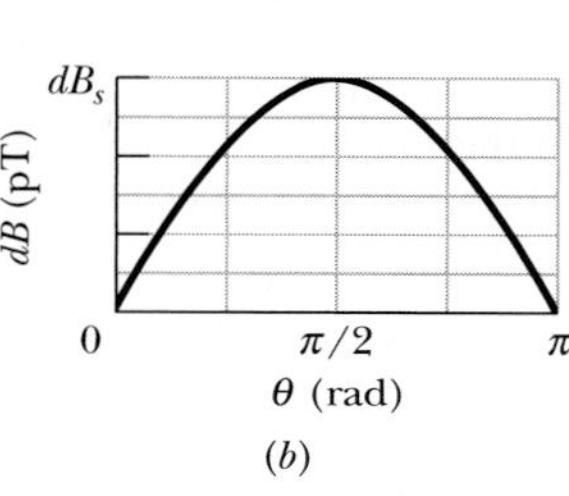

Fig. 29-34 Problem 2.

•3 SSM At a certain location in the Philippines, Earth's magnetic field of 39 μT is horizontal and directed due north. Suppose the net field is zero exactly 8.0 cm above a long, straight, horizontal wire that carries a constant current. What are the (a) magnitude and (b) direction of the current?

•4 A straight conductor carrying current i = 5.0 A splits into identical semicircular arcs as shown in Fig. 29-35. What is the magnetic field at the center C of the resulting circular loop?

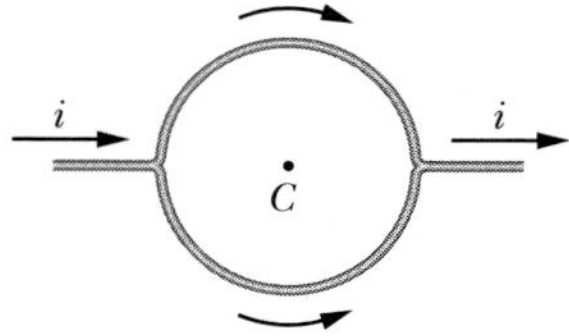

Fig. 29-35 Problem 4.

•5 In Fig. 29-36, a current i = 10 A is set up in a long hairpin conductor formed by bending a wire into a semicircle of radius R = 5.0 mm. Point b is midway between the straight sections and so distant from the semicircle that each straight section can be approximated as being an infinite wire. What are the (a) magnitude and (b) direction (into or out of the page) of $\vec{B}$ at a and the (c) magnitude and (d) direction of $\vec{B}$ at b?

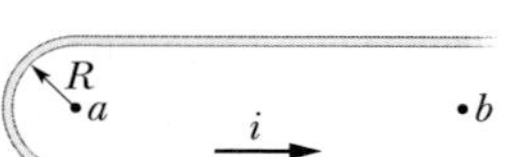

Fig. 29-36 Problem 5.

•6 In Fig. 29-37, point P is at perpendicular distance R = 2.00 cm from a very long straight wire carrying a current. The magnetic field $\vec{B}$ set up at point P is due to contributions from all the identical current-length elements $i\,d\vec{s}$ along the wire. What is the distance s to the element making (a) the greatest contribution to field $\vec{B}$ and (b) 10.0% of the greatest contribution?

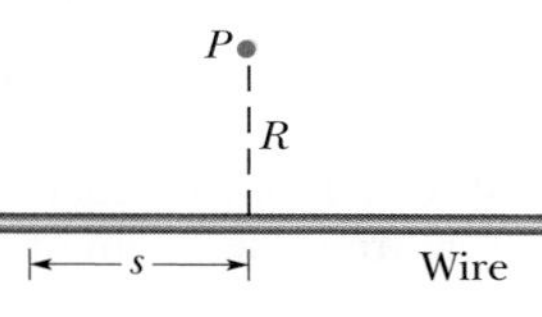

Fig. 29-37 Problem 6.

•7 GO In Fig. 29-38, two circular arcs have radii a = 13.5 cm and b = 10.7 cm, subtend angle θ = 74.0°, carry current i = 0.411 A, and share the same center of curvature P. What are the (a) magnitude and (b) direction (into or out of the page) of the net magnetic field at P?

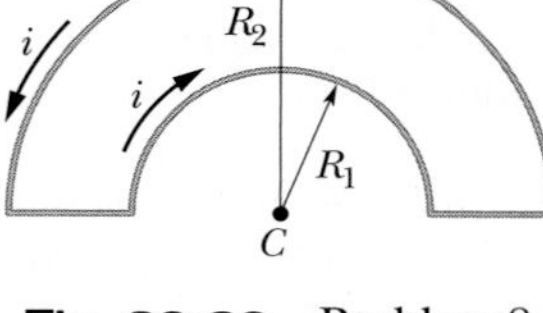

Fig. 29-38 Problem 7.

•8 In Fig. 29-39, two semicircular arcs have radii R_2 = 7.80 cm and R_1 = 3.15 cm, carry current i = 0.281 A, and share the same center of curvature C. What are the (a) magnitude and (b) direction (into or out of the page) of the net magnetic field at C?

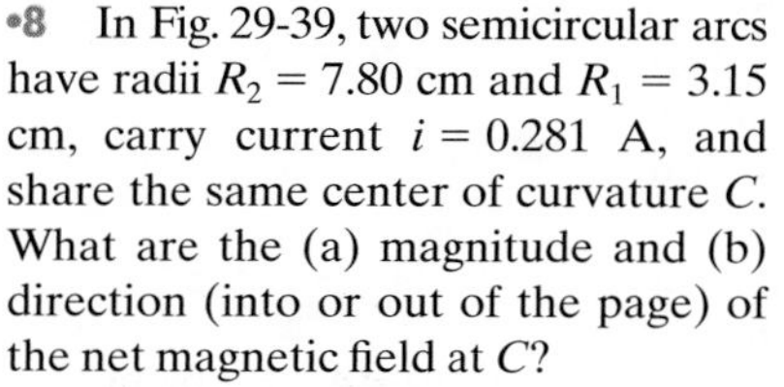

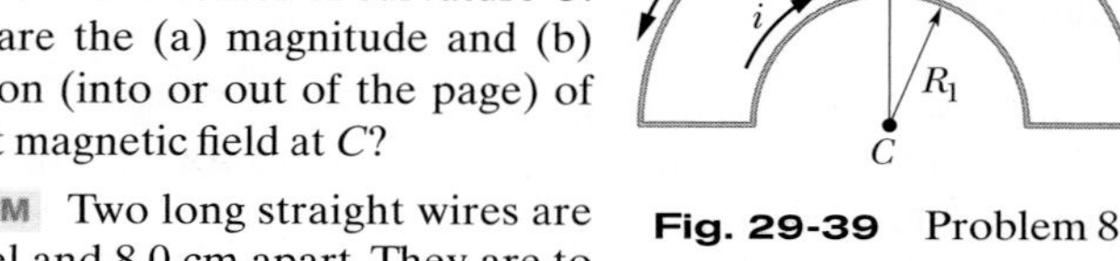

Fig. 29-39 Problem 8.

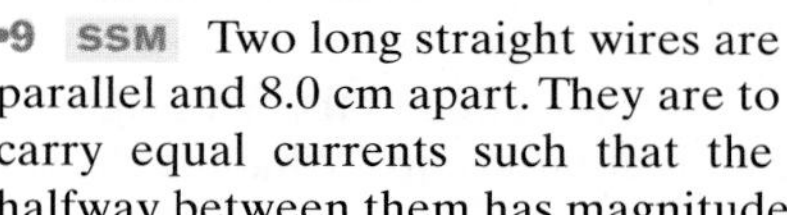

•9 SSM Two long straight wires are parallel and 8.0 cm apart. They are to carry equal currents such that the magnetic field at a point halfway between them has magnitude 300 μT. (a) Should the currents be in the same or opposite directions? (b) How much current is needed?

•10 In Fig. 29-40, a wire forms a semicircle of radius R = 9.26 cm and two (radial) straight segments each of length L = 13.1 cm. The wire carries current i = 34.8 mA. What are the (a) magnitude and (b) direction (into or out of the page) of the net magnetic field at the semicircle's center of curvature C?

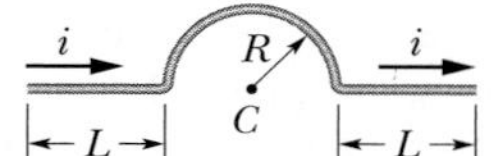

Fig. 29-40 Problem 10.

•11 In Fig. 29-41, two long straight wires are perpendicular to the page and separated by distance d_1 = 0.75 cm. Wire 1 carries 6.5 A into the page. What are the (a) magnitude and (b) direction (into or out of the page) of the current in wire 2 if the net magnetic field due to the two currents is zero at point P located at distance d_2 = 1.50 cm from wire 2?

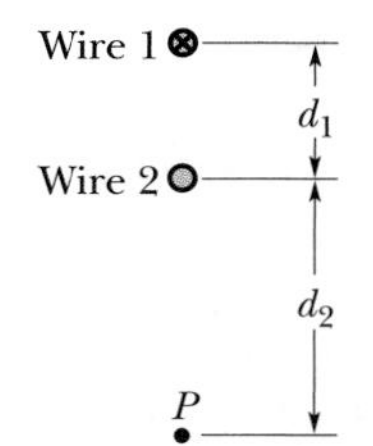

Fig. 29-41 Problem 11.

•12 In Fig. 29-42, two long straight wires at separation d = 16.0 cm carry currents i_1 = 3.61 mA and i_2 = 3.00i_1 out of the page. (a) Where on the x axis is the net magnetic field equal to zero? (b) If the two currents are doubled, is the zero-field point shifted toward wire 1, shifted toward wire 2, or unchanged?

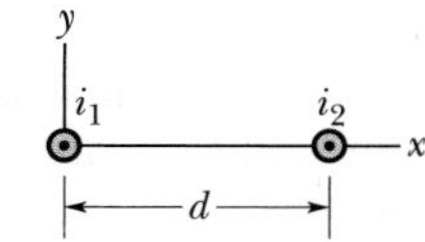

Fig. 29-42 Problem 12.

••13 In Fig. 29-43, point P_1 is at distance $R = 13.1$ cm on the perpendicular bisector of a straight wire of length $L = 18.0$ cm carrying current $i = 58.2$ mA. (Note that the wire is *not* long.) What is the magnitude of the magnetic field at P_1 due to i?

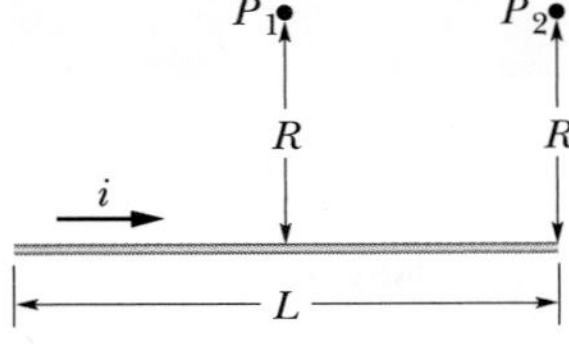

Fig. 29-43 Problems 13 and 17.

••14 Equation 29-4 gives the magnitude B of the magnetic field set up by a current in an *infinitely long* straight wire, at a point P at perpendicular distance R from the wire. Suppose that point P is actually at perpendicular distance R from the midpoint of a wire with a *finite* length L. Using Eq. 29-4 to calculate B then results in a certain percentage error. What value must the ratio L/R exceed if the percentage error is to be less than 1.00%? That is, what L/R gives

$$\frac{(B \text{ from Eq. 29-4}) - (B \text{ actual})}{(B \text{ actual})}(100\%) = 1.00\%?$$

••15 Figure 29-44 shows two current segments. The lower segment carries a current of $i_1 = 0.40$ A and includes a semicircular arc with radius 5.0 cm, angle 180°, and center point P. The upper segment carries current $i_2 = 2i_1$ and includes a circular arc with radius 4.0 cm, angle 120°, and the same center point P. What are the (a) magnitude and (b) direction of the net magnetic field $\vec{B}$ at P for the indicated current directions? What are the (c) magnitude and (d) direction of $\vec{B}$ if i_1 is reversed?

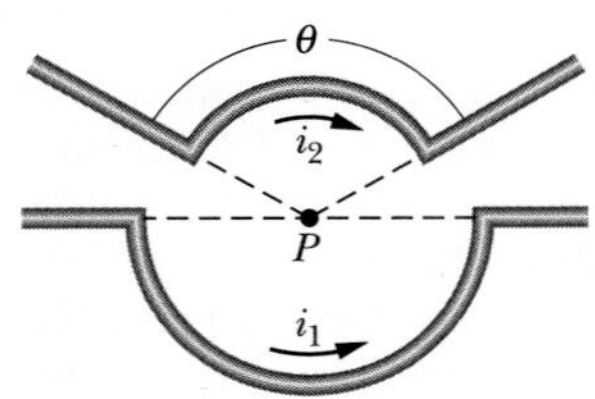

Fig. 29-44 Problem 15.

••16 GO In Fig. 29-45, two concentric circular loops of wire carrying current in the same direction lie in the same plane. Loop 1 has radius 1.50 cm and carries 4.00 mA. Loop 2 has radius 2.50 cm and carries 6.00 mA. Loop 2 is to be rotated about a diameter while the net magnetic field $\vec{B}$ set up by the two loops at their common center is measured. Through what angle must loop 2 be rotated so that the magnitude of that net field is 100 nT?

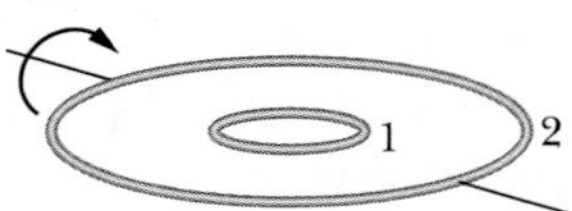

Fig. 29-45 Problem 16.

••17 SSM In Fig. 29-43, point P_2 is at perpendicular distance $R = 25.1$ cm from one end of a straight wire of length $L = 13.6$ cm carrying current $i = 0.693$ A. (Note that the wire is *not* long.) What is the magnitude of the magnetic field at P_2?

••18 A current is set up in a wire loop consisting of a semicircle of radius 4.00 cm, a smaller concentric semicircle, and two radial straight lengths, all in the same plane. Figure 29-46*a* shows the arrangement but is not drawn to scale. The magnitude of the magnetic field produced at the center of curvature is 47.25 μT. The smaller semicircle is then flipped over (rotated) until the loop is again entirely in the same plane (Fig. 29-46*b*). The magnetic field produced at the (same) center of curvature now has magnitude 15.75 μT, and its direction is reversed. What is the radius of the smaller semicircle?

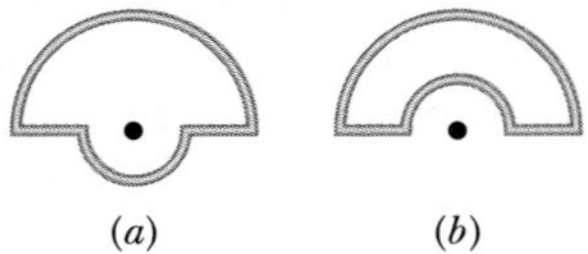

Fig. 29-46 Problem 18.

••19 One long wire lies along an x axis and carries a current of 30 A in the positive x direction. A second long wire is perpendicular to the xy plane, passes through the point (0, 4.0 m, 0), and carries a current of 40 A in the positive z direction. What is the magnitude of the resulting magnetic field at the point (0, 2.0 m, 0)?

••20 In Fig. 29-47, part of a long insulated wire carrying current $i = 5.78$ mA is bent into a circular section of radius $R = 1.89$ cm. In unit-vector notation, what is the magnetic field at the center of curvature C if the circular section (a) lies in the plane of the page as shown and (b) is perpendicular to the plane of the page after being rotated 90° counterclockwise as indicated?

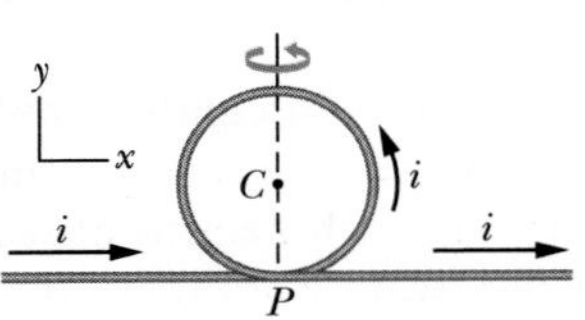

Fig. 29-47 Problem 20.

••21 GO Figure 29-48 shows two very long straight wires (in cross section) that each carry a current of 4.00 A directly out of the page. Distance $d_1 = 6.00$ m and distance $d_2 = 4.00$ m. What is the magnitude of the net magnetic field at point P, which lies on a perpendicular bisector to the wires?

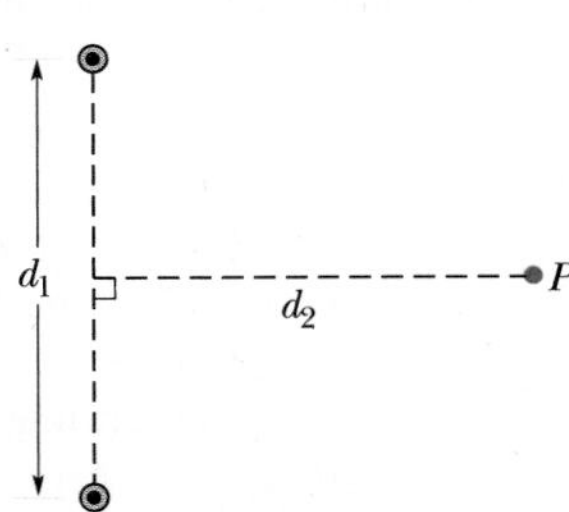

Fig. 29-48 Problem 21.

••22 Figure 29-49*a* shows, in cross section, two long, parallel wires carrying current and separated by distance L. The ratio i_1/i_2 of their currents is 4.00; the directions of the currents are not indicated. Figure 29-49*b* shows the y component B_y of their net magnetic field along the x axis to the right of wire 2. The vertical scale is set by $B_{ys} = 4.0$ nT, and the horizontal scale is set by $x_s = 20.0$ cm. (a) At what value of $x > 0$ is B_y maximum? (b) If $i_2 = 3$ mA, what is the value of that maximum? What is the direction (into or out of the page) of (c) i_1 and (d) i_2?

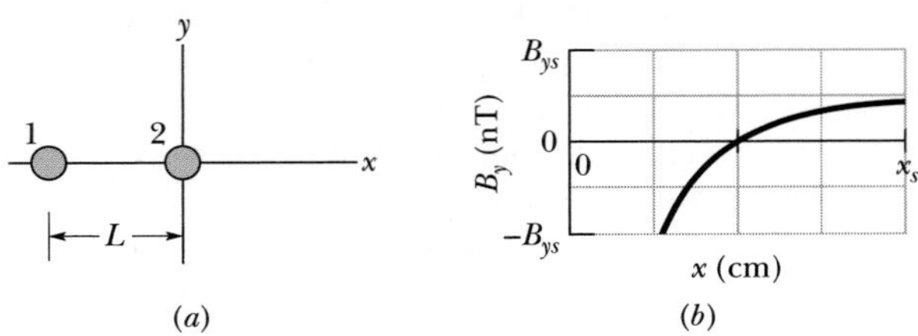

Fig. 29-49 Problem 22.

••23 ILW Figure 29-50 shows a snapshot of a proton moving at velocity $\vec{v} = (-200 \text{ m/s})\hat{j}$ toward a long straight wire with current $i = 350$ mA. At the instant shown, the proton's distance from the wire is $d = 2.89$ cm. In unit-vector notation, what is the magnetic force on the proton due to the current?

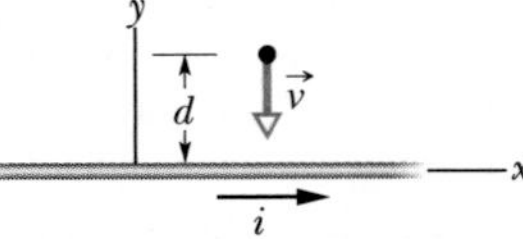

Fig. 29-50 Problem 23.

••24 GO Figure 29-51 shows, in cross section, four thin wires that are parallel, straight, and very long. They carry identical currents in the directions indicated. Initially all four wires are at distance $d = 15.0$ cm from the origin of the coordinate system, where they create a net magnetic field $\vec{B}$. (a) To what value of x must you move wire 1 along the x axis in order to rotate $\vec{B}$ counterclockwise by 30°? (b) With wire 1 in that new position, to what value of x must you move wire 3 along the x axis to rotate $\vec{B}$ by 30° back to its initial orientation?

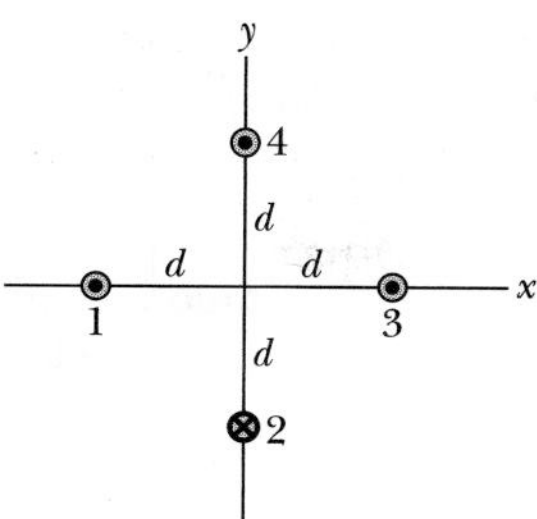

Fig. 29-51 Problem 24.

••25 SSM A wire with current $i = 3.00$ A is shown in Fig. 29-52. Two semi-infinite straight sections, both tangent to the same circle, are connected by a circular arc that has a central angle θ and runs along the circumference of the circle. The arc and the two straight sections all lie in the same plane. If $B = 0$ at the circle's center, what is θ?

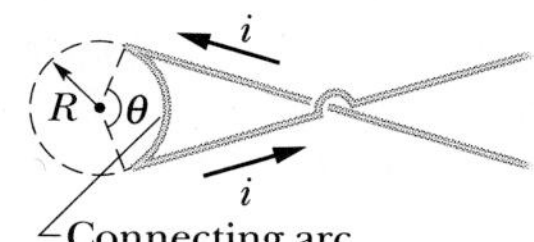

Fig. 29-52 Problem 25.

••26 In Fig. 29-53a, wire 1 consists of a circular arc and two radial lengths; it carries current $i_1 = 0.50$ A in the direction indicated. Wire 2, shown in cross section, is long, straight, and perpendicular to the plane of the figure. Its distance from the center of the arc is equal to the radius R of the arc, and it carries a current i_2 that can be varied. The two currents set up a net magnetic field $\vec{B}$ at the center of the arc. Figure 29-53b gives the square of the field's magnitude B^2 plotted versus the square of the current i_2^2. The vertical scale is set by $B_s^2 = 10.0 \times 10^{-10}\,\text{T}^2$. What angle is subtended by the arc?

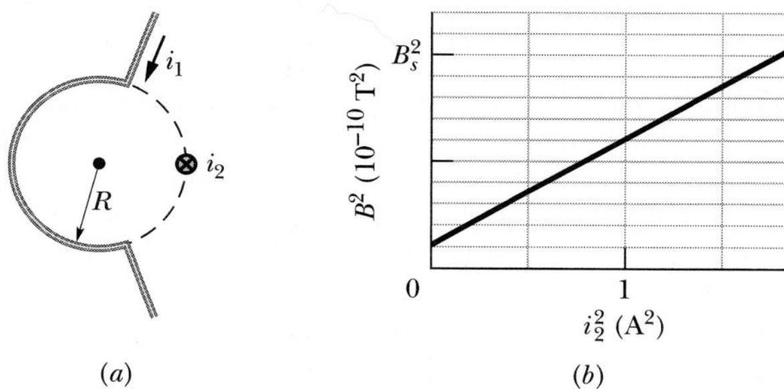

Fig. 29-53 Problem 26.

••27 In Fig. 29-54, two long straight wires (shown in cross section) carry currents $i_1 = 30.0$ mA and $i_2 = 40.0$ mA directly out of the page. They are equal distances from the origin, where they set up a magnetic field $\vec{B}$. To what value must current i_1 be changed in order to rotate $\vec{B}$ 20.0° clockwise?

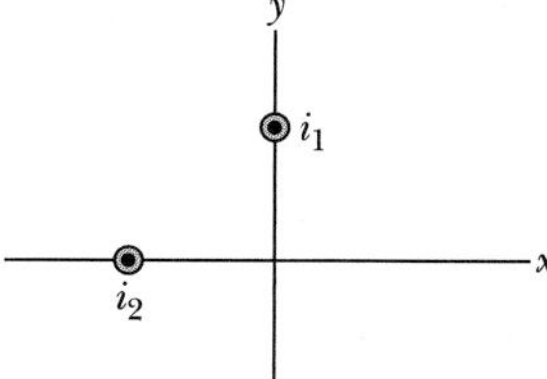

Fig. 29-54 Problem 27.

••28 GO Figure 29-55a shows two wires, each carrying a current. Wire 1 consists of a circular arc of radius R and two radial lengths; it carries current $i_1 = 2.0$ A in the direction indicated. Wire 2 is long and straight; it carries a current i_2 that can be varied; and it is at distance $R/2$ from the center of the arc. The net magnetic field $\vec{B}$ due to the two currents is measured at the center of curvature of the arc. Figure 29-55b is a plot of the component of $\vec{B}$ in the direction perpendicular to the figure as a function of current i_2. The horizontal scale is set by $i_{2s} = 1.00$ A. What is the angle subtended by the arc?

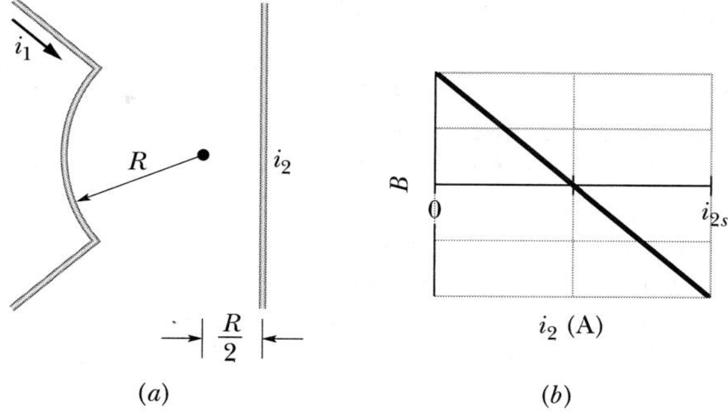

Fig. 29-55 Problem 28.

••29 SSM In Fig. 29-56, four long straight wires are perpendicular to the page, and their cross sections form a square of edge length $a = 20$ cm. The currents are out of the page in wires 1 and 4 and into the page in wires 2 and 3, and each wire carries 20 A. In unit-vector notation, what is the net magnetic field at the square's center?

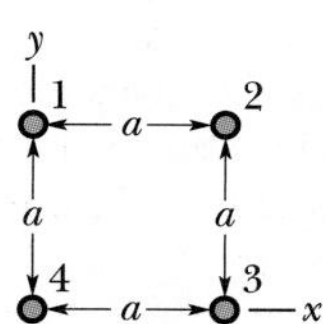

Fig. 29-56 Problems 29, 37, and 40.

•••30 Two long straight thin wires with current lie against an equally long plastic cylinder, at radius $R = 20.0$ cm from the cylinder's central axis. Figure 29-57a shows, in cross section, the cylinder and wire 1 but not wire 2. With wire 2 fixed in place, wire 1 is moved around the cylinder, from angle $\theta_1 = 0°$ to angle $\theta_1 = 180°$, through the first and second quadrants of the xy coordinate system. The net magnetic field

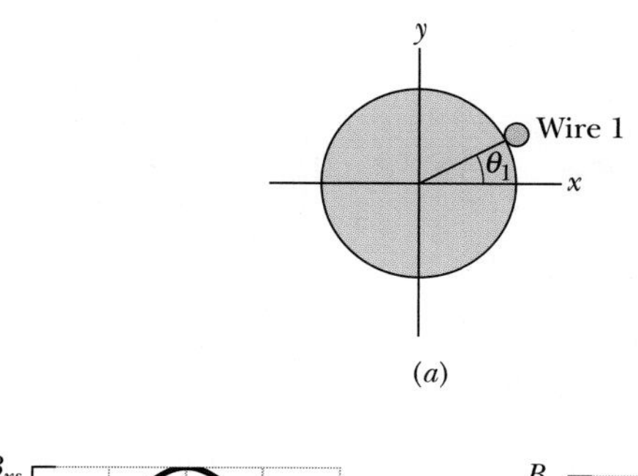

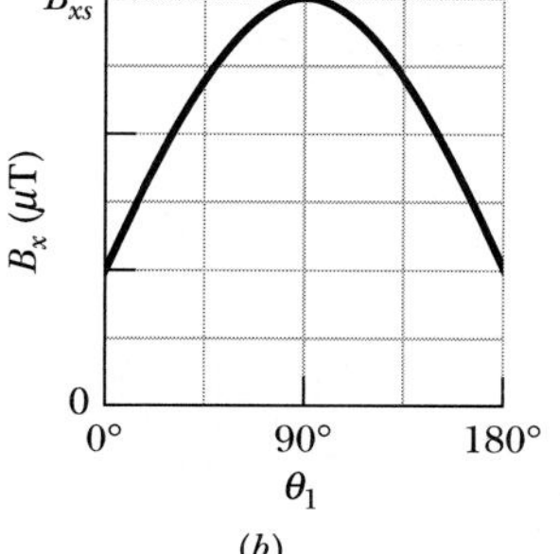

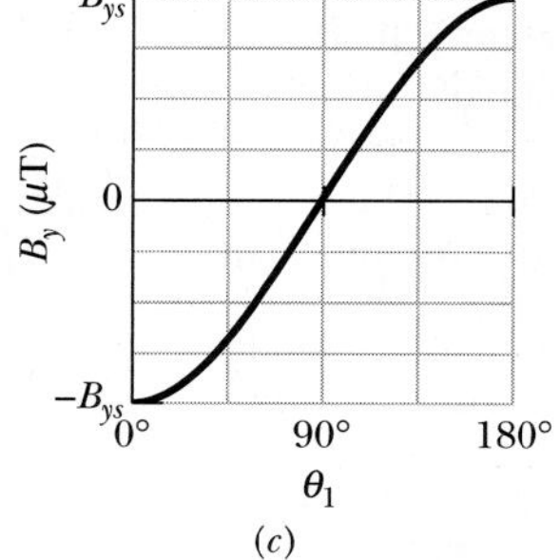

Fig. 29-57 Problem 30.

$\vec{B}$ at the center of the cylinder is measured as a function of θ_1. Figure 29-57b gives the x component B_x of that field as a function of θ_1 (the vertical scale is set by $B_{xs} = 6.0\ \mu$T), and Fig. 29-57c gives the y component B_y (the vertical scale is set by $B_{ys} = 4.0\ \mu$T). (a) At what angle θ_2 is wire 2 located? What are the (b) size and (c) direction (into or out of the page) of the current in wire 1 and the (d) size and (e) direction of the current in wire 2?

•••31 In Fig. 29-58, length a is 4.7 cm (short) and current i is 13 A. What are the (a) magnitude and (b) direction (into or out of the page) of the magnetic field at point P?

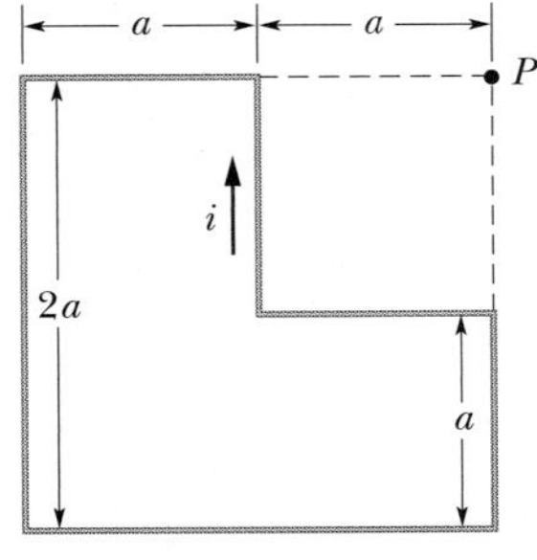

Fig. 29-58 Problem 31.

•••32 The current-carrying wire loop in Fig. 29-59a lies all in one plane and consists of a semicircle of radius 10.0 cm, a smaller semicircle with the same center, and two radial lengths. The smaller semicircle is rotated out of that plane by angle θ, until it is perpendicular to the plane (Fig. 29-59b). Figure 29-59c gives the magnitude of the net magnetic field at the center of curvature versus angle θ. The vertical scale is set by $B_a = 10.0\ \mu$T and $B_b = 12.0\ \mu$T. What is the radius of the smaller semicircle?

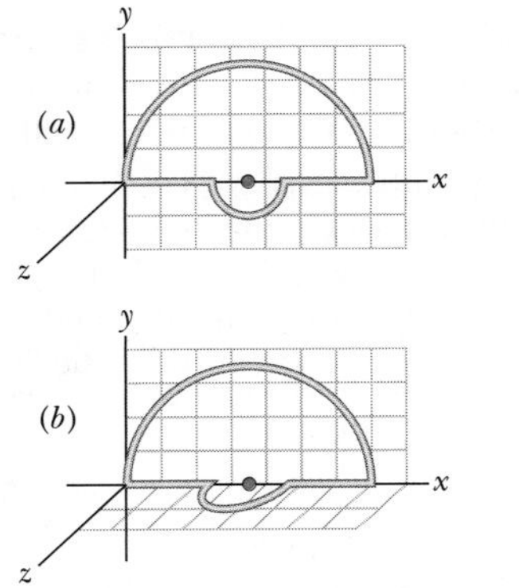

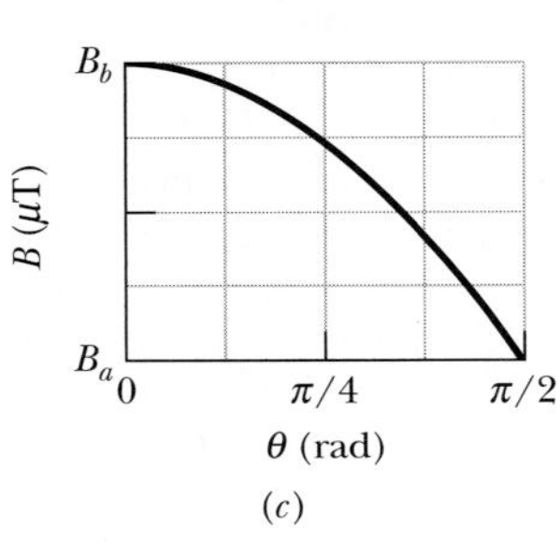

Fig. 29-59 Problem 32.

•••33 SSM ILW Figure 29-60 shows a cross section of a long thin ribbon of width $w = 4.91$ cm that is carrying a uniformly distributed total current $i = 4.61\ \mu$A into the page. In unit-vector notation, what is the magnetic field $\vec{B}$ at a point P in the plane of the ribbon at a distance $d = 2.16$ cm from its edge? (*Hint:* Imagine the ribbon as being constructed from many long, thin, parallel wires.)

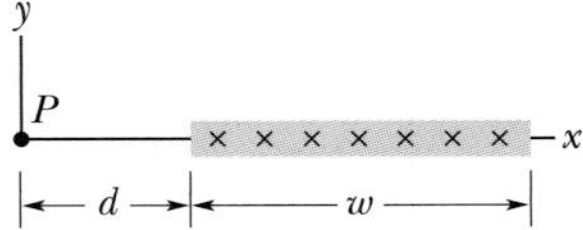

Fig. 29-60 Problem 33.

•••34 Figure 29-61 shows, in cross section, two long straight wires held against a plastic cylinder of radius 20.0 cm. Wire 1 carries current $i_1 = 60.0$ mA out of the page and is fixed in place at the left side of the cylinder. Wire 2 carries current $i_2 = 40.0$ mA out of the page and can be moved around the cylinder. At what (positive) angle θ_2 should wire 2 be positioned such that, at the origin, the net magnetic field due to the two currents has magnitude 80.0 nT?

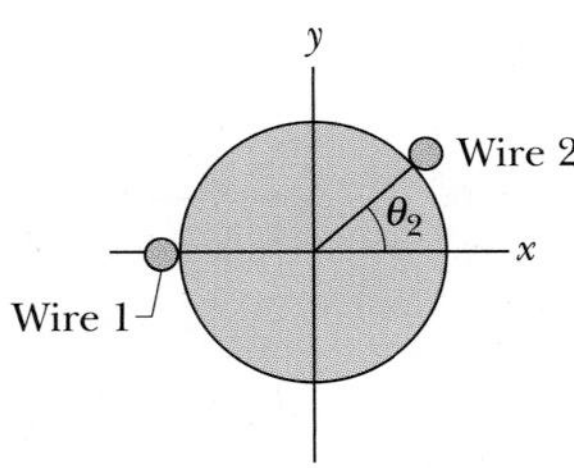

Fig. 29-61 Problem 34.

sec. 29-3 Force Between Two Parallel Currents

•35 SSM Figure 29-62 shows wire 1 in cross section; the wire is long and straight, carries a current of 4.00 mA out of the page, and is at distance $d_1 = 2.40$ cm from a surface. Wire 2, which is parallel to wire 1 and also long, is at horizontal distance $d_2 = 5.00$ cm from wire 1 and carries a current of 6.80 mA into the page. What is the x component of the magnetic force *per unit length* on wire 2 due to wire 1?

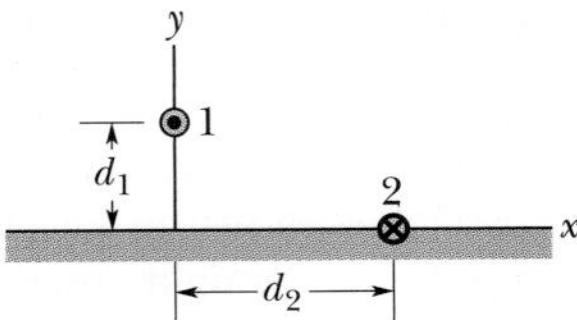

Fig. 29-62 Problem 35.

••36 In Fig. 29-63, five long parallel wires in an xy plane are separated by distance $d = 8.00$ cm, have lengths of 10.0 m, and carry identical currents of 3.00 A out of the page. Each wire experiences a magnetic force due to the other wires. In unit-vector notation, what is the net magnetic force on (a) wire 1, (b) wire 2, (c) wire 3, (d) wire 4, and (e) wire 5?

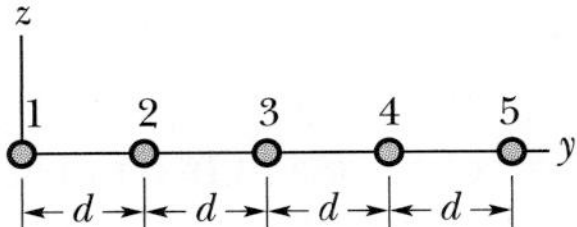

Fig. 29-63 Problems 36 and 39.

••37 GO In Fig. 29-56, four long straight wires are perpendicular to the page, and their cross sections form a square of edge length $a = 13.5$ cm. Each wire carries 7.50 A, and the currents are out of the page in wires 1 and 4 and into the page in wires 2 and 3. In unit-vector notation, what is the net magnetic force *per meter of wire length* on wire 4?

••38 Figure 29-64a shows, in cross section, three current-carrying wires that are long, straight, and parallel to one another. Wires 1 and 2 are fixed in place on an x axis, with separation d. Wire 1 has a current of 0.750 A, but the direction of the current is not given. Wire 3, with a current of 0.250 A out of the page, can be moved along the x axis to the right of wire 2. As wire 3 is moved, the magnitude of the net magnetic force $\vec{F}_2$ on wire 2 due to the currents in wires 1 and 3 changes. The x component of that force is F_{2x} and the value per unit length of wire 2 is F_{2x}/L_2. Figure 29-64b gives F_{2x}/L_2 versus the position x of wire 3. The plot has an asymptote $F_{2x}/L_2 = -0.627\ \mu$N/m as $x \to \infty$. The horizontal scale is set by $x_s = 12.0$ cm. What are the (a) size and (b) direction (into or out of the page) of the current in wire 2?

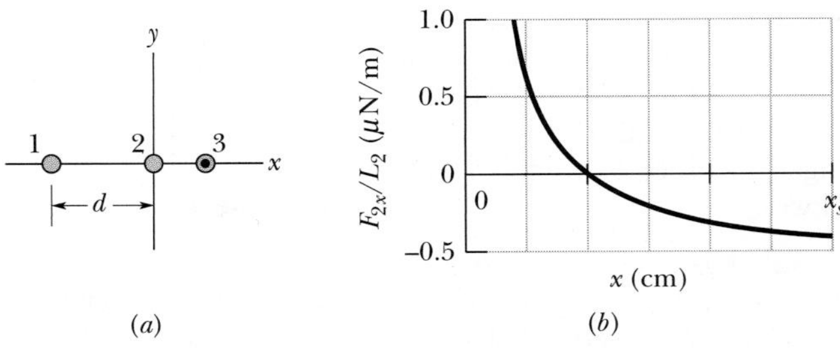

Fig. 29-64 Problem 38.

••39 GO In Fig. 29-63, five long parallel wires in an xy plane are separated by distance $d = 50.0$ cm. The currents into the page are

$i_1 = 2.00$ A, $i_3 = 0.250$ A, $i_4 = 4.00$ A, and $i_5 = 2.00$ A; the current out of the page is $i_2 = 4.00$ A. What is the magnitude of the net force *per unit length* acting on wire 3 due to the currents in the other wires?

••40 In Fig. 29-56, four long straight wires are perpendicular to the page, and their cross sections form a square of edge length $a = 8.50$ cm. Each wire carries 15.0 A, and all the currents are out of the page. In unit-vector notation, what is the net magnetic force *per meter of wire length* on wire 1?

•••41 ILW In Fig. 29-65, a long straight wire carries a current $i_1 =$ 30.0 A and a rectangular loop carries current $i_2 = 20.0$ A. Take $a =$ 1.00 cm, $b = 8.00$ cm, and $L = 30.0$ cm. In unit-vector notation, what is the net force on the loop due to i_1?

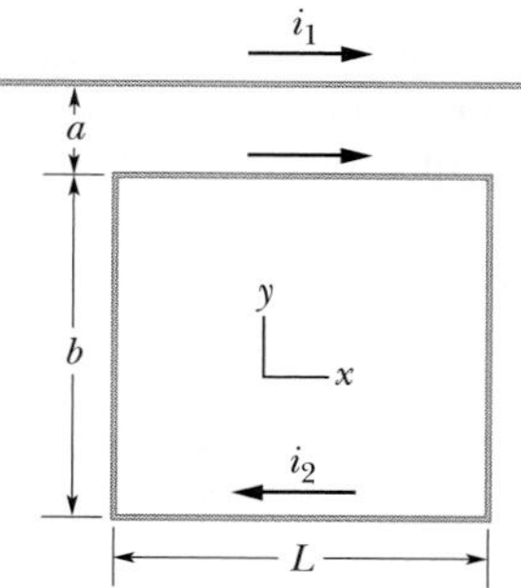

Fig. 29-65 Problem 41.

sec. 29-4 Ampere's Law

•42 In a particular region there is a uniform current density of 15 A/m² in the positive z direction. What is the value of $\oint \vec{B} \cdot d\vec{s}$ when that line integral is calculated along the three straight-line segments from (x, y, z) coordinates $(4d, 0, 0)$ to $(4d, 3d, 0)$ to $(0, 0, 0)$ to $(4d, 0, 0)$, where $d = 20$ cm?

•43 Figure 29-66 shows a cross section across a diameter of a long cylindrical conductor of radius $a = 2.00$ cm carrying uniform current 170 A. What is the magnitude of the current's magnetic field at radial distance (a) 0, (b) 1.00 cm, (c) 2.00 cm (wire's surface), and (d) 4.00 cm?

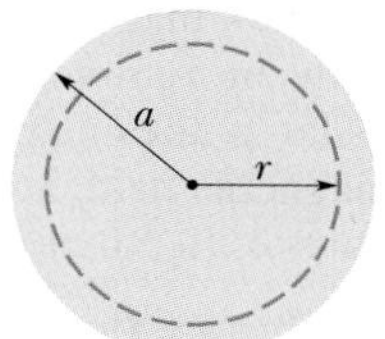

Fig. 29-66 Problem 43.

•44 Figure 29-67 shows two closed paths wrapped around two conducting loops carrying currents $i_1 = 5.0$ A and $i_2 = 3.0$ A. What is the value of the integral $\oint \vec{B} \cdot d\vec{s}$ for (a) path 1 and (b) path 2?

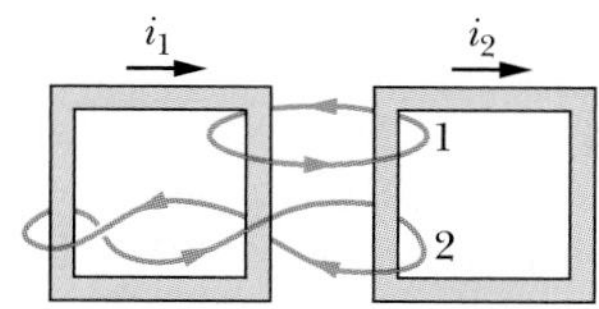

Fig. 29-67 Problem 44.

•45 SSM Each of the eight conductors in Fig. 29-68 carries 2.0 A of current into or out of the page. Two paths are indicated for the line integral $\oint \vec{B} \cdot d\vec{s}$. What is the value of the integral for (a) path 1 and (b) path 2?

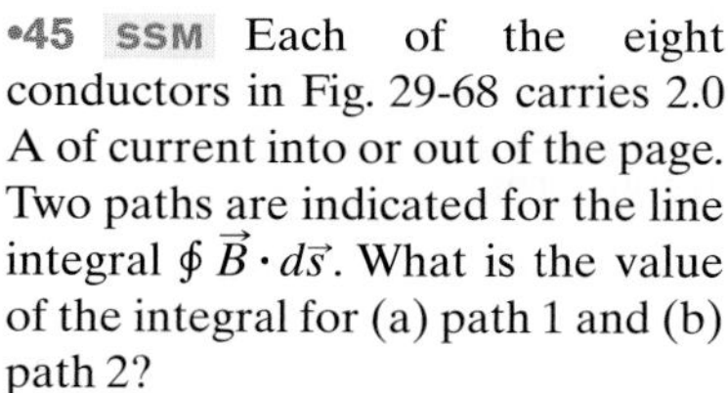
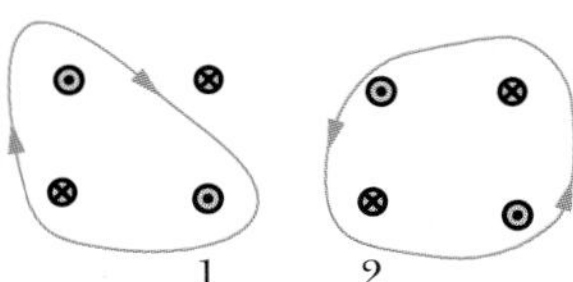

Fig. 29-68 Problem 45.

•46 Eight wires cut the page perpendicularly at the points shown in Fig. 29-69. A wire labeled with the integer k ($k = 1, 2, \ldots, 8$) carries the current ki, where $i = 4.50$ mA. For those wires with odd k, the current is out of the page; for those with even k, it is into the page. Evaluate $\oint \vec{B} \cdot d\vec{s}$ along the closed path in the direction shown.

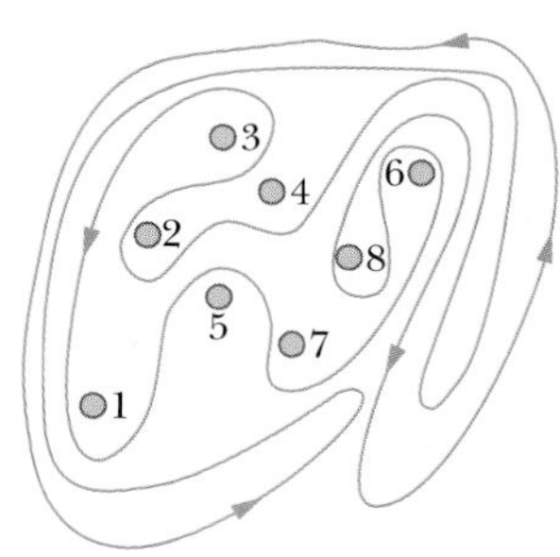

Fig. 29-69 Problem 46.

••47 ILW The current density $\vec{J}$ inside a long, solid, cylindrical wire of radius $a = 3.1$ mm is in the direction of the central axis, and its magnitude varies linearly with radial distance r from the axis according to $J = J_0 r/a$, where $J_0 = 310$ A/m². Find the magnitude of the magnetic field at (a) $r = 0$, (b) $r = a/2$, and (c) $r = a$.

••48 In Fig. 29-70, a long circular pipe with outside radius $R = 2.6$ cm carries a (uniformly distributed) current $i =$ 8.00 mA into the page. A wire runs parallel to the pipe at a distance of $3.00R$ from center to center. Find the (a) magnitude and (b) direction (into or out of the page) of the current in the wire such that the net magnetic field at point P has the same magnitude as the net magnetic field at the center of the pipe but is in the opposite direction.

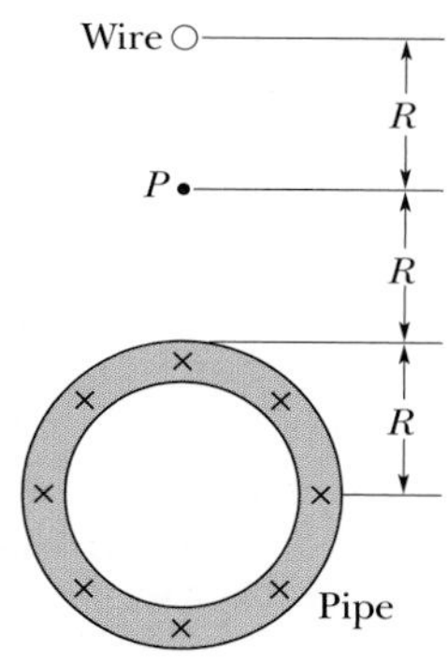

Fig. 29-70 Problem 48.

sec. 29-5 Solenoids and Toroids

•49 A toroid having a square cross section, 5.00 cm on a side, and an inner radius of 15.0 cm has 500 turns and carries a current of 0.800 A. (It is made up of a square solenoid—instead of a round one as in Fig. 29-16—bent into a doughnut shape.) What is the magnetic field inside the toroid at (a) the inner radius and (b) the outer radius?

•50 A solenoid that is 95.0 cm long has a radius of 2.00 cm and a winding of 1200 turns; it carries a current of 3.60 A. Calculate the magnitude of the magnetic field inside the solenoid.

•51 A 200-turn solenoid having a length of 25 cm and a diameter of 10 cm carries a current of 0.29 A. Calculate the magnitude of the magnetic field $\vec{B}$ inside the solenoid.

•52 A solenoid 1.30 m long and 2.60 cm in diameter carries a current of 18.0 A. The magnetic field inside the solenoid is 23.0 mT. Find the length of the wire forming the solenoid.

••53 A long solenoid has 100 turns/cm and carries current i. An electron moves within the solenoid in a circle of radius 2.30 cm perpendicular to the solenoid axis. The speed of the electron is $0.0460c$ (c = speed of light). Find the current i in the solenoid.

••54 An electron is shot into one end of a solenoid. As it enters the uniform magnetic field within the solenoid, its speed is 800 m/s and its velocity vector makes an angle of 30° with the central axis of the solenoid. The solenoid carries 4.0 A and has 8000 turns along its length. How many revolutions does the electron make along its helical path within the solenoid by the time it emerges from the solenoid's opposite end? (In a real solenoid, where the field is not uniform at the two ends, the number of revolutions would be slightly less than the answer here.)

••55 SSM ILW WWW A long solenoid with 10.0 turns/cm and a radius of 7.00 cm carries a current of 20.0 mA. A current of 6.00 A exists in a straight conductor located along the central axis of the solenoid. (a) At what radial distance from the axis will the direction of the resulting magnetic field be at 45.0° to the axial direction? (b) What is the magnitude of the magnetic field there?

sec. 29-6 A Current-Carrying Coil as a Magnetic Dipole

•56 Figure 29-71 shows an arrangement known as a Helmholtz coil. It consists of two circular coaxial coils, each of 200 turns and radius

$R = 25.0$ cm, separated by a distance $s = R$. The two coils carry equal currents $i = 12.2$ mA in the same direction. Find the magnitude of the net magnetic field at P, midway between the coils.

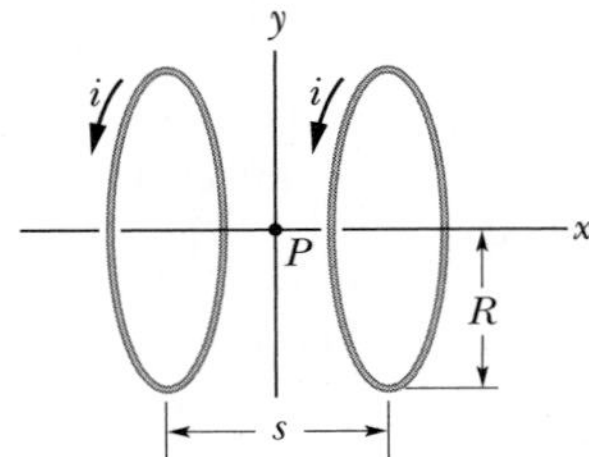

Fig. 29-71 Problems 56 and 90.

•**57** SSM A student makes a short electromagnet by winding 300 turns of wire around a wooden cylinder of diameter $d = 5.0$ cm. The coil is connected to a battery producing a current of 4.0 A in the wire. (a) What is the magnitude of the magnetic dipole moment of this device? (b) At what axial distance $z \gg d$ will the magnetic field have the magnitude 5.0 μT (approximately one-tenth that of Earth's magnetic field)?

•**58** Figure 29-72*a* shows a length of wire carrying a current i and bent into a circular coil of one turn. In Fig. 29-72*b* the same length of wire has been bent to give a coil of two turns, each of half the original radius. (a) If B_a and B_b are the magnitudes of the magnetic fields at the centers of the two coils, what is the ratio B_b/B_a? (b) What is the ratio μ_b/μ_a of the dipole moment magnitudes of the coils?

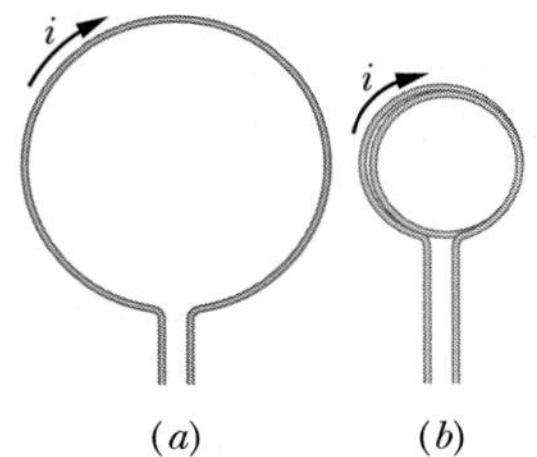

Fig. 29-72 Problem 58.

•**59** SSM What is the magnitude of the magnetic dipole moment $\vec{\mu}$ of the solenoid described in Problem 51?

••**60** GO In Fig. 29-73*a*, two circular loops, with different currents but the same radius of 4.0 cm, are centered on a y axis. They are initially separated by distance $L = 3.0$ cm, with loop 2 positioned at the origin of the axis. The currents in the two loops produce a net magnetic field at the origin, with y component B_y. That component is to be measured as loop 2 is gradually moved in the positive direction of the y axis. Figure 29-73*b* gives B_y as a function of the position y of loop 2. The curve approaches an asymptote of $B_y = 7.20$ μT as $y \to \infty$. The horizontal scale is set by $y_s = 10.0$ cm. What are (a) current i_1 in loop 1 and (b) current i_2 in loop 2?

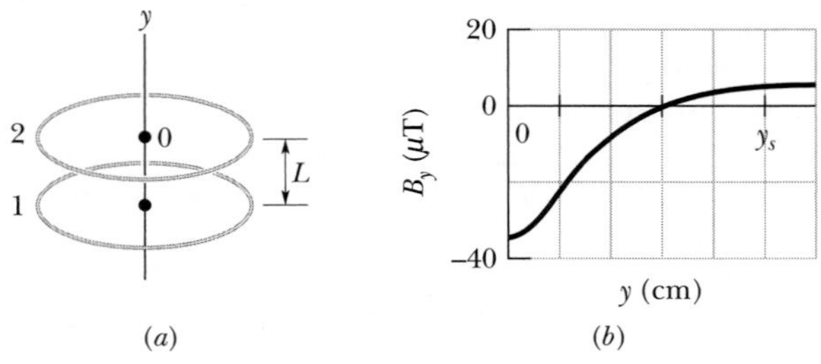

Fig. 29-73 Problem 60.

••**61** A circular loop of radius 12 cm carries a current of 15 A. A flat coil of radius 0.82 cm, having 50 turns and a current of 1.3 A, is concentric with the loop. The plane of the loop is perpendicular to the plane of the coil. Assume the loop's magnetic field is uniform across the coil. What is the magnitude of (a) the magnetic field produced by the loop at its center and (b) the torque on the coil due to the loop?

••**62** In Fig. 29-74, current $i =$ 56.2 mA is set up in a loop having two radial lengths and two semicircles of radii $a = 5.72$ cm and $b =$ 9.36 cm with a common center P. What are the (a) magnitude and (b) direction (into or out of the page) of the magnetic field at P and the (c) magnitude and (d) direction of the loop's magnetic dipole moment?

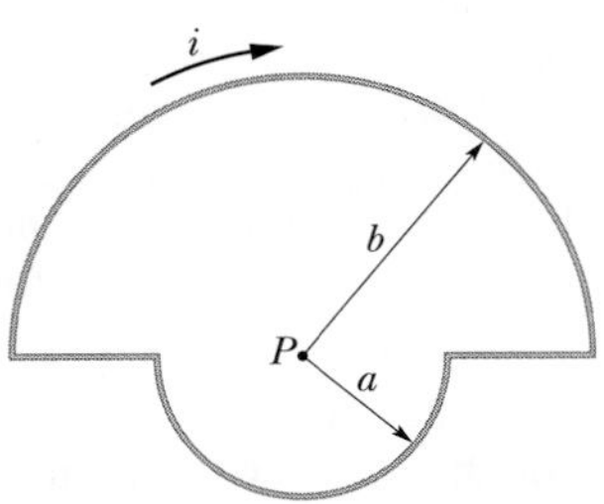

Fig. 29-74 Problem 62.

••**63** In Fig. 29-75, a conductor carries 6.0 A along the closed path *abcdefgha* running along 8 of the 12 edges of a cube of edge length 10 cm. (a) Taking the path to be a combination of three square current loops (*bcfgb*, *abgha*, and *cdefc*), find the net magnetic moment of the path in unit-vector notation. (b) What is the magnitude of the net magnetic field at the xyz coordinates of (0, 5.0 m, 0)?

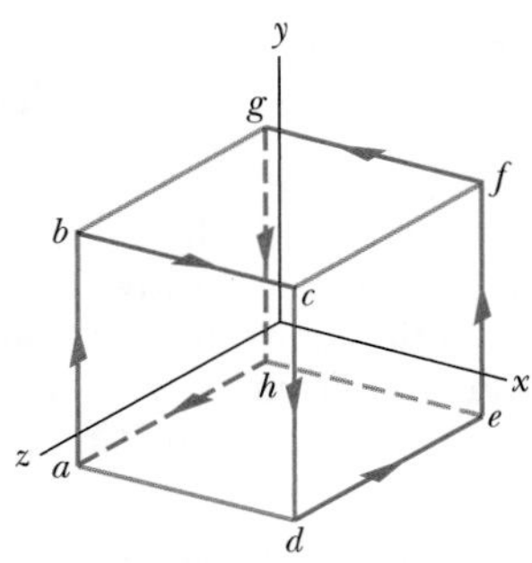

Fig. 29-75 Problem 63.

Additional Problems

64 In Fig. 29-76, a closed loop carries current $i = 200$ mA. The loop consists of two radial straight wires and two concentric circular arcs of radii 2.00 m and 4.00 m. The angle θ is $\pi/4$ rad. What are the (a) magnitude and (b) direction (into or out of the page) of the net magnetic field at the center of curvature P?

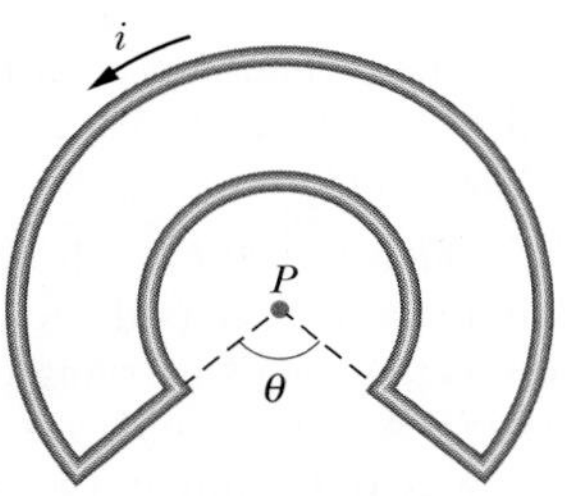

Fig. 29-76 Problem 64.

65 A cylindrical cable of radius 8.00 mm carries a current of 25.0 A, uniformly spread over its cross-sectional area. At what distance from the center of the wire is there a point within the wire where the magnetic field magnitude is 0.100 mT?

66 Two long wires lie in an xy plane, and each carries a current in the positive direction of the x axis. Wire 1 is at $y = 10.0$ cm and carries 6.00 A; wire 2 is at $y = 5.00$ cm and carries 10.0 A. (a) In unit-vector notation, what is the net magnetic field $\vec{B}$ at the origin? (b) At what value of y does $\vec{B} = 0$? (c) If the current in wire 1 is reversed, at what value of y does $\vec{B} = 0$?

67 Two wires, both of length L, are formed into a circle and a square, and each carries current i. Show that the square produces a greater magnetic field at its center than the circle produces at its center.

68 A long straight wire carries a current of 50 A. An electron, traveling at 1.0×10^7 m/s, is 5.0 cm from the wire. What is the magnitude of the magnetic force on the electron if the electron velocity is directed (a) toward the wire, (b) parallel to the wire in the direction of the current, and (c) perpendicular to the two directions defined by (a) and (b)?

69 Three long wires are parallel to a z axis, and each carries a current of 10 A in the positive z direction. Their points of intersection with the xy plane form an equilateral triangle with sides of 50 cm, as shown in Fig. 29-77. A fourth wire (wire b) passes through the midpoint of the base of the triangle and is parallel to the other three wires. If the net magnetic force on wire a is zero, what are the (a) size and (b) direction ($+z$ or $-z$) of the current in wire b?

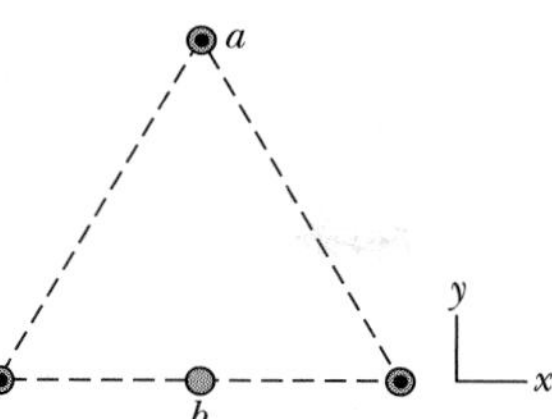

Fig. 29-77 Problem 69.

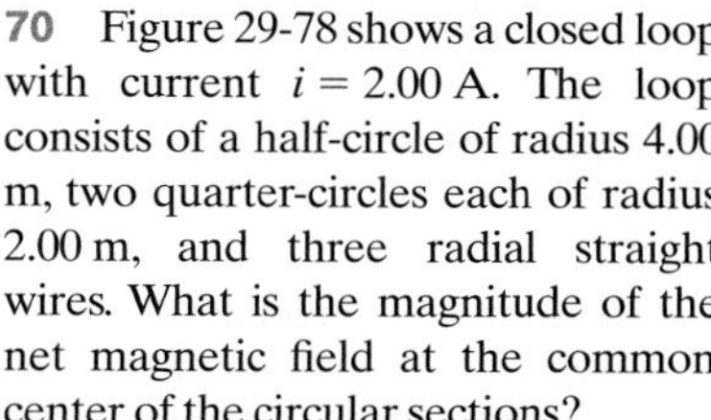

70 Figure 29-78 shows a closed loop with current $i = 2.00$ A. The loop consists of a half-circle of radius 4.00 m, two quarter-circles each of radius 2.00 m, and three radial straight wires. What is the magnitude of the net magnetic field at the common center of the circular sections?

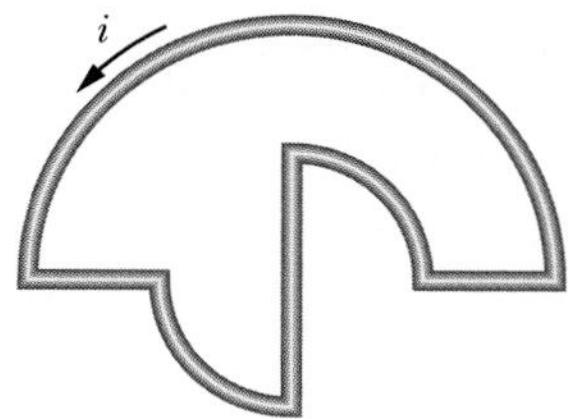

Fig. 29-78 Problem 70.

71 A 10-gauge bare copper wire (2.6 mm in diameter) can carry a current of 50 A without overheating. For this current, what is the magnitude of the magnetic field at the surface of the wire?

72 A long vertical wire carries an unknown current. Coaxial with the wire is a long, thin, cylindrical conducting surface that carries a current of 30 mA upward. The cylindrical surface has a radius of 3.0 mm. If the magnitude of the magnetic field at a point 5.0 mm from the wire is 1.0 μT, what are the (a) size and (b) direction of the current in the wire?

73 Figure 29-79 shows a cross section of a long cylindrical conductor of radius $a = 4.00$ cm containing a long cylindrical hole of radius $b = 1.50$ cm. The central axes of the cylinder and hole are parallel and are distance $d = 2.00$ cm apart; current $i = 5.25$ A is uniformly distributed over the tinted area. (a) What is the magnitude of the magnetic field at the center of the hole? (b) Discuss the two special cases $b = 0$ and $d = 0$.

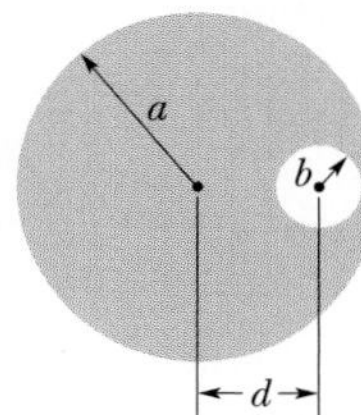

Fig. 29-79 Problem 73.

74 The magnitude of the magnetic field 88.0 cm from the axis of a long straight wire is 7.30 μT. What is the current in the wire?

75 SSM Figure 29-80 shows a wire segment of length $\Delta s = 3.0$ cm, centered at the origin, carrying current $i = 2.0$ A in the positive y direction (as part of some complete circuit). To calculate the magnitude of the magnetic field $\vec{B}$ produced by the segment at a point several meters from the origin, we can use $B = (\mu_0/4\pi)i\,\Delta s\,(\sin\theta)/r^2$ as the Biot–Savart law. This is because r and θ are essentially constant over the segment. Calculate $\vec{B}$ (in unit-vector notation) at the (x, y, z) coordinates (a) (0, 0, 5.0 m), (b) (0, 6.0 m, 0), (c) (7.0 m, 7.0 m, 0), and (d) (−3.0 m, −4.0 m, 0).

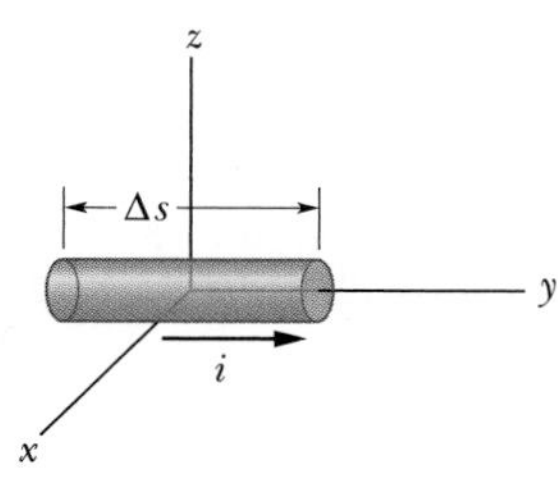

Fig. 29-80 Problem 75.

76 GO Figure 29-81 shows, in cross section, two long parallel wires spaced by distance $d = 10.0$ cm; each carries 100 A, out of the page in wire 1. Point P is on a perpendicular bisector of the line connecting the wires. In unit-vector notation, what is the net magnetic field at P if the current in wire 2 is (a) out of the page and (b) into the page?

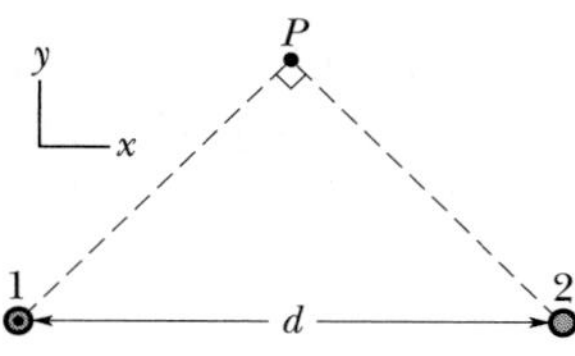

Fig. 29-81 Problem 76.

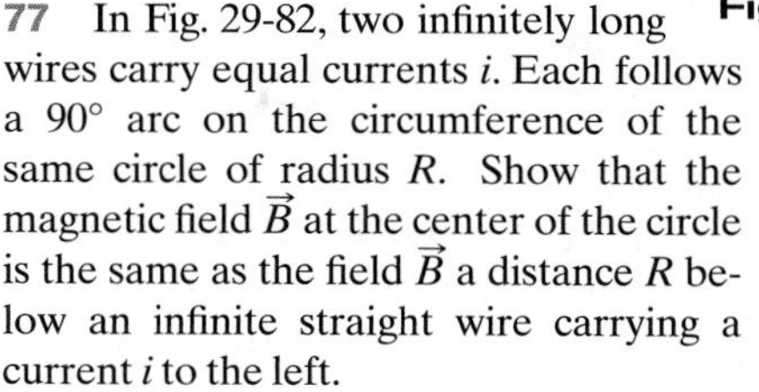

77 In Fig. 29-82, two infinitely long wires carry equal currents i. Each follows a 90° arc on the circumference of the same circle of radius R. Show that the magnetic field $\vec{B}$ at the center of the circle is the same as the field $\vec{B}$ a distance R below an infinite straight wire carrying a current i to the left.

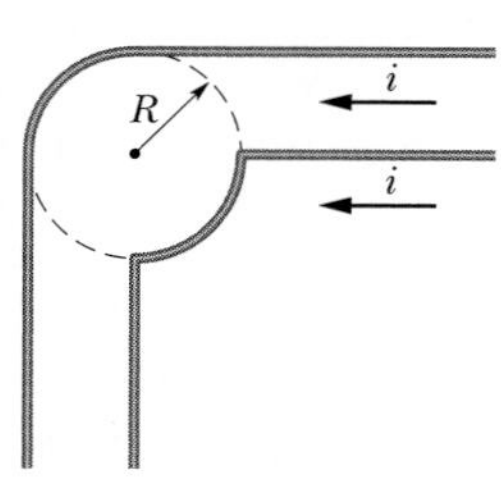

Fig. 29-82 Problem 77.

78 A long wire carrying 100 A is perpendicular to the magnetic field lines of a uniform magnetic field of magnitude 5.0 mT. At what distance from the wire is the net magnetic field equal to zero?

79 A long, hollow, cylindrical conductor (with inner radius 2.0 mm and outer radius 4.0 mm) carries a current of 24 A distributed uniformly across its cross section. A long thin wire that is coaxial with the cylinder carries a current of 24 A in the opposite direction. What is the magnitude of the magnetic field (a) 1.0 mm, (b) 3.0 mm, and (c) 5.0 mm from the central axis of the wire and cylinder?

80 A long wire is known to have a radius greater than 4.0 mm and to carry a current that is uniformly distributed over its cross section. The magnitude of the magnetic field due to that current is 0.28 mT at a point 4.0 mm from the axis of the wire, and 0.20 mT at a point 10 mm from the axis of the wire. What is the radius of the wire?

81 SSM Figure 29-83 shows a cross section of an infinite conducting sheet carrying a current per unit x-length of λ; the current emerges perpendicularly out of the page. (a) Use the Biot–Savart law and symmetry to show that for all points P above the sheet and all points P' below it, the magnetic field $\vec{B}$ is parallel to the sheet and directed as shown. (b) Use Ampere's law to prove that $B = \frac{1}{2}\mu_0\lambda$ at all points P and P'.

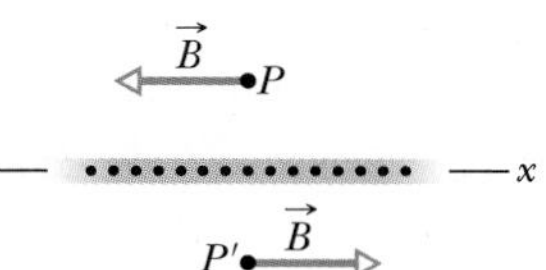

Fig. 29-83 Problem 81.

82 Figure 29-84 shows, in cross section, two long parallel wires that are separated by distance $d = 18.6$ cm. Each carries 4.23 A, out of the page in wire 1 and into the page in wire 2. In unit-vector notation, what is the net magnetic field at point P at distance $R = 34.2$ cm, due to the two currents?

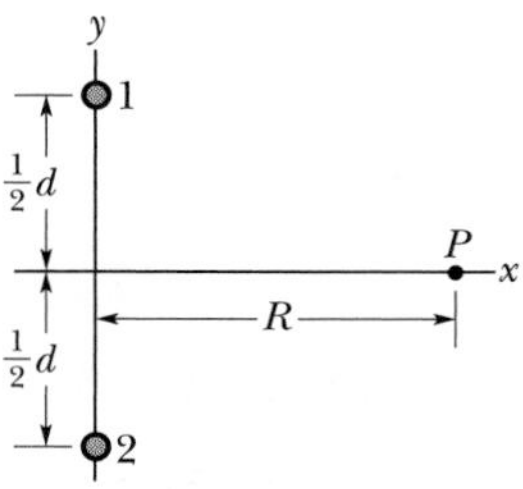

Fig. 29-84 Problem 82.

83 SSM In unit-vector notation, what is the magnetic field at point P in Fig. 29-85 if $i = 10$ A and $a = 8.0$ cm? (Note that the wires are *not* long.)

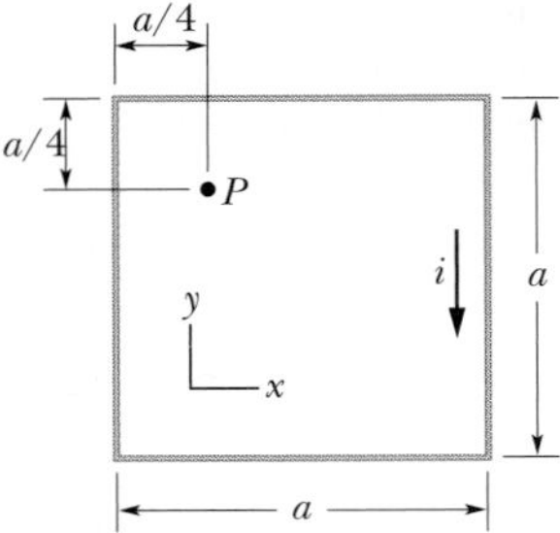

Fig. 29-85 Problem 83.

84 Three long wires all lie in an xy plane parallel to the x axis. They are spaced equally, 10 cm apart. The two outer wires each carry a current of 5.0 A in the positive x direction. What is the magnitude of the force on a 3.0 m section of either of the outer wires if the current in the center wire is 3.2 A (a) in the positive x direction and (b) in the negative x direction?

85 SSM Figure 29-86 shows a cross section of a hollow cylindrical conductor of radii a and b, carrying a uniformly distributed current i. (a) Show that the magnetic field magnitude $B(r)$ for the radial distance r in the range $b < r < a$ is given by

$$B = \frac{\mu_0 i}{2\pi(a^2 - b^2)} \frac{r^2 - b^2}{r}.$$

(b) Show that when $r = a$, this equation gives the magnetic field magnitude B at the surface of a long straight wire carrying current i; when $r = b$, it gives zero magnetic field; and when $b = 0$, it gives the magnetic field inside a solid conductor of radius a carrying current i. (c) Assume that $a = 2.0$ cm, $b = 1.8$ cm, and $i = 100$ A, and then plot $B(r)$ for the range $0 < r < 6$ cm.

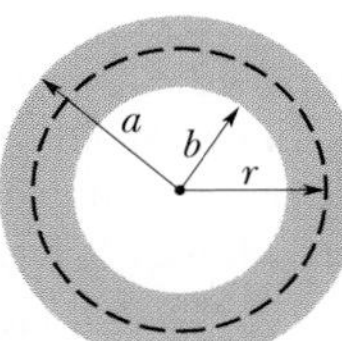

Fig. 29-86 Problem 85.

86 Show that the magnitude of the magnetic field produced at the center of a rectangular loop of wire of length L and width W, carrying a current i, is

$$B = \frac{2\mu_0 i}{\pi} \frac{(L^2 + W^2)^{1/2}}{LW}.$$

87 Figure 29-87 shows a cross section of a long conducting coaxial cable and gives its radii (a, b, c). Equal but opposite currents i are uniformly distributed in the two conductors. Derive expressions for $B(r)$ with radial distance r in the ranges (a) $r < c$, (b) $c < r < b$, (c) $b < r < a$, and (d) $r > a$. (e) Test these expressions for all the special cases that occur to you. (f) Assume that $a = 2.0$ cm, $b = 1.8$ cm, $c = 0.40$ cm, and $i = 120$ A and plot the function $B(r)$ over the range $0 < r < 3$ cm.

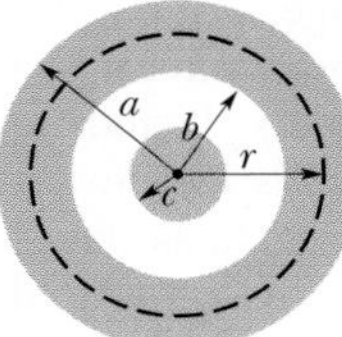

Fig. 29-87 Problem 87.

88 Figure 29-88 is an idealized schematic drawing of a rail gun. Projectile P sits between two wide rails of circular cross section; a source of current sends current through the rails and through the (conducting) projectile (a fuse is not used). (a) Let w be the distance between the rails, R the radius of each rail, and i the current. Show that the force on the projectile is directed to the right along the rails and is given approximately by

$$F = \frac{i^2 \mu_0}{2\pi} \ln \frac{w + R}{R}.$$

(b) If the projectile starts from the left end of the rails at rest, find the speed v at which it is expelled at the right. Assume that $i = 450$ kA, $w = 12$ mm, $R = 6.7$ cm, $L = 4.0$ m, and the projectile mass is 10 g.

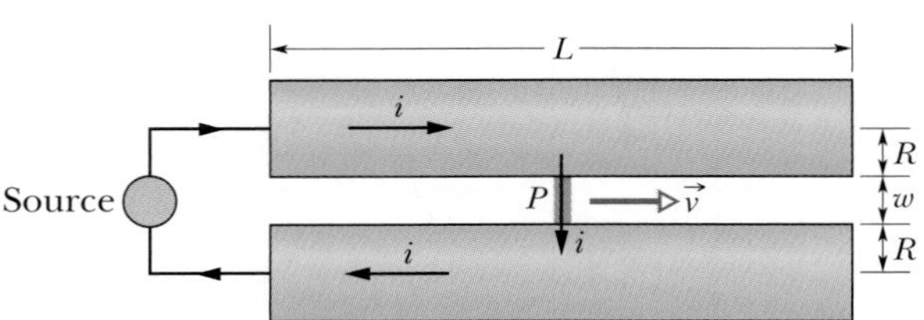

Fig. 29-88 Problem 88.

89 A square loop of wire of edge length a carries current i. Show that, at the center of the loop, the magnitude of the magnetic field produced by the current is

$$B = \frac{2\sqrt{2}\mu_0 i}{\pi a}.$$

90 In Fig. 29-71, an arrangement known as Helmholtz coils consists of two circular coaxial coils, each of N turns and radius R, separated by distance s. The two coils carry equal currents i in the same direction. (a) Show that the first derivative of the magnitude of the net magnetic field of the coils (dB/dx) vanishes at the midpoint P regardless of the value of s. Why would you expect this to be true from symmetry? (b) Show that the second derivative (d^2B/dx^2) also vanishes at P, provided $s = R$. This accounts for the uniformity of B near P for this particular coil separation.

91 SSM A square loop of wire of edge length a carries current i. Show that the magnitude of the magnetic field produced at a point on the central perpendicular axis of the loop and a distance x from its center is

$$B(x) = \frac{4\mu_0 i a^2}{\pi(4x^2 + a^2)(4x^2 + 2a^2)^{1/2}}.$$

Prove that this result is consistent with the result shown in Problem 89.

92 Show that if the thickness of a toroid is much smaller than its radius of curvature (a very skinny toroid), then Eq. 29-24 for the field inside a toroid reduces to Eq. 29-23 for the field inside a solenoid. Explain why this result is to be expected.

93 SSM Show that a uniform magnetic field $\vec{B}$ cannot drop abruptly to zero (as is suggested by the lack of field lines to the right of point a in Fig. 29-89) as one moves perpendicular to $\vec{B}$, say along the horizontal arrow in the figure. (*Hint:* Apply Ampere's law to the rectangular path shown by the dashed lines.) In actual magnets, "fringing" of the magnetic field lines always occurs, which means that $\vec{B}$ approaches zero in a gradual manner. Modify the field lines in the figure to indicate a more realistic situation.

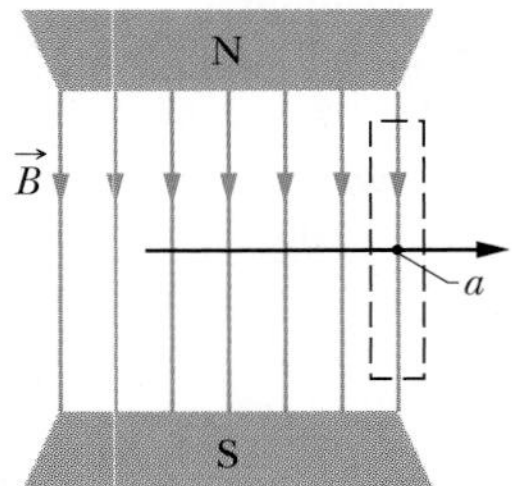

Fig. 29-89 Problem 93.

CHAPTER

INDUCTION AND INDUCTANCE

30

30-1 WHAT IS PHYSICS?

In Chapter 29 we discussed the fact that a current produces a magnetic field. That fact came as a surprise to the scientists who discovered the effect. Perhaps even more surprising was the discovery of the reverse effect: A magnetic field can produce an electric field that can drive a current. This link between a magnetic field and the electric field it produces (*induces*) is now called *Faraday's law of induction.*

The observations by Michael Faraday and other scientists that led to this law were at first just basic science. Today, however, applications of that basic science are almost everywhere. For example, induction is the basis of the electric guitars that revolutionized early rock and still drive heavy metal and punk today. It is also the basis of the electric generators that power cities and transportation lines and of the huge induction furnaces that are commonplace in foundries where large amounts of metal must be melted rapidly.

Before we get to applications like the electric guitar, we must examine two simple experiments about Faraday's law of induction.

30-2 Two Experiments

Let us examine two simple experiments to prepare for our discussion of Faraday's law of induction.

First Experiment. Figure 30-1 shows a conducting loop connected to a sensitive ammeter. Because there is no battery or other source of emf included, there is no current in the circuit. However, if we move a bar magnet toward the loop, a current suddenly appears in the circuit. The current disappears when the magnet stops. If we then move the magnet away, a current again suddenly appears, but now in the opposite direction. If we experimented for a while, we would discover the following:

1. A current appears only if there is relative motion between the loop and the magnet (one must move relative to the other); the current disappears when the relative motion between them ceases.
2. Faster motion produces a greater current.
3. If moving the magnet's north pole toward the loop causes, say, clockwise current, then moving the north pole away causes counterclockwise current. Moving the south pole toward or away from the loop also causes currents, but in the reversed directions.

The current produced in the loop is called an **induced current;** the work done per unit charge to produce that current (to move the conduction electrons that

The magnet's motion creates a current in the loop.

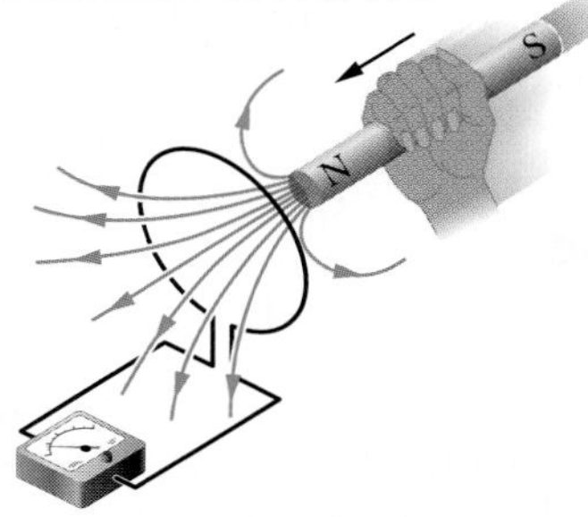

Fig. 30-1 An ammeter registers a current in the wire loop when the magnet is moving with respect to the loop.

constitute the current) is called an **induced emf;** and the process of producing the current and emf is called **induction.**

Second Experiment. For this experiment we use the apparatus of Fig. 30-2, with the two conducting loops close to each other but not touching. If we close switch S, to turn on a current in the right-hand loop, the meter suddenly and briefly registers a current—an induced current—in the left-hand loop. If we then open the switch, another sudden and brief induced current appears in the left-hand loop, but in the opposite direction. We get an induced current (and thus an induced emf) only when the current in the right-hand loop is changing (either turning on or turning off) and not when it is constant (even if it is large).

The induced emf and induced current in these experiments are apparently caused when something changes—but what is that "something"? Faraday knew.

30-3 Faraday's Law of Induction

Faraday realized that an emf and a current can be induced in a loop, as in our two experiments, by changing the *amount of magnetic field* passing through the loop. He further realized that the "amount of magnetic field" can be visualized in terms of the magnetic field lines passing through the loop. **Faraday's law of induction,** stated in terms of our experiments, is this:

> An emf is induced in the loop at the left in Figs. 30-1 and 30-2 when the number of magnetic field lines that pass through the loop is changing.

The actual number of field lines passing through the loop does not matter; the values of the induced emf and induced current are determined by the *rate* at which that number changes.

In our first experiment (Fig. 30-1), the magnetic field lines spread out from the north pole of the magnet. Thus, as we move the north pole closer to the loop, the number of field lines passing through the loop increases. That increase apparently causes conduction electrons in the loop to move (the induced current) and provides energy (the induced emf) for their motion. When the magnet stops moving, the number of field lines through the loop no longer changes and the induced current and induced emf disappear.

In our second experiment (Fig. 30-2), when the switch is open (no current), there are no field lines. However, when we turn on the current in the right-hand loop, the increasing current builds up a magnetic field around that loop and at the left-hand loop. While the field builds, the number of magnetic field lines through the left-hand loop increases. As in the first experiment, the increase in field lines through that loop apparently induces a current and an emf there. When the current in the right-hand loop reaches a final, steady value, the number of field lines through the left-hand loop no longer changes, and the induced current and induced emf disappear.

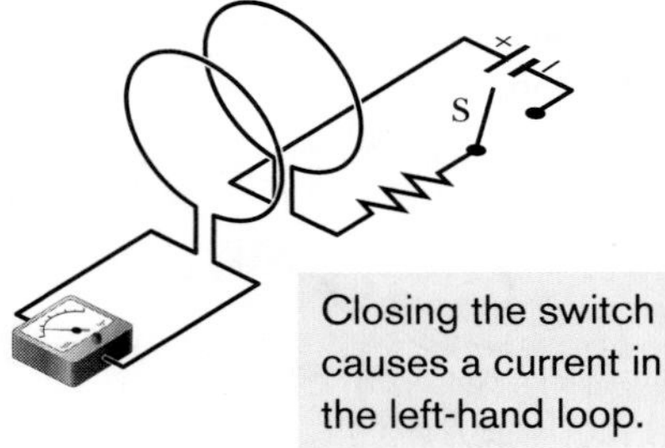

Fig. 30-2 An ammeter registers a current in the left-hand wire loop just as switch S is closed (to turn on the current in the right-hand wire loop) or opened (to turn off the current in the right-hand loop). No motion of the coils is involved.

A Quantitative Treatment

To put Faraday's law to work, we need a way to calculate the *amount of magnetic field* that passes through a loop. In Chapter 23, in a similar situation, we needed to calculate the amount of electric field that passes through a surface. There we defined an electric flux $\Phi_E = \int \vec{E} \cdot d\vec{A}$. Here we define a *magnetic flux:* Suppose a loop enclosing an area A is placed in a magnetic field $\vec{B}$. Then the **magnetic flux** through the loop is

$$\Phi_B = \int \vec{B} \cdot d\vec{A} \quad \text{(magnetic flux through area } A\text{).} \tag{30-1}$$

As in Chapter 23, $d\vec{A}$ is a vector of magnitude dA that is perpendicular to a differential area dA.

As a special case of Eq. 30-1, suppose that the loop lies in a plane and that the magnetic field is perpendicular to the plane of the loop. Then we can write the dot product in Eq. 30-1 as $B\,dA\cos 0° = B\,dA$. If the magnetic field is also uniform, then B can be brought out in front of the integral sign. The remaining $\int dA$ then gives just the area A of the loop. Thus, Eq. 30-1 reduces to

$$\Phi_B = BA \qquad (\vec{B} \perp \text{area } A,\ \vec{B} \text{ uniform}). \tag{30-2}$$

From Eqs. 30-1 and 30-2, we see that the SI unit for magnetic flux is the tesla–square meter, which is called the *weber* (abbreviated Wb):

$$1 \text{ weber} = 1 \text{ Wb} = 1 \text{ T} \cdot \text{m}^2. \tag{30-3}$$

With the notion of magnetic flux, we can state Faraday's law in a more quantitative and useful way:

The magnitude of the emf $\mathscr{E}$ induced in a conducting loop is equal to the rate at which the magnetic flux Φ_B through that loop changes with time.

As you will see in the next section, the induced emf $\mathscr{E}$ tends to oppose the flux change, so Faraday's law is formally written as

$$\mathscr{E} = -\frac{d\Phi_B}{dt} \qquad \text{(Faraday's law)}, \tag{30-4}$$

with the minus sign indicating that opposition. We often neglect the minus sign in Eq. 30-4, seeking only the magnitude of the induced emf.

If we change the magnetic flux through a coil of N turns, an induced emf appears in every turn and the total emf induced in the coil is the sum of these individual induced emfs. If the coil is tightly wound (*closely packed*), so that the same magnetic flux Φ_B passes through all the turns, the total emf induced in the coil is

$$\mathscr{E} = -N\frac{d\Phi_B}{dt} \qquad \text{(coil of } N \text{ turns)}. \tag{30-5}$$

Here are the general means by which we can change the magnetic flux through a coil:

1. Change the magnitude B of the magnetic field within the coil.
2. Change either the total area of the coil or the portion of that area that lies within the magnetic field (for example, by expanding the coil or sliding it into or out of the field).
3. Change the angle between the direction of the magnetic field $\vec{B}$ and the plane of the coil (for example, by rotating the coil so that field $\vec{B}$ is first perpendicular to the plane of the coil and then is along that plane).

CHECKPOINT 1

The graph gives the magnitude $B(t)$ of a uniform magnetic field that exists throughout a conducting loop, with the direction of the field perpendicular to the plane of the loop. Rank the five regions of the graph according to the magnitude of the emf induced in the loop, greatest first.

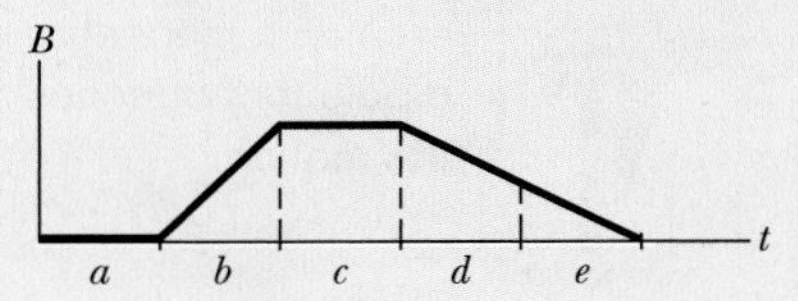

Sample Problem

Induced emf in coil due to a solenoid

The long solenoid S shown (in cross section) in Fig. 30-3 has 220 turns/cm and carries a current $i = 1.5$ A; its diameter D is 3.2 cm. At its center we place a 130-turn closely packed coil C of diameter $d = 2.1$ cm. The current in the solenoid is reduced to zero at a steady rate in 25 ms. What is the magnitude of the emf that is induced in coil C while the current in the solenoid is changing?

KEY IDEAS

1. Because it is located in the interior of the solenoid, coil C lies within the magnetic field produced by current i in the solenoid; thus, there is a magnetic flux Φ_B through coil C.
2. Because current i decreases, flux Φ_B also decreases.
3. As Φ_B decreases, emf $\mathscr{E}$ is induced in coil C.

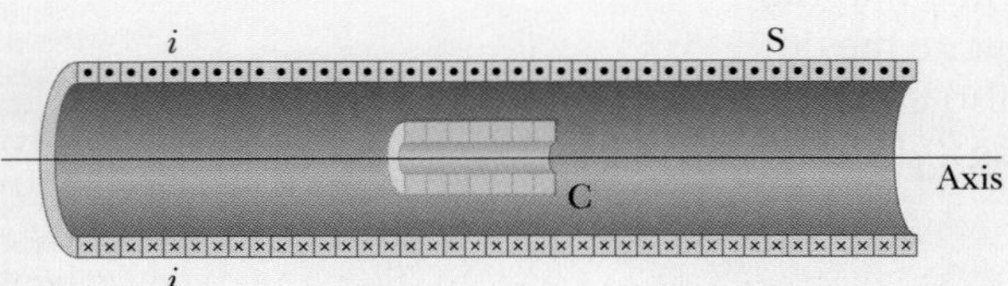

Fig. 30-3 A coil C is located inside a solenoid S, which carries current i.

4. The flux through each turn of coil C depends on the area A and orientation of that turn in the solenoid's magnetic field $\vec{B}$. Because $\vec{B}$ is uniform and directed perpendicular to area A, the flux is given by Eq. 30-2 ($\Phi_B = BA$).
5. The magnitude B of the magnetic field in the interior of a solenoid depends on the solenoid's current i and its number n of turns per unit length, according to Eq. 29-23 ($B = \mu_0 in$).

Calculations: Because coil C consists of more than one turn, we apply Faraday's law in the form of Eq. 30-5 ($\mathscr{E} = -N\, d\Phi_B/dt$), where the number of turns N is 130 and $d\Phi_B/dt$ is the rate at which the flux changes.

Because the current in the solenoid decreases at a steady rate, flux Φ_B also decreases at a steady rate, and so we can write $d\Phi_B/dt$ as $\Delta\Phi_B/\Delta t$. Then, to evaluate $\Delta\Phi_B$, we need the final and initial flux values. The final flux $\Phi_{B,f}$ is zero because the final current in the solenoid is zero. To find the initial flux $\Phi_{B,i}$, we note that area A is $\frac{1}{4}\pi d^2$ ($= 3.464 \times 10^{-4}$ m^2) and the number n is 220 turns/cm, or 22 000 turns/m. Substituting Eq. 29-23 into Eq. 30-2 then leads to

$$\begin{aligned}\Phi_{B,i} &= BA = (\mu_0 in)A \\ &= (4\pi \times 10^{-7}\,\mathrm{T \cdot m/A})(1.5\,\mathrm{A})(22\,000\ \mathrm{turns/m}) \\ &\quad \times (3.464 \times 10^{-4}\,\mathrm{m^2}) \\ &= 1.44 \times 10^{-5}\,\mathrm{Wb}.\end{aligned}$$

Now we can write

$$\begin{aligned}\frac{d\Phi_B}{dt} &= \frac{\Delta\Phi_B}{\Delta t} = \frac{\Phi_{B,f} - \Phi_{B,i}}{\Delta t} \\ &= \frac{(0 - 1.44 \times 10^{-5}\,\mathrm{Wb})}{25 \times 10^{-3}\,\mathrm{s}} \\ &= -5.76 \times 10^{-4}\,\mathrm{Wb/s} = -5.76 \times 10^{-4}\,\mathrm{V}.\end{aligned}$$

We are interested only in magnitudes; so we ignore the minus signs here and in Eq. 30-5, writing

$$\begin{aligned}\mathscr{E} &= N\frac{d\Phi_B}{dt} = (130\ \mathrm{turns})(5.76 \times 10^{-4}\,\mathrm{V}) \\ &= 7.5 \times 10^{-2}\,\mathrm{V} = 75\ \mathrm{mV}. \qquad \text{(Answer)}\end{aligned}$$

Additional examples, video, and practice available at *WileyPLUS*

30-4 Lenz's Law

Soon after Faraday proposed his law of induction, Heinrich Friedrich Lenz devised a rule for determining the direction of an induced current in a loop:

An induced current has a direction such that the magnetic field due to *the current* opposes the change in the magnetic flux that induces the current.

Furthermore, the direction of an induced emf is that of the induced current. To get a feel for **Lenz's law**, let us apply it in two different but equivalent ways to Fig. 30-4, where the north pole of a magnet is being moved toward a conducting loop.

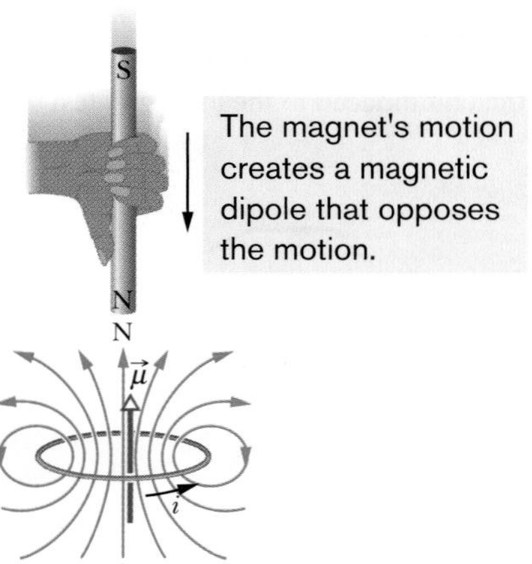

Fig. 30-4 Lenz's law at work. As the magnet is moved toward the loop, a current is induced in the loop. The current produces its own magnetic field, with magnetic dipole moment $\vec{\mu}$ oriented so as to oppose the motion of the magnet. Thus, the induced current must be counterclockwise as shown.

1. ***Opposition to Pole Movement.*** The approach of the magnet's north pole in Fig. 30-4 increases the magnetic flux through the loop and thereby induces a current in the loop. From Fig. 29-21, we know that the loop then acts as a magnetic dipole with a south pole and a north pole, and that its magnetic dipole moment $\vec{\mu}$ is directed from south to north. To *oppose* the magnetic flux increase being caused by the approaching magnet, the loop's north pole (and thus $\vec{\mu}$) must face *toward* the approaching north pole so as to repel it (Fig. 30-4). Then the curled–straight right-hand rule for $\vec{\mu}$ (Fig. 29-21) tells us that the current induced in the loop must be counterclockwise in Fig. 30-4.

   If we next pull the magnet away from the loop, a current will again be induced in the loop. Now, however, the loop will have a south pole facing the retreating north pole of the magnet, so as to oppose the retreat. Thus, the induced current will be clockwise.
2. ***Opposition to Flux Change.*** In Fig. 30-4, with the magnet initially distant, no magnetic flux passes through the loop. As the north pole of the magnet then

nears the loop with its magnetic field $\vec{B}$ directed *downward*, the flux through the loop increases. To oppose this increase in flux, the induced current i must set up its own field $\vec{B}_{ind}$ directed *upward* inside the loop, as shown in Fig. 30-5*a*; then the upward flux of field $\vec{B}_{ind}$ opposes the increasing downward flux of field $\vec{B}$. The curled–straight right-hand rule of Fig. 29-21 then tells us that i must be counterclockwise in Fig. 30-5*a*.

Note carefully that the flux of $\vec{B}_{ind}$ always opposes the *change* in the flux of $\vec{B}$, but that does not always mean that $\vec{B}_{ind}$ points opposite $\vec{B}$. For example, if we next pull the magnet away from the loop in Fig. 30-4, the flux Φ_B from the magnet is still directed downward through the loop, but it is now decreasing. The flux of $\vec{B}_{ind}$ must now be downward inside the loop, to oppose the *decrease* in Φ_B, as shown in Fig. 30-5*b*. Thus, $\vec{B}_{ind}$ and $\vec{B}$ are now in the same direction.

In Figs. 30-5*c* and *d*, the south pole of the magnet approaches and retreats from the loop, respectively.

CHECKPOINT 2

The figure shows three situations in which identical circular conducting loops are in uniform magnetic fields that are either increasing (Inc) or decreasing (Dec) in magnitude at identical rates. In each, the dashed line coincides with a diameter. Rank the situations according to the magnitude of the current induced in the loops, greatest first.

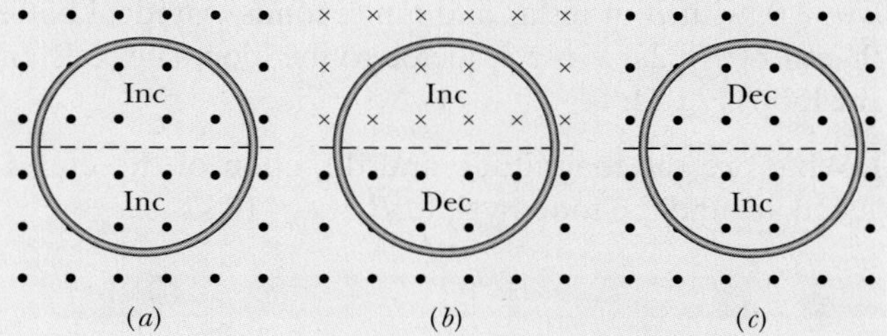

Increasing the external field $\vec{B}$ induces a current with a field $\vec{B}_{ind}$ that *opposes the change.*

Decreasing the external field $\vec{B}$ induces a current with a field $\vec{B}_{ind}$ that *opposes the change.*

Increasing the external field $\vec{B}$ induces a current with a field $\vec{B}_{ind}$ that *opposes the change.*

Decreasing the external field $\vec{B}$ induces a current with a field $\vec{B}_{ind}$ that *opposes the change.*

The induced current creates this field, trying to offset the change.

The fingers are in the current's direction; the thumb is in the induced field's direction.

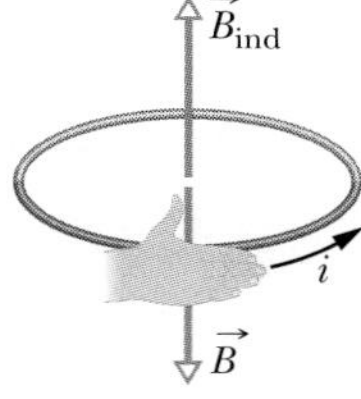

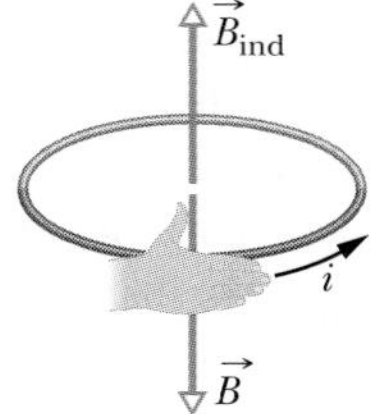

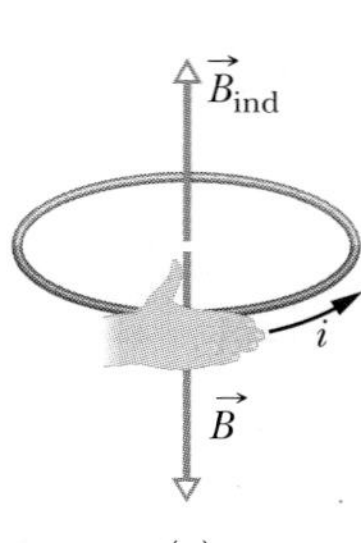

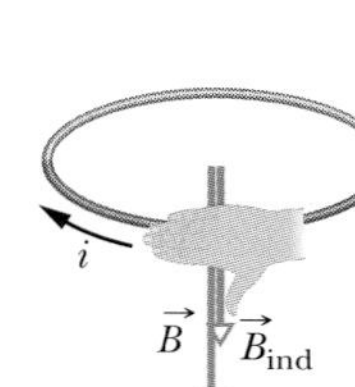

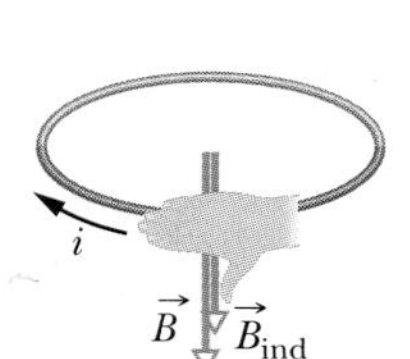

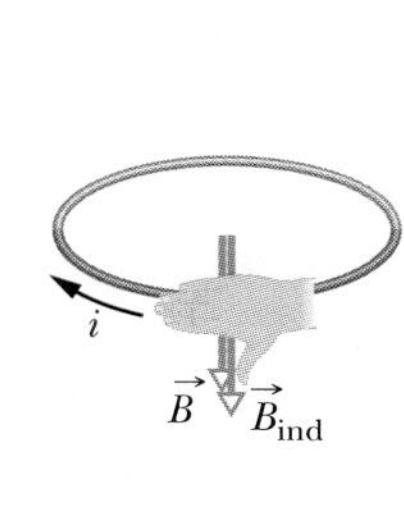

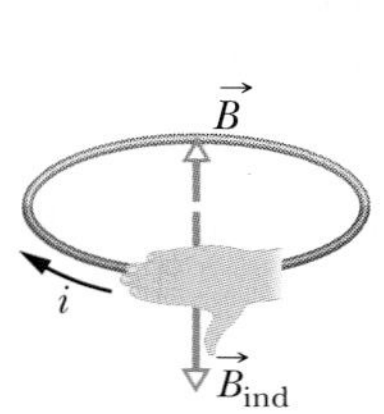

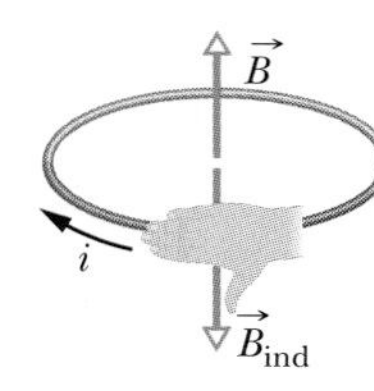

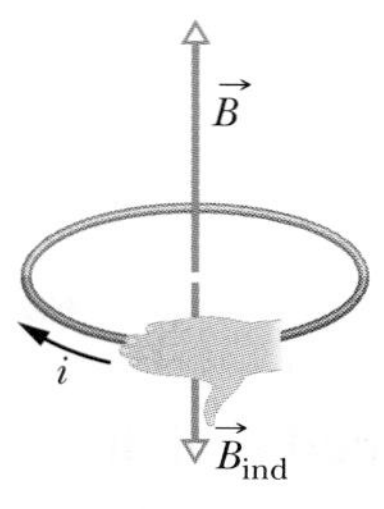

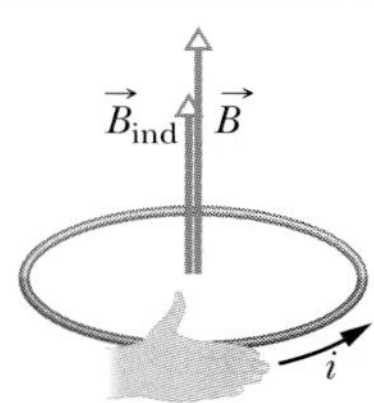

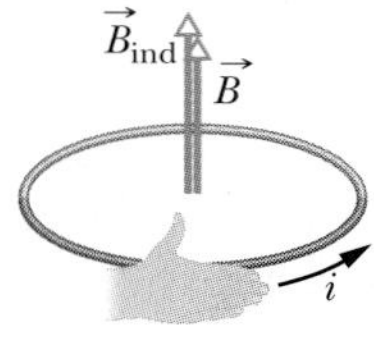

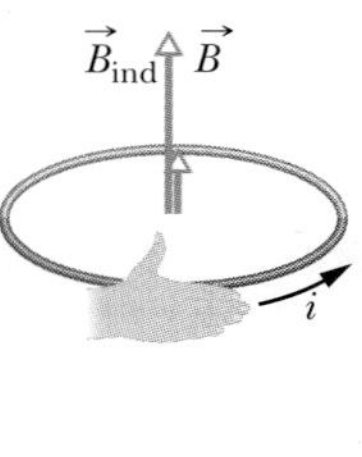

Fig. 30-5 The direction of the current i induced in a loop is such that the current's magnetic field $\vec{B}_{ind}$ opposes the *change* in the magnetic field $\vec{B}$ inducing i. The field $\vec{B}_{ind}$ is always directed opposite an increasing field $\vec{B}$ (a, c) and in the same direction as a decreasing field $\vec{B}$ (b, d). The curled–straight right-hand rule gives the direction of the induced current based on the direction of the induced field.

Sample Problem

Induced emf and current due to a changing uniform *B* field

Figure 30-6 shows a conducting loop consisting of a half-circle of radius $r = 0.20$ m and three straight sections. The half-circle lies in a uniform magnetic field $\vec{B}$ that is directed out of the page; the field magnitude is given by $B = 4.0t^2 + 2.0t + 3.0$, with B in teslas and t in seconds. An ideal battery with emf $\mathscr{E}_{bat} = 2.0$ V is connected to the loop. The resistance of the loop is 2.0 Ω.

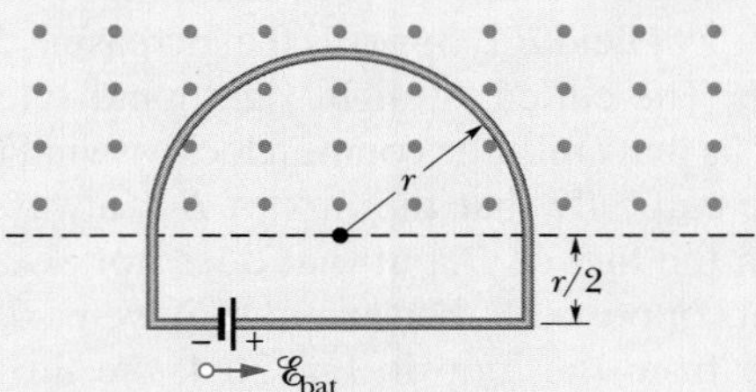

Fig. 30-6 A battery is connected to a conducting loop that includes a half-circle of radius r lying in a uniform magnetic field. The field is directed out of the page; its magnitude is changing.

(a) What are the magnitude and direction of the emf $\mathscr{E}_{ind}$ induced around the loop by field $\vec{B}$ at $t = 10$ s?

KEY IDEAS

1. According to Faraday's law, the magnitude of $\mathscr{E}_{ind}$ is equal to the rate $d\Phi_B/dt$ at which the magnetic flux through the loop changes.
2. The flux through the loop depends on how much of the loop's area lies within the flux and how the area is oriented in the magnetic field $\vec{B}$.
3. Because $\vec{B}$ is uniform and is perpendicular to the plane of the loop, the flux is given by Eq. 30-2 ($\Phi_B = BA$). (We don't need to integrate B over the area to get the flux.)
4. The induced field B_{ind} (due to the induced current) must always oppose the *change* in the magnetic flux.

Magnitude: Using Eq. 30-2 and realizing that only the field magnitude B changes in time (not the area A), we rewrite Faraday's law, Eq. 30-4, as

$$\mathscr{E}_{ind} = \frac{d\Phi_B}{dt} = \frac{d(BA)}{dt} = A\frac{dB}{dt}.$$

Because the flux penetrates the loop only within the half-circle, the area A in this equation is $\frac{1}{2}\pi r^2$. Substituting this and the given expression for B yields

$$\mathscr{E}_{ind} = A\frac{dB}{dt} = \frac{\pi r^2}{2}\frac{d}{dt}(4.0t^2 + 2.0t + 3.0)$$

$$= \frac{\pi r^2}{2}(8.0t + 2.0).$$

At $t = 10$ s, then,

$$\mathscr{E}_{ind} = \frac{\pi(0.20 \text{ m})^2}{2}[8.0(10) + 2.0]$$

$$= 5.152 \text{ V} \approx 5.2 \text{ V}. \qquad \text{(Answer)}$$

Direction: To find the direction of $\mathscr{E}_{ind}$, we first note that in Fig. 30-6 the flux through the loop is out of the page and increasing. Because the induced field B_{ind} (due to the induced current) must oppose that increase, it must be *into* the page. Using the curled–straight right-hand rule (Fig. 30-5*c*), we find that the induced current is clockwise around the loop, and thus so is the induced emf $\mathscr{E}_{ind}$.

(b) What is the current in the loop at $t = 10$ s?

KEY IDEA

The point here is that *two* emfs tend to move charges around the loop.

Calculation: The induced emf $\mathscr{E}_{ind}$ tends to drive a current clockwise around the loop; the battery's emf $\mathscr{E}_{bat}$ tends to drive a current counterclockwise. Because $\mathscr{E}_{ind}$ is greater than $\mathscr{E}_{bat}$, the net emf $\mathscr{E}_{net}$ is clockwise, and thus so is the current. To find the current at $t = 10$ s, we use Eq. 27-2 ($i = \mathscr{E}/R$):

$$i = \frac{\mathscr{E}_{net}}{R} = \frac{\mathscr{E}_{ind} - \mathscr{E}_{bat}}{R}$$

$$= \frac{5.152 \text{ V} - 2.0 \text{ V}}{2.0\ \Omega} = 1.58 \text{ A} \approx 1.6 \text{ A}. \qquad \text{(Answer)}$$

Sample Problem

Induced emf due to a changing nonuniform *B* field

Figure 30-7 shows a rectangular loop of wire immersed in a nonuniform and varying magnetic field $\vec{B}$ that is perpendicular to and directed into the page. The field's magnitude is given by $B = 4t^2x^2$, with B in teslas, t in seconds, and x in meters. (Note that the function depends on *both* time and position.) The loop has width $W = 3.0$ m and height $H = 2.0$ m. What are the magnitude and direction of the induced emf $\mathscr{E}$ around the loop at $t = 0.10$ s?

KEY IDEAS

1. Because the magnitude of the magnetic field $\vec{B}$ is changing with time, the magnetic flux Φ_B through the loop is also changing.
2. The changing flux induces an emf $\mathscr{E}$ in the loop according to Faraday's law, which we can write as $\mathscr{E} = d\Phi_B/dt$.
3. To use that law, we need an expression for the flux Φ_B at any time t. However, because B is *not* uniform over the area enclosed by the loop, we *cannot* use Eq. 30-2 ($\Phi_B = BA$) to find that expression; instead we must use Eq. 30-1 ($\Phi_B = \int \vec{B}\cdot d\vec{A}$).

Calculations: In Fig. 30-7, $\vec{B}$ is perpendicular to the plane of the loop (and hence parallel to the differential area vector $d\vec{A}$); so the dot product in Eq. 30-1 gives $B\ dA$. Because the magnetic field varies with the coordinate x but not with the coordinate y, we can take the differential area dA to be the area of a vertical strip of height H and width dx (as shown in Fig. 30-7). Then $dA = H\ dx$, and the flux through the loop is

$$\Phi_B = \int \vec{B}\cdot d\vec{A} = \int B\ dA = \int BH\ dx = \int 4t^2x^2H\ dx.$$

Treating t as a constant for this integration and inserting the integration limits $x = 0$ and $x = 3.0$ m, we obtain

$$\Phi_B = 4t^2H \int_0^{3.0} x^2\ dx = 4t^2H\left[\frac{x^3}{3}\right]_0^{3.0} = 72t^2,$$

where we have substituted $H = 2.0$ m and Φ_B is in webers. Now we can use Faraday's law to find the magnitude of $\mathscr{E}$ at any time t:

$$\mathscr{E} = \frac{d\Phi_B}{dt} = \frac{d(72t^2)}{dt} = 144t,$$

in which $\mathscr{E}$ is in volts. At $t = 0.10$ s,

$$\mathscr{E} = (144\ \text{V/s})(0.10\ \text{s}) \approx 14\ \text{V}. \qquad \text{(Answer)}$$

The flux of $\vec{B}$ through the loop is into the page in Fig. 30-7 and is increasing in magnitude because B is increasing in magnitude with time. By Lenz's law, the field B_{ind} of the induced current opposes this increase and so is directed out of the page. The curled–straight right-hand rule in Fig. 30-5*a* then tells us that the induced current is counterclockwise around the loop, and thus so is the induced emf $\mathscr{E}$.

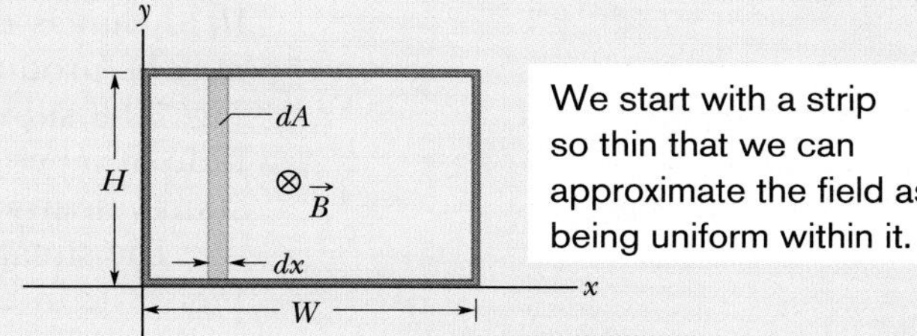

Fig. 30-7 A closed conducting loop, of width W and height H, lies in a nonuniform, varying magnetic field that points directly into the page. To apply Faraday's law, we use the vertical strip of height H, width dx, and area dA.

Additional examples, video, and practice available at *WileyPLUS*

30-5 Induction and Energy Transfers

By Lenz's law, whether you move the magnet toward or away from the loop in Fig. 30-1, a magnetic force resists the motion, requiring your applied force to do positive work. At the same time, thermal energy is produced in the material of the loop because of the material's electrical resistance to the current that is induced by the motion. The energy you transfer to the closed *loop + magnet* system via your applied force ends up in this thermal energy. (For now, we neglect energy that is radiated away from the loop as electromagnetic waves during the induction.) The faster you move the magnet, the more rapidly your applied force does work and the greater the rate at which your energy is transferred to thermal energy in the loop; that is, the power of the transfer is greater.

Regardless of how current is induced in a loop, energy is always transferred to thermal energy during the process because of the electrical resistance of the loop (unless the loop is superconducting). For example, in Fig. 30-2, when switch S is closed and a current is briefly induced in the left-hand loop, energy is transferred from the battery to thermal energy in that loop.

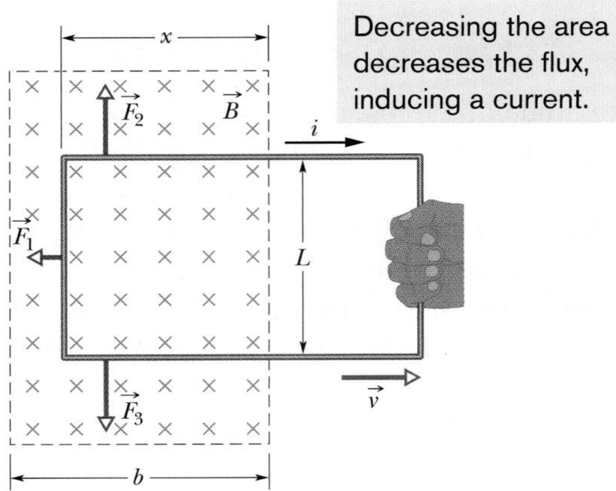

Fig. 30-8 You pull a closed conducting loop out of a magnetic field at constant velocity $\vec{v}$. While the loop is moving, a clockwise current i is induced in the loop, and the loop segments still within the magnetic field experience forces $\vec{F}_1$, $\vec{F}_2$, and $\vec{F}_3$.

Figure 30-8 shows another situation involving induced current. A rectangular loop of wire of width L has one end in a uniform external magnetic field that is directed perpendicularly into the plane of the loop. This field may be produced, for example, by a large electromagnet. The dashed lines in Fig. 30-8 show the assumed limits of the magnetic field; the fringing of the field at its edges is neglected. You are to pull this loop to the right at a constant velocity $\vec{v}$.

The situation of Fig. 30-8 does not differ in any essential way from that of Fig. 30-1. In each case a magnetic field and a conducting loop are in relative motion; in each case the flux of the field through the loop is changing with time. It is true that in Fig. 30-1 the flux is changing because $\vec{B}$ is changing and in Fig. 30-8 the flux is changing because the area of the loop still in the magnetic field is changing, but that difference is not important. The important difference between the two arrangements is that the arrangement of Fig. 30-8 makes calculations easier. Let us now calculate the rate at which you do mechanical work as you pull steadily on the loop in Fig. 30-8.

As you will see, to pull the loop at a constant velocity $\vec{v}$, you must apply a constant force $\vec{F}$ to the loop because a magnetic force of equal magnitude but opposite direction acts on the loop to oppose you. From Eq. 7-48, the rate at which you do work—that is, the power—is then

$$P = Fv, \tag{30-6}$$

where F is the magnitude of your force. We wish to find an expression for P in terms of the magnitude B of the magnetic field and the characteristics of the loop—namely, its resistance R to current and its dimension L.

As you move the loop to the right in Fig. 30-8, the portion of its area within the magnetic field decreases. Thus, the flux through the loop also decreases and, according to Faraday's law, a current is produced in the loop. It is the presence of this current that causes the force that opposes your pull.

To find the current, we first apply Faraday's law. When x is the length of the loop still in the magnetic field, the area of the loop still in the field is Lx. Then from Eq. 30-2, the magnitude of the flux through the loop is

$$\Phi_B = BA = BLx. \tag{30-7}$$

As x decreases, the flux decreases. Faraday's law tells us that with this flux decrease, an emf is induced in the loop. Dropping the minus sign in Eq. 30-4 and using Eq. 30-7, we can write the magnitude of this emf as

$$\mathscr{E} = \frac{d\Phi_B}{dt} = \frac{d}{dt} BLx = BL\frac{dx}{dt} = BLv, \tag{30-8}$$

in which we have replaced dx/dt with v, the speed at which the loop moves.

Figure 30-9 shows the loop as a circuit: induced emf $\mathscr{E}$ is represented on the left, and the collective resistance R of the loop is represented on the right. The direction of the induced current i is obtained with a right-hand rule as in Fig. 30-5*b* for decreasing flux; applying the rule tells us that the current must be clockwise, and $\mathscr{E}$ must have the same direction.

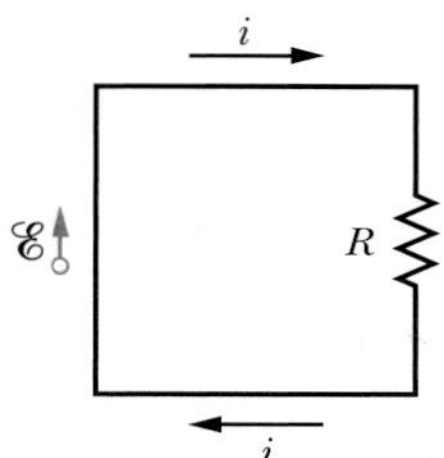

Fig. 30-9 A circuit diagram for the loop of Fig. 30-8 while the loop is moving.

To find the magnitude of the induced current, we cannot apply the loop rule for potential differences in a circuit because, as you will see in Section 30-6, we cannot define a potential difference for an induced emf. However, we can apply the equation $i = \mathscr{E}/R$. With Eq. 30-8, this becomes

$$i = \frac{BLv}{R}. \tag{30-9}$$

Because three segments of the loop in Fig. 30-8 carry this current through the magnetic field, sideways deflecting forces act on those segments. From Eq. 28-26 we know that such a deflecting force is, in general notation,

$$\vec{F}_d = i\vec{L} \times \vec{B}. \tag{30-10}$$

In Fig. 30-8, the deflecting forces acting on the three segments of the loop are marked $\vec{F}_1$, $\vec{F}_2$, and $\vec{F}_3$. Note, however, that from the symmetry, forces $\vec{F}_2$ and $\vec{F}_3$ are equal in magnitude and cancel. This leaves only force $\vec{F}_1$, which is directed opposite your force $\vec{F}$ on the loop and thus is the force opposing you. So, $\vec{F} = -\vec{F}_1$.

Using Eq. 30-10 to obtain the magnitude of $\vec{F}_1$ and noting that the angle between $\vec{B}$ and the length vector $\vec{L}$ for the left segment is 90°, we write

$$F = F_1 = iLB \sin 90° = iLB. \tag{30-11}$$

Substituting Eq. 30-9 for i in Eq. 30-11 then gives us

$$F = \frac{B^2L^2v}{R}. \tag{30-12}$$

Because B, L, and R are constants, the speed v at which you move the loop is constant if the magnitude F of the force you apply to the loop is also constant.

By substituting Eq. 30-12 into Eq. 30-6, we find the rate at which you do work on the loop as you pull it from the magnetic field:

$$P = Fv = \frac{B^2L^2v^2}{R} \quad \text{(rate of doing work)}. \tag{30-13}$$

To complete our analysis, let us find the rate at which thermal energy appears in the loop as you pull it along at constant speed. We calculate it from Eq. 26-27,

$$P = i^2R. \tag{30-14}$$

Substituting for i from Eq. 30-9, we find

$$P = \left(\frac{BLv}{R}\right)^2 R = \frac{B^2L^2v^2}{R} \quad \text{(thermal energy rate)}, \tag{30-15}$$

which is exactly equal to the rate at which you are doing work on the loop (Eq. 30-13). Thus, the work that you do in pulling the loop through the magnetic field appears as thermal energy in the loop.

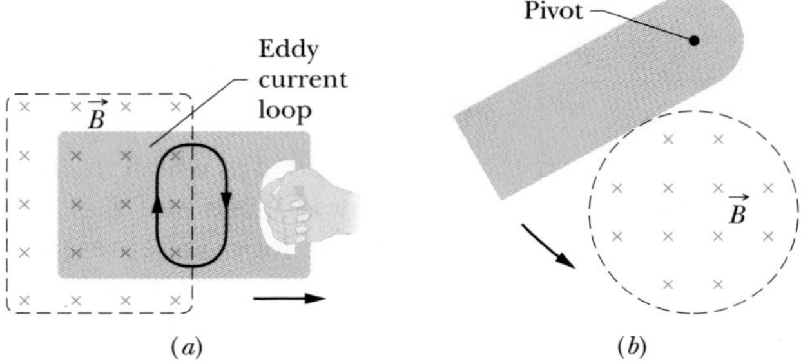

Fig. 30-10 (*a*) As you pull a solid conducting plate out of a magnetic field, *eddy currents* are induced in the plate. A typical loop of eddy current is shown. (*b*) A conducting plate is allowed to swing like a pendulum about a pivot and into a region of magnetic field. As it enters and leaves the field, eddy currents are induced in the plate.

Eddy Currents

Suppose we replace the conducting loop of Fig. 30-8 with a solid conducting plate. If we then move the plate out of the magnetic field as we did the loop (Fig. 30-10*a*), the relative motion of the field and the conductor again induces a current in the conductor. Thus, we again encounter an opposing force and must do work because of the induced current. With the plate, however, the conduction electrons making up the induced current do not follow one path as they do with the loop. Instead, the electrons swirl about within the plate as if they were caught in an eddy (whirlpool) of water. Such a current is called an *eddy current* and can be represented, as it is in Fig. 30-10*a*, *as if* it followed a single path.

As with the conducting loop of Fig. 30-8, the current induced in the plate results in mechanical energy being dissipated as thermal energy. The dissipation is more apparent in the arrangement of Fig. 30-10*b*; a conducting plate, free to rotate about a pivot, is allowed to swing down through a magnetic field like a pendulum. Each time the plate enters and leaves the field, a portion of its mechanical energy is transferred to its thermal energy. After several swings, no mechanical energy remains and the warmed-up plate just hangs from its pivot.

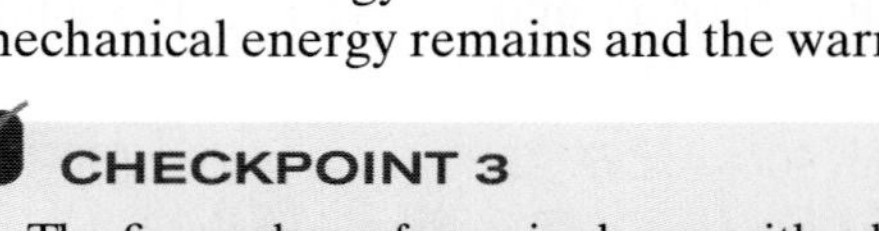

CHECKPOINT 3

The figure shows four wire loops, with edge lengths of either L or $2L$. All four loops will move through a region of uniform magnetic field $\vec{B}$ (directed out of the page) at the same constant velocity. Rank the four loops according to the maximum magnitude of the emf induced as they move through the field, greatest first.

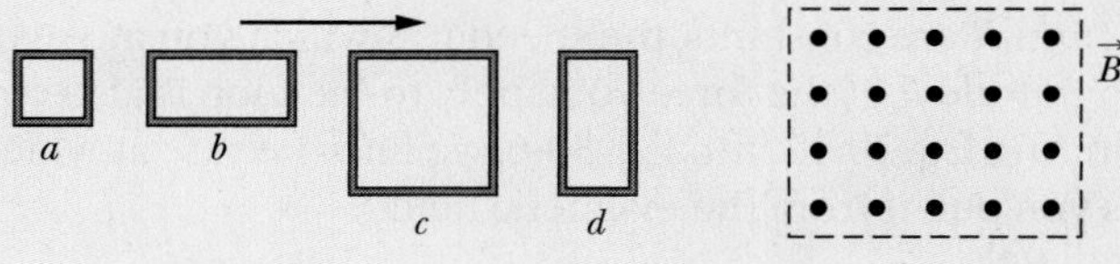

30-6 Induced Electric Fields

Let us place a copper ring of radius r in a uniform external magnetic field, as in Fig. 30-11*a*. The field—neglecting fringing—fills a cylindrical volume of radius R. Suppose that we increase the strength of this field at a steady rate, perhaps by increasing—in an appropriate way—the current in the windings of the electromagnet that produces the field. The magnetic flux through the ring will then change at a steady rate and—by Faraday's law—an induced emf and thus an induced current will appear in the ring. From Lenz's law we can deduce that the direction of the induced current is counterclockwise in Fig. 30-11*a*.

If there is a current in the copper ring, an electric field must be present along the ring because an electric field is needed to do the work of moving the conduction electrons. Moreover, the electric field must have been produced by the changing

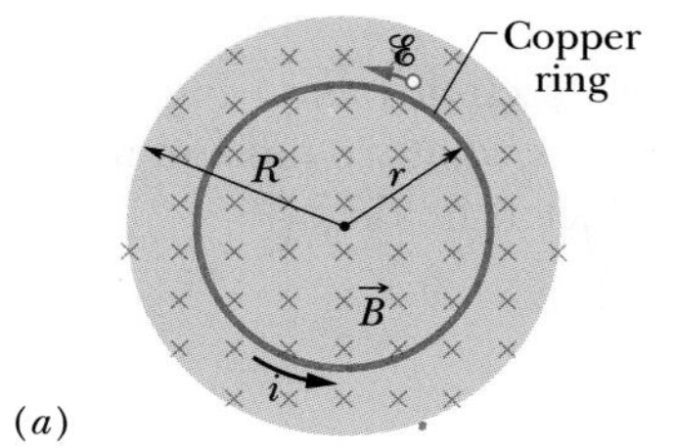

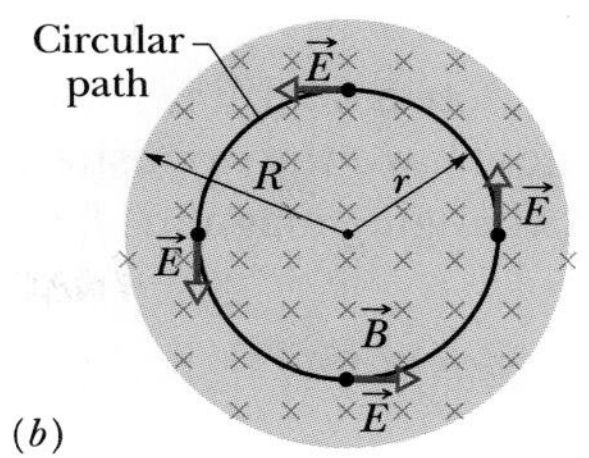

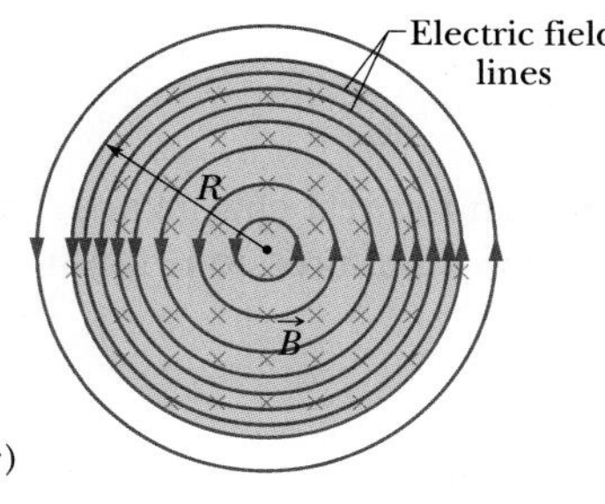

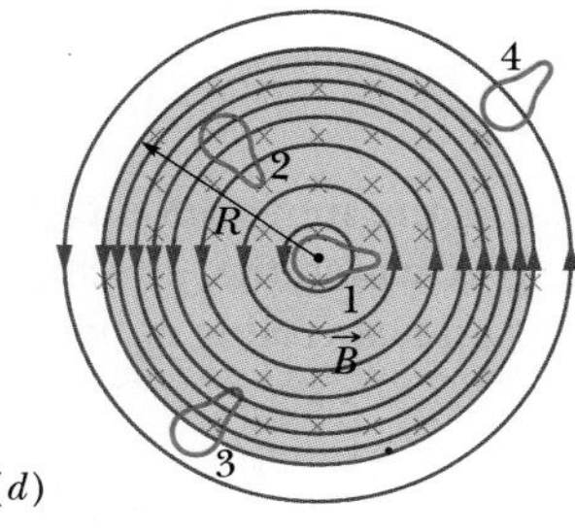

Fig. 30-11 (*a*) If the magnetic field increases at a steady rate, a constant induced current appears, as shown, in the copper ring of radius r.(*b*) An induced electric field exists even when the ring is removed; the electric field is shown at four points. (*c*) The complete picture of the induced electric field, displayed as field lines. (*d*) Four similar closed paths that enclose identical areas. Equal emfs are induced around paths 1 and 2, which lie entirely within the region of changing magnetic field. A smaller emf is induced around path 3, which only partially lies in that region. No net emf is induced around path 4, which lies entirely outside the magnetic field.

magnetic flux. This **induced electric field** $\vec{E}$ is just as real as an electric field produced by static charges; either field will exert a force $q_0\vec{E}$ on a particle of charge q_0.

By this line of reasoning, we are led to a useful and informative restatement of Faraday's law of induction:

A changing magnetic field produces an electric field.

The striking feature of this statement is that the electric field is induced even if there is no copper ring. Thus, the electric field would appear even if the changing magnetic field were in a vacuum.

To fix these ideas, consider Fig. 30-11*b*, which is just like Fig. 30-11*a* except the copper ring has been replaced by a hypothetical circular path of radius r. We assume, as previously, that the magnetic field $\vec{B}$ is increasing in magnitude at a constant rate dB/dt. The electric field induced at various points around the circular path must—from the symmetry—be tangent to the circle, as Fig. 30-11*b* shows.* Hence, the circular path is an electric field line. There is nothing special about the circle of radius r, so the electric field lines produced by the changing magnetic field must be a set of concentric circles, as in Fig. 30-11*c*.

As long as the magnetic field is *increasing* with time, the electric field represented by the circular field lines in Fig. 30-11*c* will be present. If the magnetic field remains *constant* with time, there will be no induced electric field and thus no electric field lines. If the magnetic field is *decreasing* with time (at a constant

*Arguments of symmetry would also permit the lines of $\vec{E}$ around the circular path to be *radial,* rather than tangential. However, such radial lines would imply that there are free charges, distributed symmetrically about the axis of symmetry, on which the electric field lines could begin or end; there are no such charges.

rate), the electric field lines will still be concentric circles as in Fig. 30-11*c*, but they will now have the opposite direction. All this is what we have in mind when we say "A changing magnetic field produces an electric field."

A Reformulation of Faraday's Law

Consider a particle of charge q_0 moving around the circular path of Fig. 30-11*b*. The work W done on it in one revolution by the induced electric field is $W = \mathscr{E}q_0$, where $\mathscr{E}$ is the induced emf—that is, the work done per unit charge in moving the test charge around the path. From another point of view, the work is

$$W = \int \vec{F} \cdot d\vec{s} = (q_0 E)(2\pi r), \tag{30-16}$$

where q_0E is the magnitude of the force acting on the test charge and $2\pi r$ is the distance over which that force acts. Setting these two expressions for W equal to each other and canceling q_0, we find that

$$\mathscr{E} = 2\pi r E. \tag{30-17}$$

Next we rewrite Eq. 30-16 to give a more general expression for the work done on a particle of charge q_0 moving along any closed path:

$$W = \oint \vec{F} \cdot d\vec{s} = q_0 \oint \vec{E} \cdot d\vec{s}. \tag{30-18}$$

(The loop on each integral sign indicates that the integral is to be taken around the closed path.) Substituting $\mathscr{E}q_0$ for W, we find that

$$\mathscr{E} = \oint \vec{E} \cdot d\vec{s}. \tag{30-19}$$

This integral reduces at once to Eq. 30-17 if we evaluate it for the special case of Fig. 30-11*b*.

With Eq. 30-19, we can expand the meaning of induced emf. Up to this point, induced emf has meant the work per unit charge done in maintaining current due to a changing magnetic flux, or it has meant the work done per unit charge on a charged particle that moves around a closed path in a changing magnetic flux. However, with Fig. 30-11*b* and Eq. 30-19, an induced emf can exist without the need of a current or particle: An induced emf is the sum—via integration—of quantities $\vec{E} \cdot d\vec{s}$ around a closed path, where $\vec{E}$ is the electric field induced by a changing magnetic flux and $d\vec{s}$ is a differential length vector along the path.

If we combine Eq. 30-19 with Faraday's law in Eq. 30-4 ($\mathscr{E} = -d\Phi_B/dt$), we can rewrite Faraday's law as

$$\oint \vec{E} \cdot d\vec{s} = -\frac{d\Phi_B}{dt} \quad \text{(Faraday's law).} \tag{30-20}$$

This equation says simply that a changing magnetic field induces an electric field. The changing magnetic field appears on the right side of this equation, the electric field on the left.

Faraday's law in the form of Eq. 30-20 can be applied to *any* closed path that can be drawn in a changing magnetic field. Figure 30-11*d*, for example, shows four such paths, all having the same shape and area but located in different positions in the changing field. The induced emfs $\mathscr{E}$ $(= \oint \vec{E} \cdot d\vec{s})$ for paths 1 and 2 are equal because these paths lie entirely in the magnetic field and thus have the same value of $d\Phi_B/dt$. This is true even though the electric field vectors at points along these paths are different, as indicated by the patterns of electric field lines in the figure. For path 3 the induced emf is smaller because the enclosed flux Φ_B (hence $d\Phi_B/dt$) is smaller, and for path 4 the induced emf is zero even though the electric field is not zero at any point on the path.

A New Look at Electric Potential

Induced electric fields are produced not by static charges but by a changing magnetic flux. Although electric fields produced in either way exert forces on charged particles, there is an important difference between them. The simplest evidence of this difference is that the field lines of induced electric fields form closed loops, as in Fig. 30-11*c*. Field lines produced by static charges never do so but must start on positive charges and end on negative charges.

In a more formal sense, we can state the difference between electric fields produced by induction and those produced by static charges in these words:

Electric potential has meaning only for electric fields that are produced by static charges; it has no meaning for electric fields that are produced by induction.

You can understand this statement qualitatively by considering what happens to a charged particle that makes a single journey around the circular path in Fig. 30-11*b*. It starts at a certain point and, on its return to that same point, has experienced an emf $\mathscr{E}$ of, let us say, 5 V; that is, work of 5 J/C has been done on the particle, and thus the particle should then be at a point that is 5 V greater in potential. However, that is impossible because the particle is back at the same point, which cannot have two different values of potential. Thus, potential has no meaning for electric fields that are set up by changing magnetic fields.

We can take a more formal look by recalling Eq. 24-18, which defines the potential difference between two points *i* and *f* in an electric field $\vec{E}$:

$$V_f - V_i = -\int_i^f \vec{E} \cdot d\vec{s}. \tag{30-21}$$

In Chapter 24 we had not yet encountered Faraday's law of induction; so the electric fields involved in the derivation of Eq. 24-18 were those due to static charges. If *i* and *f* in Eq. 30-21 are the same point, the path connecting them is a closed loop, V_i and V_f are identical, and Eq. 30-21 reduces to

$$\oint \vec{E} \cdot d\vec{s} = 0. \tag{30-22}$$

However, when a changing magnetic flux is present, this integral is *not* zero but is $-d\Phi_B/dt$, as Eq. 30-20 asserts. Thus, assigning electric potential to an induced electric field leads us to a contradiction. We must conclude that electric potential has no meaning for electric fields associated with induction.

CHECKPOINT 4

The figure shows five lettered regions in which a uniform magnetic field extends either directly out of the page or into the page, with the direction indicated only for region *a*. The field is increasing in magnitude at the same steady rate in all five regions; the regions are identical in area. Also shown are four numbered paths along which $\oint \vec{E} \cdot d\vec{s}$ has the magnitudes given below in terms of a quantity "mag." Determine whether the magnetic field is directed into or out of the page for regions *b* through *e*.

Path	1	2	3	4
$\oint \vec{E} \cdot d\vec{s}$	mag	2(mag)	3(mag)	0

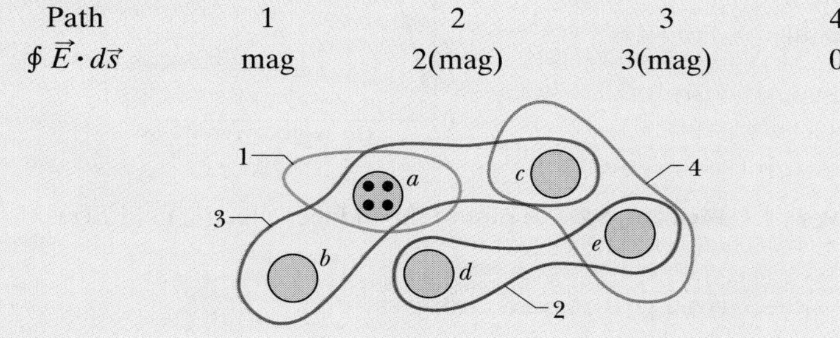

Sample Problem

Induced electric field due to changing *B* field, inside and outside

In Fig. 30-11*b*, take $R = 8.5$ cm and $dB/dt = 0.13$ T/s.

(a) Find an expression for the magnitude E of the induced electric field at points within the magnetic field, at radius r from the center of the magnetic field. Evaluate the expression for $r = 5.2$ cm.

KEY IDEA

An electric field is induced by the changing magnetic field, according to Faraday's law.

Calculations: To calculate the field magnitude E, we apply Faraday's law in the form of Eq. 30-20. We use a circular path of integration with radius $r \leq R$ because we want E for points within the magnetic field. We assume from the symmetry that $\vec{E}$ in Fig. 30-11*b* is tangent to the circular path at all points. The path vector $d\vec{s}$ is also always tangent to the circular path; so the dot product $\vec{E} \cdot d\vec{s}$ in Eq. 30-20 must have the magnitude $E\,ds$ at all points on the path. We can also assume from the symmetry that E has the same value at all points along the circular path. Then the left side of Eq. 30-20 becomes

$$\oint \vec{E} \cdot d\vec{s} = \oint E\,ds = E \oint ds = E(2\pi r). \tag{30-23}$$

(The integral $\oint ds$ is the circumference $2\pi r$ of the circular path.)

Next, we need to evaluate the right side of Eq. 30-20. Because $\vec{B}$ is uniform over the area A encircled by the path of integration and is directed perpendicular to that area, the magnetic flux is given by Eq. 30-2:

$$\Phi_B = BA = B(\pi r^2). \tag{30-24}$$

Substituting this and Eq. 30-23 into Eq. 30-20 and dropping the minus sign, we find that

$$E(2\pi r) = (\pi r^2)\frac{dB}{dt}$$

or

$$E = \frac{r}{2}\frac{dB}{dt}. \qquad \text{(Answer)} \tag{30-25}$$

Equation 30-25 gives the magnitude of the electric field at any point for which $r \leq R$ (that is, within the magnetic field). Substituting given values yields, for the magnitude of $\vec{E}$ at $r = 5.2$ cm,

$$E = \frac{(5.2 \times 10^{-2}\text{ m})}{2}(0.13\text{ T/s})$$

$$= 0.0034\text{ V/m} = 3.4\text{ mV/m}. \qquad \text{(Answer)}$$

(b) Find an expression for the magnitude E of the induced electric field at points that are outside the magnetic field, at radius r from the center of the magnetic field. Evaluate the expression for $r = 12.5$ cm.

KEY IDEAS

Here again an electric field is induced by the changing magnetic field, according to Faraday's law, except that now we use a circular path of integration with radius $r \geq R$ because we want to evaluate E for points outside the magnetic field. Proceeding as in (a), we again obtain Eq. 30-23. However, we do not then obtain Eq. 30-24 because the new path of integration is now outside the magnetic field, and so the magnetic flux encircled by the new path is only that in the area πR^2 of the magnetic field region.

Calculations: We can now write

$$\Phi_B = BA = B(\pi R^2). \tag{30-26}$$

Substituting this and Eq. 30-23 into Eq. 30-20 (without the minus sign) and solving for E yield

$$E = \frac{R^2}{2r}\frac{dB}{dt}. \qquad \text{(Answer)} \tag{30-27}$$

Because E is not zero here, we know that an electric field is induced even at points that are outside the changing magnetic field, an important result that (as you will see in Section 31-11) makes transformers possible.

With the given data, Eq. 30-27 yields the magnitude of $\vec{E}$ at $r = 12.5$ cm:

$$E = \frac{(8.5 \times 10^{-2}\text{ m})^2}{(2)(12.5 \times 10^{-2}\text{ m})}(0.13\text{ T/s})$$

$$= 3.8 \times 10^{-3}\text{ V/m} = 3.8\text{ mV/m}. \qquad \text{(Answer)}$$

Equations 30-25 and 30-27 give the same result for $r = R$. Figure 30-12 shows a plot of $E(r)$. Note that the inside and outside plots meet at $r = R$.

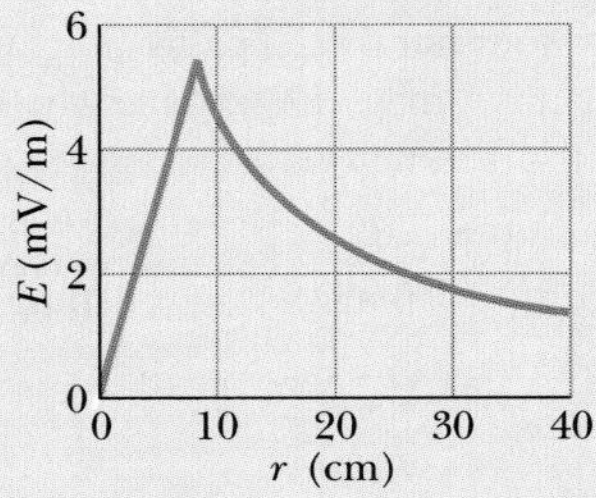

Fig. 30-12 A plot of the induced electric field $E(r)$.

Additional examples, video, and practice available at *WileyPLUS*

30-7 Inductors and Inductance

We found in Chapter 25 that a capacitor can be used to produce a desired electric field. We considered the parallel-plate arrangement as a basic type of capacitor. Similarly, an **inductor** (symbol) can be used to produce a desired magnetic field. We shall consider a long solenoid (more specifically, a short length near the middle of a long solenoid) as our basic type of inductor.

If we establish a current i in the windings (turns) of the solenoid we are taking as our inductor, the current produces a magnetic flux Φ_B through the central region of the inductor. The **inductance** of the inductor is then

$$L = \frac{N\Phi_B}{i} \quad \text{(inductance defined)}, \tag{30-28}$$

in which N is the number of turns. The windings of the inductor are said to be *linked* by the shared flux, and the product $N\Phi_B$ is called the *magnetic flux linkage.* The inductance L is thus a measure of the flux linkage produced by the inductor per unit of current.

The crude inductors with which Michael Faraday discovered the law of induction. In those days amenities such as insulated wire were not commercially available. It is said that Faraday insulated his wires by wrapping them with strips cut from one of his wife's petticoats. *(The Royal Institution/Bridgeman Art Library/NY)*

Because the SI unit of magnetic flux is the tesla–square meter, the SI unit of inductance is the tesla–square meter per ampere ($\mathrm{T \cdot m^2/A}$). We call this the **henry** (H), after American physicist Joseph Henry, the codiscoverer of the law of induction and a contemporary of Faraday. Thus,

$$1 \text{ henry} = 1 \text{ H} = 1 \text{ T} \cdot \text{m}^2/\text{A}. \tag{30-29}$$

Through the rest of this chapter we assume that all inductors, no matter what their geometric arrangement, have no magnetic materials such as iron in their vicinity. Such materials would distort the magnetic field of an inductor.

Inductance of a Solenoid

Consider a long solenoid of cross-sectional area A. What is the inductance per unit length near its middle? To use the defining equation for inductance (Eq. 30-28), we must calculate the flux linkage set up by a given current in the solenoid windings. Consider a length l near the middle of this solenoid. The flux linkage there is

$$N\Phi_B = (nl)(BA),$$

in which n is the number of turns per unit length of the solenoid and B is the magnitude of the magnetic field within the solenoid.

The magnitude B is given by Eq. 29-23,

$$B = \mu_0 in,$$

and so from Eq. 30-28,

$$L = \frac{N\Phi_B}{i} = \frac{(nl)(BA)}{i} = \frac{(nl)(\mu_0 in)(A)}{i}$$
$$= \mu_0 n^2 lA. \tag{30-30}$$

Thus, the inductance per unit length near the center of a long solenoid is

$$\frac{L}{l} = \mu_0 n^2 A \quad \text{(solenoid)}. \tag{30-31}$$

Inductance—like capacitance—depends only on the geometry of the device. The dependence on the square of the number of turns per unit length is to be expected. If you, say, triple n, you not only triple the number of turns (N) but you also triple the flux ($\Phi_B = BA = \mu_0 inA$) through each turn, multiplying the flux linkage $N\Phi_B$ and thus the inductance L by a factor of 9.

If the solenoid is very much longer than its radius, then Eq. 30-30 gives its inductance to a good approximation. This approximation neglects the spreading of the magnetic field lines near the ends of the solenoid, just as the parallel-plate capacitor formula ($C = \varepsilon_0 A/d$) neglects the fringing of the electric field lines near the edges of the capacitor plates.

From Eq. 30-30, and recalling that n is a number per unit length, we can see that an inductance can be written as a product of the permeability constant μ_0 and a quantity with the dimensions of a length. This means that μ_0 can be expressed in the unit henry per meter:

$$\begin{aligned}\mu_0 &= 4\pi \times 10^{-7}\ \text{T}\cdot\text{m/A}\\ &= 4\pi \times 10^{-7}\ \text{H/m}. \qquad (30\text{-}32)\end{aligned}$$

30-8 Self-Induction

If two coils—which we can now call inductors—are near each other, a current i in one coil produces a magnetic flux Φ_B through the second coil. We have seen that if we change this flux by changing the current, an induced emf appears in the second coil according to Faraday's law. An induced emf appears in the first coil as well.

An induced emf $\mathscr{E}_L$ appears in any coil in which the current is changing.

This process (see Fig. 30-13) is called **self-induction,** and the emf that appears is called a **self-induced emf.** It obeys Faraday's law of induction just as other induced emfs do.

For any inductor, Eq. 30-28 tells us that

$$N\Phi_B = Li. \qquad (30\text{-}33)$$

Faraday's law tells us that

$$\mathscr{E}_L = -\frac{d(N\Phi_B)}{dt}. \qquad (30\text{-}34)$$

By combining Eqs. 30-33 and 30-34 we can write

$$\mathscr{E}_L = -L\frac{di}{dt} \quad \text{(self-induced emf).} \qquad (30\text{-}35)$$

Thus, in any inductor (such as a coil, a solenoid, or a toroid) a self-induced emf appears whenever the current changes with time. The magnitude of the current has no influence on the magnitude of the induced emf; only the rate of change of the current counts.

You can find the *direction* of a self-induced emf from Lenz's law. The minus sign in Eq. 30-35 indicates that—as the law states—the self-induced emf $\mathscr{E}_L$ has the orientation such that it opposes the change in current i. We can drop the minus sign when we want only the magnitude of $\mathscr{E}_L$.

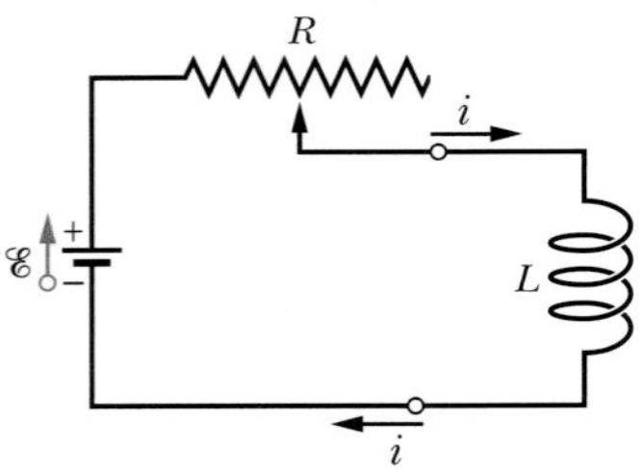

Fig. 30-13 If the current in a coil is changed by varying the contact position on a variable resistor, a self-induced emf $\mathscr{E}_L$ will appear in the coil *while the current is changing.*

Suppose that, as in Fig. 30-14*a*, you set up a current i in a coil and arrange to have the current increase with time at a rate di/dt. In the language of Lenz's law, this increase in the current is the "change" that the self-induction must oppose. For such opposition to occur, a self-induced emf must appear in the coil, pointing—as the figure shows—so as to oppose the increase in the current. If you cause the current to decrease with time, as in Fig. 30-14*b*, the self-induced emf must point in a direction that tends to oppose the decrease in the current, as the figure shows. In both cases, the emf attempts to maintain the initial condition.

In Section 30-6 we saw that we cannot define an electric potential for an electric field (and thus for an emf) that is induced by a changing magnetic flux.

This means that when a self-induced emf is produced in the inductor of Fig. 30-13, we cannot define an electric potential within the inductor itself, where the flux is changing. However, potentials can still be defined at points of the circuit that are not within the inductor—points where the electric fields are due to charge distributions and their associated electric potentials.

Moreover, we can define a self-induced potential difference V_L *across an inductor* (between its terminals, which we assume to be outside the region of changing flux). For an *ideal inductor* (its wire has negligible resistance), the magnitude of V_L is equal to the magnitude of the self-induced emf $\mathscr{E}_L$.

If, instead, the wire in the inductor has resistance r, we mentally separate the inductor into a resistance r (which we take to be outside the region of changing flux) and an ideal inductor of self-induced emf $\mathscr{E}_L$. As with a real battery of emf $\mathscr{E}$ and internal resistance r, the potential difference across the terminals of a real inductor then differs from the emf. Unless otherwise indicated, we assume here that inductors are ideal.

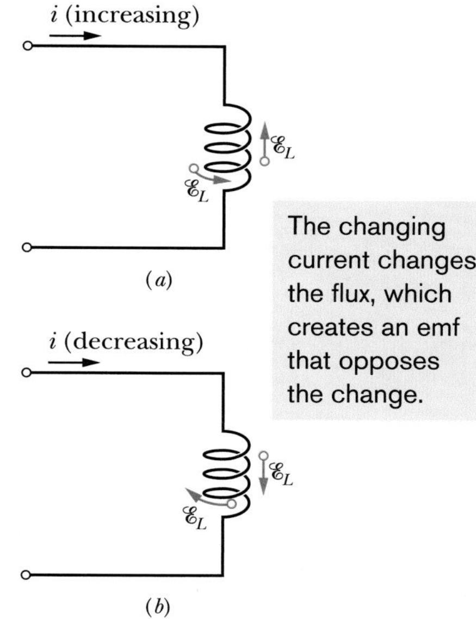

Fig. 30-14 (*a*) The current i is increasing, and the self-induced emf $\mathscr{E}_L$ appears along the coil in a direction such that it opposes the increase. The arrow representing $\mathscr{E}_L$ can be drawn along a turn of the coil or alongside the coil. Both are shown. (*b*) The current i is decreasing, and the self-induced emf appears in a direction such that it opposes the decrease.

CHECKPOINT 5

The figure shows an emf $\mathscr{E}_L$ induced in a coil. Which of the following can describe the current through the coil: (a) constant and rightward, (b) constant and leftward, (c) increasing and rightward, (d) decreasing and rightward, (e) increasing and leftward, (f) decreasing and leftward?

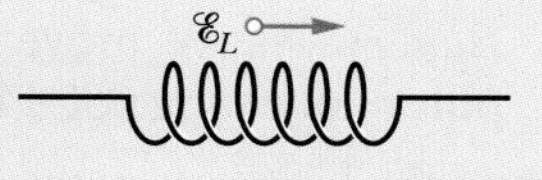

30-9 *RL* Circuits

In Section 27-9 we saw that if we suddenly introduce an emf $\mathscr{E}$ into a single-loop circuit containing a resistor R and a capacitor C, the charge on the capacitor does not build up immediately to its final equilibrium value $C\mathscr{E}$ but approaches it in an exponential fashion:

$$q = C\mathscr{E}(1 - e^{-t/\tau_C}). \tag{30-36}$$

The rate at which the charge builds up is determined by the capacitive time constant τ_C, defined in Eq. 27-36 as

$$\tau_C = RC. \tag{30-37}$$

If we suddenly remove the emf from this same circuit, the charge does not immediately fall to zero but approaches zero in an exponential fashion:

$$q = q_0 e^{-t/\tau_C}. \tag{30-38}$$

The time constant τ_C describes the fall of the charge as well as its rise.

An analogous slowing of the rise (or fall) of the current occurs if we introduce an emf $\mathscr{E}$ into (or remove it from) a single-loop circuit containing a resistor R and an inductor L. When the switch S in Fig. 30-15 is closed on a, for example, the current in the resistor starts to rise. If the inductor were not present, the current would rise rapidly to a steady value $\mathscr{E}/R$. Because of the inductor, however, a self-induced emf $\mathscr{E}_L$ appears in the circuit; from Lenz's law, this emf opposes the rise of the current, which means that it opposes the battery emf $\mathscr{E}$ in polarity. Thus, the current in the resistor responds to the difference between two emfs, a constant $\mathscr{E}$ due to the battery and a variable $\mathscr{E}_L$ $(= -L\,di/dt)$ due to self-induction. As long as $\mathscr{E}_L$ is present, the current will be less than $\mathscr{E}/R$.

As time goes on, the rate at which the current increases becomes less rapid and the magnitude of the self-induced emf, which is proportional to di/dt, becomes smaller. Thus, the current in the circuit approaches $\mathscr{E}/R$ asymptotically.

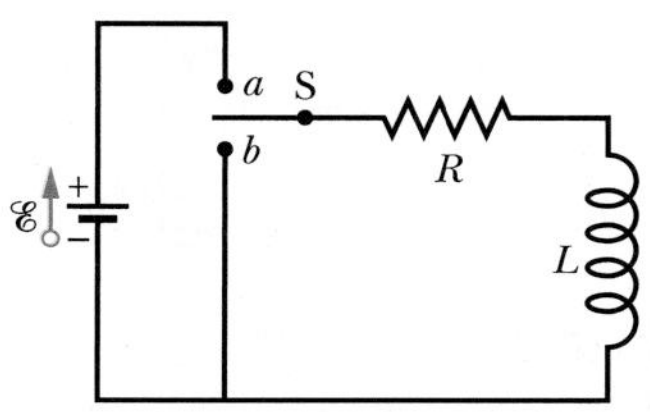

Fig. 30-15 An *RL* circuit. When switch S is closed on a, the current rises and approaches a limiting value $\mathscr{E}/R$.

We can generalize these results as follows:

> Initially, an inductor acts to oppose changes in the current through it. A long time later, it acts like ordinary connecting wire.

Now let us analyze the situation quantitatively. With the switch S in Fig. 30-15 thrown to a, the circuit is equivalent to that of Fig. 30-16. Let us apply the loop rule, starting at point x in this figure and moving clockwise around the loop along with current i.

1. *Resistor.* Because we move through the resistor in the direction of current i, the electric potential decreases by iR. Thus, as we move from point x to point y, we encounter a potential change of $-iR$.
2. *Inductor.* Because current i is changing, there is a self-induced emf $\mathscr{E}_L$ in the inductor. The magnitude of $\mathscr{E}_L$ is given by Eq. 30-35 as $L\,di/dt$. The direction of $\mathscr{E}_L$ is upward in Fig. 30-16 because current i is downward through the inductor *and* increasing. Thus, as we move from point y to point z, opposite the direction of $\mathscr{E}_L$, we encounter a potential change of $-L\,di/dt$.
3. *Battery.* As we move from point z back to starting point x, we encounter a potential change of $+\mathscr{E}$ due to the battery's emf.

Thus, the loop rule gives us

$$-iR - L\frac{di}{dt} + \mathscr{E} = 0$$

or

$$L\frac{di}{dt} + Ri = \mathscr{E} \quad (RL \text{ circuit}). \tag{30-39}$$

Equation 30-39 is a differential equation involving the variable i and its first derivative di/dt. To solve it, we seek the function $i(t)$ such that when $i(t)$ and its first derivative are substituted in Eq. 30-39, the equation is satisfied and the initial condition $i(0) = 0$ is satisfied.

Equation 30-39 and its initial condition are of exactly the form of Eq. 27-32 for an RC circuit, with i replacing q, L replacing R, and R replacing $1/C$. The solution of Eq. 30-39 must then be of exactly the form of Eq. 27-33 with the same replacements. That solution is

$$i = \frac{\mathscr{E}}{R}(1 - e^{-Rt/L}), \tag{30-40}$$

which we can rewrite as

$$i = \frac{\mathscr{E}}{R}(1 - e^{-t/\tau_L}) \quad \text{(rise of current)}. \tag{30-41}$$

Here τ_L, the **inductive time constant,** is given by

$$\tau_L = \frac{L}{R} \quad \text{(time constant)}. \tag{30-42}$$

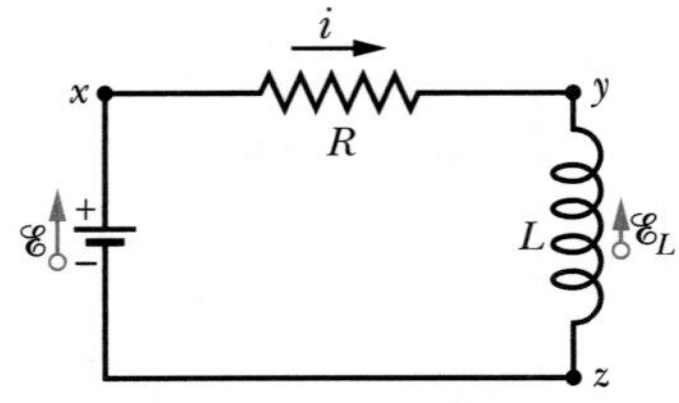

Fig. 30-16 The circuit of Fig. 30-15 with the switch closed on a. We apply the loop rule for the circuit clockwise, starting at x.

Let's examine Eq. 30-41 for just after the switch is closed (at time $t = 0$) and for a time long after the switch is closed ($t \to \infty$). If we substitute $t = 0$ into Eq. 30-41, the exponential becomes $e^{-0} = 1$. Thus, Eq. 30-41 tells us that the current is initially $i = 0$, as we expected. Next, if we let t go to ∞, then the exponential goes to $e^{-\infty} = 0$. Thus, Eq. 30-41 tells us that the current goes to its equilibrium value of $\mathscr{E}/R$.

We can also examine the potential differences in the circuit. For example, Fig. 30-17 shows how the potential differences V_R $(= iR)$ across the resistor and

V_L ($= L\,di/dt$) across the inductor vary with time for particular values of $\mathscr{E}$, L, and R. Compare this figure carefully with the corresponding figure for an RC circuit (Fig. 27-16).

To show that the quantity τ_L ($= L/R$) has the dimension of time, we convert from henries per ohm as follows:

$$1\,\frac{\text{H}}{\Omega} = 1\,\frac{\text{H}}{\Omega}\left(\frac{1\ \text{V}\cdot\text{s}}{1\ \text{H}\cdot\text{A}}\right)\left(\frac{1\ \Omega\cdot\text{A}}{1\ \text{V}}\right) = 1\ \text{s}.$$

The first quantity in parentheses is a conversion factor based on Eq. 30-35, and the second one is a conversion factor based on the relation $V = iR$.

The physical significance of the time constant follows from Eq. 30-41. If we put $t = \tau_L = L/R$ in this equation, it reduces to

$$i = \frac{\mathscr{E}}{R}(1 - e^{-1}) = 0.63\,\frac{\mathscr{E}}{R}. \tag{30-43}$$

Thus, the time constant τ_L is the time it takes the current in the circuit to reach about 63% of its final equilibrium value $\mathscr{E}/R$. Since the potential difference V_R across the resistor is proportional to the current i, a graph of the increasing current versus time has the same shape as that of V_R in Fig. 30-17*a*.

If the switch S in Fig. 30-15 is closed on *a* long enough for the equilibrium current $\mathscr{E}/R$ to be established and then is thrown to *b*, the effect will be to remove the battery from the circuit. (The connection to *b* must actually be made an instant before the connection to *a* is broken. A switch that does this is called a *make-before-break* switch.) With the battery gone, the current through the resistor will decrease. However, it cannot drop immediately to zero but must decay to zero over time. The differential equation that governs the decay can be found by putting $\mathscr{E} = 0$ in Eq. 30-39:

$$L\frac{di}{dt} + iR = 0. \tag{30-44}$$

By analogy with Eqs. 27-38 and 27-39, the solution of this differential equation that satisfies the initial condition $i(0) = i_0 = \mathscr{E}/R$ is

$$i = \frac{\mathscr{E}}{R}e^{-t/\tau_L} = i_0 e^{-t/\tau_L} \quad \text{(decay of current)}. \tag{30-45}$$

We see that both current rise (Eq. 30-41) and current decay (Eq. 30-45) in an RL circuit are governed by the same inductive time constant, τ_L.

We have used i_0 in Eq. 30-45 to represent the current at time $t = 0$. In our case that happened to be $\mathscr{E}/R$, but it could be any other initial value.

The resistor's potential difference turns on. The inductor's potential difference turns off.

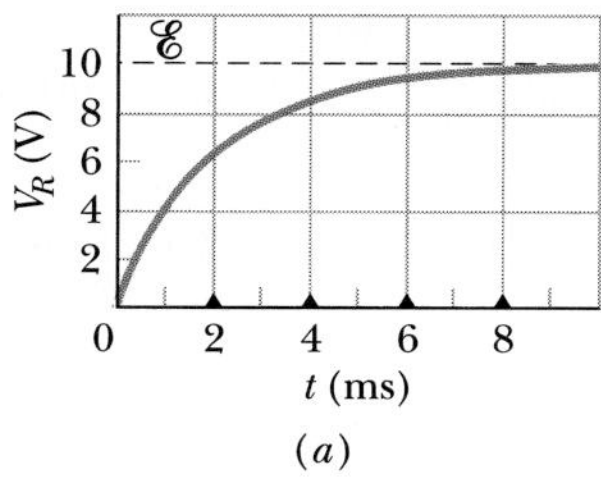

(*a*)

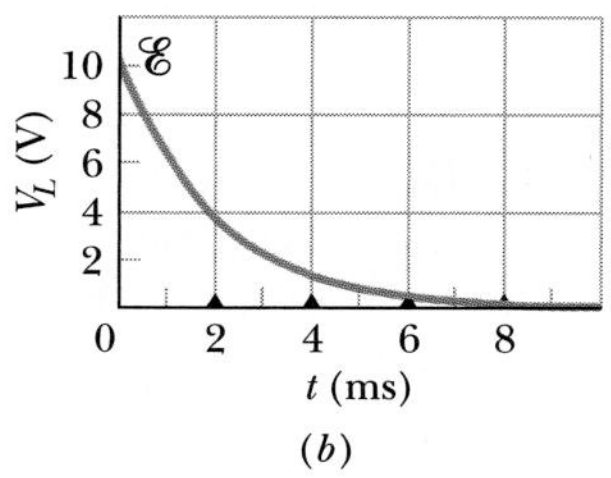

(*b*)

Fig. 30-17 The variation with time of (*a*) V_R, the potential difference across the resistor in the circuit of Fig. 30-16, and (*b*) V_L, the potential difference across the inductor in that circuit. The small triangles represent successive intervals of one inductive time constant $\tau_L = L/R$. The figure is plotted for $R = 2000\ \Omega$, $L = 4.0$ H, and $\mathscr{E} = 10$ V.

CHECKPOINT 6

The figure shows three circuits with identical batteries, inductors, and resistors. Rank the circuits according to the current through the battery (a) just after the switch is closed and (b) a long time later, greatest first. (If you have trouble here, work through the next sample problem and then try again.)

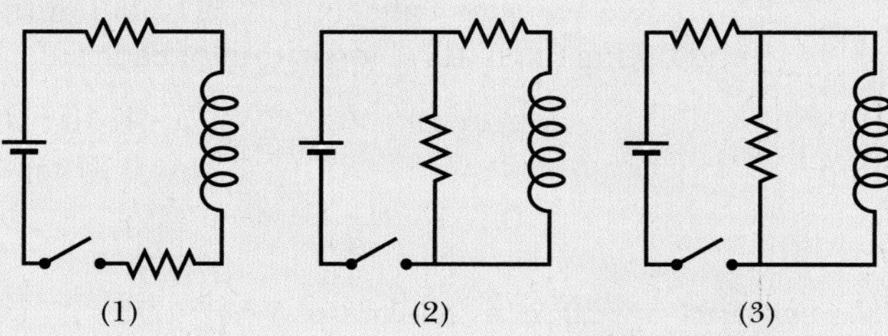

Sample Problem

RL circuit, immediately after switching and after a long time

Figure 30-18*a* shows a circuit that contains three identical resistors with resistance $R = 9.0\ \Omega$, two identical inductors with inductance $L = 2.0$ mH, and an ideal battery with emf $\mathscr{E} = 18$ V.

(a) What is the current i through the battery just after the switch is closed?

KEY IDEA

Just after the switch is closed, the inductor acts to oppose a change in the current through it.

Calculations: Because the current through each inductor is zero before the switch is closed, it will also be zero just afterward. Thus, immediately after the switch is closed, the inductors act as broken wires, as indicated in Fig. 30-18*b*. We then have a single-loop circuit for which the loop rule gives us

$$\mathscr{E} - iR = 0.$$

Substituting given data, we find that

$$i = \frac{\mathscr{E}}{R} = \frac{18\text{ V}}{9.0\ \Omega} = 2.0\text{ A}. \qquad \text{(Answer)}$$

(b) What is the current i through the battery long after the switch has been closed?

KEY IDEA

Long after the switch has been closed, the currents in the circuit have reached their equilibrium values, and the inductors act as simple connecting wires, as indicated in Fig. 30-18*c*.

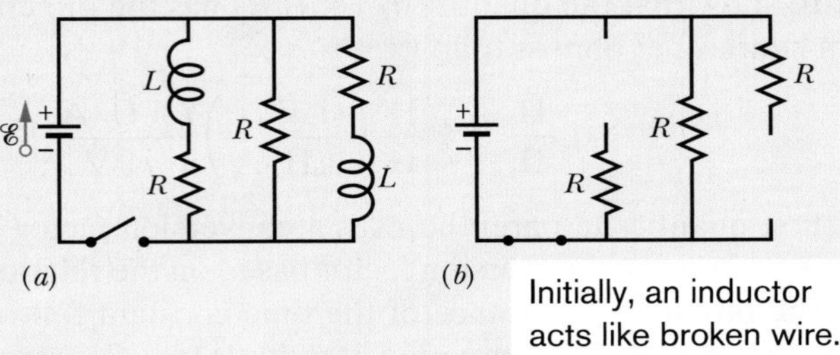

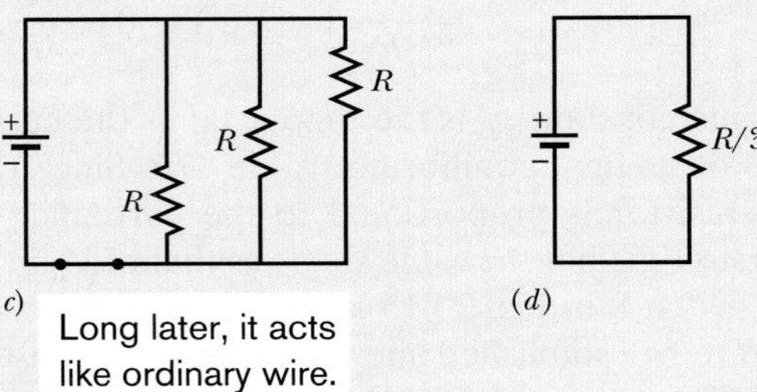

Fig. 30-18 (*a*) A multiloop *RL* circuit with an open switch. (*b*) The equivalent circuit just after the switch has been closed. (*c*) The equivalent circuit a long time later. (*d*) The single-loop circuit that is equivalent to circuit (*c*).

Calculations: We now have a circuit with three identical resistors in parallel; from Eq. 27-23, their equivalent resistance is $R_{eq} = R/3 = (9.0\ \Omega)/3 = 3.0\ \Omega$. The equivalent circuit shown in Fig. 30-18*d* then yields the loop equation $\mathscr{E} - iR_{eq} = 0$, or

$$i = \frac{\mathscr{E}}{R_{eq}} = \frac{18\text{ V}}{3.0\ \Omega} = 6.0\text{ A}. \qquad \text{(Answer)}$$

Sample Problem

RL circuit, current during the transition

A solenoid has an inductance of 53 mH and a resistance of 0.37 Ω. If the solenoid is connected to a battery, how long will the current take to reach half its final equilibrium value? (This is a *real solenoid* because we are considering its small, but nonzero, internal resistance.)

KEY IDEA

We can mentally separate the solenoid into a resistance and an inductance that are wired in series with a battery, as in Fig. 30-16. Then application of the loop rule leads to Eq. 30-39, which has the solution of Eq. 30-41 for the current i in the circuit.

Calculations: According to that solution, current i increases exponentially from zero to its final equilibrium value of $\mathscr{E}/R$. Let t_0 be the time that current i takes to reach half its equilibrium value. Then Eq. 30-41 gives us

$$\frac{1}{2}\frac{\mathscr{E}}{R} = \frac{\mathscr{E}}{R}(1 - e^{-t_0/\tau_L}).$$

We solve for t_0 by canceling $\mathscr{E}/R$, isolating the exponential, and taking the natural logarithm of each side. We find

$$t_0 = \tau_L \ln 2 = \frac{L}{R}\ln 2 = \frac{53 \times 10^{-3}\text{ H}}{0.37\ \Omega}\ln 2$$
$$= 0.10\text{ s}. \qquad \text{(Answer)}$$

Additional examples, video, and practice available at *WileyPLUS*

30-10 Energy Stored in a Magnetic Field

When we pull two charged particles of opposite signs away from each other, we say that the resulting electric potential energy is stored in the electric field of the particles. We get it back from the field by letting the particles move closer together again. In the same way we say energy is stored in a magnetic field, but now we deal with current instead of electric charges.

To derive a quantitative expression for that stored energy, consider again Fig. 30-16, which shows a source of emf $\mathscr{E}$ connected to a resistor R and an inductor L. Equation 30-39, restated here for convenience,

$$\mathscr{E} = L\frac{di}{dt} + iR, \tag{30-46}$$

is the differential equation that describes the growth of current in this circuit. Recall that this equation follows immediately from the loop rule and that the loop rule in turn is an expression of the principle of conservation of energy for single-loop circuits. If we multiply each side of Eq. 30-46 by i, we obtain

$$\mathscr{E}i = Li\frac{di}{dt} + i^2R, \tag{30-47}$$

which has the following physical interpretation in terms of the work done by the battery and the resulting energy transfers:

1. If a differential amount of charge dq passes through the battery of emf $\mathscr{E}$ in Fig. 30-16 in time dt, the battery does work on it in the amount $\mathscr{E}\,dq$. The rate at which the battery does work is $(\mathscr{E}\,dq)/dt$, or $\mathscr{E}i$. Thus, the left side of Eq. 30-47 represents the rate at which the emf device delivers energy to the rest of the circuit.
2. The rightmost term in Eq. 30-47 represents the rate at which energy appears as thermal energy in the resistor.
3. Energy that is delivered to the circuit but does not appear as thermal energy must, by the conservation-of-energy hypothesis, be stored in the magnetic field of the inductor. Because Eq. 30-47 represents the principle of conservation of energy for RL circuits, the middle term must represent the rate dU_B/dt at which magnetic potential energy U_B is stored in the magnetic field.

Thus

$$\frac{dU_B}{dt} = Li\frac{di}{dt}. \tag{30-48}$$

We can write this as

$$dU_B = Li\,di.$$

Integrating yields

$$\int_0^{U_B} dU_B = \int_0^i Li\,di$$

or

$$U_B = \tfrac{1}{2}Li^2 \quad \text{(magnetic energy)}, \tag{30-49}$$

which represents the total energy stored by an inductor L carrying a current i. Note the similarity in form between this expression and the expression for the energy stored by a capacitor with capacitance C and charge q; namely,

$$U_E = \frac{q^2}{2C}. \tag{30-50}$$

(The variable i^2 corresponds to q^2, and the constant L corresponds to $1/C$.)

Sample Problem

Energy stored in a magnetic field

A coil has an inductance of 53 mH and a resistance of 0.35 Ω.

(a) If a 12 V emf is applied across the coil, how much energy is stored in the magnetic field after the current has built up to its equilibrium value?

KEY IDEA

The energy stored in the magnetic field of a coil at any time depends on the current through the coil at that time, according to Eq. 30-49 ($U_B = \frac{1}{2}Li^2$).

Calculations: Thus, to find the energy $U_{B\infty}$ stored at equilibrium, we must first find the equilibrium current. From Eq. 30-41, the equilibrium current is

$$i_\infty = \frac{\mathscr{E}}{R} = \frac{12\text{ V}}{0.35\ \Omega} = 34.3\text{ A}. \tag{30-51}$$

Then substitution yields

$$U_{B\infty} = \tfrac{1}{2}Li_\infty^2 = (\tfrac{1}{2})(53 \times 10^{-3}\text{ H})(34.3\text{ A})^2$$
$$= 31\text{ J}. \qquad \text{(Answer)}$$

(b) After how many time constants will half this equilibrium energy be stored in the magnetic field?

Calculations: Now we are being asked: At what time t will the relation

$$U_B = \tfrac{1}{2}U_{B\infty}$$

be satisfied? Using Eq. 30-49 twice allows us to rewrite this energy condition as

$$\tfrac{1}{2}Li^2 = (\tfrac{1}{2})\tfrac{1}{2}Li_\infty^2$$

or

$$i = \left(\frac{1}{\sqrt{2}}\right)i_\infty. \tag{30-52}$$

This equation tells us that, as the current increases from its initial value of 0 to its final value of i_∞, the magnetic field will have half its final stored energy when the current has increased to this value. In general, we know that i is given by Eq. 30-41, and here i_∞ (see Eq. 30-51) is $\mathscr{E}/R$; so Eq. 30-52 becomes

$$\frac{\mathscr{E}}{R}(1 - e^{-t/\tau_L}) = \frac{\mathscr{E}}{\sqrt{2}R}.$$

By canceling $\mathscr{E}/R$ and rearranging, we can write this as

$$e^{-t/\tau_L} = 1 - \frac{1}{\sqrt{2}} = 0.293,$$

which yields

$$\frac{t}{\tau_L} = -\ln 0.293 = 1.23$$

or

$$t \approx 1.2\tau_L. \qquad \text{(Answer)}$$

Thus, the energy stored in the magnetic field of the coil by the current will reach half its equilibrium value 1.2 time constants after the emf is applied.

Additional examples, video, and practice available at *WileyPLUS*

30-11 Energy Density of a Magnetic Field

Consider a length l near the middle of a long solenoid of cross-sectional area A carrying current i; the volume associated with this length is Al. The energy U_B stored by the length l of the solenoid must lie entirely within this volume because the magnetic field outside such a solenoid is approximately zero. Moreover, the stored energy must be uniformly distributed within the solenoid because the magnetic field is (approximately) uniform everywhere inside.

Thus, the energy stored per unit volume of the field is

$$u_B = \frac{U_B}{Al}$$

or, since

$$U_B = \tfrac{1}{2}Li^2,$$

we have

$$u_B = \frac{Li^2}{2Al} = \frac{L}{l}\frac{i^2}{2A}. \tag{30-53}$$

Here L is the inductance of length l of the solenoid.

Substituting for L/l from Eq. 30-31, we find

$$u_B = \tfrac{1}{2}\mu_0 n^2 i^2, \tag{30-54}$$

where n is the number of turns per unit length. From Eq. 29-23 ($B = \mu_0 in$) we can write this *energy density* as

$$u_B = \frac{B^2}{2\mu_0} \quad \text{(magnetic energy density).} \tag{30-55}$$

This equation gives the density of stored energy at any point where the magnitude of the magnetic field is B. Even though we derived it by considering the special case of a solenoid, Eq. 30-55 holds for all magnetic fields, no matter how they are generated. The equation is comparable to Eq. 25-25,

$$u_E = \tfrac{1}{2}\varepsilon_0 E^2, \tag{30-56}$$

which gives the energy density (in a vacuum) at any point in an electric field. Note that both u_B and u_E are proportional to the square of the appropriate field magnitude, B or E.

CHECKPOINT 7

The table lists the number of turns per unit length, current, and cross-sectional area for three solenoids. Rank the solenoids according to the magnetic energy density within them, greatest first.

Solenoid	Turns per Unit Length	Current	Area
a	$2n_1$	i_1	$2A_1$
b	n_1	$2i_1$	A_1
c	n_1	i_1	$6A_1$

30-12 Mutual Induction

In this section we return to the case of two interacting coils, which we first discussed in Section 30-2, and we treat it in a somewhat more formal manner. We saw earlier that if two coils are close together as in Fig. 30-2, a steady current i in one coil will set up a magnetic flux Φ through the other coil (*linking* the other coil). If we change i with time, an emf $\mathscr{E}$ given by Faraday's law appears in the second coil; we called this process *induction.* We could better have called it **mutual induction,** to suggest the mutual interaction of the two coils and to distinguish it from *self-induction,* in which only one coil is involved.

Let us look a little more quantitatively at mutual induction. Figure 30-19*a* shows two circular close-packed coils near each other and sharing a common central axis. With the variable resistor set at a particular resistance R, the battery produces a steady current i_1 in coil 1. This current creates a magnetic field represented by the lines of $\vec{B}_1$ in the figure. Coil 2 is connected to a sensitive meter but contains no battery; a magnetic flux Φ_{21} (the flux through coil 2 associated with the current in coil 1) links the N_2 turns of coil 2.

We define the mutual inductance M_{21} of coil 2 with respect to coil 1 as

$$M_{21} = \frac{N_2\Phi_{21}}{i_1}, \tag{30-57}$$

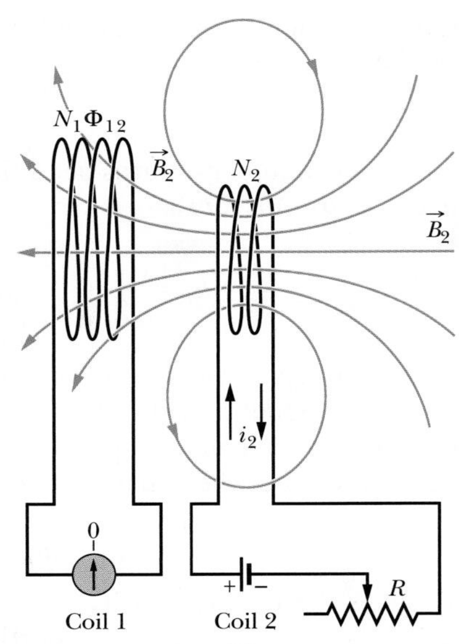

Fig. 30-19 Mutual induction. (*a*) The magnetic field $\vec{B}_1$ produced by current i_1 in coil 1 extends through coil 2. If i_1 is varied (by varying resistance R), an emf is induced in coil 2 and current registers on the meter connected to coil 2. (*b*) The roles of the coils interchanged.

which has the same form as Eq. 30-28,

$$L = N\Phi/i, \tag{30-58}$$

the definition of inductance. We can recast Eq. 30-57 as

$$M_{21}i_1 = N_2\Phi_{21}. \tag{30-59}$$

If we cause i_1 to vary with time by varying R, we have

$$M_{21}\frac{di_1}{dt} = N_2\frac{d\Phi_{21}}{dt}. \tag{30-60}$$

The right side of this equation is, according to Faraday's law, just the magnitude of the emf $\mathscr{E}_2$ appearing in coil 2 due to the changing current in coil 1. Thus, with a minus sign to indicate direction,

$$\mathscr{E}_2 = -M_{21}\frac{di_1}{dt}, \tag{30-61}$$

which you should compare with Eq. 30-35 for self-induction ($\mathscr{E} = -L\,di/dt$).

Let us now interchange the roles of coils 1 and 2, as in Fig. 30-19*b*; that is, we set up a current i_2 in coil 2 by means of a battery, and this produces a magnetic flux Φ_{12} that links coil 1. If we change i_2 with time by varying R, we then have, by the argument given above,

$$\mathscr{E}_1 = -M_{12}\frac{di_2}{dt}. \tag{30-62}$$

Thus, we see that the emf induced in either coil is proportional to the rate of change of current in the other coil. The proportionality constants M_{21} and M_{12} seem to be different. We assert, without proof, that they are in fact the same so that no subscripts are needed. (This conclusion is true but is in no way obvious.) Thus, we have

$$M_{21} = M_{12} = M, \tag{30-63}$$

and we can rewrite Eqs. 30-61 and 30-62 as

$$\mathscr{E}_2 = -M\frac{di_1}{dt} \tag{30-64}$$

and

$$\mathscr{E}_1 = -M\frac{di_2}{dt}. \tag{30-65}$$

Sample Problem

Mutual inductance of two parallel coils

Figure 30-20 shows two circular close-packed coils, the smaller (radius R_2, with N_2 turns) being coaxial with the larger (radius R_1, with N_1 turns) and in the same plane.

(a) Derive an expression for the mutual inductance M for this arrangement of these two coils, assuming that $R_1 \gg R_2$.

KEY IDEA

The mutual inductance M for these coils is the ratio of the flux linkage ($N\Phi$) through one coil to the current i in the other coil, which produces that flux linkage. Thus, we need to assume that currents exist in the coils; then we need to calculate the flux linkage in one of the coils.

Calculations: The magnetic field through the larger coil due to the smaller coil is nonuniform in both magnitude and direction; so the flux through the larger coil due to the smaller coil is nonuniform and difficult to calculate. However, the smaller coil is small enough for us to assume that the magnetic field through it due to the larger coil is approximately uniform. Thus, the flux through it due to the larger coil is also approximately uniform. Hence, to find M we shall assume a current i_1 in the larger coil and calculate the flux linkage $N_2\Phi_{21}$ in the smaller coil:

$$M = \frac{N_2\Phi_{21}}{i_1}. \qquad (30\text{-}66)$$

The flux Φ_{21} through each turn of the smaller coil is, from Eq. 30-2,

$$\Phi_{21} = B_1 A_2,$$

where B_1 is the magnitude of the magnetic field at points within the small coil due to the larger coil and A_2 ($= \pi R_2^2$) is the area enclosed by the turn. Thus, the flux linkage in the smaller coil (with its N_2 turns) is

$$N_2\Phi_{21} = N_2 B_1 A_2. \qquad (30\text{-}67)$$

To find B_1 at points within the smaller coil, we can use Eq. 29-26,

$$B(z) = \frac{\mu_0 i R^2}{2(R^2 + z^2)^{3/2}},$$

with z set to 0 because the smaller coil is in the plane of the larger coil. That equation tells us that each turn of the larger coil produces a magnetic field of magnitude $\mu_0 i_1/2R_1$ at points within the smaller coil. Thus, the larger coil (with its N_1 turns) produces a total magnetic field of magnitude

$$B_1 = N_1\frac{\mu_0 i_1}{2R_1} \qquad (30\text{-}68)$$

at points within the smaller coil.

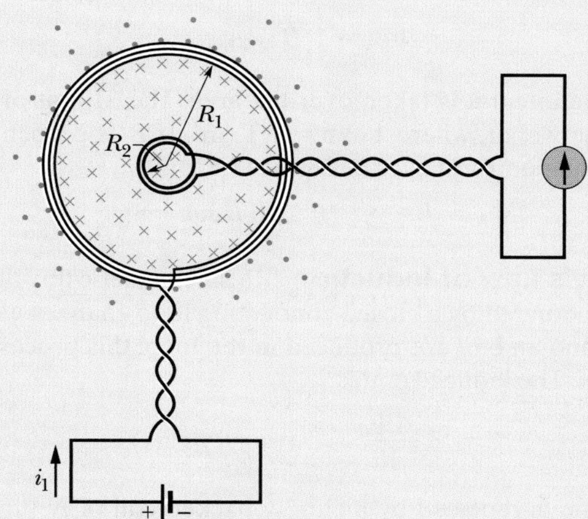

Fig. 30-20 A small coil is located at the center of a large coil. The mutual inductance of the coils can be determined by sending current i_1 through the large coil.

Substituting Eq. 30-68 for B_1 and πR_2^2 for A_2 in Eq. 30-67 yields

$$N_2\Phi_{21} = \frac{\pi\mu_0 N_1 N_2 R_2^2 i_1}{2R_1}.$$

Substituting this result into Eq. 30-66, we find

$$M = \frac{N_2\Phi_{21}}{i_1} = \frac{\pi\mu_0 N_1 N_2 R_2^2}{2R_1}. \qquad \text{(Answer)} \qquad (30\text{-}69)$$

(b) What is the value of M for $N_1 = N_2 = 1200$ turns, $R_2 = 1.1$ cm, and $R_1 = 15$ cm?

Calculations: Equation 30-69 yields

$$M = \frac{(\pi)(4\pi \times 10^{-7}\ \text{H/m})(1200)(1200)(0.011\ \text{m})^2}{(2)(0.15\ \text{m})}$$

$$= 2.29 \times 10^{-3}\ \text{H} \approx 2.3\ \text{mH}. \qquad \text{(Answer)}$$

Consider the situation if we reverse the roles of the two coils—that is, if we produce a current i_2 in the smaller coil and try to calculate M from Eq. 30-57 in the form

$$M = \frac{N_1\Phi_{12}}{i_2}.$$

The calculation of Φ_{12} (the nonuniform flux of the smaller coil's magnetic field encompassed by the larger coil) is not simple. If we were to do the calculation numerically using a computer, we would find M to be 2.3 mH, as above! This emphasizes that Eq. 30-63 ($M_{21} = M_{12} = M$) is not obvious.

Additional examples, video, and practice available at *WileyPLUS*

REVIEW & SUMMARY

Magnetic Flux The *magnetic flux* Φ_B through an area A in a magnetic field $\vec{B}$ is defined as

$$\Phi_B = \int \vec{B} \cdot d\vec{A}, \qquad (30\text{-}1)$$

where the integral is taken over the area. The SI unit of magnetic flux is the weber, where 1 Wb = 1 T · m². If $\vec{B}$ is perpendicular to the area and uniform over it, Eq. 30-1 becomes

$$\Phi_B = BA \qquad (\vec{B} \perp A, \vec{B} \text{ uniform}). \qquad (30\text{-}2)$$

Faraday's Law of Induction If the magnetic flux Φ_B through an area bounded by a closed conducting loop changes with time, a current and an emf are produced in the loop; this process is called *induction.* The induced emf is

$$\mathscr{E} = -\frac{d\Phi_B}{dt} \qquad \text{(Faraday's law).} \qquad (30\text{-}4)$$

If the loop is replaced by a closely packed coil of N turns, the induced emf is

$$\mathscr{E} = -N\frac{d\Phi_B}{dt}. \qquad (30\text{-}5)$$

Lenz's Law An induced current has a direction such that the magnetic field *due to the current* opposes the change in the magnetic flux that induces the current. The induced emf has the same direction as the induced current.

Emf and the Induced Electric Field An emf is induced by a changing magnetic flux even if the loop through which the flux is changing is not a physical conductor but an imaginary line. The changing magnetic field induces an electric field $\vec{E}$ at every point of such a loop; the induced emf is related to $\vec{E}$ by

$$\mathscr{E} = \oint \vec{E} \cdot d\vec{s}, \qquad (30\text{-}19)$$

where the integration is taken around the loop. From Eq. 30-19 we can write Faraday's law in its most general form,

$$\oint \vec{E} \cdot d\vec{s} = -\frac{d\Phi_B}{dt} \qquad \text{(Faraday's law).} \qquad (30\text{-}20)$$

A changing magnetic field induces an electric field $\vec{E}$.

Inductors An **inductor** is a device that can be used to produce a known magnetic field in a specified region. If a current i is established through each of the N windings of an inductor, a magnetic flux Φ_B links those windings. The **inductance** L of the inductor is

$$L = \frac{N\Phi_B}{i} \qquad \text{(inductance defined).} \qquad (30\text{-}28)$$

The SI unit of inductance is the **henry** (H), where 1 henry = 1 H = 1 T · m²/A. The inductance per unit length near the middle of a long solenoid of cross-sectional area A and n turns per unit length is

$$\frac{L}{l} = \mu_0 n^2 A \qquad \text{(solenoid).} \qquad (30\text{-}31)$$

Self-Induction If a current i in a coil changes with time, an emf is induced in the coil. This self-induced emf is

$$\mathscr{E}_L = -L\frac{di}{dt}. \qquad (30\text{-}35)$$

The direction of $\mathscr{E}_L$ is found from Lenz's law: The self-induced emf acts to oppose the change that produces it.

Series *RL* Circuits If a constant emf $\mathscr{E}$ is introduced into a single-loop circuit containing a resistance R and an inductance L, the current rises to an equilibrium value of $\mathscr{E}/R$ according to

$$i = \frac{\mathscr{E}}{R}(1 - e^{-t/\tau_L}) \qquad \text{(rise of current).} \qquad (30\text{-}41)$$

Here τ_L (= L/R) governs the rate of rise of the current and is called the **inductive time constant** of the circuit. When the source of constant emf is removed, the current decays from a value i_0 according to

$$i = i_0 e^{-t/\tau_L} \qquad \text{(decay of current).} \qquad (30\text{-}45)$$

Magnetic Energy If an inductor L carries a current i, the inductor's magnetic field stores an energy given by

$$U_B = \tfrac{1}{2}Li^2 \qquad \text{(magnetic energy).} \qquad (30\text{-}49)$$

If B is the magnitude of a magnetic field at any point (in an inductor or anywhere else), the density of stored magnetic energy at that point is

$$u_B = \frac{B^2}{2\mu_0} \qquad \text{(magnetic energy density).} \qquad (30\text{-}55)$$

Mutual Induction If coils 1 and 2 are near each other, a changing current in either coil can induce an emf in the other. This mutual induction is described by

$$\mathscr{E}_2 = -M\frac{di_1}{dt} \qquad (30\text{-}64)$$

and

$$\mathscr{E}_1 = -M\frac{di_2}{dt}, \qquad (30\text{-}65)$$

where M (measured in henries) is the mutual inductance.

QUESTIONS

1 If the circular conductor in Fig. 30-21 undergoes thermal expansion while it is in a uniform magnetic field, a current is induced clockwise around it. Is the magnetic field directed into or out of the page?

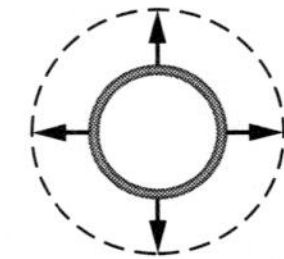

Fig. 30-21 Question 1.

2 The wire loop in Fig. 30-22*a* is subjected, in turn, to six uniform magnetic fields, each directed parallel to the z axis, which is directed out of the plane of the figure. Figure 30-22*b* gives the z components B_z of the fields versus time t. (Plots 1 and 3 are parallel; so are plots 4 and 6. Plots 2 and 5 are parallel to the time axis.) Rank the six plots according to the emf induced in the loop, greatest clockwise emf first, greatest counterclockwise emf last.

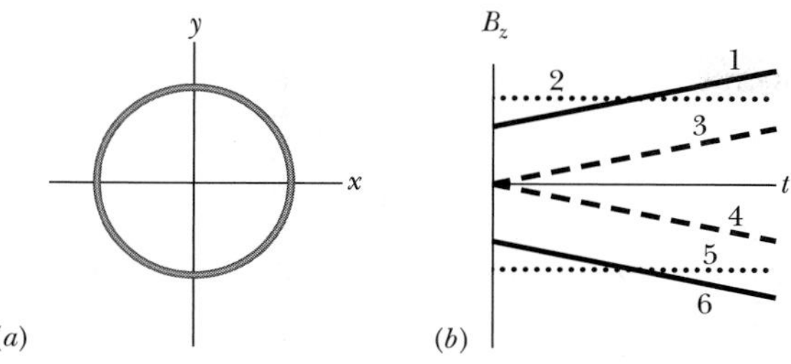

Fig. 30-22 Question 2.

3 In Fig. 30-23, a long straight wire with current i passes (without touching) three rectangular wire loops with edge lengths L, $1.5L$, and $2L$. The loops are widely spaced (so as not to affect one another). Loops 1 and 3 are symmetric about the long wire. Rank the loops according to the size of the current induced in them if current i is (a) constant and (b) increasing, greatest first.

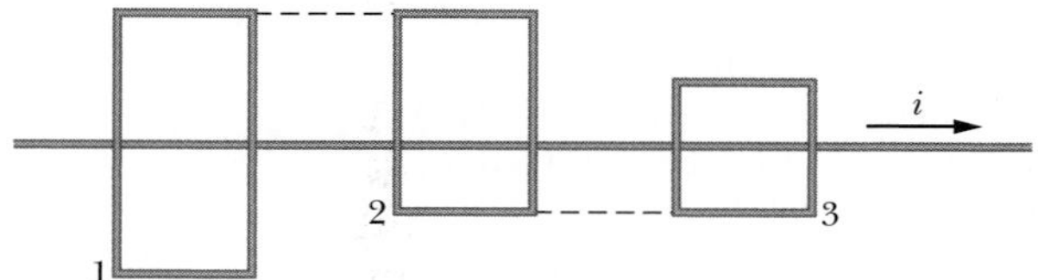

Fig. 30-23 Question 3.

4 Figure 30-24 shows two circuits in which a conducting bar is slid at the same speed v through the same uniform magnetic field and along a **U**-shaped wire. The parallel lengths of the wire are separated by $2L$ in circuit 1 and by L in circuit 2. The current induced in circuit 1 is counterclockwise. (a) Is the magnetic field into or out of the page? (b) Is the current induced in circuit 2 clockwise or counterclockwise? (c) Is the emf induced in circuit 1 larger than, smaller than, or the same as that in circuit 2?

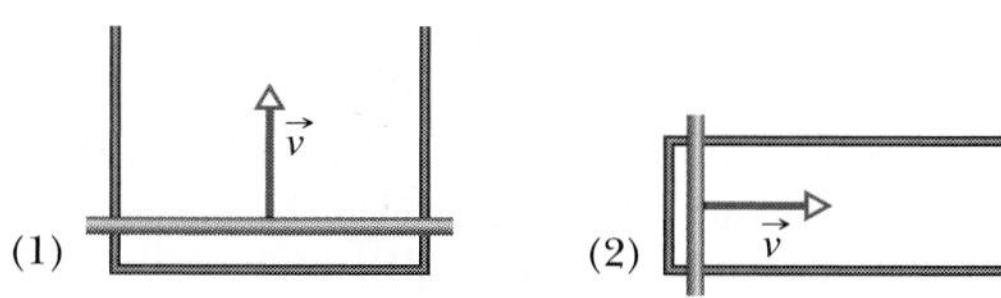

Fig. 30-24 Question 4.

5 Figure 30-25 shows a circular region in which a decreasing uniform magnetic field is directed out of the page, as well as four concentric circular paths. Rank the paths according to the magnitude of $\oint \vec{E} \cdot d\vec{s}$ evaluated along them, greatest first.

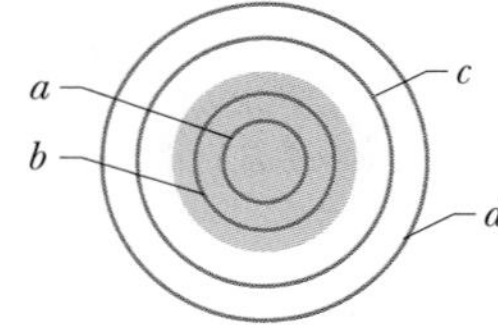

Fig. 30-25 Question 5.

6 In Fig. 30-26, a wire loop has been bent so that it has three segments: segment bc (a quarter-circle), ac (a square corner), and ab (straight). Here are three choices for a magnetic field through the loop:

(1) $\vec{B}_1 = 3\hat{\text{i}} + 7\hat{\text{j}} - 5t\hat{\text{k}}$,

(2) $\vec{B}_2 = 5t\hat{\text{i}} - 4\hat{\text{j}} - 15\hat{\text{k}}$,

(3) $\vec{B}_3 = 2\hat{\text{i}} - 5t\hat{\text{j}} - 12\hat{\text{k}}$,

where $\vec{B}$ is in milliteslas and t is in seconds. Without written calculation, rank the choices according to (a) the work done per unit charge in setting up the induced current and (b) that induced current, greatest first. (c) For each choice, what is the direction of the induced current in the figure?

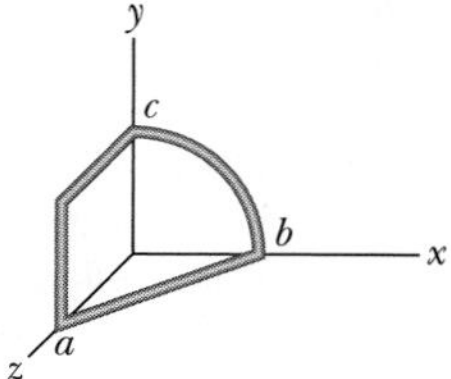

Fig. 30-26 Question 6.

7 Figure 30-27 shows a circuit with two identical resistors and an ideal inductor. Is the current through the central resistor more than, less than, or the same as that through the other resistor (a) just after the closing of switch S, (b) a long time after that, (c) just after S is reopened a long time later, and (d) a long time after that?

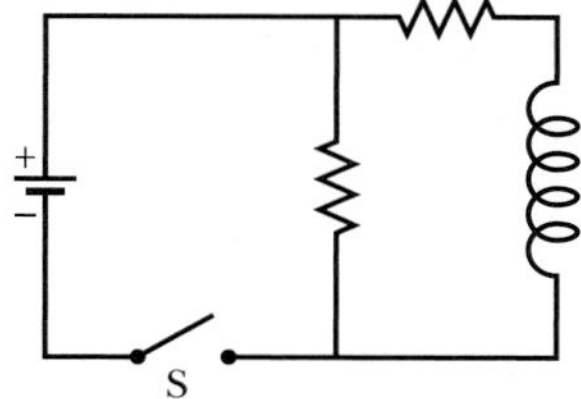

Fig. 30-27 Question 7.

8 The switch in the circuit of Fig. 30-15 has been closed on a for a very long time when it is then thrown to b. The resulting current through the inductor is indicated in Fig. 30-28 for four sets of values for the resistance R and inductance L: (1) R_0 and L_0, (2) $2R_0$ and L_0, (3) R_0 and $2L_0$, (4) $2R_0$ and $2L_0$. Which set goes with which curve?

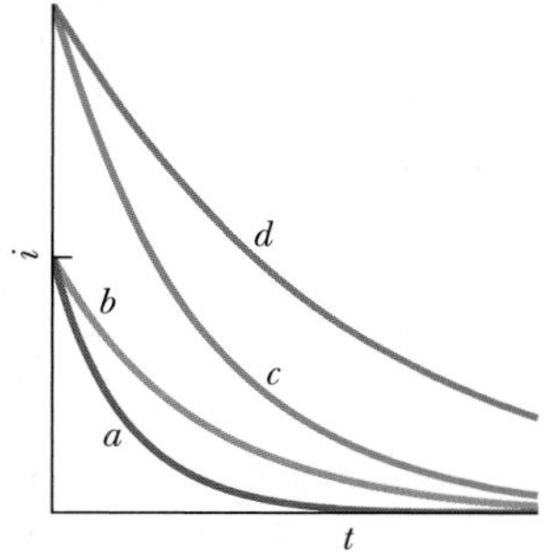

Fig. 30-28 Question 8.

9 Figure 30-29 shows three circuits with identical batteries, inductors, and resistors. Rank the circuits, greatest first, according to the current through the resistor labeled R (a) long after the switch is closed, (b) just after the switch is reopened a long time later, and (c) long after it is reopened.

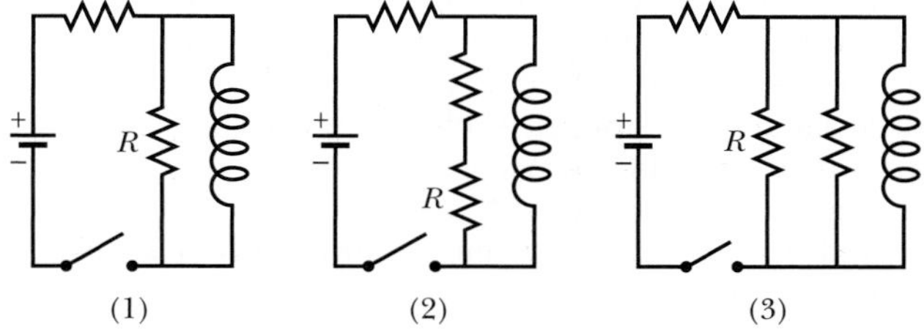

Fig. 30-29 Question 9.

10 Figure 30-30 gives the variation with time of the potential difference V_R across a resistor in three circuits wired as shown in Fig. 30-16. The circuits contain the same resistance R and emf $\mathscr{E}$ but differ in the inductance L. Rank the circuits according to the value of L, greatest first.

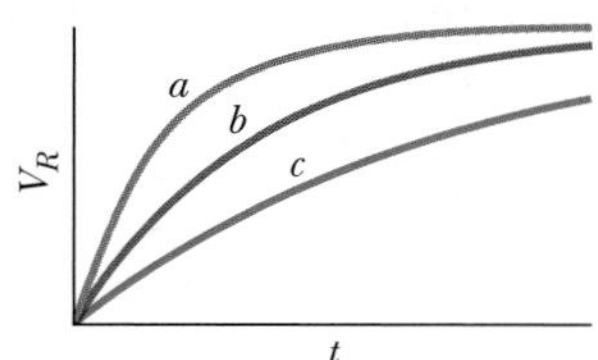

Fig. 30-30 Question 10.

PROBLEMS

GO Tutoring problem available (at instructor's discretion) in *WileyPLUS* and WebAssign
SSM Worked-out solution available in Student Solutions Manual
WWW Worked-out solution is at
• – ••• Number of dots indicates level of problem difficulty
ILW Interactive solution is at
http://www.wiley.com/college/halliday
Additional information available in *The Flying Circus of Physics* and at flyingcircusofphysics.com

sec. 30-4 Lenz's Law

•1 In Fig. 30-31, a circular loop of wire 10 cm in diameter (seen edge-on) is placed with its normal $\vec{N}$ at an angle $\theta = 30°$ with the direction of a uniform magnetic field $\vec{B}$ of magnitude 0.50 T. The loop is then rotated such that $\vec{N}$ rotates in a cone about the field direction at the rate 100 rev/min; angle θ remains unchanged during the process. What is the emf induced in the loop?

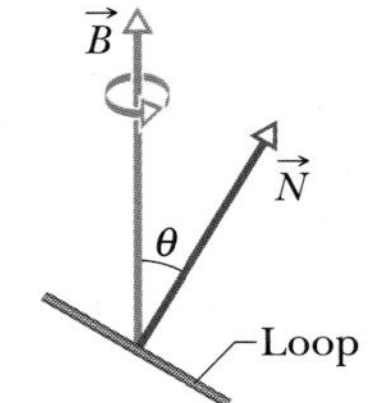

Fig. 30-31 Problem 1.

•2 A certain elastic conducting material is stretched into a circular loop of 12.0 cm radius. It is placed with its plane perpendicular to a uniform 0.800 T magnetic field. When released, the radius of the loop starts to shrink at an instantaneous rate of 75.0 cm/s. What emf is induced in the loop at that instant?

•3 SSM WWW In Fig. 30-32, a 120-turn coil of radius 1.8 cm and resistance 5.3 Ω is coaxial with a solenoid of 220 turns/cm and diameter 3.2 cm. The solenoid current drops from 1.5 A to zero in time interval $\Delta t = 25$ ms. What current is induced in the coil during Δt?

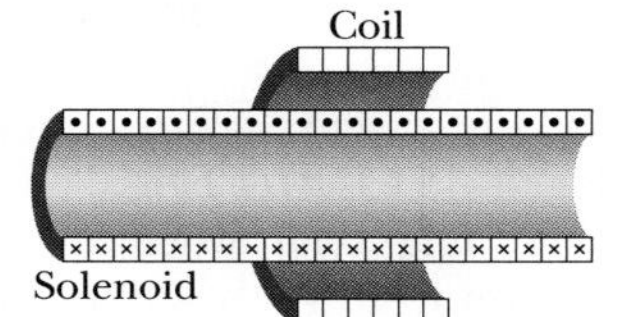

Fig. 30-32 Problem 3.

•4 A wire loop of radius 12 cm and resistance 8.5 Ω is located in a uniform magnetic field $\vec{B}$ that changes in magnitude as given in Fig. 30-33. The vertical axis scale is set by B_s = 0.50 T, and the horizontal axis scale is set by t_s = 6.00 s. The loop's plane is perpendicular to $\vec{B}$. What emf is induced in the loop during time intervals (a) 0 to 2.0 s, (b) 2.0 s to 4.0 s, and (c) 4.0 s to 6.0 s?

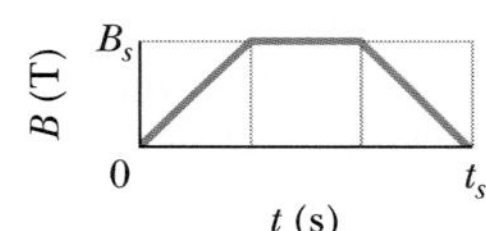

Fig. 30-33 Problem 4.

•5 In Fig. 30-34, a wire forms a closed circular loop, of radius R = 2.0 m and resistance 4.0 Ω. The circle is centered on a long straight wire; at time $t = 0$, the current in the long straight wire is 5.0 A rightward. Thereafter, the current changes according to $i = 5.0\text{ A} - (2.0\text{ A/s}^2)t^2$. (The straight wire is insulated; so there is no electrical contact between it and the wire of the loop.) What is the magnitude of the current induced in the loop at times $t > 0$?

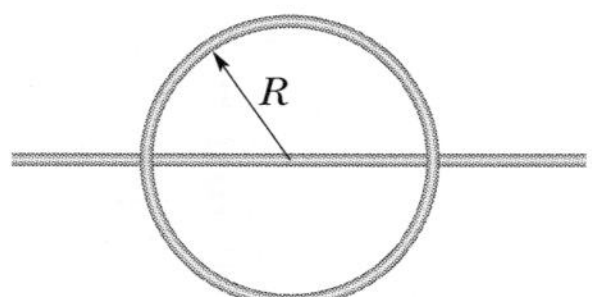

Fig. 30-34 Problem 5.

•6 Figure 30-35*a* shows a circuit consisting of an ideal battery with emf $\mathscr{E} = 6.00\ \mu$V, a resistance R, and a small wire loop of area 5.0 cm². For the time interval t = 10 s to t = 20 s, an external magnetic field is set up throughout the loop. The field is uniform, its direction is into the page in Fig. 30-35*a*, and the field magnitude is given by $B = at$, where B is in teslas, a is a constant, and t is in seconds. Figure 30-35*b* gives the current i in the circuit before, during, and after the external field is set up. The vertical axis scale is set by i_s = 2.0 mA. Find the constant a in the equation for the field magnitude.

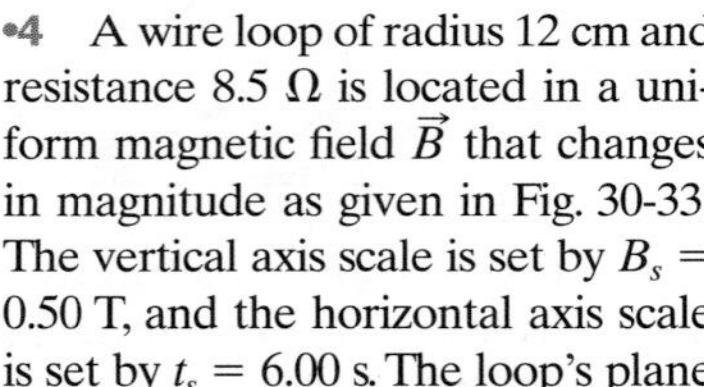

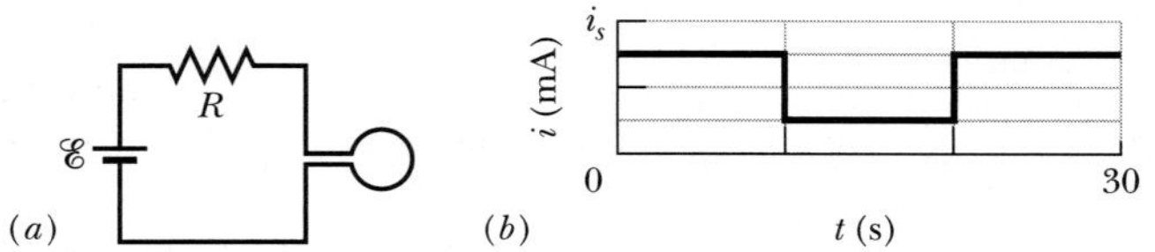

Fig. 30-35 Problem 6.

•7 In Fig. 30-36, the magnetic flux through the loop increases according to the relation $\Phi_B = 6.0t^2 + 7.0t$, where Φ_B is in milliwebers and t is in seconds. (a) What is the magnitude of the emf induced in the loop when t = 2.0 s? (b) Is the direction of the current through R to the right or left?

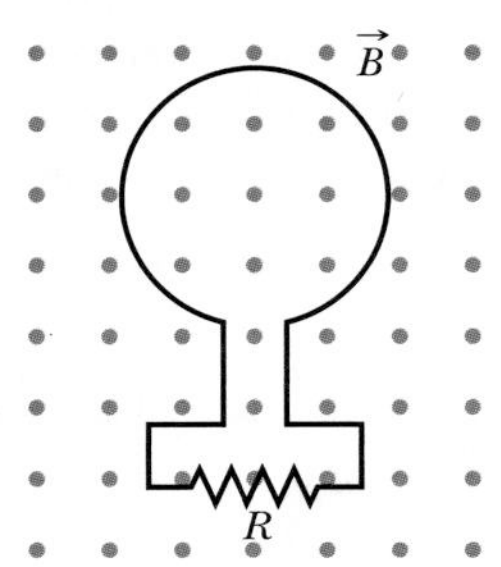

Fig. 30-36 Problem 7.

•8 A uniform magnetic field $\vec{B}$ is perpendicular to the plane of a circular loop of diameter 10 cm formed from wire of diameter 2.5 mm and resistivity $1.69 \times 10^{-8}\ \Omega \cdot$ m. At what rate must the magnitude of $\vec{B}$ change to induce a 10 A current in the loop?

•9 A small loop of area 6.8 mm² is placed inside a long solenoid that has 854 turns/cm and carries a sinusoidally varying current i of amplitude 1.28 A and angular frequency 212 rad/s. The central axes of the loop and solenoid coincide. What is the amplitude of the emf induced in the loop?

••10 Figure 30-37 shows a closed loop of wire that consists of a pair of equal semicircles, of radius 3.7 cm, lying in mutually perpendicular planes. The loop was formed by folding a flat circular loop along a diameter until the two halves became perpendicular to each other. A uniform magnetic field $\vec{B}$ of magnitude 76 mT is directed perpendicular to the fold diameter and makes equal angles (of 45°) with the planes of the semicircles. The magnetic field is reduced to zero at a uniform rate during a time interval of 4.5 ms. During this interval, what are the (a) magnitude and (b) direction (clockwise or counterclockwise when viewed along the direction of $\vec{B}$) of the emf induced in the loop?

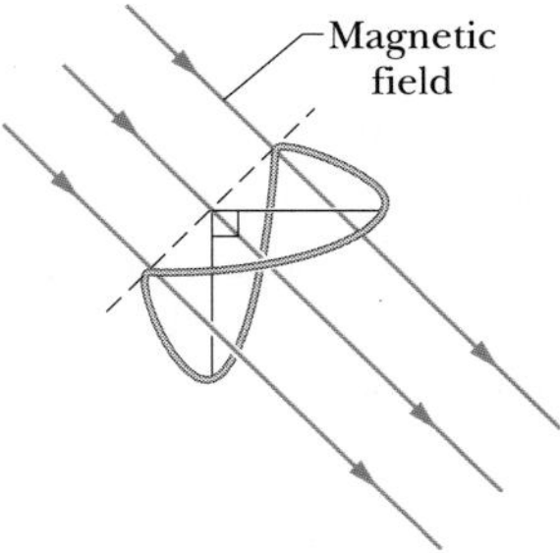

Fig. 30-37 Problem 10.

••11 A rectangular coil of N turns and of length a and width b is rotated at frequency f in a uniform magnetic field $\vec{B}$, as indicated in Fig. 30-38. The coil is connected to co-rotating cylinders, against which metal brushes slide to make contact. (a) Show that the emf induced in the coil is given (as a function of time t) by

$$\mathscr{E} = 2\pi f NabB \sin(2\pi ft) = \mathscr{E}_0 \sin(2\pi ft).$$

This is the principle of the commercial alternating-current generator. (b) What value of Nab gives an emf with $\mathscr{E}_0 = 150$ V when the loop is rotated at 60.0 rev/s in a uniform magnetic field of 0.500 T?

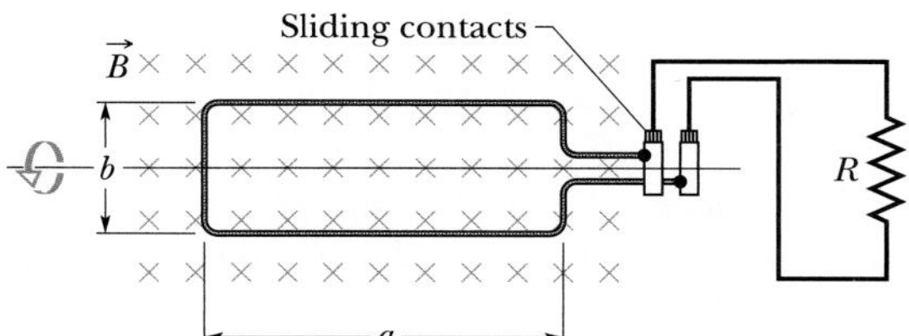

Fig. 30-38 Problem 11.

••12 In Fig. 30-39, a wire loop of lengths $L = 40.0$ cm and $W = 25.0$ cm lies in a magnetic field $\vec{B}$. What are the (a) magnitude $\mathscr{E}$ and (b) direction (clockwise or counterclockwise—or "none" if $\mathscr{E} = 0$) of the emf induced in the loop if $\vec{B} = (4.00 \times 10^{-2}\ \text{T/m})y\hat{k}$? What are (c) $\mathscr{E}$ and (d) the direction if $\vec{B} = (6.00 \times 10^{-2}\ \text{T/s})t\hat{k}$? What are (e) $\mathscr{E}$ and (f) the direction if $\vec{B} = (8.00 \times 10^{-2}\ \text{T/m}\cdot\text{s})yt\hat{k}$? What are (g) $\mathscr{E}$ and (h) the direction if $\vec{B} = (3.00 \times 10^{-2}\ \text{T/m}\cdot\text{s})xt\hat{j}$? What are (i) $\mathscr{E}$ and (j) the direction if $\vec{B} = (5.00 \times 10^{-2}\ \text{T/m}\cdot\text{s})yt\hat{i}$?

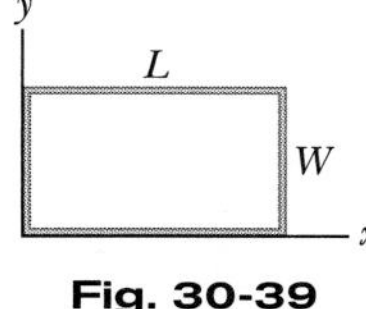

Fig. 30-39 Problem 12.

••13 ILW One hundred turns of (insulated) copper wire are wrapped around a wooden cylindrical core of cross-sectional area $1.20 \times 10^{-3}\ \text{m}^2$. The two ends of the wire are connected to a resistor. The total resistance in the circuit is 13.0 Ω. If an externally applied uniform longitudinal magnetic field in the core changes from 1.60 T in one direction to 1.60 T in the opposite direction, how much charge flows through a point in the circuit during the change?

••14 In Fig. 30-40a, a uniform magnetic field $\vec{B}$ increases in magnitude with time t as given by Fig. 30-40b, where the vertical axis scale is set by $B_s = 9.0$ mT and the horizontal scale is set by $t_s = 3.0$ s. A circular conducting loop of area $8.0 \times 10^{-4}\ \text{m}^2$ lies in the field, in the plane of the page. The amount of charge q passing point A on the loop is given in Fig. 30-40c as a function of t, with the vertical axis scale set by $q_s = 6.0$ mC and the horizontal axis scale again set by $t_s = 3.0$ s. What is the loop's resistance?

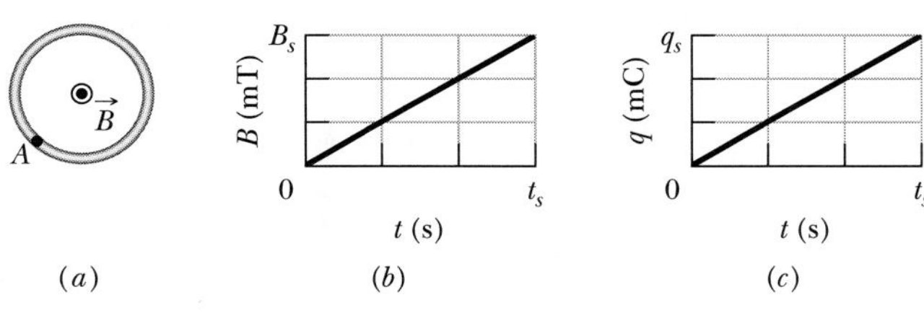

Fig. 30-40 Problem 14.

••15 GO A square wire loop with 2.00 m sides is perpendicular to a uniform magnetic field, with half the area of the loop in the field as shown in Fig. 30-41. The loop contains an ideal battery with emf $\mathscr{E} = 20.0$ V. If the magnitude of the field varies with time according to $B = 0.0420 - 0.870t$, with B in teslas and t in seconds, what are (a) the net emf in the circuit and (b) the direction of the (net) current around the loop?

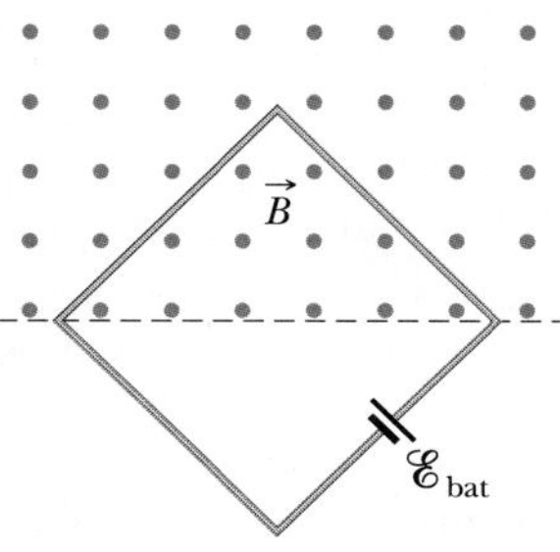

Fig. 30-41 Problem 15.

••16 GO Figure 30-42a shows a wire that forms a rectangle ($W = 20$ cm, $H = 30$ cm) and has a resistance of 5.0 mΩ. Its interior is split into three equal areas, with magnetic fields $\vec{B}_1$, $\vec{B}_2$, and $\vec{B}_3$. The fields are uniform within each region and directly out of or into the page as indicated. Figure 30-42b gives the change in the z components B_z of the three fields with time t; the vertical axis scale is set by $B_s = 4.0\ \mu\text{T}$ and $B_b = -2.5B_s$, and the horizontal axis scale is set by $t_s = 2.0$ s. What are the (a) magnitude and (b) direction of the current induced in the wire?

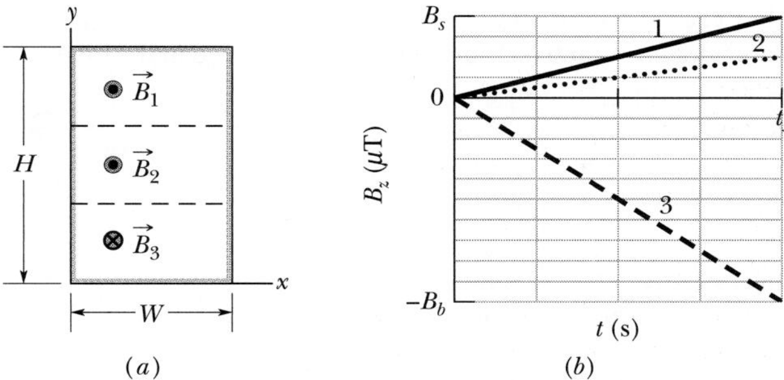

Fig. 30-42 Problem 16.

••17 A small circular loop of area $2.00\ \text{cm}^2$ is placed in the plane of, and concentric with, a large circular loop of radius 1.00 m. The current in the large loop is changed at a constant rate from 200 A to −200 A (a change in direction) in a time of 1.00 s, starting at $t = 0$. What is the magnitude of the magnetic field $\vec{B}$ at the center of the small loop due to the current in the large loop at (a) $t = 0$, (b) $t = 0.500$ s, and (c) $t = 1.00$ s? (d) From $t = 0$ to $t = 1.00$ s, is $\vec{B}$ reversed? Because the inner loop is small, assume $\vec{B}$ is uniform over its area. (e) What emf is induced in the small loop at $t = 0.500$ s?

••18 In Fig. 30-43, two straight conducting rails form a right angle. A conducting bar in contact with the rails starts at the vertex at time $t = 0$ and moves with a constant velocity of 5.20 m/s along them. A magnetic field with $B = 0.350$ T is directed out of the page. Calculate (a) the flux through the triangle formed by the rails and bar at $t = 3.00$ s and (b) the emf around the triangle at that time. (c) If the emf is $\mathscr{E} = at^n$, where a and n are constants, what is the value of n?

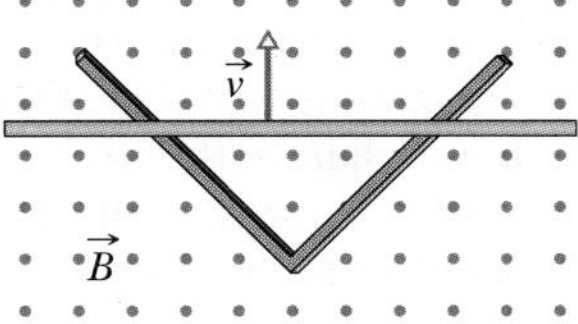

Fig. 30-43 Problem 18.

••19 ILW An electric generator contains a coil of 100 turns of wire, each forming a rectangular loop 50.0 cm by 30.0 cm. The coil is placed entirely in a uniform magnetic field with magnitude $B = 3.50$ T and with $\vec{B}$ initially perpendicular to the coil's plane. What is the maximum value of the emf produced when the coil is spun at 1000 rev/min about an axis perpendicular to $\vec{B}$?

••20 At a certain place, Earth's magnetic field has magnitude $B = 0.590$ gauss and is inclined downward at an angle of 70.0° to the horizontal. A flat horizontal circular coil of wire with a radius of 10.0 cm has 1000 turns and a total resistance of 85.0 Ω. It is connected in series to a meter with 140 Ω resistance. The coil is flipped through a half-revolution about a diameter, so that it is again horizontal. How much charge flows through the meter during the flip?

••21 In Fig. 30-44, a stiff wire bent into a semicircle of radius $a = 2.0$ cm is rotated at constant angular speed 40 rev/s in a uniform 20 mT magnetic field. What are the (a) frequency and (b) amplitude of the emf induced in the loop?

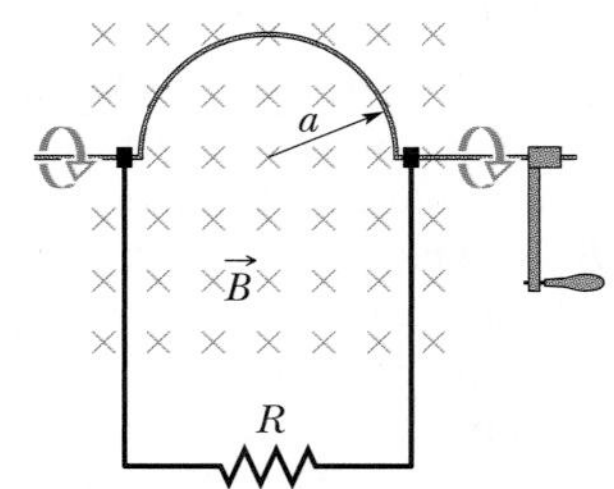

Fig. 30-44 Problem 21.

••22 A rectangular loop (area = 0.15 m^2) turns in a uniform magnetic field, $B = 0.20$ T. When the angle between the field and the normal to the plane of the loop is $\pi/2$ rad and increasing at 0.60 rad/s, what emf is induced in the loop?

••23 SSM Figure 30-45 shows two parallel loops of wire having a common axis. The smaller loop (radius r) is above the larger loop (radius R) by a distance $x \gg R$. Consequently, the magnetic field due to the counterclockwise current i in the larger loop is nearly uniform throughout the smaller loop. Suppose that x is increasing at the constant rate $dx/dt = v$. (a) Find an expression for the magnetic flux through the area of the smaller loop as a function of x. (*Hint:* See Eq. 29-27.) In the smaller loop, find (b) an expression for the induced emf and (c) the direction of the induced current.

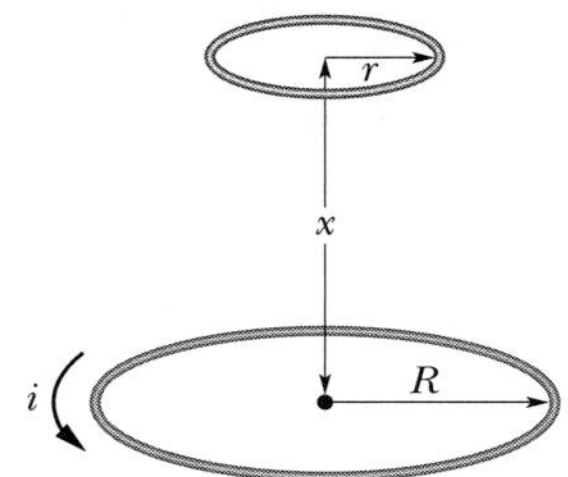

Fig. 30-45 Problem 23.

••24 A wire is bent into three circular segments, each of radius $r = 10$ cm, as shown in Fig. 30-46. Each segment is a quadrant of a circle, ab lying in the xy plane, bc lying in the yz plane, and ca lying in the zx plane. (a) If a uniform magnetic field $\vec{B}$ points in the positive x direction, what is the magnitude of the emf developed in the wire when B increases at the rate of 3.0 mT/s? (b) What is the direction of the current in segment bc?

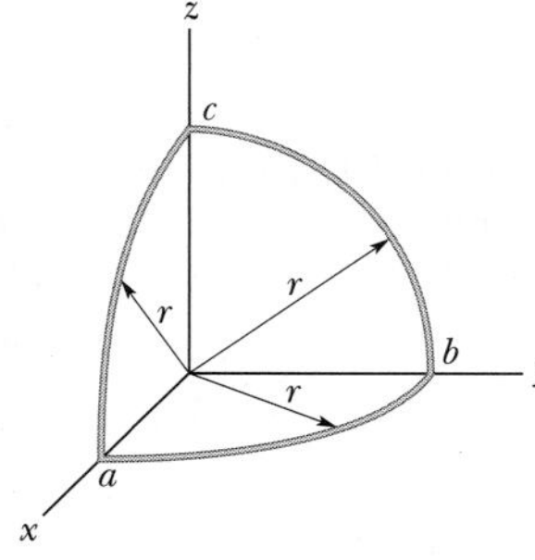

Fig. 30-46 Problem 24.

•••25 GO Two long, parallel copper wires of diameter 2.5 mm carry currents of 10 A in opposite directions. (a) Assuming that their central axes are 20 mm apart, calculate the magnetic flux per meter of wire that exists in the space between those axes. (b) What percentage of this flux lies inside the wires? (c) Repeat part (a) for parallel currents.

•••26 For the wire arrangement in Fig. 30-47, $a = 12.0$ cm and $b = 16.0$ cm. The current in the long straight wire is $i = 4.50t^2 - 10.0t$, where i is in amperes and t is in seconds. (a) Find the emf in the square loop at $t = 3.00$ s. (b) What is the direction of the induced current in the loop?

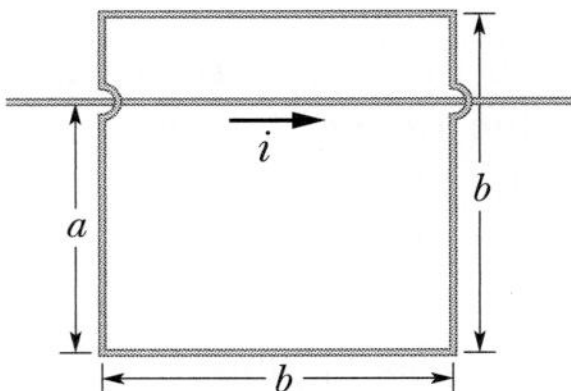

Fig. 30-47 Problem 26.

•••27 ILW As seen in Fig. 30-48, a square loop of wire has sides of length 2.0 cm. A magnetic field is directed out of the page; its magnitude is given by $B = 4.0t^2y$, where B is in teslas, t is in seconds, and y is in meters. At $t = 2.5$ s, what are the (a) magnitude and (b) direction of the emf induced in the loop?

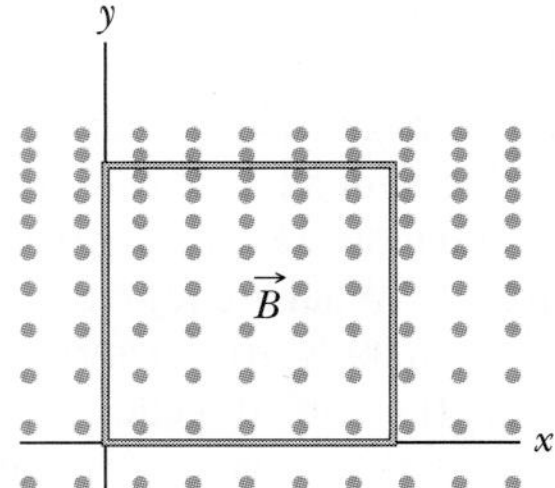

Fig. 30-48 Problem 27.

•••28 In Fig. 30-49, a rectangular loop of wire with length $a = 2.2$ cm, width $b = 0.80$ cm, and resistance $R = 0.40$ mΩ is placed near an infinitely long wire carrying current $i = 4.7$ A. The loop is then moved away from the wire at constant speed $v = 3.2$ mm/s. When the center of the loop is at distance $r = 1.5b$, what are (a) the magnitude of the magnetic flux through the loop and (b) the current induced in the loop?

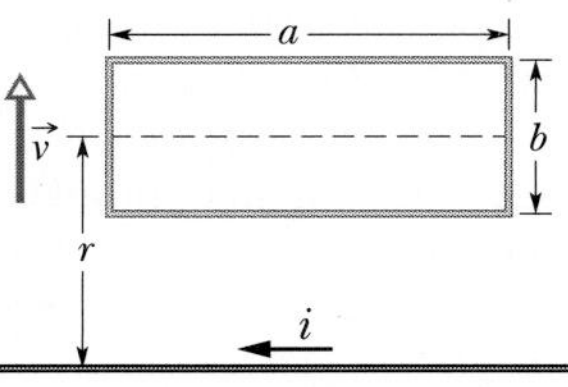

Fig. 30-49 Problem 28.

sec. 30-5 Induction and Energy Transfers

•29 In Fig. 30-50, a metal rod is forced to move with constant velocity $\vec{v}$ along two parallel metal rails, connected with a strip of metal at one end. A magnetic field of magnitude $B = 0.350$ T points out of the page. (a) If the rails are separated by $L = 25.0$ cm and the speed of the rod is 55.0 cm/s, what emf is generated? (b) If the rod has a resistance of 18.0 Ω and the rails and connector have

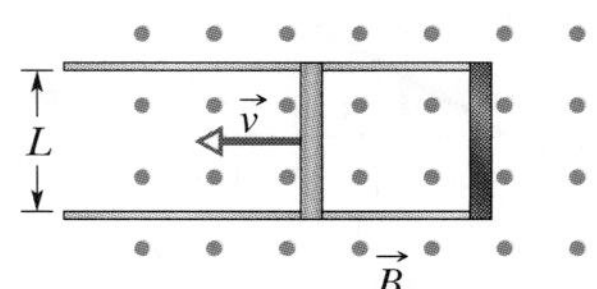

Fig. 30-50 Problems 29 and 35.

negligible resistance, what is the current in the rod? (c) At what rate is energy being transferred to thermal energy?

•30 In Fig. 30-51a, a circular loop of wire is concentric with a solenoid and lies in a plane perpendicular to the solenoid's central axis. The loop has radius 6.00 cm. The solenoid has radius 2.00 cm, consists of 8000 turns/m, and has a current i_{sol} varying with time t as given in Fig. 30-51b, where the vertical axis scale is set by i_s = 1.00 A and the horizontal axis scale is set by t_s = 2.0 s. Figure 30-51c shows, as a function of time, the energy E_{th} that is transferred to thermal energy of the loop; the vertical axis scale is set by E_s = 100.0 nJ. What is the loop's resistance?

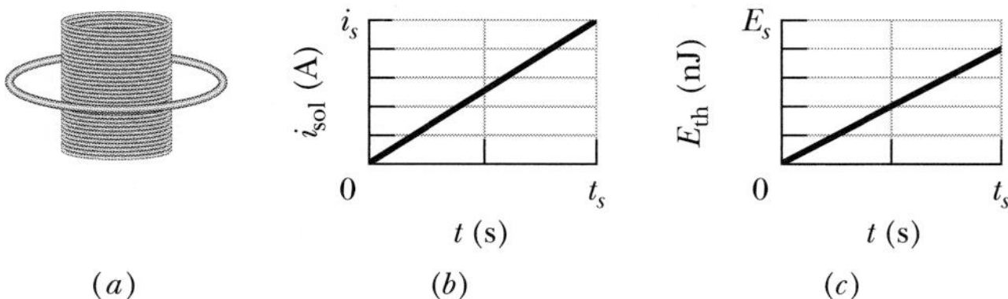

Fig. 30-51 Problem 30.

•31 SSM ILW If 50.0 cm of copper wire (diameter = 1.00 mm) is formed into a circular loop and placed perpendicular to a uniform magnetic field that is increasing at the constant rate of 10.0 mT/s, at what rate is thermal energy generated in the loop?

•32 A loop antenna of area 2.00 cm² and resistance 5.21 $\mu\Omega$ is perpendicular to a uniform magnetic field of magnitude 17.0 μT. The field magnitude drops to zero in 2.96 ms. How much thermal energy is produced in the loop by the change in field?

••33 Figure 30-52 shows a rod of length L = 10.0 cm that is forced to move at constant speed v = 5.00 m/s along horizontal rails. The rod, rails, and connecting strip at the right form a conducting loop. The rod has resistance 0.400 Ω; the rest of the loop has negligible resistance. A current i = 100 A through the long straight wire at distance a = 10.0 mm from the loop sets up a (nonuniform) magnetic field through the loop. Find the (a) emf and (b) current induced in the loop. (c) At what rate is thermal energy generated in the rod? (d) What is the magnitude of the force that must be applied to the rod to make it move at constant speed? (e) At what rate does this force do work on the rod?

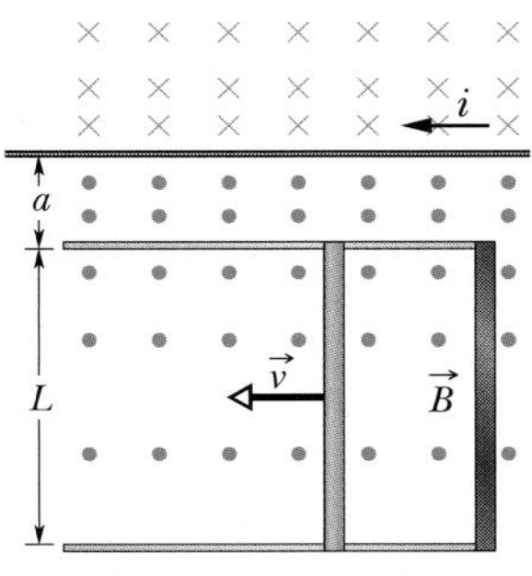

Fig. 30-52 Problem 33.

••34 In Fig. 30-53, a long rectangular conducting loop, of width L, resistance R, and mass m, is hung in a horizontal, uniform magnetic field $\vec{B}$ that is directed into the page and that exists only above line aa. The loop is then dropped; during its fall, it accelerates until it reaches a certain terminal speed v_t. Ignoring air drag, find an expression for v_t.

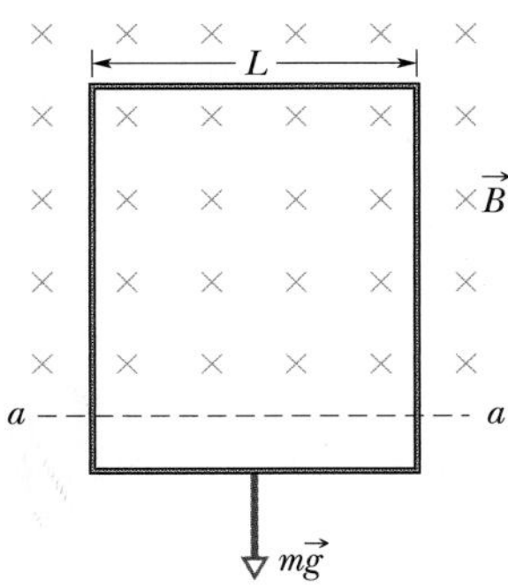

Fig. 30-53 Problem 34.

••35 The conducting rod shown in Fig. 30-50 has length L and is being pulled along horizontal, frictionless conducting rails at a constant velocity $\vec{v}$. The rails are connected at one end with a metal strip. A uniform magnetic field $\vec{B}$, directed out of the page, fills the region in which the rod moves. Assume that L = 10 cm, v = 5.0 m/s, and B = 1.2 T. What are the (a) magnitude and (b) direction (up or down the page) of the emf induced in the rod? What are the (c) size and (d) direction of the current in the conducting loop? Assume that the resistance of the rod is 0.40 Ω and that the resistance of the rails and metal strip is negligibly small. (e) At what rate is thermal energy being generated in the rod? (f) What external force on the rod is needed to maintain $\vec{v}$? (g) At what rate does this force do work on the rod?

sec. 30-6 Induced Electric Fields

•36 Figure 30-54 shows two circular regions R_1 and R_2 with radii r_1 = 20.0 cm and r_2 = 30.0 cm. In R_1 there is a uniform magnetic field of magnitude B_1 = 50.0 mT directed into the page, and in R_2 there is a uniform magnetic field of magnitude B_2 = 75.0 mT directed out of the page (ignore fringing). Both fields are decreasing at the rate of 8.50 mT/s. Calculate $\oint \vec{E} \cdot d\vec{s}$ for (a) path 1, (b) path 2, and (c) path 3.

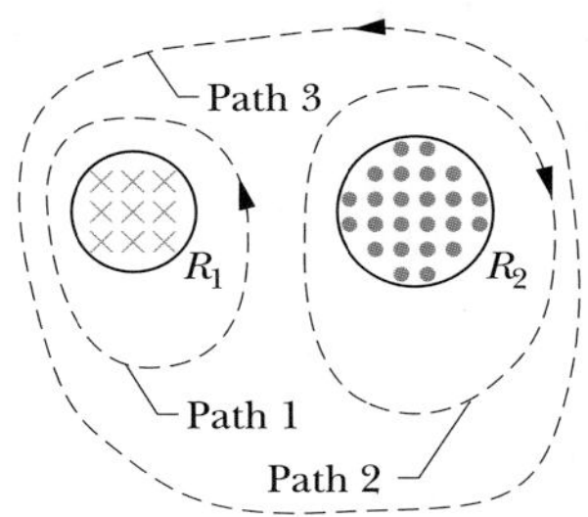

Fig. 30-54 Problem 36.

•37 SSM ILW A long solenoid has a diameter of 12.0 cm. When a current i exists in its windings, a uniform magnetic field of magnitude B = 30.0 mT is produced in its interior. By decreasing i, the field is caused to decrease at the rate of 6.50 mT/s. Calculate the magnitude of the induced electric field (a) 2.20 cm and (b) 8.20 cm from the axis of the solenoid.

••38 GO A circular region in an xy plane is penetrated by a uniform magnetic field in the positive direction of the z axis. The field's magnitude B (in teslas) increases with time t (in seconds) according to $B = at$, where a is a constant. The magnitude E of the electric field set up by that increase in the magnetic field is given by Fig. 30-55 versus radial distance r; the vertical axis scale is set by E_s = 300 μN/C, and the horizontal axis scale is set by r_s = 4.00 cm. Find a.

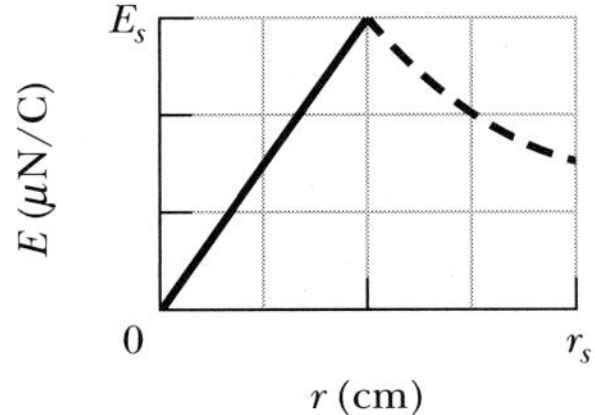

Fig. 30-55 Problem 38.

••39 The magnetic field of a cylindrical magnet that has a pole-face diameter of 3.3 cm can be varied sinusoidally between 29.6 T and 30.0 T at a frequency of 15 Hz. (The current in a wire wrapped around a permanent magnet is varied to give this variation in the net field.) At a radial distance of 1.6 cm, what is the amplitude of the electric field induced by the variation?

sec. 30-7 Inductors and Inductance

•40 The inductance of a closely packed coil of 400 turns is 8.0 mH. Calculate the magnetic flux through the coil when the current is 5.0 mA.

•41 A circular coil has a 10.0 cm radius and consists of 30.0 closely wound turns of wire. An externally produced magnetic field of magnitude 2.60 mT is perpendicular to the coil. (a) If no current is in the coil, what magnetic flux links its turns? (b) When the current in the coil is 3.80 A in a certain direction, the net flux through the coil is found to vanish. What is the inductance of the coil?

••42 Figure 30-56 shows a copper strip of width $W = 16.0$ cm that has been bent to form a shape that consists of a tube of radius $R = 1.8$ cm plus two parallel flat extensions. Current $i = 35$ mA is distributed uniformly across the width so that the tube is effectively a one-turn solenoid. Assume that the magnetic field outside the tube is negligible and the field inside the tube is uniform. What are (a) the magnetic field magnitude inside the tube and (b) the inductance of the tube (excluding the flat extensions)?

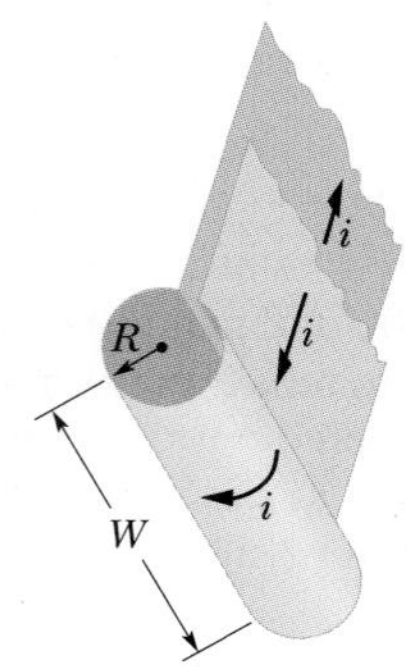

Fig. 30-56 Problem 42.

••43 GO Two identical long wires of radius $a = 1.53$ mm are parallel and carry identical currents in opposite directions. Their center-to-center separation is $d = 14.2$ cm. Neglect the flux within the wires but consider the flux in the region between the wires. What is the inductance per unit length of the wires?

sec. 30-8 Self-Induction

•44 A 12 H inductor carries a current of 2.0 A. At what rate must the current be changed to produce a 60 V emf in the inductor?

•45 At a given instant the current and self-induced emf in an inductor are directed as indicated in Fig. 30-57. (a) Is the current increasing or decreasing? (b) The induced emf is 17 V, and the rate of change of the current is 25 kA/s; find the inductance.

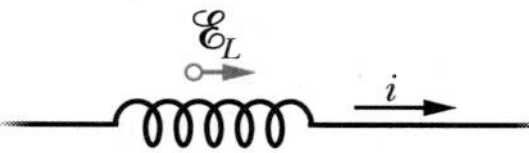

Fig. 30-57 Problem 45.

••46 The current i through a 4.6 H inductor varies with time t as shown by the graph of Fig. 30-58, where the vertical axis scale is set by $i_s = 8.0$ A and the horizontal axis scale is set by $t_s = 6.0$ ms. The inductor has a resistance of 12 Ω. Find the magnitude of the induced emf $\mathscr{E}$ during time intervals (a) 0 to 2 ms, (b) 2 ms to 5 ms, and (c) 5 ms to 6 ms. (Ignore the behavior at the ends of the intervals.)

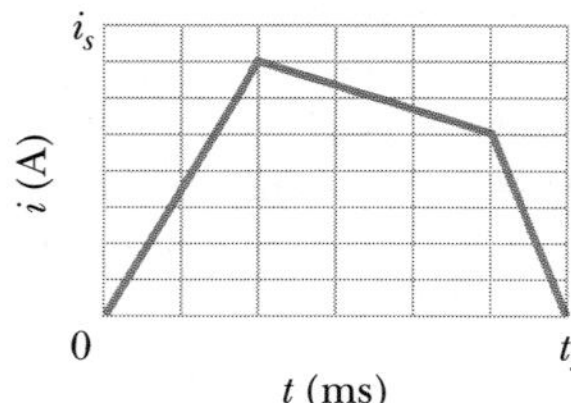

Fig. 30-58 Problem 46.

••47 *Inductors in series.* Two inductors L_1 and L_2 are connected in series and are separated by a large distance so that the magnetic field of one cannot affect the other. (a) Show that the equivalent inductance is given by

$$L_{eq} = L_1 + L_2.$$

(*Hint:* Review the derivations for resistors in series and capacitors in series. Which is similar here?) (b) What is the generalization of (a) for N inductors in series?

••48 *Inductors in parallel.* Two inductors L_1 and L_2 are connected in parallel and separated by a large distance so that the magnetic field of one cannot affect the other. (a) Show that the equivalent inductance is given by

$$\frac{1}{L_{eq}} = \frac{1}{L_1} + \frac{1}{L_2}.$$

(*Hint:* Review the derivations for resistors in parallel and capacitors in parallel. Which is similar here?) (b) What is the generalization of (a) for N inductors in parallel?

••49 The inductor arrangement of Fig. 30-59, with $L_1 = 30.0$ mH, $L_2 = 50.0$ mH, $L_3 = 20.0$ mH, and $L_4 = 15.0$ mH, is to be connected to a varying current source. What is the equivalent inductance of the arrangement? (First see Problems 47 and 48.)

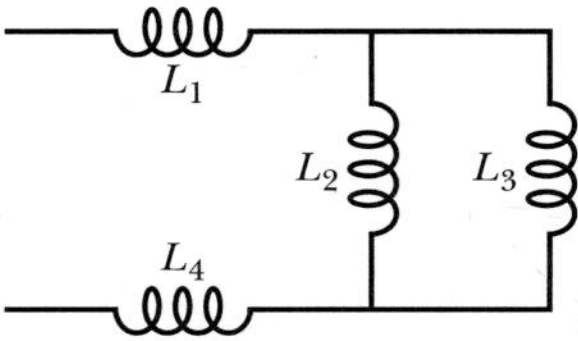

Fig. 30-59 Problem 49.

sec. 30-9 *RL* Circuits

•50 The current in an RL circuit builds up to one-third of its steady-state value in 5.00 s. Find the inductive time constant.

•51 ILW The current in an RL circuit drops from 1.0 A to 10 mA in the first second following removal of the battery from the circuit. If L is 10 H, find the resistance R in the circuit.

•52 The switch in Fig. 30-15 is closed on a at time $t = 0$. What is the ratio $\mathscr{E}_L/\mathscr{E}$ of the inductor's self-induced emf to the battery's emf (a) just after $t = 0$ and (b) at $t = 2.00\tau_L$? (c) At what multiple of τ_L will $\mathscr{E}_L/\mathscr{E} = 0.500$?

•53 SSM A solenoid having an inductance of 6.30 μH is connected in series with a 1.20 kΩ resistor. (a) If a 14.0 V battery is connected across the pair, how long will it take for the current through the resistor to reach 80.0% of its final value? (b) What is the current through the resistor at time $t = 1.0\tau_L$?

•54 In Fig. 30-60, $\mathscr{E} = 100$ V, $R_1 = 10.0$ Ω, $R_2 = 20.0$ Ω, $R_3 = 30.0$ Ω, and $L = 2.00$ H. Immediately after switch S is closed, what are (a) i_1 and (b) i_2? (Let currents in the indicated directions have positive values and currents in the opposite directions have negative values.) A long time later, what are (c) i_1 and (d) i_2? The switch is then reopened. Just then, what are (e) i_1 and (f) i_2? A long time later, what are (g) i_1 and (h) i_2?

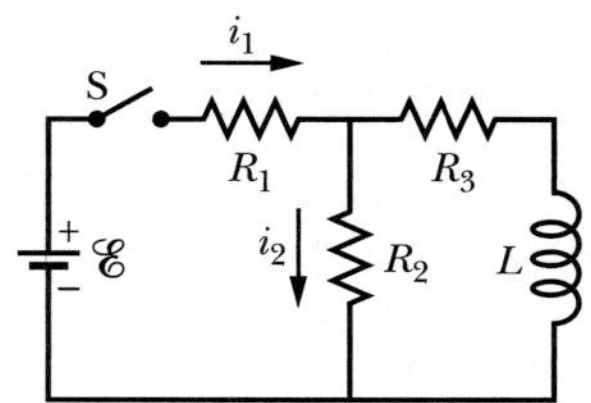

Fig. 30-60 Problem 54.

•55 SSM A battery is connected to a series RL circuit at time $t = 0$. At what multiple of τ_L will the current be 0.100% less than its equilibrium value?

•56 In Fig. 30-61, the inductor has 25 turns and the ideal battery has an emf of 16 V. Figure 30-62 gives the magnetic flux Φ through each turn versus the current i through the inductor. The vertical axis scale is set by $\Phi_s = 4.0 \times 10^{-4}\ \text{T}\cdot\text{m}^2$, and the horizontal axis scale is set by $i_s = 2.00$ A. If switch S is closed at time $t = 0$, at what rate di/dt will the current be changing at $t = 1.5\tau_L$?

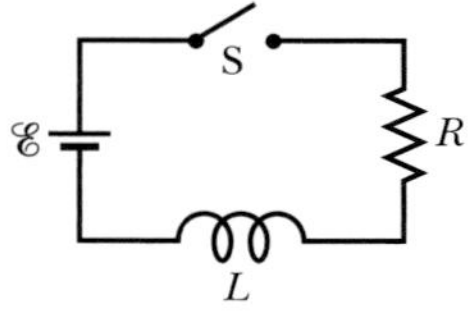

Fig. 30-61 Problems 56, 80, 83, and 93.

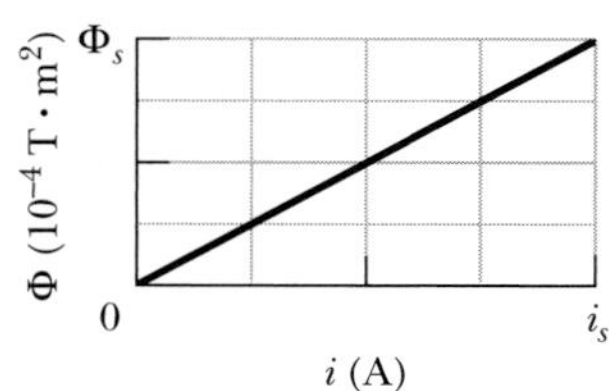

Fig. 30-62 Problem 56.

••57 GO In Fig. 30-63, $R = 15\ \Omega$, $L = 5.0$ H, the ideal battery has $\mathscr{E} = 10$ V, and the fuse in the upper branch is an ideal 3.0 A fuse. It has zero resistance as long as the current through it remains less than 3.0 A. If the current reaches 3.0 A, the fuse "blows" and thereafter has infinite resistance. Switch S is closed at time $t = 0$. (a) When does the fuse blow? (*Hint:* Equation 30-41 does not apply. Rethink Eq. 30-39.) (b) Sketch a graph of the current i through the inductor as a function of time. Mark the time at which the fuse blows.

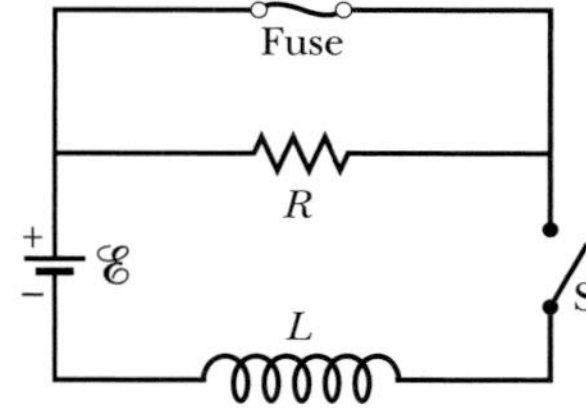

Fig. 30-63 Problem 57.

••58 GO Suppose the emf of the battery in the circuit shown in Fig. 30-16 varies with time t so that the current is given by $i(t) = 3.0 + 5.0t$, where i is in amperes and t is in seconds. Take $R = 4.0\ \Omega$ and $L = 6.0$ H, and find an expression for the battery emf as a function of t. (*Hint:* Apply the loop rule.)

•••59 SSM WWW In Fig. 30-64, after switch S is closed at time $t = 0$, the emf of the source is automatically adjusted to maintain a constant current i through S. (a) Find the current through the inductor as a function of time. (b) At what time is the current through the resistor equal to the current through the inductor?

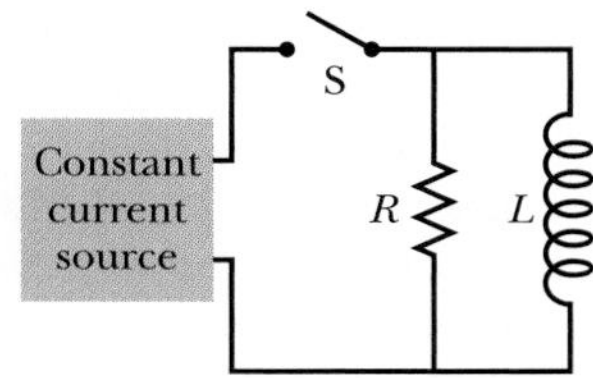

Fig. 30-64 Problem 59.

•••60 A wooden toroidal core with a square cross section has an inner radius of 10 cm and an outer radius of 12 cm. It is wound with one layer of wire (of diameter 1.0 mm and resistance per meter 0.020 Ω/m). What are (a) the inductance and (b) the inductive time constant of the resulting toroid? Ignore the thickness of the insulation on the wire.

sec. 30-10 Energy Stored in a Magnetic Field

•61 SSM A coil is connected in series with a 10.0 kΩ resistor. An ideal 50.0 V battery is applied across the two devices, and the current reaches a value of 2.00 mA after 5.00 ms. (a) Find the inductance of the coil. (b) How much energy is stored in the coil at this same moment?

•62 A coil with an inductance of 2.0 H and a resistance of 10 Ω is suddenly connected to an ideal battery with $\mathscr{E} = 100$ V. At 0.10 s after the connection is made, what is the rate at which (a) energy is being stored in the magnetic field, (b) thermal energy is appearing in the resistance, and (c) energy is being delivered by the battery?

•63 ILW At $t = 0$, a battery is connected to a series arrangement of a resistor and an inductor. If the inductive time constant is 37.0 ms, at what time is the rate at which energy is dissipated in the resistor equal to the rate at which energy is stored in the inductor's magnetic field?

•64 At $t = 0$, a battery is connected to a series arrangement of a resistor and an inductor. At what multiple of the inductive time constant will the energy stored in the inductor's magnetic field be 0.500 its steady-state value?

••65 GO For the circuit of Fig. 30-16, assume that $\mathscr{E} = 10.0$ V, $R = 6.70\ \Omega$, and $L = 5.50$ H. The ideal battery is connected at time $t = 0$. (a) How much energy is delivered by the battery during the first 2.00 s? (b) How much of this energy is stored in the magnetic field of the inductor? (c) How much of this energy is dissipated in the resistor?

sec. 30-11 Energy Density of a Magnetic Field

•66 A circular loop of wire 50 mm in radius carries a current of 100 A. Find the (a) magnetic field strength and (b) energy density at the center of the loop.

•67 SSM A solenoid that is 85.0 cm long has a cross-sectional area of 17.0 cm². There are 950 turns of wire carrying a current of 6.60 A. (a) Calculate the energy density of the magnetic field inside the solenoid. (b) Find the total energy stored in the magnetic field there (neglect end effects).

•68 A toroidal inductor with an inductance of 90.0 mH encloses a volume of 0.0200 m³. If the average energy density in the toroid is 70.0 J/m³, what is the current through the inductor?

•69 ILW What must be the magnitude of a uniform electric field if it is to have the same energy density as that possessed by a 0.50 T magnetic field?

••70 GO Figure 30-65a shows, in cross section, two wires that are straight, parallel, and very long. The ratio i_1/i_2 of the current carried by wire 1 to that carried by wire 2 is 1/3. Wire 1 is fixed in place. Wire 2 can be moved along the positive side of the x axis so as to change the magnetic energy density u_B set up by the two currents at the origin. Figure 30-65b gives u_B as a function of the position x of wire 2. The curve has an asymptote of $u_B = 1.96\ \text{nJ/m}^3$ as $x \to \infty$, and the horizontal axis scale is set by $x_s = 60.0$ cm. What is the value of (a) i_1 and (b) i_2?

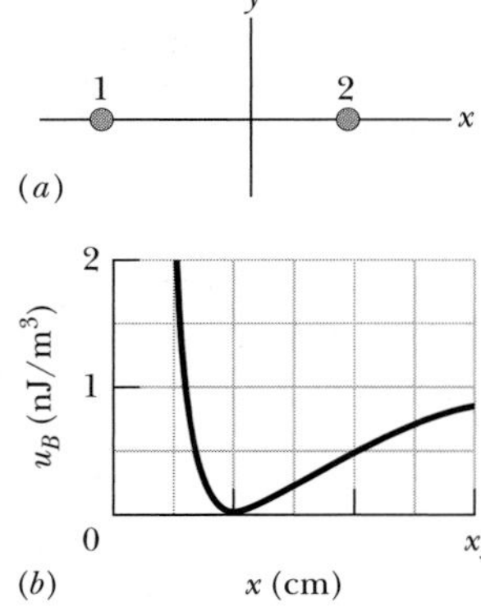

Fig. 30-65 Problem 70.

••71 A length of copper wire carries a current of 10 A uniformly distributed through its cross section. Calculate the energy density of (a) the magnetic field and (b) the electric field at the surface of the wire. The wire diameter is 2.5 mm, and its resistance per unit length is 3.3 Ω/km.

sec. 30-12 Mutual Induction

•72 Coil 1 has $L_1 = 25$ mH and $N_1 = 100$ turns. Coil 2 has $L_2 = 40$ mH and $N_2 = 200$ turns. The coils are fixed in place; their mutual inductance M is 3.0 mH. A 6.0 mA current in coil 1 is changing at the rate of 4.0 A/s. (a) What magnetic flux Φ_{12} links coil 1, and (b) what self-induced emf appears in that coil? (c) What magnetic flux Φ_{21} links coil 2, and (d) what mutually induced emf appears in that coil?

•73 SSM Two coils are at fixed locations. When coil 1 has no current and the current in coil 2 increases at the rate 15.0 A/s, the emf in coil 1 is 25.0 mV. (a) What is their mutual inductance? (b) When coil 2 has no current and coil 1 has a current of 3.60 A, what is the flux linkage in coil 2?

•74 Two solenoids are part of the spark coil of an automobile. When the current in one solenoid falls from 6.0 A to zero in 2.5 ms, an emf of 30 kV is induced in the other solenoid. What is the mutual inductance M of the solenoids?

••75 ILW A rectangular loop of N closely packed turns is positioned near a long straight wire as shown in Fig. 30-66. What is the mutual inductance M for the loop–wire combination if $N = 100$, $a = 1.0$ cm, $b = 8.0$ cm, and $l = 30$ cm?

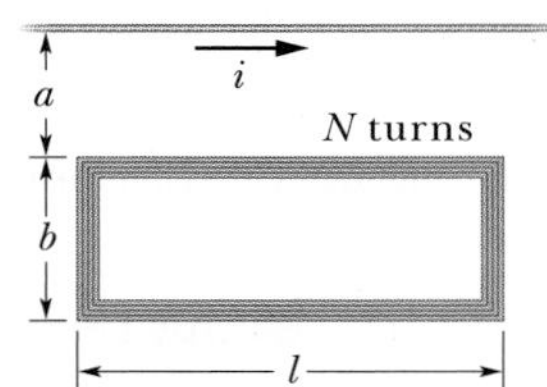

Fig. 30-66 Problem 75.

••76 A coil C of N turns is placed around a long solenoid S of radius R and n turns per unit length, as in Fig. 30-67. (a) Show that the mutual inductance for the coil–solenoid combination is given by $M = \mu_0 \pi R^2 nN$. (b) Explain why M does not depend on the shape, size, or possible lack of close packing of the coil.

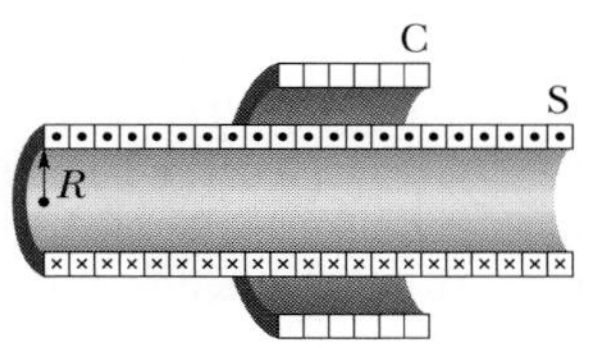

Fig. 30-67 Problem 76.

••77 SSM Two coils connected as shown in Fig. 30-68 separately have inductances L_1 and L_2. Their mutual inductance is M. (a) Show that this combination can be replaced by a single coil of equivalent inductance given by

$$L_{eq} = L_1 + L_2 + 2M.$$

(b) How could the coils in Fig. 30-68 be reconnected to yield an equivalent inductance of

$$L_{eq} = L_1 + L_2 - 2M?$$

(This problem is an extension of Problem 47, but the requirement that the coils be far apart has been removed.)

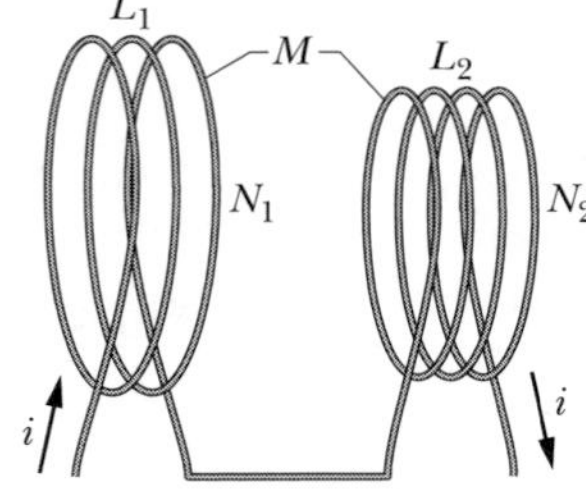

Fig. 30-68 Problem 77.

Additional Problems

78 At time $t = 0$, a 12.0 V potential difference is suddenly applied to the leads of a coil of inductance 23.0 mH and a certain resistance R. At time $t = 0.150$ ms, the current through the inductor is changing at the rate of 280 A/s. Evaluate R.

79 SSM In Fig. 30-69, the battery is ideal and $\mathscr{E} = 10$ V, $R_1 = 5.0\ \Omega$, $R_2 = 10\ \Omega$, and $L = 5.0$ H. Switch S is closed at time $t = 0$. Just afterwards, what are (a) i_1, (b) i_2, (c) the current i_S through the switch, (d) the potential difference V_2 across resistor 2, (e) the potential difference V_L across the inductor, and (f) the rate of change di_2/dt? A long time later, what are (g) i_1, (h) i_2, (i) i_S, (j) V_2, (k) V_L, and (l) di_2/dt?

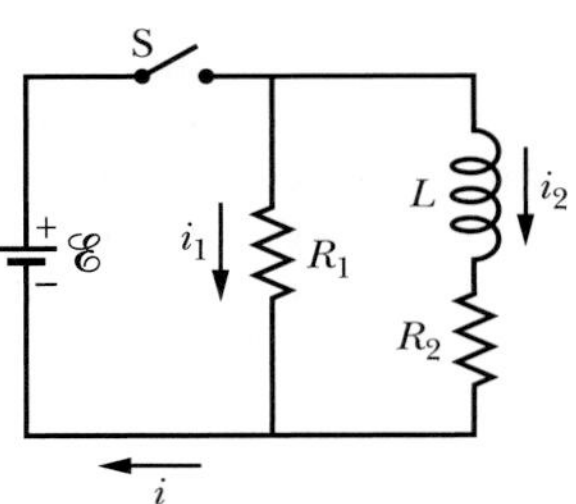

Fig. 30-69 Problem 79.

80 In Fig. 30-61, $R = 4.0$ kΩ, $L = 8.0\ \mu$H, and the ideal battery has $\mathscr{E} = 20$ V. How long after switch S is closed is the current 2.0 mA?

81 SSM Figure 30-70a shows a rectangular conducting loop of resistance $R = 0.020\ \Omega$, height $H = 1.5$ cm, and length $D = 2.5$ cm being pulled at constant speed $v = 40$ cm/s through two regions of uniform magnetic field. Figure 30-70b gives the current i induced in the loop as a function of the position x of the right side of the loop. The vertical axis scale is set by $i_s = 3.0\ \mu$A. For example, a current equal to i_s is induced clockwise as the loop enters region 1. What are the (a) magnitude and (b) direction (into or out of the page) of the magnetic field in region 1? What are the (c) magnitude and (d) direction of the magnetic field in region 2?

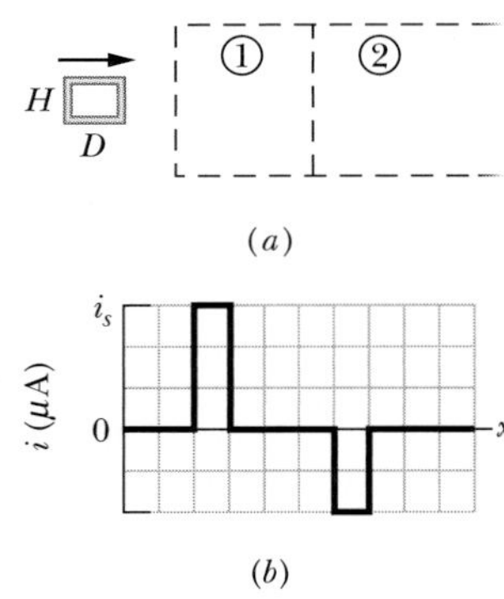

Fig. 30-70 Problem 81.

82 A uniform magnetic field $\vec{B}$ is perpendicular to the plane of a circular wire loop of radius r. The magnitude of the field varies with time according to $B = B_0 e^{-t/\tau}$, where B_0 and τ are constants. Find an expression for the emf in the loop as a function of time.

83 Switch S in Fig. 30-61 is closed at time $t = 0$, initiating the buildup of current in the 15.0 mH inductor and the 20.0 Ω resistor. At what time is the emf across the inductor equal to the potential difference across the resistor?

84 Figure 30-71a shows two concentric circular regions in which uniform magnetic fields can change. Region 1, with radius $r_1 = 1.0$ cm, has an outward magnetic field $\vec{B}_1$ that is increasing in magnitude. Region 2, with radius $r_2 = 2.0$ cm, has an outward magnetic field $\vec{B}_2$ that may also be changing. Imagine that a conducting ring of radius R is centered on the two regions and then the emf $\mathscr{E}$ around the ring is determined. Figure 30-71b gives emf $\mathscr{E}$ as a

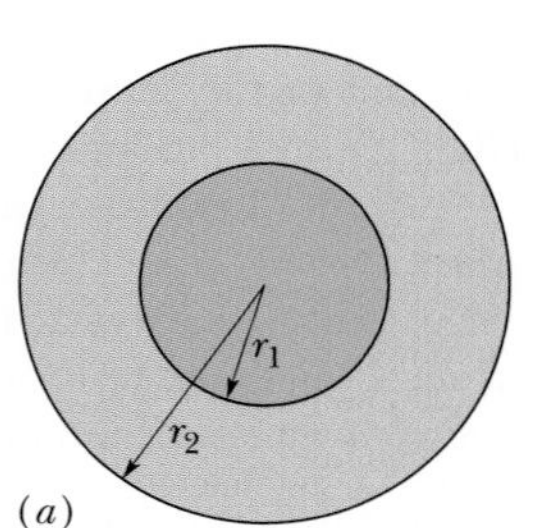

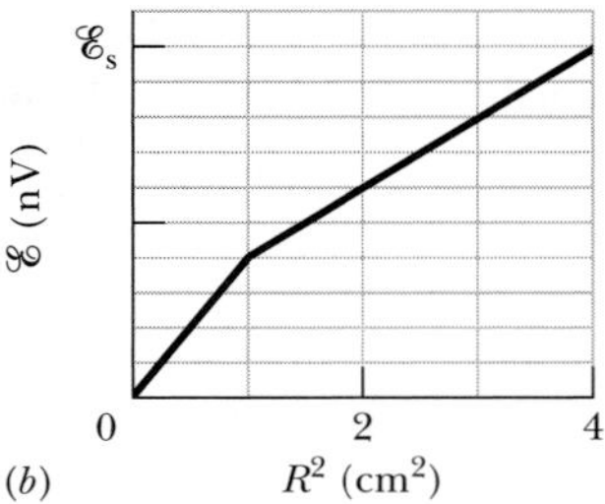

Fig. 30-71 Problem 84.

function of the square R^2 of the ring's radius, to the outer edge of region 2. The vertical axis scale is set by $\mathscr{E}_s = 20.0$ nV. What are the rates (a) dB_1/dt and (b) dB_2/dt? (c) Is the magnitude of $\vec{B}_2$ increasing, decreasing, or remaining constant?

85 SSM Figure 30-72 shows a uniform magnetic field $\vec{B}$ confined to a cylindrical volume of radius R. The magnitude of $\vec{B}$ is decreasing at a constant rate of 10 mT/s. In unit-vector notation, what is the initial acceleration of an electron released at (a) point a (radial distance $r = 5.0$ cm), (b) point b ($r = 0$), and (c) point c ($r = 5.0$ cm)?

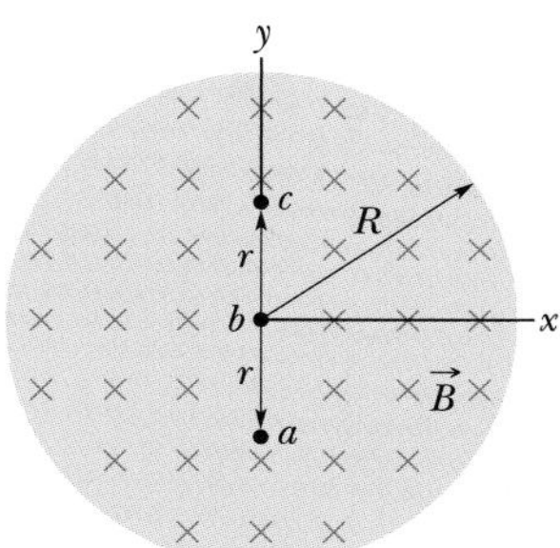

Fig. 30-72 Problem 85.

86 GO In Fig. 30-73a, switch S has been closed on A long enough to establish a steady current in the inductor of inductance $L_1 = 5.00$ mH and the resistor of resistance $R_1 = 25.0\ \Omega$. Similarly, in Fig. 30-73b, switch S has been closed on A long enough to establish a steady current in the inductor of inductance $L_2 = 3.00$ mH and the resistor of resistance $R_2 = 30.0\ \Omega$. The ratio Φ_{02}/Φ_{01} of the magnetic flux through a turn in inductor 2 to that in inductor 1 is 1.50. At time $t = 0$, the two switches are closed on B. At what time t is the flux through a turn in the two inductors equal?

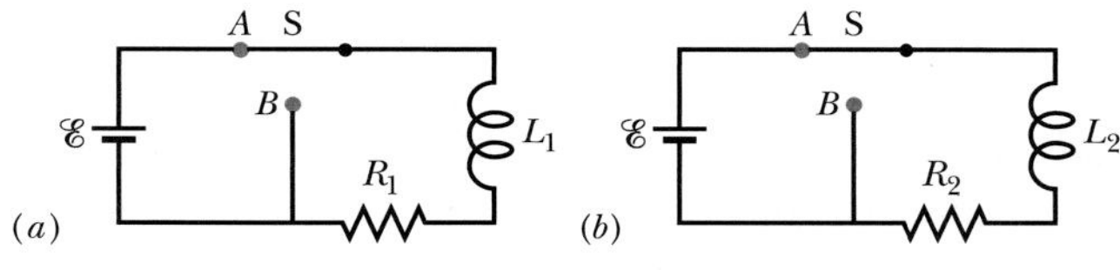

Fig. 30-73 Problem 86.

87 SSM A square wire loop 20 cm on a side, with resistance 20 mΩ, has its plane normal to a uniform magnetic field of magnitude $B = 2.0$ T. If you pull two opposite sides of the loop away from each other, the other two sides automatically draw toward each other, reducing the area enclosed by the loop. If the area is reduced to zero in time $\Delta t = 0.20$ s, what are (a) the average emf and (b) the average current induced in the loop during Δt?

88 A coil with 150 turns has a magnetic flux of 50.0 nT·m^2 through each turn when the current is 2.00 mA. (a) What is the inductance of the coil? What are the (b) inductance and (c) flux through each turn when the current is increased to 4.00 mA? (d) What is the maximum emf $\mathscr{E}$ across the coil when the current through it is given by $i = (3.00 \text{ mA}) \cos(377t)$, with t in seconds?

89 A coil with an inductance of 2.0 H and a resistance of 10 Ω is suddenly connected to an ideal battery with $\mathscr{E} = 100$ V. (a) What is the equilibrium current? (b) How much energy is stored in the magnetic field when this current exists in the coil?

90 How long would it take, following the removal of the battery, for the potential difference across the resistor in an RL circuit (with $L = 2.00$ H, $R = 3.00\ \Omega$) to decay to 10.0% of its initial value?

91 SSM In the circuit of Fig. 30-74, $R_1 = 20$ kΩ, $R_2 = 20\ \Omega$, $L = 50$ mH, and the ideal battery has $\mathscr{E} = 40$ V. Switch S has been open for a long time when it is closed at time $t = 0$. Just after the switch is closed, what are (a) the current i_{bat} through the battery and (b) the rate di_{bat}/dt? At $t = 3.0\ \mu$s, what are (c) i_{bat} and (d) di_{bat}/dt? A long time later, what are (e) i_{bat} and (f) di_{bat}/dt?

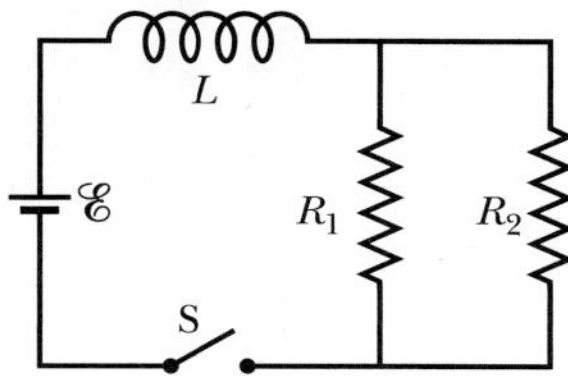

Fig. 30-74 Problem 91.

92 The flux linkage through a certain coil of 0.75 Ω resistance would be 26 mWb if there were a current of 5.5 A in it. (a) Calculate the inductance of the coil. (b) If a 6.0 V ideal battery were suddenly connected across the coil, how long would it take for the current to rise from 0 to 2.5 A?

93 In Fig. 30-61, a 12.0 V ideal battery, a 20.0 Ω resistor, and an inductor are connected by a switch at time $t = 0$. At what rate is the battery transferring energy to the inductor's field at $t = 1.61\tau_L$?

94 A long cylindrical solenoid with 100 turns/cm has a radius of 1.6 cm. Assume that the magnetic field it produces is parallel to its axis and is uniform in its interior. (a) What is its inductance per meter of length? (b) If the current changes at the rate of 13 A/s, what emf is induced per meter?

95 In Fig. 30-75, $R_1 = 8.0\ \Omega$, $R_2 = 10\ \Omega$, $L_1 = 0.30$ H, $L_2 = 0.20$ H, and the ideal battery has $\mathscr{E} = 6.0$ V. (a) Just after switch S is closed, at what rate is the current in inductor 1 changing? (b) When the circuit is in the steady state, what is the current in inductor 1?

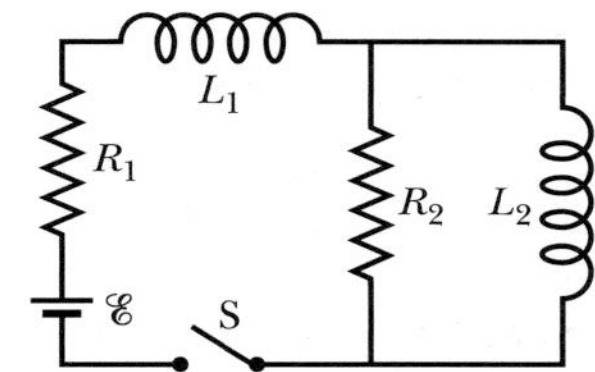

Fig. 30-75 Problem 95.

96 A square loop of wire is held in a uniform 0.24 T magnetic field directed perpendicular to the plane of the loop. The length of each side of the square is decreasing at a constant rate of 5.0 cm/s. What emf is induced in the loop when the length is 12 cm?

97 At time $t = 0$, a 45 V potential difference is suddenly applied to the leads of a coil with inductance $L = 50$ mH and resistance $R = 180\ \Omega$. At what rate is the current through the coil increasing at $t = 1.2$ ms?

98 The inductance of a closely wound coil is such that an emf of 3.00 mV is induced when the current changes at the rate of 5.00 A/s. A steady current of 8.00 A produces a magnetic flux of 40.0 μWb through each turn. (a) Calculate the inductance of the coil. (b) How many turns does the coil have?

CHAPTER

31 ELECTROMAGNETIC OSCILLATIONS AND ALTERNATING CURRENT

31-1 WHAT IS PHYSICS?

We have explored the basic physics of electric and magnetic fields and how energy can be stored in capacitors and inductors. We next turn to the associated applied physics, in which the energy stored in one location can be transferred to another location so that it can be put to use. For example, energy produced at a power plant can show up at your home to run a computer. The total worth of this applied physics is now so high that its estimation is almost impossible. Indeed, modern civilization would be impossible without this applied physics.

In most parts of the world, electrical energy is transferred not as a direct current but as a sinusoidally oscillating current (alternating current, or ac). The challenge to both physicists and engineers is to design ac systems that transfer energy efficiently and to build appliances that make use of that energy.

In our discussion of electrically oscillating systems in this chapter, our first step is to examine oscillations in a simple circuit consisting of inductance L and capacitance C.

31-2 *LC* Oscillations, Qualitatively

Of the three circuit elements resistance R, capacitance C, and inductance L, we have so far discussed the series combinations RC (in Section 27-9) and RL (in Section 30-9). In these two kinds of circuit we found that the charge, current, and potential difference grow and decay exponentially. The time scale of the growth or decay is given by a *time constant* τ, which is either capacitive or inductive.

We now examine the remaining two-element circuit combination LC. You will see that in this case the charge, current, and potential difference do not decay exponentially with time but vary sinusoidally (with period T and angular frequency ω). The resulting oscillations of the capacitor's electric field and the inductor's magnetic field are said to be **electromagnetic oscillations.** Such a circuit is said to oscillate.

Parts a through h of Fig. 31-1 show succeeding stages of the oscillations in a simple LC circuit. From Eq. 25-21, the energy stored in the electric field of the capacitor at any time is

$$U_E = \frac{q^2}{2C}, \qquad (31\text{-}1)$$

where q is the charge on the capacitor at that time. From Eq. 30-49, the energy stored in the magnetic field of the inductor at any time is

$$U_B = \frac{Li^2}{2}, \tag{31-2}$$

where i is the current through the inductor at that time.

We now adopt the convention of representing *instantaneous values* of the electrical quantities of a sinusoidally oscillating circuit with small letters, such as q, and the *amplitudes* of those quantities with capital letters, such as Q. With this convention in mind, let us assume that initially the charge q on the capacitor in Fig. 31-1 is at its maximum value Q and that the current i through the inductor is zero. This initial state of the circuit is shown in Fig. 31-1*a*. The bar graphs for energy included there indicate that at this instant, with zero current through the inductor and maximum charge on the capacitor, the energy U_B of the magnetic field is zero and the energy U_E of the electric field is a maximum. As the circuit oscillates, energy shifts back and forth from one type of stored energy to the other, but the total amount is conserved.

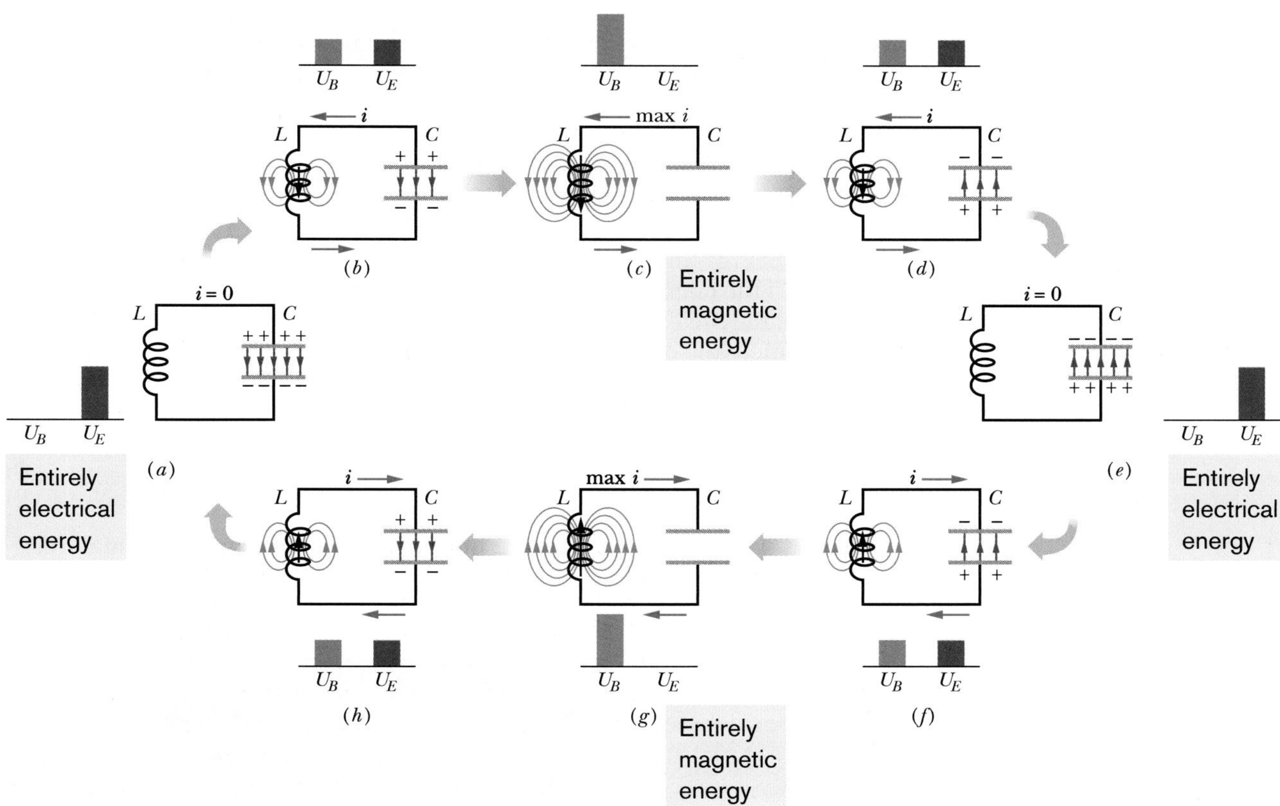

Fig. 31-1 Eight stages in a single cycle of oscillation of a resistanceless *LC* circuit. The bar graphs by each figure show the stored magnetic and electrical energies. The magnetic field lines of the inductor and the electric field lines of the capacitor are shown. (*a*) Capacitor with maximum charge, no current. (*b*) Capacitor discharging, current increasing. (*c*) Capacitor fully discharged, current maximum. (*d*) Capacitor charging but with polarity opposite that in (*a*), current decreasing. (*e*) Capacitor with maximum charge having polarity opposite that in (*a*), no current. (*f*) Capacitor discharging, current increasing with direction opposite that in (*b*). (*g*) Capacitor fully discharged, current maximum. (*h*) Capacitor charging, current decreasing.

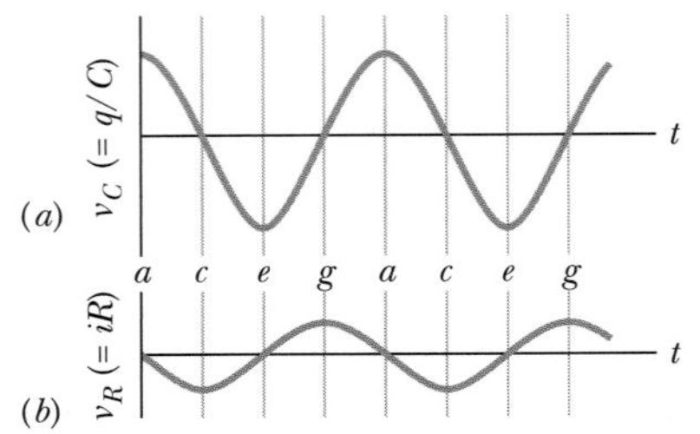

Fig. 31-2 (*a*) The potential difference across the capacitor of the circuit of Fig. 31-1 as a function of time. This quantity is proportional to the charge on the capacitor. (*b*) A potential proportional to the current in the circuit of Fig. 31-1. The letters refer to the correspondingly labeled oscillation stages in Fig. 31-1.

The capacitor now starts to discharge through the inductor, positive charge carriers moving counterclockwise, as shown in Fig. 31-1*b*. This means that a current i, given by dq/dt and pointing down in the inductor, is established. As the capacitor's charge decreases, the energy stored in the electric field within the capacitor also decreases. This energy is transferred to the magnetic field that appears around the inductor because of the current i that is building up there. Thus, the electric field decreases and the magnetic field builds up as energy is transferred from the electric field to the magnetic field.

The capacitor eventually loses all its charge (Fig. 31-1*c*) and thus also loses its electric field and the energy stored in that field. The energy has then been fully transferred to the magnetic field of the inductor. The magnetic field is then at its maximum magnitude, and the current through the inductor is then at its maximum value I.

Although the charge on the capacitor is now zero, the counterclockwise current must continue because the inductor does not allow it to change suddenly to zero. The current continues to transfer positive charge from the top plate to the bottom plate through the circuit (Fig. 31-1*d*). Energy now flows from the inductor back to the capacitor as the electric field within the capacitor builds up again. The current gradually decreases during this energy transfer. When, eventually, the energy has been transferred completely back to the capacitor (Fig. 31-1*e*), the current has decreased to zero (momentarily). The situation of Fig. 31-1*e* is like the initial situation, except that the capacitor is now charged oppositely.

The capacitor then starts to discharge again but now with a clockwise current (Fig. 31-1*f*). Reasoning as before, we see that the clockwise current builds to a maximum (Fig. 31-1*g*) and then decreases (Fig. 31-1*h*), until the circuit eventually returns to its initial situation (Fig. 31-1*a*). The process then repeats at some frequency f and thus at an angular frequency $\omega = 2\pi f$. In the ideal LC circuit with no resistance, there are no energy transfers other than that between the electric field of the capacitor and the magnetic field of the inductor. Because of the conservation of energy, the oscillations continue indefinitely. The oscillations need not begin with the energy all in the electric field; the initial situation could be any other stage of the oscillation.

To determine the charge q on the capacitor as a function of time, we can put in a voltmeter to measure the time-varying potential difference (or *voltage*) v_C that exists across the capacitor C. From Eq. 25-1 we can write

$$v_C = \left(\frac{1}{C}\right) q,$$

which allows us to find q. To measure the current, we can connect a small resistance R in series with the capacitor and inductor and measure the time-varying

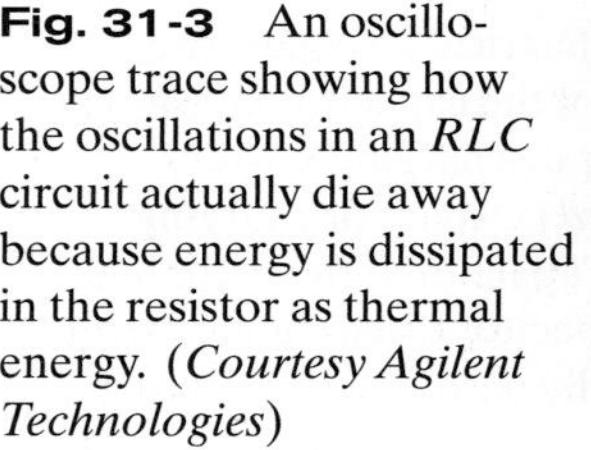

Fig. 31-3 An oscilloscope trace showing how the oscillations in an *RLC* circuit actually die away because energy is dissipated in the resistor as thermal energy. (*Courtesy Agilent Technologies*)

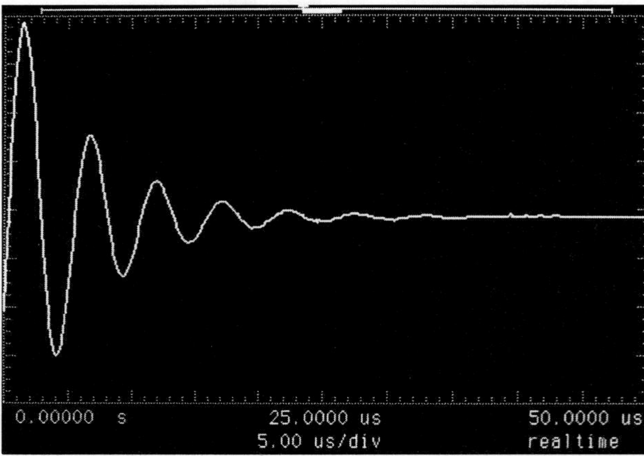

potential difference v_R across it; v_R is proportional to i through the relation

$$v_R = iR.$$

We assume here that R is so small that its effect on the behavior of the circuit is negligible. The variations in time of v_C and v_R, and thus of q and i, are shown in Fig. 31-2. All four quantities vary sinusoidally.

In an actual LC circuit, the oscillations will not continue indefinitely because there is always some resistance present that will drain energy from the electric and magnetic fields and dissipate it as thermal energy (the circuit may become warmer). The oscillations, once started, will die away as Fig. 31-3 suggests. Compare this figure with Fig. 15-15, which shows the decay of mechanical oscillations caused by frictional damping in a block–spring system.

CHECKPOINT 1

A charged capacitor and an inductor are connected in series at time $t = 0$. In terms of the period T of the resulting oscillations, determine how much later the following reach their maximum value: (a) the charge on the capacitor; (b) the voltage across the capacitor, with its original polarity; (c) the energy stored in the electric field; and (d) the current.

31-3 The Electrical-Mechanical Analogy

Let us look a little closer at the analogy between the oscillating LC system of Fig. 31-1 and an oscillating block–spring system. Two kinds of energy are involved in the block–spring system. One is potential energy of the compressed or extended spring; the other is kinetic energy of the moving block. These two energies are given by the formulas in the first energy column in Table 31-1.

The table also shows, in the second energy column, the two kinds of energy involved in LC oscillations. By looking across the table, we can see an analogy between the forms of the two pairs of energies—the mechanical energies of the block–spring system and the electromagnetic energies of the LC oscillator. The equations for v and i at the bottom of the table help us see the details of the analogy. They tell us that q corresponds to x and i corresponds to v (in both equations, the former is differentiated to obtain the latter). These correspondences then suggest that, in the energy expressions, $1/C$ corresponds to k and L corresponds to m. Thus,

q corresponds to x, $1/C$ corresponds to k,
i corresponds to v, and L corresponds to m.

These correspondences suggest that in an LC oscillator, the capacitor is mathematically like the spring in a block–spring system and the inductor is like the block.

In Section 15-3 we saw that the angular frequency of oscillation of a (frictionless) block–spring system is

$$\omega = \sqrt{\frac{k}{m}} \quad \text{(block–spring system).} \tag{31-3}$$

The correspondences listed above suggest that to find the angular frequency of oscillation for an ideal (resistanceless) LC circuit, k should be replaced by $1/C$ and m by L, yielding

$$\omega = \frac{1}{\sqrt{LC}} \quad (LC \text{ circuit}). \tag{31-4}$$

Table 31-1

Comparison of the Energy in Two Oscillating Systems

Block–Spring System		LC Oscillator	
Element	Energy	Element	Energy
Spring	Potential, $\frac{1}{2}kx^2$	Capacitor	Electrical, $\frac{1}{2}(1/C)q^2$
Block	Kinetic, $\frac{1}{2}mv^2$	Inductor	Magnetic, $\frac{1}{2}Li^2$
$v = dx/dt$		$i = dq/dt$	

31-4 *LC* Oscillations, Quantitatively

Here we want to show explicitly that Eq. 31-4 for the angular frequency of *LC* oscillations is correct. At the same time, we want to examine even more closely the analogy between *LC* oscillations and block–spring oscillations. We start by extending somewhat our earlier treatment of the mechanical block–spring oscillator.

The Block-Spring Oscillator

We analyzed block–spring oscillations in Chapter 15 in terms of energy transfers and did not—at that early stage—derive the fundamental differential equation that governs those oscillations. We do so now.

We can write, for the total energy U of a block–spring oscillator at any instant,

$$U = U_b + U_s = \tfrac{1}{2}mv^2 + \tfrac{1}{2}kx^2, \tag{31-5}$$

where U_b and U_s are, respectively, the kinetic energy of the moving block and the potential energy of the stretched or compressed spring. If there is no friction—which we assume—the total energy U remains constant with time, even though v and x vary. In more formal language, $dU/dt = 0$. This leads to

$$\frac{dU}{dt} = \frac{d}{dt}\left(\tfrac{1}{2}mv^2 + \tfrac{1}{2}kx^2\right) = mv\frac{dv}{dt} + kx\frac{dx}{dt} = 0. \tag{31-6}$$

However, $v = dx/dt$ and $dv/dt = d^2x/dt^2$. With these substitutions, Eq. 31-6 becomes

$$m\frac{d^2x}{dt^2} + kx = 0 \qquad \text{(block–spring oscillations)}. \tag{31-7}$$

Equation 31-7 is the fundamental *differential equation* that governs the frictionless block–spring oscillations.

The general solution to Eq. 31-7—that is, the function $x(t)$ that describes the block–spring oscillations—is (as we saw in Eq. 15-3)

$$x = X\cos(\omega t + \phi) \qquad \text{(displacement)}, \tag{31-8}$$

in which X is the amplitude of the mechanical oscillations (x_m in Chapter 15), ω is the angular frequency of the oscillations, and ϕ is a phase constant.

The *LC* Oscillator

Now let us analyze the oscillations of a resistanceless *LC* circuit, proceeding exactly as we just did for the block–spring oscillator. The total energy U present at any instant in an oscillating *LC* circuit is given by

$$U = U_B + U_E = \frac{Li^2}{2} + \frac{q^2}{2C}, \tag{31-9}$$

in which U_B is the energy stored in the magnetic field of the inductor and U_E is the energy stored in the electric field of the capacitor. Since we have assumed the circuit resistance to be zero, no energy is transferred to thermal energy and U remains constant with time. In more formal language, dU/dt must be zero. This leads to

$$\frac{dU}{dt} = \frac{d}{dt}\left(\frac{Li^2}{2} + \frac{q^2}{2C}\right) = Li\frac{di}{dt} + \frac{q}{C}\frac{dq}{dt} = 0. \tag{31-10}$$

However, $i = dq/dt$ and $di/dt = d^2q/dt^2$. With these substitutions, Eq. 31-10 becomes

$$L\frac{d^2q}{dt^2} + \frac{1}{C}q = 0 \qquad (LC \text{ oscillations}). \tag{31-11}$$

This is the *differential equation* that describes the oscillations of a resistanceless LC circuit. Equations 31-11 and 31-7 are exactly of the same mathematical form.

Charge and Current Oscillations

Since the differential equations are mathematically identical, their solutions must also be mathematically identical. Because q corresponds to x, we can write the general solution of Eq. 31-11, by analogy to Eq. 31-8, as

$$q = Q\cos(\omega t + \phi) \qquad (\text{charge}), \tag{31-12}$$

where Q is the amplitude of the charge variations, ω is the angular frequency of the electromagnetic oscillations, and ϕ is the phase constant.

Taking the first derivative of Eq. 31-12 with respect to time gives us the current of the LC oscillator:

$$i = \frac{dq}{dt} = -\omega Q\sin(\omega t + \phi) \qquad (\text{current}). \tag{31-13}$$

The amplitude I of this sinusoidally varying current is

$$I = \omega Q, \tag{31-14}$$

and so we can rewrite Eq. 31-13 as

$$i = -I\sin(\omega t + \phi). \tag{31-15}$$

Angular Frequencies

We can test whether Eq. 31-12 is a solution of Eq. 31-11 by substituting Eq. 31-12 and its second derivative with respect to time into Eq. 31-11. The first derivative of Eq. 31-12 is Eq. 31-13. The second derivative is then

$$\frac{d^2q}{dt^2} = -\omega^2 Q\cos(\omega t + \phi).$$

Substituting for q and d^2q/dt^2 in Eq. 31-11, we obtain

$$-L\omega^2 Q\cos(\omega t + \phi) + \frac{1}{C}Q\cos(\omega t + \phi) = 0.$$

Canceling $Q\cos(\omega t + \phi)$ and rearranging lead to

$$\omega = \frac{1}{\sqrt{LC}}.$$

Thus, Eq. 31-12 is indeed a solution of Eq. 31-11 if ω has the constant value $1/\sqrt{LC}$. Note that this expression for ω is exactly that given by Eq. 31-4, which we arrived at by examining correspondences.

The phase constant ϕ in Eq. 31-12 is determined by the conditions that exist at any certain time—say, $t = 0$. If the conditions yield $\phi = 0$ at $t = 0$, Eq. 31-12 requires that $q = Q$ and Eq. 31-13 requires that $i = 0$; these are the initial conditions represented by Fig. 31-1*a*.

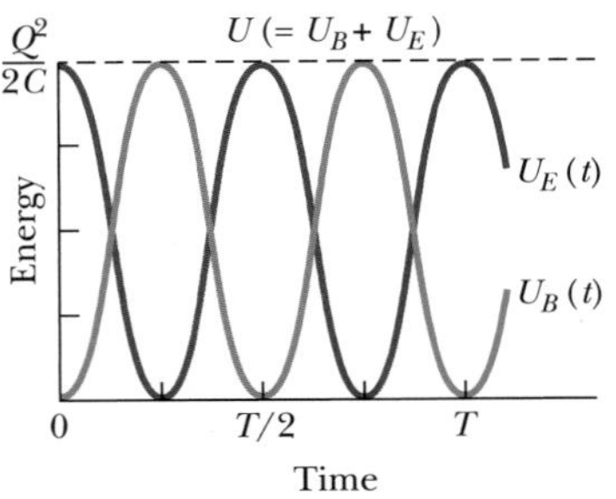

Fig. 31-4 The stored magnetic energy and electrical energy in the circuit of Fig. 31-1 as a function of time. Note that their sum remains constant. T is the period of oscillation.

Electrical and Magnetic Energy Oscillations

The electrical energy stored in the LC circuit at time t is, from Eqs. 31-1 and 31-12,

$$U_E = \frac{q^2}{2C} = \frac{Q^2}{2C}\cos^2(\omega t + \phi). \tag{31-16}$$

The magnetic energy is, from Eqs. 31-2 and 31-13,

$$U_B = \tfrac{1}{2}Li^2 = \tfrac{1}{2}L\omega^2Q^2\sin^2(\omega t + \phi).$$

Substituting for ω from Eq. 31-4 then gives us

$$U_B = \frac{Q^2}{2C}\sin^2(\omega t + \phi). \tag{31-17}$$

Figure 31-4 shows plots of $U_E(t)$ and $U_B(t)$ for the case of $\phi = 0$. Note that

1. The maximum values of U_E and U_B are both $Q^2/2C$.
2. At any instant the sum of U_E and U_B is equal to $Q^2/2C$, a constant.
3. When U_E is maximum, U_B is zero, and conversely.

CHECKPOINT 2

A capacitor in an LC oscillator has a maximum potential difference of 17 V and a maximum energy of 160 μJ. When the capacitor has a potential difference of 5 V and an energy of 10 μJ, what are (a) the emf across the inductor and (b) the energy stored in the magnetic field?

Sample Problem

LC oscillator: potential change, rate of current change

A 1.5 μF capacitor is charged to 57 V by a battery, which is then removed. At time $t = 0$, a 12 mH coil is connected in series with the capacitor to form an LC oscillator (Fig. 31-1).

(a) What is the potential difference $v_L(t)$ across the inductor as a function of time?

KEY IDEAS

(1) The current and potential differences of the circuit (both the potential difference of the capacitor and the potential difference of the coil) undergo sinusoidal oscillations. (2) We can still apply the loop rule to these oscillating potential differences, just as we did for the nonoscillating circuits of Chapter 27.

Calculations: At any time t during the oscillations, the loop rule and Fig. 31-1 give us

$$v_L(t) = v_C(t); \tag{31-18}$$

that is, the potential difference v_L across the inductor must always be equal to the potential difference v_C across the capacitor, so that the net potential difference around the circuit is zero. Thus, we will find $v_L(t)$ if we can find $v_C(t)$, and we can find $v_C(t)$ from $q(t)$ with Eq. 25-1 ($q = CV$).

Because the potential difference $v_C(t)$ is maximum when the oscillations begin at time $t = 0$, the charge q on the capacitor must also be maximum then. Thus, phase constant ϕ must be zero; so Eq. 31-12 gives us

$$q = Q\cos\omega t. \tag{31-19}$$

(Note that this cosine function does indeed yield maximum q ($= Q$) when $t = 0$.) To get the potential difference $v_C(t)$, we divide both sides of Eq. 31-19 by C to write

$$\frac{q}{C} = \frac{Q}{C}\cos\omega t,$$

and then use Eq. 25-1 to write

$$v_C = V_C\cos\omega t. \tag{31-20}$$

Here, V_C is the amplitude of the oscillations in the potential difference v_C across the capacitor.

Next, substituting $v_C = v_L$ from Eq. 31-18, we find

$$v_L = V_C\cos\omega t. \tag{31-21}$$

We can evaluate the right side of this equation by first noting that the amplitude V_C is equal to the initial (maximum) potential difference of 57 V across the capacitor. Then we find ω with Eq. 31-4:

$$\omega = \frac{1}{\sqrt{LC}} = \frac{1}{[(0.012\text{ H})(1.5 \times 10^{-6}\text{ F})]^{0.5}}$$
$$= 7454 \text{ rad/s} \approx 7500 \text{ rad/s}.$$

Thus, Eq. 31-21 becomes

$$v_L = (57\text{ V})\cos(7500\text{ rad/s})t. \qquad \text{(Answer)}$$

(b) What is the maximum rate $(di/dt)_{max}$ at which the current i changes in the circuit?

KEY IDEA

With the charge on the capacitor oscillating as in Eq. 31-12, the current is in the form of Eq. 31-13. Because $\phi = 0$, that equation gives us

$$i = -\omega Q \sin \omega t.$$

Calculations: Taking the derivative, we have

$$\frac{di}{dt} = \frac{d}{dt}(-\omega Q \sin \omega t) = -\omega^2 Q \cos \omega t.$$

We can simplify this equation by substituting CV_C for Q (because we know C and V_C but not Q) and $1/\sqrt{LC}$ for ω according to Eq. 31-4. We get

$$\frac{di}{dt} = -\frac{1}{LC} CV_C \cos \omega t = -\frac{V_C}{L} \cos \omega t.$$

This tells us that the current changes at a varying (sinusoidal) rate, with its maximum rate of change being

$$\frac{V_C}{L} = \frac{57\text{ V}}{0.012\text{ H}} = 4750 \text{ A/s} \approx 4800 \text{ A/s}. \qquad \text{(Answer)}$$

Additional examples, video, and practice available at *WileyPLUS*

31-5 Damped Oscillations in an *RLC* Circuit

A circuit containing resistance, inductance, and capacitance is called an *RLC circuit.* We shall here discuss only *series RLC circuits* like that shown in Fig. 31-5. With a resistance R present, the total *electromagnetic energy* U of the circuit (the sum of the electrical energy and magnetic energy) is no longer constant; instead, it decreases with time as energy is transferred to thermal energy in the resistance. Because of this loss of energy, the oscillations of charge, current, and potential difference continuously decrease in amplitude, and the oscillations are said to be *damped,* just as with the damped block–spring oscillator of Section 15-8.

To analyze the oscillations of this circuit, we write an equation for the total electromagnetic energy U in the circuit at any instant. Because the resistance does not store electromagnetic energy, we can use Eq. 31-9:

$$U = U_B + U_E = \frac{Li^2}{2} + \frac{q^2}{2C}. \qquad (31\text{-}22)$$

Now, however, this total energy decreases as energy is transferred to thermal energy. The rate of that transfer is, from Eq. 26-27,

$$\frac{dU}{dt} = -i^2 R, \qquad (31\text{-}23)$$

where the minus sign indicates that U decreases. By differentiating Eq. 31-22 with respect to time and then substituting the result in Eq. 31-23, we obtain

$$\frac{dU}{dt} = Li\frac{di}{dt} + \frac{q}{C}\frac{dq}{dt} = -i^2 R.$$

Substituting dq/dt for i and d^2q/dt^2 for di/dt, we obtain

$$L\frac{d^2q}{dt^2} + R\frac{dq}{dt} + \frac{1}{C}q = 0 \qquad (RLC\text{ circuit}), \qquad (31\text{-}24)$$

which is the differential equation for damped oscillations in an *RLC* circuit.

The solution to Eq. 31-24 is

$$q = Qe^{-Rt/2L}\cos(\omega' t + \phi), \qquad (31\text{-}25)$$

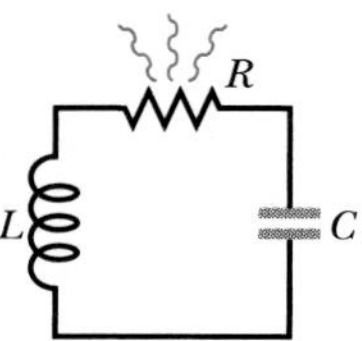

Fig. 31-5 A series *RLC* circuit. As the charge contained in the circuit oscillates back and forth through the resistance, electromagnetic energy is dissipated as thermal energy, damping (decreasing the amplitude of) the oscillations.

in which

$$\omega' = \sqrt{\omega^2 - (R/2L)^2}, \tag{31-26}$$

where $\omega = 1/\sqrt{LC}$, as with an undamped oscillator. Equation 31-25 tells us how the charge on the capacitor oscillates in a damped RLC circuit; that equation is the electromagnetic counterpart of Eq. 15-42, which gives the displacement of a damped block–spring oscillator.

Equation 31-25 describes a sinusoidal oscillation (the cosine function) with an *exponentially decaying amplitude* $Qe^{-Rt/2L}$ (the factor that multiplies the cosine). The angular frequency ω' of the damped oscillations is always less than the angular frequency ω of the undamped oscillations; however, we shall here consider only situations in which R is small enough for us to replace ω' with ω.

Let us next find an expression for the total electromagnetic energy U of the circuit as a function of time. One way to do so is to monitor the energy of the electric field in the capacitor, which is given by Eq. 31-1 ($U_E = q^2/2C$). By substituting Eq. 31-25 into Eq. 31-1, we obtain

$$U_E = \frac{q^2}{2C} = \frac{[Qe^{-Rt/2L}\cos(\omega' t + \phi)]^2}{2C} = \frac{Q^2}{2C}e^{-Rt/L}\cos^2(\omega' t + \phi). \tag{31-27}$$

Thus, the energy of the electric field oscillates according to a cosine-squared term, and the amplitude of that oscillation decreases exponentially with time.

Sample Problem

Damped *RLC* circuit: charge amplitude

A series RLC circuit has inductance $L = 12$ mH, capacitance $C = 1.6\ \mu$F, and resistance $R = 1.5\ \Omega$ and begins to oscillate at time $t = 0$.

(a) At what time t will the amplitude of the charge oscillations in the circuit be 50% of its initial value? (Note that we do not know that initial value.)

KEY IDEA

The amplitude of the charge oscillations decreases exponentially with time t: According to Eq. 31-25, the charge amplitude at any time t is $Qe^{-Rt/2L}$, in which Q is the amplitude at time $t = 0$.

Calculations: We want the time when the charge amplitude has decreased to $0.50Q$, that is, when

$$Qe^{-Rt/2L} = 0.50Q.$$

We can now cancel Q (which also means that we can answer the question without knowing the initial charge). Taking the natural logarithms of both sides (to eliminate the exponential function), we have

$$-\frac{Rt}{2L} = \ln 0.50.$$

Solving for t and then substituting given data yield

$$t = -\frac{2L}{R}\ln 0.50 = -\frac{(2)(12 \times 10^{-3}\ \text{H})(\ln 0.50)}{1.5\ \Omega}$$

$$= 0.0111\ \text{s} \approx 11\ \text{ms}. \quad \text{(Answer)}$$

(b) How many oscillations are completed within this time?

KEY IDEA

The time for one complete oscillation is the period $T = 2\pi/\omega$, where the angular frequency for LC oscillations is given by Eq. 31-4 ($\omega = 1/\sqrt{LC}$).

Calculation: In the time interval $\Delta t = 0.0111$ s, the number of complete oscillations is

$$\frac{\Delta t}{T} = \frac{\Delta t}{2\pi\sqrt{LC}}$$

$$= \frac{0.0111\ \text{s}}{2\pi[(12 \times 10^{-3}\ \text{H})(1.6 \times 10^{-6}\ \text{F})]^{1/2}} \approx 13.$$

(Answer)

Thus, the amplitude decays by 50% in about 13 complete oscillations. This damping is less severe than that shown in Fig. 31-3, where the amplitude decays by a little more than 50% in one oscillation.

Additional examples, video, and practice available at *WileyPLUS*

31-6 Alternating Current

The oscillations in an *RLC* circuit will not damp out if an external emf device supplies enough energy to make up for the energy dissipated as thermal energy in the resistance *R*. Circuits in homes, offices, and factories, including countless *RLC* circuits, receive such energy from local power companies. In most countries the energy is supplied via oscillating emfs and currents—the current is said to be an **alternating current,** or **ac** for short. (The nonoscillating current from a battery is said to be a **direct current,** or **dc**.) These oscillating emfs and currents vary sinusoidally with time, reversing direction (in North America) 120 times per second and thus having frequency $f = 60$ Hz.

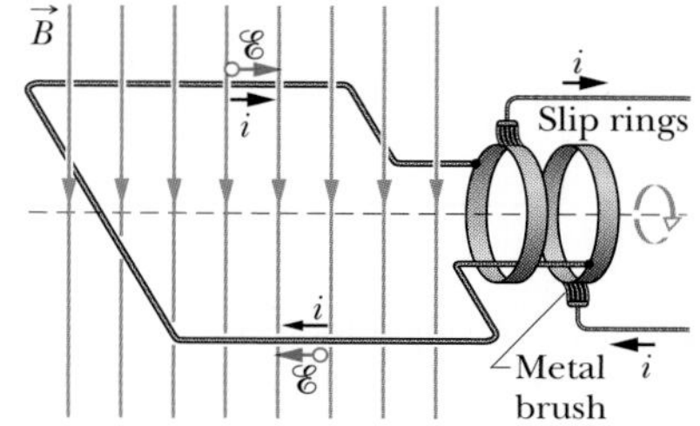

Fig. 31-6 The basic mechanism of an alternating-current generator is a conducting loop rotated in an external magnetic field. In practice, the alternating emf induced in a coil of many turns of wire is made accessible by means of slip rings attached to the rotating loop. Each ring is connected to one end of the loop wire and is electrically connected to the rest of the generator circuit by a conducting brush against which the ring slips as the loop (and it) rotates.

At first sight this may seem to be a strange arrangement. We have seen that the drift speed of the conduction electrons in household wiring may typically be 4×10^{-5} m/s. If we now reverse their direction every $\frac{1}{120}$ s, such electrons can move only about 3×10^{-7} m in a half-cycle. At this rate, a typical electron can drift past no more than about 10 atoms in the wiring before it is required to reverse its direction. How, you may wonder, can the electron ever get anywhere?

Although this question may be worrisome, it is a needless concern. The conduction electrons do not have to "get anywhere." When we say that the current in a wire is one ampere, we mean that charge passes through any plane cutting across that wire at the rate of one coulomb per second. The speed at which the charge carriers cross that plane does not matter directly; one ampere may correspond to many charge carriers moving very slowly or to a few moving very rapidly. Furthermore, the signal to the electrons to reverse directions—which originates in the alternating emf provided by the power company's generator—is propagated along the conductor at a speed close to that of light. All electrons, no matter where they are located, get their reversal instructions at about the same instant. Finally, we note that for many devices, such as lightbulbs and toasters, the direction of motion is unimportant as long as the electrons do move so as to transfer energy to the device via collisions with atoms in the device.

The basic advantage of alternating current is this: *As the current alternates, so does the magnetic field that surrounds the conductor.* This makes possible the use of Faraday's law of induction, which, among other things, means that we can step up (increase) or step down (decrease) the magnitude of an alternating potential difference at will, using a device called a transformer, as we shall discuss later. Moreover, alternating current is more readily adaptable to rotating machinery such as generators and motors than is (nonalternating) direct current.

Figure 31-6 shows a simple model of an ac generator. As the conducting loop is forced to rotate through the external magnetic field $\vec{B}$, a sinusoidally oscillating emf $\mathscr{E}$ is induced in the loop:

$$\mathscr{E} = \mathscr{E}_m \sin \omega_d t. \tag{31-28}$$

The *angular frequency* ω_d of the emf is equal to the angular speed with which the loop rotates in the magnetic field, the *phase* of the emf is $\omega_d t$, and the *amplitude* of the emf is $\mathscr{E}_m$ (where the subscript stands for maximum). When the rotating loop is part of a closed conducting path, this emf produces (*drives*) a sinusoidal (alternating) current along the path with the same angular frequency ω_d, which then is called the **driving angular frequency.** We can write the current as

$$i = I \sin(\omega_d t - \phi), \tag{31-29}$$

in which I is the amplitude of the driven current. (The phase $\omega_d t - \phi$ of the current is traditionally written with a minus sign instead of as $\omega_d t + \phi$.) We include a phase constant ϕ in Eq. 31-29 because the current i may not be in phase with the emf $\mathscr{E}$. (As you will see, the phase constant depends on the circuit to which the generator is connected.) We can also write the current i in terms of the **driving frequency** f_d of the emf, by substituting $2\pi f_d$ for ω_d in Eq. 31-29.

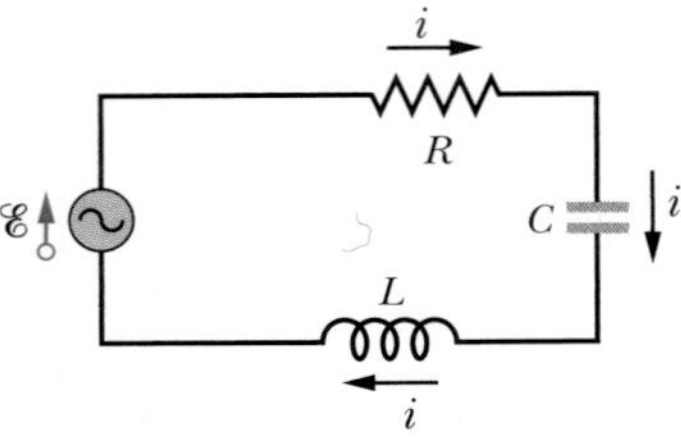

Fig. 31-7 A single-loop circuit containing a resistor, a capacitor, and an inductor. A generator, represented by a sine wave in a circle, produces an alternating emf that establishes an alternating current; the directions of the emf and current are indicated here at only one instant.

31-7 Forced Oscillations

We have seen that once started, the charge, potential difference, and current in both undamped LC circuits and damped RLC circuits (with small enough R) oscillate at angular frequency $\omega = 1/\sqrt{LC}$. Such oscillations are said to be *free oscillations* (free of any external emf), and the angular frequency ω is said to be the circuit's **natural angular frequency.**

When the external alternating emf of Eq. 31-28 is connected to an RLC circuit, the oscillations of charge, potential difference, and current are said to be *driven oscillations* or *forced oscillations.* These oscillations always occur at the driving angular frequency ω_d:

> Whatever the natural angular frequency ω of a circuit may be, forced oscillations of charge, current, and potential difference in the circuit always occur at the driving angular frequency ω_d.

However, as you will see in Section 31-9, the amplitudes of the oscillations very much depend on how close ω_d is to ω. When the two angular frequencies match—a condition known as **resonance**—the amplitude I of the current in the circuit is maximum.

31-8 Three Simple Circuits

Later in this chapter, we shall connect an external alternating emf device to a series RLC circuit as in Fig. 31-7. We shall then find expressions for the amplitude I and phase constant ϕ of the sinusoidally oscillating current in terms of the amplitude $\mathscr{E}_m$ and angular frequency ω_d of the external emf. First, let's consider three simpler circuits, each having an external emf and only one other circuit element: R, C, or L. We start with a resistive element (a purely *resistive load*).

A Resistive Load

Figure 31-8 shows a circuit containing a resistance element of value R and an ac generator with the alternating emf of Eq. 31-28. By the loop rule, we have

$$\mathscr{E} - v_R = 0.$$

With Eq. 31-28, this gives us

$$v_R = \mathscr{E}_m \sin \omega_d t.$$

Because the amplitude V_R of the alternating potential difference (or voltage) across the resistance is equal to the amplitude $\mathscr{E}_m$ of the alternating emf, we can write this as

$$v_R = V_R \sin \omega_d t. \tag{31-30}$$

From the definition of resistance ($R = V/i$), we can now write the current i_R in the resistance as

$$i_R = \frac{v_R}{R} = \frac{V_R}{R} \sin \omega_d t. \tag{31-31}$$

From Eq. 31-29, we can also write this current as

$$i_R = I_R \sin(\omega_d t - \phi), \tag{31-32}$$

where I_R is the amplitude of the current i_R in the resistance. Comparing Eqs. 31-31 and 31-32, we see that for a purely resistive load the phase constant $\phi = 0°$.

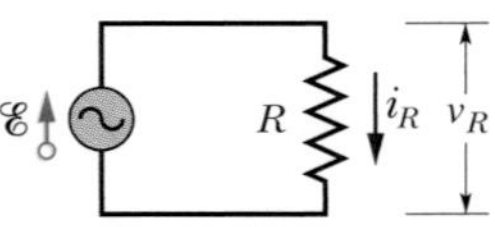

Fig. 31-8 A resistor is connected across an alternating-current generator.

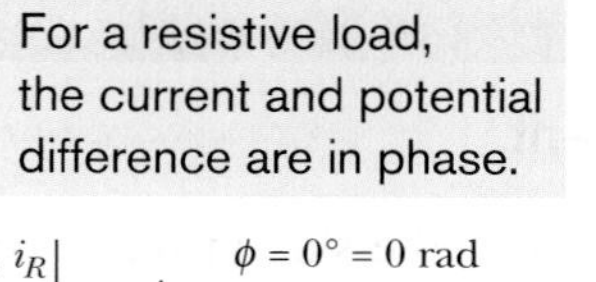

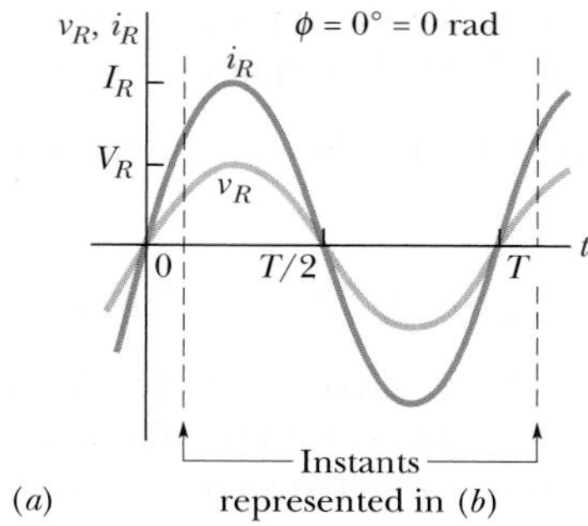

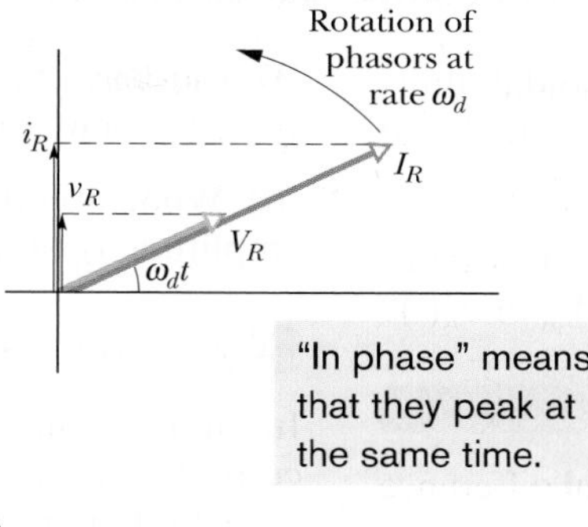

Fig. 31-9 (*a*) The current i_R and the potential difference v_R across the resistor are plotted on the same graph, both versus time t. They are in phase and complete one cycle in one period T. (*b*) A phasor diagram shows the same thing as (*a*).

We also see that the voltage amplitude and current amplitude are related by

$$V_R = I_R R \quad \text{(resistor)}. \tag{31-33}$$

Although we found this relation for the circuit of Fig. 31-8, it applies to any resistance in any ac circuit.

By comparing Eqs. 31-30 and 31-31, we see that the time-varying quantities v_R and i_R are both functions of $\sin \omega_d t$ with $\phi = 0°$. Thus, these two quantities are *in phase,* which means that their corresponding maxima (and minima) occur at the same times. Figure 31-9*a*, which is a plot of $v_R(t)$ and $i_R(t)$, illustrates this fact. Note that v_R and i_R do not decay here because the generator supplies energy to the circuit to make up for the energy dissipated in R.

The time-varying quantities v_R and i_R can also be represented geometrically by *phasors.* Recall from Section 16-11 that phasors are vectors that rotate around an origin. Those that represent the voltage across and current in the resistor of Fig. 31-8 are shown in Fig. 31-9*b* at an arbitrary time t. Such phasors have the following properties:

Angular speed: Both phasors rotate counterclockwise about the origin with an angular speed equal to the angular frequency ω_d of v_R and i_R.

Length: The length of each phasor represents the amplitude of the alternating quantity: V_R for the voltage and I_R for the current.

Projection: The projection of each phasor on the *vertical* axis represents the value of the alternating quantity at time t: v_R for the voltage and i_R for the current.

Rotation angle: The rotation angle of each phasor is equal to the phase of the alternating quantity at time t. In Fig. 31-9*b*, the voltage and current are in phase; so their phasors always have the same phase $\omega_d t$ and the same rotation angle, and thus they rotate together.

Mentally follow the rotation. Can you see that when the phasors have rotated so that $\omega_d t = 90°$ (they point vertically upward), they indicate that just then $v_R = V_R$ and $i_R = I_R$? Equations 31-30 and 31-32 give the same results.

CHECKPOINT 3

If we increase the driving frequency in a circuit with a purely resistive load, do (a) amplitude V_R and (b) amplitude I_R increase, decrease, or remain the same?

Sample Problem

Purely resistive load: potential difference and current

In Fig. 31-8, resistance R is 200 Ω and the sinusoidal alternating emf device operates at amplitude $\mathscr{E}_m = 36.0$ V and frequency $f_d = 60.0$ Hz.

(a) What is the potential difference $v_R(t)$ across the resistance as a function of time t, and what is the amplitude V_R of $v_R(t)$?

KEY IDEA

In a circuit with a purely resistive load, the potential difference $v_R(t)$ across the resistance is always equal to the potential difference $\mathscr{E}(t)$ across the emf device.

Calculations: Here we have $v_R(t) = \mathscr{E}(t)$ and $V_R = \mathscr{E}_m$. Since $\mathscr{E}_m$ is given, we can write

$$V_R = \mathscr{E}_m = 36.0 \text{ V.} \quad \text{(Answer)}$$

To find $v_R(t)$, we use Eq. 31-28 to write

$$v_R(t) = \mathscr{E}(t) = \mathscr{E}_m \sin \omega_d t \quad (31\text{-}34)$$

and then substitute $\mathscr{E}_m = 36.0$ V and

$$\omega_d = 2\pi f_d = 2\pi(60 \text{ Hz}) = 120\pi$$

to obtain

$$v_R = (36.0 \text{ V}) \sin(120\pi t). \quad \text{(Answer)}$$

We can leave the argument of the sine in this form for convenience, or we can write it as (377 rad/s)t or as (377 s^{-1})t.

(b) What are the current $i_R(t)$ in the resistance and the amplitude I_R of $i_R(t)$?

KEY IDEA

In an ac circuit with a purely resistive load, the alternating current $i_R(t)$ in the resistance is *in phase* with the alternating potential difference $v_R(t)$ across the resistance; that is, the phase constant ϕ for the current is zero.

Calculations: Here we can write Eq. 31-29 as

$$i_R = I_R \sin(\omega_d t - \phi) = I_R \sin \omega_d t. \quad (31\text{-}35)$$

From Eq. 31-33, the amplitude I_R is

$$I_R = \frac{V_R}{R} = \frac{36.0 \text{ V}}{200 \ \Omega} = 0.180 \text{ A.} \quad \text{(Answer)}$$

Substituting this and $\omega_d = 2\pi f_d = 120\pi$ into Eq. 31-35, we have

$$i_R = (0.180 \text{ A}) \sin(120\pi t). \quad \text{(Answer)}$$

Additional examples, video, and practice available at *WileyPLUS*

A Capacitive Load

Figure 31-10 shows a circuit containing a capacitance and a generator with the alternating emf of Eq. 31-28. Using the loop rule and proceeding as we did when we obtained Eq. 31-30, we find that the potential difference across the capacitor is

$$v_C = V_C \sin \omega_d t, \quad (31\text{-}36)$$

where V_C is the amplitude of the alternating voltage across the capacitor. From the definition of capacitance we can also write

$$q_C = C v_C = C V_C \sin \omega_d t. \quad (31\text{-}37)$$

Our concern, however, is with the current rather than the charge. Thus, we differentiate Eq. 31-37 to find

$$i_C = \frac{dq_C}{dt} = \omega_d C V_C \cos \omega_d t. \quad (31\text{-}38)$$

We now modify Eq. 31-38 in two ways. First, for reasons of symmetry of notation, we introduce the quantity X_C, called the **capacitive reactance** of a capacitor, defined as

$$X_C = \frac{1}{\omega_d C} \quad \text{(capacitive reactance).} \quad (31\text{-}39)$$

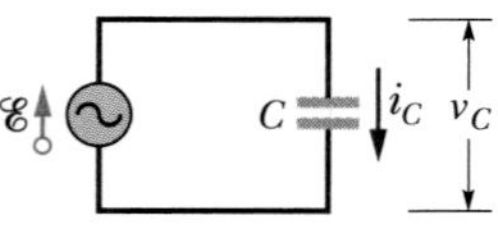

Fig. 31-10 A capacitor is connected across an alternating-current generator.

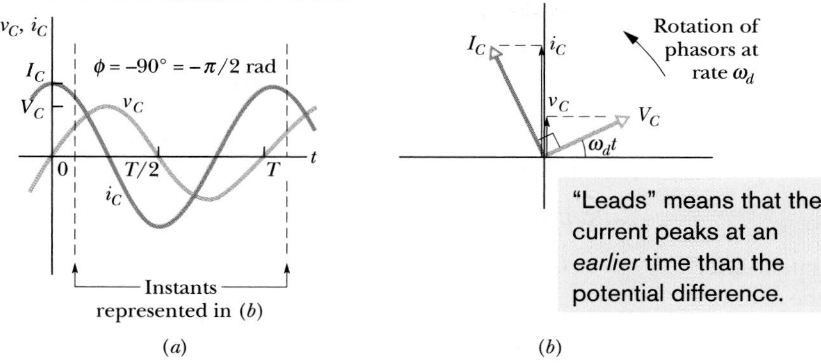

Fig. 31-11 (*a*) The current in the capacitor leads the voltage by 90° (= $\pi/2$ rad). (*b*) A phasor diagram shows the same thing.

Its value depends not only on the capacitance but also on the driving angular frequency ω_d. We know from the definition of the capacitive time constant ($\tau = RC$) that the SI unit for C can be expressed as seconds per ohm. Applying this to Eq. 31-39 shows that the SI unit of X_C is the *ohm,* just as for resistance R.

Second, we replace $\cos \omega_d t$ in Eq. 31-38 with a phase-shifted sine:

$$\cos \omega_d t = \sin(\omega_d t + 90°).$$

You can verify this identity by shifting a sine curve 90° in the negative direction.

With these two modifications, Eq. 31-38 becomes

$$i_C = \left(\frac{V_C}{X_C}\right) \sin(\omega_d t + 90°). \tag{31-40}$$

From Eq. 31-29, we can also write the current i_C in the capacitor of Fig. 31-10 as

$$i_C = I_C \sin(\omega_d t - \phi), \tag{31-41}$$

where I_C is the amplitude of i_C. Comparing Eqs. 31-40 and 31-41, we see that for a purely capacitive load the phase constant ϕ for the current is $-90°$. We also see that the voltage amplitude and current amplitude are related by

$$V_C = I_C X_C \quad \text{(capacitor).} \tag{31-42}$$

Although we found this relation for the circuit of Fig. 31-10, it applies to any capacitance in any ac circuit.

Comparison of Eqs. 31-36 and 31-40, or inspection of Fig. 31-11*a*, shows that the quantities v_C and i_C are 90°, $\pi/2$ rad, or one-quarter cycle, out of phase. Furthermore, we see that i_C *leads* v_C, which means that, if you monitored the current i_C and the potential difference v_C in the circuit of Fig. 31-10, you would find that i_C reaches its maximum *before* v_C does, by one-quarter cycle.

This relation between i_C and v_C is illustrated by the phasor diagram of Fig. 31-11*b*. As the phasors representing these two quantities rotate counterclockwise together, the phasor labeled I_C does indeed lead that labeled V_C, and by an angle of 90°; that is, the phasor I_C coincides with the vertical axis one-quarter cycle before the phasor V_C does. Be sure to convince yourself that the phasor diagram of Fig. 31-11*b* is consistent with Eqs. 31-36 and 31-40.

CHECKPOINT 4

The figure shows, in (*a*), a sine curve $S(t) = \sin(\omega_d t)$ and three other sinusoidal curves $A(t)$, $B(t)$, and $C(t)$, each of the form $\sin(\omega_d t - \phi)$. (a) Rank the three other curves according to the value of ϕ, most positive first and most negative last. (b) Which curve corresponds to which phasor in (*b*) of the figure? (c) Which curve leads the others?

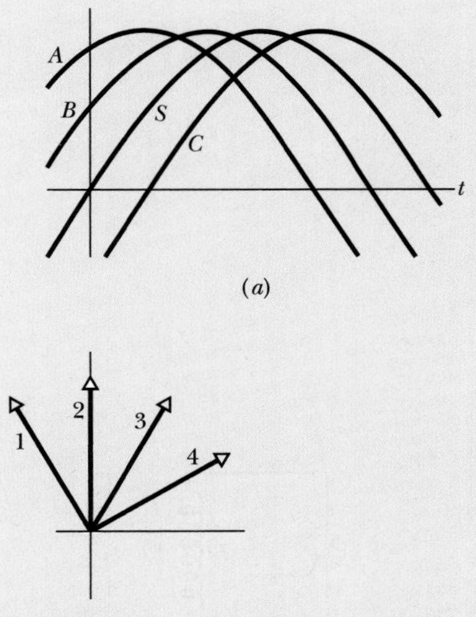

Sample Problem

Purely capacitive load: potential difference and current

In Fig. 31-10, capacitance C is 15.0 μF and the sinusoidal alternating emf device operates at amplitude $\mathscr{E}_m = 36.0$ V and frequency $f_d = 60.0$ Hz.

(a) What are the potential difference $v_C(t)$ across the capacitance and the amplitude V_C of $v_C(t)$?

KEY IDEA

In a circuit with a purely capacitive load, the potential difference $v_C(t)$ across the capacitance is always equal to the potential difference $\mathscr{E}(t)$ across the emf device.

Calculations: Here we have $v_C(t) = \mathscr{E}(t)$ and $V_C = \mathscr{E}_m$. Since $\mathscr{E}_m$ is given, we have

$$V_C = \mathscr{E}_m = 36.0 \text{ V}. \qquad \text{(Answer)}$$

To find $v_C(t)$, we use Eq. 31-28 to write

$$v_C(t) = \mathscr{E}(t) = \mathscr{E}_m \sin \omega_d t. \qquad (31\text{-}43)$$

Then, substituting $\mathscr{E}_m = 36.0$ V and $\omega_d = 2\pi f_d = 120\pi$ into Eq. 31-43, we have

$$v_C = (36.0 \text{ V}) \sin(120\pi t). \qquad \text{(Answer)}$$

(b) What are the current $i_C(t)$ in the circuit as a function of time and the amplitude I_C of $i_C(t)$?

KEY IDEA

In an ac circuit with a purely capacitive load, the alternating current $i_C(t)$ in the capacitance leads the alternating potential difference $v_C(t)$ by 90°; that is, the phase constant ϕ for the current is $-90°$, or $-\pi/2$ rad.

Calculations: Thus, we can write Eq. 31-29 as

$$i_C = I_C \sin(\omega_d t - \phi) = I_C \sin(\omega_d t + \pi/2). \qquad (31\text{-}44)$$

We can find the amplitude I_C from Eq. 31-42 ($V_C = I_C X_C$) if we first find the capacitive reactance X_C. From Eq. 31-39 ($X_C = 1/\omega_d C$), with $\omega_d = 2\pi f_d$, we can write

$$X_C = \frac{1}{2\pi f_d C} = \frac{1}{(2\pi)(60.0 \text{ Hz})(15.0 \times 10^{-6} \text{ F})} = 177\ \Omega.$$

Then Eq. 31-42 tells us that the current amplitude is

$$I_C = \frac{V_C}{X_C} = \frac{36.0 \text{ V}}{177\ \Omega} = 0.203 \text{ A}. \qquad \text{(Answer)}$$

Substituting this and $\omega_d = 2\pi f_d = 120\pi$ into Eq. 31-44, we have

$$i_C = (0.203 \text{ A}) \sin(120\pi t + \pi/2). \qquad \text{(Answer)}$$

Additional examples, video, and practice available at *WileyPLUS*

An Inductive Load

Figure 31-12 shows a circuit containing an inductance and a generator with the alternating emf of Eq. 31-28. Using the loop rule and proceeding as we did to obtain Eq. 31-30, we find that the potential difference across the inductance is

$$v_L = V_L \sin \omega_d t, \qquad (31\text{-}45)$$

where V_L is the amplitude of v_L. From Eq. 30-35 ($\mathscr{E}_L = -L\, di/dt$), we can write the potential difference across an inductance L in which the current is changing at the rate di_L/dt as

$$v_L = L \frac{di_L}{dt}. \qquad (31\text{-}46)$$

If we combine Eqs. 31-45 and 31-46, we have

$$\frac{di_L}{dt} = \frac{V_L}{L} \sin \omega_d t. \qquad (31\text{-}47)$$

Our concern, however, is with the current rather than with its time derivative. We find the former by integrating Eq. 31-47, obtaining

$$i_L = \int di_L = \frac{V_L}{L} \int \sin \omega_d t\, dt = -\left(\frac{V_L}{\omega_d L}\right) \cos \omega_d t. \qquad (31\text{-}48)$$

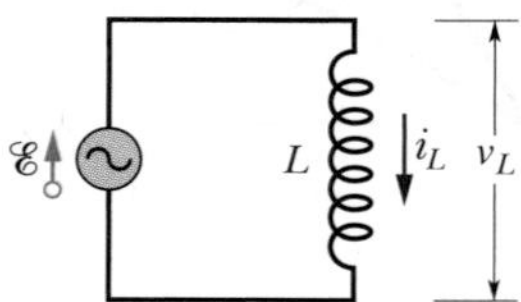

Fig. 31-12 An inductor is connected across an alternating-current generator.

We now modify this equation in two ways. First, for reasons of symmetry of notation, we introduce the quantity X_L, called the **inductive reactance** of an inductor, which is defined as

$$X_L = \omega_d L \qquad \text{(inductive reactance).} \tag{31-49}$$

The value of X_L depends on the driving angular frequency ω_d. The unit of the inductive time constant τ_L indicates that the SI unit of X_L is the *ohm*, just as it is for X_C and for R.

Second, we replace $-\cos \omega_d t$ in Eq. 31-48 with a phase-shifted sine:

$$-\cos \omega_d t = \sin(\omega_d t - 90°).$$

You can verify this identity by shifting a sine curve 90° in the positive direction.

With these two changes, Eq. 31-48 becomes

$$i_L = \left(\frac{V_L}{X_L}\right) \sin(\omega_d t - 90°). \tag{31-50}$$

From Eq. 31-29, we can also write this current in the inductance as

$$i_L = I_L \sin(\omega_d t - \phi), \tag{31-51}$$

where I_L is the amplitude of the current i_L. Comparing Eqs. 31-50 and 31-51, we see that for a purely inductive load the phase constant ϕ for the current is $+90°$. We also see that the voltage amplitude and current amplitude are related by

$$V_L = I_L X_L \qquad \text{(inductor).} \tag{31-52}$$

Although we found this relation for the circuit of Fig. 31-12, it applies to any inductance in any ac circuit.

Comparison of Eqs. 31-45 and 31-50, or inspection of Fig. 31-13*a*, shows that the quantities i_L and v_L are 90° out of phase. In this case, however, i_L *lags* v_L; that is, monitoring the current i_L and the potential difference v_L in the circuit of Fig. 31-12 shows that i_L reaches its maximum value *after* v_L does, by one-quarter cycle.

The phasor diagram of Fig. 31-13*b* also contains this information. As the phasors rotate counterclockwise in the figure, the phasor labeled I_L does indeed lag that labeled V_L, and by an angle of 90°. Be sure to convince yourself that Fig. 31-13*b* represents Eqs. 31-45 and 31-50.

For an inductive load, the current lags the potential difference by 90°.

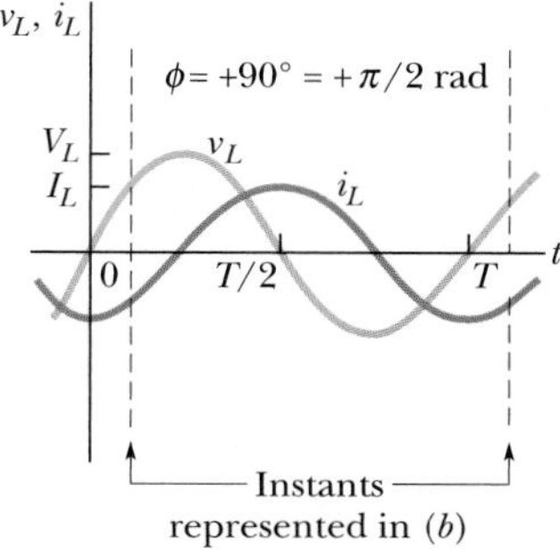

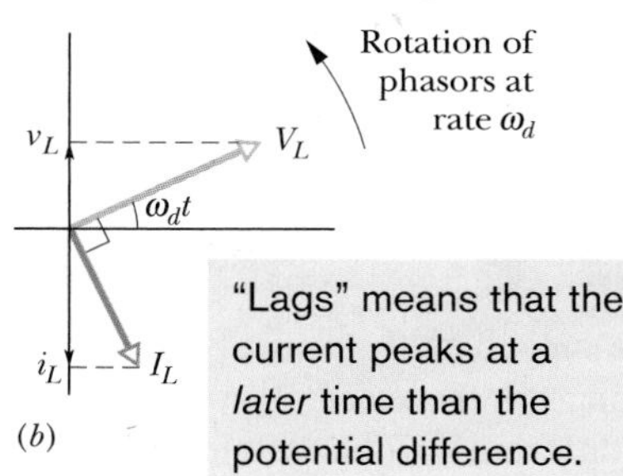

"Lags" means that the current peaks at a *later* time than the potential difference.

Fig. 31-13 (*a*) The current in the inductor lags the voltage by 90° (= $\pi/2$ rad). (*b*) A phasor diagram shows the same thing.

CHECKPOINT 5

If we increase the driving frequency in a circuit with a purely capacitive load, do (a) amplitude V_C and (b) amplitude I_C increase, decrease, or remain the same? If, instead, the circuit has a purely inductive load, do (c) amplitude V_L and (d) amplitude I_L increase, decrease, or remain the same?

Problem-Solving Tactics

Leading and Lagging in AC Circuits Table 31-2 summarizes the relations between the current i and the voltage v for each of the three kinds of circuit elements we have considered. When an applied alternating voltage produces an alternating current in these elements, the current is always in phase with the voltage across a resistor, always leads the voltage across a capacitor, and always lags the voltage across an inductor.

Many students remember these results with the mnemonic "*ELI* the *ICE* man." *ELI* contains the letter *L* (for inductor), and in it the letter *I* (for current) comes *after* the letter *E* (for emf or voltage). Thus, for an inductor, the current *lags* (comes after) the voltage. Similarly *ICE* (which contains a *C* for capacitor) means that the current *leads* (comes before) the voltage. You might also use the modified mnemonic "*ELI positively* is the *ICE* man" to remember that the phase constant ϕ is positive for an inductor.

If you have difficulty in remembering whether X_C is equal to $\omega_d C$ (wrong) or $1/\omega_d C$ (right), try remembering that C is in the "cellar"—that is, in the denominator.

Table 31-2

Phase and Amplitude Relations for Alternating Currents and Voltages

Circuit Element	Symbol	Resistance or Reactance	Phase of the Current	Phase Constant (or Angle) ϕ	Amplitude Relation
Resistor	R	R	In phase with v_R	$0°$ $(= 0 \text{ rad})$	$V_R = I_R R$
Capacitor	C	$X_C = 1/\omega_d C$	Leads v_C by $90°$ $(= \pi/2 \text{ rad})$	$-90°$ $(= -\pi/2 \text{ rad})$	$V_C = I_C X_C$
Inductor	L	$X_L = \omega_d L$	Lags v_L by $90°$ $(= \pi/2 \text{ rad})$	$+90°$ $(= +\pi/2 \text{ rad})$	$V_L = I_L X_L$

Sample Problem

Purely inductive load: potential difference and current

In Fig. 31-12, inductance L is 230 mH and the sinusoidal alternating emf device operates at amplitude $\mathscr{E}_m = 36.0$ V and frequency $f_d = 60.0$ Hz.

(a) What are the potential difference $v_L(t)$ across the inductance and the amplitude V_L of $v_L(t)$?

KEY IDEA

In a circuit with a purely inductive load, the potential difference $v_L(t)$ across the inductance is always equal to the potential difference $\mathscr{E}(t)$ across the emf device.

Calculations: Here we have $v_L(t) = \mathscr{E}(t)$ and $V_L = \mathscr{E}_m$. Since $\mathscr{E}_m$ is given, we know that

$$V_L = \mathscr{E}_m = 36.0 \text{ V}. \qquad \text{(Answer)}$$

To find $v_L(t)$, we use Eq. 31-28 to write

$$v_L(t) = \mathscr{E}(t) = \mathscr{E}_m \sin \omega_d t. \qquad (31\text{-}53)$$

Then, substituting $\mathscr{E}_m = 36.0$ V and $\omega_d = 2\pi f_d = 120\pi$ into Eq. 31-53, we have

$$v_L = (36.0 \text{ V}) \sin(120\pi t). \qquad \text{(Answer)}$$

(b) What are the current $i_L(t)$ in the circuit as a function of time and the amplitude I_L of $i_L(t)$?

KEY IDEA

In an ac circuit with a purely inductive load, the alternating current $i_L(t)$ in the inductance lags the alternating potential difference $v_L(t)$ by 90°. (In the mnemonic of the problem-solving tactic, this circuit is "positively an *ELI* circuit," which tells us that the emf E leads the current I and that ϕ is *positive.*)

Calculations: Because the phase constant ϕ for the current is $+90°$, or $+\pi/2$ rad, we can write Eq. 31-29 as

$$i_L = I_L \sin(\omega_d t - \phi) = I_L \sin(\omega_d t - \pi/2). \qquad (31\text{-}54)$$

We can find the amplitude I_L from Eq. 31-52 ($V_L = I_L X_L$) if we first find the inductive reactance X_L. From Eq. 31-49 ($X_L = \omega_d L$), with $\omega_d = 2\pi f_d$, we can write

$$X_L = 2\pi f_d L = (2\pi)(60.0 \text{ Hz})(230 \times 10^{-3} \text{ H})$$
$$= 86.7 \ \Omega.$$

Then Eq. 31-52 tells us that the current amplitude is

$$I_L = \frac{V_L}{X_L} = \frac{36.0 \text{ V}}{86.7 \ \Omega} = 0.415 \text{ A}. \qquad \text{(Answer)}$$

Substituting this and $\omega_d = 2\pi f_d = 120\pi$ into Eq. 31-54, we have

$$i_L = (0.415 \text{ A}) \sin(120\pi t - \pi/2). \qquad \text{(Answer)}$$

Additional examples, video, and practice available at *WileyPLUS*

31-9 The Series *RLC* Circuit

We are now ready to apply the alternating emf of Eq. 31-28,

$$\mathscr{E} = \mathscr{E}_m \sin \omega_d t \qquad \text{(applied emf)}, \qquad (31\text{-}55)$$

to the full RLC circuit of Fig. 31-7. Because R, L, and C are in series, the same current

$$i = I \sin(\omega_d t - \phi) \qquad (31\text{-}56)$$

is driven in all three of them. We wish to find the current amplitude I and the phase constant ϕ. The solution is simplified by the use of phasor diagrams.

The Current Amplitude

We start with Fig. 31-14*a*, which shows the phasor representing the current of Eq. 31-56 at an arbitrary time t. The length of the phasor is the current amplitude I, the projection of the phasor on the vertical axis is the current i at time t, and the angle of rotation of the phasor is the phase $\omega_d t - \phi$ of the current at time t.

Figure 31-14*b* shows the phasors representing the voltages across R, L, and C at the same time t. Each phasor is oriented relative to the angle of rotation of current phasor I in Fig. 31-14*a*, based on the information in Table 31-2:

Resistor: Here current and voltage are in phase; so the angle of rotation of voltage phasor V_R is the same as that of phasor I.

Capacitor: Here current leads voltage by 90°; so the angle of rotation of voltage phasor V_C is 90° less than that of phasor I.

Inductor: Here current lags voltage by 90°; so the angle of rotation of voltage phasor v_L is 90° greater than that of phasor I.

Figure 31-14*b* also shows the instantaneous voltages v_R, v_C, and v_L across R, C, and L at time t; those voltages are the projections of the corresponding phasors on the vertical axis of the figure.

Figure 31-14*c* shows the phasor representing the applied emf of Eq. 31-55. The length of the phasor is the emf amplitude $\mathscr{E}_m$, the projection of the phasor on the vertical axis is the emf $\mathscr{E}$ at time t, and the angle of rotation of the phasor is the phase $\omega_d t$ of the emf at time t.

From the loop rule we know that at any instant the sum of the voltages v_R, v_C, and v_L is equal to the applied emf $\mathscr{E}$:

$$\mathscr{E} = v_R + v_C + v_L. \tag{31-57}$$

Thus, at time t the projection $\mathscr{E}$ in Fig. 31-14*c* is equal to the algebraic sum of the projections v_R, v_C, and v_L in Fig. 31-14*b*. In fact, as the phasors rotate together, this equality always holds. This means that phasor $\mathscr{E}_m$ in Fig. 31-14*c* must be equal to the vector sum of the three voltage phasors V_R, V_C, and V_L in Fig. 31-14*b*.

That requirement is indicated in Fig. 31-14*d*, where phasor $\mathscr{E}_m$ is drawn as the sum of phasors V_R, V_L, and V_C. Because phasors V_L and V_C have opposite directions in the figure, we simplify the vector sum by first combining V_L and V_C to form the single phasor $V_L - V_C$. Then we combine that single phasor with V_R to find the net phasor. Again, the net phasor must coincide with phasor $\mathscr{E}_m$, as shown.

Both triangles in Fig. 31-14*d* are right triangles. Applying the Pythagorean theorem to either one yields

$$\mathscr{E}_m^2 = V_R^2 + (V_L - V_C)^2. \tag{31-58}$$

From the voltage amplitude information displayed in the rightmost column of Table 31-2, we can rewrite this as

$$\mathscr{E}_m^2 = (IR)^2 + (IX_L - IX_C)^2, \tag{31-59}$$

and then rearrange it to the form

$$I = \frac{\mathscr{E}_m}{\sqrt{R^2 + (X_L - X_C)^2}}. \tag{31-60}$$

The denominator in Eq. 31-60 is called the **impedance** Z of the circuit for the driving angular frequency ω_d:

$$Z = \sqrt{R^2 + (X_L - X_C)^2} \quad \text{(impedance defined).} \tag{31-61}$$

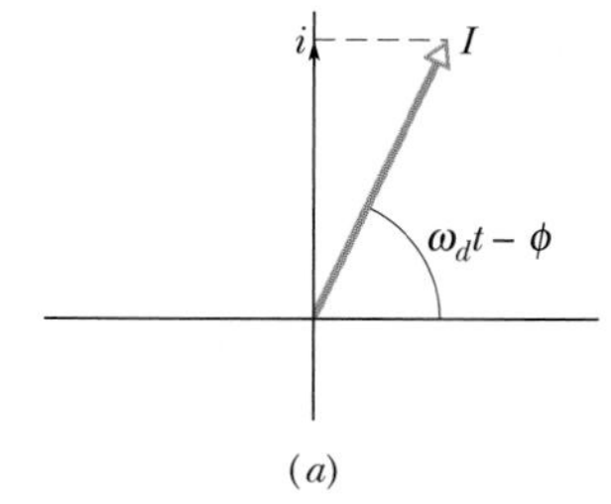

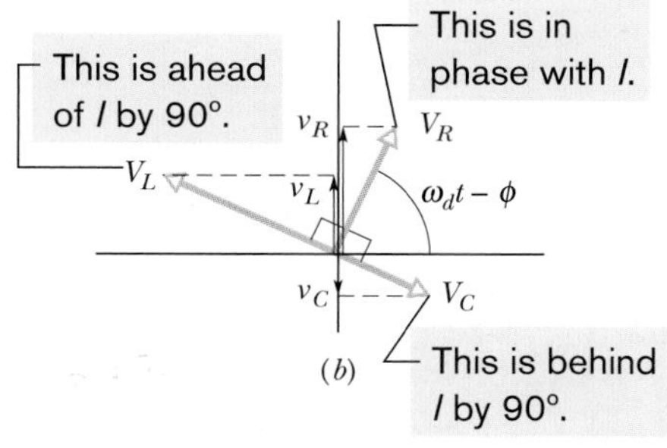

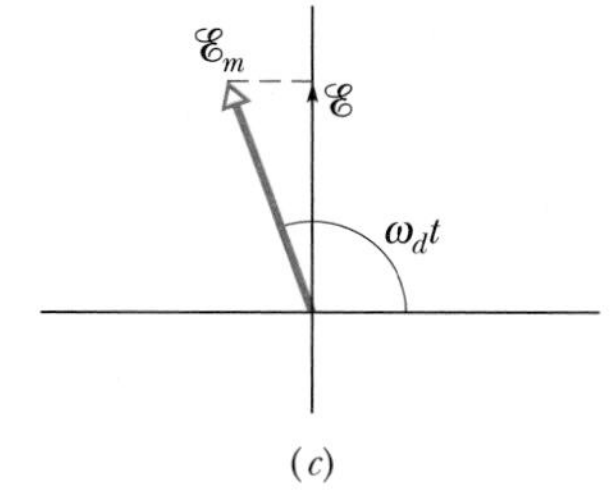

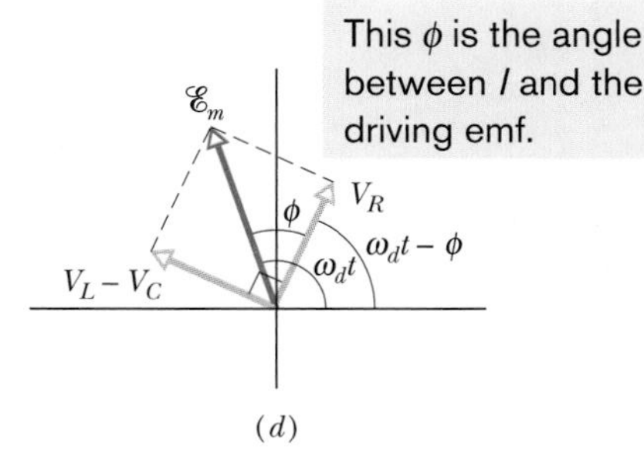

Fig. 31-14 (*a*) A phasor representing the alternating current in the driven *RLC* circuit of Fig. 31-7 at time t. The amplitude I, the instantaneous value i, and the phase $(\omega_d t - \phi)$ are shown. (*b*) Phasors representing the voltages across the inductor, resistor, and capacitor, oriented with respect to the current phasor in (*a*). (*c*) A phasor representing the alternating emf that drives the current of (*a*). (*d*) The emf phasor is equal to the vector sum of the three voltage phasors of (*b*). Here, voltage phasors V_L and V_C have been added vectorially to yield their net phasor $(V_L - V_C)$.

We can then write Eq. 31-60 as

$$I = \frac{\mathscr{E}_m}{Z}. \tag{31-62}$$

If we substitute for X_C and X_L from Eqs. 31-39 and 31-49, we can write Eq. 31-60 more explicitly as

$$I = \frac{\mathscr{E}_m}{\sqrt{R^2 + (\omega_d L - 1/\omega_d C)^2}} \quad \text{(current amplitude).} \tag{31-63}$$

We have now accomplished half our goal: We have obtained an expression for the current amplitude I in terms of the sinusoidal driving emf and the circuit elements in a series RLC circuit.

The value of I depends on the difference between $\omega_d L$ and $1/\omega_d C$ in Eq. 31-63 or, equivalently, the difference between X_L and X_C in Eq. 31-60. In either equation, it does not matter which of the two quantities is greater because the difference is always squared.

The current that we have been describing in this section is the *steady-state current* that occurs after the alternating emf has been applied for some time. When the emf is first applied to a circuit, a brief *transient current* occurs. Its duration (before settling down into the steady-state current) is determined by the time constants $\tau_L = L/R$ and $\tau_C = RC$ as the inductive and capacitive elements "turn on." This transient current can, for example, destroy a motor on start-up if it is not properly taken into account in the motor's circuit design.

The Phase Constant

From the right-hand phasor triangle in Fig. 31-14*d* and from Table 31-2 we can write

$$\tan \phi = \frac{V_L - V_C}{V_R} = \frac{IX_L - IX_C}{IR}, \tag{31-64}$$

which gives us

$$\tan \phi = \frac{X_L - X_C}{R} \quad \text{(phase constant).} \tag{31-65}$$

This is the other half of our goal: an equation for the phase constant ϕ in the sinusoidally driven series RLC circuit of Fig. 31-7. In essence, it gives us three different results for the phase constant, depending on the relative values of the reactances X_L and X_C:

$\boldsymbol{X_L > X_C}$**:** The circuit is said to be *more inductive than capacitive.* Equation 31-65 tells us that ϕ is positive for such a circuit, which means that phasor I rotates behind phasor $\mathscr{E}_m$ (Fig. 31-15*a*). A plot of $\mathscr{E}$ and i versus time is like that in Fig. 31-15*b*. (Figures 31-14*c* and *d* were drawn assuming $X_L > X_C$.)

$\boldsymbol{X_C > X_L}$**:** The circuit is said to be *more capacitive than inductive.* Equation 31-65 tells us that ϕ is negative for such a circuit, which means that phasor I rotates ahead of phasor $\mathscr{E}_m$ (Fig. 31-15*c*). A plot of $\mathscr{E}$ and i versus time is like that in Fig. 31-15*d*.

$\boldsymbol{X_C = X_L}$**:** The circuit is said to be in *resonance,* a state that is discussed next. Equation 31-65 tells us that $\phi = 0°$ for such a circuit, which means that phasors $\mathscr{E}_m$ and I rotate together (Fig. 31-15*e*). A plot of $\mathscr{E}$ and i versus time is like that in Fig. 31-15*f*.

As illustration, let us reconsider two extreme circuits: In the *purely inductive circuit* of Fig. 31-12, where X_L is nonzero and $X_C = R = 0$, Eq. 31-65 tells us that the circuit's phase constant is $\phi = +90°$ (the greatest value of ϕ), consistent with Fig. 31-13*b*. In the *purely capacitive circuit* of Fig. 31-10, where X_C is nonzero and $X_L = R = 0$, Eq. 31-65 tells us that the circuit's phase constant is $\phi = -90°$ (the least value of ϕ), consistent with Fig. 31-11*b*.

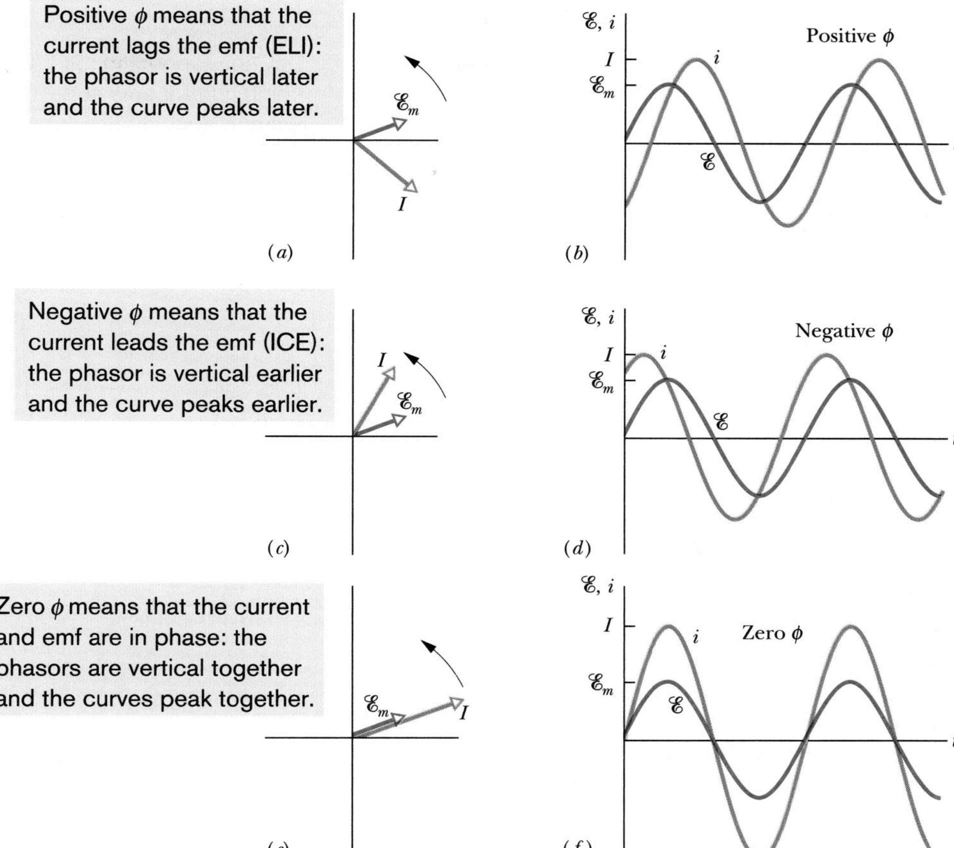

Fig. 31-15 Phasor diagrams and graphs of the alternating emf $\mathscr{E}$ and current i for the driven *RLC* circuit of Fig. 31-7. In the phasor diagram of (*a*) and the graph of (*b*), the current i lags the driving emf $\mathscr{E}$ and the current's phase constant ϕ is positive. In (*c*) and (*d*), the current i leads the driving emf $\mathscr{E}$ and its phase constant ϕ is negative. In (*e*) and (*f*), the current i is in phase with the driving emf $\mathscr{E}$ and its phase constant ϕ is zero.

Resonance

Equation 31-63 gives the current amplitude I in an *RLC* circuit as a function of the driving angular frequency ω_d of the external alternating emf. For a given resistance R, that amplitude is a maximum when the quantity $\omega_d L - 1/\omega_d C$ in the denominator is zero—that is, when

$$\omega_d L = \frac{1}{\omega_d C}$$

or

$$\omega_d = \frac{1}{\sqrt{LC}} \quad \text{(maximum } I\text{)}. \tag{31-66}$$

Because the natural angular frequency ω of the *RLC* circuit is also equal to $1/\sqrt{LC}$, the maximum value of I occurs when the driving angular frequency matches the natural angular frequency—that is, at resonance. Thus, in an *RLC* circuit, resonance and maximum current amplitude I occur when

$$\omega_d = \omega = \frac{1}{\sqrt{LC}} \quad \text{(resonance)}. \tag{31-67}$$

Figure 31-16 shows three *resonance curves* for sinusoidally driven oscillations in three series *RLC* circuits differing only in R. Each curve peaks at its maximum current amplitude I when the ratio ω_d/ω is 1.00, but the maximum value of I decreases with increasing R. (The maximum I is always $\mathscr{E}_m/R$; to see why, combine Eqs. 31-61 and 31-62.) In addition, the curves increase in width (measured in Fig. 31-16 at half the maximum value of I) with increasing R.

To make physical sense of Fig. 31-16, consider how the reactances X_L and X_C change as we increase the driving angular frequency ω_d, starting with a value

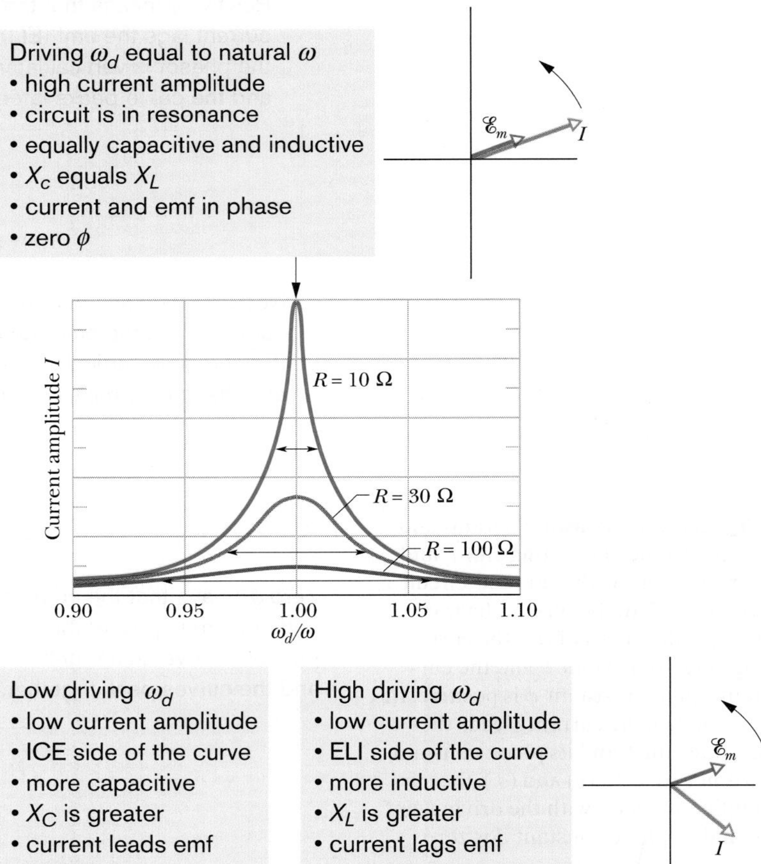

Fig. 31-16 *Resonance curves* for the driven RLC circuit of Fig. 31-7 with $L = 100\ \mu$H, $C = 100$ pF, and three values of R. The current amplitude I of the alternating current depends on how close the driving angular frequency ω_d is to the natural angular frequency ω. The horizontal arrow on each curve measures the curve's *half-width,* which is the width at the half-maximum level and is a measure of the sharpness of the resonance. To the left of $\omega_d/\omega = 1.00$, the circuit is mainly capacitive, with $X_C > X_L$; to the right, it is mainly inductive, with $X_L > X_C$.

much less than the natural frequency ω. For small ω_d, reactance X_L $(= \omega_d L)$ is small and reactance X_C $(= 1/\omega_d C)$ is large. Thus, the circuit is mainly capacitive and the impedance is dominated by the large X_C, which keeps the current low.

As we increase ω_d, reactance X_C remains dominant but decreases while reactance X_L increases. The decrease in X_C decreases the impedance, allowing the current to increase, as we see on the left side of any resonance curve in Fig. 31-16. When the increasing X_L and the decreasing X_C reach equal values, the current is greatest and the circuit is in resonance, with $\omega_d = \omega$.

As we continue to increase ω_d, the increasing reactance X_L becomes progressively more dominant over the decreasing reactance X_C. The impedance increases because of X_L and the current decreases, as on the right side of any resonance curve in Fig. 31-16. In summary, then: The low-angular-frequency side of a resonance curve is dominated by the capacitor's reactance, the high-angular-frequency side is dominated by the inductor's reactance, and resonance occurs in the middle.

CHECKPOINT 6

Here are the capacitive reactance and inductive reactance, respectively, for three sinusoidally driven series RLC circuits: (1) 50 Ω, 100 Ω; (2) 100 Ω, 50 Ω; (3) 50 Ω, 50 Ω. (a) For each, does the current lead or lag the applied emf, or are the two in phase? (b) Which circuit is in resonance?

Sample Problem

Current amplitude, impedance, and phase constant

In Fig. 31-7, let $R = 200\ \Omega$, $C = 15.0\ \mu F$, $L = 230$ mH, $f_d = 60.0$ Hz, and $\mathscr{E}_m = 36.0$ V. (These parameters are those used in the earlier sample problems above.)

(a) What is the current amplitude I?

KEY IDEA

The current amplitude I depends on the amplitude $\mathscr{E}_m$ of the driving emf and on the impedance Z of the circuit, according to Eq. 31-62 ($I = \mathscr{E}_m/Z$).

Calculations: So, we need to find Z, which depends on resistance R, capacitive reactance X_C, and inductive reactance X_L. The circuit's resistance is the given resistance R. Its capacitive reactance is due to the given capacitance and, from an earlier sample problem, $X_C = 177\ \Omega$. Its inductive reactance is due to the given inductance and, from another sample problem, $X_L = 86.7\ \Omega$. Thus, the circuit's impedance is

$$\begin{aligned} Z &= \sqrt{R^2 + (X_L - X_C)^2} \\ &= \sqrt{(200\ \Omega)^2 + (86.7\ \Omega - 177\ \Omega)^2} \\ &= 219\ \Omega. \end{aligned}$$

We then find

$$I = \frac{\mathscr{E}_m}{Z} = \frac{36.0\ \text{V}}{219\ \Omega} = 0.164\ \text{A}. \qquad \text{(Answer)}$$

(b) What is the phase constant ϕ of the current in the circuit relative to the driving emf?

KEY IDEA

The phase constant depends on the inductive reactance, the capacitive reactance, and the resistance of the circuit, according to Eq. 31-65.

Calculation: Solving Eq. 31-65 for ϕ leads to

$$\begin{aligned} \phi &= \tan^{-1}\frac{X_L - X_C}{R} = \tan^{-1}\frac{86.7\ \Omega - 177\ \Omega}{200\ \Omega} \\ &= -24.3° = -0.424\ \text{rad}. \qquad \text{(Answer)} \end{aligned}$$

The negative phase constant is consistent with the fact that the load is mainly capacitive; that is, $X_C > X_L$. In the common mnemonic for driven series RLC circuits, this circuit is an *ICE* circuit—the current *leads* the driving emf.

Additional examples, video, and practice available at *WileyPLUS*

31-10 Power in Alternating-Current Circuits

In the RLC circuit of Fig. 31-7, the source of energy is the alternating-current generator. Some of the energy that it provides is stored in the electric field in the capacitor, some is stored in the magnetic field in the inductor, and some is dissipated as thermal energy in the resistor. In steady-state operation, the average stored energy remains constant. The net transfer of energy is thus from the generator to the resistor, where energy is dissipated.

The instantaneous rate at which energy is dissipated in the resistor can be written, with the help of Eqs. 26-27 and 31-29, as

$$P = i^2R = [I\sin(\omega_d t - \phi)]^2R = I^2R\sin^2(\omega_d t - \phi). \qquad (31\text{-}68)$$

The *average* rate at which energy is dissipated in the resistor, however, is the average of Eq. 31-68 over time. Over one complete cycle, the average value of $\sin\theta$, where θ is any variable, is zero (Fig. 31-17*a*) but the average value of $\sin^2\theta$ is $\frac{1}{2}$ (Fig. 31-17*b*). (Note in Fig. 31-17*b* how the shaded areas under the curve but above the horizontal line marked $+\frac{1}{2}$ exactly fill in the unshaded spaces below that line.) Thus, we can write, from Eq. 31-68,

$$P_{avg} = \frac{I^2R}{2} = \left(\frac{I}{\sqrt{2}}\right)^2 R. \qquad (31\text{-}69)$$

The quantity $I/\sqrt{2}$ is called the **root-mean-square,** or **rms,** value of the current i:

$$I_{rms} = \frac{I}{\sqrt{2}} \quad \text{(rms current).} \qquad (31\text{-}70)$$

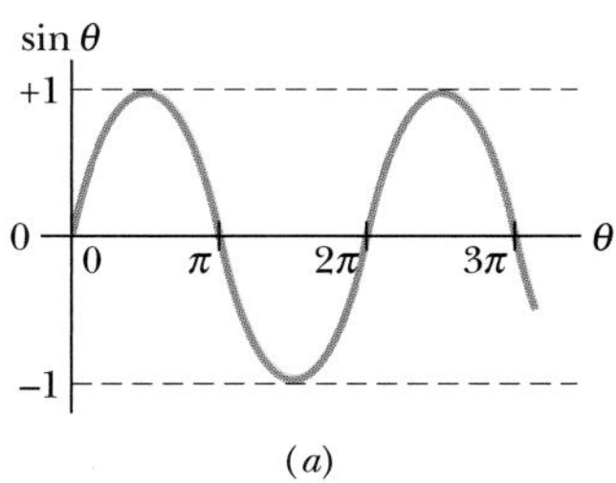

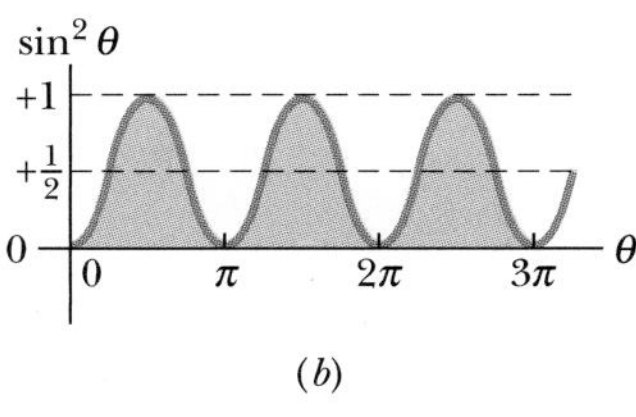

Fig. 31-17 (*a*) A plot of $\sin\theta$ versus θ. The average value over one cycle is zero. (*b*) A plot of $\sin^2\theta$ versus θ. The average value over one cycle is $\frac{1}{2}$.

We can now rewrite Eq. 31-69 as

$$P_{avg} = I_{rms}^2 R \quad \text{(average power).} \tag{31-71}$$

Equation 31-71 looks much like Eq. 26-27 ($P = i^2R$); the message is that if we switch to the rms current, we can compute the average rate of energy dissipation for alternating-current circuits just as for direct-current circuits.

We can also define rms values of voltages and emfs for alternating-current circuits:

$$V_{rms} = \frac{V}{\sqrt{2}} \quad \text{and} \quad \mathscr{E}_{rms} = \frac{\mathscr{E}_m}{\sqrt{2}} \quad \text{(rms voltage; rms emf).} \tag{31-72}$$

Alternating-current instruments, such as ammeters and voltmeters, are usually calibrated to read I_{rms}, V_{rms}, and $\mathscr{E}_{rms}$. Thus, if you plug an alternating-current voltmeter into a household electrical outlet and it reads 120 V, that is an rms voltage. The *maximum* value of the potential difference at the outlet is $\sqrt{2} \times (120\text{ V})$, or 170 V.

Because the proportionality factor $1/\sqrt{2}$ in Eqs. 31-70 and 31-72 is the same for all three variables, we can write Eqs. 31-62 and 31-60 as

$$I_{rms} = \frac{\mathscr{E}_{rms}}{Z} = \frac{\mathscr{E}_{rms}}{\sqrt{R^2 + (X_L - X_C)^2}}, \tag{31-73}$$

and, indeed, this is the form that we almost always use.

We can use the relationship $I_{rms} = \mathscr{E}_{rms}/Z$ to recast Eq. 31-71 in a useful equivalent way. We write

$$P_{avg} = \frac{\mathscr{E}_{rms}}{Z} I_{rms} R = \mathscr{E}_{rms} I_{rms} \frac{R}{Z}. \tag{31-74}$$

From Fig. 31-14*d*, Table 31-2, and Eq. 31-62, however, we see that R/Z is just the cosine of the phase constant ϕ:

$$\cos\phi = \frac{V_R}{\mathscr{E}_m} = \frac{IR}{IZ} = \frac{R}{Z}. \tag{31-75}$$

Equation 31-74 then becomes

$$P_{avg} = \mathscr{E}_{rms} I_{rms} \cos\phi \quad \text{(average power),} \tag{31-76}$$

in which the term $\cos\phi$ is called the **power factor.** Because $\cos\phi = \cos(-\phi)$, Eq. 31-76 is independent of the sign of the phase constant ϕ.

To maximize the rate at which energy is supplied to a resistive load in an *RLC* circuit, we should keep the power factor $\cos\phi$ as close to unity as possible. This is equivalent to keeping the phase constant ϕ in Eq. 31-29 as close to zero as possible. If, for example, the circuit is highly inductive, it can be made less so by putting more capacitance in the circuit, connected in series. (Recall that putting an additional capacitance into a series of capacitances decreases the equivalent capacitance C_{eq} of the series.) Thus, the resulting decrease in C_{eq} in the circuit reduces the phase constant and increases the power factor in Eq. 31-76. Power companies place series-connected capacitors throughout their transmission systems to get these results.

CHECKPOINT 7

(a) If the current in a sinusoidally driven series *RLC* circuit leads the emf, would we increase or decrease the capacitance to increase the rate at which energy is supplied to the resistance? (b) Would this change bring the resonant angular frequency of the circuit closer to the angular frequency of the emf or put it farther away?

Sample Problem

Driven *RLC* circuit: power factor and average power

A series *RLC* circuit, driven with $\mathscr{E}_{rms} = 120$ V at frequency $f_d = 60.0$ Hz, contains a resistance $R = 200\ \Omega$, an inductance with inductive reactance $X_L = 80.0\ \Omega$, and a capacitance with capacitive reactance $X_C = 150\ \Omega$.

(a) What are the power factor cos ϕ and phase constant ϕ of the circuit?

KEY IDEA

The power factor cos ϕ can be found from the resistance R and impedance Z via Eq. 31-75 ($\cos\phi = R/Z$).

Calculations: To calculate Z, we use Eq. 31-61:

$$Z = \sqrt{R^2 + (X_L - X_C)^2}$$
$$= \sqrt{(200\ \Omega)^2 + (80.0\ \Omega - 150\ \Omega)^2} = 211.90\ \Omega.$$

Equation 31-75 then gives us

$$\cos\phi = \frac{R}{Z} = \frac{200\ \Omega}{211.90\ \Omega} = 0.9438 \approx 0.944. \quad \text{(Answer)}$$

Taking the inverse cosine then yields

$$\phi = \cos^{-1} 0.944 = \pm 19.3°.$$

Both +19.3° and −19.3° have a cosine of 0.944. To determine which sign is correct, we must consider whether the current leads or lags the driving emf. Because $X_C > X_L$, this circuit is mainly capacitive, with the current leading the emf. Thus, ϕ must be negative:

$$\phi = -19.3°. \quad \text{(Answer)}$$

We could, instead, have found ϕ with Eq. 31-65. A calculator would then have given us the answer with the minus sign.

(b) What is the average rate P_{avg} at which energy is dissipated in the resistance?

KEY IDEAS

There are two ways and two ideas to use: (1) Because the circuit is assumed to be in steady-state operation, the rate at which energy is dissipated in the resistance is equal to the rate at which energy is supplied to the circuit, as given by Eq. 31-76 ($P_{avg} = \mathscr{E}_{rms} I_{rms} \cos\phi$). (2) The rate at which energy is dissipated in a resistance R depends on the square of the rms current I_{rms} through it, according to Eq. 31-71 ($P_{avg} = I^2_{rms} R$).

First way: We are given the rms driving emf $\mathscr{E}_{rms}$ and we already know cos ϕ from part (a). The rms current I_{rms} is determined by the rms value of the driving emf and the circuit's impedance Z (which we know), according to Eq. 31-73:

$$I_{rms} = \frac{\mathscr{E}_{rms}}{Z}.$$

Substituting this into Eq. 31-76 then leads to

$$P_{avg} = \mathscr{E}_{rms} I_{rms} \cos\phi = \frac{\mathscr{E}^2_{rms}}{Z} \cos\phi$$
$$= \frac{(120\ \text{V})^2}{211.90\ \Omega}(0.9438) = 64.1\ \text{W}. \quad \text{(Answer)}$$

Second way: Instead, we can write

$$P_{avg} = I^2_{rms} R = \frac{\mathscr{E}^2_{rms}}{Z^2} R$$
$$= \frac{(120\ \text{V})^2}{(211.90\ \Omega)^2}(200\ \Omega) = 64.1\ \text{W}. \quad \text{(Answer)}$$

(c) What new capacitance C_{new} is needed to maximize P_{avg} if the other parameters of the circuit are not changed?

KEY IDEAS

(1) The average rate P_{avg} at which energy is supplied and dissipated is maximized if the circuit is brought into resonance with the driving emf. (2) Resonance occurs when $X_C = X_L$.

Calculations: From the given data, we have $X_C > X_L$. Thus, we must decrease X_C to reach resonance. From Eq. 31-39 ($X_C = 1/\omega_d C$), we see that this means we must increase C to the new value C_{new}.

Using Eq. 31-39, we can write the resonance condition $X_C = X_L$ as

$$\frac{1}{\omega_d C_{new}} = X_L.$$

Substituting $2\pi f_d$ for ω_d (because we are given f_d and not ω_d) and then solving for C_{new}, we find

$$C_{new} = \frac{1}{2\pi f_d X_L} = \frac{1}{(2\pi)(60\ \text{Hz})(80.0\ \Omega)}$$
$$= 3.32 \times 10^{-5}\ \text{F} = 33.2\ \mu\text{F}. \quad \text{(Answer)}$$

Following the procedure of part (b), you can show that with C_{new}, the average power of energy dissipation P_{avg} would then be at its maximum value of

$$P_{avg,\,max} = 72.0\ \text{W}.$$

Additional examples, video, and practice available at *WileyPLUS*

31-11 Transformers

Energy Transmission Requirements

When an ac circuit has only a resistive load, the power factor in Eq. 31-76 is $\cos 0° = 1$ and the applied rms emf $\mathscr{E}_{rms}$ is equal to the rms voltage V_{rms} across the load. Thus, with an rms current I_{rms} in the load, energy is supplied and dissipated at the average rate of

$$P_{avg} = \mathscr{E}I = IV. \qquad (31\text{-}77)$$

(In Eq. 31-77 and the rest of this section, we follow conventional practice and drop the subscripts identifying rms quantities. Engineers and scientists assume that all time-varying currents and voltages are reported as rms values; that is what the meters read.) Equation 31-77 tells us that, to satisfy a given power requirement, we have a range of choices for I and V, provided only that the product IV is as required.

In electrical power distribution systems it is desirable for reasons of safety and for efficient equipment design to deal with relatively low voltages at both the generating end (the electrical power plant) and the receiving end (the home or factory). Nobody wants an electric toaster or a child's electric train to operate at, say, 10 kV. On the other hand, in the transmission of electrical energy from the generating plant to the consumer, we want the lowest practical current (hence the largest practical voltage) to minimize I^2R losses (often called *ohmic losses*) in the transmission line.

As an example, consider the 735 kV line used to transmit electrical energy from the La Grande 2 hydroelectric plant in Quebec to Montreal, 1000 km away. Suppose that the current is 500 A and the power factor is close to unity. Then from Eq. 31-77, energy is supplied at the average rate

$$P_{avg} = \mathscr{E}I = (7.35 \times 10^5 \text{ V})(500 \text{ A}) = 368 \text{ MW}.$$

The resistance of the transmission line is about 0.220 Ω/km; thus, there is a total resistance of about 220 Ω for the 1000 km stretch. Energy is dissipated due to that resistance at a rate of about

$$P_{avg} = I^2R = (500 \text{ A})^2(220\ \Omega) = 55.0 \text{ MW},$$

which is nearly 15% of the supply rate.

Imagine what would happen if we doubled the current and halved the voltage. Energy would be supplied by the plant at the same average rate of 368 MW as previously, but now energy would be dissipated at the rate of about

$$P_{avg} = I^2R = (1000 \text{ A})^2(220\ \Omega) = 220 \text{ MW},$$

which is *almost 60% of the supply rate.* Hence the general energy transmission rule: Transmit at the highest possible voltage and the lowest possible current.

The Ideal Transformer

The transmission rule leads to a fundamental mismatch between the requirement for efficient high-voltage transmission and the need for safe low-voltage generation and consumption. We need a device with which we can raise (for transmission) and lower (for use) the ac voltage in a circuit, keeping the product current × voltage essentially constant. The **transformer** is such a device. It has no moving parts, operates by Faraday's law of induction, and has no simple direct-current counterpart.

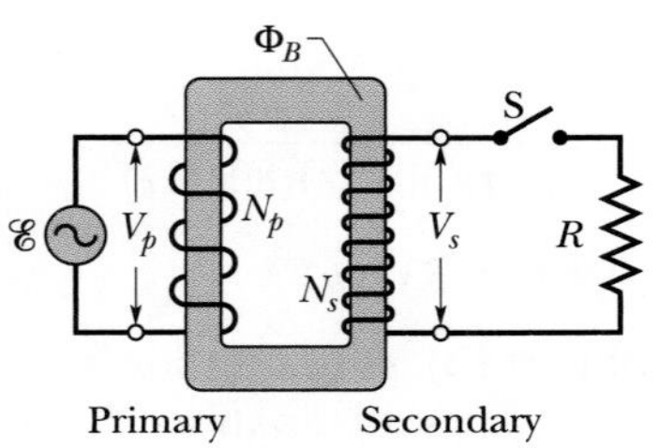

Fig. 31-18 An ideal transformer (two coils wound on an iron core) in a basic transformer circuit. An ac generator produces current in the coil at the left (the *primary*). The coil at the right (the *secondary*) is connected to the resistive load R when switch S is closed.

The *ideal transformer* in Fig. 31-18 consists of two coils, with different numbers of turns, wound around an iron core. (The coils are insulated from the core.) In use, the primary winding, of N_p turns, is connected to an alternating-current generator whose emf $\mathscr{E}$ at any time t is given by

$$\mathscr{E} = \mathscr{E}_m \sin \omega t. \qquad (31\text{-}78)$$

The secondary winding, of N_s turns, is connected to load resistance R, but its circuit is an open circuit as long as switch S is open (which we assume for the present). Thus, there can be no current through the secondary coil. We assume further for this ideal transformer that the resistances of the primary and secondary windings are negligible. Well-designed, high-capacity transformers can have energy losses as low as 1%; so our assumptions are reasonable.

For the assumed conditions, the primary winding (or *primary*) is a pure inductance and the primary circuit is like that in Fig. 31-12. Thus, the (very small) primary current, also called the *magnetizing current* I_{mag}, lags the primary voltage V_p by 90°; the primary's power factor ($= \cos\phi$ in Eq. 31-76) is zero; so no power is delivered from the generator to the transformer.

However, the small sinusoidally changing primary current I_{mag} produces a sinusoidally changing magnetic flux Φ_B in the iron core. The core acts to strengthen the flux and to bring it through the secondary winding (or *secondary*). Because Φ_B varies, it induces an emf $\mathscr{E}_{turn}$ ($= d\Phi_B/dt$) in each turn of the secondary. In fact, this emf per turn $\mathscr{E}_{turn}$ is the same in the primary and the secondary. Across the primary, the voltage V_p is the product of $\mathscr{E}_{turn}$ and the number of turns N_p; that is, $V_p = \mathscr{E}_{turn}N_p$. Similarly, across the secondary the voltage is $V_s = \mathscr{E}_{turn}N_s$. Thus, we can write

$$\mathscr{E}_{turn} = \frac{V_p}{N_p} = \frac{V_s}{N_s},$$

or
$$V_s = V_p \frac{N_s}{N_p} \quad \text{(transformation of voltage).} \qquad (31\text{-}79)$$

If $N_s > N_p$, the device is a *step-up transformer* because it steps the primary's voltage V_p *up* to a higher voltage V_s. Similarly, if $N_s < N_p$, it is a *step-down transformer.*

With switch S open, no energy is transferred from the generator to the rest of the circuit, but when we close S to connect the secondary to the resistive load R, energy *is* transferred. (In general, the load would also contain inductive and capacitive elements, but here we consider just resistance R.) Here is the process:

1. An alternating current I_s appears in the secondary circuit, with corresponding energy dissipation rate I_s^2R ($= V_s^2/R$) in the resistive load.
2. This current produces its own alternating magnetic flux in the iron core, and this flux induces an opposing emf in the primary windings.
3. The voltage V_p of the primary, however, cannot change in response to this opposing emf because it must always be equal to the emf $\mathscr{E}$ that is provided by the generator; closing switch S cannot change this fact.
4. To maintain V_p, the generator now produces (in addition to I_{mag}) an alternating current I_p in the primary circuit; the magnitude and phase constant of I_p are just those required for the emf induced by I_p in the primary to exactly cancel the emf induced there by I_s. Because the phase constant of I_p is not 90° like that of I_{mag}, this current I_p can transfer energy to the primary.

We want to relate I_s to I_p. However, rather than analyze the foregoing complex process in detail, let us just apply the principle of conservation of energy. The rate at which the generator transfers energy to the primary is equal to I_pV_p. The rate at which the primary then transfers energy to the secondary (via the alternating magnetic field linking the two coils) is I_sV_s. Because we assume that no energy is lost along the way, conservation of energy requires that

$$I_pV_p = I_sV_s.$$

Substituting for V_s from Eq. 31-79, we find that

$$I_s = I_p \frac{N_p}{N_s} \quad \text{(transformation of currents).} \qquad (31\text{-}80)$$

This equation tells us that the current I_s in the secondary can differ from the current I_p in the primary, depending on the *turns ratio* N_p/N_s.

Current I_p appears in the primary circuit because of the resistive load R in the secondary circuit. To find I_p, we substitute $I_s = V_s/R$ into Eq. 31-80 and then we substitute for V_s from Eq. 31-79. We find

$$I_p = \frac{1}{R}\left(\frac{N_s}{N_p}\right)^2 V_p. \tag{31-81}$$

This equation has the form $I_p = V_p/R_{eq}$, where equivalent resistance R_{eq} is

$$R_{eq} = \left(\frac{N_p}{N_s}\right)^2 R. \tag{31-82}$$

This R_{eq} is the value of the load resistance as "seen" by the generator; the generator produces the current I_p and voltage V_p as if the generator were connected to a resistance R_{eq}.

CHECKPOINT 8

An alternating-current emf device in a certain circuit has a smaller resistance than that of the resistive load in the circuit; to increase the transfer of energy from the device to the load, a transformer will be connected between the two. (a) Should N_s be greater than or less than N_p? (b) Will that make it a step-up or step-down transformer?

Impedance Matching

Equation 31-82 suggests still another function for the transformer. For maximum transfer of energy from an emf device to a resistive load, the resistance of the emf device must equal the resistance of the load. The same relation holds for ac circuits except that the *impedance* (rather than just the resistance) of the generator must equal that of the load. Often this condition is not met. For example, in a music-playing system, the amplifier has high impedance and the speaker set has low impedance. We can match the impedances of the two devices by coupling them through a transformer that has a suitable turns ratio N_p/N_s.

Sample Problem

Transformer: turns ratio, average power, rms currents

A transformer on a utility pole operates at $V_p = 8.5$ kV on the primary side and supplies electrical energy to a number of nearby houses at $V_s = 120$ V, both quantities being rms values. Assume an ideal step-down transformer, a purely resistive load, and a power factor of unity.

(a) What is the turns ratio N_p/N_s of the transformer?

KEY IDEA

The turns ratio N_p/N_s is related to the (given) rms primary and secondary voltages via Eq. 31-79 ($V_s = V_p N_s/N_p$).

Calculation: We can write Eq. 31-79 as

$$\frac{V_s}{V_p} = \frac{N_s}{N_p}. \tag{31-83}$$

(Note that the right side of this equation is the *inverse* of the turns ratio.) Inverting both sides of Eq. 31-83 gives us

$$\frac{N_p}{N_s} = \frac{V_p}{V_s} = \frac{8.5 \times 10^3 \text{ V}}{120 \text{ V}} = 70.83 \approx 71. \quad \text{(Answer)}$$

(b) The average rate of energy consumption (or dissipation) in the houses served by the transformer is 78 kW. What are the rms currents in the primary and secondary of the transformer?

KEY IDEA

For a purely resistive load, the power factor cos ϕ is unity; thus, the average rate at which energy is supplied and dissipated is given by Eq. 31-77 ($P_{avg} = \mathscr{E}I = IV$).

Calculations: In the primary circuit, with $V_p = 8.5$ kV, Eq. 31-77 yields

$$I_p = \frac{P_{avg}}{V_p} = \frac{78 \times 10^3 \text{ W}}{8.5 \times 10^3 \text{ V}} = 9.176 \text{ A} \approx 9.2 \text{ A}. \quad \text{(Answer)}$$

Similarly, in the secondary circuit,

$$I_s = \frac{P_{avg}}{V_s} = \frac{78 \times 10^3 \text{ W}}{120 \text{ V}} = 650 \text{ A}. \quad \text{(Answer)}$$

You can check that $I_s = I_p(N_p/N_s)$ as required by Eq. 31-80.

(c) What is the resistive load R_s in the secondary circuit? What is the corresponding resistive load R_p in the primary circuit?

One way: We can use $V = IR$ to relate the resistive load to the rms voltage and current. For the secondary circuit, we find

$$R_s = \frac{V_s}{I_s} = \frac{120\text{ V}}{650\text{ A}} = 0.1846\ \Omega \approx 0.18\ \Omega. \qquad \text{(Answer)}$$

Similarly, for the primary circuit we find

$$R_p = \frac{V_p}{I_p} = \frac{8.5 \times 10^3\text{ V}}{9.176\text{ A}} = 926\ \Omega \approx 930\ \Omega. \qquad \text{(Answer)}$$

Second way: We use the fact that R_p equals the equivalent resistive load "seen" from the primary side of the transformer, which is a resistance modified by the turns ratio and given by Eq. 31-82 ($R_{eq} = (N_p/N_s)^2R$). If we substitute R_p for R_{eq} and R_s for R, that equation yields

$$R_p = \left(\frac{N_p}{N_s}\right)^2 R_s = (70.83)^2(0.1846\ \Omega)$$

$$= 926\ \Omega \approx 930\ \Omega. \qquad \text{(Answer)}$$

Additional examples, video, and practice available at *WileyPLUS*

REVIEW & SUMMARY

_LC_ Energy Transfers In an oscillating LC circuit, energy is shuttled periodically between the electric field of the capacitor and the magnetic field of the inductor; instantaneous values of the two forms of energy are

$$U_E = \frac{q^2}{2C} \quad \text{and} \quad U_B = \frac{Li^2}{2}, \qquad (31\text{-}1, 31\text{-}2)$$

where q is the instantaneous charge on the capacitor and i is the instantaneous current through the inductor. The total energy $U\,(= U_E + U_B)$ remains constant.

_LC_ Charge and Current Oscillations The principle of conservation of energy leads to

$$L\frac{d^2q}{dt^2} + \frac{1}{C}q = 0 \quad (LC\text{ oscillations}) \qquad (31\text{-}11)$$

as the differential equation of LC oscillations (with no resistance). The solution of Eq. 31-11 is

$$q = Q\cos(\omega t + \phi) \quad \text{(charge)}, \qquad (31\text{-}12)$$

in which Q is the *charge amplitude* (maximum charge on the capacitor) and the angular frequency ω of the oscillations is

$$\omega = \frac{1}{\sqrt{LC}}. \qquad (31\text{-}4)$$

The phase constant ϕ in Eq. 31-12 is determined by the initial conditions (at $t = 0$) of the system.

The current i in the system at any time t is

$$i = -\omega Q\sin(\omega t + \phi) \quad \text{(current)}, \qquad (31\text{-}13)$$

in which ωQ is the *current amplitude I*.

Damped Oscillations Oscillations in an LC circuit are damped when a dissipative element R is also present in the circuit. Then

$$L\frac{d^2q}{dt^2} + R\frac{dq}{dt} + \frac{1}{C}q = 0 \quad (RLC\text{ circuit}). \qquad (31\text{-}24)$$

The solution of this differential equation is

$$q = Qe^{-Rt/2L}\cos(\omega' t + \phi), \qquad (31\text{-}25)$$

where

$$\omega' = \sqrt{\omega^2 - (R/2L)^2}. \qquad (31\text{-}26)$$

We consider only situations with small R and thus small damping; then $\omega' \approx \omega$.

Alternating Currents; Forced Oscillations A series RLC circuit may be set into *forced oscillation* at a *driving angular frequency* ω_d by an external alternating emf

$$\mathscr{E} = \mathscr{E}_m \sin \omega_d t. \qquad (31\text{-}28)$$

The current driven in the circuit is

$$i = I\sin(\omega_d t - \phi), \qquad (31\text{-}29)$$

where ϕ is the phase constant of the current.

Resonance The current amplitude I in a series RLC circuit driven by a sinusoidal external emf is a maximum ($I = \mathscr{E}_m/R$) when the driving angular frequency ω_d equals the natural angular frequency ω of the circuit (that is, at *resonance*). Then $X_C = X_L$, $\phi = 0$, and the current is in phase with the emf.

Single Circuit Elements The alternating potential difference across a resistor has amplitude $V_R = IR$; the current is in phase with the potential difference.

For a *capacitor,* $V_C = IX_C$, in which $X_C = 1/\omega_d C$ is the **capacitive reactance;** the current here leads the potential difference by 90° ($\phi = -90° = -\pi/2$ rad).

For an *inductor,* $V_L = IX_L$, in which $X_L = \omega_d L$ is the **inductive reactance;** the current here lags the potential difference by 90° ($\phi = +90° = +\pi/2$ rad).

Series _RLC_ Circuits For a series RLC circuit with an alternating external emf given by Eq. 31-28 and a resulting alternating current given by Eq. 31-29,

$$I = \frac{\mathscr{E}_m}{\sqrt{R^2 + (X_L - X_C)^2}} = \frac{\mathscr{E}_m}{\sqrt{R^2 + (\omega_d L - 1/\omega_d C)^2}} \quad \text{(current amplitude)} \quad (31\text{-}60, 31\text{-}63)$$

and

$$\tan\phi = \frac{X_L - X_C}{R} \quad \text{(phase constant)}. \qquad (31\text{-}65)$$

Defining the impedance Z of the circuit as

$$Z = \sqrt{R^2 + (X_L - X_C)^2} \quad \text{(impedance)} \qquad (31\text{-}61)$$

allows us to write Eq. 31-60 as $I = \mathscr{E}_m/Z$.

Power In a series RLC circuit, the **average power** P_{avg} of the generator is equal to the production rate of thermal energy in the resistor:

$$P_{avg} = I^2_{rms}R = \mathscr{E}_{rms}I_{rms}\cos\phi. \qquad (31\text{-}71, 31\text{-}76)$$

Here rms stands for **root-mean-square;** the rms quantities are related to the maximum quantities by $I_{rms} = I/\sqrt{2}$, $V_{rms} = V/\sqrt{2}$, and $\mathscr{E}_{rms} = \mathscr{E}_m/\sqrt{2}$. The term $\cos\phi$ is called the **power factor** of the circuit.

Transformers A *transformer* (assumed to be ideal) is an iron core on which are wound a primary coil of N_p turns and a secondary coil of N_s turns. If the primary coil is connected across an alternating-current generator, the primary and secondary voltages are related by

$$V_s = V_p\frac{N_s}{N_p} \quad \text{(transformation of voltage).} \qquad (31\text{-}79)$$

The currents through the coils are related by

$$I_s = I_p\frac{N_p}{N_s} \quad \text{(transformation of currents),} \qquad (31\text{-}80)$$

and the equivalent resistance of the secondary circuit, as seen by the generator, is

$$R_{eq} = \left(\frac{N_p}{N_s}\right)^2 R, \qquad (31\text{-}82)$$

where R is the resistive load in the secondary circuit. The ratio N_p/N_s is called the transformer's *turns ratio.*

QUESTIONS

1 Figure 31-19 shows three oscillating LC circuits with identical inductors and capacitors. Rank the circuits according to the time taken to fully discharge the capacitors during the oscillations, greatest first.

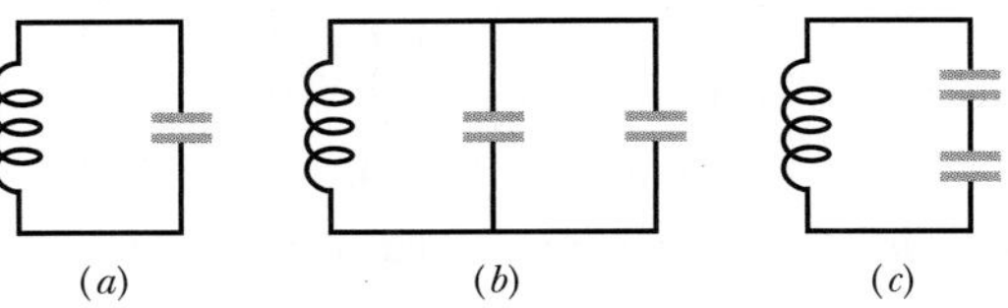

Fig. 31-19 Question 1.

2 Figure 31-20 shows graphs of capacitor voltage v_C for LC circuits 1 and 2, which contain identical capacitances and have the same maximum charge Q. Are (a) the inductance L and (b) the maximum current I in circuit 1 greater than, less than, or the same as those in circuit 2?

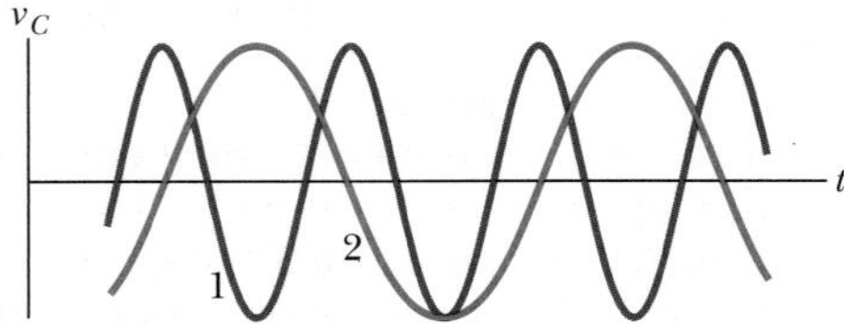

Fig. 31-20 Question 2.

3 A charged capacitor and an inductor are connected at time $t = 0$. In terms of the period T of the resulting oscillations, what is the first later time at which the following reach a maximum: (a) U_B, (b) the magnetic flux through the inductor, (c) di/dt, and (d) the emf of the inductor?

4 What values of phase constant ϕ in Eq. 31-12 allow situations *(a)*, *(c)*, *(e)*, and *(g)* of Fig. 31-1 to occur at $t = 0$?

5 Curve a in Fig. 31-21 gives the impedance Z of a driven RC circuit versus the driving angular frequency ω_d. The other two curves are similar but for different values of resistance R and capacitance C. Rank the three curves according to the corresponding value of R, greatest first.

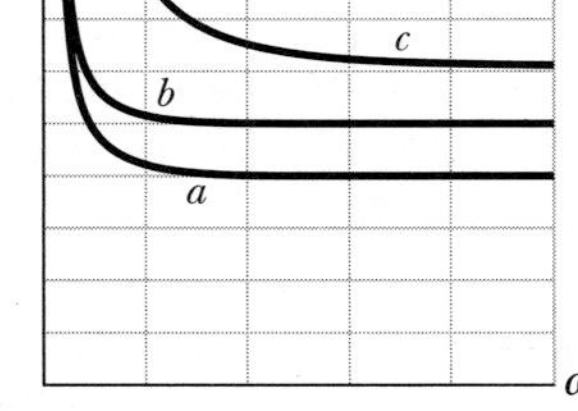

Fig. 31-21 Question 5.

6 Charges on the capacitors in three oscillating LC circuits vary as: (1) $q = 2\cos 4t$, (2) $q = 4\cos t$, (3) $q = 3\cos 4t$ (with q in coulombs and t in seconds). Rank the circuits according to (a) the current amplitude and (b) the period, greatest first.

7 An alternating emf source with a certain emf amplitude is connected, in turn, to a resistor, a capacitor, and then an inductor. Once connected to one of the devices, the driving frequency f_d is varied and the amplitude I of the resulting current through the device is measured and plotted. Which of the three plots in Fig. 31-22 corresponds to which of the three devices?

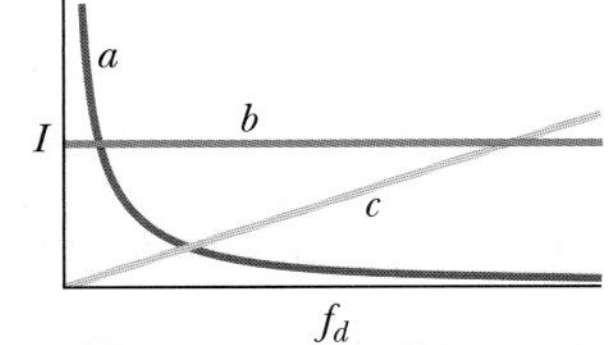

Fig. 31-22 Question 7.

8 The values of the phase constant ϕ for four sinusoidally driven series RLC circuits are (1) $-15°$, (2) $+35°$, (3) $\pi/3$ rad, and (4) $-\pi/6$ rad. (a) In which is the load primarily capacitive? (b) In which does the current lag the alternating emf?

9 Figure 31-23 shows the current i and driving emf $\mathscr{E}$ for a series RLC circuit. (a) Is the phase constant positive or negative? (b) To increase the rate at which energy is transferred to the resistive load, should L be increased or decreased? (c) Should, instead, C be increased or decreased?

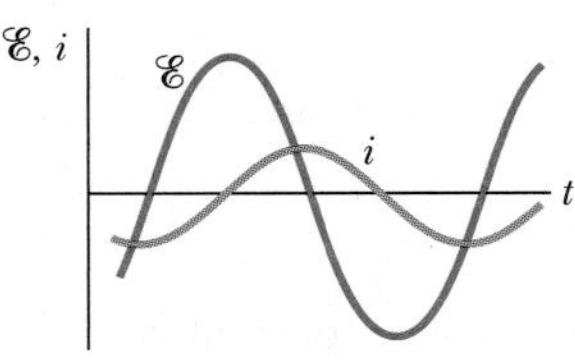

Fig. 31-23 Question 9.

10 Figure 31-24 shows three situations like those of Fig. 31-15. Is the driving angular frequency greater than, less than, or equal to the resonant angular frequency of the circuit in (a) situation 1, (b) situation 2, and (c) situation 3?

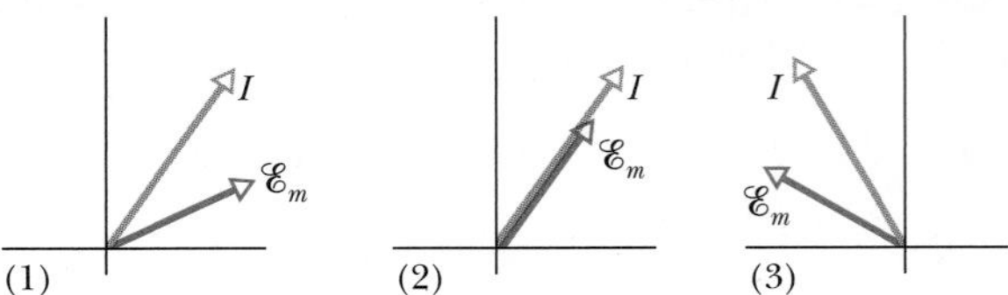

Fig. 31-24 Question 10.

11 Figure 31-25 shows the current i and driving emf $\mathscr{E}$ for a series RLC circuit. Relative to the emf curve, does the current curve shift leftward or rightward and does the amplitude of that curve increase or decrease if we slightly increase (a) L, (b) C, and (c) ω_d?

12 Figure 31-25 shows the current i and driving emf $\mathscr{E}$ for a series RLC circuit. (a) Does the current lead or lag the emf? (b) Is the circuit's load mainly capacitive or mainly inductive? (c) Is the angular frequency ω_d of the emf greater than or less than the natural angular frequency ω?

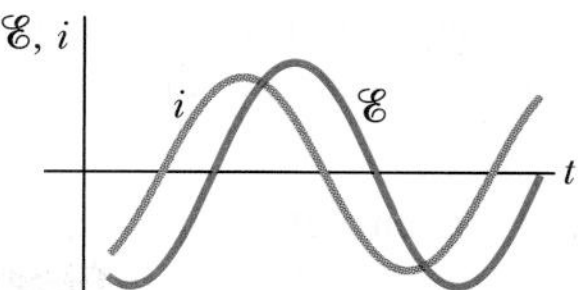

Fig. 31-25 Questions 11 and 12.

PROBLEMS

GO Tutoring problem available (at instructor's discretion) in *WileyPLUS* and WebAssign
SSM Worked-out solution available in Student Solutions Manual
• – ••• Number of dots indicates level of problem difficulty
Additional information available in *The Flying Circus of Physics* and at flyingcircusofphysics.com
WWW Worked-out solution is at
ILW Interactive solution is at
http://www.wiley.com/college/halliday

sec. 31-2 *LC* Oscillations, Qualitatively

•**1** An oscillating LC circuit consists of a 75.0 mH inductor and a 3.60 μF capacitor. If the maximum charge on the capacitor is 2.90 μC, what are (a) the total energy in the circuit and (b) the maximum current?

•**2** The frequency of oscillation of a certain LC circuit is 200 kHz. At time $t = 0$, plate A of the capacitor has maximum positive charge. At what earliest time $t > 0$ will (a) plate A again have maximum positive charge, (b) the other plate of the capacitor have maximum positive charge, and (c) the inductor have maximum magnetic field?

•**3** In a certain oscillating LC circuit, the total energy is converted from electrical energy in the capacitor to magnetic energy in the inductor in 1.50 μs. What are (a) the period of oscillation and (b) the frequency of oscillation? (c) How long after the magnetic energy is a maximum will it be a maximum again?

•**4** What is the capacitance of an oscillating LC circuit if the maximum charge on the capacitor is 1.60 μC and the total energy is 140 μJ?

•**5** In an oscillating LC circuit, $L = 1.10$ mH and $C = 4.00$ μF. The maximum charge on the capacitor is 3.00 μC. Find the maximum current.

sec. 31-3 The Electrical–Mechanical Analogy

•**6** A 0.50 kg body oscillates in SHM on a spring that, when extended 2.0 mm from its equilibrium position, has an 8.0 N restoring force. What are (a) the angular frequency of oscillation, (b) the period of oscillation, and (c) the capacitance of an LC circuit with the same period if L is 5.0 H?

••**7** SSM The energy in an oscillating LC circuit containing a 1.25 H inductor is 5.70 μJ. The maximum charge on the capacitor is 175 μC. For a mechanical system with the same period, find the (a) mass, (b) spring constant, (c) maximum displacement, and (d) maximum speed.

sec. 31-4 *LC* Oscillations, Quantitatively

•**8** A single loop consists of inductors (L_1, L_2, . . .), capacitors (C_1, C_2, . . .), and resistors (R_1, R_2, . . .) connected in series as shown, for example, in Fig. 31-26*a*. Show that regardless of the sequence of these circuit elements in the loop, the behavior of this circuit is identical to that of the simple LC circuit shown in Fig. 31-26*b*. (*Hint:* Consider the loop rule and see Problem 47 in Chapter 30.)

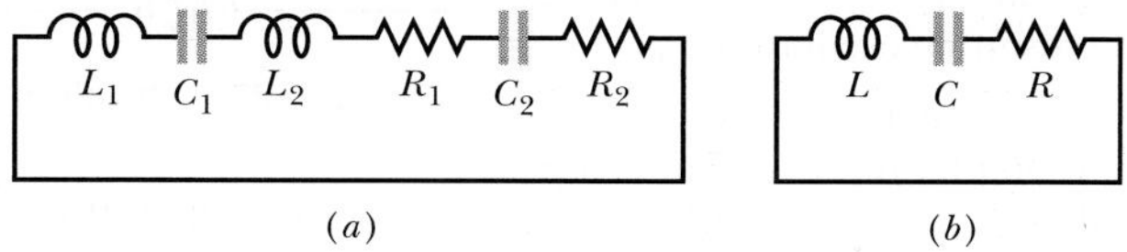

Fig. 31-26 Problem 8.

•**9** ILW In an oscillating LC circuit with $L = 50$ mH and $C = 4.0$ μF, the current is initially a maximum. How long will it take before the capacitor is fully charged for the first time?

•**10** LC oscillators have been used in circuits connected to loudspeakers to create some of the sounds of electronic music. What inductance must be used with a 6.7 μF capacitor to produce a frequency of 10 kHz, which is near the middle of the audible range of frequencies?

••**11** SSM WWW A variable capacitor with a range from 10 to 365 pF is used with a coil to form a variable-frequency LC circuit to tune the input to a radio. (a) What is the ratio of maximum frequency to minimum frequency that can be obtained with such a capacitor? If this circuit is to obtain frequencies from 0.54 MHz to 1.60 MHz, the ratio computed in (a) is too large. By adding a capacitor in parallel to the variable capacitor, this range can be adjusted. To obtain the desired frequency range, (b) what capacitance should be added and (c) what inductance should the coil have?

••**12** In an oscillating LC circuit, when 75.0% of the total energy is stored in the inductor's magnetic field, (a) what multiple of the maximum charge is on the capacitor and (b) what multiple of the maximum current is in the inductor?

••**13** In an oscillating LC circuit, $L = 3.00$ mH and $C = 2.70$ μF. At $t = 0$ the charge on the capacitor is zero and the current is 2.00 A. (a) What is the maximum charge that will appear on the capacitor? (b) At what earliest time $t > 0$ is the rate at which energy is stored in the capacitor greatest, and (c) what is that greatest rate?

••**14** To construct an oscillating LC system, you can choose from a 10 mH inductor, a 5.0 μF capacitor, and a 2.0 μF capacitor. What are the (a) smallest, (b) second smallest, (c) second largest, and (d) largest oscillation frequency that can be set up by these elements in various combinations?

••15 ILW An oscillating LC circuit consisting of a 1.0 nF capacitor and a 3.0 mH coil has a maximum voltage of 3.0 V. What are (a) the maximum charge on the capacitor, (b) the maximum current through the circuit, and (c) the maximum energy stored in the magnetic field of the coil?

••16 An inductor is connected across a capacitor whose capacitance can be varied by turning a knob. We wish to make the frequency of oscillation of this LC circuit vary linearly with the angle of rotation of the knob, going from 2×10^5 to 4×10^5 Hz as the knob turns through 180°. If $L = 1.0$ mH, plot the required capacitance C as a function of the angle of rotation of the knob.

••17 ILW GO In Fig. 31-27, $R = 14.0\ \Omega$, $C = 6.20\ \mu$F, and $L = 54.0$ mH, and the ideal battery has emf $\mathscr{E} = 34.0$ V. The switch is kept at a for a long time and then thrown to position b. What are the (a) frequency and (b) current amplitude of the resulting oscillations?

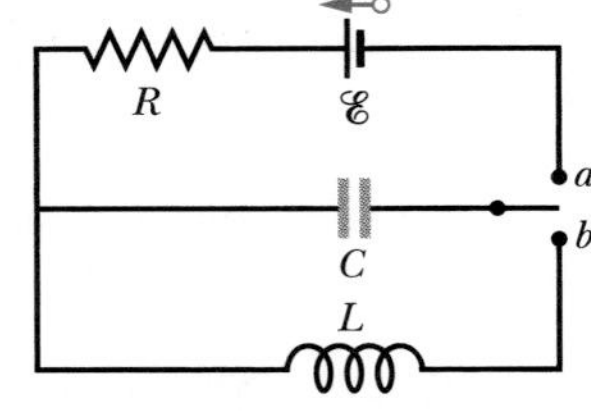

Fig. 31-27 Problem 17.

••18 An oscillating LC circuit has a current amplitude of 7.50 mA, a potential amplitude of 250 mV, and a capacitance of 220 nF. What are (a) the period of oscillation, (b) the maximum energy stored in the capacitor, (c) the maximum energy stored in the inductor, (d) the maximum rate at which the current changes, and (e) the maximum rate at which the inductor gains energy?

••19 Using the loop rule, derive the differential equation for an LC circuit (Eq. 31-11).

••20 GO In an oscillating LC circuit in which $C = 4.00\ \mu$F, the maximum potential difference across the capacitor during the oscillations is 1.50 V and the maximum current through the inductor is 50.0 mA. What are (a) the inductance L and (b) the frequency of the oscillations? (c) How much time is required for the charge on the capacitor to rise from zero to its maximum value?

••21 ILW In an oscillating LC circuit with $C = 64.0\ \mu$F, the current is given by $i = (1.60) \sin(2500t + 0.680)$, where t is in seconds, i in amperes, and the phase constant in radians. (a) How soon after $t = 0$ will the current reach its maximum value? What are (b) the inductance L and (c) the total energy?

••22 A series circuit containing inductance L_1 and capacitance C_1 oscillates at angular frequency ω. A second series circuit, containing inductance L_2 and capacitance C_2, oscillates at the same angular frequency. In terms of ω, what is the angular frequency of oscillation of a series circuit containing all four of these elements? Neglect resistance. (*Hint:* Use the formulas for equivalent capacitance and equivalent inductance; see Section 25-4 and Problem 47 in Chapter 30.)

••23 In an oscillating LC circuit, $L = 25.0$ mH and $C = 7.80\ \mu$F. At time $t = 0$ the current is 9.20 mA, the charge on the capacitor is 3.80 μC, and the capacitor is charging. What are (a) the total energy in the circuit, (b) the maximum charge on the capacitor, and (c) the maximum current? (d) If the charge on the capacitor is given by $q = Q \cos(\omega t + \phi)$, what is the phase angle ϕ? (e) Suppose the data are the same, except that the capacitor is discharging at $t = 0$. What then is ϕ?

sec. 31-5 Damped Oscillations in an *RLC* Circuit

••24 GO A single-loop circuit consists of a 7.20 Ω resistor, a 12.0 H inductor, and a 3.20 μF capacitor. Initially the capacitor has a charge of 6.20 μC and the current is zero. Calculate the charge on the capacitor N complete cycles later for (a) $N = 5$, (b) $N = 10$, and (c) $N = 100$.

••25 ILW What resistance R should be connected in series with an inductance $L = 220$ mH and capacitance $C = 12.0\ \mu$F for the maximum charge on the capacitor to decay to 99.0% of its initial value in 50.0 cycles? (Assume $\omega' \approx \omega$.)

••26 In an oscillating series RLC circuit, find the time required for the maximum energy present in the capacitor during an oscillation to fall to half its initial value. Assume $q = Q$ at $t = 0$.

•••27 SSM In an oscillating series RLC circuit, show that $\Delta U/U$, the fraction of the energy lost per cycle of oscillation, is given to a close approximation by $2\pi R/\omega L$. The quantity $\omega L/R$ is often called the Q of the circuit (for *quality*). A high-Q circuit has low resistance and a low fractional energy loss ($= 2\pi/Q$) per cycle.

sec. 31-8 Three Simple Circuits

•28 A 1.50 μF capacitor is connected as in Fig. 31-10 to an ac generator with $\mathscr{E}_m = 30.0$ V. What is the amplitude of the resulting alternating current if the frequency of the emf is (a) 1.00 kHz and (b) 8.00 kHz?

•29 ILW A 50.0 mH inductor is connected as in Fig. 31-12 to an ac generator with $\mathscr{E}_m = 30.0$ V. What is the amplitude of the resulting alternating current if the frequency of the emf is (a) 1.00 kHz and (b) 8.00 kHz?

•30 A 50.0 Ω resistor is connected as in Fig. 31-8 to an ac generator with $\mathscr{E}_m = 30.0$ V. What is the amplitude of the resulting alternating current if the frequency of the emf is (a) 1.00 kHz and (b) 8.00 kHz?

•31 (a) At what frequency would a 6.0 mH inductor and a 10 μF capacitor have the same reactance? (b) What would the reactance be? (c) Show that this frequency would be the natural frequency of an oscillating circuit with the same L and C.

••32 GO An ac generator has emf $\mathscr{E} = \mathscr{E}_m \sin \omega_d t$, with $\mathscr{E}_m = 25.0$ V and $\omega_d = 377$ rad/s. It is connected to a 12.7 H inductor. (a) What is the maximum value of the current? (b) When the current is a maximum, what is the emf of the generator? (c) When the emf of the generator is -12.5 V and increasing in magnitude, what is the current?

••33 SSM An ac generator has emf $\mathscr{E} = \mathscr{E}_m \sin(\omega_d t - \pi/4)$, where $\mathscr{E}_m = 30.0$ V and $\omega_d = 350$ rad/s. The current produced in a connected circuit is $i(t) = I \sin(\omega_d t - 3\pi/4)$, where $I = 620$ mA. At what time after $t = 0$ does (a) the generator emf first reach a maximum and (b) the current first reach a maximum? (c) The circuit contains a single element other than the generator. Is it a capacitor, an inductor, or a resistor? Justify your answer. (d) What is the value of the capacitance, inductance, or resistance, as the case may be?

••34 GO An ac generator with emf $\mathscr{E} = \mathscr{E}_m \sin \omega_d t$, where $\mathscr{E}_m = 25.0$ V and $\omega_d = 377$ rad/s, is connected to a 4.15 μF capacitor. (a) What is the maximum value of the current? (b) When the current is a maximum, what is the emf of the generator? (c) When the emf of the generator is -12.5 V and increasing in magnitude, what is the current?

sec. 31-9 The Series *RLC* Circuit

•35 ILW A coil of inductance 88 mH and unknown resistance and a 0.94 μF capacitor are connected in series with an alternating emf of frequency 930 Hz. If the phase constant between the applied voltage and the current is 75°, what is the resistance of the coil?

•36 An alternating source with a variable frequency, a capacitor with capacitance C, and a resistor with resistance R are connected in series. Figure 31-28 gives the impedance Z of the circuit versus the driving angular frequency ω_d; the curve reaches an asymptote of 500 Ω, and the horizontal scale is set by $\omega_{ds} = 300$ rad/s. The figure also gives the reactance X_C for the capacitor versus ω_d. What are (a) R and (b) C?

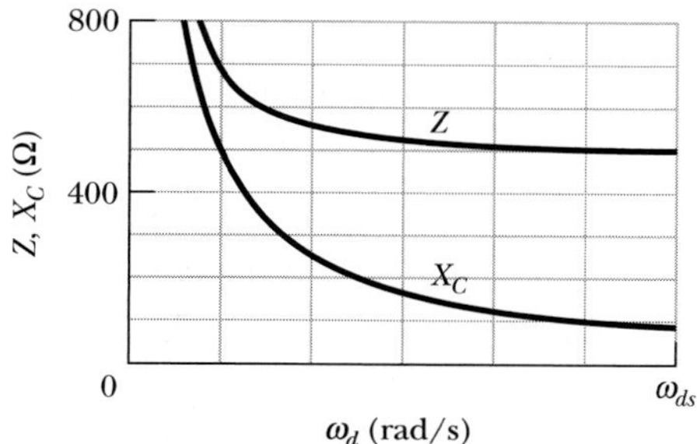

Fig. 31-28 Problem 36.

•37 An electric motor has an effective resistance of 32.0 Ω and an inductive reactance of 45.0 Ω when working under load. The rms voltage across the alternating source is 420 V. Calculate the rms current.

•38 The current amplitude I versus driving angular frequency ω_d for a driven RLC circuit is given in Fig. 31-29, where the vertical axis scale is set by $I_s = 4.00$ A. The inductance is 200 μH, and the emf amplitude is 8.0 V. What are (a) C and (b) R?

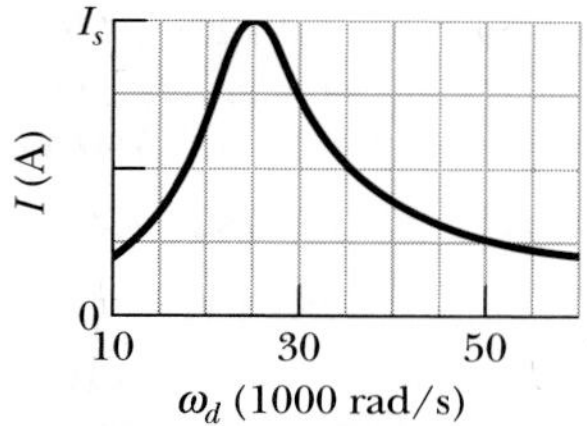

Fig. 31-29 Problem 38.

•39 Remove the inductor from the circuit in Fig. 31-7 and set $R = 200$ Ω, $C = 15.0$ μF, $f_d = 60.0$ Hz, and $\mathscr{E}_m = 36.0$ V. What are (a) Z, (b) ϕ, and (c) I? (d) Draw a phasor diagram.

•40 An alternating source drives a series RLC circuit with an emf amplitude of 6.00 V, at a phase angle of +30.0°. When the potential difference across the capacitor reaches its maximum positive value of +5.00 V, what is the potential difference across the inductor (sign included)?

•41 SSM In Fig. 31-7, set $R = 200$ Ω, $C = 70.0$ μF, $L = 230$ mH, $f_d = 60.0$ Hz, and $\mathscr{E}_m = 36.0$ V. What are (a) Z, (b) ϕ, and (c) I? (d) Draw a phasor diagram.

•42 An alternating source with a variable frequency, an inductor with inductance L, and a resistor with resistance R are connected in series. Figure 31-30 gives the impedance Z of the circuit versus the driving angular frequency ω_d, with the horizontal axis scale set by $\omega_{ds} = 1600$ rad/s. The figure also gives the reactance X_L for the inductor versus ω_d. What are (a) R and (b) L?

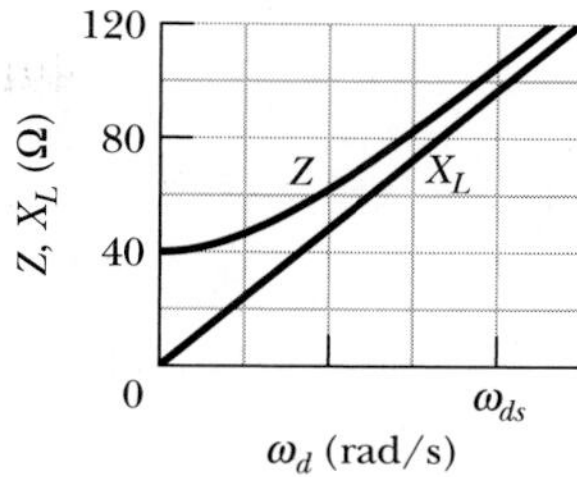

Fig. 31-30 Problem 42.

•43 Remove the capacitor from the circuit in Fig. 31-7 and set $R = 200$ Ω, $L = 230$ mH, $f_d = 60.0$ Hz, and $\mathscr{E}_m = 36.0$ V. What are (a) Z, (b) ϕ, and (c) I? (d) Draw a phasor diagram.

••44 GO An ac generator with $\mathscr{E}_m = 220$ V and operating at 400 Hz causes oscillations in a series RLC circuit having $R = 220$ Ω, $L = 150$ mH, and $C = 24.0$ μF. Find (a) the capacitive reactance X_C, (b) the impedance Z, and (c) the current amplitude I. A second capacitor of the same capacitance is then connected in series with the other components. Determine whether the values of (d) X_C, (e) Z, and (f) I increase, decrease, or remain the same.

••45 ILW GO (a) In an RLC circuit, can the amplitude of the voltage across an inductor be greater than the amplitude of the generator emf? (b) Consider an RLC circuit with $\mathscr{E}_m = 10$ V, $R = 10$ Ω, $L = 1.0$ H, and $C = 1.0$ μF. Find the amplitude of the voltage across the inductor at resonance.

••46 An alternating emf source with a variable frequency f_d is connected in series with a 50.0 Ω resistor and a 20.0 μF capacitor. The emf amplitude is 12.0 V. (a) Draw a phasor diagram for phasor V_R (the potential across the resistor) and phasor V_C (the potential across the capacitor). (b) At what driving frequency f_d do the two phasors have the same length? At that driving frequency, what are (c) the phase angle in degrees, (d) the angular speed at which the phasors rotate, and (e) the current amplitude?

••47 SSM WWW An RLC circuit such as that of Fig. 31-7 has $R = 5.00$ Ω, $C = 20.0$ μF, $L = 1.00$ H, and $\mathscr{E}_m = 30.0$ V. (a) At what angular frequency ω_d will the current amplitude have its maximum value, as in the resonance curves of Fig. 31-16? (b) What is this maximum value? At what (c) lower angular frequency ω_{d1} and (d) higher angular frequency ω_{d2} will the current amplitude be half this maximum value? (e) For the resonance curve for this circuit, what is the fractional half-width $(\omega_{d1} - \omega_{d2})/\omega$?

••48 GO Figure 31-31 shows a driven RLC circuit that contains two identical capacitors and two switches. The emf amplitude is set at 12.0 V, and the driving frequency is set at 60.0 Hz. With both switches open, the current leads the emf by 30.9°. With switch S_1 closed and switch S_2 still open, the emf leads the current by 15.0°. With both switches closed, the current amplitude is 447 mA. What are (a) R, (b) C, and (c) L?

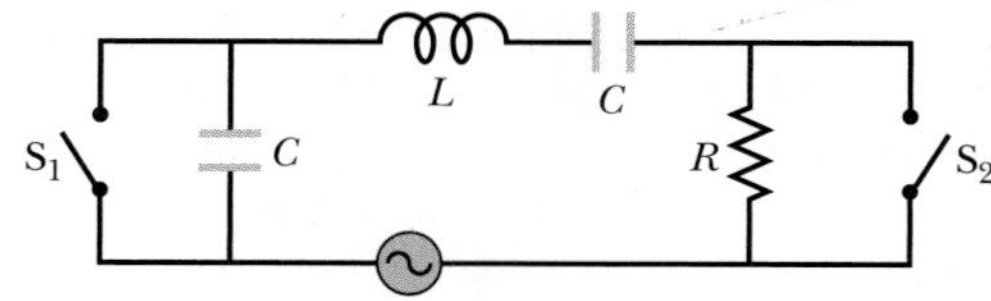

Fig. 31-31 Problem 48.

••49 In Fig. 31-32, a generator with an adjustable frequency of oscillation is connected to resistance $R = 100\ \Omega$, inductances $L_1 = 1.70$ mH and $L_2 = 2.30$ mH, and capacitances $C_1 = 4.00\ \mu$F, $C_2 = 2.50\ \mu$F, and $C_3 = 3.50\ \mu$F. (a) What is the resonant frequency of the circuit? (*Hint:* See Problem 47 in Chapter 30.) What happens to the resonant frequency if (b) R is increased, (c) L_1 is increased, and (d) C_3 is removed from the circuit?

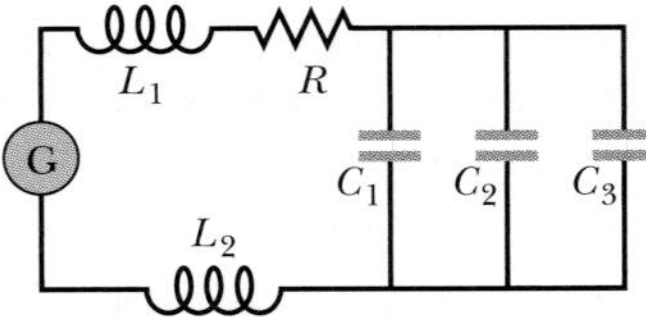

Fig. 31-32 Problem 49.

••50 An alternating emf source with a variable frequency f_d is connected in series with an 80.0 Ω resistor and a 40.0 mH inductor. The emf amplitude is 6.00 V. (a) Draw a phasor diagram for phasor V_R (the potential across the resistor) and phasor V_L (the potential across the inductor). (b) At what driving frequency f_d do the two phasors have the same length? At that driving frequency, what are (c) the phase angle in degrees, (d) the angular speed at which the phasors rotate, and (e) the current amplitude?

••51 SSM The fractional half-width $\Delta\omega_d$ of a resonance curve, such as the ones in Fig. 31-16, is the width of the curve at half the maximum value of I. Show that $\Delta\omega_d/\omega = R(3C/L)^{1/2}$, where ω is the angular frequency at resonance. Note that the ratio $\Delta\omega_d/\omega$ increases with R, as Fig. 31-16 shows.

sec. 31-10 Power in Alternating-Current Circuits

•52 An ac voltmeter with large impedance is connected in turn across the inductor, the capacitor, and the resistor in a series circuit having an alternating emf of 100 V (rms); the meter gives the same reading in volts in each case. What is this reading?

•53 SSM An air conditioner connected to a 120 V rms ac line is equivalent to a 12.0 Ω resistance and a 1.30 Ω inductive reactance in series. Calculate (a) the impedance of the air conditioner and (b) the average rate at which energy is supplied to the appliance.

•54 What is the maximum value of an ac voltage whose rms value is 100 V?

•55 What direct current will produce the same amount of thermal energy, in a particular resistor, as an alternating current that has a maximum value of 2.60 A?

••56 A typical light dimmer used to dim the stage lights in a theater consists of a variable inductor L (whose inductance is adjustable between zero and L_{max}) connected in series with a lightbulb B, as shown in Fig. 31-33. The electrical supply is 120 V (rms) at 60.0 Hz; the lightbulb is rated at 120 V, 1000 W. (a) What L_{max} is required if the rate of energy dissipation in the lightbulb is to be varied by a factor of 5 from its upper limit of 1000 W? Assume that the resistance of the lightbulb is independent of its temperature. (b) Could one use a variable resistor (adjustable between zero and R_{max}) instead of an inductor? (c) If so, what R_{max} is required? (d) Why isn't this done?

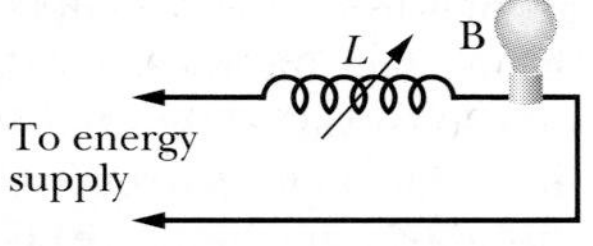

Fig. 31-33 Problem 56.

••57 In an RLC circuit such as that of Fig. 31-7 assume that $R = 5.00\ \Omega$, $L = 60.0$ mH, $f_d = 60.0$ Hz, and $\mathscr{E}_m = 30.0$ V. For what values of the capacitance would the average rate at which energy is dissipated in the resistance be (a) a maximum and (b) a minimum? What are (c) the maximum dissipation rate and the corresponding (d) phase angle and (e) power factor? What are (f) the minimum dissipation rate and the corresponding (g) phase angle and (h) power factor?

••58 For Fig. 31-34, show that the average rate at which energy is dissipated in resistance R is a maximum when R is equal to the internal resistance r of the ac generator. (In the text discussion we tacitly assumed that $r = 0$.)

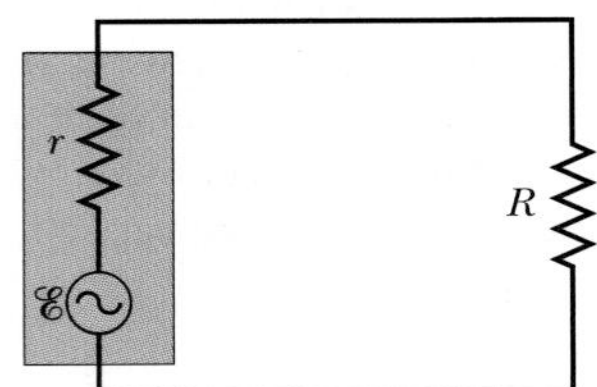

Fig. 31-34 Problems 58 and 66.

••59 In Fig. 31-7, $R = 15.0\ \Omega$, $C = 4.70\ \mu$F, and $L = 25.0$ mH. The generator provides an emf with rms voltage 75.0 V and frequency 550 Hz. (a) What is the rms current? What is the rms voltage across (b) R, (c) C, (d) L, (e) C and L together, and (f) R, C, and L together? At what average rate is energy dissipated by (g) R, (h) C, and (i) L?

••60 GO In a series oscillating RLC circuit, $R = 16.0\ \Omega$, $C = 31.2\ \mu$F, $L = 9.20$ mH, and $\mathscr{E}_m = \mathscr{E}_m \sin \omega_d t$ with $\mathscr{E}_m = 45.0$ V and $\omega_d = 3000$ rad/s. For time $t = 0.442$ ms find (a) the rate P_g at which energy is being supplied by the generator, (b) the rate P_C at which the energy in the capacitor is changing, (c) the rate P_L at which the energy in the inductor is changing, and (d) the rate P_R at which energy is being dissipated in the resistor. (e) Is the sum of P_C, P_L, and P_R greater than, less than, or equal to P_g?

••61 SSM WWW Figure 31-35 shows an ac generator connected to a "black box" through a pair of terminals. The box contains an RLC circuit, possibly even a multiloop circuit, whose elements and connections we do not know. Measurements outside the box reveal that

$$\mathscr{E}(t) = (75.0\text{ V})\sin \omega_d t$$

and

$$i(t) = (1.20\text{ A})\sin(\omega_d t + 42.0°).$$

(a) What is the power factor? (b) Does the current lead or lag the emf? (c) Is the circuit in the box largely inductive or largely capacitive? (d) Is the circuit in the box in resonance? (e) Must there be a capacitor in the box? (f) An inductor? (g) A resistor? (h) At what average rate is energy delivered to the box by the generator? (i) Why don't you need to know ω_d to answer all these questions?

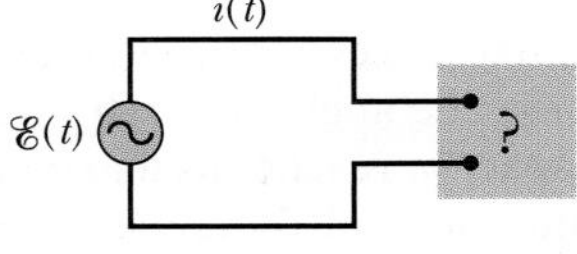

Fig. 31-35 Problem 61.

sec. 31-11 Transformers

•62 A generator supplies 100 V to a transformer's primary coil, which has 50 turns. If the secondary coil has 500 turns, what is the secondary voltage?

•63 SSM ILW A transformer has 500 primary turns and 10 secondary turns. (a) If V_p is 120 V (rms), what is V_s with an open circuit? If the secondary now has a resistive load of 15 Ω, what is the current in the (b) primary and (c) secondary?

•64 Figure 31-36 shows an "autotransformer." It consists of a single coil (with an iron core). Three taps T_i are provided. Between taps T_1 and T_2 there are 200 turns, and between taps T_2 and T_3 there are 800 turns. Any two taps can be chosen as the primary terminals, and any two taps can be chosen as the secondary terminals. For choices producing a step-up transformer, what are the (a) smallest, (b) second smallest, and (c) largest values of the ratio V_s/V_p? For a step-down transformer, what are the (d) smallest, (e) second smallest, and (f) largest values of V_s/V_p?

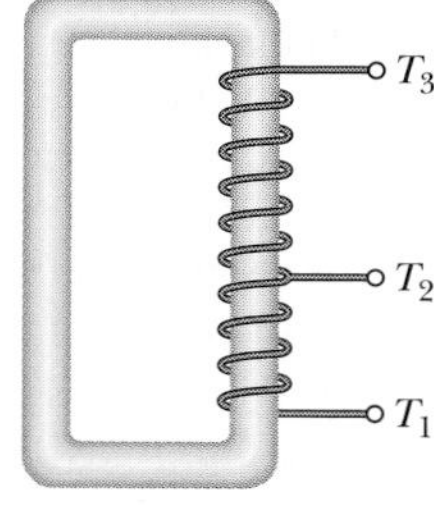

Fig. 31-36 Problem 64.

••65 An ac generator provides emf to a resistive load in a remote factory over a two-cable transmission line. At the factory a step-down transformer reduces the voltage from its (rms) transmission value V_t to a much lower value that is safe and convenient for use in the factory. The transmission line resistance is 0.30 Ω/cable, and the power of the generator is 250 kW. If V_t = 80 kV, what are (a) the voltage decrease ΔV along the transmission line and (b) the rate P_d at which energy is dissipated in the line as thermal energy? If V_t = 8.0 kV, what are (c) ΔV and (d) P_d? If V_t = 0.80 kV, what are (e) ΔV and (f) P_d?

Additional Problems

66 In Fig. 31-34, let the rectangular box on the left represent the (high-impedance) output of an audio amplifier, with r = 1000 Ω. Let R = 10 Ω represent the (low-impedance) coil of a loudspeaker. For maximum transfer of energy to the load R we must have $R = r$, and that is not true in this case. However, a transformer can be used to "transform" resistances, making them behave electrically as if they were larger or smaller than they actually are. (a) Sketch the primary and secondary coils of a transformer that can be introduced between the amplifier and the speaker in Fig. 31-34 to match the impedances. (b) What must be the turns ratio?

67 An ac generator produces emf $\mathscr{E} = \mathscr{E}_m \sin(\omega_d t - \pi/4)$, where $\mathscr{E}_m$ = 30.0 V and ω_d = 350 rad/s. The current in the circuit attached to the generator is $i(t) = I \sin(\omega_d t + \pi/4)$, where I = 620 mA. (a) At what time after $t = 0$ does the generator emf first reach a maximum? (b) At what time after $t = 0$ does the current first reach a maximum? (c) The circuit contains a single element other than the generator. Is it a capacitor, an inductor, or a resistor? Justify your answer. (d) What is the value of the capacitance, inductance, or resistance, as the case may be?

68 A series RLC circuit is driven by a generator at a frequency of 2000 Hz and an emf amplitude of 170 V. The inductance is 60.0 mH, the capacitance is 0.400 μF, and the resistance is 200 Ω. (a) What is the phase constant in radians? (b) What is the current amplitude?

69 A generator of frequency 3000 Hz drives a series RLC circuit with an emf amplitude of 120 V. The resistance is 40.0 Ω, the capacitance is 1.60 μF, and the inductance is 850 μH. What are (a) the phase constant in radians and (b) the current amplitude? (c) Is the circuit capacitive, inductive, or in resonance?

70 A 45.0 mH inductor has a reactance of 1.30 kΩ. (a) What is its operating frequency? (b) What is the capacitance of a capacitor with the same reactance at that frequency? If the frequency is doubled, what is the new reactance of (c) the inductor and (d) the capacitor?

71 An RLC circuit is driven by a generator with an emf amplitude of 80.0 V and a current amplitude of 1.25 A. The current leads the emf by 0.650 rad. What are the (a) impedance and (b) resistance of the circuit? (c) Is the circuit inductive, capacitive, or in resonance?

72 A series RLC circuit is driven in such a way that the maximum voltage across the inductor is 1.50 times the maximum voltage across the capacitor and 2.00 times the maximum voltage across the resistor. (a) What is ϕ for the circuit? (b) Is the circuit inductive, capacitive, or in resonance? The resistance is 49.9 Ω, and the current amplitude is 200 mA. (c) What is the amplitude of the driving emf?

73 A capacitor of capacitance 158 μF and an inductor form an LC circuit that oscillates at 8.15 kHz, with a current amplitude of 4.21 mA. What are (a) the inductance, (b) the total energy in the circuit, and (c) the maximum charge on the capacitor?

74 An oscillating LC circuit has an inductance of 3.00 mH and a capacitance of 10.0 μF. Calculate the (a) angular frequency and (b) period of the oscillation. (c) At time $t = 0$, the capacitor is charged to 200 μC and the current is zero. Roughly sketch the charge on the capacitor as a function of time.

75 For a certain driven series RLC circuit, the maximum generator emf is 125 V and the maximum current is 3.20 A. If the current leads the generator emf by 0.982 rad, what are the (a) impedance and (b) resistance of the circuit? (c) Is the circuit predominantly capacitive or inductive?

76 A 1.50 μF capacitor has a capacitive reactance of 12.0 Ω. (a) What must be its operating frequency? (b) What will be the capacitive reactance if the frequency is doubled?

77 SSM In Fig. 31-37, a three-phase generator G produces electrical power that is transmitted by means of three wires. The electric potentials (each relative to a common reference level) are $V_1 = A \sin \omega_d t$ for wire 1, $V_2 = A \sin(\omega_d t - 120°)$ for wire 2, and $V_3 = A \sin(\omega_d t - 240°)$ for wire 3. Some types of industrial equipment (for example, motors) have three terminals and are designed to be connected directly to these three wires. To use a more conventional two-terminal device (for example, a lightbulb), one connects it to any two of the three wires. Show that the potential difference between *any two* of the wires (a) oscillates sinusoidally with angular frequency ω_d and (b) has an amplitude of $A\sqrt{3}$.

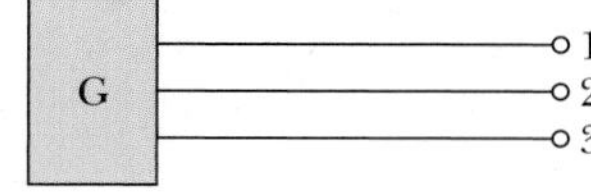

Fig. 31-37 Problem 77.

78 An electric motor connected to a 120 V, 60.0 Hz ac outlet does mechanical work at the rate of 0.100 hp (1 hp = 746 W). (a) If the motor draws an rms current of 0.650 A, what is its effective resistance, relative to power transfer? (b) Is this the same as the resistance of the motor's coils, as measured with an ohmmeter with the motor disconnected from the outlet?

79 SSM (a) In an oscillating LC circuit, in terms of the maximum charge Q on the capacitor, what is the charge there when the energy in the electric field is 50.0% of that in the magnetic field? (b) What fraction of a period must elapse following the time the capacitor is fully charged for this condition to occur?

80 A series *RLC* circuit is driven by an alternating source at a frequency of 400 Hz and an emf amplitude of 90.0 V. The resistance is 20.0 Ω, the capacitance is 12.1 μF, and the inductance is 24.2 mH. What is the rms potential difference across (a) the resistor, (b) the capacitor, and (c) the inductor? (d) What is the average rate at which energy is dissipated?

81 SSM In a certain series *RLC* circuit being driven at a frequency of 60.0 Hz, the maximum voltage across the inductor is 2.00 times the maximum voltage across the resistor and 2.00 times the maximum voltage across the capacitor. (a) By what angle does the current lag the generator emf? (b) If the maximum generator emf is 30.0 V, what should be the resistance of the circuit to obtain a maximum current of 300 mA?

82 A 1.50 mH inductor in an oscillating *LC* circuit stores a maximum energy of 10.0 μJ. What is the maximum current?

83 A generator with an adjustable frequency of oscillation is wired in series to an inductor of $L = 2.50$ mH and a capacitor of $C = 3.00$ μF. At what frequency does the generator produce the largest possible current amplitude in the circuit?

84 A series *RLC* circuit has a resonant frequency of 6.00 kHz. When it is driven at 8.00 kHz, it has an impedance of 1.00 kΩ and a phase constant of 45°. What are (a) *R*, (b) *L*, and (c) *C* for this circuit?

85 SSM An *LC* circuit oscillates at a frequency of 10.4 kHz. (a) If the capacitance is 340 μF, what is the inductance? (b) If the maximum current is 7.20 mA, what is the total energy in the circuit? (c) What is the maximum charge on the capacitor?

86 When under load and operating at an rms voltage of 220 V, a certain electric motor draws an rms current of 3.00 A. It has a resistance of 24.0 Ω and no capacitive reactance. What is its inductive reactance?

87 The ac generator in Fig. 31-38 supplies 120 V at 60.0 Hz. With the switch open as in the diagram, the current leads the generator emf by 20.0°. With the switch in position 1, the current lags the generator emf by 10.0°. When the switch is in position 2, the current amplitude is 2.00 A. What are (a) *R*, (b) *L*, and (c) *C*?

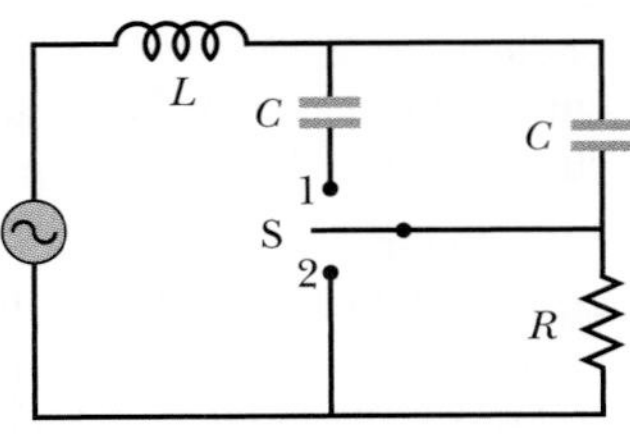

Fig. 31-38 Problem 87.

88 In an oscillating *LC* circuit, $L = 8.00$ mH and $C = 1.40$ μF. At time $t = 0$, the current is maximum at 12.0 mA. (a) What is the maximum charge on the capacitor during the oscillations? (b) At what earliest time $t > 0$ is the rate of change of energy in the capacitor maximum? (c) What is that maximum rate of change?

89 SSM For a sinusoidally driven series *RLC* circuit, show that over one complete cycle with period *T* (a) the energy stored in the capacitor does not change; (b) the energy stored in the inductor does not change; (c) the driving emf device supplies energy $(\frac{1}{2}T)\mathscr{E}_m I \cos\phi$; and (d) the resistor dissipates energy $(\frac{1}{2}T)RI^2$. (e) Show that the quantities found in (c) and (d) are equal.

90 What capacitance would you connect across a 1.30 mH inductor to make the resulting oscillator resonate at 3.50 kHz?

91 A series circuit with resistor–inductor–capacitor combination R_1, L_1, C_1 has the same resonant frequency as a second circuit with a different combination R_2, L_2, C_2. You now connect the two combinations in series. Show that this new circuit has the same resonant frequency as the separate circuits.

92 Consider the circuit shown in Fig. 31-39. With switch S_1 closed and the other two switches open, the circuit has a time constant τ_C. With switch S_2 closed and the other two switches open, the circuit has a time constant τ_L. With switch S_3 closed and the other two switches open, the circuit oscillates with a period *T*. Show that $T = 2\pi\sqrt{\tau_C \tau_L}$.

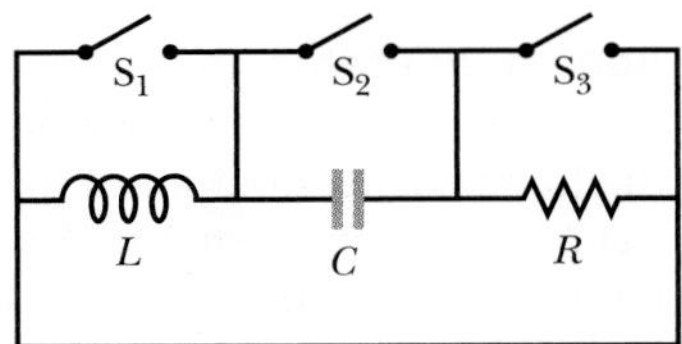

Fig. 31-39 Problem 92.

CHAPTER

32

MAXWELL'S EQUATIONS; MAGNETISM OF MATTER

32-1 WHAT IS PHYSICS?

This chapter reveals some of the breadth of physics because it ranges from the basic science of electric and magnetic fields to the applied science and engineering of magnetic materials. First, we conclude our basic discussion of electric and magnetic fields, finding that most of the physics principles in the last 11 chapters can be summarized in only *four* equations, known as Maxwell's equations.

Second, we examine the science and engineering of magnetic materials. The careers of many scientists and engineers are focused on understanding why some materials are magnetic and others are not and on how existing magnetic materials can be improved. These researchers wonder why Earth has a magnetic field but you do not. They find countless applications for inexpensive magnetic materials in cars, kitchens, offices, and hospitals, and magnetic materials often show up in unexpected ways. For example, if you have a tattoo (Fig. 32-1) and undergo an MRI (magnetic resonance imaging) scan, the large magnetic field used in the scan may noticeably tug on your tattooed skin because some tattoo inks contain magnetic particles. In another example, some breakfast cereals are advertised as being "iron fortified" because they contain small bits of iron for you to ingest. Because these iron bits are magnetic, you can collect them by passing a magnet over a slurry of water and cereal.

Fig. 32-1 Some of the inks used for tattoos contain magnetic particles. *(Oliver Strewe/Getty Images, Inc.)*

Our first step here is to revisit Gauss' law, but this time for magnetic fields.

32-2 Gauss' Law for Magnetic Fields

Figure 32-2 shows iron powder that has been sprinkled onto a transparent sheet placed above a bar magnet. The powder grains, trying to align themselves with the magnet's magnetic field, have fallen into a pattern that reveals the field. One end of the magnet is a *source* of the field (the field lines diverge from it) and the other end is a *sink* of the field (the field lines converge toward it). By convention, we call the source the *north pole* of the magnet and the sink the *south pole,* and we say that the magnet, with its two poles, is an example of a **magnetic dipole.**

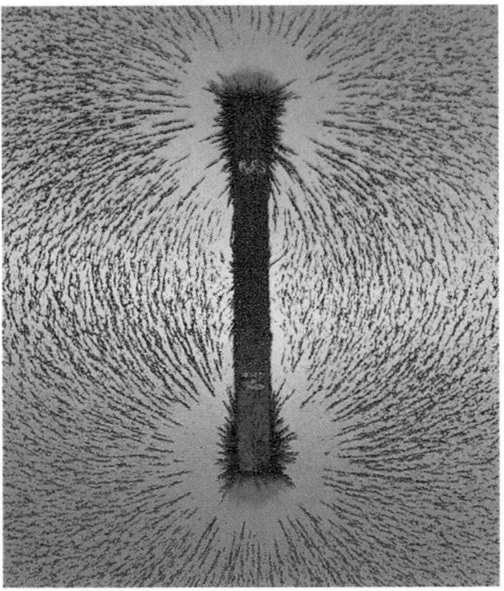

Fig. 32-2 A bar magnet is a magnetic dipole. The iron filings suggest the magnetic field lines. (Colored light fills the background.) *(Runk/Schoenberger/Grant Heilman Photography)*

Suppose we break a bar magnet into pieces the way we can break a piece of chalk (Fig. 32-3). We should, it seems, be able to isolate a single magnetic pole, called a *magnetic monopole.* However, we cannot—not even if we break the magnet down to its individual atoms and then to its electrons and nuclei. Each fragment has a north pole and a south pole. Thus:

The simplest magnetic structure that can exist is a magnetic dipole. Magnetic monopoles do not exist (as far as we know).

Gauss' law for magnetic fields is a formal way of saying that magnetic monopoles do not exist. The law asserts that the net magnetic flux Φ_B through any closed Gaussian surface is zero:

$$\Phi_B = \oint \vec{B} \cdot d\vec{A} = 0 \qquad \text{(Gauss' law for magnetic fields).} \qquad (32\text{-}1)$$

Contrast this with Gauss' law for electric fields,

$$\Phi_E = \oint \vec{E} \cdot d\vec{A} = \frac{q_{enc}}{\varepsilon_0} \qquad \text{(Gauss' law for electric fields).}$$

In both equations, the integral is taken over a *closed* Gaussian surface. Gauss' law for electric fields says that this integral (the net electric flux through the surface) is proportional to the net electric charge q_{enc} enclosed by the surface. Gauss' law for magnetic fields says that there can be no net magnetic flux through the surface because there can be no net "magnetic charge" (individual magnetic poles) enclosed by the surface. The simplest magnetic structure that can exist and thus be enclosed by a Gaussian surface is a dipole, which consists of both a source and a sink for the field lines. Thus, there must always be as much magnetic flux into the surface as out of it, and the net magnetic flux must always be zero.

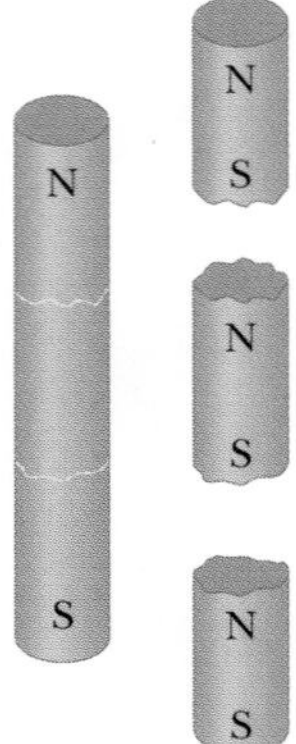

Fig. 32-3 If you break a magnet, each fragment becomes a separate magnet, with its own north and south poles.

Gauss' law for magnetic fields holds for structures more complicated than a magnetic dipole, and it holds even if the Gaussian surface does not enclose the entire structure. Gaussian surface II near the bar magnet of Fig. 32-4 encloses no poles, and we can easily conclude that the net magnetic flux through it is zero. Gaussian surface I is more difficult. It may seem to enclose only the north pole of the magnet because it encloses the label N and not the label S. However, a south pole must be associated with the lower boundary of the surface because magnetic field lines enter the surface there. (The enclosed section is like one piece of the broken bar magnet in Fig. 32-3.) Thus, Gaussian surface I encloses a magnetic dipole, and the net flux through the surface is zero.

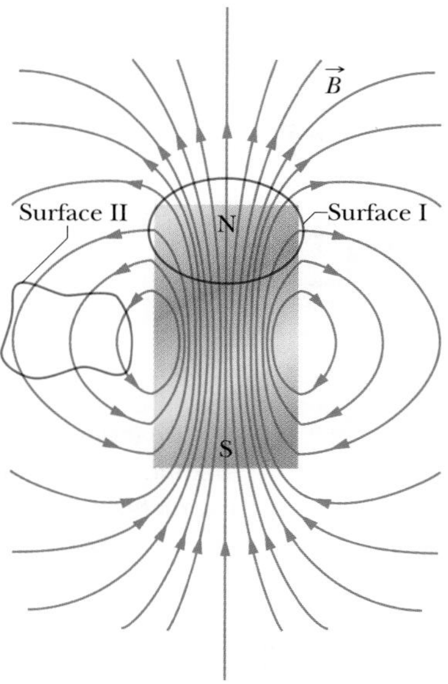

Fig. 32-4 The field lines for the magnetic field $\vec{B}$ of a short bar magnet. The red curves represent cross sections of closed, three-dimensional Gaussian surfaces.

CHECKPOINT 1

The figure here shows four closed surfaces with flat top and bottom faces and curved sides. The table gives the areas A of the faces and the magnitudes B of the uniform and perpendicular magnetic fields through those faces; the units of A and B are arbitrary but consistent. Rank the surfaces according to the magnitudes of the magnetic flux through their curved sides, greatest first.

Surface	A_{top}	B_{top}	A_{bot}	B_{bot}
a	2	6, outward	4	3, inward
b	2	1, inward	4	2, inward
c	2	6, inward	2	8, outward
d	2	3, outward	3	2, outward

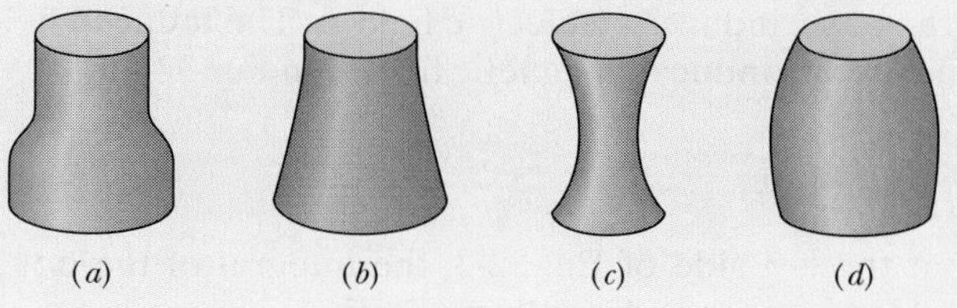

32-3 Induced Magnetic Fields

In Chapter 30 you saw that a changing magnetic flux induces an electric field, and we ended up with Faraday's law of induction in the form

$$\oint \vec{E} \cdot d\vec{s} = -\frac{d\Phi_B}{dt} \quad \text{(Faraday's law of induction).} \tag{32-2}$$

Here $\vec{E}$ is the electric field induced along a closed loop by the changing magnetic flux Φ_B encircled by that loop. Because symmetry is often so powerful in physics, we should be tempted to ask whether induction can occur in the opposite sense; that is, can a changing electric flux induce a magnetic field?

The answer is that it can; furthermore, the equation governing the induction of a magnetic field is almost symmetric with Eq. 32-2. We often call it Maxwell's law of induction after James Clerk Maxwell, and we write it as

$$\oint \vec{B} \cdot d\vec{s} = \mu_0\varepsilon_0 \frac{d\Phi_E}{dt} \quad \text{(Maxwell's law of induction).} \tag{32-3}$$

Here $\vec{B}$ is the magnetic field induced along a closed loop by the changing electric flux Φ_E in the region encircled by that loop.

As an example of this sort of induction, we consider the charging of a parallel-plate capacitor with circular plates. (Although we shall focus on this arrangement,

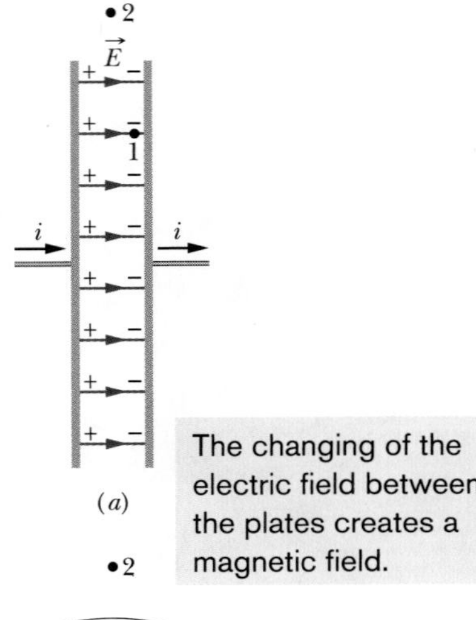

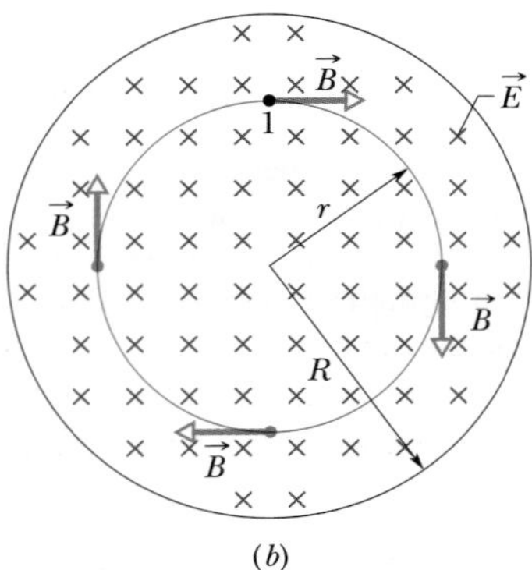

Fig. 32-5 (*a*) A circular parallel-plate capacitor, shown in side view, is being charged by a constant current i. (*b*) A view from within the capacitor, looking toward the plate at the right in (*a*). The electric field $\vec{E}$ is uniform, is directed into the page (toward the plate), and grows in magnitude as the charge on the capacitor increases. The magnetic field $\vec{B}$ induced by this changing electric field is shown at four points on a circle with a radius r less than the plate radius R.

a changing electric flux will always induce a magnetic field whenever it occurs.) We assume that the charge on our capacitor (Fig. 32-5*a*) is being increased at a steady rate by a constant current i in the connecting wires. Then the electric field magnitude between the plates must also be increasing at a steady rate.

Figure 32-5*b* is a view of the right-hand plate of Fig. 32-5*a* from between the plates. The electric field is directed into the page. Let us consider a circular loop through point 1 in Figs. 32-5*a* and *b*, a loop that is concentric with the capacitor plates and has a radius smaller than that of the plates. Because the electric field through the loop is changing, the electric flux through the loop must also be changing. According to Eq. 32-3, this changing electric flux induces a magnetic field around the loop.

Experiment proves that a magnetic field $\vec{B}$ *is* indeed induced around such a loop, directed as shown. This magnetic field has the same magnitude at every point around the loop and thus has circular symmetry about the *central axis* of the capacitor plates (the axis extending from one plate center to the other).

If we now consider a larger loop—say, through point 2 outside the plates in Figs. 32-5*a* and *b*—we find that a magnetic field is induced around that loop as well. Thus, while the electric field is changing, magnetic fields are induced between the plates, both inside and outside the gap. When the electric field stops changing, these induced magnetic fields disappear.

Although Eq. 32-3 is similar to Eq. 32-2, the equations differ in two ways. First, Eq. 32-3 has the two extra symbols μ_0 and ε_0, but they appear only because we employ SI units. Second, Eq. 32-3 lacks the minus sign of Eq. 32-2, meaning that the induced electric field $\vec{E}$ and the induced magnetic field $\vec{B}$ have opposite directions when they are produced in otherwise similar situations. To see this opposition, examine Fig. 32-6, in which an increasing magnetic field $\vec{B}$, directed into the page, induces an electric field $\vec{E}$. The induced field $\vec{E}$ is counterclockwise, opposite the induced magnetic field $\vec{B}$ in Fig. 32-5*b*.

Ampere-Maxwell Law

Now recall that the left side of Eq. 32-3, the integral of the dot product $\vec{B} \cdot d\vec{s}$ around a closed loop, appears in another equation—namely, Ampere's law:

$$\oint \vec{B} \cdot d\vec{s} = \mu_0 i_{\text{enc}} \quad \text{(Ampere's law)}, \tag{32-4}$$

where i_{enc} is the current encircled by the closed loop. Thus, our two equations that specify the magnetic field $\vec{B}$ produced by means other than a magnetic material (that is, by a current and by a changing electric field) give the field in exactly the same form. We can combine the two equations into the single equation

$$\oint \vec{B} \cdot d\vec{s} = \mu_0 \varepsilon_0 \frac{d\Phi_E}{dt} + \mu_0 i_{\text{enc}} \quad \text{(Ampere–Maxwell law)}. \tag{32-5}$$

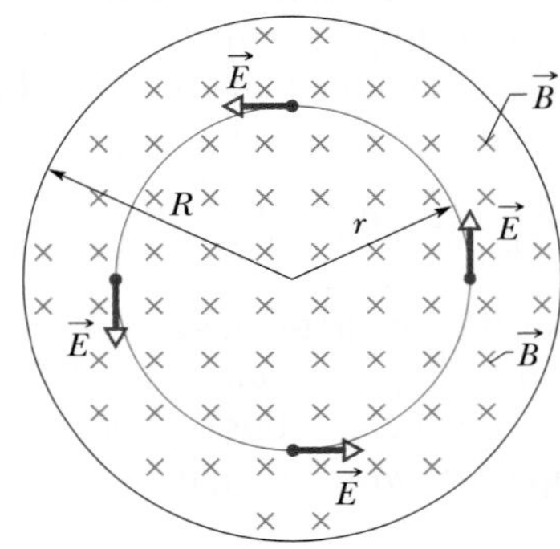

Fig. 32-6 A uniform magnetic field $\vec{B}$ in a circular region. The field, directed into the page, is increasing in magnitude. The electric field $\vec{E}$ induced by the changing magnetic field is shown at four points on a circle concentric with the circular region. Compare this situation with that of Fig. 32-5*b*.

When there is a current but no change in electric flux (such as with a wire carrying a constant current), the first term on the right side of Eq. 32-5 is zero, and so Eq. 32-5 reduces to Eq. 32-4, Ampere's law. When there is a change in electric flux but no current (such as inside or outside the gap of a charging capacitor), the second term on the right side of Eq. 32-5 is zero, and so Eq. 32-5 reduces to Eq. 32-3, Maxwell's law of induction.

CHECKPOINT 2

The figure shows graphs of the electric field magnitude E versus time t for four uniform electric fields, all contained within identical circular regions as in Fig. 32-5*b*. Rank the fields according to the magnitudes of the magnetic fields they induce at the edge of the region, greatest first.

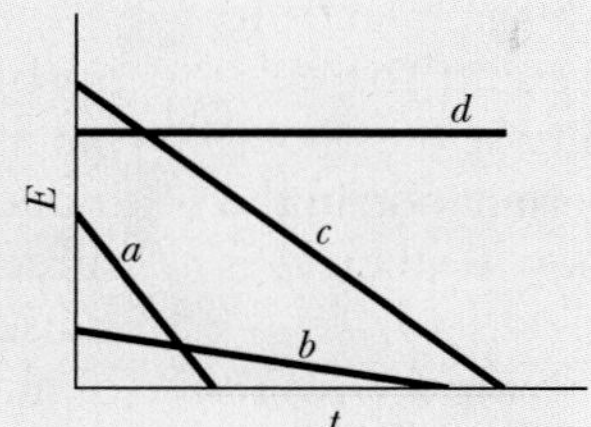

Sample Problem

Magnetic field induced by changing electric field

A parallel-plate capacitor with circular plates of radius R is being charged as in Fig. 32-5*a*.

(a) Derive an expression for the magnetic field at radius r for the case $r \leq R$.

KEY IDEAS

A magnetic field can be set up by a current and by induction due to a changing electric flux; both effects are included in Eq. 32-5. There is no current between the capacitor plates of Fig. 32-5, but the electric flux there is changing. Thus, Eq. 32-5 reduces to

$$\oint \vec{B} \cdot d\vec{s} = \mu_0 \varepsilon_0 \frac{d\Phi_E}{dt}. \qquad (32\text{-}6)$$

We shall separately evaluate the left and right sides of this equation.

Left side of Eq. 32-6: We choose a circular Amperian loop with a radius $r \leq R$ as shown in Fig. 32-5*b* because we want to evaluate the magnetic field for $r \leq R$—that is, inside the capacitor. The magnetic field $\vec{B}$ at all points along the loop is tangent to the loop, as is the path element $d\vec{s}$. Thus, $\vec{B}$ and $d\vec{s}$ are either parallel or antiparallel at each point of the loop. For simplicity, assume they are parallel (the choice does not alter our outcome here). Then

$$\oint \vec{B} \cdot d\vec{s} = \oint B\, ds \cos 0° = \oint B\, ds.$$

Due to the circular symmetry of the plates, we can also assume that $\vec{B}$ has the same magnitude at every point around the loop. Thus, B can be taken outside the integral on the right side of the above equation. The integral that remains is $\oint ds$, which simply gives the circumference $2\pi r$ of the loop. The left side of Eq. 32-6 is then $(B)(2\pi r)$.

Right side of Eq. 32-6: We assume that the electric field $\vec{E}$ is uniform between the capacitor plates and directed perpendicular to the plates. Then the electric flux Φ_E through the Amperian loop is EA, where A is the area encircled by the loop within the electric field. Thus, the right side of Eq. 32-6 is $\mu_0\varepsilon_0\, d(EA)/dt$.

Combining results: Substituting our results for the left and right sides into Eq. 32-6, we get

$$(B)(2\pi r) = \mu_0\varepsilon_0 \frac{d(EA)}{dt}.$$

Because A is a constant, we write $d(EA)$ as $A\, dE$; so we have

$$(B)(2\pi r) = \mu_0\varepsilon_0 A \frac{dE}{dt}. \qquad (32\text{-}7)$$

The area A that is encircled by the Amperian loop within the electric field is the *full* area πr^2 of the loop because the loop's radius r is less than (or equal to) the plate radius R. Substituting πr^2 for A in Eq. 32-7 leads to, for $r \leq R$,

$$B = \frac{\mu_0\varepsilon_0 r}{2} \frac{dE}{dt}. \qquad \text{(Answer)} \quad (32\text{-}8)$$

This equation tells us that, inside the capacitor, B increases linearly with increased radial distance r, from 0 at the central axis to a maximum value at plate radius R.

(b) Evaluate the field magnitude B for $r = R/5 = 11.0$ mm and $dE/dt = 1.50 \times 10^{12}$ V/m · s.

Calculation: From the answer to (a), we have

$$B = \frac{1}{2}\mu_0\varepsilon_0 r \frac{dE}{dt}$$
$$= \tfrac{1}{2}(4\pi \times 10^{-7}\ \text{T}\cdot\text{m/A})(8.85 \times 10^{-12}\ \text{C}^2/\text{N}\cdot\text{m}^2)$$
$$\times (11.0 \times 10^{-3}\ \text{m})(1.50 \times 10^{12}\ \text{V/m}\cdot\text{s})$$
$$= 9.18 \times 10^{-8}\ \text{T}. \qquad \text{(Answer)}$$

(c) Derive an expression for the induced magnetic field for the case $r \geq R$.

Calculation: Our procedure is the same as in (a) except we now use an Amperian loop with a radius r that is greater than the plate radius R, to evaluate B outside the capacitor. Evaluating the left and right sides of Eq. 32-6 again leads to Eq. 32-7. However, we then need this subtle point: The electric field exists only between the plates, not outside the plates. Thus, the area A that is encircled by the Amperian loop in the electric field is *not* the full area πr^2 of the loop. Rather, A is only the plate area πR^2.

Substituting πR^2 for A in Eq. 32-7 and solving the result for B give us, for $r \geq R$,

$$B = \frac{\mu_0 \varepsilon_0 R^2}{2r}\frac{dE}{dt}. \qquad \text{(Answer)} \quad (32\text{-}9)$$

This equation tells us that, outside the capacitor, B decreases with increased radial distance r, from a maximum value at the plate edges (where $r = R$). By substituting $r = R$ into Eqs. 32-8 and 32-9, you can show that these equations are consistent; that is, they give the same maximum value of B at the plate radius.

The magnitude of the induced magnetic field calculated in (b) is so small that it can scarcely be measured with simple apparatus. This is in sharp contrast to the magnitudes of induced electric fields (Faraday's law), which can be measured easily. This experimental difference exists partly because induced emfs can easily be multiplied by using a coil of many turns. No technique of comparable simplicity exists for multiplying induced magnetic fields. In any case, the experiment suggested by this sample problem has been done, and the presence of the induced magnetic fields has been verified quantitatively.

Additional examples, video, and practice available at *WileyPLUS*

32-4 Displacement Current

If you compare the two terms on the right side of Eq. 32-5, you will see that the product $\varepsilon_0(d\Phi_E/dt)$ must have the dimension of a current. In fact, that product has been treated as being a fictitious current called the **displacement current** i_d:

$$i_d = \varepsilon_0 \frac{d\Phi_E}{dt} \quad \text{(displacement current).} \qquad (32\text{-}10)$$

"Displacement" is poorly chosen in that nothing is being displaced, but we are stuck with the word. Nevertheless, we can now rewrite Eq. 32-5 as

$$\oint \vec{B} \cdot d\vec{s} = \mu_0 i_{d,\text{enc}} + \mu_0 i_{\text{enc}} \quad \text{(Ampere–Maxwell law),} \qquad (32\text{-}11)$$

in which $i_{d,\text{enc}}$ is the displacement current that is encircled by the integration loop.

Let us again focus on a charging capacitor with circular plates, as in Fig. 32-7*a*. The real current i that is charging the plates changes the electric field $\vec{E}$ between the plates. The fictitious displacement current i_d between the plates is associated with that changing field $\vec{E}$. Let us relate these two currents.

The charge q on the plates at any time is related to the magnitude E of the field between the plates at that time by Eq. 25-4:

$$q = \varepsilon_0 A E, \qquad (32\text{-}12)$$

in which A is the plate area. To get the real current i, we differentiate Eq. 32-12 with respect to time, finding

$$\frac{dq}{dt} = i = \varepsilon_0 A \frac{dE}{dt}. \qquad (32\text{-}13)$$

To get the displacement current i_d, we can use Eq. 32-10. Assuming that the electric field $\vec{E}$ between the two plates is uniform (we neglect any fringing), we can replace the electric flux Φ_E in that equation with EA. Then Eq. 32-10

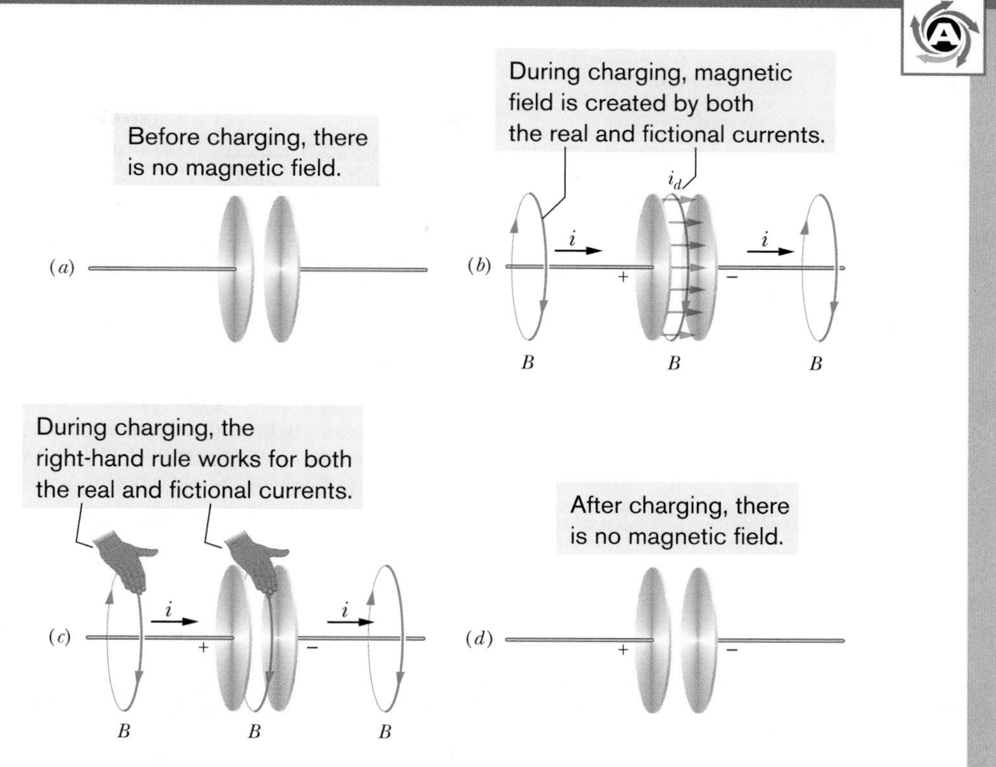

Fig. 32-7 (*a*) Before and (*d*) after the plates are charged, there is no magnetic field. (*b*) During the charging, magnetic field is created by both the real current and the (fictional) displacement current. (*c*) The same right-hand rule works for both currents to give the direction of the magnetic field.

becomes

$$i_d = \varepsilon_0 \frac{d\Phi_E}{dt} = \varepsilon_0 \frac{d(EA)}{dt} = \varepsilon_0 A \frac{dE}{dt}. \tag{32-14}$$

Comparing Eqs. 32-13 and 32-14, we see that the real current i charging the capacitor and the fictitious displacement current i_d between the plates have the same magnitude:

$$i_d = i \quad \text{(displacement current in a capacitor)}. \tag{32-15}$$

Thus, we can consider the fictitious displacement current i_d to be simply a continuation of the real current i from one plate, across the capacitor gap, to the other plate. Because the electric field is uniformly spread over the plates, the same is true of this fictitious displacement current i_d, as suggested by the spread of current arrows in Fig. 32-7*b*. Although no charge actually moves across the gap between the plates, the idea of the fictitious current i_d can help us to quickly find the direction and magnitude of an induced magnetic field, as follows.

Finding the Induced Magnetic Field

In Chapter 29 we found the direction of the magnetic field produced by a real current i by using the right-hand rule of Fig. 29-4. We can apply the same rule to find the direction of an induced magnetic field produced by a fictitious displacement current i_d, as is shown in the center of Fig. 32-7*c* for a capacitor.

We can also use i_d to find the magnitude of the magnetic field induced by a charging capacitor with parallel circular plates of radius R. We simply consider the space between the plates to be an imaginary circular wire of radius R carrying the imaginary current i_d. Then, from Eq. 29-20, the magnitude of the magnetic field at a point inside the capacitor at radius r from the center is

$$B = \left(\frac{\mu_0 i_d}{2\pi R^2}\right) r \quad \text{(inside a circular capacitor).} \tag{32-16}$$

Similarly, from Eq. 29-17, the magnitude of the magnetic field at a point outside the capacitor at radius r is

$$B = \frac{\mu_0 i_d}{2\pi r} \quad \text{(outside a circular capacitor).} \tag{32-17}$$

CHECKPOINT 3

The figure is a view of one plate of a parallel-plate capacitor from within the capacitor. The dashed lines show four integration paths (path b follows the edge of the plate). Rank the paths according to the magnitude of $\oint \vec{B} \cdot d\vec{s}$ along the paths during the discharging of the capacitor, greatest first.

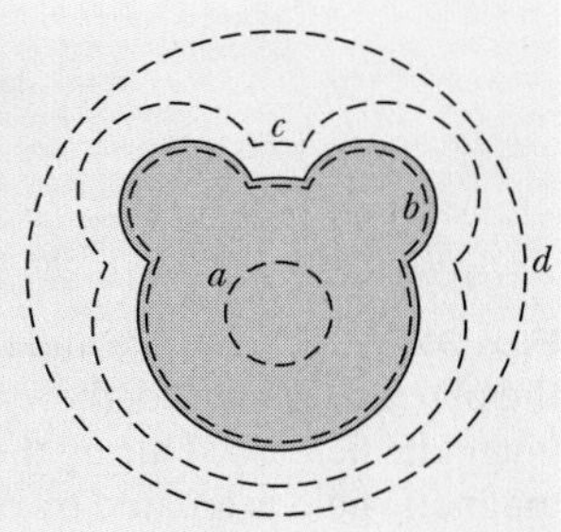

Sample Problem

Treating a changing electric field as a displacement current

A circular parallel-plate capacitor with plate radius R is being charged with a current i.

(a) Between the plates, what is the magnitude of $\oint \vec{B} \cdot d\vec{s}$, in terms of μ_0 and i, at a radius $r = R/5$ from their center?

KEY IDEA

A magnetic field can be set up by a current and by induction due to a changing electric flux (Eq. 32-5). Between the plates in Fig. 32-5, the current is zero and we can account for the changing electric flux with a fictitious displacement current i_d. Then integral $\oint \vec{B} \cdot d\vec{s}$ is given by Eq. 32-11, but because there is no real current i between the capacitor plates, the equation reduces to

$$\oint \vec{B} \cdot d\vec{s} = \mu_0 i_{d,\text{enc}}. \tag{32-18}$$

Calculations: Because we want to evaluate $\oint \vec{B} \cdot d\vec{s}$ at radius $r = R/5$ (within the capacitor), the integration loop encircles only a portion $i_{d,\text{enc}}$ of the total displacement current i_d. Let's assume that i_d is uniformly spread over the full plate area. Then the portion of the displacement current encircled by the loop is proportional to the area encircled by the loop:

$$\frac{\left(\begin{array}{c}\text{encircled displacement}\\ \text{current } i_{d,\text{enc}}\end{array}\right)}{\left(\begin{array}{c}\text{total displacement}\\ \text{current } i_d\end{array}\right)} = \frac{\text{encircled area } \pi r^2}{\text{full plate area } \pi R^2}.$$

This gives us

$$i_{d,\text{enc}} = i_d \frac{\pi r^2}{\pi R^2}.$$

Substituting this into Eq. 32-18, we obtain

$$\oint \vec{B} \cdot d\vec{s} = \mu_0 i_d \frac{\pi r^2}{\pi R^2}. \tag{32-19}$$

Now substituting $i_d = i$ (from Eq. 32-15) and $r = R/5$ into Eq. 32-19 leads to

$$\oint \vec{B} \cdot d\vec{s} = \mu_0 i \frac{(R/5)^2}{R^2} = \frac{\mu_0 i}{25}. \quad \text{(Answer)}$$

(b) In terms of the maximum induced magnetic field, what is the magnitude of the magnetic field induced at $r = R/5$, inside the capacitor?

KEY IDEA

Because the capacitor has parallel circular plates, we can treat the space between the plates as an imaginary wire of radius R carrying the imaginary current i_d. Then we can use Eq. 32-16 to find the induced magnetic field magnitude B at any point inside the capacitor.

Calculations: At $r = R/5$, Eq. 32-16 yields

$$B = \left(\frac{\mu_0 i_d}{2\pi R^2}\right) r = \frac{\mu_0 i_d (R/5)}{2\pi R^2} = \frac{\mu_0 i_d}{10\pi R}. \tag{32-20}$$

From Eq. 32-16, the maximum field magnitude B_{max} within the capacitor occurs at $r = R$. It is

$$B_{max} = \left(\frac{\mu_0 i_d}{2\pi R^2}\right)R = \frac{\mu_0 i_d}{2\pi R}. \quad (32\text{-}21)$$

Dividing Eq. 32-20 by Eq. 32-21 and rearranging the result, we find that the field magnitude at $r = R/5$ is

$$B = \tfrac{1}{5}B_{max}. \quad \text{(Answer)}$$

We should be able to obtain this result with a little reasoning and less work. Equation 32-16 tells us that inside the capacitor, B increases linearly with r. Therefore, a point $\frac{1}{5}$ the distance out to the full radius R of the plates, where B_{max} occurs, should have a field B that is $\frac{1}{5}B_{max}$.

Additional examples, video, and practice available at *WileyPLUS*

32-5 Maxwell's Equations

Equation 32-5 is the last of the four fundamental equations of electromagnetism, called *Maxwell's equations* and displayed in Table 32-1. These four equations explain a diverse range of phenomena, from why a compass needle points north to why a car starts when you turn the ignition key. They are the basis for the functioning of such electromagnetic devices as electric motors, television transmitters and receivers, telephones, fax machines, radar, and microwave ovens.

Maxwell's equations are the basis from which many of the equations you have seen since Chapter 21 can be derived. They are also the basis of many of the equations you will see in Chapters 33 through 36 concerning optics.

Table 32-1

Maxwell's Equations[a]

Name	Equation	
Gauss' law for electricity	$\oint \vec{E}\cdot d\vec{A} = q_{enc}/\varepsilon_0$	Relates net electric flux to net enclosed electric charge
Gauss' law for magnetism	$\oint \vec{B}\cdot d\vec{A} = 0$	Relates net magnetic flux to net enclosed magnetic charge
Faraday's law	$\oint \vec{E}\cdot d\vec{s} = -\frac{d\Phi_B}{dt}$	Relates induced electric field to changing magnetic flux
Ampere–Maxwell law	$\oint \vec{B}\cdot d\vec{s} = \mu_0\varepsilon_0\frac{d\Phi_E}{dt} + \mu_0 i_{enc}$	Relates induced magnetic field to changing electric flux and to current

[a]Written on the assumption that no dielectric or magnetic materials are present.

32-6 Magnets

The first known magnets were *lodestones,* which are stones that have been *magnetized* (made magnetic) naturally. When the ancient Greeks and ancient Chinese discovered these rare stones, they were amused by the stones' ability to attract metal over a short distance, as if by magic. Only much later did they learn to use lodestones (and artificially magnetized pieces of iron) in compasses to determine direction.

Today, magnets and magnetic materials are ubiquitous. Their magnetic properties can be traced to their atoms and electrons. In fact, the inexpensive magnet you might use to hold a note on the refrigerator door is a direct result of the quantum physics taking place in the atomic and subatomic material within the magnet. Before we explore some of this physics, let's briefly discuss the largest magnet we commonly use—namely, Earth itself.

For Earth, the south pole of the dipole is actually in the north.

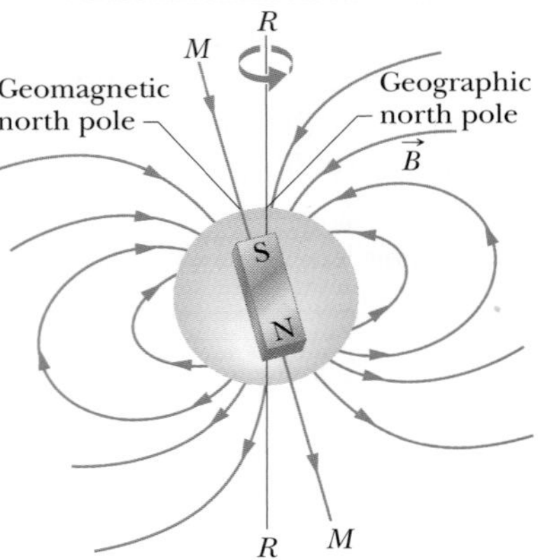

Fig. 32-8 Earth's magnetic field represented as a dipole field. The dipole axis MM makes an angle of 11.5° with Earth's rotational axis RR. The south pole of the dipole is in Earth's Northern Hemisphere.

The Magnetism of Earth

Earth is a huge magnet; for points near Earth's surface, its magnetic field can be approximated as the field of a huge bar magnet—a magnetic dipole—that straddles the center of the planet. Figure 32-8 is an idealized symmetric depiction of the dipole field, without the distortion caused by passing charged particles from the Sun.

Because Earth's magnetic field is that of a magnetic dipole, a magnetic dipole moment $\vec{\mu}$ is associated with the field. For the idealized field of Fig. 32-8, the magnitude of $\vec{\mu}$ is 8.0×10^{22} J/T and the direction of $\vec{\mu}$ makes an angle of 11.5° with the rotation axis (RR) of Earth. The *dipole axis* (MM in Fig. 32-8) lies along $\vec{\mu}$ and intersects Earth's surface at the *geomagnetic north pole* off the northwest coast of Greenland and the *geomagnetic south pole* in Antarctica. The lines of the magnetic field $\vec{B}$ generally emerge in the Southern Hemisphere and reenter Earth in the Northern Hemisphere. Thus, the magnetic pole that is in Earth's Northern Hemisphere and known as a "north magnetic pole" *is really the south pole of Earth's magnetic dipole.*

The direction of the magnetic field at any location on Earth's surface is commonly specified in terms of two angles. The **field declination** is the angle (left or right) between geographic north (which is toward 90° latitude) and the horizontal component of the field. The **field inclination** is the angle (up or down) between a horizontal plane and the field's direction.

Magnetometers measure these angles and determine the field with much precision. However, you can do reasonably well with just a *compass* and a *dip meter.* A compass is simply a needle-shaped magnet that is mounted so it can rotate freely about a vertical axis. When it is held in a horizontal plane, the north-pole end of the needle points, generally, toward the geomagnetic north pole (really a south magnetic pole, remember). The angle between the needle and geographic north is the field declination. A dip meter is a similar magnet that can rotate freely about a horizontal axis. When its vertical plane of rotation is aligned with the direction of the compass, the angle between the meter's needle and the horizontal is the field inclination.

At any point on Earth's surface, the measured magnetic field may differ appreciably, in both magnitude and direction, from the idealized dipole field of Fig. 32-8. In fact, the point where the field is actually perpendicular to Earth's surface and inward is not located at the geomagnetic north pole off Greenland as we would expect; instead, this so-called *dip north pole* is located in the Queen Elizabeth Islands in northern Canada, far from Greenland.

In addition, the field observed at any location on the surface of Earth varies with time, by measurable amounts over a period of a few years and by substantial amounts over, say, 100 years. For example, between 1580 and 1820 the direction indicated by compass needles in London changed by 35°.

In spite of these local variations, the average dipole field changes only slowly over such relatively short time periods. Variations over longer periods can be studied by measuring the weak magnetism of the ocean floor on either side of the Mid-Atlantic Ridge (Fig. 32-9). This floor has been formed by molten magma

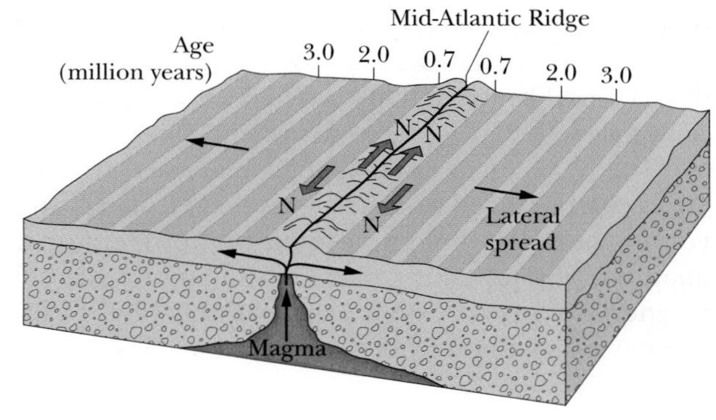

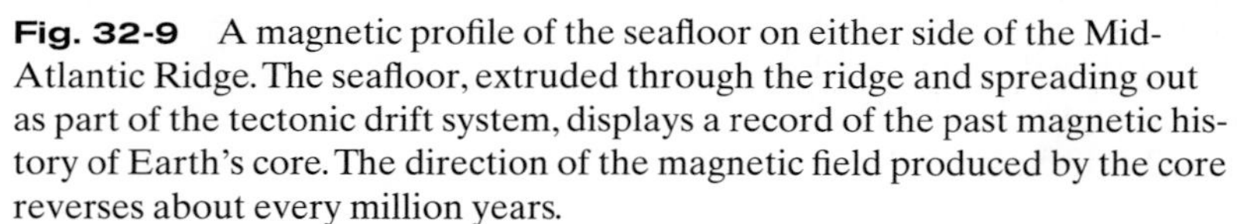
Fig. 32-9 A magnetic profile of the seafloor on either side of the Mid-Atlantic Ridge. The seafloor, extruded through the ridge and spreading out as part of the tectonic drift system, displays a record of the past magnetic history of Earth's core. The direction of the magnetic field produced by the core reverses about every million years.

that oozed up through the ridge from Earth's interior, solidified, and was pulled away from the ridge (by the drift of tectonic plates) at the rate of a few centimeters per year. As the magma solidified, it became weakly magnetized with its magnetic field in the direction of Earth's magnetic field at the time of solidification. Study of this solidified magma across the ocean floor reveals that Earth's field has reversed its *polarity* (directions of the north pole and south pole) about every million years. The reason for the reversals is not known. In fact, the mechanism that produces Earth's magnetic field is only vaguely understood.

32-7 Magnetism and Electrons

Magnetic materials, from lodestones to videotapes, are magnetic because of the electrons within them. We have already seen one way in which electrons can generate a magnetic field: Send them through a wire as an electric current, and their motion produces a magnetic field around the wire. There are two more ways, each involving a magnetic dipole moment that produces a magnetic field in the surrounding space. However, their explanation requires quantum physics that is beyond the physics presented in this book, and so here we shall only outline the results.

Spin Magnetic Dipole Moment

An electron has an intrinsic angular momentum called its **spin angular momentum** (or just **spin**) $\vec{S}$; associated with this spin is an intrinsic **spin magnetic dipole moment** $\vec{\mu}_s$. (By *intrinsic,* we mean that $\vec{S}$ and $\vec{\mu}_s$ are basic characteristics of an electron, like its mass and electric charge.) Vectors $\vec{S}$ and $\vec{\mu}_s$ are related by

$$\vec{\mu}_s = -\frac{e}{m}\vec{S}, \qquad (32\text{-}22)$$

in which e is the elementary charge (1.60×10^{-19} C) and m is the mass of an electron (9.11×10^{-31} kg). The minus sign means that $\vec{\mu}_s$ and $\vec{S}$ are oppositely directed.

Spin $\vec{S}$ is different from the angular momenta of Chapter 11 in two respects:

1. Spin $\vec{S}$ itself cannot be measured. However, its component along any axis can be measured.
2. A measured component of $\vec{S}$ is *quantized,* which is a general term that means it is restricted to certain values. A measured component of $\vec{S}$ can have only two values, which differ only in sign.

Let us assume that the component of spin $\vec{S}$ is measured along the z axis of a coordinate system. Then the measured component S_z can have only the two values given by

$$S_z = m_s\frac{h}{2\pi}, \qquad \text{for } m_s = \pm\tfrac{1}{2}, \qquad (32\text{-}23)$$

where m_s is called the *spin magnetic quantum number* and h ($= 6.63 \times 10^{-34}$ J·s) is the Planck constant, the ubiquitous constant of quantum physics. The signs given in Eq. 32-23 have to do with the direction of S_z along the z axis. When S_z is parallel to the z axis, m_s is $+\frac{1}{2}$ and the electron is said to be *spin up*. When S_z is antiparallel to the z axis, m_s is $-\frac{1}{2}$ and the electron is said to be *spin down.*

The spin magnetic dipole moment $\vec{\mu}_s$ of an electron also cannot be measured; only its component along any axis can be measured, and that component too is quantized, with two possible values of the same magnitude but different signs. We can relate the component $\mu_{s,z}$ measured on the z axis to S_z by rewriting Eq. 32-22 in component form for the z axis as

$$\mu_{s,z} = -\frac{e}{m}S_z.$$

For an electron, the spin is opposite the magnetic dipole moment.

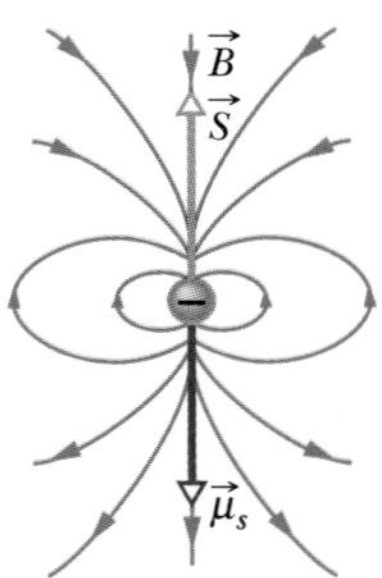

Fig. 32-10 The spin $\vec{S}$, spin magnetic dipole moment $\vec{\mu}_s$, and magnetic dipole field $\vec{B}$ of an electron represented as a microscopic sphere.

Substituting for S_z from Eq. 32-23 then gives us

$$\mu_{s,z} = \pm\frac{eh}{4\pi m}, \tag{32-24}$$

where the plus and minus signs correspond to $\mu_{s,z}$ being parallel and antiparallel to the z axis, respectively.

The quantity on the right side of Eq. 32-24 is called the *Bohr magneton* μ_B:

$$\mu_B = \frac{eh}{4\pi m} = 9.27 \times 10^{-24} \text{ J/T} \quad \text{(Bohr magneton).} \tag{32-25}$$

Spin magnetic dipole moments of electrons and other elementary particles can be expressed in terms of μ_B. For an electron, the magnitude of the measured z component of $\vec{\mu}_s$ is

$$|\mu_{s,z}| = 1\mu_B. \tag{32-26}$$

(The quantum physics of the electron, called *quantum electrodynamics,* or QED, reveals that $\mu_{s,z}$ is actually slightly greater than $1\mu_B$, but we shall neglect that fact.)

When an electron is placed in an external magnetic field $\vec{B}_{ext}$, an energy U can be associated with the orientation of the electron's spin magnetic dipole moment $\vec{\mu}_s$ just as an energy can be associated with the orientation of the magnetic dipole moment $\vec{\mu}$ of a current loop placed in $\vec{B}_{ext}$. From Eq. 28-38, the orentation energy for the electron is

$$U = -\vec{\mu}_s \cdot \vec{B}_{ext} = -\mu_{s,z}B_{ext}, \tag{32-27}$$

where the z axis is taken to be in the direction of $\vec{B}_{ext}$.

If we imagine an electron to be a microscopic sphere (which it is not), we can represent the spin $\vec{S}$, the spin magnetic dipole moment $\vec{\mu}_s$, and the associated magnetic dipole field as in Fig. 32-10. Although we use the word "spin" here, electrons do not spin like tops. How, then, can something have angular momentum without actually rotating? Again, we would need quantum physics to provide the answer.

Protons and neutrons also have an intrinsic angular momentum called spin and an associated intrinsic spin magnetic dipole moment. For a proton those two vectors have the same direction, and for a neutron they have opposite directions. We shall not examine the contributions of these dipole moments to the magnetic fields of atoms because they are about a thousand times smaller than that due to an electron.

CHECKPOINT 4

The figure here shows the spin orientations of two particles in an external magnetic field $\vec{B}_{ext}$. (a) If the particles are electrons, which spin orientation is at lower energy? (b) If, instead, the particles are protons, which spin orientation is at lower energy?

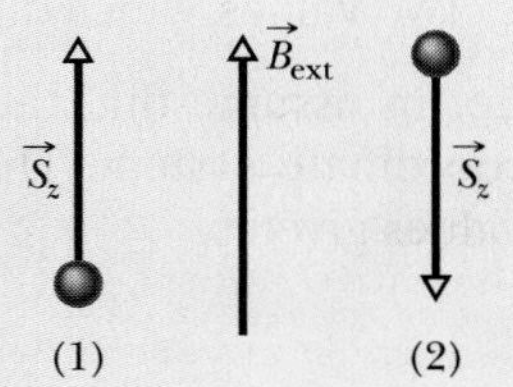

Orbital Magnetic Dipole Moment

When it is in an atom, an electron has an additional angular momentum called its **orbital angular momentum** $\vec{L}_{orb}$. Associated with $\vec{L}_{orb}$ is an **orbital magnetic dipole moment** $\vec{\mu}_{orb}$; the two are related by

$$\vec{\mu}_{orb} = -\frac{e}{2m}\vec{L}_{orb}. \tag{32-28}$$

The minus sign means that $\vec{\mu}_{orb}$ and $\vec{L}_{orb}$ have opposite directions.

Orbital angular momentum $\vec{L}_{orb}$ cannot be measured; only its component along any axis can be measured, and that component is quantized. The component along, say, a z axis can have only the values given by

$$L_{orb,z} = m_\ell \frac{h}{2\pi}, \qquad \text{for } m_\ell = 0, \pm 1, \pm 2, \ldots, \pm(\text{limit}), \tag{32-29}$$

in which m_ℓ is called the *orbital magnetic quantum number* and "limit" refers to some largest allowed integer value for m_ℓ. The signs in Eq. 32-29 have to do with the direction of $L_{orb,z}$ along the z axis.

The orbital magnetic dipole moment $\vec{\mu}_{orb}$ of an electron also cannot itself be measured; only its component along an axis can be measured, and that component is quantized. By writing Eq. 32-28 for a component along the same z axis as above and then substituting for $L_{orb,z}$ from Eq. 32-29, we can write the z component $\mu_{orb,z}$ of the orbital magnetic dipole moment as

$$\mu_{orb,z} = -m_\ell \frac{eh}{4\pi m} \tag{32-30}$$

and, in terms of the Bohr magneton, as

$$\mu_{orb,z} = -m_\ell \mu_B. \tag{32-31}$$

When an atom is placed in an external magnetic field $\vec{B}_{ext}$, an energy U can be associated with the orientation of the orbital magnetic dipole moment of each electron in the atom. Its value is

$$U = -\vec{\mu}_{orb} \cdot \vec{B}_{ext} = -\mu_{orb,z} B_{ext}, \tag{32-32}$$

where the z axis is taken in the direction of $\vec{B}_{ext}$.

Although we have used the words "orbit" and "orbital" here, electrons do not orbit the nucleus of an atom like planets orbiting the Sun. How can an electron have an orbital angular momentum without orbiting in the common meaning of the term? Once again, this can be explained only with quantum physics.

Loop Model for Electron Orbits

We can obtain Eq. 32-28 with the nonquantum derivation that follows, in which we assume that an electron moves along a circular path with a radius that is much larger than an atomic radius (hence the name "loop model"). However, the derivation does not apply to an electron within an atom (for which we need quantum physics).

We imagine an electron moving at constant speed v in a circular path of radius r, counterclockwise as shown in Fig. 32-11. The motion of the negative charge of the electron is equivalent to a conventional current i (of positive charge) that is clockwise, as also shown in Fig. 32-11. The magnitude of the orbital magnetic dipole moment of such a *current loop* is obtained from Eq. 28-35 with $N = 1$:

$$\mu_{orb} = iA, \tag{32-33}$$

where A is the area enclosed by the loop. The direction of this magnetic dipole moment is, from the right-hand rule of Fig. 29-21, downward in Fig. 32-11.

To evaluate Eq. 32-33, we need the current i. Current is, generally, the rate at which charge passes some point in a circuit. Here, the charge of magnitude e takes a time $T = 2\pi r/v$ to circle from any point back through that point, so

$$i = \frac{\text{charge}}{\text{time}} = \frac{e}{2\pi r/v}. \tag{32-34}$$

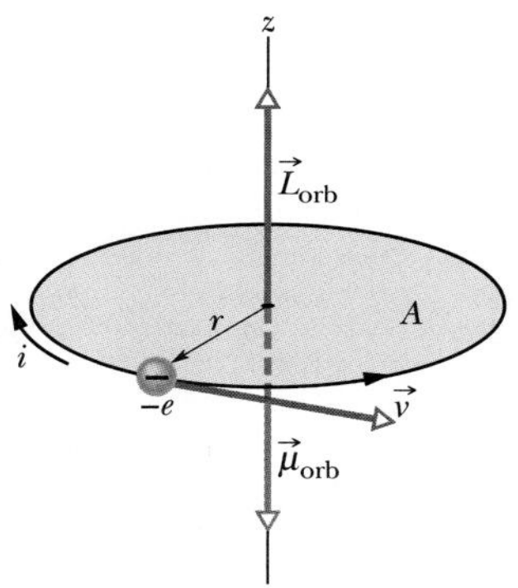

Fig. 32-11 An electron moving at constant speed v in a circular path of radius r that encloses an area A. The electron has an orbital angular momentum $\vec{L}_{orb}$ and an associated orbital magnetic dipole moment $\vec{\mu}_{orb}$. A clockwise current i (of positive charge) is equivalent to the counterclockwise circulation of the negatively charged electron.

Substituting this and the area $A = \pi r^2$ of the loop into Eq. 32-33 gives us

$$\mu_{orb} = \frac{e}{2\pi r/v}\,\pi r^2 = \frac{evr}{2}. \tag{32-35}$$

To find the electron's orbital angular momentum $\vec{L}_{orb}$, we use Eq. 11-18, $\vec{\ell} = m(\vec{r} \times \vec{v})$. Because $\vec{r}$ and $\vec{v}$ are perpendicular, $\vec{L}_{orb}$ has the magnitude

$$L_{orb} = mrv \sin 90° = mrv. \tag{32-36}$$

The vector $\vec{L}_{orb}$ is directed upward in Fig. 32-11 (see Fig. 11-12). Combining Eqs. 32-35 and 32-36, generalizing to a vector formulation, and indicating the opposite directions of the vectors with a minus sign yield

$$\vec{\mu}_{orb} = -\frac{e}{2m}\vec{L}_{orb},$$

which is Eq. 32-28. Thus, by "classical" (nonquantum) analysis we have obtained the same result, in both magnitude and direction, given by quantum physics. You might wonder, seeing as this derivation gives the correct result for an electron within an atom, why the derivation is invalid for that situation. The answer is that this line of reasoning yields other results that are contradicted by experiments.

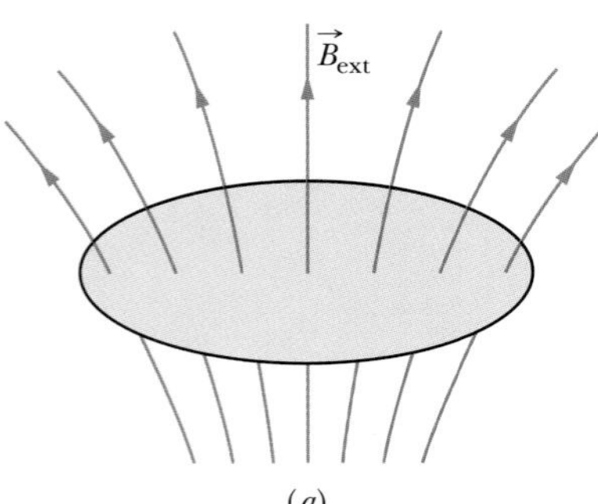

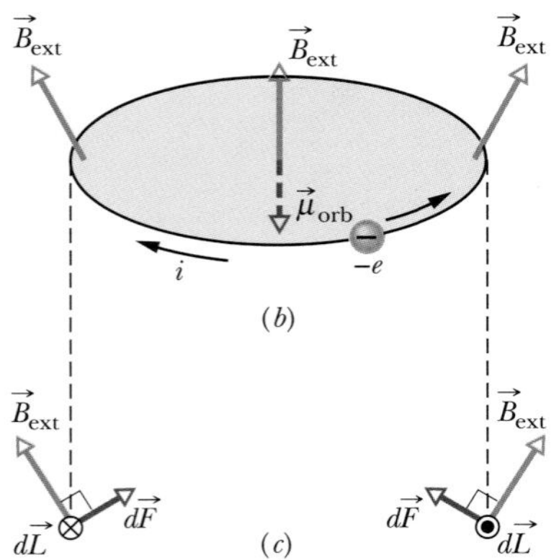

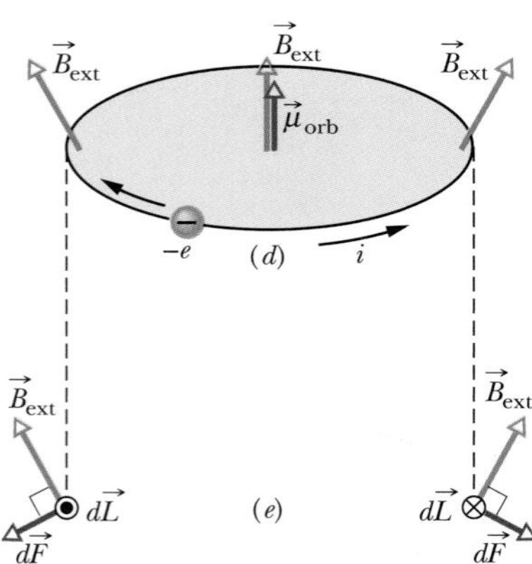

Fig. 32-12 (*a*) A loop model for an electron orbiting in an atom while in a nonuniform magnetic field $\vec{B}_{ext}$. (*b*) Charge $-e$ moves counterclockwise; the associated conventional current i is clockwise. (*c*) The magnetic forces $d\vec{F}$ on the left and right sides of the loop, as seen from the plane of the loop. The net force on the loop is upward. (*d*) Charge $-e$ now moves clockwise. (*e*) The net force on the loop is now downward.

Loop Model in a Nonuniform Field

We continue to consider an electron orbit as a current loop, as we did in Fig. 32-11. Now, however, we draw the loop in a nonuniform magnetic field $\vec{B}_{ext}$ as shown in Fig. 32-12*a*. (This field could be the diverging field near the north pole of the magnet in Fig. 32-4.) We make this change to prepare for the next several sections, in which we shall discuss the forces that act on magnetic materials when the materials are placed in a nonuniform magnetic field. We shall discuss these forces by assuming that the electron orbits in the materials are tiny current loops like that in Fig. 32-12*a*.

Here we assume that the magnetic field vectors all around the electron's circular path have the same magnitude and form the same angle with the vertical, as shown in Figs. 32-12*b* and *d*. We also assume that all the electrons in an atom move either counterclockwise (Fig. 32-12*b*) or clockwise (Fig. 32-12*d*). The associated conventional current i around the current loop and the orbital magnetic dipole moment $\vec{\mu}_{orb}$ produced by i are shown for each direction of motion.

Figures 32-12*c* and *e* show diametrically opposite views of a length element $d\vec{L}$ of the loop that has the same direction as i, as seen from the plane of the orbit. Also shown are the field $\vec{B}_{ext}$ and the resulting magnetic force $d\vec{F}$ on $d\vec{L}$. Recall that a current along an element $d\vec{L}$ in a magnetic field $\vec{B}_{ext}$ experiences a magnetic force $d\vec{F}$ as given by Eq. 28-28:

$$d\vec{F} = i\,d\vec{L} \times \vec{B}_{ext}. \tag{32-37}$$

On the left side of Fig. 32-12*c*, Eq. 32-37 tells us that the force $d\vec{F}$ is directed upward and rightward. On the right side, the force $d\vec{F}$ is just as large and is directed upward and leftward. Because their angles are the same, the horizontal components of these two forces cancel and the vertical components add. The same is true at any other two symmetric points on the loop. Thus, the net force on the current loop of Fig. 32-12*b* must be upward. The same reasoning leads to a downward net force on the loop in Fig. 32-12*d*. We shall use these two results shortly when we examine the behavior of magnetic materials in nonuniform magnetic fields.

32-8 Magnetic Materials

Each electron in an atom has an orbital magnetic dipole moment and a spin magnetic dipole moment that combine vectorially. The resultant of these two vector quantities combines vectorially with similar resultants for all other electrons in the atom, and the resultant for each atom combines with those for all the other atoms in a sample of a material. If the combination of all these magnetic dipole moments produces a magnetic field, then the material is magnetic. There are three general types of magnetism: diamagnetism, paramagnetism, and ferromagnetism.

1. ***Diamagnetism*** is exhibited by all common materials but is so feeble that it is masked if the material also exhibits magnetism of either of the other two types. In diamagnetism, weak magnetic dipole moments are produced in the atoms of the material when the material is placed in an external magnetic field $\vec{B}_{ext}$; the combination of all those induced dipole moments gives the material as a whole only a feeble net magnetic field. The dipole moments and thus their net field disappear when $\vec{B}_{ext}$ is removed. The term *diamagnetic material* usually refers to materials that exhibit only diamagnetism.
2. ***Paramagnetism*** is exhibited by materials containing transition elements, rare earth elements, and actinide elements (see Appendix G). Each atom of such a material has a permanent resultant magnetic dipole moment, but the moments are randomly oriented in the material and the material as a whole lacks a net magnetic field. However, an external magnetic field $\vec{B}_{ext}$ can partially align the atomic magnetic dipole moments to give the material a net magnetic field. The alignment and thus its field disappear when $\vec{B}_{ext}$ is removed. The term *paramagnetic material* usually refers to materials that exhibit primarily paramagnetism.
3. ***Ferromagnetism*** is a property of iron, nickel, and certain other elements (and of compounds and alloys of these elements). Some of the electrons in these materials have their resultant magnetic dipole moments aligned, which produces regions with strong magnetic dipole moments. An external field $\vec{B}_{ext}$ can then align the magnetic moments of such regions, producing a strong magnetic field for a sample of the material; the field partially persists when $\vec{B}_{ext}$ is removed. We usually use the terms *ferromagnetic material* and *magnetic material* to refer to materials that exhibit primarily ferromagnetism.

The next three sections explore these three types of magnetism.

32-9 Diamagnetism

We cannot yet discuss the quantum physical explanation of diamagnetism, but we can provide a classical explanation with the loop model of Figs. 32-11 and 32-12. To begin, we assume that in an atom of a diamagnetic material each electron can orbit only clockwise as in Fig. 32-12*d* or counterclockwise as in Fig. 32-12*b*. To account for the lack of magnetism in the absence of an external magnetic field $\vec{B}_{ext}$, we assume the atom lacks a net magnetic dipole moment. This implies that before $\vec{B}_{ext}$ is applied, the number of electrons orbiting in one direction is the same as that orbiting in the opposite direction, with the result that the net upward magnetic dipole moment of the atom equals the net downward magnetic dipole moment.

Now let's turn on the nonuniform field $\vec{B}_{ext}$ of Fig. 32-12*a*, in which $\vec{B}_{ext}$ is directed upward but is diverging (the magnetic field lines are diverging). We could do this by increasing the current through an electromagnet or by moving

Fig. 32-13 An overhead view of a frog that is being levitated in a magnetic field produced by current in a vertical solenoid below the frog. *(Courtesy A. K. Gein, High Field Magnet Laboratory, University of Nijmegen, The Netherlands)*

the north pole of a bar magnet closer to, and below, the orbits. As the magnitude of $\vec{B}_{ext}$ increases from zero to its final maximum, steady-state value, a clockwise electric field is induced around each electron's orbital loop according to Faraday's law and Lenz's law. Let us see how this induced electric field affects the orbiting electrons in Figs. 32-12*b* and *d*.

In Fig. 32-12*b*, the counterclockwise electron is accelerated by the clockwise electric field. Thus, as the magnetic field $\vec{B}_{ext}$ increases to its maximum value, the electron speed increases to a maximum value. This means that the associated conventional current *i* and the downward magnetic dipole moment $\vec{\mu}$ due to *i* also *increase.*

In Fig. 32-12*d*, the clockwise electron is decelerated by the clockwise electric field. Thus, here, the electron speed, the associated current *i*, and the upward magnetic dipole moment $\vec{\mu}$ due to *i* all *decrease.* By turning on field $\vec{B}_{ext}$, we have given the atom a *net* magnetic dipole moment that is downward. This would also be so if the magnetic field were uniform.

The nonuniformity of field $\vec{B}_{ext}$ also affects the atom. Because the current *i* in Fig. 32-12*b* increases, the upward magnetic forces $d\vec{F}$ in Fig. 32-12*c* also increase, as does the net upward force on the current loop. Because current *i* in Fig. 32-12*d* decreases, the downward magnetic forces $d\vec{F}$ in Fig. 32-12*e* also decrease, as does the net downward force on the current loop. Thus, by turning on the *nonuniform* field $\vec{B}_{ext}$, we have produced a net force on the atom; moreover, that force is directed *away* from the region of greater magnetic field.

We have argued with fictitious electron orbits (current loops), but we have ended up with exactly what happens to a diamagnetic material: If we apply the magnetic field of Fig. 32-12, the material develops a downward magnetic dipole moment and experiences an upward force. When the field is removed, both the dipole moment and the force disappear. The external field need not be positioned as shown in Fig. 32-12; similar arguments can be made for other orientations of $\vec{B}_{ext}$. In general,

> A diamagnetic material placed in an external magnetic field $\vec{B}_{ext}$ develops a magnetic dipole moment directed opposite $\vec{B}_{ext}$. If the field is nonuniform, the diamagnetic material is repelled *from* a region of greater magnetic field *toward* a region of lesser field.

The frog in Fig. 32-13 is diamagnetic (as is any other animal). When the frog was placed in the diverging magnetic field near the top end of a vertical current-carrying solenoid, every atom in the frog was repelled upward, away from the region of stronger magnetic field at that end of the solenoid. The frog moved upward into weaker and weaker magnetic field until the upward magnetic force balanced the gravitational force on it, and there it hung in midair. The frog is not in discomfort because *every* atom is subject to the same forces and thus there is no force variation within the frog. The sensation is similar to the "weightless" situation of floating in water, which frogs like very much. If we went to the expense of building a much larger solenoid, we could similarly levitate a person in midair due to the person's diamagnetism.

CHECKPOINT 5

The figure shows two diamagnetic spheres located near the south pole of a bar magnet. Are (a) the magnetic forces on the spheres and (b) the magnetic dipole moments of the spheres directed toward or away from the bar magnet? (c) Is the magnetic force on sphere 1 greater than, less than, or equal to that on sphere 2?

32-10 Paramagnetism

In paramagnetic materials, the spin and orbital magnetic dipole moments of the electrons in each atom do not cancel but add vectorially to give the atom a net (and permanent) magnetic dipole moment $\vec{\mu}$. In the absence of an external magnetic field, these atomic dipole moments are randomly oriented, and the net magnetic dipole moment of the material is zero. However, if a sample of the material is placed in an external magnetic field $\vec{B}_{ext}$, the magnetic dipole moments tend to line up with the field, which gives the sample a net magnetic dipole moment. This alignment with the external field is the opposite of what we saw with diamagnetic materials.

> A paramagnetic material placed in an external magnetic field $\vec{B}_{ext}$ develops a magnetic dipole moment in the direction of $\vec{B}_{ext}$. If the field is nonuniform, the paramagnetic material is attracted *toward* a region of greater magnetic field *from* a region of lesser field.

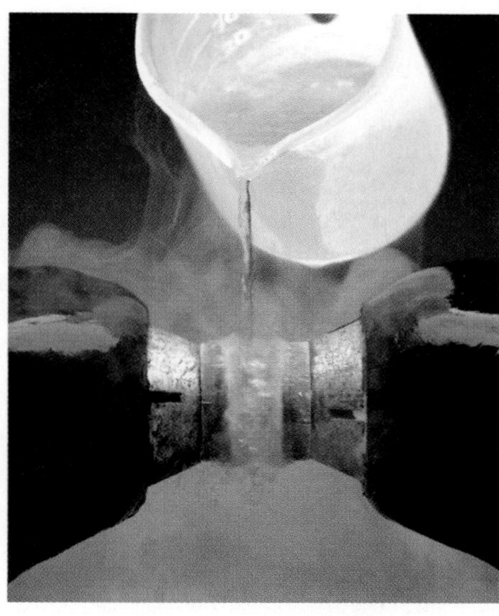

Liquid oxygen is suspended between the two pole faces of a magnet because the liquid is paramagnetic and is magnetically attracted to the magnet. *(Richard Megna/Fundamental Photographs)*

A paramagnetic sample with N atoms would have a magnetic dipole moment of magnitude $N\mu$ if alignment of its atomic dipoles were complete. However, random collisions of atoms due to their thermal agitation transfer energy among the atoms, disrupting their alignment and thus reducing the sample's magnetic dipole moment.

The importance of thermal agitation may be measured by comparing two energies. One, given by Eq. 19-24, is the mean translational kinetic energy $K\,(=\frac{3}{2}kT)$ of an atom at temperature T, where k is the Boltzmann constant (1.38×10^{-23} J/K) and T is in kelvins (not Celsius degrees). The other, derived from Eq. 28-38, is the difference in energy $\Delta U_B\,(=2\mu B_{ext})$ between parallel alignment and antiparallel alignment of the magnetic dipole moment of an atom and the external field. (The lower energy state is $-\mu B_{ext}$ and the higher energy state is $+\mu B_{ext}$.) As we shall show below, $K \gg \Delta U_B$, even for ordinary temperatures and field magnitudes. Thus, energy transfers during collisions among atoms can significantly disrupt the alignment of the atomic dipole moments, keeping the magnetic dipole moment of a sample much less than $N\mu$.

We can express the extent to which a given paramagnetic sample is magnetized by finding the ratio of its magnetic dipole moment to its volume V. This vector quantity, the magnetic dipole moment per unit volume, is the **magnetization** $\vec{M}$ of the sample, and its magnitude is

$$M = \frac{\text{measured magnetic moment}}{V}. \tag{32-38}$$

The unit of $\vec{M}$ is the ampere–square meter per cubic meter, or ampere per meter (A/m). Complete alignment of the atomic dipole moments, called *saturation* of the sample, corresponds to the maximum value $M_{max} = N\mu/V$.

In 1895 Pierre Curie discovered experimentally that the magnetization of a paramagnetic sample is directly proportional to the magnitude of the external magnetic field $\vec{B}_{ext}$ and inversely proportional to the temperature T in kelvins:

$$M = C\frac{B_{ext}}{T}. \tag{32-39}$$

Equation 32-39 is known as *Curie's law,* and C is called the *Curie constant.* Curie's law is reasonable in that increasing B_{ext} tends to align the atomic dipole moments in a sample and thus to increase M, whereas increasing T tends to disrupt the alignment via thermal agitation and thus to decrease M. However, the law is actually an approximation that is valid only when the ratio B_{ext}/T is not too large.

Fig. 32-14 A *magnetization curve* for potassium chromium sulfate, a paramagnetic salt. The ratio of magnetization M of the salt to the maximum possible magnetization M_{max} is plotted versus the ratio of the applied magnetic field magnitude B_{ext} to the temperature T. Curie's law fits the data at the left; quantum theory fits all the data. After W. E. Henry.

Figure 32-14 shows the ratio M/M_{max} as a function of B_{ext}/T for a sample of the salt potassium chromium sulfate, in which chromium ions are the paramagnetic substance. The plot is called a *magnetization curve.* The straight line for Curie's law fits the experimental data at the left, for B_{ext}/T below about 0.5 T/K. The curve that fits all the data points is based on quantum physics. The data on the right side, near saturation, are very difficult to obtain because they require very strong magnetic fields (about 100 000 times Earth's field), even at very low temperatures.

CHECKPOINT 6

The figure here shows two paramagnetic spheres located near the south pole of a bar magnet. Are (a) the magnetic forces on the spheres and (b) the magnetic dipole moments of the spheres directed toward or away from the bar magnet? (c) Is the magnetic force on sphere 1 greater than, less than, or equal to that on sphere 2?

Sample Problem

Orientation energy of a magnetic field in a paramagnetic gas

A paramagnetic gas at room temperature ($T = 300$ K) is placed in an external uniform magnetic field of magnitude $B = 1.5$ T; the atoms of the gas have magnetic dipole moment $\mu = 1.0\mu_B$. Calculate the mean translational kinetic energy K of an atom of the gas and the energy difference ΔU_B between parallel alignment and antiparallel alignment of the atom's magnetic dipole moment with the external field.

KEY IDEAS

(1) The mean translational kinetic energy K of an atom in a gas depends on the temperature of the gas. (2) The energy U_B of a magnetic dipole $\vec{\mu}$ in an external magnetic field $\vec{B}$ depends on the angle θ between the directions of $\vec{\mu}$ and $\vec{B}$.

Calculations: From Eq. 19-24, we have

$$K = \tfrac{3}{2}kT = \tfrac{3}{2}(1.38 \times 10^{-23}\text{ J/K})(300\text{ K})$$
$$= 6.2 \times 10^{-21}\text{ J} = 0.039\text{ eV}. \qquad \text{(Answer)}$$

From Eq. 28-38 ($U_B = -\vec{\mu}\cdot\vec{B}$), we can write the difference ΔU_B between parallel alignment ($\theta = 0°$) and antiparallel alignment ($\theta = 180°$) as

$$\Delta U_B = -\mu B\cos 180° - (-\mu B\cos 0°) = 2\mu B$$
$$= 2\mu_B B = 2(9.27 \times 10^{-24}\text{ J/T})(1.5\text{ T})$$
$$= 2.8 \times 10^{-23}\text{ J} = 0.000\,17\text{ eV}. \qquad \text{(Answer)}$$

Here K is about 230 times ΔU_B; so energy exchanges among the atoms during their collisions with one another can easily reorient any magnetic dipole moments that might be aligned with the external magnetic field. That is, as soon as a magnetic dipole moment happens to become aligned with the external field, in the dipole's low energy state, chances are very good that a neighboring atom will hit the atom, transferring enough energy to put the dipole in a higher energy state. Thus, the magnetic dipole moment exhibited by the paramagnetic gas must be due to fleeting partial alignments of the atomic dipole moments.

Additional examples, video, and practice available at *WileyPLUS*

32-11 Ferromagnetism

When we speak of magnetism in everyday conversation, we almost always have a mental picture of a bar magnet or a disk magnet (probably clinging to a refrigerator door). That is, we picture a ferromagnetic material having strong, permanent magnetism, and not a diamagnetic or paramagnetic material having weak, temporary magnetism.

Iron, cobalt, nickel, gadolinium, dysprosium, and alloys containing these elements exhibit ferromagnetism because of a quantum physical effect called *exchange coupling* in which the electron spins of one atom interact with those of neighboring atoms. The result is alignment of the magnetic dipole moments of the atoms, in spite of the randomizing tendency of atomic collisions due to thermal agitation. This persistent alignment is what gives ferromagnetic materials their permanent magnetism.

If the temperature of a ferromagnetic material is raised above a certain critical value, called the *Curie temperature,* the exchange coupling ceases to be effective. Most such materials then become simply paramagnetic; that is, the dipoles still tend to align with an external field but much more weakly, and thermal agitation can now more easily disrupt the alignment. The Curie temperature for iron is 1043 K (= 770°C).

The magnetization of a ferromagnetic material such as iron can be studied with an arrangement called a *Rowland ring* (Fig. 32-15). The material is formed into a thin toroidal core of circular cross section. A primary coil P having n turns per unit length is wrapped around the core and carries current i_P. (The coil is essentially a long solenoid bent into a circle.) If the iron core were not present, the magnitude of the magnetic field inside the coil would be, from Eq. 29-23,

$$B_0 = \mu_0 i_P n. \tag{32-40}$$

However, with the iron core present, the magnetic field $\vec{B}$ inside the coil is greater than $\vec{B}_0$, usually by a large amount. We can write the magnitude of this field as

$$B = B_0 + B_M, \tag{32-41}$$

where B_M is the magnitude of the magnetic field contributed by the iron core. This contribution results from the alignment of the atomic dipole moments within the iron, due to exchange coupling and to the applied magnetic field B_0, and is proportional to the magnetization M of the iron. That is, the contribution B_M is proportional to the magnetic dipole moment per unit volume of the iron. To determine B_M we use a secondary coil S to measure B, compute B_0 with Eq. 32-40, and subtract as suggested by Eq. 32-41.

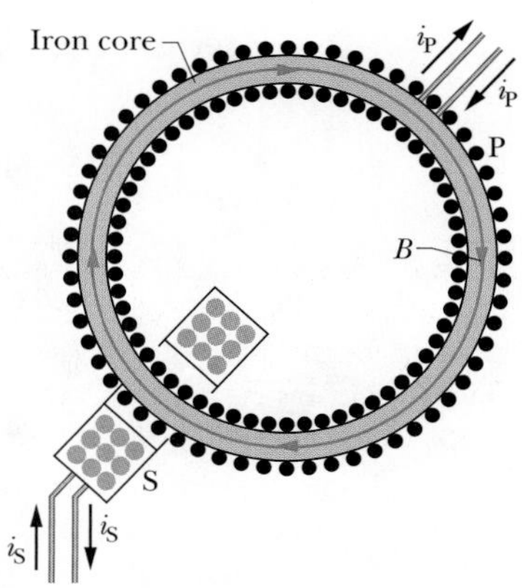

Fig. 32-15 A Rowland ring. A primary coil P has a core made of the ferromagnetic material to be studied (here iron). The core is magnetized by a current i_P sent through coil P. (The turns of the coil are represented by dots.) The extent to which the core is magnetized determines the total magnetic field $\vec{B}$ within coil P. Field $\vec{B}$ can be measured by means of a secondary coil S.

Figure 32-16 shows a magnetization curve for a ferromagnetic material in a Rowland ring: The ratio $B_M/B_{M,\max}$, where $B_{M,\max}$ is the maximum possible value of B_M, corresponding to saturation, is plotted versus B_0. The curve is like Fig. 32-14, the magnetization curve for a paramagnetic substance: Both curves show the extent to which an applied magnetic field can align the atomic dipole moments of a material.

For the ferromagnetic core yielding Fig. 32-16, the alignment of the dipole moments is about 70% complete for $B_0 \approx 1 \times 10^{-3}$ T. If B_0 were increased to 1 T, the alignment would be almost complete (but $B_0 = 1$ T, and thus almost complete saturation, is quite difficult to obtain).

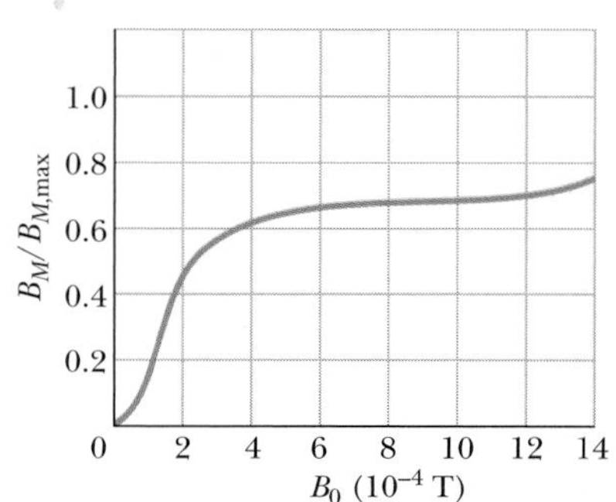

Fig. 32-16 A magnetization curve for a ferromagnetic core material in the Rowland ring of Fig. 32-15. On the vertical axis, 1.0 corresponds to complete alignment (saturation) of the atomic dipoles within the material.

Magnetic Domains

Exchange coupling produces strong alignment of adjacent atomic dipoles in a ferromagnetic material at a temperature below the Curie temperature. Why, then, isn't the material naturally at saturation even when there is no applied magnetic field B_0? Why isn't every piece of iron a naturally strong magnet?

To understand this, consider a specimen of a ferromagnetic material such as iron that is in the form of a single crystal; that is, the arrangement of the atoms that

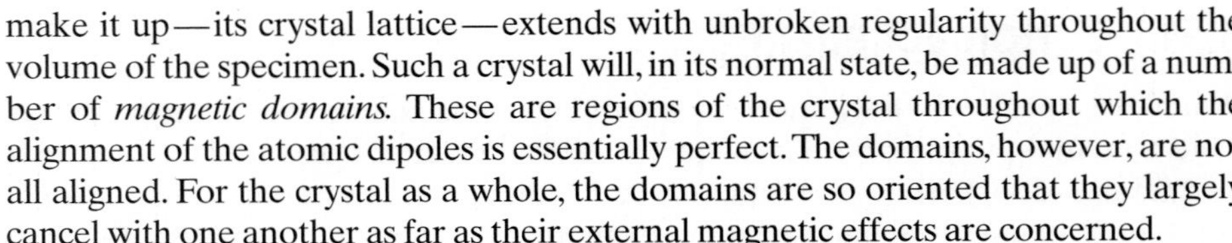

make it up—its crystal lattice—extends with unbroken regularity throughout the volume of the specimen. Such a crystal will, in its normal state, be made up of a number of *magnetic domains.* These are regions of the crystal throughout which the alignment of the atomic dipoles is essentially perfect. The domains, however, are not all aligned. For the crystal as a whole, the domains are so oriented that they largely cancel with one another as far as their external magnetic effects are concerned.

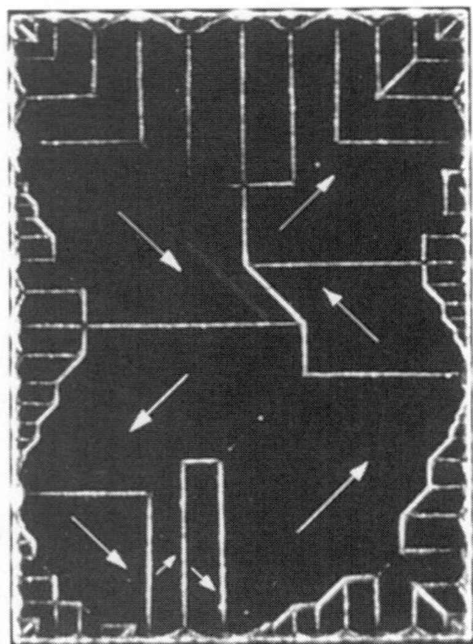

Fig. 32-17 A photograph of domain patterns within a single crystal of nickel; white lines reveal the boundaries of the domains. The white arrows superimposed on the photograph show the orientations of the magnetic dipoles within the domains and thus the orientations of the net magnetic dipoles of the domains. The crystal as a whole is unmagnetized if the net magnetic field (the vector sum over all the domains) is zero. *(Courtesy Ralph W. DeBlois)*

Figure 32-17 is a magnified photograph of such an assembly of domains in a single crystal of nickel. It was made by sprinkling a colloidal suspension of finely powdered iron oxide on the surface of the crystal. The domain boundaries, which are thin regions in which the alignment of the elementary dipoles changes from a certain orientation in one of the domains forming the boundary to a different orientation in the other domain, are the sites of intense, but highly localized and nonuniform, magnetic fields. The suspended colloidal particles are attracted to these boundaries and show up as the white lines (not all the domain boundaries are apparent in Fig. 32-17). Although the atomic dipoles in each domain are completely aligned as shown by the arrows, the crystal as a whole may have only a very small resultant magnetic moment.

Actually, a piece of iron as we ordinarily find it is not a single crystal but an assembly of many tiny crystals, randomly arranged; we call it a *polycrystalline solid.* Each tiny crystal, however, has its array of variously oriented domains, just as in Fig. 32-17. If we magnetize such a specimen by placing it in an external magnetic field of gradually increasing strength, we produce two effects; together they produce a magnetization curve of the shape shown in Fig. 32-16. One effect is a growth in size of the domains that are oriented along the external field at the expense of those that are not. The second effect is a shift of the orientation of the dipoles within a domain, as a unit, to become closer to the field direction.

Exchange coupling and domain shifting give us the following result:

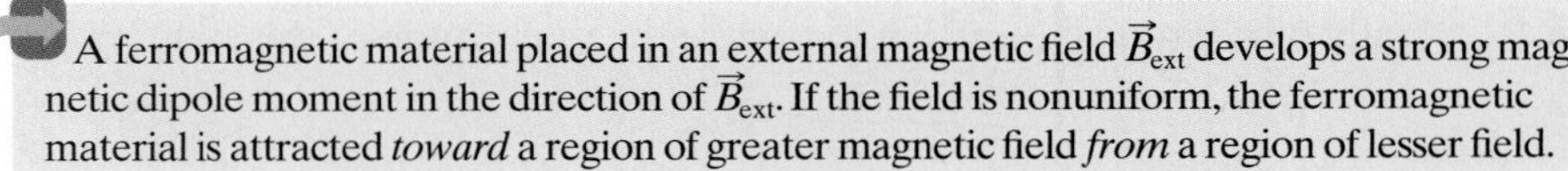

A ferromagnetic material placed in an external magnetic field $\vec{B}_{ext}$ develops a strong magnetic dipole moment in the direction of $\vec{B}_{ext}$. If the field is nonuniform, the ferromagnetic material is attracted *toward* a region of greater magnetic field *from* a region of lesser field.

Hysteresis

Magnetization curves for ferromagnetic materials are not retraced as we increase and then decrease the external magnetic field B_0. Figure 32-18 is a plot of B_M versus B_0 during the following operations with a Rowland ring: (1) Starting with the iron unmagnetized (point a), increase the current in the toroid until B_0 ($= \mu_0 in$) has the value corresponding to point b; (2) reduce the current in the toroid winding (and thus B_0) back to zero (point c); (3) reverse the toroid current and increase it in magnitude until B_0 has the value corresponding to point d; (4) reduce the current to zero again (point e); (5) reverse the current once more until point b is reached again.

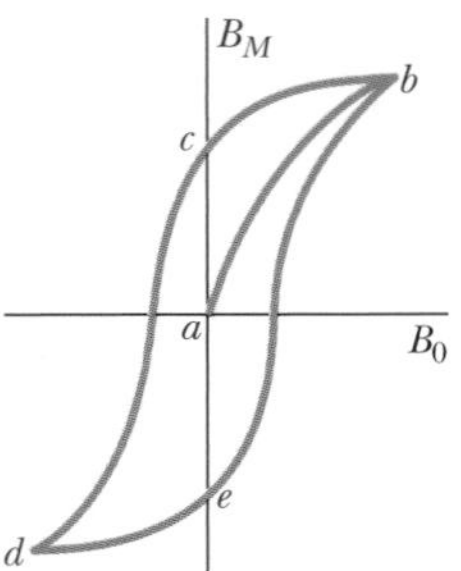

Fig. 32-18 A magnetization curve (ab) for a ferromagnetic specimen and an associated hysteresis loop ($bcdeb$).

The lack of retraceability shown in Fig. 32-18 is called **hysteresis,** and the curve $bcdeb$ is called a *hysteresis loop.* Note that at points c and e the iron core is magnetized, even though there is no current in the toroid windings; this is the familiar phenomenon of permanent magnetism.

Hysteresis can be understood through the concept of magnetic domains. Evidently the motions of the domain boundaries and the reorientations of the domain directions are not totally reversible. When the applied magnetic field B_0 is increased and then decreased back to its initial value, the domains do not return completely to their original configuration but retain some "memory" of their alignment after the initial increase. This memory of magnetic materials is essential for the magnetic storage of information.

This memory of the alignment of domains can also occur naturally. When lightning sends currents along multiple tortuous paths through the ground, the currents produce intense magnetic fields that can suddenly magnetize any ferromagnetic material in nearby rock. Because of hysteresis, such rock material retains some of that magnetization after the lightning strike (after the currents disappear). Pieces of the rock—later exposed, broken, and loosened by weathering—are then lodestones.

Sample Problem

Magnetic dipole moment of a compass needle

A compass needle made of pure iron (density 7900 kg/m^3) has a length L of 3.0 cm, a width of 1.0 mm, and a thickness of 0.50 mm. The magnitude of the magnetic dipole moment of an iron atom is $\mu_{Fe} = 2.1 \times 10^{-23}$ J/T. If the magnetization of the needle is equivalent to the alignment of 10% of the atoms in the needle, what is the magnitude of the needle's magnetic dipole moment $\vec{\mu}$?

KEY IDEAS

(1) Alignment of all N atoms in the needle would give a magnitude of $N\mu_{Fe}$ for the needle's magnetic dipole moment $\vec{\mu}$. However, the needle has only 10% alignment (the random orientation of the rest does not give any net contribution to $\vec{\mu}$). Thus,

$$\mu = 0.10N\mu_{Fe}. \quad (32\text{-}42)$$

(2) We can find the number of atoms N in the needle from the needle's mass:

$$N = \frac{\text{needle's mass}}{\text{iron's atomic mass}}. \quad (32\text{-}43)$$

Finding N: Iron's atomic mass is not listed in Appendix F, but its molar mass M is. Thus, we write

$$\text{iron's atomic mass} = \frac{\text{iron's molar mass } M}{\text{Avogadro's number } N_A}. \quad (32\text{-}44)$$

Next, we can rewrite Eq. 32-43 in terms of the needle's mass m, the molar mass M, and Avogadro's number N_A:

$$N = \frac{mN_A}{M}. \quad (32\text{-}45)$$

The needle's mass m is the product of its density and its volume. The volume works out to be 1.5×10^{-8} m^3; so

$$\begin{aligned}\text{needle's mass } m &= (\text{needle's density})(\text{needle's volume}) \\ &= (7900 \text{ kg/m}^3)(1.5 \times 10^{-8} \text{ m}^3) \\ &= 1.185 \times 10^{-4} \text{ kg}.\end{aligned}$$

Substituting into Eq. 32-45 with this value for m, and also 55.847 g/mol (= 0.055 847 kg/mol) for M and 6.02×10^{23} for N_A, we find

$$\begin{aligned}N &= \frac{(1.185 \times 10^{-4} \text{ kg})(6.02 \times 10^{23})}{0.055\,847 \text{ kg/mol}} \\ &= 1.2774 \times 10^{21}.\end{aligned}$$

Finding μ: Substituting our value of N and the value of μ_{Fe} into Eq. 32-42 then yields

$$\begin{aligned}\mu &= (0.10)(1.2774 \times 10^{21})(2.1 \times 10^{-23} \text{ J/T}) \\ &= 2.682 \times 10^{-3} \text{ J/T} \approx 2.7 \times 10^{-3} \text{ J/T}. \quad \text{(Answer)}\end{aligned}$$

Additional examples, video, and practice available at *WileyPLUS*

REVIEW & SUMMARY

Gauss' Law for Magnetic Fields The simplest magnetic structures are magnetic dipoles. Magnetic monopoles do not exist (as far as we know). **Gauss' law** for magnetic fields,

$$\Phi_B = \oint \vec{B} \cdot d\vec{A} = 0, \quad (32\text{-}1)$$

states that the net magnetic flux through any (closed) Gaussian surface is zero. It implies that magnetic monopoles do not exist.

Maxwell's Extension of Ampere's Law A changing electric flux induces a magnetic field $\vec{B}$. Maxwell's law,

$$\oint \vec{B} \cdot d\vec{s} = \mu_0\varepsilon_0 \frac{d\Phi_E}{dt} \quad \text{(Maxwell's law of induction)}, \quad (32\text{-}3)$$

relates the magnetic field induced along a closed loop to the changing electric flux Φ_E through the loop. Ampere's law, $\oint \vec{B} \cdot d\vec{s} = \mu_0 i_{enc}$ (Eq. 32-4), gives the magnetic field generated by a current i_{enc} encircled by a closed loop. Maxwell's law and Ampere's law can be written as the single equation

$$\oint \vec{B} \cdot d\vec{s} = \mu_0\varepsilon_0 \frac{d\Phi_E}{dt} + \mu_0 i_{enc} \quad \text{(Ampere–Maxwell law)}. \quad (32\text{-}5)$$

Displacement Current We define the fictitious *displacement current* due to a changing electric field as

$$i_d = \varepsilon_0 \frac{d\Phi_E}{dt}. \quad (32\text{-}10)$$

Equation 32-5 then becomes

$$\oint \vec{B} \cdot d\vec{s} = \mu_0 i_{d,\text{enc}} + \mu_0 i_{\text{enc}} \quad \text{(Ampere–Maxwell law)}, \qquad (32\text{-}11)$$

where $i_{d,\text{enc}}$ is the displacement current encircled by the integration loop. The idea of a displacement current allows us to retain the notion of continuity of current through a capacitor. However, displacement current is *not* a transfer of charge.

Maxwell's Equations Maxwell's equations, displayed in Table 32-1, summarize electromagnetism and form its foundation, including optics.

Earth's Magnetic Field Earth's magnetic field can be approximated as being that of a magnetic dipole whose dipole moment makes an angle of 11.5° with Earth's rotation axis, and with the south pole of the dipole in the Northern Hemisphere. The direction of the local magnetic field at any point on Earth's surface is given by the *field declination* (the angle left or right from geographic north) and the *field inclination* (the angle up or down from the horizontal).

Spin Magnetic Dipole Moment An electron has an intrinsic angular momentum called *spin angular momentum* (or *spin*) $\vec{S}$, with which an intrinsic *spin magnetic dipole moment* $\vec{\mu}_s$ is associated:

$$\vec{\mu}_s = -\frac{e}{m}\vec{S}. \qquad (32\text{-}22)$$

Spin $\vec{S}$ cannot itself be measured, but any component can be measured. Assuming that the measurement is along the z axis of a coordinate system, the component S_z can have only the values given by

$$S_z = m_s \frac{h}{2\pi}, \qquad \text{for } m_s = \pm\tfrac{1}{2}, \qquad (32\text{-}23)$$

where h ($= 6.63 \times 10^{-34}$ J·s) is the Planck constant. Similarly, the electron's spin magnetic dipole moment $\vec{\mu}_s$ cannot itself be measured but its component can be measured. Along a z axis, the component is

$$\mu_{s,z} = \pm\frac{eh}{4\pi m} = \pm\mu_B, \qquad (32\text{-}24, 32\text{-}26)$$

where μ_B is the *Bohr magneton:*

$$\mu_B = \frac{eh}{4\pi m} = 9.27 \times 10^{-24} \text{ J/T}. \qquad (32\text{-}25)$$

The energy U associated with the orientation of the spin magnetic dipole moment in an external magnetic field $\vec{B}_{\text{ext}}$ is

$$U = -\vec{\mu}_s \cdot \vec{B}_{\text{ext}} = -\mu_{s,z}B_{\text{ext}}. \qquad (32\text{-}27)$$

Orbital Magnetic Dipole Moment An electron in an atom has an additional angular momentum called its *orbital angular momentum* $\vec{L}_{\text{orb}}$, with which an *orbital magnetic dipole moment* $\vec{\mu}_{\text{orb}}$ is associated:

$$\vec{\mu}_{\text{orb}} = -\frac{e}{2m}\vec{L}_{\text{orb}}. \qquad (32\text{-}28)$$

Orbital angular momentum is quantized and can have only values given by

$$L_{\text{orb},z} = m_\ell \frac{h}{2\pi},$$

$$\text{for } m_\ell = 0, \pm1, \pm2, \ldots, \pm(\text{limit}). \qquad (32\text{-}29)$$

This means that the associated magnetic dipole moment measured along a z axis is given by

$$\mu_{\text{orb},z} = -m_\ell \frac{eh}{4\pi m} = -m_\ell \mu_B. \qquad (32\text{-}30, 32\text{-}31)$$

The energy U associated with the orientation of the orbital magnetic dipole moment in an external magnetic field $\vec{B}_{\text{ext}}$ is

$$U = -\vec{\mu}_{\text{orb}} \cdot \vec{B}_{\text{ext}} = -\mu_{\text{orb},z}B_{\text{ext}}. \qquad (32\text{-}32)$$

Diamagnetism *Diamagnetic materials* do not exhibit magnetism until they are placed in an external magnetic field $\vec{B}_{\text{ext}}$. They then develop a magnetic dipole moment directed opposite $\vec{B}_{\text{ext}}$. If the field is nonuniform, the diamagnetic material is repelled from regions of greater magnetic field. This property is called *diamagnetism.*

Paramagnetism In a *paramagnetic material,* each atom has a permanent magnetic dipole moment $\vec{\mu}$, but the dipole moments are randomly oriented and the material as a whole lacks a magnetic field. However, an external magnetic field $\vec{B}_{\text{ext}}$ can partially align the atomic dipole moments to give the material a net magnetic dipole moment in the direction of $\vec{B}_{\text{ext}}$. If $\vec{B}_{\text{ext}}$ is nonuniform, the material is attracted to regions of greater magnetic field. These properties are called *paramagnetism.*

The alignment of the atomic dipole moments increases with an increase in $\vec{B}_{\text{ext}}$ and decreases with an increase in temperature T. The extent to which a sample of volume V is magnetized is given by its *magnetization* $\vec{M}$, whose magnitude is

$$M = \frac{\text{measured magnetic moment}}{V}. \qquad (32\text{-}38)$$

Complete alignment of all N atomic magnetic dipoles in a sample, called *saturation* of the sample, corresponds to the maximum magnetization value $M_{\text{max}} = N\mu/V$. For low values of the ratio B_{ext}/T, we have the approximation

$$M = C\frac{B_{\text{ext}}}{T} \qquad \text{(Curie's law)}, \qquad (32\text{-}39)$$

where C is called the *Curie constant.*

Ferromagnetism In the absence of an external magnetic field, some of the electrons in a ferromagnetic material have their magnetic dipole moments aligned by means of a quantum physical interaction called *exchange coupling,* producing regions (domains) within the material with strong magnetic dipole moments. An external field $\vec{B}_{\text{ext}}$ can align the magnetic dipole moments of those regions, producing a strong net magnetic dipole moment for the material as a whole, in the direction of $\vec{B}_{\text{ext}}$. This net magnetic dipole moment can partially persist when field $\vec{B}_{\text{ext}}$ is removed. If $\vec{B}_{\text{ext}}$ is nonuniform, the ferromagnetic material is attracted to regions of greater magnetic field. These properties are called *ferromagnetism.* Exchange coupling disappears when a sample's temperature exceeds its *Curie temperature.*

QUESTIONS

1 Figure 32-19*a* shows a capacitor, with circular plates, that is being charged. Point *a* (near one of the connecting wires) and point *b* (inside the capacitor gap) are equidistant from the central axis, as are point *c* (not so near the wire) and point *d* (between the plates but outside the gap). In Fig. 32-19*b*, one curve gives the variation with distance *r* of the magnitude of the magnetic field inside and outside the wire. The other curve gives the variation with distance *r* of the magnitude of the magnetic field inside and outside the gap. The two curves partially overlap. Which of the three points on the curves correspond to which of the four points of Fig. 32-19*a*?

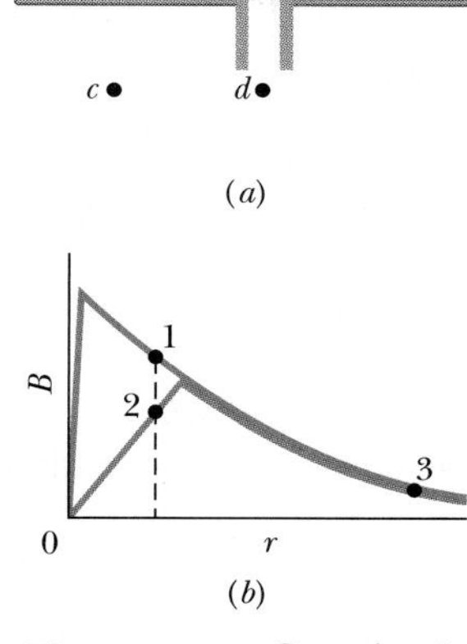

Fig. 32-19 Question 1.

2 Figure 32-20 shows a parallel-plate capacitor and the current in the connecting wires that is discharging the capacitor. Are the directions of (a) electric field $\vec{E}$ and (b) displacement current i_d leftward or rightward between the plates? (c) Is the magnetic field at point *P* into or out of the page?

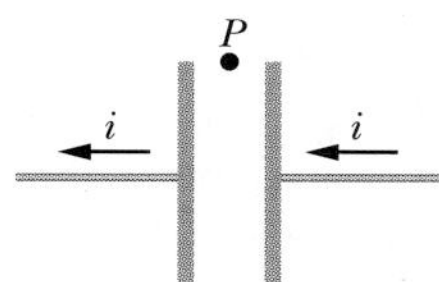

Fig. 32-20 Question 2.

3 Figure 32-21 shows, in two situations, an electric field vector $\vec{E}$ and an induced magnetic field line. In each, is the magnitude of $\vec{E}$ increasing or decreasing?

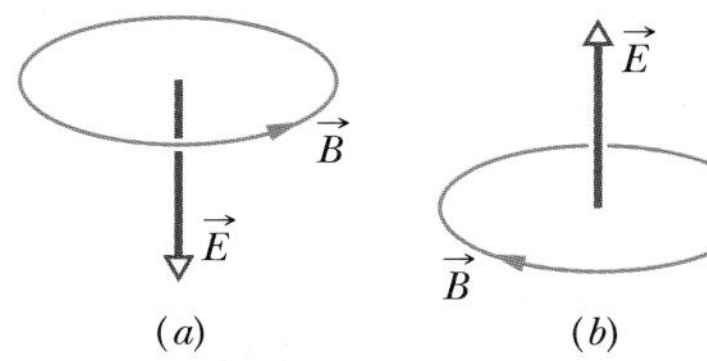

Fig. 32-21 Question 3.

4 Figure 32-22*a* shows a pair of opposite spin orientations for an electron in an external magnetic field $\vec{B}_{ext}$. Figure 32-22*b* gives three choices for the graph of the potential energies associated with those orientations as a function of the magnitude of $\vec{B}_{ext}$. Choices *b* and *c* consist of intersecting lines, choice *a* of parallel lines. Which is the correct choice?

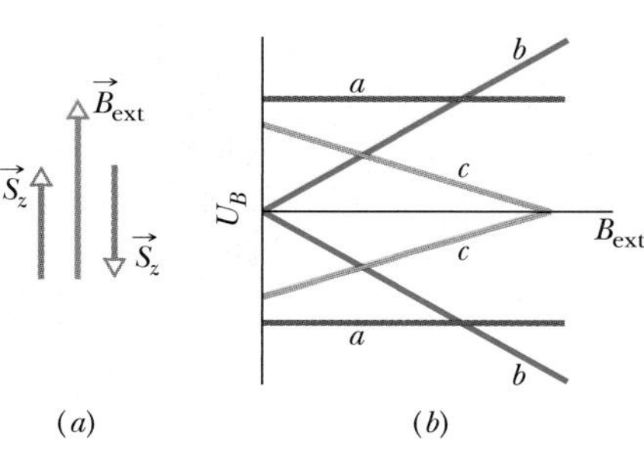

Fig. 32-22 Question 4.

5 An electron in an external magnetic field $\vec{B}_{ext}$ has its spin angular momentum S_z antiparallel to $\vec{B}_{ext}$. If the electron undergoes a *spin-flip* so that S_z is then parallel with $\vec{B}_{ext}$, must energy be supplied to or lost by the electron?

6 Does the magnitude of the net force on the current loop of Figs. 32-12*a* and *b* increase, decrease, or remain the same if we increase (a) the magnitude of $\vec{B}_{ext}$ and (b) the divergence of $\vec{B}_{ext}$?

7 Figure 32-23 shows a face-on view of one of the two square plates of a parallel-plate capacitor, as well as four loops that are located between the plates. The capacitor is being discharged. (a) Neglecting fringing of the magnetic field, rank the loops according to the magnitude of $\oint \vec{B} \cdot d\vec{s}$ along them, greatest first. (b) Along which loop, if any, is the angle between the directions of $\vec{B}$ and $d\vec{s}$ constant (so that their dot product can easily be evaluated)? (c) Along which loop, if any, is *B* constant (so that *B* can be brought in front of the integral sign in Eq. 32-3)?

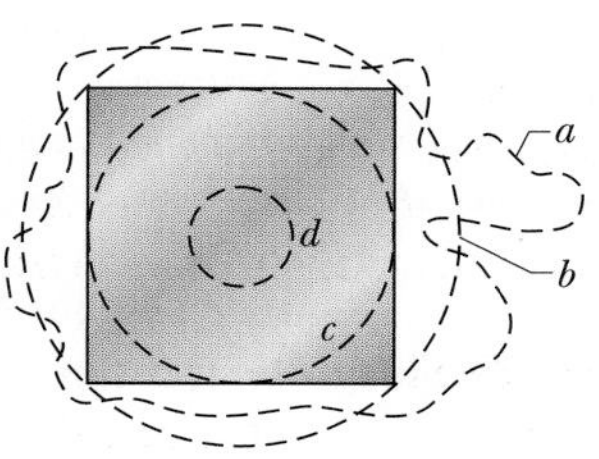

Fig. 32-23 Question 7.

8 Figure 32-24 shows three loop models of an electron orbiting counterclockwise within a magnetic field. The fields are nonuniform for models 1 and 2 and uniform for model 3. For each model, are (a) the magnetic dipole moment of the loop and (b) the magnetic force on the loop directed up, directed down, or zero?

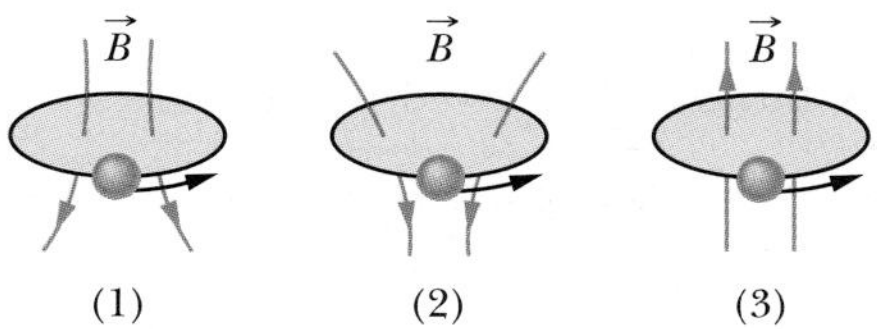

Fig. 32-24 Questions 8, 9, and 10.

9 Replace the current loops of Question 8 and Fig. 32-24 with diamagnetic spheres. For each field, are (a) the magnetic dipole moment of the sphere and (b) the magnetic force on the sphere directed up, directed down, or zero?

10 Replace the current loops of Question 8 and Fig. 32-24 with paramagnetic spheres. For each field, are (a) the magnetic dipole moment of the sphere and (b) the magnetic force on the sphere directed up, directed down, or zero?

11 Figure 32-25 represents three rectangular samples of a ferromagnetic material in which the magnetic dipoles of the domains have been directed out of the page (encircled dot) by a very strong applied field B_0. In each sample, an island domain still has its magnetic field directed into the page (encircled ×). Sample 1 is one (pure) crystal. The other samples contain impurities collected along lines; domains cannot easily spread across such lines.

The applied field is now to be reversed and its magnitude kept moderate. The change causes the island domain to grow. (a) Rank the three samples according to the success of that growth, greatest growth first. Ferromagnetic materials in which the magnetic dipoles are easily changed are said to be *magnetically soft;* when the changes are difficult, requiring strong applied fields, the materials are said to be *magnetically hard.* (b) Of the three samples, which is the most magnetically hard?

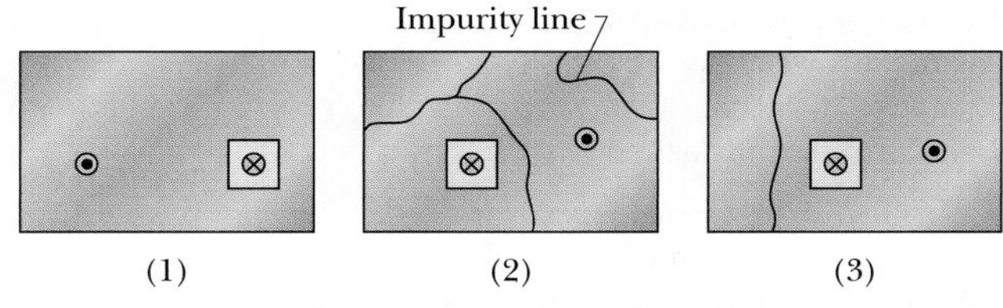

Fig. 32-25 Question 11.

PROBLEMS

 Tutoring problem available (at instructor's discretion) in *WileyPLUS* and WebAssign

SSM Worked-out solution available in Student Solutions Manual

• – ••• Number of dots indicates level of problem difficulty

Additional information available in *The Flying Circus of Physics* and at flyingcircusofphysics.com

WWW Worked-out solution is at http://www.wiley.com/college/halliday

ILW Interactive solution is at http://www.wiley.com/college/halliday

sec. 32-2 Gauss' Law for Magnetic Fields

•1 The magnetic flux through each of five faces of a die (singular of "dice") is given by $\Phi_B = \pm N$ Wb, where N (= 1 to 5) is the number of spots on the face. The flux is positive (outward) for N even and negative (inward) for N odd. What is the flux through the sixth face of the die?

•2 Figure 32-26 shows a closed surface. Along the flat top face, which has a radius of 2.0 cm, a perpendicular magnetic field $\vec{B}$ of magnitude 0.30 T is directed outward. Along the flat bottom face, a magnetic flux of 0.70 mWb is directed outward. What are the (a) magnitude and (b) direction (inward or outward) of the magnetic flux through the curved part of the surface?

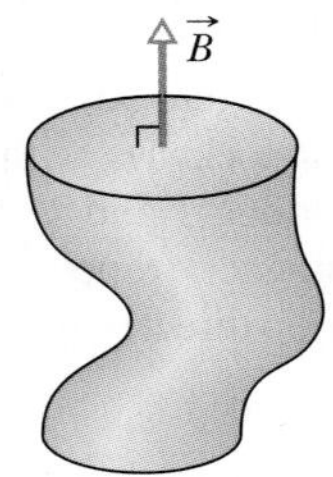

Fig. 32-26 Problem 2.

••3 SSM ILW A Gaussian surface in the shape of a right circular cylinder with end caps has a radius of 12.0 cm and a length of 80.0 cm. Through one end there is an inward magnetic flux of 25.0 μWb. At the other end there is a uniform magnetic field of 1.60 mT, normal to the surface and directed outward. What are the (a) magnitude and (b) direction (inward or outward) of the net magnetic flux through the curved surface?

•••4 Two wires, parallel to a z axis and a distance $4r$ apart, carry equal currents i in opposite directions, as shown in Fig. 32-27. A circular cylinder of radius r and length L has its axis on the z axis, midway between the wires. Use Gauss' law for magnetism to derive an expression for the net outward magnetic flux through the half of the cylindrical surface above the x axis. (*Hint:* Find the flux through the portion of the xz plane that lies within the cylinder.)

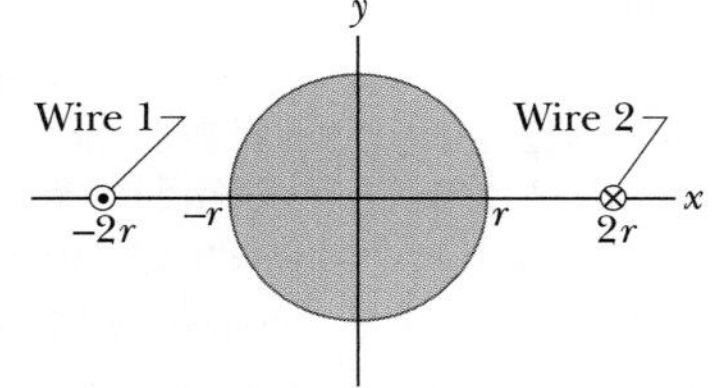

Fig. 32-27 Problem 4.

sec. 32-3 Induced Magnetic Fields

•5 SSM The induced magnetic field at radial distance 6.0 mm from the central axis of a circular parallel-plate capacitor is 2.0×10^{-7} T. The plates have radius 3.0 mm. At what rate $d\vec{E}/dt$ is the electric field between the plates changing?

•6 A capacitor with square plates of edge length L is being discharged by a current of 0.75 A. Figure 32-28 is a head-on view of one of the plates from inside the capacitor. A dashed rectangular path is shown. If $L = 12$ cm, $W = 4.0$ cm, and $H = 2.0$ cm, what is the value of $\oint \vec{B} \cdot d\vec{s}$ around the dashed path?

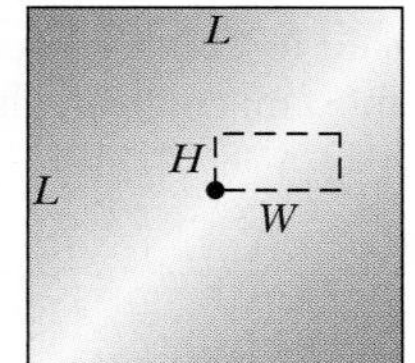

Fig. 32-28 Problem 6.

••7 GO *Uniform electric flux.* Figure 32-29 shows a circular region of radius $R = 3.00$ cm in which a uniform electric flux is directed out of the plane of the page. The total electric flux through the region is given by $\Phi_E = (3.00 \text{ mV} \cdot \text{m/s})t$, where t is in seconds. What is the magnitude of the magnetic field that is induced at radial distances (a) 2.00 cm and (b) 5.00 cm?

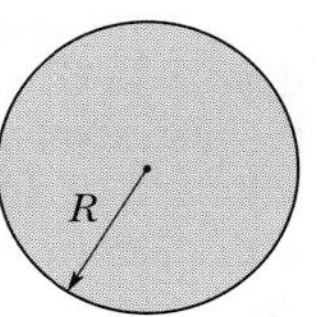

Fig. 32-29 Problems 7 to 10 and 19 to 22.

••8 GO *Nonuniform electric flux.* Figure 32-29 shows a circular region of radius $R = 3.00$ cm in which an electric flux is directed out of the plane of the page. The flux encircled by a concentric circle of radius r is given by $\Phi_{E,\text{enc}} = (0.600 \text{ V} \cdot \text{m/s})(r/R)t$, where $r \le R$ and t is in seconds. What is the magnitude of the induced magnetic field at radial distances (a) 2.00 cm and (b) 5.00 cm?

••9 GO *Uniform electric field.* In Fig. 32-29, a uniform electric field is directed out of the page within a circular region of radius $R = 3.00$ cm. The field magnitude is given by $E = (4.50 \times 10^{-3} \text{ V/m} \cdot \text{s})t$, where t is in seconds. What is the magnitude of the induced magnetic field at radial distances (a) 2.00 cm and (b) 5.00 cm?

••10 GO *Nonuniform electric field.* In Fig. 32-29, an electric field is directed out of the page within a circular region of radius R = 3.00 cm. The field magnitude is $E = (0.500 \text{ V/m} \cdot \text{s})(1 - r/R)t$, where t is in seconds and r is the radial distance ($r \le R$). What is the magnitude of the induced magnetic field at radial distances (a) 2.00 cm and (b) 5.00 cm?

••11 Suppose that a parallel-plate capacitor has circular plates with radius $R = 30$ mm and a plate separation of 5.0 mm. Suppose also that a sinusoidal potential difference with a maximum value of 150 V and a frequency of 60 Hz is applied across the plates; that is,

$$V = (150 \text{ V}) \sin[2\pi(60 \text{ Hz})t].$$

(a) Find $B_{max}(R)$, the maximum value of the induced magnetic field that occurs at $r = R$. (b) Plot $B_{max}(r)$ for $0 < r < 10$ cm.

••12 A parallel-plate capacitor with circular plates of radius 40 mm is being discharged by a current of 6.0 A. At what radius (a) inside and (b) outside the capacitor gap is the magnitude of the induced magnetic field equal to 75% of its maximum value? (c) What is that maximum value?

sec. 32-4 Displacement Current

•13 At what rate must the potential difference between the plates of a parallel-plate capacitor with a 2.0 μF capacitance be changed to produce a displacement current of 1.5 A?

•14 A parallel-plate capacitor with circular plates of radius R is being charged. Show that the magnitude of the current density of the displacement current is $J_d = \varepsilon_0(dE/dt)$ for $r \le R$.

•15 SSM Prove that the displacement current in a parallel-plate capacitor of capacitance C can be written as $i_d = C(dV/dt)$, where V is the potential difference between the plates.

•16 A parallel-plate capacitor with circular plates of radius 0.10 m is being discharged. A circular loop of radius 0.20 m is concentric with the capacitor and halfway between the plates. The displacement current through the loop is 2.0 A. At what rate is the electric field between the plates changing?

••17 A silver wire has resistivity $\rho = 1.62 \times 10^{-8}\ \Omega \cdot \text{m}$ and a cross-sectional area of 5.00 mm^2. The current in the wire is uniform and changing at the rate of 2000 A/s when the current is 100 A. (a) What is the magnitude of the (uniform) electric field in the wire when the current in the wire is 100 A? (b) What is the displacement current in the wire at that time? (c) What is the ratio of the magnitude of the magnetic field due to the displacement current to that due to the current at a distance r from the wire?

••18 The circuit in Fig. 32-30 consists of switch S, a 12.0 V ideal battery, a 20.0 MΩ resistor, and an air-filled capacitor. The capacitor has parallel circular plates of radius 5.00 cm, separated by 3.00 mm. At time $t = 0$, switch S is closed to begin charging the capacitor. The electric field between the plates is uniform. At $t = 250\ \mu$s, what is the magnitude of the magnetic field within the capacitor, at radial distance 3.00 cm?

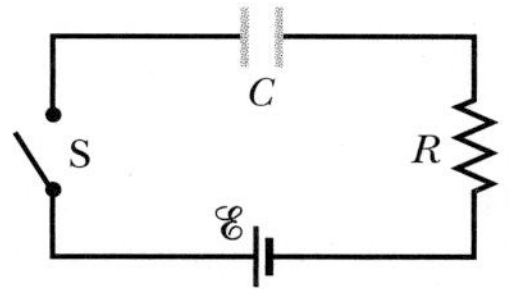

Fig. 32-30 Problem 18.

••19 *Uniform displacement-current density.* Figure 32-29 shows a circular region of radius $R = 3.00$ cm in which a displacement current is directed out of the page. The displacement current has a uniform density of magnitude $J_d = 6.00\ \text{A/m}^2$. What is the magnitude of the magnetic field due to the displacement current at radial distances (a) 2.00 cm and (b) 5.00 cm?

••20 *Uniform displacement current.* Figure 32-29 shows a circular region of radius $R = 3.00$ cm in which a uniform displacement current $i_d = 0.500$ A is out of the page. What is the magnitude of the magnetic field due to the displacement current at radial distances (a) 2.00 cm and (b) 5.00 cm?

••21 GO *Nonuniform displacement-current density.* Figure 32-29 shows a circular region of radius $R = 3.00$ cm in which a displacement current is directed out of the page. The magnitude of the density of this displacement current is $J_d = (4.00\ \text{A/m}^2)(1 - r/R)$, where r is the radial distance ($r \leq R$). What is the magnitude of the magnetic field due to the displacement current at (a) $r = 2.00$ cm and (b) $r = 5.00$ cm?

••22 GO *Nonuniform displacement current.* Figure 32-29 shows a circular region of radius $R = 3.00$ cm in which a displacement current i_d is directed out of the page. The magnitude of the displacement current is given by $i_d = (3.00\ \text{A})(r/R)$, where r is the radial distance ($r \leq R$). What is the magnitude of the magnetic field due to i_d at radial distances (a) 2.00 cm and (b) 5.00 cm?

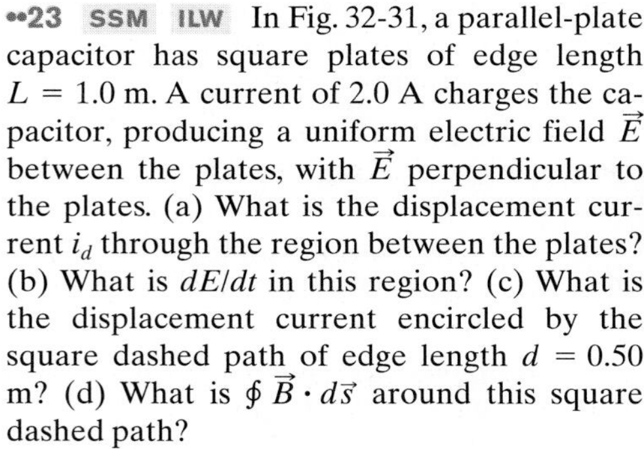

••23 SSM ILW In Fig. 32-31, a parallel-plate capacitor has square plates of edge length $L = 1.0$ m. A current of 2.0 A charges the capacitor, producing a uniform electric field $\vec{E}$ between the plates, with $\vec{E}$ perpendicular to the plates. (a) What is the displacement current i_d through the region between the plates? (b) What is dE/dt in this region? (c) What is the displacement current encircled by the square dashed path of edge length $d = 0.50$ m? (d) What is $\oint \vec{B} \cdot d\vec{s}$ around this square dashed path?

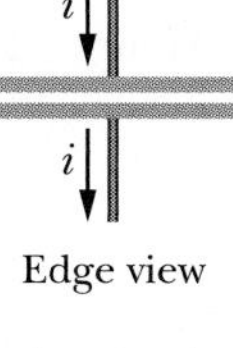

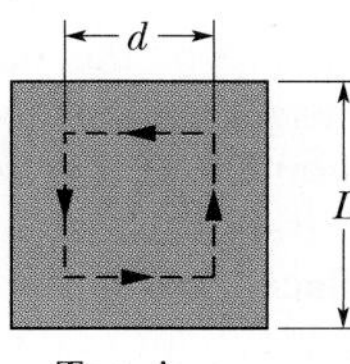

Fig. 32-31 Problem 23.

••24 The magnitude of the electric field between the two circular parallel plates in Fig. 32-32 is $E = (4.0 \times 10^5) - (6.0 \times 10^4 t)$, with E in volts per meter and t in seconds. At $t = 0$, $\vec{E}$ is upward. The plate area is 4.0×10^{-2} m^2. For $t \geq 0$, what are the (a) magnitude and (b) direction (up or down) of the displacement current between the plates and (c) is the direction of the induced magnetic field clockwise or counterclockwise in the figure?

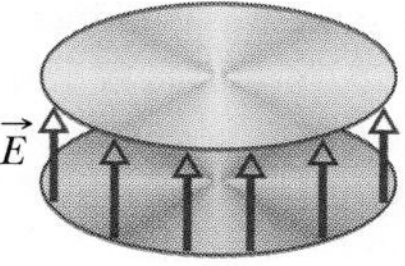

Fig. 32-32 Problem 24.

••25 ILW As a parallel-plate capacitor with circular plates 20 cm in diameter is being charged, the current density of the displacement current in the region between the plates is uniform and has a magnitude of 20 A/m^2. (a) Calculate the magnitude B of the magnetic field at a distance $r = 50$ mm from the axis of symmetry of this region. (b) Calculate dE/dt in this region.

••26 A capacitor with parallel circular plates of radius $R = 1.20$ cm is discharging via a current of 12.0 A. Consider a loop of radius $R/3$ that is centered on the central axis between the plates. (a) How much displacement current is encircled by the loop? The maximum induced magnetic field has a magnitude of 12.0 mT. At what radius (b) inside and (c) outside the capacitor gap is the magnitude of the induced magnetic field 3.00 mT?

••27 ILW In Fig. 32-33, a uniform electric field $\vec{E}$ collapses. The vertical axis scale is set by $E_s = 6.0 \times 10^5$ N/C, and the horizontal axis scale is set by $t_s = 12.0\ \mu$s. Calculate the magnitude of the displacement current through a 1.6 m^2 area perpendicular to the field during each of the time intervals a, b, and c shown on the graph. (Ignore the behavior at the ends of the intervals.)

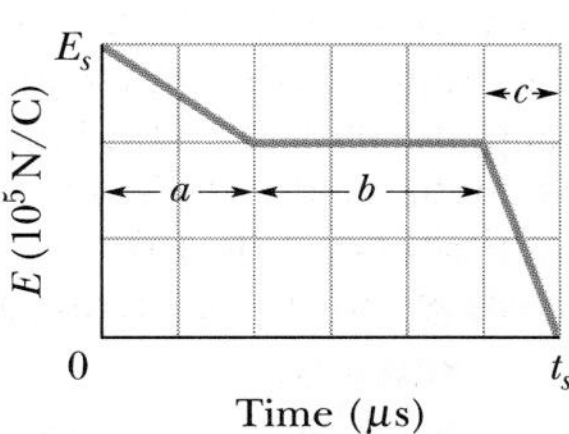

Fig. 32-33 Problem 27.

••28 GO Figure 32-34a shows the current i that is produced in a wire of resistivity $1.62 \times 10^{-8}\ \Omega \cdot \text{m}$. The magnitude of the current versus time t is shown in Fig. 32-34b. The vertical axis scale is set by $i_s = 10.0$ A, and the horizontal axis scale is set by $t_s = 50.0$ ms. Point P is at radial distance 9.00 mm from the wire's center. Determine the magnitude of the magnetic field $\vec{B}_i$ at point P due to the actual current i in the wire at (a) $t = 20$ ms, (b) $t = 40$ ms, and (c) $t = 60$ ms. Next, assume that the electric field driving the current is confined to the wire. Then determine the magnitude of the magnetic field $\vec{B}_{id}$ at point P due to the displacement current i_d in the wire at (d) $t = 20$ ms, (e) $t = 40$ ms, and (f) $t = 60$ ms. At point P at $t = 20$ s, what is the direction (into or out of the page) of (g) $\vec{B}_i$ and (h) $\vec{B}_{id}$?

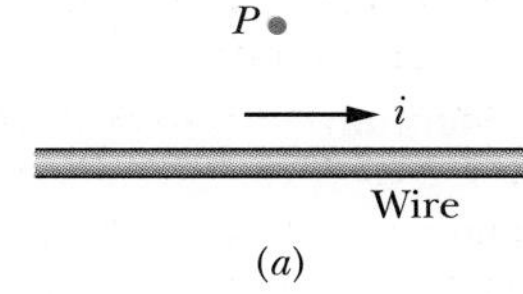

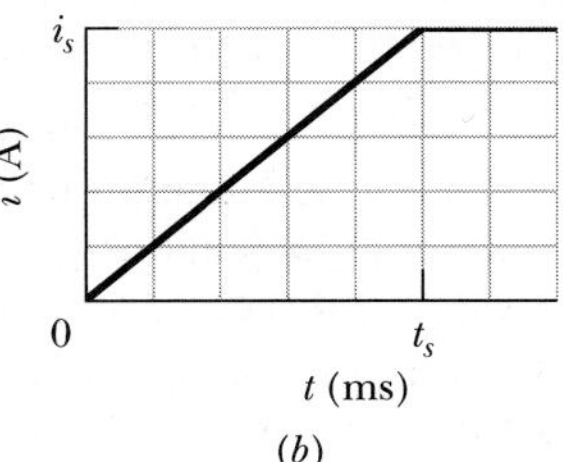

Fig. 32-34 Problem 28.

•••29 In Fig. 32-35, a capacitor with circular plates of radius $R =$

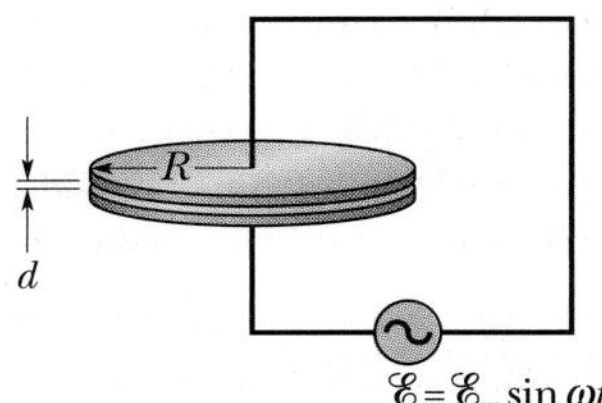

Fig. 32-35 Problem 29.

18.0 cm is connected to a source of emf $\mathscr{E} = \mathscr{E}_m \sin \omega t$, where $\mathscr{E}_m =$ 220 V and $\omega = 130$ rad/s. The maximum value of the displacement current is $i_d = 7.60\ \mu$A. Neglect fringing of the electric field at the edges of the plates. (a) What is the maximum value of the current i in the circuit? (b) What is the maximum value of $d\Phi_E/dt$, where Φ_E is the electric flux through the region between the plates? (c) What is the separation d between the plates? (d) Find the maximum value of the magnitude of $\vec{B}$ between the plates at a distance $r =$ 11.0 cm from the center.

sec. 32-6 Magnets

•**30** Assume the average value of the vertical component of Earth's magnetic field is 43 μT (downward) for all of Arizona, which has an area of 2.95×10^5 km^2. What then are the (a) magnitude and (b) direction (inward or outward) of the net magnetic flux through the rest of Earth's surface (the entire surface excluding Arizona)?

•**31** In New Hampshire the average horizontal component of Earth's magnetic field in 1912 was 16 μT, and the average inclination or "dip" was 73°. What was the corresponding magnitude of Earth's magnetic field?

sec. 32-7 Magnetism and Electrons

•**32** Figure 32-36a is a one-axis graph along which two of the allowed energy values (*levels*) of an atom are plotted. When the atom is placed in a magnetic field of 0.500 T, the graph changes to that of Fig. 32-36b because of the energy associated with $\vec{\mu}_{orb} \cdot \vec{B}$. (We neglect $\vec{\mu}_s$.) Level E_1 is unchanged, but level E_2 splits into a (closely spaced) triplet of levels. What are the allowed values of m_ℓ associated with (a) energy level E_1 and (b) energy level E_2? (c) In joules, what amount of energy is represented by the spacing between the triplet levels?

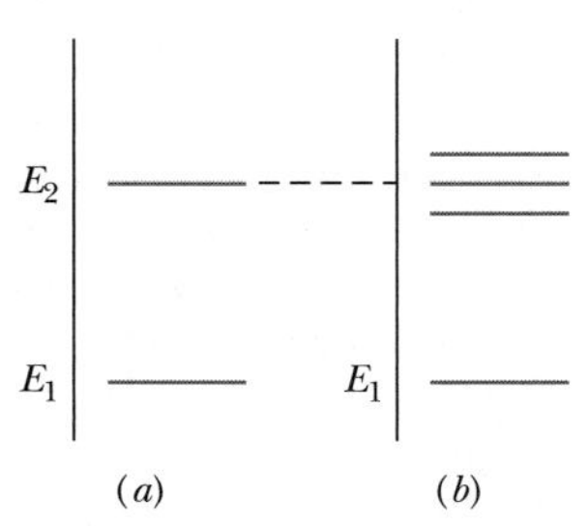

Fig. 32-36 Problem 32.

•**33** SSM WWW If an electron in an atom has an orbital angular momentum with $m = 0$, what are the components (a) $L_{orb,z}$ and (b) $\mu_{orb,z}$? If the atom is in an external magnetic field $\vec{B}$ that has magnitude 35 mT and is directed along the z axis, what are (c) the energy U_{orb} associated with $\vec{\mu}_{orb}$ and (d) the energy U_{spin} associated with $\vec{\mu}_s$? If, instead, the electron has $m = -3$, what are (e) $L_{orb,z}$, (f) $\mu_{orb,z}$, (g) U_{orb}, and (h) U_{spin}?

•**34** What is the energy difference between parallel and antiparallel alignment of the z component of an electron's spin magnetic dipole moment with an external magnetic field of magnitude 0.25 T, directed parallel to the z axis?

•**35** What is the measured component of the orbital magnetic dipole moment of an electron with (a) $m_\ell = 1$ and (b) $m_\ell = -2$?

•**36** An electron is placed in a magnetic field $\vec{B}$ that is directed along a z axis. The energy difference between parallel and antiparallel alignments of the z component of the electron's spin magnetic moment with $\vec{B}$ is 6.00×10^{-25} J. What is the magnitude of $\vec{B}$?

sec. 32-9 Diamagnetism

•**37** Figure 32-37 shows a loop model (loop L) for a diamagnetic material. (a) Sketch the magnetic field lines within and about the material due to the bar magnet. What is the direction of (b) the loop's net magnetic dipole moment $\vec{\mu}$, (c) the conventional current i in the loop (clockwise or counterclockwise in the figure), and (d) the magnetic force on the loop?

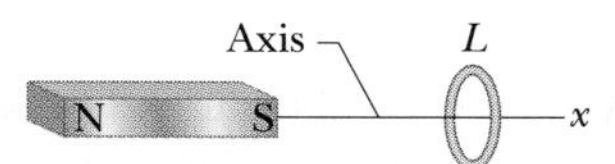

Fig. 32-37 Problems 37 and 71.

•••**38** Assume that an electron of mass m and charge magnitude e moves in a circular orbit of radius r about a nucleus. A uniform magnetic field $\vec{B}$ is then established perpendicular to the plane of the orbit. Assuming also that the radius of the orbit does not change and that the change in the speed of the electron due to field $\vec{B}$ is small, find an expression for the change in the orbital magnetic dipole moment of the electron due to the field.

sec. 32-10 Paramagnetism

•**39** A sample of the paramagnetic salt to which the magnetization curve of Fig. 32-14 applies is to be tested to see whether it obeys Curie's law. The sample is placed in a uniform 0.50 T magnetic field that remains constant throughout the experiment. The magnetization M is then measured at temperatures ranging from 10 to 300 K. Will it be found that Curie's law is valid under these conditions?

•**40** A sample of the paramagnetic salt to which the magnetization curve of Fig. 32-14 applies is held at room temperature (300 K). At what applied magnetic field will the degree of magnetic saturation of the sample be (a) 50% and (b) 90%? (c) Are these fields attainable in the laboratory?

•**41** SSM ILW A magnet in the form of a cylindrical rod has a length of 5.00 cm and a diameter of 1.00 cm. It has a uniform magnetization of 5.30×10^3 A/m. What is its magnetic dipole moment?

•**42** A 0.50 T magnetic field is applied to a paramagnetic gas whose atoms have an intrinsic magnetic dipole moment of 1.0×10^{-23} J/T. At what temperature will the mean kinetic energy of translation of the atoms equal the energy required to reverse such a dipole end for end in this magnetic field?

••**43** An electron with kinetic energy K_e travels in a circular path that is perpendicular to a uniform magnetic field, which is in the positive direction of a z axis. The electron's motion is subject only to the force due to the field. (a) Show that the magnetic dipole moment of the electron due to its orbital motion has magnitude $\mu = K_e/B$ and that it is in the direction opposite that of $\vec{B}$. What are the (b) magnitude and (c) direction of the magnetic dipole moment of a positive ion with kinetic energy K_i under the same circumstances? (d) An ionized gas consists of 5.3×10^{21} electrons/m^3 and the same number density of ions. Take the average electron kinetic energy to be 6.2×10^{-20} J and the average ion kinetic energy to be 7.6×10^{-21} J. Calculate the magnetization of the gas when it is in a magnetic field of 1.2 T.

••**44** Figure 32-38 gives the magnetization curve for a paramagnetic material. The vertical axis scale is set by $a = 0.15$, and the horizontal axis scale is set by $b = 0.2$ T/K. Let μ_{sam} be the measured net magnetic moment of a sample of the material and μ_{max} be the maximum possible net magnetic moment of that sample. According to Curie's law, what would be the ratio μ_{sam}/μ_{max} were the sample placed in a uniform magnetic field of magnitude 0.800 T, at a temperature of 2.00 K?

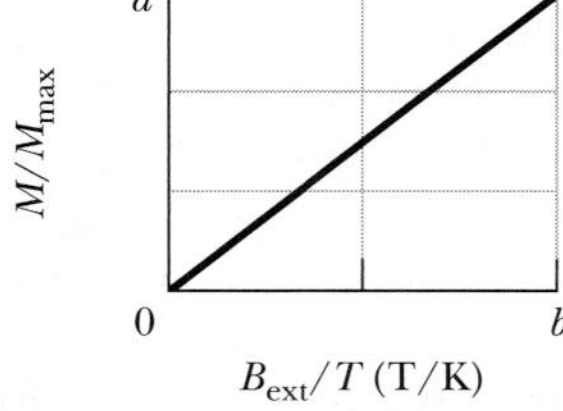

Fig. 32-38 Problem 44.

•••**45** SSM Consider a solid containing N atoms per unit volume, each atom having a magnetic dipole moment $\vec{\mu}$. Suppose the direction of $\vec{\mu}$ can be only parallel or antiparallel to an externally applied magnetic field $\vec{B}$ (this will be the case if $\vec{\mu}$ is due to the spin of a single electron). According to statistical mechanics, the probability of an atom being in a state with energy U is proportional to $e^{-U/kT}$, where T is the temperature and k is Boltzmann's constant. Thus, because energy U is $-\vec{\mu} \cdot \vec{B}$, the fraction of atoms whose dipole moment is parallel to $\vec{B}$ is proportional to $e^{\mu B/kT}$ and the fraction of atoms whose dipole moment is antiparallel to $\vec{B}$ is proportional to $e^{-\mu B/kT}$. (a) Show that the magnitude of the magnetization of this solid is $M = N\mu \tanh(\mu B/kT)$. Here tanh is the hyperbolic tangent function: $\tanh(x) = (e^x - e^{-x})/(e^x + e^{-x})$. (b) Show that the result given in (a) reduces to $M = N\mu^2 B/kT$ for $\mu B \ll kT$. (c) Show that the result of (a) reduces to $M = N\mu$ for $\mu B \gg kT$. (d) Show that both (b) and (c) agree qualitatively with Fig. 32-14.

sec. 32-11 Ferromagnetism

••46 GO You place a magnetic compass on a horizontal surface, allow the needle to settle, and then give the compass a gentle wiggle to cause the needle to oscillate about its equilibrium position. The oscillation frequency is 0.312 Hz. Earth's magnetic field at the location of the compass has a horizontal component of 18.0 μT. The needle has a magnetic moment of 0.680 mJ/T. What is the needle's rotational inertia about its (vertical) axis of rotation?

••47 SSM ILW WWW The magnitude of the magnetic dipole moment of Earth is 8.0×10^{22} J/T. (a) If the origin of this magnetism were a magnetized iron sphere at the center of Earth, what would be its radius? (b) What fraction of the volume of Earth would such a sphere occupy? Assume complete alignment of the dipoles. The density of Earth's inner core is 14 g/cm^3. The magnetic dipole moment of an iron atom is 2.1×10^{-23} J/T. (*Note:* Earth's inner core is in fact thought to be in both liquid and solid forms and partly iron, but a permanent magnet as the source of Earth's magnetism has been ruled out by several considerations. For one, the temperature is certainly above the Curie point.)

••48 The magnitude of the dipole moment associated with an atom of iron in an iron bar is 2.1×10^{-23} J/T. Assume that all the atoms in the bar, which is 5.0 cm long and has a cross-sectional area of 1.0 cm^2, have their dipole moments aligned. (a) What is the dipole moment of the bar? (b) What torque must be exerted to hold this magnet perpendicular to an external field of magnitude 1.5 T? (The density of iron is 7.9 g/cm^3.)

••49 SSM The exchange coupling mentioned in Section 32-11 as being responsible for ferromagnetism is *not* the mutual magnetic interaction between two elementary magnetic dipoles. To show this, calculate (a) the magnitude of the magnetic field a distance of 10 nm away, along the dipole axis, from an atom with magnetic dipole moment 1.5×10^{-23} J/T (cobalt), and (b) the minimum energy required to turn a second identical dipole end for end in this field. (c) By comparing the latter with the mean translational kinetic energy of 0.040 eV, what can you conclude?

••50 A magnetic rod with length 6.00 cm, radius 3.00 mm, and (uniform) magnetization 2.70×10^3 A/m can turn about its center like a compass needle. It is placed in a uniform magnetic field $\vec{B}$ of magnitude 35.0 mT, such that the directions of its dipole moment and $\vec{B}$ make an angle of 68.0°. (a) What is the magnitude of the torque on the rod due to $\vec{B}$? (b) What is the change in the orientation energy of the rod if the angle changes to 34.0°?

••51 The saturation magnetization M_{max} of the ferromagnetic metal nickel is 4.70×10^5 A/m. Calculate the magnetic dipole moment of a single nickel atom. (The density of nickel is 8.90 g/cm^3, and its molar mass is 58.71 g/mol.)

••52 Measurements in mines and boreholes indicate that Earth's interior temperature increases with depth at the average rate of 30 C°/km. Assuming a surface temperature of 10°C, at what depth does iron cease to be ferromagnetic? (The Curie temperature of iron varies very little with pressure.)

••53 A Rowland ring is formed of ferromagnetic material. It is circular in cross section, with an inner radius of 5.0 cm and an outer radius of 6.0 cm, and is wound with 400 turns of wire. (a) What current must be set up in the windings to attain a toroidal field of magnitude $B_0 = 0.20$ mT? (b) A secondary coil wound around the toroid has 50 turns and resistance 8.0 Ω. If, for this value of B_0, we have $B_M = 800B_0$, how much charge moves through the secondary coil when the current in the toroid windings is turned on?

Additional Problems

54 Using the approximations given in Problem 61, find (a) the altitude above Earth's surface where the magnitude of its magnetic field is 50.0% of the surface value at the same latitude; (b) the maximum magnitude of the magnetic field at the core–mantle boundary, 2900 km below Earth's surface; and the (c) magnitude and (d) inclination of Earth's magnetic field at the north geographic pole. (e) Suggest why the values you calculated for (c) and (d) differ from measured values.

55 Earth has a magnetic dipole moment of 8.0×10^{22} J/T. (a) What current would have to be produced in a single turn of wire extending around Earth at its geomagnetic equator if we wished to set up such a dipole? Could such an arrangement be used to cancel out Earth's magnetism (b) at points in space well above Earth's surface or (c) on Earth's surface?

56 A charge q is distributed uniformly around a thin ring of radius r. The ring is rotating about an axis through its center and perpendicular to its plane, at an angular speed ω. (a) Show that the magnetic moment due to the rotating charge has magnitude $\mu = \frac{1}{2}q\omega r^2$. (b) What is the direction of this magnetic moment if the charge is positive?

57 A magnetic compass has its needle, of mass 0.050 kg and length 4.0 cm, aligned with the horizontal component of Earth's magnetic field at a place where that component has the value B_h = 16 μT. After the compass is given a momentary gentle shake, the needle oscillates with angular frequency $\omega = 45$ rad/s. Assuming that the needle is a uniform thin rod mounted at its center, find the magnitude of its magnetic dipole moment.

58 The capacitor in Fig. 32-7 is being charged with a 2.50 A current. The wire radius is 1.50 mm, and the plate radius is 2.00 cm. Assume that the current i in the wire and the displacement current i_d in the capacitor gap are both uniformly distributed. What is the magnitude of the magnetic field due to i at the following radial distances from the wire's center: (a) 1.00 mm (inside the wire), (b) 3.00 mm (outside the wire), and (c) 2.20 cm (outside the wire)? What is the magnitude of the magnetic field due to i_d at the following radial distances from the central axis between the plates: (d) 1.00 mm (inside the gap), (e) 3.00 mm (inside the gap), and (f) 2.20 cm (outside the gap)? (g) Explain why the fields at the two smaller radii are so different for the wire and the gap but the fields at the largest radius are not.

59 A parallel-plate capacitor with circular plates of radius $R = 16$ mm and gap width $d = 5.0$ mm has a uniform electric field between the plates. Starting at time $t = 0$, the potential difference between the two plates is $V = (100\text{ V})e^{-t/\tau}$, where the

time constant $\tau = 12$ ms. At radial distance $r = 0.80R$ from the central axis, what is the magnetic field magnitude (a) as a function of time for $t \geq 0$ and (b) at time $t = 3\tau$?

60 A magnetic flux of 7.0 mWb is directed outward through the flat bottom face of the closed surface shown in Fig. 32-39. Along the flat top face (which has a radius of 4.2 cm) there is a 0.40 T magnetic field $\vec{B}$ directed perpendicular to the face. What are the (a) magnitude and (b) direction (inward or outward) of the magnetic flux through the curved part of the surface?

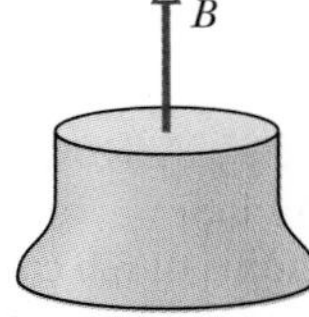

Fig. 32-39 Problem 60.

61 SSM The magnetic field of Earth can be approximated as the magnetic field of a dipole. The horizontal and vertical components of this field at any distance r from Earth's center are given by

$$B_h = \frac{\mu_0 \mu}{4\pi r^3} \cos \lambda_m, \qquad B_v = \frac{\mu_0 \mu}{2\pi r^3} \sin \lambda_m,$$

where λ_m is the *magnetic latitude* (this type of latitude is measured from the geomagnetic equator toward the north or south geomagnetic pole). Assume that Earth's magnetic dipole moment has magnitude $\mu = 8.00 \times 10^{22}\ \text{A}\cdot\text{m}^2$. (a) Show that the magnitude of Earth's field at latitude λ_m is given by

$$B = \frac{\mu_0 \mu}{4\pi r^3} \sqrt{1 + 3 \sin^2 \lambda_m}.$$

(b) Show that the inclination ϕ_i of the magnetic field is related to the magnetic latitude λ_m by $\tan \phi_i = 2 \tan \lambda_m$.

62 Use the results displayed in Problem 61 to predict the (a) magnitude and (b) inclination of Earth's magnetic field at the geomagnetic equator, the (c) magnitude and (d) inclination at geomagnetic latitude 60.0°, and the (e) magnitude and (f) inclination at the north geomagnetic pole.

63 A parallel-plate capacitor with circular plates of radius 55.0 mm is being charged. At what radius (a) inside and (b) outside the capacitor gap is the magnitude of the induced magnetic field equal to 50.0% of its maximum value?

64 A sample of the paramagnetic salt to which the magnetization curve of Fig. 32-14 applies is immersed in a uniform magnetic field of 2.0 T. At what temperature will the degree of magnetic saturation of the sample be (a) 50% and (b) 90%?

65 A parallel-plate capacitor with circular plates of radius R is being discharged. The displacement current through a central circular area, parallel to the plates and with radius $R/2$, is 2.0 A. What is the discharging current?

66 Figure 32-40 gives the variation of an electric field that is perpendicular to a circular area of 2.0 m². During the time period shown, what is the greatest displacement current through the area?

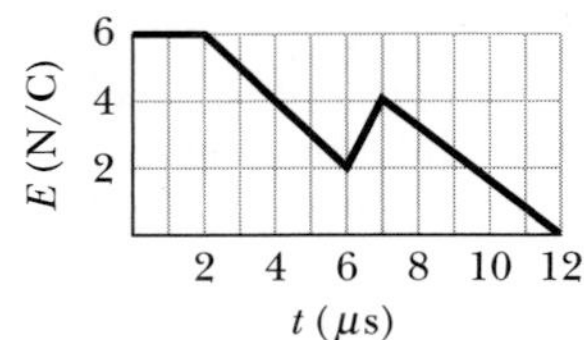

Fig. 32-40 Problem 66.

67 In Fig. 32-41, a parallel-plate capacitor is being discharged by a current $i = 5.0$ A. The plates are square with edge length $L = 8.0$ mm. (a) What is the rate at which the electric field between the plates is changing? (b) What is the value of $\oint \vec{B} \cdot d\vec{s}$ around the dashed path, where $H = 2.0$ mm and $W = 3.0$ mm?

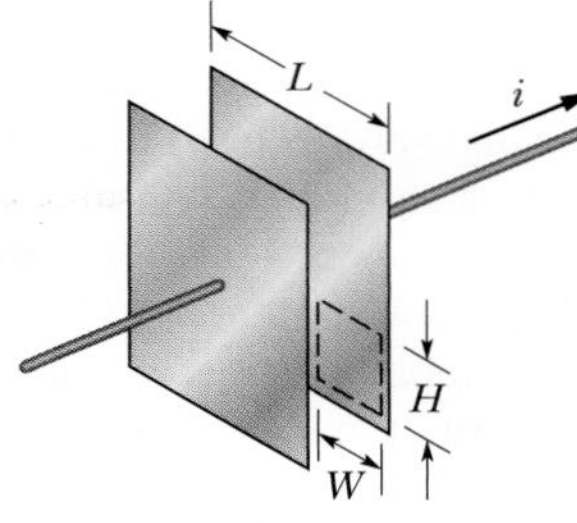

Fig. 32-41 Problem 67.

68 What is the measured component of the orbital magnetic dipole moment of an electron with the values (a) $m_\ell = 3$ and (b) $m_\ell = -4$?

69 In Fig. 32-42, a bar magnet lies near a paper cylinder. (a) Sketch the magnetic field lines that pass through the surface of the cylinder. (b) What is the sign of $\vec{B} \cdot d\vec{A}$ for every area $d\vec{A}$ on the surface? (c) Does this contradict Gauss' law for magnetism? Explain.

Fig. 32-42 Problem 69.

70 In the lowest energy state of the hydrogen atom, the most probable distance of the single electron from the central proton (the nucleus) is 5.2×10^{-11} m. (a) Compute the magnitude of the proton's electric field at that distance. The component $\mu_{s,z}$ of the proton's spin magnetic dipole moment measured on a z axis is 1.4×10^{-26} J/T. (b) Compute the magnitude of the proton's magnetic field at the distance 5.2×10^{-11} m on the z axis. (*Hint:* Use Eq. 29-27.) (c) What is the ratio of the spin magnetic dipole moment of the electron to that of the proton?

71 Figure 32-37 shows a loop model (loop L) for a paramagnetic material. (a) Sketch the field lines through and about the material due to the magnet. What is the direction of (b) the loop's net magnetic dipole moment $\vec{\mu}$, (c) the conventional current i in the loop (clockwise or counterclockwise in the figure), and (d) the magnetic force acting on the loop?

72 Two plates (as in Fig. 32-7) are being discharged by a constant current. Each plate has a radius of 4.00 cm. During the discharging, at a point between the plates at radial distance 2.00 cm from the central axis, the magnetic field has a magnitude of 12.5 nT. (a) What is the magnitude of the magnetic field at radial distance 6.00 cm? (b) What is the current in the wires attached to the plates?

73 SSM If an electron in an atom has orbital angular momentum with m_ℓ values limited by ±3, how many values of (a) $L_{orb,z}$ and (b) $\mu_{orb,z}$ can the electron have? In terms of h, m, and e, what is the greatest allowed magnitude for (c) $L_{orb,z}$ and (d) $\mu_{orb,z}$? (e) What is the greatest allowed magnitude for the z component of the electron's *net* angular momentum (orbital plus spin)? (f) How many values (signs included) are allowed for the z component of its net angular momentum?

74 A parallel-plate capacitor with circular plates is being charged. Consider a circular loop centered on the central axis and located between the plates. If the loop radius of 3.00 cm is greater than the plate radius, what is the displacement current between the plates when the magnetic field along the loop has magnitude 2.00 μT?

75 Suppose that ±4 are the limits to the values of m_ℓ for an electron in an atom. (a) How many different values of the electron's $\mu_{orb,z}$ are possible? (b) What is the greatest magnitude of those possible values? Next, if the atom is in a magnetic field of magnitude 0.250 T, in the positive direction of the z axis, what are (c) the maximum energy and (d) the minimum energy associated with those possible values of $\mu_{orb,z}$?

The International System of Units (SI)*

TABLE 1

The SI Base Units

Quantity	Name	Symbol	Definition
length	meter	m	". . . the length of the path traveled by light in vacuum in 1/299,792,458 of a second." (1983)
mass	kilogram	kg	". . . this prototype [a certain platinum–iridium cylinder] shall henceforth be considered to be the unit of mass." (1889)
time	second	s	". . . the duration of 9,192,631,770 periods of the radiation corresponding to the transition between the two hyperfine levels of the ground state of the cesium-133 atom." (1967)
electric current	ampere	A	". . . that constant current which, if maintained in two straight parallel conductors of infinite length, of negligible circular cross section, and placed 1 meter apart in vacuum, would produce between these conductors a force equal to 2×10^{-7} newton per meter of length." (1946)
thermodynamic temperature	kelvin	K	". . . the fraction 1/273.16 of the thermodynamic temperature of the triple point of water." (1967)
amount of substance	mole	mol	". . . the amount of substance of a system which contains as many elementary entities as there are atoms in 0.012 kilogram of carbon-12." (1971)
luminous intensity	candela	cd	". . . the luminous intensity, in a given direction, of a source that emits monochromatic radiation of frequency 540×10^{12} hertz and that has a radiant intensity in that direction of 1/683 watt per steradian." (1979)

*Adapted from "The International System of Units (SI)," National Bureau of Standards Special Publication 330, 1972 edition. The definitions above were adopted by the General Conference of Weights and Measures, an international body, on the dates shown. In this book we do not use the candela.

TABLE 2

Some SI Derived Units

Quantity	Name of Unit	Symbol	
area	square meter	m^2	
volume	cubic meter	m^3	
frequency	hertz	Hz	s^{-1}
mass density (density)	kilogram per cubic meter	kg/m^3	
speed, velocity	meter per second	m/s	
angular velocity	radian per second	rad/s	
acceleration	meter per second per second	m/s^2	
angular acceleration	radian per second per second	rad/s^2	
force	newton	N	$kg \cdot m/s^2$
pressure	pascal	Pa	N/m^2
work, energy, quantity of heat	joule	J	$N \cdot m$
power	watt	W	J/s
quantity of electric charge	coulomb	C	$A \cdot s$
potential difference, electromotive force	volt	V	W/A
electric field strength	volt per meter (or newton per coulomb)	V/m	N/C
electric resistance	ohm	Ω	V/A
capacitance	farad	F	$A \cdot s/V$
magnetic flux	weber	Wb	$V \cdot s$
inductance	henry	H	$V \cdot s/A$
magnetic flux density	tesla	T	Wb/m^2
magnetic field strength	ampere per meter	A/m	
entropy	joule per kelvin	J/K	
specific heat	joule per kilogram kelvin	$J/(kg \cdot K)$	
thermal conductivity	watt per meter kelvin	$W/(m \cdot K)$	
radiant intensity	watt per steradian	W/sr	

TABLE 3

The SI Supplementary Units

Quantity	Name of Unit	Symbol
plane angle	radian	rad
solid angle	steradian	sr

Some Fundamental Constants of Physics*

Constant	Symbol	Computational Value	Best (1998) Value	
			Value[a]	Uncertainty[b]
Speed of light in a vacuum	c	3.00×10^{8} m/s	2.997 924 58	exact
Elementary charge	e	1.60×10^{-19} C	1.602 176 487	0.025
Gravitational constant	G	6.67×10^{-11} m^3/s$^2\cdot$kg	6.674 28	100
Universal gas constant	R	8.31 J/mol$\cdot$K	8.314 472	1.7
Avogadro constant	N_A	6.02×10^{23} mol^{-1}	6.022 141 79	0.050
Boltzmann constant	k	1.38×10^{-23} J/K	1.380 650 4	1.7
Stefan–Boltzmann constant	σ	5.67×10^{-8} W/m$^2\cdot$K^4	5.670 400	7.0
Molar volume of ideal gas at STP[d]	V_m	2.27×10^{-2} m^3/mol	2.271 098 1	1.7
Permittivity constant	ϵ_0	8.85×10^{-12} F/m	8.854 187 817 62	exact
Permeability constant	μ_0	1.26×10^{-6} H/m	1.256 637 061 43	exact
Planck constant	h	6.63×10^{-34} J$\cdot$s	6.626 068 96	0.050
Electron mass[c]	m_e	9.11×10^{-31} kg	9.109 382 15	0.050
		5.49×10^{-4} u	5.485 799 094 3	4.2×10^{-4}
Proton mass[c]	m_p	1.67×10^{-27} kg	1.672 621 637	0.050
		1.0073 u	1.007 276 466 77	1.0×10^{-4}
Ratio of proton mass to electron mass	m_p/m_e	1840	1836.152 672 47	4.3×10^{-4}
Electron charge-to-mass ratio	e/m_e	1.76×10^{11} C/kg	1.758 820 150	0.025
Neutron mass[c]	m_n	1.68×10^{-27} kg	1.674 927 211	0.050
		1.0087 u	1.008 664 915 97	4.3×10^{-4}
Hydrogen atom mass[c]	$m_{^1H}$	1.0078 u	1.007 825 031 6	0.0005
Deuterium atom mass[c]	$m_{^2H}$	2.0136 u	2.013 553 212 724	3.9×10^{-5}
Helium atom mass[c]	$m_{^4He}$	4.0026 u	4.002 603 2	0.067
Muon mass	m_μ	1.88×10^{-28} kg	1.883 531 30	0.056
Electron magnetic moment	μ_e	9.28×10^{-24} J/T	9.284 763 77	0.025
Proton magnetic moment	μ_p	1.41×10^{-26} J/T	1.410 606 662	0.026
Bohr magneton	μ_B	9.27×10^{-24} J/T	9.274 009 15	0.025
Nuclear magneton	μ_N	5.05×10^{-27} J/T	5.050 783 24	0.025
Bohr radius	a	5.29×10^{-11} m	5.291 772 085 9	6.8×10^{-4}
Rydberg constant	R	1.10×10^{7} m^{-1}	1.097 373 156 852 7	6.6×10^{-6}
Electron Compton wavelength	λ_C	2.43×10^{-12} m	2.426 310 217 5	0.0014

[a]Values given in this column should be given the same unit and power of 10 as the computational value.

[b]Parts per million.

[c]Masses given in u are in unified atomic mass units, where 1 u = $1.660\ 538\ 782 \times 10^{-27}$ kg.

[d]STP means standard temperature and pressure: 0°C and 1.0 atm (0.1 MPa).

*The values in this table were selected from the 1998 CODATA recommended values (www.physics.nist.gov).

Some Astronomical Data

Some Distances from Earth

To the Moon*	3.82×10^8 m	To the center of our galaxy	2.2×10^{20} m
To the Sun*	1.50×10^{11} m	To the Andromeda Galaxy	2.1×10^{22} m
To the nearest star (Proxima Centauri)	4.04×10^{16} m	To the edge of the observable universe	$\sim 10^{26}$ m

*Mean distance.

The Sun, Earth, and the Moon

Property	Unit	Sun		Earth	Moon
Mass	kg	1.99×10^{30}		5.98×10^{24}	7.36×10^{22}
Mean radius	m	6.96×10^8		6.37×10^6	1.74×10^6
Mean density	kg/m^3	1410		5520	3340
Free-fall acceleration at the surface	m/s^2	274		9.81	1.67
Escape velocity	km/s	618		11.2	2.38
Period of rotation[a]	—	37 d at poles[b]	26 d at equator[b]	23 h 56 min	27.3 d
Radiation power[c]	W	3.90×10^{26}			

[a]Measured with respect to the distant stars.
[b]The Sun, a ball of gas, does not rotate as a rigid body.
[c]Just outside Earth's atmosphere solar energy is received, assuming normal incidence, at the rate of 1340 W/m^2.

Some Properties of the Planets

	Mercury	Venus	Earth	Mars	Jupiter	Saturn	Uranus	Neptune	Pluto
Mean distance from Sun, 10^6 km	57.9	108	150	228	778	1430	2870	4500	5900
Period of revolution, y	0.241	0.615	1.00	1.88	11.9	29.5	84.0	165	248
Period of rotation,[a] d	58.7	−243[b]	0.997	1.03	0.409	0.426	−0.451[b]	0.658	6.39
Orbital speed, km/s	47.9	35.0	29.8	24.1	13.1	9.64	6.81	5.43	4.74
Inclination of axis to orbit	<28°	≈3°	23.4°	25.0°	3.08°	26.7°	97.9°	29.6°	57.5°
Inclination of orbit to Earth's orbit	7.00°	3.39°		1.85°	1.30°	2.49°	0.77°	1.77°	17.2°
Eccentricity of orbit	0.206	0.0068	0.0167	0.0934	0.0485	0.0556	0.0472	0.0086	0.250
Equatorial diameter, km	4880	12 100	12 800	6790	143 000	120 000	51 800	49 500	2300
Mass (Earth = 1)	0.0558	0.815	1.000	0.107	318	95.1	14.5	17.2	0.002
Density (water = 1)	5.60	5.20	5.52	3.95	1.31	0.704	1.21	1.67	2.03
Surface value of g,[c] m/s^2	3.78	8.60	9.78	3.72	22.9	9.05	7.77	11.0	0.5
Escape velocity,[c] km/s	4.3	10.3	11.2	5.0	59.5	35.6	21.2	23.6	1.3
Known satellites	0	0	1	2	63 + ring	60 + rings	27 + rings	13 + rings	3

[a]Measured with respect to the distant stars.
[b]Venus and Uranus rotate opposite their orbital motion.
[c]Gravitational acceleration measured at the planet's equator.

Conversion Factors

Conversion factors may be read directly from these tables. For example, 1 degree = 2.778×10^{-3} revolutions, so $16.7° = 16.7 \times 2.778 \times 10^{-3}$ rev. The SI units are fully capitalized. Adapted in part from G. Shortley and D. Williams, *Elements of Physics,* 1971, Prentice-Hall, Englewood Cliffs, NJ.

Plane Angle

°	′	″	RADIAN	rev
1 degree = 1	60	3600	1.745×10^{-2}	2.778×10^{-3}
1 minute = 1.667×10^{-2}	1	60	2.909×10^{-4}	4.630×10^{-5}
1 second = 2.778×10^{-4}	1.667×10^{-2}	1	4.848×10^{-6}	7.716×10^{-7}
1 RADIAN = 57.30	3438	2.063×10^{5}	1	0.1592
1 revolution = 360	2.16×10^{4}	1.296×10^{6}	6.283	1

Solid Angle

1 sphere = 4π steradians = 12.57 steradians

Length

cm	METER	km	in.	ft	mi
1 centimeter = 1	10^{-2}	10^{-5}	0.3937	3.281×10^{-2}	6.214×10^{-6}
1 METER = 100	1	10^{-3}	39.37	3.281	6.214×10^{-4}
1 kilometer = 10^{5}	1000	1	3.937×10^{4}	3281	0.6214
1 inch = 2.540	2.540×10^{-2}	2.540×10^{-5}	1	8.333×10^{-2}	1.578×10^{-5}
1 foot = 30.48	0.3048	3.048×10^{-4}	12	1	1.894×10^{-4}
1 mile = 1.609×10^{5}	1609	1.609	6.336×10^{4}	5280	1

1 angström = 10^{-10} m
1 nautical mile = 1852 m
= 1.151 miles = 6076 ft

1 fermi = 10^{-15} m
1 light-year = 9.461×10^{12} km
1 parsec = 3.084×10^{13} km

1 fathom = 6 ft
1 Bohr radius = 5.292×10^{-11} m
1 yard = 3 ft

1 rod = 16.5 ft
1 mil = 10^{-3} in.
1 nm = 10^{-9} m

Area

METER^2	cm^2	ft^2	in.^{2}
1 SQUARE METER = 1	10^{4}	10.76	1550
1 square centimeter = 10^{-4}	1	1.076×10^{-3}	0.1550
1 square foot = 9.290×10^{-2}	929.0	1	144
1 square inch = 6.452×10^{-4}	6.452	6.944×10^{-3}	1

1 square mile = 2.788×10^{7} ft^2 = 640 acres
1 barn = 10^{-28} m^2

1 acre = 43 560 ft^2
1 hectare = 10^{4} m^2 = 2.471 acres

Volume

METER^3	cm^3	L	ft^3	in.^{3}
1 CUBIC METER = 1	10^6	1000	35.31	6.102×10^4
1 cubic centimeter = 10^{-6}	1	1.000×10^{-3}	3.531×10^{-5}	6.102×10^{-2}
1 liter = 1.000×10^{-3}	1000	1	3.531×10^{-2}	61.02
1 cubic foot = 2.832×10^{-2}	2.832×10^4	28.32	1	1728
1 cubic inch = 1.639×10^{-5}	16.39	1.639×10^{-2}	5.787×10^{-4}	1

1 U.S. fluid gallon = 4 U.S. fluid quarts = 8 U.S. pints = 128 U.S. fluid ounces = 231 in.^{3}

1 British imperial gallon = 277.4 in.^{3} = 1.201 U.S. fluid gallons

Mass

Quantities in the colored areas are not mass units but are often used as such. For example, when we write 1 kg "=" 2.205 lb, this means that a kilogram is a *mass* that *weighs* 2.205 pounds at a location where g has the standard value of 9.80665 m/s^2.

g	KILOGRAM	slug	u	oz	lb	ton
1 gram = 1	0.001	6.852×10^{-5}	6.022×10^{23}	3.527×10^{-2}	2.205×10^{-3}	1.102×10^{-6}
1 KILOGRAM = 1000	1	6.852×10^{-2}	6.022×10^{26}	35.27	2.205	1.102×10^{-3}
1 slug = 1.459×10^4	14.59	1	8.786×10^{27}	514.8	32.17	1.609×10^{-2}
1 atomic mass unit = 1.661×10^{-24}	1.661×10^{-27}	1.138×10^{-28}	1	5.857×10^{-26}	3.662×10^{-27}	1.830×10^{-30}
1 ounce = 28.35	2.835×10^{-2}	1.943×10^{-3}	1.718×10^{25}	1	6.250×10^{-2}	3.125×10^{-5}
1 pound = 453.6	0.4536	3.108×10^{-2}	2.732×10^{26}	16	1	0.0005
1 ton = 9.072×10^5	907.2	62.16	5.463×10^{29}	3.2×10^4	2000	1

1 metric ton = 1000 kg

Density

Quantities in the colored areas are weight densities and, as such, are dimensionally different from mass densities. See the note for the mass table.

slug/ft^3	KILOGRAM/ METER^3	g/cm^3	lb/ft^3	lb/in.^{3}
1 slug per foot^3 = 1	515.4	0.5154	32.17	1.862×10^{-2}
1 KILOGRAM per METER^3 = 1.940×10^{-3}	1	0.001	6.243×10^{-2}	3.613×10^{-5}
1 gram per centimeter^3 = 1.940	1000	1	62.43	3.613×10^{-2}
1 pound per foot^3 = 3.108×10^{-2}	16.02	16.02×10^{-2}	1	5.787×10^{-4}
1 pound per inch^3 = 53.71	2.768×10^4	27.68	1728	1

Time

y	d	h	min	SECOND
1 year = 1	365.25	8.766×10^3	5.259×10^5	3.156×10^7
1 day = 2.738×10^{-3}	1	24	1440	8.640×10^4
1 hour = 1.141×10^{-4}	4.167×10^{-2}	1	60	3600
1 minute = 1.901×10^{-6}	6.944×10^{-4}	1.667×10^{-2}	1	60
1 SECOND = 3.169×10^{-8}	1.157×10^{-5}	2.778×10^{-4}	1.667×10^{-2}	1

Speed

	ft/s	km/h	METER/SECOND	mi/h	cm/s
1 foot per second =	1	1.097	0.3048	0.6818	30.48
1 kilometer per hour =	0.9113	1	0.2778	0.6214	27.78
1 METER per SECOND =	3.281	3.6	1	2.237	100
1 mile per hour =	1.467	1.609	0.4470	1	44.70
1 centimeter per second =	3.281×10^{-2}	3.6×10^{-2}	0.01	2.237×10^{-2}	1

1 knot = 1 nautical mi/h = 1.688 ft/s   1 mi/min = 88.00 ft/s = 60.00 mi/h

Force

Force units in the colored areas are now little used. To clarify: 1 gram-force (= 1 gf) is the force of gravity that would act on an object whose mass is 1 gram at a location where g has the standard value of 9.80665 m/s^2.

	dyne	NEWTON	lb	pdl	gf	kgf
1 dyne =	1	10^{-5}	2.248×10^{-6}	7.233×10^{-5}	1.020×10^{-3}	1.020×10^{-6}
1 NEWTON =	10^5	1	0.2248	7.233	102.0	0.1020
1 pound =	4.448×10^5	4.448	1	32.17	453.6	0.4536
1 poundal =	1.383×10^4	0.1383	3.108×10^{-2}	1	14.10	1.410×10^2
1 gram-force =	980.7	9.807×10^{-3}	2.205×10^{-3}	7.093×10^{-2}	1	0.001
1 kilogram-force =	9.807×10^5	9.807	2.205	70.93	1000	1

1 ton = 2000 lb

Pressure

	atm	dyne/cm^2	inch of water	cm Hg	PASCAL	lb/in.^{2}	lb/ft^2
1 atmosphere =	1	1.013×10^6	406.8	76	1.013×10^5	14.70	2116
1 dyne per centimeter^2 =	9.869×10^{-7}	1	4.015×10^{-4}	7.501×10^{-5}	0.1	1.405×10^{-5}	2.089×10^{-3}
1 inch of water[a] at 4°C =	2.458×10^{-3}	2491	1	0.1868	249.1	3.613×10^{-2}	5.202
1 centimeter of mercury[a] at 0°C =	1.316×10^{-2}	1.333×10^4	5.353	1	1333	0.1934	27.85
1 PASCAL =	9.869×10^{-6}	10	4.015×10^{-3}	7.501×10^{-4}	1	1.450×10^{-4}	2.089×10^{-2}
1 pound per inch^2 =	6.805×10^{-2}	6.895×10^4	27.68	5.171	6.895×10^3	1	144
1 pound per foot^2 =	4.725×10^{-4}	478.8	0.1922	3.591×10^{-2}	47.88	6.944×10^{-3}	1

[a]Where the acceleration of gravity has the standard value of 9.80665 m/s^2.

1 bar = 10^6 dyne/cm^2 = 0.1 MPa   1 millibar = 10^3 dyne/cm^2 = 10^2 Pa   1 torr = 1 mm Hg

Energy, Work, Heat

Quantities in the colored areas are not energy units but are included for convenience. They arise from the relativistic mass–energy equivalence formula $E = mc^2$ and represent the energy released if a kilogram or unified atomic mass unit (u) is completely converted to energy (bottom two rows) or the mass that would be completely converted to one unit of energy (rightmost two columns).

	Btu	erg	ft · lb	hp · h	JOULE	cal	kW · h	eV	MeV	kg	u
1 British thermal unit =	1	1.055×10^{10}	777.9	3.929×10^{-4}	1055	252.0	2.930×10^{-4}	6.585×10^{21}	6.585×10^{15}	1.174×10^{-14}	7.070×10^{12}
1 erg =	9.481×10^{-11}	1	7.376×10^{-8}	3.725×10^{-14}	10^{-7}	2.389×10^{-8}	2.778×10^{-14}	6.242×10^{11}	6.242×10^{5}	1.113×10^{-24}	670.2
1 foot-pound =	1.285×10^{-3}	1.356×10^{7}	1	5.051×10^{-7}	1.356	0.3238	3.766×10^{-7}	8.464×10^{18}	8.464×10^{12}	1.509×10^{-17}	9.037×10^{9}
1 horsepower-hour =	2545	2.685×10^{13}	1.980×10^{6}	1	2.685×10^{6}	6.413×10^{5}	0.7457	1.676×10^{25}	1.676×10^{19}	2.988×10^{-11}	1.799×10^{16}
1 JOULE =	9.481×10^{-4}	10^{7}	0.7376	3.725×10^{-7}	1	0.2389	2.778×10^{-7}	6.242×10^{18}	6.242×10^{12}	1.113×10^{-17}	6.702×10^{9}
1 calorie =	3.968×10^{-3}	4.1868×10^{7}	3.088	1.560×10^{-6}	4.1868	1	1.163×10^{-6}	2.613×10^{19}	2.613×10^{13}	4.660×10^{-17}	2.806×10^{10}
1 kilowatt-hour =	3413	3.600×10^{13}	2.655×10^{6}	1.341	3.600×10^{6}	8.600×10^{5}	1	2.247×10^{25}	2.247×10^{19}	4.007×10^{-11}	2.413×10^{16}
1 electron-volt =	1.519×10^{-22}	1.602×10^{-12}	1.182×10^{-19}	5.967×10^{-26}	1.602×10^{-19}	3.827×10^{-20}	4.450×10^{-26}	1	10^{-6}	1.783×10^{-36}	1.074×10^{-9}
1 million electron-volts =	1.519×10^{-16}	1.602×10^{-6}	1.182×10^{-13}	5.967×10^{-20}	1.602×10^{-13}	3.827×10^{-14}	4.450×10^{-20}	10^{-6}	1	1.783×10^{-30}	1.074×10^{-3}
1 kilogram =	8.521×10^{13}	8.987×10^{23}	6.629×10^{16}	3.348×10^{10}	8.987×10^{16}	2.146×10^{16}	2.497×10^{10}	5.610×10^{35}	5.610×10^{29}	1	6.022×10^{26}
1 unified atomic mass unit =	1.415×10^{-13}	1.492×10^{-3}	1.101×10^{-10}	5.559×10^{-17}	1.492×10^{-10}	3.564×10^{-11}	4.146×10^{-17}	9.320×10^{8}	932.0	1.661×10^{-27}	1

Power

	Btu/h	ft · lb/s	hp	cal/s	kW	WATT
1 British thermal unit per hour =	1	0.2161	3.929×10^{-4}	6.998×10^{-2}	2.930×10^{-4}	0.2930
1 foot-pound per second =	4.628	1	1.818×10^{-3}	0.3239	1.356×10^{-3}	1.356
1 horsepower =	2545	550	1	178.1	0.7457	745.7
1 calorie per second =	14.29	3.088	5.615×10^{-3}	1	4.186×10^{-3}	4.186
1 kilowatt =	3413	737.6	1.341	238.9	1	1000
1 WATT =	3.413	0.7376	1.341×10^{-3}	0.2389	0.001	1

Magnetic Field

	gauss	TESLA	milligauss
1 gauss =	1	10^{-4}	1000
1 TESLA =	10^4	1	10^7
1 milligauss =	0.001	10^{-7}	1

1 tesla = 1 weber/meter^2

Magnetic Flux

	maxwell	WEBER
1 maxwell =	1	10^{-8}
1 WEBER =	10^8	1

Mathematical Formulas

Geometry

Circle of radius r: circumference $= 2\pi r$; area $= \pi r^2$.

Sphere of radius r: area $= 4\pi r^2$; volume $= \frac{4}{3}\pi r^3$.

Right circular cylinder of radius r and height h:
area $= 2\pi r^2 + 2\pi rh$; volume $= \pi r^2 h$.

Triangle of base a and altitude h: area $= \frac{1}{2}ah$.

Quadratic Formula

If $ax^2 + bx + c = 0$, then $x = \dfrac{-b \pm \sqrt{b^2 - 4ac}}{2a}$.

Trigonometric Functions of Angle θ

$\sin\theta = \dfrac{y}{r}$ $\quad \cos\theta = \dfrac{x}{r}$

$\tan\theta = \dfrac{y}{x}$ $\quad \cot\theta = \dfrac{x}{y}$

$\sec\theta = \dfrac{r}{x}$ $\quad \csc\theta = \dfrac{r}{y}$

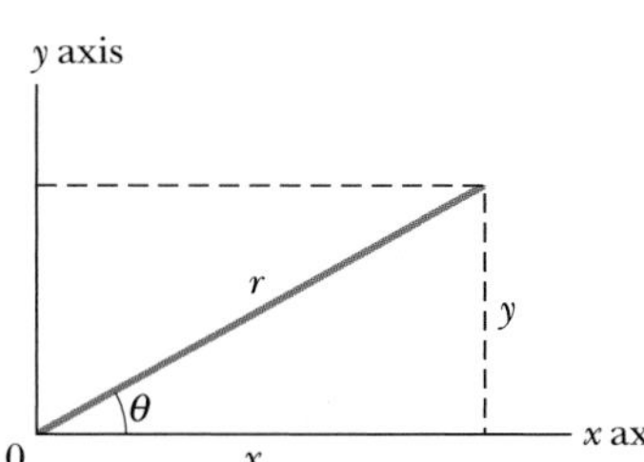

Pythagorean Theorem

In this right triangle,
$a^2 + b^2 = c^2$

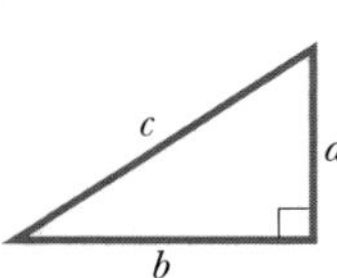

Triangles

Angles are A, B, C

Opposite sides are a, b, c

Angles $A + B + C = 180°$

$\dfrac{\sin A}{a} = \dfrac{\sin B}{b} = \dfrac{\sin C}{c}$

$c^2 = a^2 + b^2 - 2ab\cos C$

Exterior angle $D = A + C$

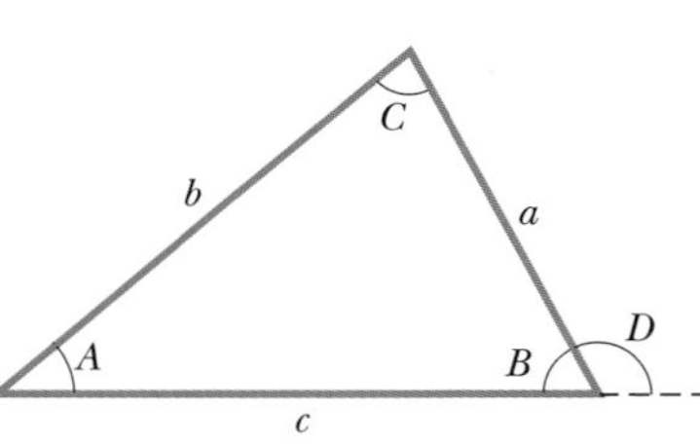

Mathematical Signs and Symbols

$=$ equals

$\approx$ equals approximately

$\sim$ is the order of magnitude of

$\neq$ is not equal to

$\equiv$ is identical to, is defined as

$>$ is greater than ($\gg$ is much greater than)

$<$ is less than ($\ll$ is much less than)

$\geq$ is greater than or equal to (or, is no less than)

$\leq$ is less than or equal to (or, is no more than)

$\pm$ plus or minus

$\propto$ is proportional to

Σ the sum of

x_{avg} the average value of x

Trigonometric Identities

$\sin(90° - \theta) = \cos\theta$

$\cos(90° - \theta) = \sin\theta$

$\sin\theta/\cos\theta = \tan\theta$

$\sin^2\theta + \cos^2\theta = 1$

$\sec^2\theta - \tan^2\theta = 1$

$\csc^2\theta - \cot^2\theta = 1$

$\sin 2\theta = 2\sin\theta\cos\theta$

$\cos 2\theta = \cos^2\theta - \sin^2\theta = 2\cos^2\theta - 1 = 1 - 2\sin^2\theta$

$\sin(\alpha \pm \beta) = \sin\alpha\cos\beta \pm \cos\alpha\sin\beta$

$\cos(\alpha \pm \beta) = \cos\alpha\cos\beta \mp \sin\alpha\sin\beta$

$\tan(\alpha \pm \beta) = \dfrac{\tan\alpha \pm \tan\beta}{1 \mp \tan\alpha\tan\beta}$

$\sin\alpha \pm \sin\beta = 2\sin\frac{1}{2}(\alpha \pm \beta)\cos\frac{1}{2}(\alpha \mp \beta)$

$\cos\alpha + \cos\beta = 2\cos\frac{1}{2}(\alpha + \beta)\cos\frac{1}{2}(\alpha - \beta)$

$\cos\alpha - \cos\beta = -2\sin\frac{1}{2}(\alpha + \beta)\sin\frac{1}{2}(\alpha - \beta)$

Binomial Theorem

$$(1 + x)^n = 1 + \frac{nx}{1!} + \frac{n(n-1)x^2}{2!} + \cdots \qquad (x^2 < 1)$$

Exponential Expansion

$$e^x = 1 + x + \frac{x^2}{2!} + \frac{x^3}{3!} + \cdots$$

Logarithmic Expansion

$$\ln(1 + x) = x - \tfrac{1}{2}x^2 + \tfrac{1}{3}x^3 - \cdots \qquad (|x| < 1)$$

Trigonometric Expansions (θ in radians)

$$\sin \theta = \theta - \frac{\theta^3}{3!} + \frac{\theta^5}{5!} - \cdots$$

$$\cos \theta = 1 - \frac{\theta^2}{2!} + \frac{\theta^4}{4!} - \cdots$$

$$\tan \theta = \theta + \frac{\theta^3}{3} + \frac{2\theta^5}{15} + \cdots$$

Cramer's Rule

Two simultaneous equations in unknowns x and y,

$$a_1x + b_1y = c_1 \quad \text{and} \quad a_2x + b_2y = c_2,$$

have the solutions

$$x = \frac{\begin{vmatrix} c_1 & b_1 \\ c_2 & b_2 \end{vmatrix}}{\begin{vmatrix} a_1 & b_1 \\ a_2 & b_2 \end{vmatrix}} = \frac{c_1b_2 - c_2b_1}{a_1b_2 - a_2b_1}$$

and

$$y = \frac{\begin{vmatrix} a_1 & c_1 \\ a_2 & c_2 \end{vmatrix}}{\begin{vmatrix} a_1 & b_1 \\ a_2 & b_2 \end{vmatrix}} = \frac{a_1c_2 - a_2c_1}{a_1b_2 - a_2b_1}.$$

Products of Vectors

Let $\hat{\text{i}}$, $\hat{\text{j}}$, and $\hat{\text{k}}$ be unit vectors in the x, y, and z directions. Then

$$\hat{\text{i}}\cdot\hat{\text{i}} = \hat{\text{j}}\cdot\hat{\text{j}} = \hat{\text{k}}\cdot\hat{\text{k}} = 1, \quad \hat{\text{i}}\cdot\hat{\text{j}} = \hat{\text{j}}\cdot\hat{\text{k}} = \hat{\text{k}}\cdot\hat{\text{i}} = 0,$$

$$\hat{\text{i}} \times \hat{\text{i}} = \hat{\text{j}} \times \hat{\text{j}} = \hat{\text{k}} \times \hat{\text{k}} = 0,$$

$$\hat{\text{i}} \times \hat{\text{j}} = \hat{\text{k}}, \quad \hat{\text{j}} \times \hat{\text{k}} = \hat{\text{i}}, \quad \hat{\text{k}} \times \hat{\text{i}} = \hat{\text{j}}$$

Any vector $\vec{a}$ with components a_x, a_y, and a_z along the x, y, and z axes can be written as

$$\vec{a} = a_x\hat{\text{i}} + a_y\hat{\text{j}} + a_z\hat{\text{k}}.$$

Let $\vec{a}$, $\vec{b}$, and $\vec{c}$ be arbitrary vectors with magnitudes a, b, and c. Then

$$\vec{a} \times (\vec{b} + \vec{c}) = (\vec{a} \times \vec{b}) + (\vec{a} \times \vec{c})$$

$$(s\vec{a}) \times \vec{b} = \vec{a} \times (s\vec{b}) = s(\vec{a} \times \vec{b}) \qquad (s = \text{a scalar}).$$

Let θ be the smaller of the two angles between $\vec{a}$ and $\vec{b}$. Then

$$\vec{a}\cdot\vec{b} = \vec{b}\cdot\vec{a} = a_xb_x + a_yb_y + a_zb_z = ab\cos\theta$$

$$\vec{a} \times \vec{b} = -\vec{b} \times \vec{a} = \begin{vmatrix} \hat{\text{i}} & \hat{\text{j}} & \hat{\text{k}} \\ a_x & a_y & a_z \\ b_x & b_y & b_z \end{vmatrix}$$

$$= \hat{\text{i}}\begin{vmatrix} a_y & a_z \\ b_y & b_z \end{vmatrix} - \hat{\text{j}}\begin{vmatrix} a_x & a_z \\ b_x & b_z \end{vmatrix} + \hat{\text{k}}\begin{vmatrix} a_x & a_y \\ b_x & b_y \end{vmatrix}$$

$$= (a_yb_z - b_ya_z)\hat{\text{i}} + (a_zb_x - b_za_x)\hat{\text{j}} + (a_xb_y - b_xa_y)\hat{\text{k}}$$

$$|\vec{a} \times \vec{b}| = ab\sin\theta$$

$$\vec{a}\cdot(\vec{b} \times \vec{c}) = \vec{b}\cdot(\vec{c} \times \vec{a}) = \vec{c}\cdot(\vec{a} \times \vec{b})$$

$$\vec{a} \times (\vec{b} \times \vec{c}) = (\vec{a}\cdot\vec{c})\vec{b} - (\vec{a}\cdot\vec{b})\vec{c}$$

Derivatives and Integrals

In what follows, the letters u and v stand for any functions of x, and a and m are constants. To each of the indefinite integrals should be added an arbitrary constant of integration. The *Handbook of Chemistry and Physics* (CRC Press Inc.) gives a more extensive tabulation.

1. $\dfrac{dx}{dx} = 1$
2. $\dfrac{d}{dx}(au) = a\dfrac{du}{dx}$
3. $\dfrac{d}{dx}(u + v) = \dfrac{du}{dx} + \dfrac{dv}{dx}$
4. $\dfrac{d}{dx}x^m = mx^{m-1}$
5. $\dfrac{d}{dx}\ln x = \dfrac{1}{x}$
6. $\dfrac{d}{dx}(uv) = u\dfrac{dv}{dx} + v\dfrac{du}{dx}$
7. $\dfrac{d}{dx}e^x = e^x$
8. $\dfrac{d}{dx}\sin x = \cos x$
9. $\dfrac{d}{dx}\cos x = -\sin x$
10. $\dfrac{d}{dx}\tan x = \sec^2 x$
11. $\dfrac{d}{dx}\cot x = -\csc^2 x$
12. $\dfrac{d}{dx}\sec x = \tan x \sec x$
13. $\dfrac{d}{dx}\csc x = -\cot x \csc x$
14. $\dfrac{d}{dx}e^u = e^u\dfrac{du}{dx}$
15. $\dfrac{d}{dx}\sin u = \cos u\dfrac{du}{dx}$
16. $\dfrac{d}{dx}\cos u = -\sin u\dfrac{du}{dx}$

1. $\displaystyle\int dx = x$
2. $\displaystyle\int au\, dx = a\int u\, dx$
3. $\displaystyle\int (u + v)\, dx = \int u\, dx + \int v\, dx$
4. $\displaystyle\int x^m\, dx = \frac{x^{m+1}}{m+1} \quad (m \neq -1)$
5. $\displaystyle\int \frac{dx}{x} = \ln|x|$
6. $\displaystyle\int u\frac{dv}{dx}\, dx = uv - \int v\frac{du}{dx}\, dx$
7. $\displaystyle\int e^x\, dx = e^x$
8. $\displaystyle\int \sin x\, dx = -\cos x$
9. $\displaystyle\int \cos x\, dx = \sin x$
10. $\displaystyle\int \tan x\, dx = \ln|\sec x|$
11. $\displaystyle\int \sin^2 x\, dx = \tfrac{1}{2}x - \tfrac{1}{4}\sin 2x$
12. $\displaystyle\int e^{-ax}\, dx = -\frac{1}{a}e^{-ax}$
13. $\displaystyle\int xe^{-ax}\, dx = -\frac{1}{a^2}(ax + 1)\, e^{-ax}$
14. $\displaystyle\int x^2e^{-ax}\, dx = -\frac{1}{a^3}(a^2x^2 + 2ax + 2)e^{-ax}$
15. $\displaystyle\int_0^\infty x^ne^{-ax}\, dx = \frac{n!}{a^{n+1}}$
16. $\displaystyle\int_0^\infty x^{2n}e^{-ax^2}\, dx = \frac{1\cdot 3\cdot 5\cdots(2n-1)}{2^{n+1}a^n}\sqrt{\frac{\pi}{a}}$
17. $\displaystyle\int \frac{dx}{\sqrt{x^2 + a^2}} = \ln(x + \sqrt{x^2 + a^2})$
18. $\displaystyle\int \frac{x\, dx}{(x^2 + a^2)^{3/2}} = -\frac{1}{(x^2 + a^2)^{1/2}}$
19. $\displaystyle\int \frac{dx}{(x^2 + a^2)^{3/2}} = \frac{x}{a^2(x^2 + a^2)^{1/2}}$
20. $\displaystyle\int_0^\infty x^{2n+1}\, e^{-ax^2}\, dx = \frac{n!}{2a^{n+1}} \quad (a > 0)$
21. $\displaystyle\int \frac{x\, dx}{x + d} = x - d\ln(x + d)$

Properties of the Elements

All physical properties are for a pressure of 1 atm unless otherwise specified.

Element	Symbol	Atomic Number Z	Molar Mass, g/mol	Density, g/cm^3 at 20°C	Melting Point, °C	Boiling Point, °C	Specific Heat, J/(g·°C) at 25°C
Actinium	Ac	89	(227)	10.06	1323	(3473)	0.092
Aluminum	Al	13	26.9815	2.699	660	2450	0.900
Americium	Am	95	(243)	13.67	1541	—	—
Antimony	Sb	51	121.75	6.691	630.5	1380	0.205
Argon	Ar	18	39.948	1.6626×10^{-3}	−189.4	−185.8	0.523
Arsenic	As	33	74.9216	5.78	817 (28 atm)	613	0.331
Astatine	At	85	(210)	—	(302)	—	—
Barium	Ba	56	137.34	3.594	729	1640	0.205
Berkelium	Bk	97	(247)	14.79	—	—	—
Beryllium	Be	4	9.0122	1.848	1287	2770	1.83
Bismuth	Bi	83	208.980	9.747	271.37	1560	0.122
Bohrium	Bh	107	262.12	—	—	—	—
Boron	B	5	10.811	2.34	2030	—	1.11
Bromine	Br	35	79.909	3.12 (liquid)	−7.2	58	0.293
Cadmium	Cd	48	112.40	8.65	321.03	765	0.226
Calcium	Ca	20	40.08	1.55	838	1440	0.624
Californium	Cf	98	(251)	—	—	—	—
Carbon	C	6	12.01115	2.26	3727	4830	0.691
Cerium	Ce	58	140.12	6.768	804	3470	0.188
Cesium	Cs	55	132.905	1.873	28.40	690	0.243
Chlorine	Cl	17	35.453	3.214×10^{-3} (0°C)	−101	−34.7	0.486
Chromium	Cr	24	51.996	7.19	1857	2665	0.448
Cobalt	Co	27	58.9332	8.85	1495	2900	0.423
Copernicium	Cp	112	(285)	—	—	—	—
Copper	Cu	29	63.54	8.96	1083.40	2595	0.385
Curium	Cm	96	(247)	13.3	—	—	—
Darmstadtium	Ds	110	(271)	—	—	—	—
Dubnium	Db	105	262.114	—	—	—	—
Dysprosium	Dy	66	162.50	8.55	1409	2330	0.172
Einsteinium	Es	99	(254)	—	—	—	—
Erbium	Er	68	167.26	9.15	1522	2630	0.167
Europium	Eu	63	151.96	5.243	817	1490	0.163
Fermium	Fm	100	(237)	—	—	—	—
Fluorine	F	9	18.9984	1.696×10^{-3} (0°C)	−219.6	−188.2	0.753
Francium	Fr	87	(223)	—	(27)	—	—
Gadolinium	Gd	64	157.25	7.90	1312	2730	0.234
Gallium	Ga	31	69.72	5.907	29.75	2237	0.377
Germanium	Ge	32	72.59	5.323	937.25	2830	0.322
Gold	Au	79	196.967	19.32	1064.43	2970	0.131

Element	Symbol	Atomic Number Z	Molar Mass, g/mol	Density, g/cm^3 at 20°C	Melting Point, °C	Boiling Point, °C	Specific Heat, J/(g · °C) at 25°C
Hafnium	Hf	72	178.49	13.31	2227	5400	0.144
Hassium	Hs	108	(265)	—	—	—	—
Helium	He	2	4.0026	0.1664×10^{-3}	−269.7	−268.9	5.23
Holmium	Ho	67	164.930	8.79	1470	2330	0.165
Hydrogen	H	1	1.00797	0.08375×10^{-3}	−259.19	−252.7	14.4
Indium	In	49	114.82	7.31	156.634	2000	0.233
Iodine	I	53	126.9044	4.93	113.7	183	0.218
Iridium	Ir	77	192.2	22.5	2447	(5300)	0.130
Iron	Fe	26	55.847	7.874	1536.5	3000	0.447
Krypton	Kr	36	83.80	3.488×10^{-3}	−157.37	−152	0.247
Lanthanum	La	57	138.91	6.189	920	3470	0.195
Lawrencium	Lr	103	(257)	—	—	—	—
Lead	Pb	82	207.19	11.35	327.45	1725	0.129
Lithium	Li	3	6.939	0.534	180.55	1300	3.58
Lutetium	Lu	71	174.97	9.849	1663	1930	0.155
Magnesium	Mg	12	24.312	1.738	650	1107	1.03
Manganese	Mn	25	54.9380	7.44	1244	2150	0.481
Meitnerium	Mt	109	(266)	—	—	—	—
Mendelevium	Md	101	(256)	—	—	—	—
Mercury	Hg	80	200.59	13.55	−38.87	357	0.138
Molybdenum	Mo	42	95.94	10.22	2617	5560	0.251
Neodymium	Nd	60	144.24	7.007	1016	3180	0.188
Neon	Ne	10	20.183	0.8387×10^{-3}	−248.597	−246.0	1.03
Neptunium	Np	93	(237)	20.25	637	—	1.26
Nickel	Ni	28	58.71	8.902	1453	2730	0.444
Niobium	Nb	41	92.906	8.57	2468	4927	0.264
Nitrogen	N	7	14.0067	1.1649×10^{-3}	−210	−195.8	1.03
Nobelium	No	102	(255)	—	—	—	—
Osmium	Os	76	190.2	22.59	3027	5500	0.130
Oxygen	O	8	15.9994	1.3318×10^{-3}	−218.80	−183.0	0.913
Palladium	Pd	46	106.4	12.02	1552	3980	0.243
Phosphorus	P	15	30.9738	1.83	44.25	280	0.741
Platinum	Pt	78	195.09	21.45	1769	4530	0.134
Plutonium	Pu	94	(244)	19.8	640	3235	0.130
Polonium	Po	84	(210)	9.32	254	—	—
Potassium	K	19	39.102	0.862	63.20	760	0.758
Praseodymium	Pr	59	140.907	6.773	931	3020	0.197
Promethium	Pm	61	(145)	7.22	(1027)	—	—
Protactinium	Pa	91	(231)	15.37 (estimated)	(1230)	—	—
Radium	Ra	88	(226)	5.0	700	—	—
Radon	Rn	86	(222)	9.96×10^{-3} (0°C)	(−71)	−61.8	0.092
Rhenium	Re	75	186.2	21.02	3180	5900	0.134
Rhodium	Rh	45	102.905	12.41	1963	4500	0.243
Roentgenium	Rg	111	(280)	—	—	—	—
Rubidium	Rb	37	85.47	1.532	39.49	688	0.364
Ruthenium	Ru	44	101.107	12.37	2250	4900	0.239
Rutherfordium	Rf	104	261.11	—	—	—	—

Element	Symbol	Atomic Number Z	Molar Mass, g/mol	Density, g/cm^3 at 20°C	Melting Point, °C	Boiling Point, °C	Specific Heat, $J/(g \cdot °C)$ at 25°C
Samarium	Sm	62	150.35	7.52	1072	1630	0.197
Scandium	Sc	21	44.956	2.99	1539	2730	0.569
Seaborgium	Sg	106	263.118	—	—	—	—
Selenium	Se	34	78.96	4.79	221	685	0.318
Silicon	Si	14	28.086	2.33	1412	2680	0.712
Silver	Ag	47	107.870	10.49	960.8	2210	0.234
Sodium	Na	11	22.9898	0.9712	97.85	892	1.23
Strontium	Sr	38	87.62	2.54	768	1380	0.737
Sulfur	S	16	32.064	2.07	119.0	444.6	0.707
Tantalum	Ta	73	180.948	16.6	3014	5425	0.138
Technetium	Tc	43	(99)	11.46	2200	—	0.209
Tellurium	Te	52	127.60	6.24	449.5	990	0.201
Terbium	Tb	65	158.924	8.229	1357	2530	0.180
Thallium	Tl	81	204.37	11.85	304	1457	0.130
Thorium	Th	90	(232)	11.72	1755	(3850)	0.117
Thulium	Tm	69	168.934	9.32	1545	1720	0.159
Tin	Sn	50	118.69	7.2984	231.868	2270	0.226
Titanium	Ti	22	47.90	4.54	1670	3260	0.523
Tungsten	W	74	183.85	19.3	3380	5930	0.134
Unnamed	Uut	113	(284)	—	—	—	—
Unnamed	Unq	114	(289)	—	—	—	—
Unnamed	Uup	115	(288)	—	—	—	—
Unnamed	Uuh	116	(293)	—	—	—	—
Unnamed	Uus	117	—	—	—	—	—
Unnamed	Uuo	118	(294)	—	—	—	—
Uranium	U	92	(238)	18.95	1132	3818	0.117
Vanadium	V	23	50.942	6.11	1902	3400	0.490
Xenon	Xe	54	131.30	5.495×10^{-3}	−111.79	−108	0.159
Ytterbium	Yb	70	173.04	6.965	824	1530	0.155
Yttrium	Y	39	88.905	4.469	1526	3030	0.297
Zinc	Zn	30	65.37	7.133	419.58	906	0.389
Zirconium	Zr	40	91.22	6.506	1852	3580	0.276

The values in parentheses in the column of molar masses are the mass numbers of the longest-lived isotopes of those elements that are radioactive. Melting points and boiling points in parentheses are uncertain.

The data for gases are valid only when these are in their usual molecular state, such as H_2, He, O_2, Ne, etc. The specific heats of the gases are the values at constant pressure.

Source: Adapted from J. Emsley, *The Elements,* 3rd ed., 1998, Clarendon Press, Oxford. See also www.webelements.com for the latest values and newest elements.

Periodic Table of the Elements

Metals
Metalloids
Nonmetals

THE HORIZONTAL PERIODS

Period	Alkali metals IA	IIA	IIIB	IVB	VB	VIB	VIIB	VIIIB	VIIIB	VIIIB	IB	IIB	IIIA	IVA	VA	VIA	VIIA	Noble gases 0
1	1 H																	2 He
2	3 Li	4 Be											5 B	6 C	7 N	8 O	9 F	10 Ne
3	11 Na	12 Mg											13 Al	14 Si	15 P	16 S	17 Cl	18 Ar
4	19 K	20 Ca	21 Sc	22 Ti	23 V	24 Cr	25 Mn	26 Fe	27 Co	28 Ni	29 Cu	30 Zn	31 Ga	32 Ge	33 As	34 Se	35 Br	36 Kr
5	37 Rb	38 Sr	39 Y	40 Zr	41 Nb	42 Mo	43 Tc	44 Ru	45 Rh	46 Pd	47 Ag	48 Cd	49 In	50 Sn	51 Sb	52 Te	53 I	54 Xe
6	55 Cs	56 Ba	57-71 *	72 Hf	73 Ta	74 W	75 Re	76 Os	77 Ir	78 Pt	79 Au	80 Hg	81 Tl	82 Pb	83 Bi	84 Po	85 At	86 Rn
7	87 Fr	88 Ra	89-103 †	104 Rf	105 Db	106 Sg	107 Bh	108 Hs	109 Mt	110 Ds	111 Rg	112 Cp	113	114	115	116	117	118

Transition metals

Inner transition metals

Series															
Lanthanide series *	57 La	58 Ce	59 Pr	60 Nd	61 Pm	62 Sm	63 Eu	64 Gd	65 Tb	66 Dy	67 Ho	68 Er	69 Tm	70 Yb	71 Lu
Actinide series †	89 Ac	90 Th	91 Pa	92 U	93 Np	94 Pu	95 Am	96 Cm	97 Bk	98 Cf	99 Es	100 Fm	101 Md	102 No	103 Lr

Evidence for the discovery of elements 113 through 118 has been reported. See www.webelements.com for the latest information and newest elements.

ANSWERS

to Checkpoints and Odd-Numbered Questions and Problems

CHAPTER 21

CP **1.** C and D attract; B and D attract **2.** (a) leftward; (b) leftward; (c) leftward **3.** (a) a, c, b; (b) less than **4.** $-15e$ (net charge of $-30e$ is equally shared)
Q **1.** 3, 1, 2, 4 (zero) **3.** a and b **5.** $2kq^2/r^2$, up the page **7.** b and c tie, then a (zero) **9.** (a) same; (b) less than; (c) cancel; (d) add; (e) adding components; (f) positive direction of y; (g) negative direction of y; (h) positive direction of x; (i) negative direction of x
P **1.** 0.500 **3.** 1.39 m **5.** 2.81 N **7.** -4.00 **9.** (a) $-1.00\ \mu C$; (b) $3.00\ \mu C$ **11.** (a) 0.17 N; (b) -0.046 N **13.** (a) -14 cm; (b) 0 **15.** (a) 35 N; (b) $-10°$; (c) -8.4 cm; (d) $+2.7$ cm **17.** (a) 1.60 N; (b) 2.77 N **19.** (a) 3.00 cm; (b) 0; (c) -0.444 **21.** 3.8×10^{-8} C **23.** (a) 0; (b) 12 cm; (c) 0; (d) 4.9×10^{-26} N **25.** 6.3×10^{11} **27.** (a) 3.2×10^{-19} C; (b) 2 **29.** (a) -6.05 cm; (b) 6.05 cm **31.** 122 mA **33.** 1.3×10^7 C **35.** (a) 0; (b) 1.9×10^{-9} N **37.** (a) ^{9}B; (b) ^{13}N; (c) ^{12}C **39.** 1.31×10^{-22} N **41.** (a) 5.7×10^{13} C; (b) cancels out; (c) 6.0×10^5 kg **43.** (b) 3.1 cm **45.** 0.19 MC **47.** $-45\ \mu C$ **49.** 3.8 N **51.** (a) 2.00×10^{10} electrons; (b) 1.33×10^{10} electrons **53.** (a) 8.99×10^9 N; (b) 8.99 kN **55.** (a) 0.5; (b) 0.15; (c) 0.85 **57.** 1.7×10^8 N **59.** -1.32×10^{13} C **61.** (a) $(0.829\ \text{N})\hat{i}$; (b) $(-0.621\ \text{N})\hat{j}$ **63.** 2.2×10^{-6} kg **65.** 4.68×10^{-19} N **67.** (a) $2.72L$; (b) 0 **69.** (a) 5.1×10^2 N; (b) $7.7 \times 10^{28}\ \text{m/s}^2$

CHAPTER 22

CP **1.** (a) rightward; (b) leftward; (c) leftward; (d) rightward (p and e have same charge magnitude, and p is farther) **2.** (a) toward positive y; (b) toward positive x; (c) toward negative y **3.** (a) leftward; (b) leftward; (c) decrease **4.** (a) all tie; (b) 1 and 3 tie, then 2 and 4 tie
Q **1.** a, b, c **3.** (a) yes; (b) toward; (c) no (the field vectors are not along the same line); (d) cancel; (e) add; (f) adding components; (g) toward negative y **5.** (a) to their left; (b) no **7.** (a) 4, 3, 1, 2; (b) 3, then 1 and 4 tie, then 2 **9.** a, b, c **11.** e, b, then a and c tie, then d (zero)
P **3.** (a) 3.07×10^{21} N/C; (b) outward **5.** 56 pC **7.** $(1.02 \times 10^5\ \text{N/C})\hat{j}$ **9.** (a) 1.38×10^{-10} N/C; (b) 180° **11.** -30 cm **13.** (a) 3.60×10^{-6} N/C; (b) 2.55×10^{-6} N/C; (c) 3.60×10^{-4} N/C; (d) 7.09×10^{-7} N/C; (e) As the proton nears the disk, the forces on it from electrons e_s more nearly cancel. **15.** (a) 160 N/C; (b) 45° **17.** (a) $-90°$; (b) $+2.0\ \mu C$; (c) $-1.6\ \mu C$ **19.** (a) $qd/4\pi\varepsilon_0 r^3$; (b) $-90°$ **23.** 0.506 **25.** (a) 1.62×10^6 N/C; (b) $-45°$ **27.** (a) 23.8 N/C; (b) $-90°$ **29.** 1.57 **31.** (a) -5.19×10^{-14} C/m; (b) 1.57×10^{-3} N/C; (c) $-180°$; (d) 1.52×10^{-8} N/C; (e) 1.52×10^{-8} N/C **35.** 0.346 m **37.** 28% **39.** $-5e$ **41.** (a) 1.5×10^3 N/C; (b) 2.4×10^{-16} N; (c) up; (d) 1.6×10^{-26} N; (e) 1.5×10^{10} **43.** $3.51 \times 10^{15}\ \text{m/s}^2$ **45.** 6.6×10^{-15} N **47.** (a) $1.92 \times 10^{12}\ \text{m/s}^2$; (b) 1.96×10^5 m/s **49.** (a) 0.245 N; (b) $-11.3°$; (c) 108 m; (d) -21.6 m **51.** 2.6×10^{-10} N; (b) 3.1×10^{-8} N; (c) moves to stigma **53.** 27 μm **55.** (a) 2.7×10^6 m/s; (b) 1.0 kN/C **57.** (a) 9.30×10^{-15} C·m; (b) 2.05×10^{-11} J **59.** 1.22×10^{-23} J **61.** $(1/2\pi)(pE/I)^{0.5}$ **63.** (a) 8.87×10^{-15} N; (b) 120 **65.** 217° **67.** 61 N/C **69.** (a) 47 N/C; (b) 27 N/C **71.** 38 N/C **73.** (a) -1.0 cm; (b) 0; (c) 10 pC **75.** $+1.00\ \mu C$ **77.** (a) 6.0 mm; (b) 180° **79.** 9:30 **81.** (a) -0.029 C; (b) repulsive forces would explode the sphere **83.** (a) -1.49×10^{-26} J; (b) $(-1.98 \times 10^{-26}\ \text{N·m})\hat{k}$; (c) 3.47×10^{-26} J **85.** (a) top row: 4, 8, 12; middle row: 5, 10, 14; bottom row: 7, 11, 16; (b) 1.63×10^{-19} C **87.** (a) $(-1.80\ \text{N/C})\hat{i}$; (b) $(43.2\ \text{N/C})\hat{i}$; (c) $(-6.29\ \text{N/C})\hat{i}$

CHAPTER 23

CP **1.** (a) $+EA$; (b) $-EA$; (c) 0; (d) 0 **2.** (a) 2; (b) 3; (c) 1 **3.** (a) equal; (b) equal; (c) equal **4.** 3 and 4 tie, then 2, 1
Q **1.** (a) 8 N·m²/C; (b) 0 **3.** all tie **5.** all tie **7.** a, c, then b and d tie (zero) **9.** (a) 2, 1, 3; (b) all tie ($+4q$)
P **1.** -0.015 N·m²/C **3.** (a) 0; (b) -3.92 N·m²/C; (c) 0; (d) 0 **5.** 3.01 nN·m²/C **7.** 2.0×10^5 N·m²/C **9.** (a) 8.23 N·m²/C; (b) 72.9 pC; (c) 8.23 N·m²/C; (d) 72.9 pC **11.** -1.70 nC **13.** $3.54\ \mu C$ **15.** (a) 0; (b) 0.0417 **17.** (a) $37\ \mu C$; (b) 4.1×10^6 N·m²/C **19.** (a) $4.5 \times 10^{-7}\ \text{C/m}^2$; (b) 5.1×10^4 N/C **21.** (a) -3.0×10^{-6} C; (b) $+1.3 \times 10^{-5}$ C **23.** (a) $0.32\ \mu C$; (b) $0.14\ \mu C$ **25.** $5.0\ \mu C/m$ **27.** $3.8 \times 10^{-8}\ \text{C/m}^2$ **29.** (a) 0.214 N/C; (b) inward; (c) 0.855 N/C; (d) outward; (e) -3.40×10^{-12} C; (f) -3.40×10^{-12} C **31.** (a) 2.3×10^6 N/C; (b) outward; (c) 4.5×10^5 N/C; (d) inward **33.** (a) 0; (b) 0; (c) $(-7.91 \times 10^{-11}\ \text{N/C})\hat{i}$ **35.** -1.5 **37.** (a) 5.3×10^7 N/C; (b) 60 N/C **39.** 5.0 nC/m² **41.** 0.44 mm **43.** (a) 0; (b) $1.31\ \mu N/C$; (c) $3.08\ \mu N/C$; (d) $3.08\ \mu N/C$ **45.** (a) 2.50×10^4 N/C; (b) 1.35×10^4 N/C **47.** -7.5 nC **49.** (a) 0; (b) 56.2 mN/C; (c) 112 mN/C; (d) 49.9 mN/C; (e) 0; (f) 0; (g) -5.00 fC; (h) 0 **51.** $1.79 \times 10^{-11}\ \text{C/m}^2$ **53.** (a) 7.78 fC; (b) 0; (c) 5.58 mN/C; (d) 22.3 mN/C **55.** $6K\varepsilon_0 r^3$ **57.** (a) 0; (b) 2.88×10^4 N/C; (c) 200 N/C **59.** (a) 5.4 N/C; (b) 6.8 N/C **61.** (a) 0; (b) $q_a/4\pi\varepsilon_0 r^2$; (c) $(q_a + q_b)/4\pi\varepsilon_0 r^2$ **63.** -1.04 nC **65.** (a) 0.125; (b) 0.500 **67.** (a) $+2.0$ nC; (b) -1.2 nC; (c) $+1.2$ nC; (d) $+0.80$ nC **69.** $(5.65 \times 10^4\ \text{N/C})\hat{j}$ **71.** (a) -2.53×10^{-2} N·m²/C; (b) $+2.53 \times 10^{-2}$ N·m²/C **75.** 3.6 nC **77.** (a) $+4.0\ \mu C$; (b) $-4.0\ \mu C$ **79.** (a) 693 kg/s; (b) 693 kg/s; (c) 347 kg/s; (d) 347 kg/s; (e) 575 kg/s **81.** (a) $0.25R$; (b) $2.0R$

CHAPTER 24

CP **1.** (a) negative; (b) increase **2.** (a) positive; (b) higher **3.** (a) rightward; (b) 1, 2, 3, 5: positive; 4, negative; (c) 3, then 1, 2, and 5 tie, then 4 **4.** all tie **5.** a, c (zero), b **6.** (a) 2, then 1 and 3 tie; (b) 3; (c) accelerate leftward
Q **1.** $-4q/4\pi\varepsilon_0 d$ **3.** (a) 1 and 2; (b) none; (c) no; (d) 1 and 2, yes; 3 and 4, no **5.** (a) higher; (b) positive; (c) negative; (d) all tie **7.** (a) 0; (b) 0; (c) 0; (d) all three quantities still 0 **9.** (a) 3 and 4 tie, then 1 and 2 tie; (b) 1 and 2, increase; 3 and 4, decrease
P **1.** (a) 3.0×10^5 C; (b) 3.6×10^6 J **3.** 2.8×10^5 **5.** 8.8 mm **7.** -32.0 V **9.** (a) 1.87×10^{-21} J; (b) -11.7 mV **11.** (a) -0.268 mV; (b) -0.681 mV **13.** (a) 3.3 nC; (b) 12 nC/m² **15.** (a) 0.54 mm; (b) 790 V **17.** 0.562 mV **19.** (a) 6.0 cm; (b) -12.0 cm **21.** $16.3\ \mu V$ **23.** (a) 24.3 mV; (b) 0 **25.** (a) -2.30 V; (b) -1.78 V **27.** 13 kV **29.** 32.4 mV **31.** $47.1\ \mu V$ **33.** 18.6 mV **35.** $(-12\ \text{V/m})\hat{i} + (12\ \text{V/m})\hat{j}$ **37.** 150 N/C **39.** $(-4.0 \times 10^{-16}\ \text{N})\hat{i} + (1.6 \times 10^{-16}\ \text{N})\hat{j}$ **41.** (a) 0.90 J; (b) 4.5 J **43.** -0.192 pJ **45.** 2.5 km/s **47.** 22 km/s **49.** 0.32 km/s **51.** (a) $+6.0 \times 10^4$ V; (b) -7.8×10^5 V; (c) 2.5 J; (d) increase; (e) same; (f) same **53.** (a) 0.225 J; (b) A 45.0 m/s², B 22.5 m/s²; (c) A 7.75 m/s, B 3.87 m/s **55.** 1.6×10^{-9} m **57.** (a) 3.0 J; (b) -8.5 m **59.** (a) proton; (b) 65.3 km/s **61.** (a) 12; (b) 2 **63.** (a) -1.8×10^2 V; (b) 2.9 kV; (c) -8.9 kV **65.** 2.5×10^{-8} C **67.** (a) 12 kN/C; (b) 1.8 kV; (c) 5.8 cm **69.** (a) 64 N/C; (b) 2.9 V; (c) 0 **71.** $p/2\pi\varepsilon_0 r^3$ **73.** (a) 3.6×10^5 V; (b) no **75.** 6.4×10^8 V **77.** 2.90 kV **79.** 7.0×10^5 m/s **81.** (a) 1.8 cm; (b) 8.4×10^5 m/s; (c) 2.1×10^{-17} N; (d) positive; (e) 1.6×10^{-17} N; (f) negative **83.** (a) $+7.19 \times 10^{-10}$ V; (b) $+2.30 \times 10^{-28}$ J; (c) $+2.43 \times 10^{-29}$ J **85.** 2.30×10^{-28} J **87.** 2.1 days **89.** 2.30×10^{-22} J **91.** 1.48×10^7 m/s **93.** -1.92 MV

95. (a) $Q/4\pi\varepsilon_0 r$; (b) $(\rho/3\varepsilon_0)(1.5r_2^2 - 0.50r^2 - r_1^3 r^{-1})$, $\rho = Q/[(4\pi/3)(r_2^3 - r_1^3)]$; (c) $(\rho/2\varepsilon_0)(r_2^2 - r_1^2)$, with ρ as in (b); (d) yes **101.** (a) 0.484 MeV; (b) 0 **103.** −1.7 **105.** (a) 38 s; (b) 280 days

CHAPTER 25

CP **1.** (a) same; (b) same **2.** (a) decreases; (b) increases; (c) decreases **3.** (a) $V, q/2$; (b) $V/2$; q
Q **1.** a, 2; b, 1; c, 3 **3.** (a) no; (b) yes; (c) all tie **5.** (a) same; (b) same; (c) more; (d) more **7.** a, series; b, parallel; c, parallel **9.** (a) increase; (b) same; (c) increase; (d) increase; (e) increase; (f) increase **11.** parallel, C_1 alone, C_2 alone, series
P **1.** (a) 3.5 pF; (b) 3.5 pF; (c) 57 V **3.** (a) 144 pF; (b)17.3 nC **5.** 0.280 pF **7.** 6.79×10^{-4} F/m² **9.** 315 mC **11.** 3.16 μF **13.** 43 pF **15.** (a) 3.00 μF; (b) 60.0 μC; (c) 10.0 V; (d) 30.0 μC; (e) 10.0 V; (f) 20.0 μC; (g) 5.00 V; (h) 20.0 μC **17.** (a) 789 μC; (b) 78.9 V **19.** (a) 4.0 μF; (b) 2.0 μF **21.** (a) 50 V; (b) 5.0×10^{-5} C; (c) 1.5×10^{-4} C **23.** (a) 4.5×10^{14}; (b) 1.5×10^{14}; (c) 3.0×10^{14}; (d) 4.5×10^{14}; (e) up; (f) up **25.** 3.6 pC **27.** (a) 9.00 μC; (b) 16.0 μC; (c) 9.00 μC; (d) 16.0 μC; (e) 8.40 μC; (f) 16.8 μC; (g) 10.8 μC; (h) 14.4 μC **29.** 72 F **31.** 0.27 J **33.** 0.11 J/m³ **35.** (a) 9.16×10^{-18} J/m³; (b) 9.16×10^{-6} J/m³; (c) 9.16×10^{6} J/m³; (d) 9.16×10^{18} J/m³; (e) ∞ **37.** (a) 16.0 V; (b) 45.1 pJ; (c) 120 pJ; (d) 75.2 pJ **39.** (a) 190 V; (b) 95 mJ **41.** 81 pF/m **43.** Pyrex **45.** 66 μJ **47.** 0.63 m² **49.** 17.3 pF **51.** (a) 10 kV/m; (b) 5.0 nC; (c) 4.1 nC **53.** (a) 89 pF; (b) 0.12 nF; (c) 11 nC; (d) 11 nC; (e) 10 kV/m; (f) 2.1 kV/m; (g) 88 V; (h) −0.17 μJ **55.** (a) 0.107 nF; (b) 7.79 nC; (c) 7.45 nC **57.** 45 μC **59.** 16 μC **61.** (a) 7.20 μC; (b) 18.0 μC; (c) Battery supplies charges only to plates to which it is connected; charges on other plates are due to electron transfers between plates, in accord with new distribution of voltages across the capacitors. So the battery does not directly supply charge on capacitor 4. **63.** (a) 10 μC; (b) 20 μC **65.** 1.06 nC **67.** (a) 2.40 μF; (b) 0.480 mC; (c) 80 V; (d) 0.480 mC; (e) 120 V **69.** 4.9% **71.** (a) 0.708 pF; (b) 0.600 ; (c) 1.02×10^{-9} J; (d) sucked in **73.** 5.3 V **75.** 40 μF **77.** (a) 200 kV/m; (b) 200 kV/m; (c) 1.77 μC/m²; (d) 4.60 μC/m²; (e) −2.83 μC/m²

CHAPTER 26

CP **1.** 8 A, rightward **2.** (a)–(c) rightward **3.** a and c tie, then b **4.** device 2 **5.** (a) and (b) tie, then (d), then (c)
Q **1.** tie of A, B, and C, then tie of $A+B$ and $B+C$, then $A+B+C$ **3.** (a) top-bottom, front-back, left-right; (b) top-bottom, front-back, left-right; (c) top-bottom, front-back, left-right; (d) top-bottom, front-back, left-right **5.** a, b, and c all tie, then d **7.** (a) B, A, C; (b) B, A, C **9.** (a) C, B, A; (b) all tie; (c) A, B, C; (d) all tie
P **1.** (a) 1.2 kC; (b) 7.5×10^{21} **3.** 6.7 μC/m² **5.** (a) 6.4 A/m²; (b) north; (c) cross-sectional area **7.** 0.38 mm **9.** 18.1 μA **11.** (a) 1.33 A; (b) 0.666 A; (c) J_a **13.** 13 min **15.** 2.4 Ω **17.** 2.0×10^{6} (Ω·m)$^{-1}$ **19.** 2.0×10^{-8} Ω·m **21.** (1.8×10^{3})°C **23.** 8.2×10^{-4} Ω·m **25.** 54 Ω **27.** 3.0 **29.** 3.35×10^{-7} C **31.** (a) 6.00 mA; (b) 1.59×10^{-8} V; (c) 21.2 nΩ **33.** (a) 38.3 mA; (b) 109 A/m²; (c) 1.28 cm/s; (d) 227 V/m **35.** 981 kΩ **39.** 150 s **41.** (a) 1.0 kW; (b) US$0.25 **43.** 0.135 W **45.** (a) 10.9 A; (b) 10.6 Ω; (c) 4.50 MJ **47.** (a) 5.85 m; (b) 10.4 m **49.** (a) US$4.46; (b) 144 Ω; (c) 0.833 A **51.** (a) 5.1 V; (b) 10 V; (c) 10 W; (d) 20 W **53.** (a) 28.8 Ω; (b) 2.60×10^{19} s^{-1} **55.** 660 W **57.** 28.8 kC **59.** (a) silver; (b) 51.6 nΩ **61.** (a) 2.3×10^{12}; (b) 5.0×10^{3}; (c) 10 MV **63.** 2.4 kW **65.** (a) 1.37; (b) 0.730 **67.** (a) −8.6%; (b) smaller **69.** 146 kJ **71.** (a) 250°C; (b) yes **73.** 3.0×10^{6} J/kg **75.** 560 W

CHAPTER 27

CP **1.** (a) rightward; (b) all tie; (c) b, then a and c tie; (d) b, then a and c tie **2.** (a) all tie; (b) R_1, R_2, R_3 **3.** (a) less; (b) greater; (c) equal **4.** (a) $V/2, i$; (b) $V, i/2$ **5.** (a) 1, 2, 4, 3; (b) 4, tie of 1 and 2, then 3
Q **1.** (a) equal; (b) more **3.** parallel, R_2, R_1, series **5.** (a) series; (b) parallel; (c) parallel **7.** (a) less; (b) less; (c) more **9.** (a) same; (b) same; (c) less; (d) more **11.** (a) all tie; (b) 1, 3, 2
P **1.** (a) 0.50 A; (b) 1.0 W; (c) 2.0 W; (d) 6.0 W; (e) 3.0 W; (f) supplied; (g) absorbed **3.** (a) 14 V; (b) 1.0×10^{2} W; (c) 6.0×10^{2} W; (d) 10 V; (e) 1.0×10^{2} W **5.** 11 kJ **7.** (a) 80 J; (b) 67 J; (c) 13 J **9.** (a) 12.0 eV; (b) 6.53 W **11.** (a) 50 V; (b) 48 V; (c) negative **13.** (a) 6.9 km; (b) 20 Ω **15.** 8.0 Ω **17.** (a) 0.004 Ω; (b) 1 **19.** (a) 4.00 Ω; (b) parallel **21.** 5.56 A **23.** (a) 50 mA; (b) 60 mA; (c) 9.0 V **25.** $3d$ **27.** 3.6×10^{3} A **29.** (a) 0.333 A; (b) right; (c) 720 J **31.** (a) −11 V; (b) −9.0 V **33.** 48.3 V **35.** (a) 5.25 V; (b) 1.50 V; (c) 5.25 V; (d) 6.75 V **37.** 1.43 Ω **39.** (a) 0.150 Ω; (b) 240 W **41.** (a) 0.709 W; (b) 0.050 W; (c) 0.346 W; (d) 1.26 W; (e) −0.158 W **43.** 9 **45.** (a) 0.67 A; (b) down; (c) 0.33 A; (d) up; (e) 0.33 A; (f) up; (g) 3.3 V **47.** (a) 1.11 A; (b) 0.893 A; (c) 126 m **49.** (a) 0.45 A **51.** (a) 55.2 mA; (b) 4.86 V; (c) 88.0 Ω; (d) decrease **53.** −3.0% **57.** 0.208 ms **59.** 4.61 **61.** (a) 2.41 μs; (b) 161 pF **63.** (a) 1.1 mA; (b) 0.55 mA; (c) 0.55 mA; (d) 0.82 mA; (e) 0.82 mA; (f) 0; (g) 4.0×10^{2} V; (h) 6.0×10^{2} V **65.** 411 μA **67.** 0.72 MΩ **69.** (a) 0.955 μC/s; (b) 1.08 μW; (c) 2.74 μW; (d) 3.82 μW **71.** (a) 3.00 A; (b) 3.75 A; (c) 3.94 A **73.** (a) 1.32×10^{7} A/m²; (b) 8.90 V; (c) copper; (d) 1.32×10^{7} A/m²; (e) 51.1 V; (f) iron **75.** (a) 3.0 kV; (b) 10 s; (c) 11 GΩ **77.** (a) 85.0 Ω; (b) 915 Ω **81.** 4.0 V **83.** (a) 24.8 Ω; (b) 14.9 kΩ **85.** the cable **87.** −13 μC **89.** 20 Ω **91.** (a) 3.00 A; (b) down; (c) 1.60 A; (d) down; (e) supply; (f) 55.2 W; (g) supply; (h) 6.40 W **93.** (a) 1.0 V; (b) 50 mΩ **95.** 3 **99.** (a) 1.5 mA; (b) 0; (c) 1.0 mA

CHAPTER 28

CP **1.** a, $+z$; b, $-x$; c, $\vec{F}_B = 0$ **2.** (a) 2, then tie of 1 and 3 (zero); (b) 4 **3.** (a) electron; (b) clockwise **4.** $-y$ **5.** (a) all tie; (b) 1 and 4 tie, then 2 and 3 tie
Q **1.** (a) no, because $\vec{v}$ and $\vec{F}_B$ must be perpendicular; (b) yes; (c) no, because $\vec{B}$ and $\vec{F}_B$ must be perpendicular **3.** (a) $+z$ and $-z$ tie, then $+y$ and $-y$ tie, then $+x$ and $-x$ tie (zero); (b) $+y$ **5.** (a) $\vec{F}_E$; (b) $\vec{F}_B$ **7.** (a) $\vec{B}_1$; (b) $\vec{B}_1$ into page, $\vec{B}_2$ out of page; (c) less **9.** (a) positive; (b) $2 \rightarrow 1$ and $2 \rightarrow 4$ tie, then $2 \rightarrow 3$ (which is zero) **11.** (a) negative; (b) equal; (c) equal; (d) half-circle
P **1.** (a) 400 km/s; (b) 835 eV **3.** (a) $(6.2 \times 10^{-14}\ \text{N})\hat{k}$; (b) $(-6.2 \times 10^{-14}\ \text{N})\hat{k}$ **5.** −2.0 T **7.** $(-11.4\ \text{V/m})\hat{i} - (6.00\ \text{V/m})\hat{j} + (4.80\ \text{V/m})\hat{k}$ **9.** $-(0.267\ \text{mT})\hat{k}$ **11.** 0.68 MV/m **13.** 7.4 μV **15.** (a) $(-600\ \text{mV/m})\hat{k}$; (b) 1.20 V **17.** (a) 2.60×10^{6} m/s; (b) 0.109 μs; (c) 0.140 MeV; (d) 70.0 kV **19.** 1.2×10^{-9} kg/C **21.** (a) 2.05×10^{7} m/s; (b) 467 μT; (c) 13.1 MHz; (d) 76.3 ns **23.** 21.1 μT **25.** (a) 0.978 MHz; (b) 96.4 cm **27.** (a) 495 mT; (b) 22.7 mA; (c) 8.17 MJ **29.** 65.3 km/s **31.** 5.07 ns **33.** (a) 0.358 ns; (b) 0.166 mm; (c) 1.51 mm **35.** (a) 200 eV; (b) 20.0 keV; (c) 0.499% **37.** 2.4×10^{2} m **39.** (a) 28.2 N; (b) horizontally west **41.** (a) 467 mA; (b) right **43.** (a) 0; (b) 0.138 N; (c) 0.138 N; (d) 0 **45.** $(-2.50\ \text{mN})\hat{j} + (0.750\ \text{mN})\hat{k}$ **47.** (a) 0.12 T; (b) 31° **49.** $(-4.3 \times 10^{-3}\ \text{N·m})\hat{j}$ **51.** 2.45 A **55.** (a) 2.86 A·m²; (b) 1.10 A·m² **57.** (a) 12.7 A; (b) 0.0805 N·m **59.** (a) 0.30 A·m²; (b) 0.024 N·m **61.** (a) −72.0 μJ; (b) $(96.0\hat{i} + 48.0\hat{k})$ μN·m **63.** (a) $-(9.7 \times 10^{-4}\ \text{N·m})\hat{i} - (7.2 \times 10^{-4}\ \text{N·m})\hat{j} + (8.0 \times 10^{-4}\ \text{N·m})\hat{k}$; (b) -6.0×10^{-4} J **65.** (a) 90°; (b) 1; (c) 1.28×10^{-7} N·m

67. (a) 20 min; (b) 5.9×10^{-2} N·m **69.** 8.2 mm **71.** 127 u **73.** (a) 6.3×10^{14} m/s^2; (b) 3.0 mm **75.** (a) 1.4; (b) 1.0 **77.** $(-500 \text{ V/m})\hat{j}$ **79.** (a) 0.50; (b) 0.50; (c) 14 cm; (d) 14 cm **81.** $(0.80\hat{j} - 1.1\hat{k})$ mN **83.** −40 mC **85.** (a) $(12.8\hat{i} + 6.41\hat{j}) \times 10^{-22}$ N; (b) 90°; (c) 173°

CHAPTER 29

CP **1.** *b, c, a* **2.** *d*, tie of *a* and *c*, then *b* **3.** *d, a*, tie of *b* and *c* (zero) **Q** **1.** *c, a, b* **3.** *c, d*, then *a* and *b* tie (zero) **5.** *a, c, b* **7.** *c* and *d* tie, then *b, a* **9.** *b, a, d, c* (zero) **11.** (a) 1, 3, 2; (b) less **P** **1.** (a) 3.3 μT; (b) yes **3.** (a) 16 A; (b) east **5.** (a) 1.0 mT; (b) out; (c) 0.80 mT; (d) out **7.** (a) 0.102 μT; (b) out **9.** (a) opposite; (b) 30 A **11.** (a) 4.3 A; (b) out **13.** 50.3 nT **15.** (a) 1.7 μT; (b) into; (c) 6.7 μT; (d) into **17.** 132 nT **19.** 5.0 μT **21.** 256 nT **23.** $(-7.75 \times 10^{-23}\text{ N})\hat{i}$ **25.** 2.00 rad **27.** 61.3 mA **29.** $(80\ \mu\text{T})\hat{j}$ **31.** (a) 20 μT; (b) into **33.** $(22.3\text{ pT})\hat{j}$ **35.** 88.4 pN/m **37.** $(-125\ \mu\text{N/m})\hat{i} + (41.7\ \mu\text{N/m})\hat{j}$ **39.** 800 nN/m **41.** $(3.20\text{ mN})\hat{j}$ **43.** (a) 0; (b) 0.850 mT; (c) 1.70 mT; (d) 0.850 mT **45.** (a) −2.5 μT·m; (b) 0 **47.** (a) 0; (b) 0.10 μT; (c) 0.40 μT **49.** (a) 533 μT; (b) 400 μT **51.** 0.30 mT **53.** 0.272 A **55.** (a) 4.77 cm; (b) 35.5 μT **57.** (a) 2.4 A·m^2; (b) 46 cm **59.** 0.47 A·m^2 **61.** (a) 79 μT; (b) 1.1×10^{-6} N·m **63.** (a) $(0.060\text{ A·m}^2)\hat{j}$; (b) $(96\text{ pT})\hat{j}$ **65.** 1.28 mm **69.** (a) 15 A; (b) $-z$ **71.** 7.7 mT **73.** (a) 15.3 μT **75.** (a) $(0.24\hat{i})$ nT; (b) 0; (c) $(-43\hat{k})$ pT; (d) $(0.14\hat{k})$ nT **79.** (a) 4.8 mT; (b) 0.93 mT; (c) 0 **83.** $(-0.20\text{ mT})\hat{k}$ **87.** (a) $\mu_0 ir/2\pi c^2$; (b) $\mu_0 i/2\pi r$; (c) $\mu_0 i(a^2 - r^2)/2\pi(a^2 - b^2)r$; (d) 0

CHAPTER 30

CP **1.** *b*, then *d* and *e* tie, and then *a* and *c* tie (zero) **2.** *a* and *b* tie, then *c* (zero) **3.** *c* and *d* tie, then *a* and *b* tie **4.** *b*, out; *c*, out; *d*, into; *e*, into **5.** d and e **6.** (a) 2, 3, 1 (zero); (b) 2, 3, 1 **7.** *a* and *b* tie, then *c* **Q** **1.** out **3.** (a) all tie (zero); (b) 2, then 1 and 3 tie (zero) **5.** *d* and *c* tie, then *b, a* **7.** (a) more; (b) same; (c) same; (d) same (zero) **9.** (a) all tie (zero); (b) 1 and 2 tie, then 3; (c) all tie (zero) **P** **1.** 0 **3.** 30 mA **5.** 0 **7.** (a) 31 mV; (b) left **9.** 0.198 mV **11.** (b) 0.796 m^2 **13.** 29.5 mC **15.** (a) 21.7 V; (b) counterclockwise **17.** (a) 1.26×10^{-4} T; (b) 0; (c) 1.26×10^{-4} T; (d) yes; (e) 5.04×10^{-8} V **19.** 5.50 kV **21.** (a) 40 Hz; (b) 3.2 mV **23.** (a) $\mu_0 iR^2\pi r^2/2x^3$; (b) $3\mu_0 i\pi R^2 r^2 v/2x^4$; (c) counterclockwise **25.** (a) 13 μWb/m; (b) 17%; (c) 0 **27.** (a) 80 μV; (b) clockwise **29.** (a) 48.1 mV; (b) 2.67 mA; (c) 0.129 mW **31.** 3.68 μW **33.** (a) 240 μV; (b) 0.600 mA; (c) 0.144 μW; (d) 2.87×10^{-8} N; (e) 0.144 μW **35.** (a) 0.60 V; (b) up; (c) 1.5 A; (d) clockwise; (e) 0.90 W; (f) 0.18 N; (g) 0.90 W **37.** (a) 71.5 μV/m; (b) 143 μV/m **39.** 0.15 V/m **41.** (a) 2.45 mWb; (b) 0.645 mH **43.** 1.81 μH/m **45.** (a) decreasing; (b) 0.68 mH **47.** (b) $L_{eq} = \Sigma L_j$, sum from $j = 1$ to $j = N$ **49.** 59.3 mH **51.** 46 Ω **53.** (a) 8.45 ns; (b) 7.37 mA **55.** 6.91 **57.** (a) 1.5 s **59.** (a) $i[1 - \exp(-Rt/L)]$; (b) $(L/R)\ln 2$ **61.** (a) 97.9 H; (b) 0.196 mJ **63.** 25.6 ms **65.** (a) 18.7 J; (b) 5.10 J; (c) 13.6 J **67.** (a) 34.2 J/m^3; (b) 49.4 mJ **69.** 1.5×10^8 V/m **71.** (a) 1.0 J/m^3; (b) 4.8×10^{-15} J/m^3 **73.** (a) 1.67 mH; (b) 6.00 mWb **75.** 13 μH **77.** (b) have the turns of the two solenoids wrapped in opposite directions **79.** (a) 2.0 A; (b) 0; (c) 2.0 A; (d) 0; (e) 10 V; (f) 2.0 A/s; (g) 2.0 A; (h) 1.0 A; (i) 3.0 A; (j) 10 V; (k) 0; (l) 0 **81.** (a) 10 μT; (b) out; (c) 3.3 μT; (d) out **83.** 0.520 ms **85.** (a) $(4.4 \times 10^7\text{ m/s}^2)\hat{i}$; (b) 0; (c) $(-4.4 \times 10^7\text{ m/s}^2)\hat{i}$ **87.** (a) 0.40 V; (b) 20 A **89.** (a) 10 A; (b) 1.0×10^2 J **91.** (a) 0; (b) 8.0×10^2 A/s; (c) 1.8 mA; (d) 4.4×10^2 A/s; (e) 4.0 mA; (f) 0 **93.** 1.15 W **95.** (a) 20 A/s; (b) 0.75 A **97.** 12 A/s

CHAPTER 31

CP **1.** (a) *T*/2; (b) *T*; (c) *T*/2; (d) *T*/4 **2.** (a) 5 V; (b) 150 μJ **3.** (a) remains the same; (b) remains the same **4.** (a) *C, B, A*; (b) 1, *A*; 2, *B*; 3, *S*; 4, *C*; (c) *A* **5.** (a) remains the same; (b) increases; (c) remains the same; (d) decreases **6.** (a) 1, lags; 2, leads; 3, in phase; (b) 3 ($\omega_d = \omega$ when $X_L = X_C$) **7.** (a) increase (circuit is mainly capacitive; increase *C* to decrease X_C to be closer to resonance for maximum P_{avg}); (b) closer **8.** (a) greater; (b) step-up **Q** **1.** *b, a, c* **3.** (a) *T*/4; (b) *T*/4; (c) *T*/2; (d) *T*/2 **5.** *c, b, a* **7.** *a* inductor; *b* resistor; *c* capacitor **9.** (a) positive; (b) decreased (to decrease X_L and get closer to resonance); (c) decreased (to increase X_C and get closer to resonance) **11.** (a) rightward, increase (X_L increases, closer to resonance); (b) rightward, increase (X_C decreases, closer to resonance); (c) rightward, increase (ω_d/ω increases, closer to resonance) **P** **1.** (a) 1.17 μJ; (b) 5.58 mA **3.** (a) 6.00 μs; (b) 167 kHz; (c) 3.00 μs **5.** 45.2 mA **7.** (a) 1.25 kg; (b) 372 N/m; (c) 1.75×10^{-4} m; (d) 3.02 mm/s **9.** 7.0×10^{-4} s **11.** (a) 6.0; (b) 36 pF; (c) 0.22 mH **13.** (a) 0.180 mC; (b) 70.7 μs; (c) 66.7 W **15.** (a) 3.0 nC; (b) 1.7 mA; (c) 4.5 nJ **17.** (a) 275 Hz; (b) 365 mA **21.** (a) 356 μs; (b) 2.50 mH; (c) 3.20 mJ **23.** (a) 1.98 μJ; (b) 5.56 μC; (c) 12.6 mA; (d) −46.9°; (e) +46.9° **25.** 8.66 mΩ **29.** (a) 95.5 mA; (b) 11.9 mA **31.** (a) 0.65 kHz; (b) 24 Ω **33.** (a) 6.73 ms; (b) 11.2 ms; (c) inductor; (d) 138 mH **35.** 89 Ω **37.** 7.61 A **39.** (a) 267 Ω; (b) −41.5°; (c) 135 mA **41.** (a) 206 Ω; (b) 13.7°; (c) 175 mA **43.** (a) 218 Ω; (b) 23.4°; (c) 165 mA **45.** (a) yes; (b) 1.0 kV **47.** (a) 224 rad/s; (b) 6.00 A; (c) 219 rad/s; (d) 228 rad/s; (e) 0.040 **49.** (a) 796 Hz; (b) no change; (c) decreased; (d) increased **53.** (a) 12.1 Ω; (b) 1.19 kW **55.** 1.84 A **57.** (a) 117 μF; (b) 0; (c) 90.0 W; (d) 0°; (e) 1; (f) 0; (g) −90°; (h) 0 **59.** (a) 2.59 A; (b) 38.8 V; (c) 159 V; (d) 224 V; (e) 64.2 V; (f) 75.0 V; (g) 100 W; (h) 0; (i) 0 **61.** (a) 0.743; (b) lead; (c) capacitive; (d) no; (e) yes; (f) no; (g) yes; (h) 33.4 W **63.** (a) 2.4 V; (b) 3.2 mA; (c) 0.16 A **65.** (a) 1.9 V; (b) 5.9 W; (c) 19 V; (d) 5.9×10^2 W; (e) 0.19 kV; (f) 59 kW **67.** (a) 6.73 ms; (b) 2.24 ms; (c) capacitor; (d) 59.0 μF **69.** (a) −0.405 rad; (b) 2.76 A; (c) capacitive **71.** (a) 64.0 Ω; (b) 50.9 Ω; (c) capacitive **73.** (a) 2.41 μH; (b) 21.4 pJ; (c) 82.2 nC **75.** (a) 39.1 Ω; (b) 21.7 Ω; (c) capacitive **79.** (a) 0.577*Q*; (b) 0.152 **81.** (a) 45.0°; (b) 70.7 Ω **83.** 1.84 kHz **85.** (a) 0.689 μH; (b) 17.9 pJ; (c) 0.110 μC **87.** (a) 165 Ω; (b) 313 mH; (c) 14.9 μF

CHAPTER 32

CP **1.** *d, b, c, a* (zero) **2.** *a, c, b, d* (zero) **3.** tie of *b, c,* and *d*, then *a* **4.** (a) 2; (b) 1 **5.** (a) away; (b) away; (c) less **6.** (a) toward; (b) toward; (c) less **Q** **1.** 1 *a*, 2 *b*, 3 *c* and *d* **3.** a, decreasing; b, decreasing **5.** supplied **7.** (a) *a* and *b* tie, then *c, d*; (b) none (because plate lacks circular symmetry, $\vec{B}$ not tangent to any circular loop); (c) none **9.** (a) 1 up, 2 up, 3 down; (b) 1 down, 2 up, 3 zero **11.** (a) 1, 3, 2; (b) 2 **P** **1.** +3 Wb **3.** (a) 47.4 μWb; (b) inward **5.** 2.4×10^{13} V/m·s **7.** (a) 1.18×10^{-19} T; (b) 1.06×10^{-19} T **9.** (a) 5.01×10^{-22} T; (b) 4.51×10^{-22} T **11.** (a) 1.9 pT **13.** 7.5×10^5 V/s **17.** (a) 0.324 V/m; (b) 2.87×10^{-16} A; (c) 2.87×10^{-18} **19.** (a) 75.4 nT; (b) 67.9 nT **21.** (a) 27.9 nT; (b) 15.1 nT **23.** (a) 2.0 A; (b) 2.3×10^{11} V/m·s; (c) 0.50 A; (d) 0.63 μT·m **25.** (a) 0.63 μT; (b) 2.3×10^{12} V/m·s **27.** (a) 0.71 A; (b) 0; (c) 2.8 A **29.** (a) 7.60 μA; (b) 859 kV·m/s; (c) 3.39 mm; (d) 5.16 pT **31.** 55 μT **33.** (a) 0; (b) 0; (c) 0; (d) $\pm 3.2 \times 10^{-25}$ J; (e) -3.2×10^{-34} J·s; (f) 2.8×10^{-23} J/T; (g) -9.7×10^{-25} J; (h) $\pm 3.2 \times 10^{-25}$ J **35.** (a) -9.3×10^{-24} J/T; (b) 1.9×10^{-23} J/T **37.** (b) $+x$;

(c) clockwise; (d) $+x$ **39.** yes **41.** 20.8 mJ/T **43.** (b) K_i/B; (c) $-z$; (d) 0.31 kA/m **47.** (a) 1.8×10^2 km; (b) 2.3×10^{-5} **49.** (a) 3.0 μT; (b) 5.6×10^{-10} eV **51.** 5.15×10^{-24} A·m^2 **53.** (a) 0.14 A; (b) 79 μC **55.** (a) 6.3×10^8 A; (b) yes; (c) no **57.** 0.84 kJ/T **59.** (a) $(1.2 \times 10^{-13}\,\mathrm{T})\exp[-t/(0.012\,\mathrm{s})]$; (b) 5.9×10^{-15} T **63.** (a) 27.5 mm; (b) 110 mm **65.** 8.0 A

67. (a) -8.8×10^{15} V/m·s; (b) 5.9×10^{-7} T·m **69.** (b) sign is minus; (c) no, because there is compensating positive flux through open end nearer to magnet **71.** (b) $-x$; (c) counterclockwise; (d) $-x$ **73.** (a) 7; (b) 7; (c) $3h/2\pi$; (d) $3eh/4\pi m$; (e) $3.5h/2\pi$; (f) 8 **75.** (a) 9; (b) 3.71×10^{-23} J/T; (c) $+9.27 \times 10^{-24}$ J; (d) -9.27×10^{-24} J

Figures are noted by page numbers in *italics*, tables are indicated by t following the page number.

G

H

I

Some Physical Constants*

Speed of light	c	2.998×10^{8} m/s
Gravitational constant	G	6.673×10^{-11} N · m²/kg²
Avogadro constant	N_A	6.022×10^{23} mol^{-1}
Universal gas constant	R	8.314 J/mol · K
Mass–energy relation	c^2	8.988×10^{16} J/kg
		931.49 MeV/u
Permittivity constant	ε_0	8.854×10^{-12} F/m
Permeability constant	μ_0	1.257×10^{-6} H/m
Planck constant	h	6.626×10^{-34} J · s
		4.136×10^{-15} eV·s
Boltzmann constant	k	1.381×10^{-23} J/K
		8.617×10^{-5} eV/K
Elementary charge	e	1.602×10^{-19} C
Electron mass	m_e	9.109×10^{-31} kg
Proton mass	m_p	1.673×10^{-27} kg
Neutron mass	m_n	1.675×10^{-27} kg
Deuteron mass	m_d	3.344×10^{-27} kg
Bohr radius	a	5.292×10^{-11} m
Bohr magneton	μ_B	9.274×10^{-24} J/T
		5.788×10^{-5} eV/T
Rydberg constant	R	$1.097\,373 \times 10^{7}$ m^{-1}

*For a more complete list, showing also the best experimental values, see Appendix B.

The Greek Alphabet

Alpha	A	α	Iota	I	ι	Rho	P	ρ
Beta	B	β	Kappa	K	κ	Sigma	Σ	σ
Gamma	Γ	γ	Lambda	Λ	λ	Tau	T	τ
Delta	Δ	δ	Mu	M	μ	Upsilon	Υ	υ
Epsilon	E	ϵ	Nu	N	ν	Phi	Φ	ϕ, φ
Zeta	Z	ζ	Xi	Ξ	ξ	Chi	X	χ
Eta	H	η	Omicron	O	o	Psi	Ψ	ψ
Theta	Θ	θ	Pi	Π	π	Omega	Ω	ω